CELL BIOLOGY

A LABORATORY HANDBOOK

Third Edition
Volume 4

CELL BIOLOGY

A LABORATORY HANDBOOK

Third Edition
Volume 4

Edited by

Julio E. Celis

Institute of Cancer Biology, Danish Cancer Society,
Copenhagen, Denmark

ELSEVIER
ACADEMIC
PRESS

AMSTERDAM • BOSTON • HEIDELBERG • LONDON
NEW YORK • OXFORD • PARIS • SAN DIEGO
SAN FRANCISCO • SINGAPORE • SYDNEY • TOKYO

Elsevier Academic Press
30 Corporate Drive, Suite 400, Burlington, MA 01803, USA
525 B Street, Suite 1900, San Diego, California 92101-4495, USA
84 Theobald's Road, London WC1X 8RR, UK

This book is printed on acid-free paper. ∞

Library of Congress Cataloging-in-Publication Data
Application Submitted

British Library Cataloguing in Publication Data
A catalogue record for this book is available from the British Library

ISBN 13: 978-0-12-164734-6
ISBN 10: 0-12-164734-X
Set ISBN 13: 978-0-12-164730-8
Set ISBN 10: 0-12-164730-7

For all information on all Elsevier Academic Press publications visit our Web site at www.books.elsevier.com

Printed in China
05 06 07 08 09 10 9 8 7 6 5 4 3 2 1

Contents of Volume 4

PART E. APPENDIX

Section 16. Appendix

Contents of other Volumes

Contributors

Numbers in parenthesis indicate the volume (bold face) and page on which the authors' contribution begins.

Mads Aaboe (**4**: 83)   Clinical Biochemical Department, Molecular Diagnostic Laboratory, Aarhus University Hospital, Skejby, Brendstrupgaardvej, Aarhus N, DK-8200, DENMARK

Tanja Aarvak (**1**: 239)   Dynal Biotech ASA, PO Box 114, Smestad, N-0309, NORWAY

Harindra R. Abeysinghe (**3**: 345)   Department of Pathology and Laboratory Medicine, University of Rochester School of Medicine, 601 Elmwood Ave., Rm 1-6337, Rochester, NY 14642

Ruedi Aebersold (**4**: 437)   The Institute for Systems Biology, 1441 North 34th Street, Seattle, WA 98103-8904

Ueli Aebi (**3**: 233, 241)   ME Muller Institute for Microscopy, Biozentrum, University of Basel, Klingelbergstr. 50/70, Basel, CH-4056, SWITZERLAND

Cheol-Hee Ahn (**4**: 29)   School of Materials Science and Engineering, Seoul National University, Seoul, 151-744, SOUTH KOREA

Natalie G. Ahn (**4**: 443)   Department of Chemistry & Biochemistry, University of Colorado, 215 UCB, Boulder, CO 80309

Ramiro Alberio (**4**: 45)   School of Biosciences, University of Nottingham, Sutton Bonington, Loughborough, Leics, LE12 5RD, UNITED KINGDOM

Donna G. Albertson (**3**: 445)   Cancer Research Institute, Department of Laboratory Medicine, The University of California, San Francisco, Box 0808, San Francisco, CA 94143-0808

Heiner Albiez (**1**: 291)   Department of Biology II, Ludwig-Maximilians University of Munich, Munich, GERMANY

Terence Allen (**3**: 325)   CRC Structural Cell Biology Group, Paterson Institute for Cancer Research, Christie Hospital NHS Trust, Wilmslow Road, Withington, Manchester, M20 4BX, UNITED KINGDOM

Noona Ambartsumian (**1**: 363)   Department of Molecular Cancer Biology, Danish Cancer Society, Institute of Cancer Biology, Strandboulevarden 49, Copenhagen, DK-2100, DENMARK

Øystein Åmellem (**1**: 239)   Immunosystems, Dynal Biotech ASA, PO Box 114, Smestad, N-0309, NORWAY

Patrick Amstutz (**1**: 497)   Department of Biochemistry, University of Zürich, Winterthurerstr. 190, Zurich, CH-8057, SWITZERLAND

Jens S. Andersen (**4**: 427)   Protein Interaction Laboratory, University of Southern Denmark—Odense, Campusvej 55, Odense M, DK-5230, DENMARK

Mads Hald Andersen (**1**: 97)   Tumor Immunology Group, Institute of Cancer Biology, Danish Cancer Society, Strandboulevarden 49, Copenhagen, DK-2100, DENMARK

Helena Andersson (**4**: 63)   Bioscience at Novum, Karolinska Institutet, Halsovagen 7-9, Huddinge, SE-141 57, SWEDEN

Peter W. Andrews (**1**: 183)   Department of Biomedical Science, The University of Sheffield, Rm B2 238, Sheffield, S10 2TN, UNITED KINGDOM

Elsa Anes (**2**: 57)   Faculdade de Farmacia, Universidade de Lisboa, Av. Forcas Armadas, Lisboa, 1649-019, PORTUGAL

James M. Angelastro (**1**: 171)   Department of Pathology and Center for Neurobiology and Behavior, Columbia University College of Physicians and Surgeons, 630 West 168th Street, New York, NY 10032

Sergey V. Anisimov (4: 103)   Molecular Cardiology Unit, National Institute on Aging, NIH, 5600 Nathan Shock Drive, Baltimore, MD 21224

Celia Antonio (2: 379)   Department of Biochemistry & Molecular Biophysics, College of Physicians & Surgeons, Columbia University, 701 W 168ST HHSC 724, New York, NY 69117

Shigehisa Aoki (1: 411)   Department of Pathology & Biodefence, Faculty of Medicine, Saga University, Nebeshima 5-1-1, Saga, 849-8501, JAPAN

Ron D. Appel (4: 207)   Swiss Institute of Bioinformatics, CMU, Rue Michel Servet 1, Geneva 4, CH-1211, SWITZERLAND

Rolf Apweiler (4: 469)   EMBL Outstation, European Bioinformatics Institute, Wellcome Trust Genome Campus, Hinxton, Cambridge, CB10 1SD, UNITED KINGDOM

Nobukazu Araki (2: 147)   Department of Histology and Cell Biology, School of Medicine, Kagawa University, Mki, Kagawa, 761-0793, JAPAN

Christopher M. Armstrong (4: 295)   Dana Faber Cancer Institute, Harvard University, 44 Binney Street, Boston, MA 02115

Anthony J. Ashford (2: 155)   Antibody Facility, Max Planck Institute of Mollecular Cell Biology and Genetics, Pfotenhauerstrsse 108, Dresden, D-01307, GERMANY

Daniel Axelrod (3: 19)   Dept of Physics & Biophysics Research Division, University of Michigan, Ann Arbor, MI 48109-1055

Sheree Bailey (1: 475)   Dept of Immunology, Allergy and Arthritis, Flinders Medical Centre and Flinders University, Bedford Park, Adelaide, SA, 5051, SOUTH AUSTRALIA

Nathalie Q. Balaban (2: 419)   Department of Physics, The Hebrew University-Givat Ram, Racah Institute, Jerusalem, 91904, ISRAEL

William E. Balch (2: 209)   Department of Cell and Molecular Biology, The Scripps Research Institute, 10550 North Torrey Pines Road, La Jolla, CA 92037

Debabrata Banerjee (1: 315)   Department of Medicine, Cancer Institute of New Jersey, 195 Little Albany Street, New Brunswick, NJ 08903

Jiri Bartek (4: 253)   Department of Cell Cycle and Cancer, Danish Cancer Society, Strandboulevarden 49, Copenhagen, DK-2100, DENMARK

Werner Baschong (3: 5)   ME Muller Institute for Microscopy, Biozentrum, University of Basel, Klingelbergstrasse 50/70, Basel, CH-4056, SWITZERLAND

Philippe I. H. Bastiaens (3: 153)   Cell Biology and Cell Biophysics Program, European Molecular Biology Laboratory, Meyerhofstrasse 1, Heidelberg, 69117, GERMANY

Jürgen C. Becker (1: 103)   Department of Dermatology, University of Würzburg, Sanderring 2, Würzburg, 97070, GERMANY

Martin Béhé (4: 149)   Department of Nuclear Medicine, Philipp's-University of Marburg, Baldingerstraße, Marburg/Lahn, D-35043, GERMANY

Thomas M. Behr (4: 149)   Department of Nuclear Medicine, Philipp's-University of Marburg, Baldingerstraße, Marburg, D-35043, GERMANY

Stefanie Benesch (2: 399)   Department of Cell Biology, Gesellschaft fur Biotechnoogische Forschung, Mascheroder Weg 1, Braunschweig, D-38124, GERMANY

Aaron Bensimon (3: 429)   Laboratoire de Biophysique de l'ADN, Departement des Biotechnologies, Institut Pasteur, 25 rue du Dr. Roux, Paris Cedex 15, F-75724, FRANCE

John J. M. Bergeron (2: 41)   Department of Anatomy and Cell Biology, Faculty of Medicine, McGill University, STRATHCONA Anatomy & Dentistry Building, Montreal, QC, H3A 2B2, CANADA

Michael W. Berns (3: 351)   Beckman Laser Institute, University of California, Irvine, 1002 Health Sciences Road E, Irvine, CA 92697-1475

Joseph R. Bertino (1: 315)   The Cancer Institute of New Jersey, 195 Little Albany Street, New Brunswick, NJ 08901

Paulo Bianco (1: 79)   Dipartimento di Medicina Sperimentale e Patologia, Universita 'La Sapienza', Viale Regina Elena 324, Roma, I-00161, ITALY

Hans Kaspar Binz (1: 497)   Department of Biochemistry, University of Zürich, Winterthurerstr. 190, Zürich, CH-8057, SWITZERLAND

R. Curtis Bird (1: 247)   Department of Pathobiology, Auburn University, Auburn, Alabama 36849

Mina J. Bissell (1: 139)   Life Sciences Division, Lawrence Berkeley National Laboratory, 1 Cyclotron Road, Bldg 83-101, Berkeley, CA 94720

Stephanie Blackwood (**3**: 445)   Cancer Research Institute, University of California San Francisco, PO Box 0808, San Francisco, CA 94143-0808

Blagoy Blagoev (**4**: 427)   Protein Interaction Laboratory, University of Southern Denmark—Odense, Campusvej 55, Odense M, DK-5230, DENMARK

Kenneth R. Boheler (**4**: 103)   Laboratory of Cardiovascular Science, National Institute on Aging, NIH, 5600 Nathan Shock Drive, Baltimore, MD 21224-6825

Michelle A. Booden (**1**: 345)   Lineberger Comprehensive Cancer Center, University of North Carolina at Chapel Hill, Chapel Hill, NC 27599-7295

Gary G. Borisy (**3**: 277)   Department of Cell and Molecular Biology, Northwestern University Medical School, Chicago, IL 6011-3072

Elliot Botvinick (**3**: 351)   Beckman Laser Institute, University of California, Irvine, 1002 Health Sciences Road, East, Irvine, CA 92697-1475

Gérard Bouchet (**4**: 207)   Swiss Institute of Bioinformatics (SIB), CMU, rue Michel-Servet 1, Genève 4, CH-1211, SWITZERLAND

Rosemary Boyle (**4**: 437)   The Institute for Systems Biology, 1441 North 34th St., Seattle, WA 98109

Susanne Brandfass (**1**: 563)   Department of Biochemistry and Cell Biology, Max Planck Institute of Biophysical Chemistry, Am Faßberg 11, Gottingen, D-37077, GERMANY

Pascal Braun (**4**: 73)   Department of Chemistry and Chemical Biology, Harvard University, 12 Oxford Street, Cambridge, MA 02138

Steven A. Braut (**4**: 121)   Department of Anatomy and Structural Biology, Golding # 601, Albert Einstein College of Medicine of Yeshiva University, 1300 Morris Park Avenue, Bronx, NY 10461

Alvis Brazma (**4**: 95)   EMBL Outstation—Hinxton, European Bioinformatics Institute, Wellcome Trust Genome Campus, Hinxton, Cambridge, CB10 1SD, UNITED KINGDOM

J. David Briley (**3**: 471)   Department of Genomic Sciences, Glaxo Wellcome Research and Development, 5 Moore Drive, Research Triangle Park, NC 27709-3398

Simon Broad (**1**: 133)   Keratinocyte Laboratory, London Research Institute, 44 Lincoln's Inn Fields, London, WC2A 3PX, UNITED KINGDOM

Nicholas H. Brown (**3**: 77)   Wellcome Trust/Cancer Research UK Institute and Departmentt of Anatomy, University of Cambridge, Tennis Court Road, Cambridge, CB2 1QR, UNITED KINGDOM

Heather L. Brownell (**2**: 329, 341)   Office of Technology Licensing and Industry Sponsored Research, Harvard Medical School, 25 Shattuck Street, Gordon Hall of Medicine, Room 414, Boston, MA 02115

Damien Brunner (**3**: 69)   Cell Biology and Cell Biophysics Programme, European Molecular Biology Laboratory, Meyerhofstrasse 1, Heidelberg, D-69117, GERMANY

Suzannah Bumpstead (**3**: 463)   Genotyping / Chr 20, The Wellcome Trust Sanger Institute, The Wellcome Trust Genome Campus, Hinxton, Cambridge, CB10 1SA, UNITED KINGDOM

Deborah C. Burford (**3**: 403)   Wellcome Trust, Sanger Institute, The Wellcome Trust Genome Campus, Hinxton, Cambridge, CB10 1SA, UNITED KINGDOM

Gerald Burgstaller (**2**: 161)   Department of Cell Biology, Institute of Molecular Biology, Austrian Academy of Sciences, Billrothstrasse 11, Salzburg, A-5020, AUSTRIA

Ian M. Caldicott (**1**: 157)

Angelique S. Camp (**1**: 457)   Gene Therapy Centre, University of North Carolina at Chapel Hill, 7119 Thurston-Bowles (G44 Wilson Hall), Chapel Hill, NC 27599-7352

Keith H. S. Campbell (**4**: 45)   School of Biosciences, Sutton Bonington, Loughborough, Leics, LE12 5RD, UNITED KINGDOM

Yihai Cao (**1**: 373)   Microbiology & Tumor Biology Center, Karolinska Institute, Room: Skrivrum (G415), Box 280, Stockholm, SE-171 77, SWEDEN

Maria Carmo-Fonseca (**2**: 277, **3**: 419)   Institute of Molecular Medicine, Faculty of Medicine, University of Lisbon, Av. Prof. Egas Moniz, Lisbon, 1649-028, PORTUGAL

T. Carneiro (**3**: 419)   Faculty of Medicine, Institute of Molecular Medicine, University of Lisbon, Av. Prof. Egas Moniz, Lisboa, 1649-028, PORTUGAL

Nigel P. Carter (**2**: 133)   The Wellcome Trust, Sanger Institute, The Wellcome Trust, Genome Campus, Hinxton, Cambridge, CB10 1SA, UNITED KINGDOM

Célia Carvalho (3: 419)   Faculty of Medicine, Institute of Molecular Medicine, University of Lisbon, Av. Prof. Egas Moniz, Lisboa, 1649-028, PORTUGAL

Lucy A. Carver (2: 11)   Cellular and Molecular Biology Program, Sidney Kimmel Cancer Center, 10835 Altman Row, San Diego, CA 92121

Doris Cassio (1: 231, 3: 387)   INSERM U-442: Signalisation cellulaire et calcium, Bat 443, Universite Paris-Sud, Street George Clemenceau Pack, 444, Orsay, Cedex, F-91405, FRANCE

Chris Catton (3: 207)   Department of Zoology, University of Oxford, South Parks Road, Oxford, OX1 3PS, UNITED KINGDOM

Julio E. Celis (1: 527, 4: 69, 165, 219, 243, 289) Danish Cancer Society, Institute of Cancer Biology and Danish Centre for Translational Breast Cancer Research, Strandboulevarden 49, Copenhagen O, DK-2100, DENMARK

Pierre Chambon (3: 501)   Institut de Génétique et de Biologie Moléculaire et Cellulaire, 1 rue Laurent Fries, B.P.10142, Illkirch CEDEX, F-67404, FRANCE

Francis Ka-Ming Chan (2: 355)   Department of Pathology, University of Massachusetts Medical School, Room S2-125, 55 Lake Avenue North, Worcester, MA 01655

Ming-Shien Chang (3: 87)   Department of Physics, Duke University, 107 Physics Bldg, Durham, NC 27708-1000

Samit Chatterjee (2: 241)   Margaret M. Dyson Vision Research Institute, Department of Ophthalmology, Weill Medical College of Cornell University, 1300 York Avenue, New York, NY 10021

Sandeep Chaudhary (1: 121)   Veterans Affairs Medical Center, San Diego (V111G), 3350 La Jolla Village Drive, San Diego, CA 92161

Jingwen Chen (3: 471)   Department of Genomic Sciences, Glaxo Wellcome Research and Development, 5 Moore Drive, Research Triangle Park, NC 27709

Yonglong Chen (1: 191)   Institute for Biochemistry and Molecular Cell Biology, University of Goettingen, Justus-von-Liebig-Weg 11, Göttingen, D-37077, GERMANY

Yong Woo Cho (4: 29)   Akina, Inc., Business & Technology Center, 1291 Cumberland Ave., #E130, West Lafayette, IN 47906

Juno Choe (1: 269)   Institute for Systems Biology, 1441 N. 34th St, Seattle, WA 98103

Claus R. L. Christensen (1: 363)   Department of Molecular Cancer Biology, Danish Cancer Society, Institute of Cancer Biology, Strandboulevarden 49, Copenhagen, DK-2100, DENMARK

Theodore Ciaraldi (1: 121)   Veterans Affairs Medical Center, University of California, San Diego, 9500 Gilman Drive, La Jolla, CA 92093-9111

Aaron Ciechanover (4: 351)   Center for Tumor and Vascular Biology, The Rappaport Faculty of Medicine and Research Institute, Technion-Israel Institute of Technology, POB 9649, Efron Street, Bat Galim, Haifa, 31096, ISRAEL

Mark S. F. Clarke (2: 233, 4: 5)   Department of Health and Human Performance, University of Houston, 3855 Holman Street, Garrison—Rm 104D, Houston, TX 77204-6015

Martin Clynes (1: 335)   National Institute for Cellular Biotechnology, Dublin City University, Glasnevin, Dublin, 9, IRELAND

Philippe Collas (1: 207)   Institute of Medical Biochemistry, University of Oslo, PO Box 1112 Blindern, Oslo, 0317, NORWAY

Kristen Correia (4: 35)   Krumlauf Lab, Stowers Institute for Medical Research, 1000 East 50th Street, Kansas City, MO 64110

Pascale Cossart (2: 407)   Unite des Interactions Bacteries-Cellules/Unité INSERM 604, Institut Pasteur, 28, rue du Docteur Roux, Paris Cedex 15, F-75724, FRANCE

Thomas Cremer (1: 291)   Department of Biology II, Ludwig-Maximilians University of Munich, Munich, 80333, GERMANY

Robert A. Cross (2: 371)   Molecular Motors Group, Marie Curie Research Institute, The Chart, Oxted, Surrey, RH8 0TE, UNITED KINGDOM

Matthew E. Cunningham (1: 171)   Hospital for Special Surgery, New York Hospital, 520 E. 70th Street, New York, NY 10021

Noélia Custódio (3: 419)   Faculty of Medicine, Institute of Molecular Medicine, University of Lisbon, Av. Prof. Egas Moniz, Lisboa, 1649-028, PORTUGAL

Zbigniew Darzynkiewwicz (1: 279)   The Cancer Research Institute, New York Medical College, 19 Bradhurst Avenue, Hawthorne, NY 10532

Ilan Davis (3: 187)   Wellcome Trust Centre for Cell Biology, Institute of Cell and Molecular Biology, The University of Edinburgh, Michael Swann Building, The King's Buildings, Mayfield Road, Edinburgh, EH9 3JR, SCOTLAND

Stephen C. De Rosa (1: 257)   Vaccine Research Center, National Institutes of Health, 40 Convent Dr., Room 5610, Bethesda, MD 20892-3015

Nicholas M. Dean (3: 523)   Functional Genomics, GeneTrove, GeneTrove (a division of Isis Isis Pharmaceuticals, Inc.), 2292 Faraday Avenue, Carlsbad, CA 92008

Anne Dell (4: 415)   Department of Biological Sciences, Biochemistry Building, Imperial College of Science, Technology & Medicine, Biochemistry Building, London, SW7 2AY, UNITED KINGDOM

Panos Deloukas (3: 463)   The Wellcome Trust, Sanger Institute, Hinxton, Cambridge, CB10 1SA, UNITED KINGDOM

Nicolas Demaurex (3: 163)   Department of Cell Physiology and Metabolism, University of Geneva Medical Center, 1 Michel-Servet, Geneva, CH-1211, SWITZERLAND

Chris Denning (4: 45)   Division of Animal Physiology, School of Biosciences, Institute of Genetics Room C15, University of Nottingham, Queens Medical Centre, Nottingham, NG7 2UH, UNITED KINGDOM

Ami Deora (2: 241)   Margaret M. Dyson Vision Research Institute, Department of Ophthalmology, Weill Medical College of Cornell University, 1300 York Avenue, New York, NY 10021

Julien Depollier (4: 13)   Centre de Recherche en Biochimie Macromoléculaire (UPR 1086), Centre National de la Recherche Scientifique (CNRS), 1919 Route de Mende, Montpellier Cedex 5, F-34293, FRANCE

Channing J. Der (1: 345)   Department of Pharmacology, University of North Carolina at Chapel Hill, Lineberger Comprehensive Cancer Center, Chapel Hill, NC 27599

Bart Devreese (4: 259)   Department of Biochemistry, Physiology and Microbiology, University of Ghent, K.L. Ledeganckstraat 35, Ghent, B-9000, BELGIUM

Alberto Diaspro (3: 201)   Department of Physics, University of Genoa, Via Dodecaneso 33, Genoa, I-16146, ITALY

James Fred Dice (4: 345)   Department Physiology, Tufts University School of Medicine, 136 Harrison Ave, Boston, MA 02111

Thomas J. Diefenbach (4: 307)   Department of Physiology, Tufts University School of Medicine, 136 Harrison Avenue, Boston, MA 02111

Chris Dinant (2: 121)   Biomolecular Sciences, UMIST, PO Box 88, Manchester, M60 1QD, UNITED KINGDOM

Da-Qiao Ding (3: 171)   Structural Biology Section and CREST Research Project, Kansai Advanced Research Center, Communications Research Laboratory, 588-2 Iwaoka, Iwaoka-cho, Nishi-ku, Kobe, 651-2492, JAPAN

Gilles Divita (4: 13)   Centre de Recherche en Biochimie Macromoléculaire (UPR 1086), Centre National de la Recherche Scientifique (CNRS), 1919 Route de Mende, Montpellier Cedex 5, F-34293, FRANCE

Eric P. Dixon (1: 483)   TriPath Oncology, 4025 Stirrup Creek Drive, Suite 400, Durham, NC 27703

Bernhard Dobberstein (2: 215)   Zentrum fur Molekulare Biologie, Universitat Heidelberg, Im Neuenheimer Feld 282, Heidelberg, D-69120, GERMANY

Lynda J. Donald (4: 457)   Department of Chemistry, University of Manitoba, Room 531 Parker Building, Winnipeg, MB, R3T 2N2, CANADA

Wolfgang R. G. Dostmann (2: 299)   Department of Pharmacology, University of Vermont, Health Science Research Facility 330, Burlington, VT 05405-0068

Adam Douglass (3: 129)   Department of Cellular and Molecular Pharmacology, The University of California, San Francisco, School of Medicine, Medical Sciences Building, Room S1210, 513 Parnassus Avenue, San Francisco, CA 94143-0450

Kate Downes (3: 463)   Genotyping / Chr 20, The Wellcome Trust, Sanger Institute, The Wellcome Trust Genome Campus, Hinxton, Cambridge, CB10 1SA, UNITED KINGDOM

Harry W. Duckworth (4: 457)   Department of Chemistry, University of Manitoba, Room 531 Parker Building, Winnipeg, MB, R3T 2N2, CANADA

Derek M. Dykxhoorn (3: 511)   CBR Institute for Biomedical Research, Harvard Medical School, 200 Longwood Ave, Boston, MA 02115

Lars Dyrskjøt (**4**: 83)   Clinical Biochemical Department, Molecular Diagnostic Laboratory, Aarhus University Hospital, Skejby, Brendstrupgaardvej, Aarhus N, DK-8200, DENMARK

Christoph Eckerskorn (**4**: 157)   Protein Analytics, Max Planck Institute for Biochemistry, Klopferspitz 18, Martinsried, D-82152, GERMANY

Glenn S. Edwards (**3**: 87)   Department of Physics, Duke University, 221 FEL Bldg, Box 90305, Durham, NC 27708-0305

Andreas A. O. Eggert (**1**: 103)   Department of Dermatology, Julius-Maximillians University, Josef-Schneider-Str. 2, Würzburg, 97080, GERMANY

Maria Ekström (**4**: 63)   Bioscience at Novum, Karolinska Institutet, Huddinge, SE-141 57, SWEDEN

Andreas Engel (**3**: 317)   Maurice E. Müller Institute for Microscopy at the Biozentrum, University of Basel, Klingelbergstrasse 70, Basel, CH-4056, SWITZERLAND

Anne-Marie Engel (**1**: 353)   Bartholin Institutte, Bartholinsgade 2, Copenhagen K, DK-1356, DENMARK

José A. Enríquez (**2**: 69)   Department of Biochemistry and Molecular and Cellular Biology, Universidad de Zaragoza, Miguel Servet, 177, Zaragoza, E-50013, SPAIN

Rachel Errington (**1**: 305)   Department of Medical Biochemistry and Immunology, University of Wales College of Medicine, Heath Park, Cardiff, CF14 4XN, UNITED KINGDOM

Virginia Espina (**3**: 339)   Microdissection Core Facility, Laboratory of Pathology, National Cancer Institute, 9000 Rockville Pike, Building 10, Room B1B53, Bethesda, MD 20892

H. Dariush Fahimi (**2**: 63)   Department of Anatomy and Cell Biology II, University of Heidelberg, Im Neuenheimer Feld 307, Heidelberg, D-69120, GERMANY

Federico Federici (**3**: 201)   Department of Physics, University of Genoa, Via Dodecaneso 33, Genoa, I-16146, ITALY

Daniel L. Feeback (**2**: 233, **4**: 5)   Space and Life Sciences Directorate, NASA-Johnson Space Center, 3600 Bay Area Blvd, Houston, TX 77058

Patricio Fernández-Silva (**2**: 69)   Dept of Biochemistry and Molecular and Cellular Biology, Univeridad de Zaragoza, Miguel Servet 177, Zaragoza, E-50013, SPAIN

Erika Fernández-Vizarra (**2**: 69)   Dept of Biochemistry and Molecular and Cellulary Biology, Universidad de Zaragoza, Miguel Servet, 177, Zaragoza, E-50013, SPAIN

Patrick F. Finn (**4**: 345)   Department of Physiology, Tufts University School of Medicine, 136 Harrison Ave, Boston, MA 02111

Kevin L. Firth (**2**: 329, **2**: 341)   ASK Science Products Inc., 487 Victoria St, Kingston, Ontario, K7L 3Z8, CANADA

Raluca Flükiger-Gagescu (**2**: 27)   Unitec— Office of Technology Transfer, University of Geneva and University of Geneva Hospitals, 24, Rue Général-Dufour, Geneva 4, CH-1211, SWITZERLAND

Leonard J. Foster (**4**: 363, 427)   Protein Interaction Laboratory, University of Southern Denmark, Odense, Campusvej 55, Odense M, DK-5230, DENMARK

Dimitrios Fotiadis (**3**: 317)   M. E. Müller Institute for Microscopy at the Biozentrum, University of Basel, Klingelbergstrasse 70, Basel, CH-4056, SWITZERLAND

Patrick L. T. M. Frederix (**3**: 317)   M. E. Müller Institute for Microscopy at the Biozentrum, University of Basel, Klingelbergstrasse 70, Basel, CH-4056, SWITZERLAND

Marcus Frohme (**4**: 113)   Functional Genome Analysis, German Cancer Research Center, Deutsches Krebsforschungszentrum, Im Neuenheimer Feld 580, Heidelberg, D-69120, GERMANY

Masanori Fujimoto (**4**: 197)   Department of Biochemistry and Biomolecular Recognition, Yamaguchi University School of Medicine, 1-1-1, Minami-kogushi, Ube, Yamaguchi, 755-8505, JAPAN

Margarida Gama-Carvalho (**2**: 277)   Faculty of Medicine, Institute of Molecular Medicine, University of Lisbon, AV. Prof. Egas Moniz, Lisbon, 1649-028, PORTUGAL

Henrik Garoff (**1**: 419, **4**: 63)   Unit for Cell Biology, Center for Biotechnology. Karolinska Institute, Huddinge, SE-141 57, SWEDEN

Susan M. Gasser (**2**: 359)   Friedrich Miescher Institute fr Biomedical Research, Maulbeerstrasse 66, Basel, CH-1211, SWITZERLAND

Kristine G. Gaustad (**1**: 207)   Institute of Medical Biochemistry, University of Oslo, PO Box 1112 Blindern, Oslo, 0317, NORWAY

Benjamin Geiger (**2**: 419)   Dept. of Molecular Cell Biology, Weizman Institute of Science, Wolfson Building, Rm 617, Rehovot, 76100, ISRAEL

Kris Gevaert (**4**: 379, **4**: 457)   Dept. Medical Protein Research, Flanders Interuniversity Institute for Biotechnology, Faculty of Medicine and Health Sciences, Ghent University, Instituut Rommelaere—Blok D, Albert Baertsoenkaai 3, Gent, B-9000, BELGIUM

Jilur Ghori (**3**: 463)   Genotyping / Chr 20, The Wellcome Trust, Sanger Institute, The Wellcome Trust, Genome Campus, Hinxton, Cambridge, CB10 1SA, UNITED KINGDOM

Alasdair J. Gibb (**1**: 395)   Department of Pharmacology, University College London, Gower Street, London, WC1E GBT, UNITED KINGDOM

Mario Gimona (**1**: 557, **2**: 161, **4**: 145)   Department of Cell Biology, Institute of Molecular Biology, Austrian Academy of Sciences, Billrothstrasse 11, Salzburg, A-5020, AUSTRIA

David A. Glesne (**1**: 165)   Biosciences Division, Argonne National Laboratory, 9700 South Cass Avenue, Argonne, IL 60439-4844

Martin Goldberg (**3**: 325)   Science Laboratories, University of Durham, South Road, Durham, DH1 3LE, UNITED KINGDOM

Kenneth N. Goldie (**3**: 267)   Structural and Computational Biology Programme, EMBL, Meyerhofstrasse 1, Heidelberg, D-69117, GERMANY

Jon W. Gordon (**3**: 487)   Geriatrics and Adult Development, Mount Sinai School of Medicine, One Gustave L. Levy Place, New York, NY 10029

Angelika Görg (**4**: 175)   Fachgebiet Proteomik, Technische Universität München, Am Forum 2, Freising Weihenstephan, D-85350, GERMANY

Martin Gotthardt (**4**: 149)   Department of Nuclear Medicine, Philipp's-University of Marburg, Baldingerstraße, Marburg/Lahn, D-35043, GERMANY

Frank L. Graham (**1**: 435)   Department of Biology, McMaster University, Life Sciences Building, Room 430, Hamilton, Ontario, L8S 4K1, CANADA

Claude Granier (**1**: 519)   UMR 5160, Faculté de Pharmacie, 15 Av. Charles Flahault, Montpellier Cedex 5, BP 14491, 34093, FRANCE

Lloyd A. Greene (**1**: 171)   Department of Pathology and Center for Neurobiology and Behavior, Columbia University, College of Physicians and Surgeons, 630 W. 168th Street, New York, NY 10032

Susan M. Gribble (**3**: 403)   Sanger Institute, The Wellcome Trust, The Wellcome Trust Genome Campus, Hinxton, Cambridge, CB10 1SA, UNITED KINGDOM

Gareth Griffiths (**2**: 57, **3**: 299)   Department of Cell Biology, EMBL, Postfach 102209, Heidelberg, D-69117, GERMANY

Sergio Grinstein (**3**: 163)   Cell Biology Program, Hospital for Sick Children, 555 University Avenue, Toronto, Ontario, M5G 1X8, CANADA

Pavel Gromov (**1**: 527, **4**: 69, 165, 243, 289)   Institute of Cancer Biology and Danish Centre for Translational Breast Cancer Research, Danish Cancer Society, Strandboulevarden 49, Copenhagen, DK-2100, DENMARK

Irina Gromova (**4**: 219)   Department of Medical Biochemistry and Danish Centre for Translational Breast Cancer Research, Danish Cancer Society, Strandboulevarden 49, Copenhagen, DK-2100, DENMARK

Dale F. Gruber (**1**: 33)   Cell Culture Research and Development, GIBCO/Invitrogen Corporation, 3175 Staley Road, Grand Island, NY 14072

Markus Grubinger (**4**: 145)   Institute of Physics and Biophysics, University of Salzburg, Hellbrunnerstr. 34, Salzburg, A-5020, AUSTRIA

Jean Gruenberg (**2**: 27, 201)   Department of Biochemistry, University of Geneva, 30, quai Ernest Ansermet, Geneva 4, CH-1211, SWITZERLAND

Stephanie L. Gupton (**3**: 137)   10550 North Torrey Pines Road, CB 163, La Jolla, CA 92037

Cemal Gurkan (**2**: 209)   Department of Cell and Molecular Biology, The Scripps Research Institute, 10550 North Torrey Pines Road, La Jolla, CA 92037

Martin Guttenberger (4: 131)   Zentrum für Molekulariologie der Pflanzen, Universitat Tübingen, Entwicklungsgenetik, Auf der Morgenstelle 3, Tübingen, D-72076, GERMANY

Thomas Haaf (3: 409)   Institute for Human Genetics, Johannes Gutenberg-Universität Mainz, 55101, Mainz, D-55131, GERMANY

Christine M. Hager-Braun (1: 511)   Health and Human Services, NIH National Institute of Environmental Health Sciences, MD F0-04, PO Box 12233, Research Triangle Park, NC 27709

Anne-Mari Håkelien (1: 207)   Institute of Medical Biochemistry, Institute of Medical Biochemistry, University of Oslo, PO Box 1112 Blindern, Oslo, 0317, NORWAY

Fiona C. Halliday (1: 395)   GlaxoSmithKline, Greenford, Middlesex, UB6 OHE, UNITED KINGDOM

Gerald Hammond (2: 223)   Molecular Neuropathobiology Laboratory, Cancer Reasearch UK London Research Institute, 44 Lincoln's Inn Fields, London, WC2A 3PX, UNITED KINGDOM

Klaus Hansen (4: 253)

Hironobu Harada (1: 367)   Department of Neurosurgery, Ehime University School of Medicine, Shitsukawa, Toon-shi, Ehime, 791-0295, JAPAN

Robert J. Hay (1: 43, 49, 573)   Viitro Enterprises Incorporated, 1113 Marsh Road, PO Box 328, Bealeton, VA 22712

Izumi Hayashi (1: 151)   National Medical Center and Beckman Research Institute, Division of Neurosciences, City of Hope, 1500 E. Duarte Rd, Duarte, CA 91010-3000

Timothy A. Haystead (4: 265)   Department of Pharmacology and Cancer Biology, Duke University Medical Center, Box 3813 Med Ctr, Durham, NC 27710

Rebecca Heald (2: 379)   Molecular and Cell Biology Department, University of California, Berkeley, Berkeley, CA 94720-3200

Florence Hediger (2: 359)   Department of Molecular Biology, University of Geneva, 30, Quai Ernest Ansermet, Geneva, CH-1211, SWITZERLAND

Rainer Heintzmann (3: 29)   Randall Division of Cell and Molecular Biophysics, King's College London, Guy's Campus, London, SE1 1UL, UNITED KINGDOM

Frederic Heitz (4: 13)   Centre de Recherche en Biochimie Macromoléculaire (UPR 1086), Centre National de la Recherche Scientifique (CNRS), 1919 Route de Mende, Montpellier Cedex 5, F-34293, FRANCE

Johannes W. Hell (2: 85)   Department of Pharmacology, University of Iowa, 2152 Bowen Science Building, Iowa City, IA 52242

Kai Hell (4: 269)   Adolf-Butenandt-Institut fur Physiologische Chemie, Lehrstuhl: Physiologische Chemie, Universitat Munchen, Butenandtstr. 5, Gebäude B, Munchen, D-81377, GERMANY

Robert R. Henry (1: 121)   Veterans Affairs Medical Center, San Diego (V111G), 3350 La Jolla Village Drive, San Diego, CA 92161

Johan Hiding (2: 45)   Göteborg University, Institute of Medical Biochemistry, PO Box 440, Göteborg, SE-403-50, SWEDEN

Yasushi Hiraoka (3: 171)   Structural Biology Section and CREST Research Project, Kansai Advanced Research Center, Communications Research Laboratory, 588-2 Iwaoka, Iwaoka-cho, Nishi-ku, Kobe, 651-2492, JAPAN

Mary M. Hitt (1: 435)   Department of Pathology & Molecular Medicine, McMaster University, 1200 Main Street West, Hamilton, Ontario, L8N 3Z5, CANADA

Julie Hodgkinson (3: 307)   School of Crystallography, Birkbeck College, Unversity of London, Malet Street, London, WC1E 7HX, UNITED KINGDOM

Klaus P. Hoeflich (2: 307)   Division of Molecular and Structural Biology, Ontario Cancer Institute, Department of Medical Biophysics, University of Toronto, 610 University Avenue, 7-707A, Toronto, Ontario, M5G 2M9, CANADA

Tracy L. Hoffman (1: 21)   ATCC, P.O. Box 1549, Manassas, VA 20108

Jörg D. Hoheisel (4: 113)   Functional Genome Analysis, German Cancer Research Center, Deutsches Krebsforschungszentrum, Im Neuenheimer Feld 580, Heidelberg, D-69120, GERMANY

Thomas Hollemann (1: 191)   Institute for Biochemistry and Molecular Cell Biology, University of Goettingen, Justus-von-Liebig-Weg 11, Göttingen, D-37077, GERMANY

Caterina Holz (4: 57)   PSF biotech AG, Huebnerweg 6, Berlin, D-14059, GERMANY

Akira Honda (**2**: 299)   Department of Pharmacology, University of Vermont, Health Science Research Facility 330, Burlington, VT 05405-0068

Masanori Honsho (**2**: 5)   Max Planck Institute of Molecular Cell Biology and Genetics, Pfotenhauerstrasse 108, Dresden, D-01307, GERMANY

Andrew N. Hoofnagle (**4**: 443)   School of Medicine, University of Colorado Health Sciences Center, Denver, CO 80262

Eliezer Huberman (**1**: 165)   Gene Expression and Function Group, Argonne National Laboratory, 9700 South Cass Avenue, Argonne, IL 60439-4844

M. Shane Hutson (**3**: 87)   Department of Physics, Duke University, 107 Physics Bldg, Durham, NC 27708-1000

Andreas Hüttmann (**1**: 115)   Abteilung für Hämatologie, Universitäkrankenhaus Essn, Hufelandstr. 55, Essen, 45122, GERMANY

Anthony A. Hyman (**2**: 155)   Max Planck Institute of Molecular Cell Biology and Gene Technology, Pfotenhauerstrasse 108, Dresden, D-01307, GERMANY

Sherrif F. Ibrahim (**1**: 269)   Institute for Systems Biology, 1441 N. 34th St, Seattle, WA 98103

Kazuo Ikeda (**1**: 151)   National Medical Center and Beckman Research Institute, Division of Neurosciences, City of Hope, 1500 East Duarte Road, Duarte, CA 91010-3000

Elina Ikonen (**2**: 181)   The LIPID Cell Biology Group, Department of Biochemistry, The Finnish National Public Health Institute, Mannerheimintie 166, Helsinki, FIN-00300, FINLAND

Pranvera Ikonomi (**1**: 49)   Director, Cell Biology, American Type Culture Collection (ATCC), 10801 University Blvd., Manassas, VA 20110-2209

Mitsuhiko Ikura (**2**: 307)   Division of Molecular and Structural Biology, Ontario Cancer Institute, Department of Medical Biophysics, University of Toronto, 610 University Avenue 7-707A, Toronto, Ontario, M5G 2M9, CANADA

Arup Kumar Indra (**3**: 501)   Institut de Génétique et de Biologie Moléculaire et Cellulaire (IGBMC), 1 rue Laurent Fries, B.P.10142, Illkirch CEDEX, F-67404, FRANCE

Takayoshi Inoue (**4**: 35)   National Institute for Neuroscience, 4-1-1 Ogawahigashi, Kodaira, Tokyo, 187-8502, JAPAN

Kumiko Ishii (**2**: 139)   Supra-Biomolecular System Research Group, RIKEN (Institute of Physical and Chemical Research), 2-1, Hirosawa, Wako-shi, Saitama, 351-0198, JAPAN

Dean A. Jackson (**2**: 121)   Department of Biomolecular Sciences, UMIST, PO Box 88, Manchester, M60 1QD, UNITED KINGDOM

Reinhard Jahn (**2**: 85)   Department of Neurobiology, Max-Planck-Institut for Biophysical Chemistry, Am Faßberg 11, Gottingen, D-37077, GERMANY

Kim D. Janda (**1**: 491)   Department of Chemistry, BCC-582, The Scripps Research Institute, 10550 N. Torrey Pines Road, La Jolla, CA 92037

Harry W. Jarrett (**4**: 335)   Department of Biochemistry, University of Tennessee Health Sciences Center, Memphis, TN 38163

Daniel G. Jay (**4**: 307)   Dept. Physiology, Tufts University School of Medicine, 136 Harrison Avenue, Boston, MA 02111

David W. Jayme (**1**: 33)   Cell Culture Research and Development, GIBCO/Invitrogen Corporation, 3175 Staley Road, Grand Island, NY 14072

Ole Nørregaard Jensen (**4**: 409)   Protein Research Group, Department of Biochemistry and Molecular Biology, University of Southern Denmark, Campusvej 55, Odense M, DK-5230, DENMARK

Jae Hyun Jeong (**4**: 29)   Department of Chemical & Biomolecular Engineering, Center for Ultramicrochemical Process Systems, Korea Advanced Institute of Science and Technology, Daejeon, 305-701, SOUTH KOREA

Jeff A. Jones (**2**: 233)   Space and Life Sciences Directorate, NASA-Johnson Space Center, TX 77058

Gloria Juan (**1**: 279)   Research Pathology Division, Room S-830, Memorial Sloan-Kettering Cancer Center, 1275 York Avenue, New York, NY 10021

Melissa S. Jurica (**2**: 109)   Molecular, Cell & Developmental Biology, Center for Molecular Biology of RNA, UC Santa Cruz, 1156 High Street, Santa Cruz, CA 95064

Eckhart Kämpgen (1: 103) Department of Dermatology, Friedrich Alexander University, Hartmannstr. 14, Erlangen, D-91052, GERMANY

Roger Karlsson (2: 165) Department of Cell Biology, The Wenner-Gren Institute, Stockholm University, Stockholm, S-10691, SWEDEN

Fredrik Kartberg (2: 45) Göteborg University, Institute of Medical Biochemistry, PO Box 440, Gothenburg, SE, 403-50, SWEDEN

Irina N. Kaverina (3: 111) Institute of Molecular Biotechnology, Austrian Academy of Sciences, Dr. Bohrgasse 3-5, Vienna, A-1030, AUSTRIA

Ralph H. Kehlenbach (2: 267) Hygiene-Institut-Abteilung Virologie, Universitat Heidelberg, Im Neuenheimer Feld 324, Heidelberg, D-69120, GERMANY

Daniel P. Kiehart (3: 87) Department of Biology, Duke University, B330g Levine Sci Bldg, Box 91000, Durham, NC 27708-1000

Katherine E. Kilpatrick (1: 483) Senior Research Investigator, TriPath Oncology, 4025 Stirrup Creek Drive, Suite 400, Durham, NC 27703

Jong-Duk Kim (4: 29) Department of Chemical & Biomolecular Engineering, Center for Ultramicrochemical Process Systems, Korea Advanced Institute of Science and Technology, Daejeon, 305-701, SOUTH KOREA

Maurice Kléber (1: 69) Institute of Cell Biology, Department of Biology, Swiss Federal Institute of Technology, ETH—Hönggerberg, Zurich, CH-8093, SWITZERLAND

Toshihide Kobayashi (2: 139) Supra-Biomolecular System Research Group, RIKEN (Institute of Physical and Chemical Research) Frontier Research System, 2-1, Hirosawa, Wako-shi, Saitama, 351-0198, JAPAN

Stefan Kochanek (1: 445) Division of Gene Therapy, University of Ulm, Helmholtz Str. 8/I, Ulm, D-89081, GERMANY

Anna Koffer (2: 223) Physiology Department, University College London, 21 University Street, London, WC1E 6JJ, UNITED KINGDOM

Antonius Koller (4: 383) Department of Cell Biology, Torrey Mesa Research Institute, 3115 Merryfield Row, San Diego, CA 92121

Erich Koller (3: 523) Functional Genomics, GeneTrove, Isis Pharmaceuticals, Inc., 2292 Faraday Ave., Carlsbad, CA 92008

Robert L. Kortum (1: 215) The Eppley Institute for Research in Cancer, The University of Nebraska Medical Center, 986805 Nebraska Medical Center, Omaha, NE 68198-6805

Irina Kratchmarova (4: 427) Protein Interaction Laboratory, University of Southern Denmark—Odense, Campusvej 55, Odense M, DK-5230, DENMARK

Geri E. Kreitzer (2: 189) Cell and Developmental Biology, Weill Medical College of Cornell University, LC-300, New York, NY 10021

Florian Kreppel (1: 445) Division of Gene Therapy, University of Ulm, Helmholtz Str. 8/I, Ulm, D-89081, GERMANY

Mogens Kruhøffer (4: 83) Molecular Diagnostic Laboratory, Clinical Biochemical Department, Aarhus University Hospital, Skejby, Brendstrupgaardvej, Aarhus N, DK-8200, DENMARK

Robb Krumlauf (4: 35) Stowers Institute for Medical Research, 1000 East 50th Street, Kansas City, MO 64110

Michael Kühl (1: 191) Development Biochemistry, University of Ulm, Albert-Einstein-Allee 11, Ulm, D-89081, GERMANY

Mark Kühnel (2: 57) Department of Cell Biology, EMBL, Postfach 102209, Heidelberg, D-69117, GERMANY

Anuj Kumar (3: 179) Dept. of Molecular, Cellular, and Developmental Biology and Life Sciences Institute, University of Michigan, 210 Washtenaw Avenue, Ann Arbor, MI 48109-2216

Thomas Küntziger (1: 207) Institute of Medical Biochemistry, Institute of Medical Biochemistry, University of Oslo, PO Box 1112 Blindern, Oslo, 0317, NORWAY

Yasuhiro Kuramitsu (4: 197) Department of Biochemistry and Biomolecular Recognition, Yamaguchi University School of Medicine, 1-1-1 Minami-kogushi, Ube, Yamaguchi, 755-8505, JAPAN

Sergei A. Kuznetsov (1: 79) Craniofacial and Skeletal Disease Branch, NIDCR, NIH, Department of Health and Human Services, 30 Convent Drive MSC 4320, Bethesda, MD 20892

Joshua Labaer (4: 73)   Harvard Institute of Proteomics, 320 Charles Street, Boston, MA 02141-2023

Frank Lafont (2: 181)   Department of Biochemistry, University of Geneva, 30, quai Ernest-Ansermet 1211, Geneva 4, CH-1211, SWITZERLAND

Yun Wah Lam (2: 103, 115)   Wellcome Trust Biocentre, MSI/WTB Complex, University of Dundee, Dow Street, Dundee, DD1 5EH, UNITED KINGDOM

Angus I. Lamond (2: 103, 115)   Wellcome Trust Biocentre, MSI/WTB Complex, University of Dundee, Dow Street, Dundee, DD1 5EH, UNITED KINGDOM

Lukas Landmann (3: 5)   Institute for Anatomy (LL), Anatomisches Institut, University of Basel, Pestalozzistrasse 20, Basel, CH-4056, SWITZERLAND

Helga B. Landsverk (1: 207)   Institute of Medical Biochemistry, Institute of Medical Biochemistry, University of Oslo, PO Box 1112 Blindern, Oslo, 0317, NORWAY

Christine Lang (4: 57)   Department of Microbiology and Genetics, Berlin University of Technology, Gustav-Meyer-Allee 25, Berlin, D-13355, GERMANY

Paul LaPointe (2: 209)   Department of Cell and Molecular Biology, The Scripps Research Institute, 10550 North Torrey Pines Road, La Jolla, CA 92037

Martin R. Larsen (4: 371)   Department of Biochemistry and Molecular Biology, University of Southern Denmark, Campusvej 55, Odense M, DK-5230, DENMARK

Pamela L. Larsen (1: 157)   Department of Cellular and Structural Biology, University of Texas Health Science Center at San Antonio, San Antonio, TX 78229-3900

Eugene Ngo-Lung Lau (1: 115)   Leukaemia Foundation of Queensland Leukaemia Research Laboratories, Queensland Institute of Medical Research, Royal Brisbane Hospital Post Office, Brisbane, Queensland, Q4029, AUSTRALIA

Sabrina Laugesen (4: 371)   Department of Biochemistry and Molecular Biology, University of Southern Denmark, Campusvej 55, Odense M, DK-5230, DENMARK

Daniel Laune (1: 519)   Centre de Pharmacologie et Biotechnologie pour la Santé, CNRS UMR 5160, Faculté de Pharmacie, Avenue Charles Flahault, Montpellier Cedex 5, F-34093, FRANCE

Andre Le Bivic (2: 241)   Groupe Morphogenese et Compartimentation Membranaire, UMR 6156, IBDM, Faculte des Sciences de Luminy, case 907, Marseille cedex 09, F-13288, FRANCE

Ronald Lebofsky (3: 429)   Laboratoire de Biophysique de l'ADN, Departement des Biotechnologies, Institut Pasteur, 25 rue du Dr. Roux, Paris Cedex 15, F-75724, FRANCE

Chuan-PU Lee (2: 259)   The Department of Biochemistry and Molecular Biology, Wayne State University School of Medicine, 4374 Scott Hall, 540 E. Canfield, Detroit, MI 48201

Eva Lee (1: 139)   Life Sciences Division, Lawrence Berkeley National Laboratory, 1 Cyclotron Road, Bldg 83-101, Berkeley, CA 94720

Joon-Hee Lee (4: 45)   School of Biosciences, University of Nottingham, Sutton Bonington, Loughborough, Leics, LE12 5RD, UNITED KINGDOM

Kwangmoon Lee (1: 215)   The Eppley Institute for Research in Cancer, The University of Nebraska Medical Center, 986805 Nebraska Medical Center, Omaha, NE 68198-6805

Thomas Lee (4: 443)   Dept of Chemistry and Biochemistry, Univ of Colorado, 215 UCB, Boulder, CO 80309-0215

Margaret Leversha (3: 395)   Memorial Sloan Kettering Cancer Center, 1275 York Avenue, New York, NY 10021

Jeffrey M. Levsky (4: 121)   Department of Anatomy and Structural Biology, Golding # 601, Albert Einstein College of Medicine of Yeshiva University, 1300 Morris Park Avenue, Bronx, NY 10461

Alexandre Lewalle (3: 37)   Randall Centre, New Hunt's House, Guy's Campus, London, SE1 1UL, UNITED KINGDOM

Chung Leung Li (1: 115)   Experimental Haematology Laboratory, Stem Cell Program, Institute of Zoology/Genomics Research Center, Academia Sinica, Nankang 115, Nankang, Taipei, 11529, R.O.C.

LiQiong Li (**3**: 345) Department of Pathology and Laboratory Medicine, University of Rochester School of Medicine, 601 Elmwood Ave., Rm 1-6337, Rochester, NY 14642

Mei Li (**3**: 501) Institut de Génétique et de Biologie Moléculaire et Cellulaire (IGBMC), 1 rue Laurent Fries, B.P.10142, Illkirch CEDEX, F-67404, FRANCE

Siming Li (**4**: 295) Dana Faber Cancer Institute, Harvard University, 44 Binney Street, Boston, MA 02115

Lih-huei Liaw (**3**: 351) Beckman Laser Institute, University of California, Irvine, 1002 Health Sciences Road E, Irvine, CA 92697-1475

Antonietta M. Lillo (**1**: 491) Department of Chemistry, BCC-582, The Scripps Research Institute, 10550 N. Torrey Pines Road, La Jolla, CA 92037

Uno Lindberg (**2**: 165) Department of Cell Biology, Stockholm University, The Wenner-Gren Institute, Stockhólm, S-10691, SWEDEN

Christian Linden (**1**: 103) Department of Virology, Julius-Maximillins University, Versbacher Str. 7, Würzburg, D-97080, GERMANY

Robert Lindner (**2**: 51) Department of Cell Biology in the Center of Anatomy, Hannover Medical School, Hannover, D-30625, GERMANY

Lance A. Liotta (**3**: 339) Chief, Laboratory of Pathology, National Cancer Institute Building 10, Room 2A33, 9000 Rockville Pike, Bethesda, MD 20892

Adam J. Liska (**4**: 399) Max Planck Institute of Molecular Cell Biology and Genetics, Pfotenhauerst 108, Dresden, D-01307, GERMANY

Hong Liu (**1**: 139) Life Sciences Division, Lawrence Berkeley National Laboratory, 1 Cyclotron Road, Bldg 83-101, Berkeley, CA 94720

Silvia Lommel (**2**: 399) Department of Cell Biology, German Research Center for Biotechnology (GBF), Mascheroder Weg 1, Braunschweig, D-38124, GERMANY

Giuseppe S. A. Longo-Sorbello (**1**: 315) Centro di Riferimento Oncologico, Ospedale "S. Vincenzo", Taormina, Contradra Sirinam, 08903, ITALY

Lovisa Lovmar (**3**: 455) Department of Medical Sciences, Uppsala University, Akademiska sjukhuset, Uppsala, SE-75185, SWEDEN

Eugene Lukanidin (**1**: 363) Department of Molecular Cancer Biology, Institute of Cancer Biology, Danish Cancer Society, Strandboulevarden 49, Copenhagen, DK-2100, DENMARK

Jiri Lukas (**4**: 253) Department of Cell Cycle and Cancer, Danish Cancer Society, Strandboulevarden 49, Copenhagen, DK-2100, DENMARK

Peter J. Macardle (**1**: 475) Department of Immunology, Allergy and Artritis, Flinders Medical Centre and Flinders University, Bedford Park, Adelaie, SA, 5051, SOUTH AUSTRALIA

Peder S. Madsen (**4**: 69) Institute of Medical Biochemistry, University of Aarhus, Ole Worms Alle, Building 170, Aarhus C, DK-8000, DENMARK

Nils E. Magnusson (**4**: 83) Clinical Biochemical Department, Molecular Diagnostic Laboratory, Aarhus University Hospital, Skejby, Brendstrupgaardvej, Aarhus N, DK-8200, DENMARK

Asami Makino (**2**: 139) Supra-Biomolecular System Research Group, RIKEN (Institute of Physical and Chemical Research) Frontier Research System, 2-1, Hirosawa, Wako-shi, Saitama, 351-0198, JAPAN

G. Mike Makrigiorgos (**3**: 477) Department of Radiation Oncology, Dana Farber-Brigham and Women's Cancer Center, 75 Francis Street, Level L2, Boston, MA 02215

Matthias Mann (**4**: 363, 427) Protein Interaction Laboratory, University of Southern Denmark, Odense, Campusvej 55, Odense M, DK-5230, DENMARK

Edward Manser (**4**: 285) Glaxo-IMCB Group, Institute of Molecular and Cell Biology, Singapore, 117609, SINGAPORE

Ahmed Mansouri (**3**: 491) Department of Molecular Cell Biology, Max-Planck-Institute of Biophysical Chemistry, Am Fassberg 11, Göttingen, D-37077, GERMANY

Alan D. Marmorstein (**2**: 241) Cole Eye Institute, Weill Medical College of Cornell Cleveland Clinic, 9500 Euclid Avenue, i31, Cleveland, OH 44195

Bruno Martoglio (**2**: 215) Institute of Biochemistry, ETH Zentrum, Building CHN, Room L32.3, Zurich, CH-8092, SWITZERLAND

Susanne E. Mason (**1**: 407) Department of Physiology, University of Maryland School of Medicine, 655 W. Baltimore St., Baltimore, MD 21201

Stephen J. Mather (**1**: 539) Dept of Nuclear Medicine, St Bartholomews Hospital, London, EC1A 7BE, UNITED KINGDOM

Arvid B. Maunsbach (3: 221, 289)   Department of Cell Biology, Institute of Anatomy, Aarhus University, Aarhus, DK-8000, DENMARK

William Hayes McDonald (4: 391)   Department of Cell Biology, The Scripps Research Institute, 10550 North Torrey Pines Rd, La Jolla, CA 92370

Kathleen M. McKenzie (1: 491)   Department of Chemistry, BCC-582, The Scripps Research Institute, 10550 N. Torrey Pines Road, La Jolla, CA 92037

Alexander D. McLellan (1: 103)   Department of Microbiology & Immunology, University of Otago, PO Box 56, 720 Cumberland St, Dunedin, NEW ZEALAND

Scott W. McPhee (1: 457)   Department of Surgery, University of Medicine and Dentistry of New Jersey, Camden, NJ 08103

Jill Meisenhelder (4: 139)   Molecular and Cell Biology Laboratory, The Salk Institute, 10010 North Torrey Pines Road, La Jolla, CA 92037

Paula Meleady (1: 13)   National Institute for Cellular Biotechnology, Dublin City University, Glasnevin, Dublin, 9, IRELAND

Nicholas T. Mesires (4: 345)   Department of Physiology, Tufts University School of Medicine, 136 Harrison Ave, Boston, MA 02111

Daniel Metzger (3: 501)   Institut de Génétique et de Biologie Moléculaire et Cellulaire (IGBMC), Institut Clinique de la Souris (ICS), 1 rue Laurent Fries, B.P.10142, Illkirch CEDEX, F-67404, FRANCE

Martina Mirlacher (3: 369)   Division of Molecular Pathology, Institue of Pathology, University of Basel, Schonbeinstrasse 40, Basel, CH-4031, SWITZERLAND

Suchareeta Mitra (4: 335)   Department of Biochemistry, University of Tennessee Health Sciences Center, Memphis, TN 38163

Atsushi Miyawaki (2: 317)   Laboratory for Cell Function and Dynamics, Advanced Technology Center, Brain Science Institute, Institute of Physical and Chemical Research (RIKEN), 2-1 Horosawa, Wako, Saitama, 351-0198, JAPAN

Dejana Mokranjac (4: 269)   Adolf-Butenandt-Institut fur Physiologische Chemie, Lehrstuhl: Physiologische Chemie, Universitat Munchen, Butenandtstr. 5, Gebäude B, Munchen, D-81377, GERMANY

Peter L. Molloy (4: 325)   CSIRO Molecular Science, PO Box 184, North Ryde, NSW, 1670, AUSTRALIA

Richard A. Moravec (1: 25)   Promega Corporation, 2800 Woods Hollow Road, Madison, WI 53711-5399

José M. A. Moreira (1: 527)   Institute of Cancer Biology and Danish Centre for Translational Breast Cancer Research, Danish Cancer Society, Strandboulevarden 49, Copenhagen O, DK-2100, DENMARK

May C. Morris (4: 13)   Centre de Recherche en Biochimie Macromoléculaire (UPR 1086), Centre National de la Recherche Scientifique (CNRS), 1919 Route de Mende, Montpellier Cedex 5, F-34293, FRANCE

Robert A. Moxley (4: 335)   Department of Biochemistry, University of Tennessee Health Sciences Center, Memphis, TN 38163

Anne Muesch (2: 189)   Margaret M. Dyson Vision Research Institute, Department of Opthalmology, Weill Medical College of Cornell University, New York, NY 10021

Peggy Müller (1: 325)   Zentrum für Angewandte Medizinische und Humanbiologische Forschung, Labor für Molekulare Hepatologie der Universitätsklinik und Polikinik für Innere Medizin I, Martin Luther University Halle-Wittenburg, Heinrich-Damerow-Street 1, Saale, Halle, D-06097, GERMANY

Steve Murray (3: 325)   CRC Structural Cell Biology Group, Paterson Institute for Cancer Research, Christie Hospital NHS Trust, Wilmslow Road, Withington, Manchester, M20 4BX, UNITED KINGDOM

Connie Myers (1: 139)   Life Sciences Division, Lawrence Berkeley National Laboratory, 1 Cyclotron Road, Bldg 83-101, Berkeley, CA 94720

Kazuyuki Nakamura (4: 197)   Department of Biochemistry and Biomolecular Recognition, Yamaguchi University School of Medicine, 1-1-1 Minami-kogushi, Ube, Yamaguchi, 755-8505, JAPAN

Maithreyi Narasimha (3: 77)   Wellcome Trust/Cancer Research UK Institute and Dept of Anatomyaraki, University of Cambridge, Tennis Court Road, Cambridge, CB2 1QR, UNITED KINGDOM

Dobrin Nedelkov (4: 279)   Intrinsic Bioprobes, Inc., 625 S. Smith Road, Suite 22, Tempe, AZ 85281

Randall W. Nelson (4: 279)   Intrinsic Bioprobes Inc., 625 S. Smith Road, Suite 22, Tempe, AZ 85281

Frank R. Neumann (2: 359)   Department of Molecular Biology, University of Geneva, 30, Quai Ernest Ansermet, Geneva, CH-1211, SWITZERLAND

Walter Neupert (4: 269)   Adolf-Butenandt-Institut fur Physiologische Chemie, Lehrstuhl: Physiologische Chemie, Universitat Munchen, Butenandtstr. 5, Gebäude B, Munchen, D-81377, GERMANY

Axl Alois Neurauter (1: 239)   Immunosystem R & D, Dynal Biotech ASA, PO Box 114, Smestad, N-0309, NORWAY

Phillip Ng (1: 435)   Dept of Molecular and Human Genetics, Baylor College of Medicine, One Baylor Plaza, Houston, TX 77030

Garth L. Nicolson (1: 359)   The Institute for Molecular Medicine, 15162 Triton Lane, Huntington Beach, CA 92649-1041

Trine Nilsen (4: 275)   Department of Biochemistry, Institute for Cancer Research, The Norwegian Radium Hospital, Montebello, Oslo, N-0310, NORWAY

Tommy Nilsson (2: 45)   Göteborg University, Institute of Medical Biochemistry, PO Box, Göteborg, SE-403 50, SWEDEN

Lars Norderhaug (1: 239)   Dynal Biotech ASA, PO Box 114, Smestad, N-0309, NORWAY

Robert O'Connor (1: 5, 13, 335)   National Institute for Cellular Biotechnology, Dublin City University, Glasnevin, Dublin, 9, IRELAND

Lorraine O'Driscoll (1: 5, 335)   National Institute for Cellular Biotechnology, Dublin City University, Glasnevin, Dublin, 9, IRELAND

Martin Offterdinger (3: 153)   Cell Biology and Cell Biophysics Program, European Molecular Biology Laboratory, Meyerhofstrasse 1, Heidelberg, D-69117, GERMANY

Philip Oh (2: 11)   Cellular and Molecular Biology Program, Sidney Kimmel Cancer Center, 10835 Altman Row, San Diego, CA 92121

Takanori Ohnishi (1: 367)   Department of Neurosurgery, Ehime University School of Medicine, Shitsukawa, Toon-shi, Ehime, 791-0295, JAPAN

Sjur Olsnes (4: 19, 275)   Department of Biochemistry, The Norwegian Radium Hospital, Montebello, Oslo, 0310, NORWAY

Shao-En Ong (4: 427)   Protein Interaction Laboratory, University of Southern Denmark— Odense, Campusvej 55, Odense M, DK-5230, DENMARK

Akifumi Ootani (1: 411)   Department of Internal Medicine, Faculty of Medicine, Saga University, Nebeshima 5-1-1, Saga, 849-8501, JAPAN

Valerio Orlando (4: 317)   Dulbecco Telethon Institute, Institute of Genetics & Biophysics CNR, Via Pietro Castellino 111, Naples, I-80131, ITALY

Torben F. Ørntoft (4: 83)   Clinical Biochemical Department, Molecular Diagnostic Laboratory, Aarhus University Hospital, Skejby, Brendstrupgaardvej 100, Aarhus N, DK-8200, DENMARK

Mary Osborn (1: 549, 563)   Department of Biochemistry and Cell Biology, Max Planck Institute of Biophysical Chemistry, Am Faßberg 11, Gottingen, D-37077, GERMANY

Lawrence E. Ostrowski (2: 99)   Cystic Fibrosis/ Pulmonary Research and Treatment Centre, University of North Carolina at Chapel Hill, Thurston-Bowles Building, Chapel Hill, NC 27599-7248

Hendrik Otto (2: 253)   Institut für Biochemie und Molekularbiologie, Universität Freiburg, Hermann-Herder-Str. 7, Freiburg, D-79104, GERMANY

Kerstin Otto (1: 103)   Department of Dermatology, Julius-Maximillians University, Josef-Schneider-Str. 2, Würzburg, 97080, GERMANY

Michel M. Ouellette (1: 215)   Department of Biochemistry and Molecular Biology, Eppley Institute for Research in Cancer, The University of Nebraska Medical Center, 986805 Nebraska Medical Center, Omaha, NE 68198-6805

Jacques Paiement (2: 41)   Département de pathologie et biologie cellulaire, Université de Montréal, Case postale 6128, Succursale "Centre-Ville", Montreal, QC, H3C 3J7, CANADA

Patricia M. Palagi (4: 207)   Swiss Institute of Bioinformatics, CMU, 1 Michel Servet, Geneva 4, CH-1211, SWITZERLAND

Kinam Park (4: 29)   Department of Pharmaceutics and Biomedical Engineering, Purdue University School of Pharmacy, 575 Stadium Mall Drive, Room G22, West Lafayette, IN 47907-2091

Helen Parkinson (4: 95)   EMBL Outstation—Hinxton, European Bioinformatics Institute, Wellcome Trust Genome Campus, Hinxton, Cambridge, CB10 1SD, UNITED KINGDOM

Richard M. Parton (3: 187)   Wellcome Trust Centre for Cell Biology, Institute of Cell and Molecular Biology The University of Edinburgh, Michael Swann Building, The King's Buildings, Mayfield Road, Edinburgh, EH9 3JR, SCOTLAND

Bryce M. Paschal (2: 267)   Center for Cell Signaling, University of Virginia, 1400 Jefferson Park Avenue, West Complex Room 7021, Charlottesville, VA 22908-0577

Wayne F. Patton (4: 225)   Perkin-Elmer LAS, Building 100-1, 549 Albany Street, Boston, MA 02118

Staffan Paulie (1: 533)   Mabtech AB, Box 1233, Nacha Strand, SE-131 28, SWEDEN

Rainer Pepperkok (3: 121)   Cell Biology and Cell Biophysics Programme, European Molecular Biology Laboratory (EMBL), Meyerhofstrasse 1, Heidelberg, D-69117, GERMANY

Xomalin G. Peralta (3: 87)   Department of Physics, Duke University, 107 Physics Bldg, Durham, NC 27708-1000

Martha Perez-Magallanes (1: 151)   National Medical Center and Beckman Research Institute, Division of Neurosciences, City of Hope, 1500 E. Duarte Rd, Duarte, CA 91010

Stephen P. Perfetto (1: 257)   Vaccine Research Center, National Institutes of Health, 40 Convent Dr., Room 5509, Bethesda, MD 20892-3015

Hedvig Perlmann (1: 533)   Department of Immunology, Stockholm University, Biology Building F5, Top floor, Svante Arrhenius väg 16, Stockholm, SE-10691, SWEDEN

Peter Perlmann (1: 533)   Department of Immunology, Stockholm University, Biology Building F5, Top floor, Svante Arrhenius väg 16, Stockholm, SE-10691, SWEDEN

Timothy W. Petersen (1: 269)   Institute for Systems Biology, 1441 N. 34th St, Seattle, WA 98103

Patti Lynn Peterson (2: 259)   Department of Neurology, Wayne State University School of Medicine, 5L26 Detroit Receiving Hospital, Detroit Medical Center, Detroit, MI 48201

Nisha Philip (4: 265)   Department of Pharmacology and Cancer Biology, Duke University, Research Dr. LSRC Rm C115, Box 3813, Durham, NC 27710

Thomas Pieler (1: 191)   Institute for Biochemistry and Molecular Cell Biology, University of Goettingen, Humboldtallee 23, Göttingen, D-37073, GERMANY

Daniel Pinkel (3: 445)   Department of Laboratory Medicine, University of California San Francisco, Box 0808, San Francisco, CA 94143-0808

Javier Pizarro Cerdá (2: 407)   Unite des Interactions Bacteries-Cellules/Unité INSERM 604, Institut Pasteur, 28, rue du Docteur Roux, Paris Cedex 15, F-75724, FRANCE

Andreas Plückthun (1: 497)   Department of Biochemistry, University of Zürich, Winterthurerstrasse 190, Zürich, CH-8057, SWITZERLAND

Helen Plutner (2: 209)   Department of Cell and Molecular Biology, The Scripps Research Institute, 10550 North Torrey Pines Road, La Jolla, CA 92037

Piotr Pozarowski (1: 279)   Brander Cancer Research Institute, New York Medical College, Valhalla, NY 10595

Johanna Prast (1: 557)   Institute of Molecular Biology, Austrian Academy of Sciences, Billrothstrasse 11, Salzburg, A-5020, AUSTRIA

Brendan D. Price (3: 477)   Department of Radiation Oncology, Dana Farber-Brigham and Women's Cancer Center, 75 Francis Street, Level L2, Boston, MA 02215

Elena Prigmore (3: 403)   Sanger Institute, The Wellcome Trust, The Wellcome Trust Genome Campus, Hinxton, Cambridge, CB10 1SA, UNITED KINGDOM

Gottfried Proess (1: 467)   Eurogentec S.A., Liege Science Park, 4102 Seraing, B-, BELGIUM

David M. Prowse (1: 133)   Centre for Cutaneous Research, Barts and The London Queen Mary's School of Medicine and Dentistry, Institute of Cell and Molecular Science, 2 Newark Street, Whitechapel London, WC2A 3PX, UNITED KINGDOM

Manuela Pruess (4: 469)   EMBL outstation—Hinxton, European Bioinformatic Institute, Welcome Trust Genome Campus, Hinxton, Cambridge, CB10 1SD, UNITED KINGDOM

Eric Quéméneur (4: 235) Life Sciences Division, CEA Valrhô, BP 17171 Bagnols-sur-Cèze, F-30207, FRANCE

Leda Helen Raptis (2: 329, 341) Department of Microbiology and Immunology, Queen's University, Room 716 Botterell Hall, Kingston, Ontario, K7L3N6, CANADA

Anne-Marie Rasmussen (1: 239) Dynal Biotech ASA, PO Box 114, Smestad, N-0309, NORWAY

Andreas S. Reichert (4: 269) Department of Physiological Chemistry, University of Munich, Butenandtstr. 5, München, D-81377, GERMANY

Siegfreid Reipert (3: 325) Ordinariat II, Institute of Biochemistry and Molecular Biology, Vienna Biocenter, Dr. Bohr-Gasse 9, Vienna, A-1030, AUSTRIA

Guenter P. Resch (3: 267) Institute of Molecular Biology, Dr. Bohrgasse 3-5, Vienna, A-1030, AUSTRIA

Katheryn A. Resing (4: 443) Dept of Chemistry and Biochemistry, University of Colorado, 215 UCB, Boulder, CO 80309-0215

Donald L Riddle (1: 157) Division of Biological Sciences, University of Missouri, 311 Tucker Hall, Columbia, MO 65211

Mara Riminucci (1: 79) Department of Experimental Medicine, Universita' dell' Aquila, Via Vetoio, Coppito II, L'Aquila, I-67100, ITALY

Terry L. Riss (1: 25) Promega Corporation, 2800 Woods Hollow Road, Madison, WI 53711-5399

Pamela Gehron Robey (1: 79) Craniofacial and Skeletal Disease Branch, NIDCR, NIH, Department of Health and Human Services 30 Convent Dr, MSC 4320, Bethesda, MD 20892-4320

Linda J. Robinson (2: 201)

Philippe Rocca-Serra (4: 95) EMBL Outstation—Hinxton, European Bioinformatics Institute, Wellcome Trust Genome Campus, Hinxton, Cambridge, CB10 1SD, UNITED KINGDOM

Alice Rodriguez (3: 87) Department of Biology, Duke University, Durham, NC 27708-1000

Enrique Rodriguez-Boulan (2: 189, 241) Margaret M Dyson Vision Research Institute, Department of Opthalmology, Weill Medical College of Cornell University, New York, NY 10021

Mario Roederer (1: 257) ImmunoTechnology Section and Flow Cytometry Core, Vaccine Research Center, National Institute for Allergy and Infectious Diseases, National Institutes of Health, 40 Convent Dr., Room 5509, Bethesda, MD 20892-3015

Peter Roepstorff (4: 371) Department of Biochemistry and Molecular Biology, University of Southern Denmark, Campusvej 55, Odense M, DK-5230, DENMARK

Manfred Rohde (2: 399) Department of Microbial Pathogenicity, Gesellschaft fur Biotechnoogische Forschung, Mascheroder Weg 1, Braunschweig, D-38124, GERMANY

Norbert Roos (3: 299) Electron Microscopical Unit for Biological Sciences, University of Oslo, Blindern, Oslo, 0316, NORWAY

Sabine Rospert (2: 253) Institut für Biochemie und Molekularbiologie, Universität Freiburg, Hermann-Herder-Str. 7, Freiburg, D-79104, GERMANY

Klemens Rottner (3: 111) Cytoskeleton Dynamics Group, German Research Centre for Biotechnology (GBF), Mascheroder Weg 1, Braunschweig, D-38124, GERMANY

Line Roy (2: 41) Department of Anatomy and Cell Biology, Faculty of Medicine, McGill University, STRATHCONA Anatomy & Dentistry Building, Montreal, QC, H3A 2B2, CANADA

Sandra Rutherford (3: 325) CRC Structural Cell Biology Group, Paterson Institute for Cancer Research, Christie Hospital NHS Trust, Wilmslow Road, Withington, Manchester, M20 4BX, UNITED KINGDOM

Beth Rycroft (1: 395) Department of Pharmacology, University College London, Gower Street, London, WC1E GBT, UNITED KINGDOM

Patrick Salmon (1: 425) Department of Genetics and Microbiology, Faculty of Medicine, University of Geneva, CMU-1 Rue Michel-Servet, Geneva 4, CH-1211, SWITZERLAND

Paul M. Salvaterra (1: 151) National Medical Center and Beckman Research Institute, Division of Neurosciences, City of Hope, 1500 E. Duarte Rd, Duarte, CA 91010-3000

R. Jude Samulski (1: 457) Gene Therapy Centre, Department of Pharmacology, University of North Carolina at Chapel Hill, 7119 Thurston Bowles, Chapel Hill, NC 27599-7352

Susanna-Assunta Sansone (**4**: 95)   EMBL
Outstation—Hinxton, European Bioinformatics
Institute, Wellcome Trust Genome Campus, Hinxton,
Cambridge, CB10 1SD, UNITED KINGDOM

Ugis Sarkan (**4**: 95)   EMBL Outstation—Hinxton,
European Bioinformatics Institute, Wellcome Trust
Genome Campus, Hinxton, Cambridge, CB10 1SD,
UNITED KINGDOM

Moritoshi Sato (**2**: 325)   Department of Chemistry,
School of Science, University of Tokyo, 7-3-1 Hongo,
Bunkyo-Ku, Tokyo, 113-0033, JAPAN

Guido Sauter (**3**: 369)   Institute of Pathology,
University of Basel, Schonbeinstrasse 40, Basel,
CH-4003, SWITZERLAND

Carolyn L. Sawyer (**2**: 299)   Department of
Pharmacology, University of Vermont, Health Science
Research Facility 330, Burlington, VT 05405-0068

Guray Saydam (**1**: 315)   Department of Medicine,
Section of Hematology, Ege University Hospital,
Bornova Izmir, 35100, TURKEY

Silvia Scaglione (**3**: 201)   BIOLab, Department of
Informatic, Systemistic and Telematic, University of
Genoa, Viale Causa 13, Genoa, I-16145, ITALY

Lothar Schermelleh (**1**: 291, 301)   Department of
Biology II, Biocenter of the Ludwig-Maximilians
University of Munich (LMU), Großhadernerstr. 2,
Planegg-Martinsried, 82152, GERMANY

Gudrun Schiedner (**1**: 445)   CEVEC
Pharmaceuticals GmbH, Gottfried-Hagen-Straße 62,
Köln, D-51105, GERMANY

David Schieltz (**4**: 383)   Department of Cell Biology,
Torrey Mesa Research Institute, 3115 Merryfield Row,
San Diego, CA 92121

Jan E. Schnitzer (**2**: 11)   Sidney Kimmel Cancer
Center, 10835 Altman Row, San Diego, CA 92121

Morten Schou (**1**: 353)   Bartholin Institute,
Bartholinsgade 2, Copenhagen K, DK-1356,
DENMARK

Sebastian Schuck (**2**: 5)   Max Planck Institute of
Molecular Cell Biology and Genetics,
Pfotenhauerstrasse 108, Dresden, D-01307,
GERMANY

Herwig Schüler (**2**: 165)   Department of Cell
Biology, The Wnner-Gren Institute, Stockholm
University, Stockholm, S-10691, SWEDEN

Michael Schuler (**3**: 501)   Institut de Génétique et de
Biologie Moléculaire et Cellulaire (IGBMC), 1 rue
Laurent Fries, B.P.10142, Illkirch CEDEX, F-67404,
FRANCE

Ulrich S. Schwarz (**2**: 419)   Theory Division, Max
Planck Institute of Colloids and Interfaces, Potsdam,
14476, GERMANY

Antonio S. Sechi (**2**: 393)   Institute for Biomedical
Technology-Cell Biology, Universitaetsklinikum
Aachen, RWTH, Pauwelsstrasse 30, Aachen, D-52057,
GERMANY

Richard L. Segraves (**3**: 445)   Comprehensive
Cancer Center, University of California San
Francisco, Box 0808, 2400 Sutter N-426, San
Francisco, CA 94143-0808

James R. Sellers (**2**: 387)   Cellular and Motility
Section, Laboratory of Molecular Cardiology,
National Heart, Lung and Blood Institute (NHLBI),
National Institutes of Health, 10 Center Drive, MSC
1762, Bethesda, MD 20892-1762

Nicholas J. Severs (**3**: 249)   Cardiac Medicine,
National Heart and Lung Institute, Imperial College,
Faculty of Medicine, Royal Brompton Hospital,
Dovehouse Street, London, SW3 6LY, UNITED
KINGDOM

Jagesh Shah (**3**: 351)   Laboratory of Cell Biology,
Ludwig Institute for Cancer Research, University of
California, 9500 Gilman Drive, MC 0660, La Jolla,
CA 92093-0660

Norman E. Sharpless (**1**: 223)   The Lineberger
Comprehensive Cancer Center, The University of
North Carolina School of Medicine, Lineberger
Cancer Center, CB# 7295, Chapel Hill, NC
27599-7295

Andrej Shevchenko (**4**: 399)   Max Planck Institute
for Molecular Cell Biology and Genetics,
Pfotenhauerstrasse 108, Dresden, D-01307,
GERMANY

David M. Shotton (**3**: 207, 249, 257)   Department of
Zoology, University of Oxford, South Parks Road,
Oxford, OX1 3PS, UNITED KINGDOM

David I. Shreiber (**1**: 379)   Department of
Biomedical Engineering, Rutgers, the State
University of New Jersey, 617 Bowser Road,
Piscataway, NJ 08854-8014

Snaevar Sigurdsson (**3**: 455)   Department of Medical Sciences, Uppsala University, Akademiska sjukhuset, Uppsala, SE-751 85, SWEDEN

Stephen Simkins (**1**: 483)   TriPath Oncology, 4025 Stirrup Creek Drive, Suite 400, Durham, NC 27703

Ronald Simon (**3**: 369)   Division of Molecular Pathology, Institute of Pathology, University of Basel, Schonbeinstrasse 40, Basel, CH-4031, SWITZERLAND

Kai Simons (**1**: 127, **2**: 5, 181)   Max Planck Institute of Molecular Cell Biology and Genetics, Pfotenhauerstrasse 108, Dresden, D-01307, GERMANY

Jeremy C. Simpson (**3**: 121)   Cell Biology and Cell Biophysics Programme, European Molecular Biology Laboratory (EMBL), Meyerhofstrasse 1, Heidelberg, D-69117, GERMANY

Robert H. Singer (**4**: 121)   Department of Anatomy and Structural Biology, Golding # 601, Albert Einstein College of Medicine of Yeshiva University, 1300 Morris Park Avenue, Bronx, NY 10461

Mathilda Sjöberg (**1**: 419)   Department of Biosciences at Novum, Karolinska Instituet, Huddinge, SE-141-57, SWEDEN

Camilla Skiple Skjerpen (**4**: 275)   Department of Biochemistry, Institute for Cancer Research, The Norwegian Radium Hospital, Montebello, Oslo, N-0310, NORWAY

John Sleep (**3**: 37)   Randall Division, Guy's Campus, New Hunt's House, London, SE1 1UL, UNITED KINGDOM

J. Victor Small (**1**: 557)   Department of Cell Biology, Institute of Molecular Biology, Austrian Academy of Sciences, Billrothstrasse 11, Salzburg, A-5020, AUSTRIA

Joël Smet (**4**: 259)   Department of Pediatrics and Medical Genetics, University Hospital, De Pintelaan 185, Ghent, B-9000, BELGIUM

Kim Smith (**3**: 381)   Director of Cytogenetic Services, Oxford Radcliffe NHS Trust, Headington, Oxford, OX3 9DU, UNITED KINGDOM

Paul J. Smith (**1**: 305)   Dept of Pathology, University of Wales College of Medicine, Heath Park, Cardiff, CF14 4XN, UNITED KINGDOM

Antoine M. Snijders (**3**: 445)   Comprehensive Cancer Center, Cancer Research Institute, The University of California, San Francisco, Box 0808, 2340 Sutter Street N-, San Fransisco, CA 94143-0808

Michael Snyder (**3**: 179)   Department of Molecular, Cellular and Developmental Biology, Yale University, P. O. Box 208103, Kline Biology Tower, 219 Prospect St., New Haven, CT 06520-8103

Irina Solovei (**1**: 291)   Department of Biology II, Anthropology & Human Genetics, Ludwig-Maximilians University of Munich, Munich, GERMANY

Marion Sölter (**1**: 191)   Institute for Biochemistry and Molecular Cell Biology, University of Goettingen, Justus-von-Liebig-Weg 11, Göttingen, D-37077, GERMANY

Lukas Sommer (**1**: 69)   Institute of Cell Biology, Department of Biology, Swiss Federal Institute of Technology, ETH—Hönggerberg, Zurich, CH-8093, SWITZERLAND

Simon Sparks (**3**: 207)   Department of Zoology, University of Oxford, South Parks Road, Oxford, OX1 3PS, UNITED KINGDOM

Kenneth G. Standing (**4**: 457)   Department of Physics and Astronomy, University of Manitoba, 510 Allen Bldg, Winnipeg, MB, R3T 2N2, CANADA

Walter Steffen (**3**: 37, 307)   Randall Division, Guy's Campus, New Hunt's House, London, SE1 1UL, UNITED KINGDOM

Theresia E. B. Stradal (**3**: 111)   Department of Cell Biology, German Research Centre for Biotechnology (GBF), Mascheroder Weg 1, Braunschweig, D-38124, GERMANY

Per Thor Straten (**1**: 97)   Tumor Immunology Group, Institute of Cancer Biology, Danish Cancer Society, Strandboulevarden 49, Copenhagen, DK-2100, DENMARK

Hajime Sugihara (**1**: 411)   Department of Pathology & Biodefence, Faculty of Medicine, Saga University, Nebeshima 5-1-1, Saga, 849-8501, JAPAN

Chung-Ho Sun (**3**: 351)   Beckman Laser Institute, University of California, Irvine, 1002 Health Sciences Road E, Irvine, CA 92697-1475

Mark Sutton-Smith (**4**: 415)   Department of Biological Sciences, Imperial College of Science, Technology and Medicine, Biochemistry Building, London, SW7 2AY, UNITED KINGDOM

Tatyana M. Svitkina (**3**: 277)   Department of Cell and Molecular Biology, Northwestern University Medical School, Chicago, IL 60611

Ann-Christine Syvänen (3: 455)   Department of Medical Sciences, Uppsala University, Forskningsavd 2, ing 70, Uppsala, SE-751 85, SWEDEN

Masako Tada (1: 199)   ReproCELL Incorporation, 1-1-1 Uchisaiwai-cho, Chiyoda-ku, Tokyo, 100-0011, JAPAN

Takashi Tada (1: 199)   Stem Cell Engineering, Stem Cell Research Center, Institute for Frontier Medical Sciences, Kyoto University, 53 Kawahara-cho Shogoin, Sakyo-ku, Kyoto, 606-8507, JAPAN

Angela Taddei (2: 359)   Department of Molecular Biology, University of Geneva, 30, Quai Ernest Ansermet, Geneva 4, CH-1211, SWITZERLAND

Tomohiko Taguchi (2: 33)   Department of Cell Biology, Yale University School of Medicine, 333 Cedar Street, PO Box 208002, New Haven, CT 06520-8002

Kazusuke Takeo (4: 197)   Department of Biochemistry and Biomolecular Recognition, Yamaguchi University School of Medicine, 1-1-1, Minami-kogushi, Ube, Yamaguchi, 755-8505, JAPAN

Nobuyuki Tanahashi (2: 91)   Laboratory of Frontier Science, Core Technology and Research Center, The Tokyo Metropolitan Institute of Medical Sciences, 3-18-22 Honkomagome, Bunkyo-ku, Tokyo, 113-8613, JAPAN

Keiji Tanaka (2: 91)   Laboratory of Frontier Science, Core Technology and Research Center, The Tokyo Metropolitan Institute of Medical Sciences, 3-18-22 Honkomagome, Bunkyo-ku, Tokyo, 133-8613, JAPAN

Chi Tang (2: 121)   Dept of Biomolecular Sciences, UMIST, PO Box 88, Manchester, M60 1QD, UNITED KINGDOM

Kirill V. Tarasov (4: 103)   Molecular Cardiology Unit, National Institute on Aging, NIH, 5600 Nathan Shock Drive, Baltimore, MD 21224

J. David Taylor (3: 471)   Department of Genomic Sciences, Glaxo Wellcome Research and Development, 5 Moore Drive, Research Triangle Park, NC 27709-3398

Nancy Smyth Templeton (4: 25)   Department of Molecular and Cellular Biology & the Center for Cell and Gene Therapy, Baylor College of Medicine, One Baylor Plaza, Alkek Bldg., Room N1010, Houston, TX 77030

Kenneth K. Teng (1: 171)   Department of Medicine, Wiell Medical of Cornell University, 1300 York Ave., Rm-A663, New York, NY 10021

Patrick Terheyden (1: 103)   Department of Dermatology, Julius-Maximillians University, Josef-Schneider-Str. 2, Würzburg, D-97080, GERMANY

Scott M. Thompson (1: 407)   Department of Physiology, University of Maryland School of Medicine, 655 W. Baltimore St., Baltimore, MD 21201

John F. Timms (4: 189)   Department of Biochemistry and Molecular Biology, Ludwig Institute of Cancer Research, Cruciform Building 1.1.09, Gower Street, London, WC1E 6BT, UNITED KINGDOM

Shuji Toda (1: 411)   Department of Pathology & Biodefence, Faculty of Medicine, Saga University, Nebeshima 5-1-1, Saga, 849-8501, JAPAN

Yoichiro Tokutake (3: 87)   Department of Physics, Duke University, 107 Physics Bldg, Durham, NC 27708-1000

Evi Tomai (2: 329)   Department of Microbiology and Immunology, Queen's University, Room 716 Botterell Hall, Kingston, Ontario, K7L3N6, CANADA

Kenneth B. Tomer (1: 511)   Mass Spectrometry, Laboratory of Structural Biology, National Institute of Environmental Health Sciences NIEH/NIH, 111 Alexander Drive, PO Box 12233, Research Triangle Park, NC 27709

Derek Toomre (3: 19)   Department of Cell Biology, Yale University School of Medicine, SHM-C227/229, PO Box 208002, 333 Cedar Street, New Haven, CT 06520-8002

Sharon A. Tooze (2: 79)   Secretory Pathways Laboratory, Cancer Research UK London Research Institute, 44 Lincoln's Inn Fields, London, WC2A 3PX, UNITED KINGDOM

David Tosh (1: 177)   Centre for Regenerative Medicine, Department of Biology and Biochemistry, University of Bath, Claverton Down, Bath, BA2 7AY, UNITED KINGDOM

Yusuke Toyama (3: 87)   Department of Physics, Duke University, 107 Physics Bldg, Box 90305, Durham, NC 27708-0305

Robert T. Tranquillo (1: 379)   Department of Biomedical Engineering and Department of Chemical Engineering and Materials Science, University of Minnesota, Biomedical Engineering, 7-112 BSBE. 312 Church St SE, Minneapolis, MN 55455

Signe Trentemølle (4: 165)   Institute of Cancer Biology and Danish Centre for Translational Breast Cancer Research, Danish Cancer Society,

Strandboulevarden 49, Copenhagen, DK-2100, DENMARK

Didier Trono (1: 425)   Department of Genetics and Microbiology, Faculty of Medicine, University of Geneva, CMU-1 Rue Michel-Servet, Geneva 4, CH-1211, SWITZERLAND

Kevin Truong (2: 307)   Division of Molecular and Structural Biology, Ontario Cancer Institute, Department of Medical Biophysics, University of Toronto, 610 University Avenue, 7-707A, Toronto, Ontario, M5G 2M9, CANADA

Jessica K. Tyler (2: 287)   Department of Biochemistry and Molecular Genetics, University of Colorado Health Sciences Center at Fitzsimons, PO Box 6511, Aurora, CO 80045

Aylin S. Ulku (1: 345)   Department of Pharmacology, University of North Carolina at Chapel Hill, Lineberger Comprehensive Cancer Center, Chapel Hill, NC 27599-7295

Yoshio Umezawa (2: 325)   Department of Chemistry, The School of Science, University of Tokyo, 7-3-1 Hongo, Bunkyo-ku, Tokyo, 113-0033, JAPAN

Ronald Vale (3: 129)   Department of Cellular and Molecular Pharmacology, The Howard Hughes Medical Institute, The University of California, San Francisco, N316, Genentech Hall, 1600 16th Street, San Francisco, CA 94107

Jozef Van Beeumen (4: 259)   Department of Biochemistry, Physiology and Microbiology, University of Ghent, K.L. Ledeganckstraat 35, Ghent, B-9000, BELGIUM

Rudy N. A. van Coster (4: 259)   Department of Pediatrics and Medical Genetics, University Hospital, University of Ghent, De Pintelaan 185, Ghent, B-9000, BELGIUM

Ger van den Engh (1: 269)   Institute for Systems Biology, 1441 North 34th Street, Seattle, WA 98103-8904

Peter van der Geer (4: 139)   Department of Chemistry and Biochemistry, University of California, San Diego, 9500 Gilman Dr., La Jolla, CA 92093-0601

Joël Vandekerckhovr (4: 379, 457)   Department of Medical Protein Research, Flanders Interuniversity Institute for Biotechnology, KL Ledeganckstraat 35, Gent, B-9000, BELGIUM

Charles R. Vanderburg (4: 5)   Department of Neurology, Massachusetts General Hospital, 114 Sixteenth Street, Charlestown, MA 02129

John Venable (4: 383)   Department of Cell Biology, Scripps Research Institute, 10550 North Torrey Pines Road, La Jolla, CA 92037

Isabelle Vernos (2: 379)   Cell Biology and Cell Biophysics Programme, European Molecular Biology Laboratory, Meyerhofstrasse 1, Heidelberg, D-69117, GERMANY

Peter J. Verveer (3: 153)   Cell Biology and Cell Biophysics Program, European Molecular Biology Laboratory, Meyerhofstrasse 1, Heidelberg, D-69117, GERMANY

Marc Vidal (4: 295)   Cancer Biology Department, Dana-Farber Cancer Institute, 44 Binney Street, Boston, MA 02115

Emmanuel Vignal (2: 427)   Department Genie, Austrian Academy of Sciences, Billrothstrasse 11, Salzbourg–Autriche, A-520, 5020, AUSTRIA

Sylvie Villard (1: 519)   Centre de Pharmacologie et Biotechnologie pour la Santé, CNRS—UMR 5094, Faculté de Pharmacie, Avenue Charles Flahault, Montpellier Cedex 5, F-34093, FRANCE

Hikka Virta (1: 127)   Department of Cell Biology, European Molecular Biology Laboratory, Cell Biology Programme, Heidelberg, D-69012, GERMANY

Alfred Völkl (2: 63)   Department of Anatomy and Cell Biology II, University of Heidelberg, Im Neuenheimer Feld 307, Heidelberg, D-69120, GERMANY

Sonja Voordijk (4: 207)   Geneva Bioinformatics SA, Avenue de Champel 25, Geneva, CH-1211, SWITZERLAND

Adina Vultur (2: 329, 341)   Department of Microbiology and Immunology, Queen's University, Room 716 Botterell Hall, Kingston, Ontario, K7L3N6, CANADA

Teruhiko Wakayama (1: 87)   Center for Developmental Biology, RIKEN, 2-2-3 Minatojima-minamimachi, Kobe, 650-0047, JAPAN

Daniel Walther (4: 207)   Swiss Institute of Bioinformatics (SIB), CMU, rue Michel-Servet 1, Genève 4, 1211, SWITZERLAND

Gang Wang (3: 477)   Department of Radiation Oncology, Dana Farber-Brigham and Women's Cancer Center, 75 Francis Street, Level L2, Boston, MA 02215

Nancy Wang (3: 345)   Department of Pathology and Laboratory Medicine, University of Rochester School of Medicine, 601 Elmwood Ave., Rm 1-6337, Rochester, NY 14642

Xiaodong Wang (2: 209)   Department of Cell and Molecular Biology, The Scripps Research Institute, 10550 North Torrey Pines Road, La Jolla, CA 92037

Yanzhuang Wang (2: 33)   Department of Cell Biology, Yale University School of Medicine, 333 Cedar Street, PO Box 208002, New Haven, CT 06520-8002

Yu-Li Wang (3: 107)   Department of Physiology, University of Massachusetts Medical School, 377 Plantation St., Rm 327, Worcester, MA 01605

Graham Warren (2: 33)   Department of Cell Biology, Yale University School of Medicine, 333 Cedar Street, PO Box 208002, New Haven, CT 06520-8002

Clare M. Waterman-Storer (3: 137)   10550 North Torrey Pines Road, CB 163, La Jolla, CA 92037

Jennifer C. Waters (3: 49)   Department of Cell Biology, Department of Systems Biology, Harvard Medical School, 240 Longwood Ave, Boston, MA 02115

Fiona M. Watt (1: 133)   Keratinocyte Laboratory, London Research Institute, 44 Lincoln's Inn Fields, London, WC2A 3PX, UNITED KINGDOM

Gerhard Weber (4: 157)   Protein Analytics, Max Planck Institute for Biochemistry, Klopferspitz 18, Martinsried, D-82152, GERMANY

Peter J. A. Weber (4: 157)   Proteomics Division, Tecan Munich GmbH, Feldkirchnerstr. 12a, Kirchheim, D-, 85551, GERMANY

Paul Webster (3: 299)   Electron Microscopy Laboratory, House Ear Institute, 2100 West Third Street, Los Angeles, CA 90057

Jürgen Wehland (2: 399)   Department of Cell Biology, Gesellschaft fur Biotechnoogische Forschung, Mascheroder Weg 1, Braunschweig, D-38124, GERMANY

Dieter G. Weiss (3: 57)   Institute of Cell Biology and Biosystems Technology, Department of Biological Sciences, Universitat Rostock, Albert-Einstein-Str. 3, Rostock, D-18051, GERMANY

Walter Weiss (4: 175)   Fachgebiet Proteomik, Technische Universität Muenchen, Am Forum 2, Freising Weihenstephan, D-85350, GERMANY

Adrienne R. Wells (3: 87)   Department of Biology, Duke University, Durham, NC 27708-1000

Jørgen Wesche (4: 19)   Department of Biochemistry, The Norwegian Radium Hospital, Montebello, Oslo, 0310, NORWAY

Pamela Whittaker (3: 463)   Genotyping / Chr 20, The Wellcome Trust Sanger Institute, The Wellcome Trust Genome Campus, Hinxton, Cambridge, CB10 1SA, UNITED KINGDOM

John Wiemann (3: 87)   Department of Biology, Duke University, B330g Levine Sci Bldg, Box 91000, Durham, NC 27708-1000

Sebastian Wiesner (2: 173)   Dynamique du Cytosquelette, Laboratoire d'Enzymologie et Biochimie Structurales, UPR A 9063 CNRS, Building 34, Bat. 34, avenue de la Terrasse, Gif-sur-Yvette, F-91198, FRANCE

Ilona Wolff (1: 325)   Prodekanat Forschung, Medizinnishce Fakultät, Martin Luther University Halle-Wittenburg, Magdeburger Str 8, Saale, Halle, D-06097, GERMANY

Ye Xiong (2: 259)   The Department of Biochemistry and Molecular Biology, Wayne State University School of Medicine, 4374 Scott Hall, 540 E. Canfield, Detroit, MI 48201

David P. Yarnall (3: 471)   Department of Metabolic Diseases, Glaxo Wellcome Inc, 5 Moore Drive, Research Triangle Park, NC 27709-3398

Hideki Yashirodas (2: 91)   Laboratory of Frontier Science, Core Technology and Research Center, The Tokyo Metropolitan Institute of Medical Sciences, 3-18-22 Honkomagome, Bunkyo-ku, Tokyo, 133-8613, JAPAN

John R. Yates III (4: 383, 391)   Department of Cell Biology, Scripps Research Institute, 10550 North Torrey Pines Road, La Jolla, CA 92037

Charles Yeaman (2: 189)   Department of Cell and Developmental Biology, Weill Medical College of Cornell University, New York, NY 10021

Robin Young (2: 41)   Département de pathologie et biologie cellulaire, Université de Montréal, Case postale 6128, Succursale "Centre-Ville", Montreal, QC, H3C 3J7, CANADA

Christian Zahnd (**1**: 497) Department of Biochemistry, Univerisity of Zürich, Winterthurerstr. 190, Zürich, CH-8057, SWITZERLAND

Zhuo-shen Zhao (**4**: 285) Glaxo-IMCB Group, Institute of Molecular and Cell Biology, Singapore, 117609, SINGAPORE

Huilin Zhou (**4**: 437) Department of Cellular and Molecular Medicine, Ludwig Institute for Cancer Research, University of California, San Diego, 9500 Gilman Drive, CMM-East, Rm 3050, La Jolla, CA 92093-0660

Timo Zimmermann (**3**: 69) Cell Biology and Cell Biophysics Programme, EMBL, Meyerhofstrasse 1, Heidelberg, D-69117, GERMANY

Chiara Zurzolo (**2**: 241) Department of Cell Biology and Infection, Pasteur Institute, 25,28 rue du Docteur Roux, Paris, 75015, FRANCE

Preface

Scientific progress often takes place when new technologies are developed, or when old procedures are improved. Today, more than ever, we are in need of complementary technology platforms to tackle complex biological problems, as we are rapidly moving from the analysis of single molecules to the study of multifaceted biological problems. The third edition of *Cell Biology: A Laboratory Handbook* brings together 236 articles covering novel techniques and procedures in cell and molecular biology, proteomics, genomics, and functional genomics. It contains 165 new articles, many of which were commissioned in response to the extraordinary feedback we received from the scientific community at large.

As in the case of the second edition, the *Handbook* has been divided in four volumes. The first volume covers tissue culture and associated techniques, viruses, antibodies, and immunohistochemistry. Volume 2 covers organelles and cellular structures as well as assays in cell biology. Volume 3 includes imaging techniques, electron microscopy, scanning probe and scanning electron microscopy, micro-dissection, tissue arrays, cytogenetics and in situ hybridization, genomics, transgenic, knockouts, and knockdown methods. The last volume includes transfer of macromolecules, expression systems, and gene expression profiling in addition to various proteomic technologies. Appendices include representative cultured cell lines and their characteristics, Internet resources in cell biology, and bioinformatic resources for in silico proteome analysis. The Handbook provides in a single source most of the classical and emerging technologies that are essential for research in the life sciences. Short of having an expert at your side, the protocols enable researchers at all stages of their career to embark on the study of biological problems using a variety of technologies and model systems. Techniques are presented in a friendly, step-by-step fashion, and gives useful tips as to potential pitfalls of the methodology.

I would like to extend my gratitude to the Associate Editors for their hard work, support, and vision in selecting new techniques. I would also like to thank the staff at Elsevier for their constant support and dedication to the project. Many people participated in the realization of the *Handbook* and I would like to thank in particular Lisa Tickner, Karen Dempsey, Angela Dooley, and Tari Paschall for coordinating and organizing the preparation of the volumes. My gratitude is also extended to all the authors for the time and energy they dedicated to the project.

Julio E. Celis
Editor

PART
A

TRANSFER OF MACROMOLECULES

Proteins

Impact-Mediated Cytoplasmic Loading of Macromolecules into Adherent Cells

Mark S. F. Clarke, Daniel L. Feeback, and Charles R. Vanderburg

I. INTRODUCTION

The advent of modern molecular biology, including the development of gene array technologies, has resulted in an explosion of information concerning the specific genes activated during normal cellular development, as well as those associated with a variety of pathological conditions. These techniques have provided a highly efficient, broad-based screening approach for those specific genes involved in regulating normal cellular physiology and identifying candidate genes directly associated with the etiology of specific disease states. However, this approach provides information at the transcriptional level only and does not necessarily indicate that the gene in question is in fact translated into a protein or whether posttranslational modification of the protein occurs.

The critical importance of posttranslational modification (i.e., phosphorylation, glycosylation, sialylation) to protein function has been recognized with regard to a number of proteins involved in a variety of important disease states. For example, altered glycosylation of the β-amyloid precursor protein results in an increase in the amount of β-amyloid peptide generated and hence secreted as insoluble extracellular amyloid deposits (Georgopoulou *et al.*, 2001; Walter *et al.*, 2001), a pathological hallmark of Alzheimer's disease. Abnormal phosphorylation of synapsin I has been linked to alterations in synaptic vesicle trafficking, leading to defective neurotransmission in Huntington's disease (Lievens *et al.*, 2002). Altered phosphorylation of the TAU protein involved in microtubule function has been linked to a number of neurodegenative diseases, such as Alzheimer's disease (Billingsley and Kincaid, 1997; Sanchez *et al.*, 2001).

Aberrant sialylation of cell surface antigens has been detected in a number of different tumor cell types and has been linked to the acquisition of a neoplastic phenotype (Sell, 1990), whereas improper sialylation of sodium channels in cardiac tissue has been linked to heart failure (Ufret-Vincenty *et al.*, 2001; Fozzard and Kyle, 2002). In addition to the limitations associated with the total lack of information provided regarding posttranslational modification of the encoded protein, gene analysis cannot provide information on the role of protein–protein interactions within the cellular milieu. For example, a combination of gene and protein analysis may indicate that a particular protein is unregulated at both mRNA and protein levels, respectively. However, the presence of a regulatory protein(s) in the cellular milieu may result in this overexpression being of no physiological importance. For example, cytokine stimulation leading to an elicited cellular response is not only associated with the action of a particular cytokine at its specific receptor, but also with a cascade of downstream signal transduction events that are under the control of a variety of negative regulatory proteins (Carra *et al.*, 2000; Murphy *et al.*, 2000). Such protein–protein interactions underscore the complexity of the interplay between proteins in modulating cell function and are an excellent example of the limitations of gene analysis techniques in probing "cause-and-effect" relationships between gene expression and cellular response. It is the study of the interaction among gene expression, modified transcript splicing, and posttranslational modification on protein function as it impacts cellular/tissue phenotype, which is now the modern research field of proteomics.

One approach to overcoming the limitations of genetic analysis is to insert a protein of interest into a cell and directly observe the effects on cell function. The direct "insertion" of a protein (in contrast to transfection of a genetic construct followed by protein expression) allows investigation of the functional role of the inserted molecule in determining cell phenotype in a fashion that also considers the role of protein–protein interactions in the cellular milieu. Insertion of individual or mixtures of purified proteins prepared in the *native* state also increases the possibility that any observed effects of the protein on cell phenotype/response are of true physiological relevance. As yet, such functional proteomic studies are limited by the availability of technologies that allow efficient insertion of native proteins directly into cells.

Microinjection (Bloom *et al.*, 2003; Bubb *et al.*, 2003), electroporation (Chow and Gawler, 1999; Ponsaerts *et al.*, 2002), and lipid vesicle-mediated protein loading (Chen *et al.*, 1993) all have been demonstrated to achieve protein insertion into cultured cells. Mechanical-based loading techniques such as biolistics (Maddelein *et al.*, 2002), syringe loading (Clarke and McNeil, 1992), and scrape loading (McNeil *et al.*, 1989) can also be used to load a variety of cell types with a range of proteins. However, apart from microinjection, which requires expensive equipment and skilled personnel to produce relatively few loaded cells, none of the aforementioned techniques have wide applicability to loading of either individual or mixtures of purified proteins into adherent primary cells in tissue culture. This article describes a technique, known as impact-mediated loading, that is capable of simultaneously loading a large number of adherent primary cells (>10,000 cells during a single procedure) with a variety of proteins at high efficiency. This approach is based on the production of transient plasma membrane wounds by particle impact with the cell membrane. We have used this technique previously in order to load dyes, antibodies, and plasmid constructs into the cytoplasm of a number of primary and established cell lines utilizing a crude, relatively uncontrolled experimental apparatus (Clarke *et al.*, 1994). In addition, we have used a more refined version of this technique to study the effects of altered gravitational conditions on the membrane-wounding response of human primary myoblasts (Clarke *et al.*, 2002). Based on observations made during these latter studies, we have developed a novel technology, known as the G-Loader (Fig. 1), which utilizes the effects of hypergravity at $200\ g$ to enhance macromolecular loading into the cytoplasm of adherent cells via impact-mediated plasma membrane wounding (Clarke *et al.*, 2001). The protocol described here specifically details the experimental approach used to load primary human skeletal myocytes with FITC-labeled IgG

immunoglobulin as an example of the suitability of G-Loader technology for loading biologically active proteins into primary adherent cells.

II. MATERIALS AND INSTRUMENTATION

Dulbecco's modified Eagle's medium–F12 medium (X1 concentration) (DMEM/F12, Cat. No. 320-1885AG), bovine fetal calf serum (CS) (Cat. No. 200-6170AG), and penicillin–streptomycin solution (Cat. No. 600-5140AG) are obtained ready to use from Gibco BRL (Grand Island, NY). Alexa Fluor 488-labeled goat anti-mouse IgG (M_r 150,000 kDa) (Cat. No. A-11001) is obtained from Molecular Probes (Eugene, OR). Tissue culture flasks (T-75, T25, and 35-mm diameter plates) (Cat. Nos. 10-126-41, 10-126-26, and 25050-35) and sterile polypropylene conical centrifuge tubes (50 ml capacity) (Cat. No. 05-538-55A) are from Fisher Scientific (Pittsburgh, PA). The G-Loader technology and associated consumables are from Peilear Technologies (Houston, TX). Living cells loaded with fluorescently labeled IgG are examined utilizing an Zeiss Axiophot inverted microscope in order to view them in the 35-mm tissue culture plate directly.

III. PROCEDURES

A. Preparation of Tissue-Cultured Cells for Impact-Mediated Loading Using the G-Loader

Stock Solutions and Media Preparation

1. *Stock IgG solution*: Stock Alexa Fluor 488 goat anti-mouse IgG (2 mg/ml) can be stored at 4°C in the dark for up to a month or aliquoted and frozen in the dark at −80°C for storage up to a year. Prior to use on living cells, it is important to remove any preservative (e.g., sodium azide) from the immunoglobulin solution, as exposure of living cells to such contaminants can result in rapid cell death. Removal of such preservatives can be carried out using either a miniature dialysis device or buffer exchange using a 30,000 MW cutoff centrifugal miniature concentrator. One milliliter of stock IgG (2 mg/ml) solution should be dialyzed against a minimum of three changes of 100 ml of sterile DMEM/F12 medium over a period of 24 h prior to use. The total volume of dialyzed IgG solution should be adjusted to original volume prior to dialysis to maintain the overall IgG concentration. Buffer exchange using a centrifugal concentrator should

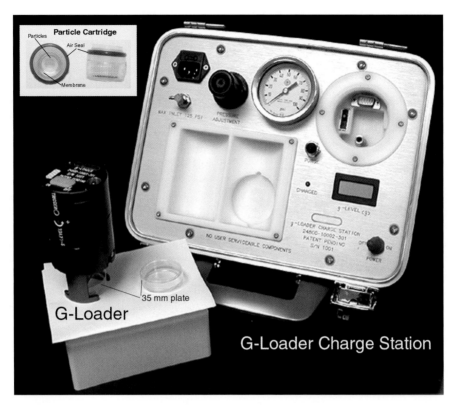

FIGURE 1 Impact-mediated loading apparatus known as the G-Loader. The G-Loader device itself is approximately 25 cm tall and 10 cm in diameter. A 35-mm tissue culture plate containing a monolayer of adherent cells is placed in the base of the G-Loader after being incubated with loading medium containing the macromolecule of interest. Prior to placing the cell sample into the G-Loader, a particle cartridge is inserted into the device consisting of a circular cartridge that supports a rupturable membrane on which is located a layer of 10-μm particles arranged in a specific pattern (see insert). In addition, the device has been charged with air to the required pressure and the on-board *g*-load accelerometer trigger has been set to the required *g* value using the G-Loader charging station (for benchtop operation there is a manual trigger button located on the side wall of the device). When the G-Loader is activated, either manually or by reaching a set *g* load, a metered volume of pressurized air is directed into the particle cartridge and the membrane ruptures and disperses particles into the airstream. The particles then impact the cell layer, inducing membrane wounding in a highly controllable and reproducible fashion.

involve a minimum of three complete exchanges of the original volume of stock IgG solution with serum-free DMEM/F12 medium.

2. *IgG loading solution*: Add 125 μl of stock IgG solution to 0.875 ml of serum-free DMEM/F12 to obtain final concentrations of 250 μg IgG/ml. Equilibrate this loading solution to the correct temperature and pH by incubating in a standard 5% CO_2 tissue culture incubator maintained at 37°C for a period of 30 min prior to loading cells with the IgG.

3. *10% FCS.DMEM/F12 tissue culture medium*: Add 5 ml of sterile penicillin/streptomycin solution and 50 ml of sterile fetal calf serum to 445 ml of sterile (X1) DMEM/F12 solution to obtain DMEM/F12 culture medium containing 10% FCS, 100 IU/ml penicillin, and 100 μg/ml streptomycin (10% FCS.DMEM). Store at 4°C for up to 21 days.

B. Culture of Human Primary Myoblasts

1. Obtain primary human myoblasts from the Clonetics Corporation (Walkersville, MD) and culture to confluence in T-75 (75 cm²) culture flasks using 15 ml of 10% FCS.DMEM/F12 maintained at 37°C in a 5% CO_2 humidified atmosphere with subculture every fourth day.

2. The day prior to the loading procedure, trypsinize, collect by centrifugation, and resuspend cells in 5 ml of 10% FCS.DMEM/F12 as a single cell suspension. Determine the number of cells in the suspension using a hemacytometer and adjust cell density in the solution to 50,000 cells/ml using 10% FCS.DMEM/F12. Place 2 ml of this cell suspension into a 35-mm tissue culture plate and incubate cells overnight to allow formation of a monolayer. Under

these culture conditions the center of the 35-mm plate is between 60 and 80% confluent after 24 h of culture prior to loading.

3. Prior to loading, wash cells three times over a period of 5 min with warm, serum-free DMEM/F12 *containing no antibiotics* to remove any unattached cells and to wash away serum components that may interact with the IgG molecules being loaded.

C. Impact-Mediated Loading of IgG Using G-Loader Technology

1. Remove serum-free medium from the cell monolayer, replace with 400 µl of loading solution, and agitate the 35-mm plate gently so that the loading solution covers all of the culture surface. Agitate the loading solution for a minimum of 30 s to ensure that the dissolved macromolecule is in contact with the surface of the cells.

2. Remove the loading solution from the 35-mm plate by tilting the plate to one side and removing the loading solution with a 1-ml sterile pipette. *Note*: The loading solution can be reused on up to five additional 35-mm plates if used immediately after first centrifuging at 10,000 *g* for 1 min to remove any particulate/cellular material.

3. The cells are now ready to be loaded using the G-Loader technology. Place the 35-mm plate in the G-Loader and operate the firing mechanism in the tissue culture hood. Alternatively, the device and cells can be placed into the centrifuge and loading performed under hypergravity conditions (i.e., 200 *g*) (Fig. 1). In the case of human primary skeletal myoblasts and IgG loading, highly efficient cytoplasmic loading of IgG can be achieved without the need for the use of

hypergravity conditions during impact-mediated loading.

4. Remove the 35-mm plate from the G-Loader and immediately place 2 ml of warm serum-free DMEM/F12 medium (without antibiotics) into the 35-mm plate to prevent drying of the cells. Gently wash the cell layer three times over 2 min with three changes of serum-free DMEM/F12 medium to remove both excess IgG and particles impacted with the cell surface.

5. Replace washing medium with 10% FCS.DME/F12 tissue culture medium and incubate for the required time period. After this time (0–24 h depending on the goal of the project), wash once with warm D-PBS (pH 7.2), replace with fresh D-PBS, and view the living cell monolayer by inverted fluorescent microscopy using UV excitation and emission at 488 nm.

IV. COMMENTS

This technology provides a highly efficient means of loading large macromolecules into living adherent cells with little or no disruption of the adherent monolayer. As can be seen in Fig. 2, there is little or no visually discernible damage to the cell monolayer 1 h after impact-mediated cytoplasmic loading of a 2×10^6-Da-sized fluorescent dextran. The protocol detailed here describes the procedure required to load fluorescently labeled IgG into cultured primary skeletal muscle cells (Fig. 3). In addition, this technique has been applied to a large number of different cell types and a wide variety of different macromolecules, including

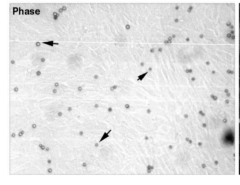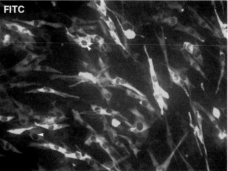

FIGURE 2   **Paired phase-contrast and FITC-fluorescent micrographs of a Swiss 3T3 fibroblast monolayer 1 h after impact-mediated cytoplasmic loading of 2×10^6-Da fluorescent dextran (FDx).** The monolayer was washed three times with warm serum-free DMEM to remove excess FDx (200 kDa) and to remove the majority of the particles (arrows). Note the clear exclusion of the 200-kDa dextran from the nuclear region of the cells (arrowheads) and little or no visual damage to the cell monolayer (left). Complete bead removal can be achieved by washing the monolayer multiple times with medium over a period of 4 h.

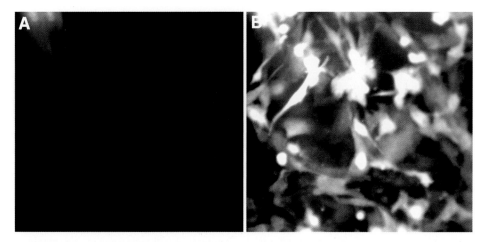

FIGURE 3   **Fluorescent micrographs of (A) control and (B) Alexa Fluor 488 goat anti-mouse IgG-loaded primary human myoblasts.** Control monolayers are treated in an identical fashion to loaded cells (including incubation with IgG loading solution) except that they are not subjected to impact-mediated loading. Micrographs were photographed under identical conditions immediately after loading and washing of monolayers. Note the somewhat "perturbed" appearance of the cells, an appearance that disappears after 20 to 60 min of additional culture in serum-containing medium (see Fig. 6).

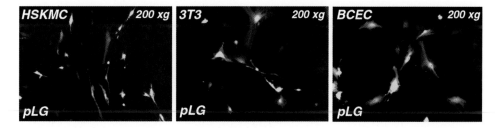

FIGURE 4   Fluorescent micrographs of primary human skeletal myoblasts (HSKMC), Swiss 3T3 cells (3T3), and primary bovine capillary endothelial cells (BCEC) 24 h after impact-mediated loading of a plasmid construct encoding for a green fluorescent protein, Lantern Green (pLG).

plasmid DNA constructs as a means of transfecting both primary and established cell lines (Fig. 4). Furthermore, this technique has the ability to simultaneously load two or more macromolecules directly into the cell cytoplasm of living cells, which are then localized to specific cellular compartments within the living cell based on size exclusion (Fig. 5). In addition, the use of hypergravity conditions during impact-mediated loading significantly increases the amount of fluorescently labeled IgG protein loaded into the cell cytoplasm (Fig. 6).

V. PITFALLS

The impact-mediated loading technology described here (i.e., The G-Loader) is a highly efficient means of simultaneously loading large numbers of living cells with large macromolecules. Under the conditions required for loading of macromolecules below 1×10^6 Da in size, the number of cells that are wounded, loaded, and survive the procedure is extremely high (approximately 70–80% of the starting population) (Fig. 1). One of the disadvantages of the technique described in this article for loading IgG is that not all cells in the 35-mm tissue culture dish are loaded. This is due to a limited particle impact area (approximately one-half of the culture area radiating from the center of the plate) generated by the device. This drawback can be overcome by growing cells only in the center region of the 35-mm plates utilizing a 20-mm-diameter cloning ring or plating cells on 20-mm-diameter glass coverslips located in the center of the 35-mm plates and removing the coverslips after loading. This approach yields a cell population in

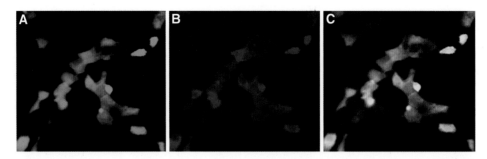

FIGURE 5   Matched fluorescent micrographs of NGF-differentiated PC-12 cells 20 min after impact-mediated loading of a mixture of fluorescently labeled dextrans. Two different-sized, different-colored dextran molecules (*green, FDx–10 kDa; red, TRITC–30 kDa*) were loaded simultaneously into monolayers of PC-12 cells using impact-mediated loading. Note exclusion of the larger dextran (red signal) from the nucleus of the cells (B), whereas the smaller dextran (green signal) is found in both the cytoplasm and the nucleus of the loaded cells (A). When a digital overlay of the images is generated, cells are colored predominantly orange in their cytoplasm (i.e., mixed red and green signal), whereas cell nuclei are green (C).

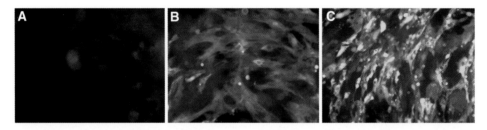

FIGURE 6   Fluorescent micrographs of Swiss 3T3 cell monolayers loaded cytoplasmically with Fluor 488 goat antimouse IgG employing impact-mediated loading at 1 *g* (B) and (C) 200 *g* in a benchtop centrifuge. Control cells (A) were exposed to IgG for the same period of time but were not impact loaded. Micrographs were taken at the same photographic conditions. Note the larger amount of IgG present in the cytoplasm of cells loaded at 200 *g* as compared to those loaded at 1 *g* and that IgG is excluded from the nuclear regions of the loaded cells.

which approximately 70–80% of the starting population is loaded. However, in those experiments where direct localization by microscopic techniques is utilized, the presence of unloaded cells in the region closest to the edge of the plate serves as an excellent internal negative control.

A second area of concern is the purity of the macromolecule that is being loaded. In initial experiments using commercially produced, fluorescently labeled IgG, IgM, and plant lectins, impact-mediated loading was observed at high efficiencies immediately after loading as determined by direct inspection of the cell layer by fluorescent microscopy. Subsequently, however, the majority of the cells died within 2 h of loading. After investigation, it was found that most commercially produced products had some form of preservative in the storage buffer that needed to be removed prior to contact with living cells during the loading procedure.

The G-Loader technology utilizes a combination of particle impact-induced membrane wounding and hypergravity conditions to load macromolecules directly into the cell cytoplasm. This technology appears to have utility for loading in a wide range of cell types, both established cell lines (e.g., COS, 3T3, C2C12 cells) and primary cell types (e.g., human myoblasts, bovine endothelial cells, chick corneal fibroblasts, rat gut epithelium), as well as multicellular microorganisms (e.g., hydra, nematodes, *Xenopus* ova). However, in order to achieve optimal loading efficiency, a mechanistic approach to determining impact pressure, macromolecular concentration in the loading solution, and whether or not hypergravity conditions are required for efficient loading is suggested. In addition, the nature of the macromolecule being loaded is also important. Those compounds that are highly charged or may interact with cell surface components may need longer contact with the cells prior to loading, a higher loading solution concentration, or higher g load (i.e., 200 g) (Fig. 5) in order to achieve the most efficient cytoplasmic loading possible.

References

Billingsley, M. L., and Kincaid, R. L. (1997). Regulated phosphorylation and dephosphorylation of tau protein: Effects on microtubule interaction, intracellular trafficking and neurodegeneration. *Biochem. J.* **323**(Pt 3), 577–591.

Bloom, O., Evergren, E., Tomilin, N., Kjaerulff, O., Low, P., Brodin, L., Pieribone, V. A., Greengard, P., and Shupliakov, O. (2003). Colocalization of synapsin and actin during synaptic vesicle recycling. *J. Cell Biol.* **161**(4), 737–747.

Bubb, M. R., Yarmola, E. G., Gibson, B. G., and Southwick, F. S. (2003). Depolymerization of actin filaments by profilin: Effects of profilin on capping protein function. *J. Biol. Chem.* **278**(27), 24629–24635.

Carra, G., Gerosa, F., and Trinchieri, G. (2000). Biosynthesis and posttranslational regulation of human IL-12. *J. Immunol.* **164**(9), 4752–4761.

Chen, W., Carbone, F. R., and McCluskey, J. (1993). Electroporation and commercial liposomes efficiently deliver soluble protein into the MHC class I presentation pathway: Priming *in vitro* and *in vivo* for class I-restricted recognition of soluble antigen. *J. Immunol. Methods* **160**(1), 49–57.

Chow, A., and Gawler, D. (1999). Mapping the site of interaction between annexin VI and the p120GAP C2 domain. *FEBS Lett.* **460**(1), 166–172.

Clarke, M. S., and McNeil, P. L. (1992). Syringe loading introduces macromolecules into living mammalian cell cytosol. *J. Cell Sci.* **102**(Pt 3), 533–541.

Clarke, M. S., Vanderburg, C. R., and Feeback, D. L. (2002). The effect of acute microgravity on mechanically-induced membrane damage and membrane-membrane fusion events. *J. Gravit. Physiol.* **8**(2), 37–47.

Clarke, M. S., Vanderburg, C. R., Hay, E. D., and McNeil, P. L. (1994). Cytoplasmic loading of dyes, protein and plasmid DNA using an impact-mediated procedure. *Biotechniques* **17**(6), 1118–1125.

Clarke, M. S. F., Feeback, D. L., and Vanderburg, C. R. (2001). U.S. Patent Number **6,221,666.** "Method and apparatus for cytoplasmic loading using an impact-mediated procedure." USA.

Fozzard, H. A., and Kyle, J. W. (2002). Do defects in ion channel glycosylation set the stage for lethal cardiac arrhythmias? *Sci. STKE* **2002**(130), PE19.

Georgopoulou, N., McLaughlin, M., McFarlane, I., and Breen, K. C. (2001). The role of post-translational modification in beta-amyloid precursor protein processing. *Biochem. Soc. Symp.* **67**, 23–36.

Lievens, J. C., Woodman, B., Mahal, A., and Bates, G. P. (2002). Abnormal phosphorylation of synapsin I predicts a neuronal transmission impairment in the R6/2 Huntington's disease transgenic mice. *Mol. Cell Neurosci.* **20**(4), 638–648.

Maddelein, M. L., Dos Reis, S., Duvezin-Caubet, S., Coulary-Salin, B., and Saupe, S. J. (2002). Amyloid aggregates of the HET-s prion protein are infectious. *Proc. Natl. Acad. Sci. USA* **99**(11), 7402–7407.

McNeil, P. L., Muthukrishnan, L., Warder, E., and D'Amore, P. A. (1989). Growth factors are released by mechanically wounded endothelial cells. *J. Cell Biol.* **109**(2), 811–822.

Murphy, F. J., Hayes, M. P., and Burd, P. R. (2000). Disparate intracellular processing of human IL-12 preprotein subunits: Atypical processing of the P35 signal peptide. *J. Immunol.* **164**(2), 839–847.

Ponsaerts, P., Van den Bosch, G., Cools, N., Van Driessche, A., Nijs, G., Lenjou, M., Lardon, F., Van Broeckhoven, C., Van Bockstaele, D. R., Berneman, Z. N., and Van Tendeloo, V. F. (2002). Messenger RNA electroporation of human monocytes, followed by rapid *in vitro* differentiation, leads to highly stimulatory antigen-loaded mature dendritic cells. *J. Immunol.* **169**(4), 1669–1675.

Sanchez, M. P., Alvarez-Tallada, V., and Avila, J. (2001). The microtubule-associated protein tau in neurodegenerative diseases. *Rev. Neurol.* **33**(2), 169–177.

Sell, S. (1990). Cancer-associated carbohydrates identified by monoclonal antibodies. *Hum. Pathol.* **21**(10), 1003–1019.

Ufret-Vincenty, C. A., Baro, D. J., Lederer, W. J., Rockman, H. A., Quinones, L. E., and Santana, L. F. (2001). Role of sodium channel deglycosylation in the genesis of cardiac arrhythmias in heart failure. *J. Biol. Chem.* **276**(30), 28197–28203.

Walter, J., Fluhrer, R., Hartung, B., Willem, M., Kaether, C., Capell, A., Lammich, S., Multhaup, G., and Haass, C. (2001). Phosphorylation regulates intracellular trafficking of beta-secretase. *J. Biol. Chem.* **276**(18), 14634–14641.

A Peptide Carrier for the Delivery of Biologically Active Proteins into Mammalian Cells: Application to the Delivery of Antibodies and Therapeutic Proteins

May C. Morris, Julien Depollier, Frederic Heitz, and Gilles Divita

I. INTRODUCTION

In order to circumvent the technological problems of gene delivery, an increasing interest is being taken in designing novel strategies to deliver full-length proteins into a large number of cells (Ford *et al.*, 2001; Wadia and Dowdy, 2002). However, the development of peptide drugs and therapeutic proteins remains limited by the poor permeability and the selectivity of the cell membrane. Substantial progress has been made in the development of cell-penetrating peptide-based drug delivery systems that are able overcome both extracellular and intracellular limitations (Morris *et al.*, 2000; Gariepy *et al.*, 2001; Langel *et al.*, 2002). A series of small protein domains, termed protein transduction domains (PTDs), have been shown to cross biological membranes efficiently, independently of transporters or specific receptors, and to promote the delivery of peptides and proteins into cells (for review see, Langel *et al.*, 2002). The use of PTD-mediated transfection has proven that "protein therapy" can have a major impact on the future of therapies in a variety of viral diseases and cancers. TAT protein from immunodeficiency virus (HIV-1) (Schwarze *et al.*, 1999; Schwarze and Dowdy, 2000), the third α helix of Antennapedia homeodomain (Derossi *et al.*, 1994), VP22 protein from herpes simplex virus (Elliott and

O'Hare, 1997), and transportan (Pooga *et al.*, 2001), as well as polyarginine sequences (Futaki *et al.*, 2000), have been used successfully to improve the delivery of covalently linked peptides or proteins into cells. However, PTDs display a certain number of limitations in that they all require cross-linking to the target peptide or protein. Moreover, protein transduction using the PTD-TAT-fusion protein system may require denaturation of the protein prior to delivery, introducing an additional delay between time of delivery and intracellular activation of the protein.

We have described a new strategy for the delivery of full-length proteins and peptides into mammalian cells based on a short amphipathic peptide carrier, Pep-1. This peptide carrier allows the delivery of several distinct proteins and peptides into different cell lines in a fully biologically active form, without the need for prior chemical covalent coupling or denaturation steps (Morris *et al.*, 2001). Pep-1 is a 21 residue peptide (KETWWETWWTEWSQPKKKRKV) consisting of three domains with specific functions: a hydrophobic tryptophan-rich motif (KETWWETWWTEW), involved in the main interactions with macromolecules and required for efficient targeting to the cell membrane; a hydrophilic lysine-rich domain (KKKRKV) derived from the nuclear localization sequence (NLS) of simian virus 40 (SV-40) large T antigen, required to improve intracellular delivery and solubility of the peptide vector; and a spacer domain (SQP), which improves the flexibility and the integrity of both hydrophobic and hydrophilic domains (Morris *et al.*, 2001). Given that the mechanism through which Pep-1 delivers macromolecules does not involve the endosomal pathway, the degradation of macromolecules delivered is significantly limited and rapid dissociation of the Pep-1/macromolecule particle is favoured as soon as it has crossed the cell membrane. This peptide-based protein delivery strategy presents several advantages, including rapid delivery of proteins into cells with very high efficiency, stability in physiological buffers, lack of toxicity, and lack of sensitivity to serum. Pep-1 technology constitutes a powerful tool for basic research, and several studies have demonstrated that Pep-1 technology is extremely powerful for studying the role of proteins and for targeting specific protein/protein interactions *in vitro* as well as *in vivo* (Morris *et al.*, 2001; Gallo *et al.*, 2002; Wu *et al.*, 2002, Pratt and Kurch, 2002, Aoshiba *et al.*, 2003).

This article describes several protocols for the use of noncovalent Pep-1 technology for the delivery of biologically active proteins into mammalian cells *in vitro*.

II. MATERIALS AND INSTRUMENTATIONS

A. Products

β-Galactosidase is from the Chariot kit (Cat. No. 30025) or the β-galactosidase staining kit (Cat. No. 35001) from Active Motif (Carsbad-USA). Phosphate-buffered saline (PBS) (Cat. No. 14190-169), cell culture reagents Dulbecco's modified Eagle's medium (DMEM) (Cat. No. 41965-062), glutamine, and streptomycin/penicillin (Cat. No. 15140-130) are from Invitrogen Life Technologies (Carsbad, CA). Foetal bovine serum (FBS) is from PERBIO (Lot 3264EHJ, Cat. No. CH30160-03). Potassium ferricyanide (Cat. No. P3667), potassium ferrocyanide (Cat. No. P9387), magnesium chloride (Cat. No. M2670), 5-bromo-4-chloro-3-indolyl-β-D-galactopyranoside (X-gal; Cat. No. B4252), *N, N*-dimethylformamide (DMF) (Cat. No. D4551), formaldehyde (Cat. No. F8775), and glutaraldehyde (Cat. No. G5882) are from Sigma (St. Louis, MO).

B. Peptide Carrier Pep-1/Chariot

Pep-1 is 21 residue peptide KETWWETW WTEWSQPKKKRKV (molecular mass 2907 Da). The peptide is acetylated at it N terminus and carries a cysteamide group at its C terminus, both of which are essential for the stability of the peptide and the transduction mechanism (Morris *et al.*, 2001). Pep-1 can be synthesized in house or obtained from commercial sources. Protocol for the synthesis and purification of Pep-1 and derivatives are described in Mery *et al.* (1993) and Morris *et al.* (1999, 2001). Also, Pep-1 is manufactured under the name of Chariot by Active Motif Inc. (Carlsbad, CA; http://www.activemotif.com, Cat. No. 30025). The protein transduction kit contains lyophilized Chariot and additional reagents required for the transduction protocol, including a positive transduction control (β-galactosidase Cat. No. 35001). All the protocols described for Pep-1 are directly transposable to Chariot.

III. PROCEDURES

A. Preparation of Pep-1/Protein or Pep-1/Peptide Complexes

This procedure is modified according to Morris *et al.* (2001).

Solutions

1. *Storage of Pep-1 powder*: Pep-1 is stable at least one year when stored at −20°C in a lyophilized form

2. *Stock solution of Pep-1*: Take the vial containing the peptide powder out of the freezer and equilibrate for 30 min at room temperature without opening the vial. This step is essential to limit hydration of the peptide powder. Resuspend the peptide at a concentration of 2 mg/ml (concentration: 0.68 mM) in water or in PBS buffer. Mix gently by tapping the tube or by vortexing at low speed for 20 s.

3. *Storage of the Pep-1 solution*: Because repeated freeze/thaw cycles can induce peptide aggregation, is recommended to aliquot the Pep-1 stock solution into tubes containing the amount you expect to use in a typical experiment prior to freezing. The Pep-1 stock solution is stable for about 2 months when stored at −20°C.

Steps

1. Dilute the amount of protein or peptide in 100 μl of PBS for each reaction. Use 0.5–1 μg of protein or 0.1–0.5 μg of peptide or low molecular weight protein per transduction reaction. The protein or peptide can be used at concentrations varying from 50 n*M* up

to 1 μM or higher depending on the biological response expected.

2. For peptide or low molecular weight protein (<10 kDa) transduction, the Pep-1 solution must be diluted 1 : 10 in sterile H_2O (concentration: 68 μM). At this stage, sonication of the peptide solution is recommended, to limit aggregation, for 5 min in a water bath sonicator. A probe sonicator can also be used: place the tube in water and sonicate for 1 min at an amplitude of 30%. In a tube dilute the appropriate volume of Pep-1 into 100 μl of sterile water or PBS for each reaction. For optimal transduction the molar ratio between Pep-1 and protein or peptide is generally 15 : 1 to 20 : 1.

3. Add 100 μl of diluted protein to 100 μl diluted Pep-1. Mix gently by tapping the tube. It is necessary to make the Pep-1/protein complex in a concentrated solution, which will be added to the cells and then be diluted to the final transduction volume (600 μl for 35-mm culture plate).

4. Incubate at 37°C for 30 min to allow the Pep-1/protein complex to form and then proceed immediately to the transduction experiments. At this stage, Pep-1/protein complexes should not be stored.

B. Protein Transduction Protocol for Adherent Cell Lines

The protocol is modified for 35-mm culture plates according to Morris *et al.* (2001). The amount of protein, Pep-1, transduction volume, and number of cells should be adjusted accordingly to the size of the culture plate used.

Steps

1. In a 6-well or a 35-mm tissue culture plate, seed 0.3×10^6 cells per well in 2 ml of complete growth medium. Incubate the cells at 37°C in a humidified atmosphere containing 5% CO_2 until the cells are 50–70% confluent. It is recommended to pass the cells the day before treatment for a better response following transduction.

2. Aspirate the medium from the cells to be transfected.

3. Wash the cells with PBS.

4. Overlay the cells with the 200 μl Pep-1/protein complex. Add 400 μl of serum-free medium to the overlay to achieve the final transduction volume of 600 μl for a 35-mm plate.

5. Incubate at 37°C in a humidified atmosphere containing 5% CO_2 for 1 h.

6. Add 1 ml of complete growth medium to the cells. Do not remove the Pep-1/protein complex. Then

incubate at 37°C in a humidified atmosphere containing 5% CO_2 for 30 min to 24 h, depending on the cellular response expected. Peptides and proteins are fully released in the cells 1 and 2 h later, respectively.

7. Process the cells for observation or detection assays. Cells may be fixed or observed directly using live imaging technology.

8. For a 6-well plate, dilute Pep-1 into 600 μl of water or PBS and the protein into 600 μl of PBS. Mix the two solutions and then proceed as described in steps 1–7. For multiple transductions, do not use a mix that exceeds the volume required for six transductions per tube, as this may cause aggregation.

C. Protein Transduction Protocol for Suspension Cells

These conditions were optimized using several cell lines for the delivery of protein of size ranging from 5 to 100 kDa. However, efficient transduction may require optimization of Pep-1 concentration, cell numbers, and exposure time of cells to the Pep-1/protein complex.

Steps

1. The same number of cells recommended for seeding adherent cells is recommended for suspension cells (confluency between 50 and 70%). Culture cells in standard medium in 35-mm dishes or 6-well plates.

2. Form the Pep-1/protein or Pep-1/peptide complexes as described for adherent cells steps 1–4 in section III,A.

3. Collect the suspension cells by centrifugation at 400 g for 5 min. Remove the supernatant.

4. Wash the cells twice with PBS.

5. Centrifuge at 400 g for 5 min to pellet the cells. Remove the supernatant.

6. Solubilize the cell pellet in the Pep-1/protein or peptide complex solution (200 μl). Add serum-free medium to achieve a final transduction volume of 600 μl.

7. Incubate at 37°C in a humidified atmosphere containing 5% CO_2 for 1 h.

8. Add complete growth medium to the cells. Do not remove the Pep-1/protein or peptide complex. Continue to incubate at 37°C in a humidified atmosphere containing 5% CO_2 for 1 to 24 h depending on the expected cellular response. As described for adherent cells, peptides and proteins are fully released in the cells 1 and 2 h later, respectively.

9. Process the cells for observation or detection assays. Cells may be fixed or observed directly using live imaging spectroscopy.

D. Application: Transduction of β-Galactosidase as a Control Protein

Solutions

1. *Preparation of β-galactosidase*: β-Galactosidase is provided in a lyophilized form and stored at –20°C. Make a stock solution at 0.25 mg/ml in water. Aliquot into several tubes to avoid repeated freeze/thaw cycles, taking into account that 0.5–1.0 μg of β-galactosidase is used per transduction reaction on 35-mm plates. Store the stock solution at –20°C.

2. *Staining solution*: Prepare stock solutions in water. Mix 400 mM potassium ferricyanide (13.2 g for 100 ml), 400 mM potassium ferrocyanide (16.9 g for 100 ml), 200 mM of magnesium chloride (4.6 g for 10 ml), and 20 mg/ml X-Gal resuspended in DMF. Store solutions at –20°C in the dark. The staining solution should be prepared freshly: 250 μl of 400 mM potassium ferricyanide (4 mM final), 250 μl of 400 mM potassium ferrocyanide (4 mM final), 250 μl ml of 200 mM magnesium chloride (2 mM final), 1.25 ml X-Gal (20 mg/ml in DMF) (1 mg/ml final), and 23 ml of PBS.

3. *Fixation solution*: Prepare a 10-fold concentrated solution in PBS (10X) containing 20% formaldehyde and 2% glutaraldehyde. Store at 20°C.

1. Transduction Protocol

Steps

1. Dilute 6 μl of Pep-1 stock solution (0.68 mM) in 100 μl of sterile water.
2. Dilute 0.5 μg of β-Gal in 100 μl of PBS.
3. Add the Pep-1 solution to the β-Gal solution and mix gently
4. Incubate at 37°C for 30 min to allow the Pep-1/β-Gal complex to form.
5. Use the transduction protocol described in Section III,B (steps 1–7) for adherent cell lines and in Section III,C (steps 1–8) for suspension cell lines.

2. Staining Protocol

Steps

1. After 2 to 3 h, remove the growth medium from the transfected cells.
2. Rinse the cells three times with PBS.
3. Add fixing Solution. (Dilute the 10X stock in sterile water to make a 1X solution.) Incubate at room temperature for 5 min and rinse extensively with PBS.
4. Add a freshly made staining solution to the cells.
5. Incubate the cells at 37°C for 30 min to 2 h. Cells can be kept overnight at room temperature before analysis.
6. Analyze the percentage of cells transfected with β-galactosidase under a microscope.

E. Application: Delivery of Fluorescently Labelled Antibodies

Pep-1 strategy has been used for the delivery of antibodies into living cells. The protocol was optimized according to Morris *et al.* (2001) and to the Chariot guideline manual (www.activemotif.com).

Steps

1. Detection of fluorescently labelled antibodies in cells requires a large amount of antibody to be delivered. Depending on the sensitivity required to detect your antibody, it may be necessary to increase the amount of Pep-1 and antibody used. Using fluorescently labelled antibodies, concentrations in the final transduction volume of 0.1 μM for antibody and 5 μM for Pep-1 are recommended.
2. Incubate Pep-1 and antibody solution for 30 min at 37°C.
3. Overlay onto cultured cells for 1 to 3 h as described in Section III,C or III,D.
4. Extensively wash and observe cells by fluorescence microscopy. There is no need to fix the cells for observation.

IV. COMMENTS

A large variety of proteins, antibodies, and peptides have been successfully introduced into different cell lines using Pep-1/Chariot strategy and have been shown to retain their biological activity (Morris *et al.*, 2001; Buster *et al.*, 2002, Jurney *et al.*, 2002; Gallo *et al.*, 2002; Wu *et al.*, 2002; Pratt and Kinch, 2002; Aoshiba *et al.*, 2003). An update of published work using Chariot strategy is available on the Web site of Active Motif.

The advantages of the Pep-1 technology are directly associated with the mechanism through which this carrier promotes the delivery of proteins and peptides into cells.

a. The fact that Pep-1-mediated protein transduction is independent of the endosomal pathway significantly limits degradation and preserves the biological activity of the internalized protein for prolonged time periods.

b. An important property of Pep-1-based protein delivery is that it allows for the relatively rapid introduction of proteins into cells. That there is no need of covalent coupling for the formation of Pep-1/macromolecule particles favours rapid release of peptides or proteins into the cytoplasm as soon as the cell membrane has been crossed. Then the final localization of the macromolecule is determined by its inherent intracellular targeting properties. Entry of Pep-1/ macromolecule complexes into the cell occurs in a short time (10 min), and release of macromolecules varies from 1 to 3 h, depending on the nature of the protein and on its affinity for the target (Morris *et al.*, 2001).

c. It is important that the process of complex formation between the Pep-1/protein was performed in the absence of serum to limit interactions with serum proteins. However, the transduction process itself is not affected by the presence of serum, which is a considerable advantage for most biological applications (Morris *et al.*, 2001).

V. PITFALLS

1. Although this protocol was tested on several cells lines, conditions for efficient protein transduction should also be optimized for each cell line, including reagent concentration, cell number, and exposure time of cells to the Pep1/macromolecule complexes. β-Galactosidase should always be used as a positive control of transduction.

2. Low protein transduction efficiency may be associated with several parameters.

    a. For adherent cells, the optimal confluence is about 50–60%; a higher confluence (90%) reduces the transduction efficiency dramatically.

    b. Conditions for the formation of Pep-1/macro-molecule complexes are critical and should be respected. Special attention should be paid to the recommended volumes, incubation times for the formation of Pep-1/macromolecule complexes, and time of exposure of these complexes to cells.

    c. ***Transduced macromolecules may be cytotoxic.*** Transfect the macromolecule at a lower concentration. Compare untransfected cells, cells incubated with Pep-1 alone, and cells with macromolecule alone. Transfect the β-galactosidase control protein.

3. Pep-1 interacts via hydrophobic and hydrophilic interactions. Each peptide or protein will have a different hydrophobicity pattern and may require a different amount of Pep-1. The association of Pep-1 with proteins depends on the structure and biophysical properties of both components of the complex, and thus complex formation may vary between different combinations of Pep-1 and macromolecules (Morris *et al.*, 2001). The size of the protein or peptide may also affect the decaging procedure, and low efficiency

may be observed for peptides shorter than 15 residues.

4. Diluting Pep-1 in dimethyl sulfoxide (DMSO) may improve the delivery of high molecular weight proteins and antibodies. Pep-1 should first be diluted in 60% DMSO and then combined with the antibody or protein dilution. The final concentration of DMSO should not exceed 20%. However, as this amount of DMSO may be toxic, it may need to be lowered for certain cell lines.

References

Aoshiba, K., Yokohori, N., and Nagai, A. (2003). Alveolar wall apoptosis causes lung destruction and amphysematous changes. *Am. J. Respir. Cell. Mol. Biol.* **28**, 555–562.

Buster, D., McNally, K., and McNally, F. J. (2002). Katanin inhibition prevents the redistribution of ã-tubulin at mitosis. *J. Cell Sci.* **115**, 1083–1092.

Derossi, D., Joliot, A. H., Chassaings, G., and Prochiantz, A. (1994). The third helix of the antennapedia homeodomain translocates through biological membranes. *J. Biol. Chem.* **269**, 10444–10450.

Elliott, G., and O'Hare, P. (1997). Intercellular trafficking and protein delivery by a Herpesvirus structural protein. *Cell* **88**, 223–233.

Ford, K. G., Souberbielle, B. E., Darling, D., and Farzaneh, F. (2001). Protein transduction: An alternative to genetic intervention? *Gene Ther.* **8**, 1–4.

Futaki, S., Suzuki, T., Ohashi, W., Yagami, T., Tanaka, S., Ueda, K., and Sugiura, Y. (2001). Arginine-rich peptides: An abundant source of membrane-permeable peptides having potential as carriers for intracellular protein delivery. *J. Biol. Chem.* **276**, 5836–5840.

Gallo, G., Yee, H. F., and Letourneau, P. C. (2002). Actin turnover is required to prevent axon retraction driven by endogenous actomyosin contractility. *J. Cell Biol.* **158**, 1219–1228.

Gariepy, J., and Kawamura, K. (2000). Vectorial delivery of macromolecules into cells using peptide-based vehicles. *Trends Biotechnol.* **19**, 21–26.

Langel, U. (2002). "*Cell Penetrating Peptides: Processes and Application.*" CRC press, Pharmacology & Toxicology series.

Mery, J., Granier, C., Juin, M., and Brugidou, J. (1993). Disulfide linkage to polyacrylic resin for automated Fmoc peptide synthesis: Immunochemical applications of peptide resins and mercaptoamide peptides. *Int. J. Pept. Protein Res.* **42**, 44–52.

Morris, M. C., Chaloin, L., Heitz, F., and Divita, G. (2000). Translocating peptides and proteins and their use for gene delivery. *Curr. Opin. Biotechnol.* **11**, 461–466.

Morris, M. C., Chaloin, L., Mery, J., Heitz, F., and Divita G. (1999). A novel potent strategy for gene delivery using a single peptide vector as a carrier. *Nucleic Acids Res.* **27**, 3510–3517.

Morris, M. C., Depollier, J., Mery, J., Heitz, F., and Divita, G. (2001). A peptide carrier for the delivery of biologically active proteins into mammalian cells. *Nature Biotechnol.* **19**, 1173.

Pooga, M., Kut, C., Kihlmark, M., Hallbrink, M., Fernaeus, S., Raid, R., Land, T., Hallberg, E., Bartfai, T., and Langel, U. (2001). Cellular translocation of proteins by transportan. *FASEB J.* **8**, 1451.

Pratt, R. L., and Kinch, M. S. (2002). Activation of the EphA2 tyrosine kinase stimulates the MAP/ERK kinase signaling cascade. *Oncogene* **21**, 7690–7699.

Schwarze, S. R., and Dowdy, S. F. (2000). *In vivo* protein transduction: Intracellular delivery of biologically active proteins, compounds and DNA. *Trend Pharmacol. Sci.* **21**, 45–48.

Schwarze, S. R., Ho, A., Vocero-Akbani, A., and Dowdy, S. F. (1999). *In vivo* protein transduction: Delivery of a biologically active protein into the mouse. *Science* **285**, 1569–1572.

Wadia, J. S., and Dowdy, S. F. (2002). Protein transduction technology. *Curr. Opin. Biotechnol.* **13**, 52–56.

Wu, Y., Wood, M. D., Yi, T., and Katagiri, F. (2002). Direct delivery of bacterial avirulence proteins into resistant *Arabidopsis* protoplasts leads to hypersensitive cell death. *Plant J.* **33**, 130–137.

3

Selective Permeabilization of the Cell-Surface Membrane by Streptolysin O

Jørgen Wesche and Sjur Olsnes

I. INTRODUCTION

Streptolysin O (SLO) is a bacterial exotoxin secreted by β-hemolytic streptococci. The toxin binds to cholesterol in the cell surface of mammalian cells where it oligomerises to form large pores in the plasma membrane (Bhakdi *et al.*, 1985; Palmer *et al.*, 1998). The pores can reach a diameter of 35 nm and allow the release of large proteins from the cytosol. Importantly, SLO binds to cholesterol on the plasma membrane at 4°C, while pore formation requires 37°C (Bhakdi *et al.*, 1993). Thus, permeabilisation can be controlled easily by prebinding the toxin to cells at 4°C and then allowing permeabilisation to occur by incubating the cells at 37°C (Fig. 1).

SLO is used as a tool in cell biology to analyse the transport of toxins with intracellular targets (e.g., ricin and diphtheria toxin) and growth factors into the cytosol (Rapak *et al.*, 1997; Wesche *et al.*, 1999, 2000). In these experiments, labelled toxin or growth factor is added to intact cells and allowed to translocate into the cytosol before SLO is used to permeabilise the cells and allow the cytosolic proteins to leak out for analysis. In this way, only the protein that has reached the cytosol is obtained and not protein in intracellular membrane-bound compartments such as endosomes and the Golgi apparatus. This is very useful for studying the translocation of toxin mutants devoid of the enzymatic activity often used to assay translocation and for translocation of growth factors without defined cytosolic targets.

The toxin is also widely used in experiments involving semi-intact cells to follow cellular processes such as exocytosis and endocytosis. In these experiments, the plasma membrane is permeabilised selectively by

SLO to release the cytosol without affecting intracellular membranes. Cytosol can then be added back either depleted for certain proteins or supplemented with antibodies or membrane-impermeable inhibitors and then cellular processes may be investigated under these conditions.

II. MATERIALS

Minimal essential medium with Earle's salts and without $NaHCO_3$ (Cat. No. 61100-087) is from GIBCO. HEPES (Cat. No. H-3375) and MESNA (2-mercaptoethanesulfonic acid sodium salt, Cat. No. M-1511) are from Sigma. SLO is from S. Bakhdi (for information on how to order SLO from Bhakdi's laboratory please consult http://www.mikrobiologie.medizin.uni-mainz.de/en/index.html). SLO can also be supplied by Sigma (Cat. No. S-5265).

III. PROCEDURES

A. Optimizing Permeabilisation

The concentration of toxin used to obtain optimal permeabilisation in the cells of interest must be determined experimentally. Increasing concentrations of the toxin should be tested for the optimal release of cytosolic components. An easy way to measure the degree of release is to measure the activity of the enzyme lactate dehydrogenase (LDH, 135 kDa). Commercial kits exist for the analysis of LDH activity (Sigma, Cat. No. TOX-7).

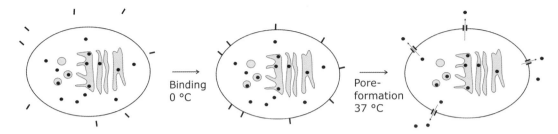

FIGURE 1   Selective permeabilisation using SLO. Radiolabelled ligand (black dots) may localise to intra-cellular membrane-bound compartments after endocytosis. Part of the ligand may also translocate to the cytosol. To discriminate between the different localisations, SLO is added (black lines) to selectively permeabilise the cell-surface membrane and allow the cytosolic proteins to leak out of the cell for analysis. SLO binds to the cell at 4°C, while insertion and pore formation require 37°C. Permeabilisation of the plasma membrane can therefore be controlled easily.

Solutions

1. *1 M MESNA*: To make 1 ml, dissolve 164.2 mg MESNA in 1 ml water

2. *HEPES medium*: Bicarbonate-free minimal essential medium with Earle's salts buffered with 20 mM HEPES to pH 7.4

Steps

1. Store SLO (~500 µg/ml) in aliquots at –80°C and thaw carefully on ice. To be active, SLO must be reduced by agents such as dithiothreitol or MESNA (MESNA is membrane impermeable and might be advantageous to avoid reduction of cellular components after addition of the toxin to cells).

2. Seed out cells growing in culture on tissue culture plates the day prior to the experiment. Before use, wash cells once in HEPES medium.

3. Dilute the toxin in HEPES medium to obtain the required concentrations (1–25 µg/ml) and add 10 mM MESNA to activate the toxin. After a 30-min incubation at 37°C with the reducing agent, add the toxin mixture to the cells and keep for 10 min at 4°C.

4. Remove nonbound toxin (and MESNA) by washing the cells once in HEPES medium.

5. Incubate the cells at 37°C for 10 min to allow permeabilisation to occur.

6. Transfer the cells to 4°C and keep for 30 min to let cytosolic proteins leak out into the medium.

7. The degree of permeabilisation can then be measured by analysing the release of the protein LDH into the medium. The optimal concentration of released LDH is normally between 60 and 80%. SLO concentrations that allow a higher release of LDH often affect intracellular membranes as well. To check if intracellular membranes have been affected, a Western blot analysis of organelle marker proteins may be used (Rapak *et al.*, 1997). Ligands that are endocytosed by

the cells, but not translocated to the cytosol (e.g., transferrin), can also be used to check if endosomes are affected by the treatment (Wesche *et al.*, 2000).

B. Experiment

In the following experiment, translocation of the growth factor FGF-1 is used as an example of how SLO can be used to monitor translocation of an exogenously added protein to the cytosol.

Steps

1. Seed out NIH/3T3 cells on 6-wells tissue culture plates at a density of 2×10^5 cells per well. The next day, change the medium to medium containing 0.5% serum for starvation of the cells. After 24 h, add radiolabelled FGF-1 to the cells and incubate further for 6 h. Wash the cells once in HEPES medium and once in high salt/low pH medium (2 M NaCl, pH 4.0, in phosphate-buffered saline) to remove surface-bound FGF-1. (Pronase may also be used to remove surface-bound proteins.) Then wash the cells an additional time with HEPES medium.

2. Add SLO to the cells and keep for 10 min at 4°C.

3. Remove nonbound toxin by washing the cells in HEPES medium.

4. Incubate the cells at 37°C for 10 min to allow permeabilisation to occur.

5. Keep the cells at 4°C for 30 min to let cytosolic proteins leak out into the medium.

6. The content of the cytosol (released into the medium) can now be analysed by SDS–PAGE and fluorography (see Fig. 2). The remaining cell components can be lysed (1% Triton X-100 in HEPES buffer) and fractionated into a membrane and nuclear fraction by centrifugation (Fig. 2).

FGF-1 appears in the cytosol after 4 h and reaches a maximum after 6 h. FGF-1 is also found in the nuclear

A

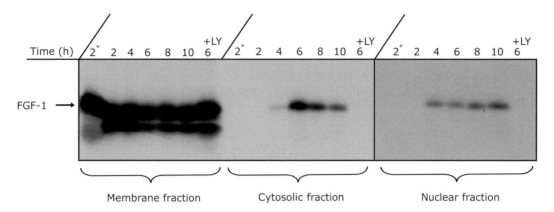

B

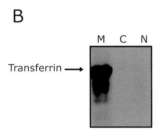

FIGURE 2 Transport of FGF-1 to the cytosol and the nucleus. Radiolabelled FGF-1 or transferrin was added to serum-starved NIH/3T3 cells and incubated for 6 h at 37°C. In lanes labelled +LY, the PI-3 kinase inhibitor LY294002 was added to block translocation to the cytosol and the nucleus. The time point labelled with an asterisk indicates that the incubation was performed at 4°C. The cells were then treated with SLO as described in the text. Cytosolic proteins were recovered from the medium and designated the cytosolic fraction. The remaining cell components were lysed with 1% Triton X-100 and fractionated into a nuclear fraction and a membrane fraction. FGF-1 was recovered from the different fractions by adsorption to heparin–Sepharose. The adsorbed material was analysed by SDS–PAGE and fluorography.

fraction. Transferrin that does not translocate to the cytosol or the nucleus is only found in the membrane fraction (Fig. 2).

IV. COMMENTS

The preparations of SLO vary a lot depending on the suppliers and there are also considerable batch variations. Each batch must therefore be tested for activity.

A mutant that does not require reduction has been constructed and has been reported to have the same activity as wild-type toxin (Pinkney *et al.*, 1989).

An alternative to SLO for plasma membrane permeabilisation is the mild nonionic detergent digitonin. As SLO, digitonin binds to cholesterol and therefore makes holes preferentially in the plasma membrane. A major difference, however, is that digitonin also permeabilises the cells at 4°C and is therefore more difficult to control.

Permeabilisation can also been done on cells in suspension by carefully centrifugating the cells between washes.

V. PITFALLS

SLO isolated from cultures of β-hemolytic streptococci may contain proteases that could cause cells growing on support to detach and may also attack intracellular proteins after permeabilisation (Bhakdi *et al.*, 1993).

Acknowledgment

J.W. is a fellow of the Norwegian Cancer Society.

References

Bhakdi, S., Tranum-Jensen, J., and Sziegoleit, A. (1985). Mechanism of membrane damage by streptolysin-O. *Infect. Immun.* **47**, 52–60.

Bhakdi, S., Weller, U., Walev, I., Martin, E., Jonas, D., and Palmer, M. (1993). A guide to the use of pore-forming toxins for controlled permeabilization of cell membranes. *Med. Microbiol. Immunol. (Berl)* **182**, 167–175.

Palmer, M., Harris, R., Freytag, C., Kehoe, M., Tranum-Jensen, J., and Bhakdi, S. (1998). Assembly mechanism of the oligomeric streptolysin O pore: The early membrane lesion is lined by a free edge of the lipid membrane and is extended gradually during oligomerization. *EMBO J.* **17**, 1598–1605.

Pinkney, M., Beachey, E., and Kehoe, M. (1989). The thiol-activated toxin streptolysin O does not require a thiol group for cytolytic activity. *Infect. Immun.* **57**, 2553–2558.

Rapak, A., Falnes, P. O., and Olsnes, S. (1997). Retrograde transport of mutant ricin to the endoplasmic reticulum with subsequent translocation to cytosol. *Proc. Natl. Acad. Sci. USA* **94**, 3783–3788.

Wesche, J., Rapak, A., and Olsnes, S. (1999). Dependence of ricin toxicity on translocation of the toxin A-chain from the endoplasmic reticulum to the cytosol. *J. Biol. Chem.* **274**, 34443–34449.

Wesche, J., Wiedlocha, A., Falnes, P. O., Choe, S., and Olsnes, S. (2000). Externally added aFGF mutants do not require extensive unfolding for transport to the cytosol and the nucleus in NIH/3T3 cells. *Biochemistry* **39**, 15091–15100.

Genes

4

New Cationic Liposomes for Gene Transfer

Nancy Smyth Templeton

I. INTRODUCTION

We developed improved liposomes that produce efficacy for the treatment of cancer (Ramesh *et al.*, 2001; Shi *et al.*, 2002; Tirone *et al.*, 2001), cardiovascular diseases, and HIV-1 related diseases in small and large animal models. These liposomes condense nucleic acids, mixtures of nucleic acids and proteins, and viruses (Yotnda *et al.*, 2002) on the interior of bilamellar invaginated vesicles (BIVs) produced by a novel extrusion procedure (Templeton *et al.*, 1997). These nucleic acid : liposome complexes have extended half-life in the circulation, are stable in physiological concentrations of serum, have broad biodistribution, efficiently encapsulate various sizes of nucleic acids and other molecules, are targetable to specific organs and cell types, penetrate through tight barriers in several organs, are fusogenic with cell membranes, are optimized for nucleic acid : lipid ratio and colloidal suspension *in vivo*, can be size fractionated to produce a totally homogeneous population of complexes prior to injection, and can be administered repeatedly. We can add specific ligands either by ionic interactions or by covalent attachments to the surface of these nucleic acid–liposome complexes to accomplish targeted delivery to specific cell surface receptors. The ligands include monoclonal antibodies, Fab fragments, peptides, peptide mimetics, small molecules and drugs, proteins, and parts of proteins. In addition, the charge on the surface of these complexes can be modified in order to avoid uptake by nontarget cells using our novel technology called "reversible masking." We have also achieved high-dose systemic delivery of these complexes without toxicity *in vivo* by further purification of plasmid DNA. We have developed proprietary technologies for the detection and removal of contaminants found at high levels in clinical grade plasmid DNA preparations produced by several different companies. For instance, these DNA contaminants preclude high-dose delivery of complexes that may be required for the treatment of metastatic cancer. Our "Super Clean DNA" also provides for elevated levels of gene expression and prolonged transient expression *in vitro* and *in vivo*. Our complexes have been injected into mice, rats, rabbits, pigs, nonhuman primates, and humans. Currently, these complexes are injected intravenously into patients in clinical trials to treat lung cancer and will be used in upcoming trials to treat breast, pancreatic, head and neck cancers, hepatitis B and C, and restenosis. This article focuses on the production of these BIV liposomes and on mixing nucleic acid–liposome complexes. Because our processes are reproducible, we have standard operating procedures for the cGMP manufacture of these reagents that have been approved by the Food and Drug Administration for use in phase I/II clinical trials.

II. MATERIALS AND INSTRUMENTATION

The following materials are necessary: ethanol, paper towels, gloves, spatulas, 1-liter round-bottomed flask, cork ring to hold the round-bottomed flask, timer, small weigh boats, sterile 5% dextrose in water (D5W; Abbott Laboratories Inc. Part No. 1522-02), cGMP grade (Cat. No. 770890) or GLP grade (Cat. No. 890890) 18:1 dimethyldioctadecylammonium bromide (DOTAP) powder form (Avanti Polar Lipids), cGMP

25

grade synthetic cholesterol (Sigma Cat. No. C1231), ACS grade chloroform (Fisher Cat. No. C298-1), 0-grade argon gas, RBS pF detergent concentrate (Pierce Cat. No. 27960), sterile glass vials with screw tops, sterile 50-ml polypropylene conical tubes (Falcon Cat. No. 352098), sterile 15-ml polypropylene conical tubes (Falcon Cat. No. 352097), sterile polysulfone syringe filters 1.0 µm pore size 13 mm diameter (Whatman Cat. No. 6780-1310), sterile polysulfone syringe filters 0.45 µm pore size 13 mm diameter (Whatman Cat. No. 6780-1304), sterile ANOTOP syringe filters 0.2 µm pore size 10 mm diameter (Whatman Cat. No. 6809-1122), sterile ANOTOP syringe filters 0.1 µm pore size 10 mm diameter (Whatman Cat. No. 6809-1112), glass pipettes, sterile polystyrene pipettes, lens paper, parafilm, sterile 10-cc syringes (Becton Dickinson Cat. No. 309604), sterile water (Baxter Cat. No. 2F7114), sterile microfuge tubes, sterile pipette tips with no barriers, latex control particles (Beckman-Coulter Part No. 6602336), plastic cuvettes with caps (Beckman-Coulter Part No. 7800091), conductivity calibration standard (Beckman-Coulter Part No. 8301355), and EMPSL7 mobility standard (Beckman-Coulter Part No. 8304073).

The following instrumentation is necessary: Class II type A/B3 laminar flow biological safety cabinet (hood), vortex, analytical balance, rotary evaporator (Buchi Model R-114), vacuum aspirator (Brinkmann Model B-169), water baths, circulating water bath (Lauda Model E100), freeze dryer (Labconco Model Freezone 4.5), low-frequency sonicator (Lab-Line TansSonic Model 820/H), pipette aid, pipetmen, spectrophotometer (Beckman Model DU640B), spectrophotometer turbidity cell holder (Beckman Part No. 517151), UV silica-masked semimicrocell (Beckman Part No. 533041), submicron particle size analyzer with computer (Coulter Model N4 Plus), and zeta potential analyzer (Coulter Model Delsa 440SX).

III. PROCEDURES

A. Liposome Preparation

This procedure prepares 30 ml of BIV liposomes. The final yield is about 95%; therefore, approximately 28.5 ml of 5× stock liposomes is produced.

1. Clean all work surfaces and the hood with 70% ethanol.
2. Allow lipids, DOTAP, and synthetic cholesterol to come to room temperature.
3. Weigh 420 mg of DOTAP and place into a 1-liter round-bottomed flask.
4. Weigh 208 mg of synthetic cholesterol and place into the 1-liter round-bottomed flask.
5. Dissolve the lipids in 30 ml of chloroform.
6. Rotate on the rotary evaporator at 30°C for 30 min to make a thin film.
7. Dry the film in the flask under vacuum for 15 min.
8. Under the hood, add 30 ml of D5W to the film, cover with Parafilm, and make small holes in the Parafilm with a syringe needle.
9. Rotate the flask in a 50°C water bath for 45 min, rotating and swirling the flask until the film is in solution.
10. Place the flask in a 37°C water bath and continue to swirl for 10 min.
11. Cover the flask with more Parafilm (over the Parafilm with holes) and allow flask to sit overnight at room temperature.
12. The next day, remove the Parafilm with no holes and sonicate the flask at low frequency (35 kHz) for 5 min at 50°C.
13. Under the hood, place the contents from the flask into a sterile 50-ml tube and aliquot liposomes evenly into three sterile 15-ml tubes.
14. Place one 15-ml tube into a circulating water bath at 50°C (nearby the hood) for 10 min.
15. Under the hood, remove the plungers from four sterile 10-cc syringes and attach one of each pore size sterile filter (1.0, 0.45, 0.2, and 0.1 µm) to the syringes.
16. Under the hood, rapidly filter the liposomes through all four filters, starting with the largest pore size (1.0 µm) and ending with the smallest pore size (0.1 µm).
17. Under the hood, collect liposomes in a sterile 50-ml tube.
18. Repeat steps 14–16 for the remaining two 15-ml tubes and collect all liposomes in the same sterile 50-ml tube (listed in step 17) under the hood.
19. Under the hood, aliquot the liposomes into sterile glass tubes with screw caps, flush with argon or nitrogen gas through a 0.2-µm filter, and store at 4°C.
20. Rinse the round-bottomed flask with hot water, then with ethanol, and fill with 1× RBS and allow to sit overnight. The next day, rinse out the flask, then rinse with hot water, then with distilled water, and finally with ethanol. Seal the flask with Parafilm after it is dry.

B. Quality Assurance/Quality Control Testing for the Liposome Stock

1. Place 10 µl of liposomes into 190 µl of sterile water. Read the sample on the spectrophotometer at

OD_{400} using the turbidity holder and corresponding cuvette. The reading should be about 4.0, and the acceptable range is any reading lower than 4.5 OD_{400}.

2. Determine the size for the L100 and L200 standards using the particle size analyzer. If the standards are within range, proceed to assessing the liposome stock. Place 12 µl of liposomes into 4 ml of sterile water in the plastic cuvette. Determine the particle size, which should be about 200 nm, and the acceptable range is between 50 and 250 nm.

3. Access the mobility for the conductivity calibration standard and the mobility standard using the zeta potential analyzer. If the standard is within range, proceed to assessing the liposome stock. Place 120 µl of liposomes into 1880 µl of sterile water. Determine the zeta potential, which should be about 65 mV, and the acceptable range is between 50 and 80 mV.

4. Assays for residual chloroform in the liposome stock should not detect any chloroform. All sterility tests should show no observable growth or no contamination.

C. Complex Preparation

Using pipetmen and microfuge tubes, final volumes ranging from 50 to 1200 µl of complexes can be prepared.

1. Bring all reagents to room temperature.
2. Under the hood, dilute the 5× stock liposomes to 2× in D5W in a microfuge tube. For example, to mix a 300-µl final volume of complexes, dilute 60 µl of 5× stock liposomes in 90 µl of D5W.
3. Under the hood in a second microfuge tube, dilute the stock nucleic acid in D5W to 1 µg/ul. The volume of diluted nucleic acid must be identical to the volume of diluted stock liposomes in step 2. The final mixed complexes have a concentration of 0.5 µg of nucleic acid/ul. *Note*: The stock nucleic acid should be vortexed well prior to removing any aliquot for mixing complexes.
4. Under the hood, pipette the diluted nucleic acid into the diluted liposomes by rapid mixing at the surface of the liposomes. Rinse up and down about twice using the pipetman.
5. Read the sample (10 µl of the complexes in 190 µl of sterile water) on the spectrophotometer at OD_{400} using the turbidity holder and corresponding cuvette. The OD_{400} should be about 0.8, and the acceptable range is between 0.65 and 0.95 OD_{400}. If the complexes fall out of this range, then the amount of nucleic acid used for mixing must be adjusted. Specifically, if the OD_{400} is too low, then more nucleic acid must be used for mixing. If the OD_{400} is too high, then less nucleic

acid must be used for mixing. If the nucleic acid is not adequately pure, then the appropriate OD_{400} of the complexes will not be obtained. *Note*: To avoid wasting material in order to establish the proper mixing conditions, we mix final volumes of complexes at 50 µl first before we begin large-scale mixing using new lots of nucleic acids.

6. When the proper mixing volumes have been established, more complexes can be mixed if needed following the steps given earlier. All complexes can be pooled after mixing, and the OD_{400} of the pooled complexes should be measured and should fall between 0.65 and 0.95 OD_{400}.

7. The particle size of the complexes can also be measured. Determine the size for the L300 and L500 standards using the particle size analyzer. If the standards are within range, proceed to assessing the complexes. Place 12 µl of complexes into 4 ml of sterile water in the plastic cuvette. Determine the particle size, which should be about 400 nm, and the acceptable range is between 200 and 500 nm.

8. Because the encapsulation of nucleic acids is spontaneous, the complexes can be administered after mixing. Complexes can also be stored in glass vials at 4°C and administered the following day.

IV. COMMENTS

The complexes can be administered and readministered into animals and humans by any delivery route. The dose and administration schedule will be determined by the specific nucleic acid therapeutic and disease model of the investigator and, therefore, must be optimized. Please also see Section V.

We have found that BIV DOTAP is best for transfection of cells in culture (Yotnda *et al.*, 2002; N. S. Templeton, unpublished data). The same protocol is used to produce these liposomes; however, cholesterol is not added. Complexes made with BIV DOTAP are mixed using the procedure given earlier.

V. PITFALLS

1. Many investigators are not set up to produce liposomes in their laboratories. Therefore, the BIV liposomes and other custom services can be purchased from our nonviral core facility at the Baylor College of Medicine (contact Nancy Smyth Templeton at **NANCYT@bcm.tmc.edu** to place orders). The BIV DOTAP:Chol liposomes described here are also sold

by Qbiogene and are listed as the *In Vivo* GeneShuttle in their catalog.

2. Using the optimal liposomes for transfection is not necessarily enough to ensure success. Other reagents must be optimized, such as expression plasmid design (Lu *et al.*, 2002) and plasmid DNA preparation (Templeton, 2002), for applications that use plasmid DNA.

References

Lu, H., Zhang, Y., Roberts, D. D., Osborne, C. K., and Templeton, N. S. (2002). Enhanced gene expression in breast cancer cells *in vitro* and tumors *in vivo*. *Mol. Ther.* **6**, 783–792.

Ramesh, R., Saeki, T., Templeton, N. S., Ji, L., Stephens, L. C., Ito, I., Wilson, D. R., Wu, Z., Branch, C. D., Minna, J. D., and Roth, J. A. (2001). Successful treatment of primary and disseminated human lung cancers by systemic delivery of tumor suppressor genes using an improved liposome vector. *Mol. Ther.* **3**, 337–350.

Shi, H. Y., Liang, R., Templeton, N. S., and Zhang, M. (2002). Inhibition of breast tumor progression by systemic delivery of the maspin gene in a syngeneic tumor model. *Mol. Ther.* **5**, 755–761.

Templeton, N. S. (2002). Liposomal delivery of nucleic acids *in vivo*. *DNA Cell Biol.* **21**, 857–867.

Templeton, N. S., Lasic, D. D., Frederik, P. M., Strey, H. H., Roberts, D. D., and Pavlakis, G. N. (1997). Improved DNA:liposome complexes for increased systemic delivery and gene expression. *Nature Biotech.* **15**, 647–652.

Tirone, T. A., Fagan, S. P., Templeton, N. S., Wang, X. P., and Brunicardi, F. C. (2001). Insulinoma induced hypoglycemic death in mice is prevented with beta cell specific gene therapy. *Ann. Surg.* **233**, 603–611.

Yotnda, P., Chen, D.-H., Chiu, W., Piedra, P. A., Davis, A., Templeton, N. S., and Brenner, M. K. (2002). Bilamellar cationic liposomes protect adenovectors from preexisting humoral immune responses. *Mol. Ther.* **5**, 233–241.

5

Cationic Polymers for Gene Delivery: Formation of Polycation–DNA Complexes and *in vitro* Transfection

Yong Woo Cho, Jae Hyun Jeong, Cheol-Hee Ahn, Jong-Duk Kim, and Kinam Park

I. INTRODUCTION

Although nonviral gene delivery systems were first introduced with cationic lipids, cationic polymers (or polycations) have gained increasing attention in recent years as a nonviral vector for gene therapy due to their nonimmunogenicity and low acute toxicity. In both systems, DNA is incorporated into a complex by electrostatic interactions between anionic phosphate groups of DNA and cationic groups of lipids or polycations under physiological conditions. A large number of polycations have been studied as nonviral vectors, such as poly-L-lysine (Wu and Wu 1987), poly(ethylenimine) (Behr *et al.*, 1999; Ahn *et al.*, 2002), poly(amidoamine) dendrimers (Haensler and Szoka, 1993), poly(2-dimethylaminoethyl methacrylate) (Cherng *et al.*, 1996), and chitosan (Lee *et al.*, 1998). Cationic polymers have flexibility in designing a carrier with well-defined structural and chemical properties on a large scale as well as the ability to introduce functional moieties (e.g., targeting moieties). In general, the ability to vary and control the physicochemical properties in cationic lipid–based system is relatively limited. This article describes methods that are used commonly in characterizing polycation-based gene delivery systems in *in vitro* transfection, focusing on complex formation between polycations and DNA.

II. MATERIALS AND INSTRUMENTATION

HEPES (Cat. No. H4034), bicinchoninic acid (BCA) protein assay kit (Cat. No. BCA-1), fetal bovine serum (FBS, Cat. No. F3885), 100× penicillin–streptomycin (Cat. No. P0781), trypan blue (Cat. No. T0076), methylthiazoletetrazolium (MTT, Cat. No. M5655), and poly(L-aspartic acid) (Cat. No. P5387) are from Sigma (St. Louis, MO). Polyethylenimine (PEI, Cat. No. 40,872-7), glycerol (Cat. No. 19,161-2), and dimethyl sulfoxide (DMSO, Cat. No. 27,043-1) from Aldrich (Milwaukee, WI). Dulbecco's modified eagle's medium (DMEM, Cat. No. 12100-046) and 10× trypsin–EDTA (Cat. No. 15400-054) are from Gibco-BRL (Carlsbad, CA). Agarose (Cat. No. 161-3101), ethidium bromide (EtBr Cat. No. 161-0433), TBE (Tris/boric acid/EDTA, Cat. No. 161-0733), bromphenol blue (Cat. No. 161-0404), and xylene cyanole FF (Cat. No. 161-0423) are from Bio-Rad (Hercules, CA). The plasmid maxi kit (Cat. No. 12162) is from QIAgen (Valencia, CA). pSV-β-galactosidase (Cat. No. E1081), β-galactosidase enzyme assay system (Cat. No. E2000), pGL3 control containing the SV40 promoter-driven luciferase reporter gene (Cat. No. E1741), and Luciferase assay system (Cat. No. E1500) are from Promega (Madison, WI).

29

The spectrofluorometer (Spex FluoroMax-2) is from JY Horiba (Edison, NJ). The gel electrophoresis (Mini-PROTEAN 3 electrophoresis cell) system is from Bio-Rad. The microtiter plate reader (SOFTmax PRO) is from Molecular Device Corp. (Sunnyvale, CA). The luminometer (Lumat LB9507) is from Berthold Technologies (Oak Ridge, TN). The dynamic light scattering (90 plus) is from Brookhaven Instruments Corp. (Holtsville, NY).

III. PROCEDURES

A. Formation of Polycation/DNA Complexes (Polyplexes)

Solutions

1. *HEPES-buffered saline:* 15 mM HEPES, 150 mM NaCl, pH 7.4. To make 500 ml, add 1.787 g HEPES and 4.383 g NaCl to 480 ml water and adjust the pH to 7.4 with 1 M NaOH. Adjust the total volume to 500 ml with water. Sterilize the buffer by filtering through a filter with a 0.22-μm pore size.

2. *Plasmid stock solution:* Plasmid DNA (e.g., pSV-β-galactosidase or pGL3 control containing the SV40 promoter-driven luciferase reporter gene) is transformed into *Escherichia coli*. The transformed cells are grown in larger quantities (0.5–1.0 liter) of Luria and Bertan (LB) broth. The plasmid DNA is isolated using the plasmid maxi kit from QIAgen according to instructions of the manufacturer. The plasmid DNA is collected in HEPES-buffered saline and stored at 4°C. The purity is confirmed by 1% agarose gel electrophoresis, and DNA concentration can be measured by UV absorption at 260 nm.

3. *Polycation stock solution, 5 mg/ml in HEPES-buffered saline:* Dissolve 50 mg of a cationic polymer in 10 ml HEPES-buffered saline. Sterilize the polymer solution through a filter with a 0.22-μm pore size.

Steps

1. DNA: Dilute the plasmid DNA stock solution to a final concentration of 20 μg/ml in HEPES-buffered saline.

2. Cationic polymers: Make a series of dilutions in HEPES-buffered saline (see Table I). The N : P ratio is defined as the molar ratio of amino groups in polycations to phosphate groups in DNA.

3. Add polycation solutions to plasmid DNA solutions at different N : P ratios and vortex gently.

4. Incubate at room temperature for 30 min to allow complex formation.

5. Store at 4°C.

B. Analysis of Polycation/DNA Complexes

1. Ethidium Bromide Displacement Assay

The degree of DNA condensation by polycations can be determined by an EtBr displacement assay using a fluorometer (Wadhwa *et al.*, 1995; Choi *et al.*, 1998). Ethidium bromide intercalates between stacked base pairs of double-stranded DNA to give a significant increase of fluorescence intensity. Addition of a polycation causes a large drop in fluorescence intensity due to displacement of ethidium bromide molecules from DNA, which indicates the condensation of DNA to form complex particles.

Solutions

1. *Ethidium bromide at 10 mg/ml:* The stock solution should be stored in a bottle wrapped in aluminum foil at 4°C and in the dark.

2. *Plasmid DNA stock solution: (See solutions in Section III,A)*

3. *Stock solutions of polycations: (See solutions in Section III,A)*

Steps

1. The spectrofluorometer is operated with an excitation wavelength (λ_{ex}) of 510 nm and an emission wavelength (λ_{em}) of 590 nm. Use slit widths set at 10 nm and an integration time of 3 s. Perform all experiments in triplicate.

2. Dilute the DNA stock solution to a final concentration of 10 μg/ml, including 0.4 μg/ml EtBr in a test curvette (total volume of 2 ml).

3. Incubate the DNA solution for 15 min to ensure interactions between DNA and EtBr.

4. Measure the fluorescence and calibrate to 100%.

5. Measure the background fluorescence with EtBr alone and set to 0%.

6. Add aliquots of the polycation stock solution sequentially to the DNA solution at various N : P ratios, mix gently, and measure the fluorescence after each addition.

7. Plot the graph the relative fluorescence (%) *vs* N : P ratio.

TABLE I   **Amount of Polyethylenimine (PEI) for Forming Polyplexes with Various N : P ratios[a]**

N : P ratio	0.2	0.5	1.0	1.5	2.0	3.0	4.0	5.0	6.0
PEI (μl)[b]	0.10	0.26	0.52	0.77	1.03	1.55	2.06	2.58	3.10

[a] Twenty micrograms of DNA is used.
[b] PEI stock solution at 5 mg/ml in HEPES-buffered saline.

8. Determine or compare the abilities of different polycations to condense DNA.

2. Agarose Gel Retardation Assay

The complex formation between polycations and DNA can be observed by a decrease of mobility of DNA in agarose gel electrophoresis (Ahn *et al.*, 2002).

Solutions

1. *Ethidium bromide solution*: *See solutions in Section III,B*

2. *Loading buffer solution:* 0.25% (w/v) bromphenol blue, 0.25% (w/v) xylene cyanole FF, 5% glycerol

3. *Electrophoresis buffer solution*: The ionic strength and pH of the buffer can play major roles in complex stability. Standard electrophoresis buffers such as TBE (90 mM Tris–borate, 2 mM EDTA, pH 8.3) and TAE (40 mM Tris–acetate, 1 mM EDTA, pH 7.9) commonly prove satisfactory, but others such as TE (10 mM Tris, 1 mM EDTA) have been used.

4. *Dissociation buffer solution*: Dissolve 100 mg poly(L-aspartic acid) in 5 ml double distilled water.

Steps

1. Prepare polycation/DNA complexes at various N : P ratios in HEPES-buffered saline and incubate at room temperature for 30 min to allow the complexes to form properly.

2. Add aliquots of the polyplexes (20 µl) in Eppendorf tubes containing 2 µl gel-loading buffer and mix gently.

3. Load the complexes onto wells of the 0.8% agarose gel containing ethidium bromide (1 µg/ml). Also, apply the controls for free DNA and a free polycation to the gel.

4. Perform electrophoresis in 0.5× TBE buffer at 100 V until the bromphenol blue has migrated 5–7 cm through the gel.

5. Visualize and photograph the electrophoresed gel on an UV illuminator to show the location of DNA and complexes.

6. Incubate the gels in the dissociation buffer for 30 min to disturb polycation/DNA complexes and rephotograph the gel to show the presence of DNA dissociated from the polymer.

3. Measurements of Particle Size
Solutions

1. *Plasmid DNA stock solution in double-distilled water*: *See solutions in Section III,A*

2. *Stock solutions of polycations in double-distilled water*: *See solutions in Section III,A*

Steps

1. Turn on dynamic light-scattering equipped with a He–Ne laser at a scattering angle of 90°.

2. Set parameters for software. Set viscosity to 0.890 centipoise (cP), refractive index (RI) medium to 1.333 in water. However, if complexes are in 150 mM NaCl, set viscosity to 1.145 cP and RI medium to 1.340; if in HEPES, set viscosity to 1.546 cP and RI medium to 1.363. Set the temperature to 25°C for all the solutions.

3. Prepare 500-µl complexes at 20 µg/ml in test tubes and filter through a 0.45-µm filter. Cap the sample and allow it to equilibrate for 30 min before initiating measurements.

4. Calculate the particle size and size distribution using nonnegative least squares (NNLS) algorithms. When the difference between the measured and calculated baselines is less than 0.2%, accept the correlation function. If not, the concentration of complexes can be controlled to give a reasonable signal.

5. Measure the mean particle size, polydispersity factor using the Stokes–Einstein equation, and the cumulant method.

C. In Vitro Transfection

This procedure describes the *in vitro* transfection of 293T cells using cationic polymers, such as PEI and PLL, as a vector. The protocol can be used for all adherent cell types with slight modifications. For suspension type cells, cells need to be spun down before changing media.

Solutions

1. *HEPES-buffered saline*: *See solutions in Section III,A*

2. *Phosphate-buffered saline (PBS)*: 137 mM NaCl, 2.7 mM KCl, 8.1 mM Na$_2$HPO$_4$, 1.47 mM KH$_2$PO$_4$, pH 7.4. To make 1 liter of 10× PBS, dissolve 80.06 g NaCl, 2.01 g KCl, 11.50 g of Na$_2$HPO$_4$, and 2.00 g KH$_2$PO$_4$ in 800 ml water. Adjust the pH to 7.4 with 1 N HCl. Adjust the total volume to 1 liter with water. To make 1 liter of 1× PBS, mix 100 ml 10× PBS (described earlier) with 900 ml of water. Sterilize the buffer through a filter with a 0.22-µm pore size.

3. *Tissue culture medium*: Tissue culture medium may vary, depending on the requirements of the cell line. Typically, Dulbecco's modified Eagle's medium (DMEM) supplemented with 10% fetal bovine serum, 100 units/ml penicillin–streptomycin, and 2 mM L-glutamine are used to maintain cell lines.

4. *Plasmid stock solution: See solutions in Section III,A*

5. *Stock solutions of polycations: See solutions in Section III,A*

6. *0.05% trypsin–0.02% EDTA solution (1×):* Add 10 ml of 10× trypsin–EDTA to 90 ml of PBS and store at 4°C

7. *0.5% trypan blue solution:* Dissolve 0.5 g trypan blue in 100 ml PBS and filtrate through a filtration paper to remove possible crystals. Store at –20°C

Steps

1. Passage 293T cells 3–4 days before the transfection experiment.

2. Detach the cells with trypsin–EDTA solution and determine the cell number and cell viability using trypan blue. Mix 50 µl cell suspension and 50 µl of 0.5% trypan blue solution. Bring the mixture into a counting chamber and count the number of uncolored (vital) and blue (dead) cells in a number of squares using a microscope. When more than 10 cells are counted per square, dilute the cell suspension and count again.

3. Seed cells in a 6-well tissue culture plate at a density of ~2 × 10^5 cells/well in 2 ml completed DMEM and incubate overnight at 37°C in a humidified atmosphere under 5% CO_2.

4. Prepare polycations/DNA complexes at various N : P ratios in HEPES-buffered saline and incubate for 20 min at room temperature (see steps in Section III,A).

5. Remove culture media from the cells and replace with 2 ml of serum-free DMEM.

6. Introduce 400 µl polyplexes to each well and incubate for 4 h at 37°C in an incubator.

7. Aspirate transfection media, replace with 2 ml of completed DMEM, and culture the cells in an incubator for 48 h at 37°C.

8. Evaluate the transfection efficiency.

D. Determination of Transfection Efficiency

1. β-Galactosidase Activity

The β-galactosidase activity in transfected cell lysates can be determined using the substrate *O*-nitrophenyl-β-ᴅ-galactopyranoside (ONPG).

Solutions

1. *Phosphate-buffered saline: See solutions in Section III,C*

2. *β-Galactosidase enzyme assay system with reporter lysis buffer (Promega) including reporter lysis buffer (5×), assay 2× buffer, β-galactosidase, 1 M sodium carbonate*

Steps

1. Remove growth media from the cells and wash twice with 2 ml PBS.

2. Add 200 µl of a lysis buffer to cover the cells and rock the 6-well plate slowly several times to ensure complete coverage of the cells.

3. Incubate at room temperature for 15 min.

4. Transfer the cell lysate to a microcentrifuge tube.

5. Centrifuge at top speed in a microcentrifuge for 2 min and transfer the supernatant to a fresh tube.

6. Pipette 50 µl of the cell lysates into wells of a 96-well plate.

7. Meanwhile, make a series of dilutions for the standard curve of β-galactosidase in 1× lysis buffer between 0 and $5.0 × 10^{-3}$ units. Prepare 50 µl of each β-galactosidase standard per well.

8. Add 50 µl of assay 2× buffer to each well of the 96-well plate and mix by pipetting.

9. Incubate at 37°C until faint yellow color has developed or the highest standard shows an absorbance of 2 or more.

10. Stop the reaction by adding 150 µl of 1 *M* sodium carbonate.

11. Measure the absorbance of the samples at 420 nm in a microtiter plate reader and calculate the β-galactosidase amount of a sample by comparing with the linear standard curve.

12. The β-galactosidase amount in each sample is normalized to milligrams of protein. Protein concentrations in cell lysates can be measured using a BCA protein assay kit according to instructions of the manufacturer.

2. Luciferase Activity

Solutions

1. *Phosphate-buffered saline: See solutions in Section III,C*

2. *Luciferase assay system (Promega) including a lysis buffer and luciferase assay reagent*

Steps

1. Remove growth media from the cells and wash twice with 2 ml PBS.

2. Add 200 µl of a lysis buffer to cover the cells and rock the 6-well plate slowly several times to ensure complete coverage of the cells.

3. Incubate at room temperature for 15 min.

4. Transfer the cell lysate to a microcentrifuge tube.

5. Centrifuge at top speed in a microcentrifuge for 2 min and transfer the supernatant to a fresh tube.

6. Dispense 20 µl of the supernatant into a luminometer tube.

7. Set the read time to 10 s (the read time can be varied).

8. Initiate reading by injecting 100 µl of Luciferase assay reagent into the tube.

9. Luciferase activity in each sample is normalized to milligrams of protein. Protein concentrations in cell lysates can be measured using a BCA protein assay kit according to instructions of the manufacturer.

E. *In Vitro* Cytotoxicity Assay

Cytotoxicity of polycation/DNA complexes can be evaluated by the MTT assay. The assay is based on the cleavage of the yellow tetrazolium salt, MTT, to form dark blue formazan crystals by active mitochondria dehydrogenases. This conversion only occurs in living cells.

Solutions

1. *MTT stock solution, 5 mg/ml in PBS*
2. *Tissue culture medium:* DMEM
3. *Polyplexes solution*

Steps

1. Seed cells in 24-well microplates at a density of ~4 × 10⁴ cells/well in 1 ml of completed DMEM and incubate overnight at 37°C in a humidified atmosphere under 5% CO_2.
2. Remove culture media from the cells and replace with 1 ml of serum-free DMEM.
3. Introduce 40 µl of polyplexes to each well and incubate for 4 h at 37°C in an incubator.
4. Aspirate transfection media, replace with 1 ml of completed DMEM, and culture cells in an incubator for 48 h at 37°C.
5. Remove old media and replace with new completed DMEM.
6. Add 50 µl of 5 mg/ml MTT solution to each well and incubate for 4 h at 37°C.
7. Remove MTT-containing media and add 750 µl of dime to each well and pipette up and down to dissolve formazan crystals formed by live cells.
8. Measure the absorbance at 570 nm using a microtiter plate reader
9. Calculate the absorbance percentage relative to that of untreated control cells.

IV. COMMENTS

Complex formation between polycations and DNA can be affected by several factors, such as the nature of the cation, the molecular weight of the polycation, the molecular architecture of the polycation, and the N : P ratio. Full complexation to form stable nanoparticles between 20 and 100 nm is necessary for effective transfection. Generally, shorter polycations need a higher N : P ratio to achieve full condensation.

V. PITFALLS

1. Large aggregates or precipitated materials can be observed in some cases. They lower the transfection efficiency and may cause toxic effects to cells. They may occur when the N : P ratio is too low or the pH of the transfection medium is too high. Generally, small cationic particles can be formed at the excess of positive groups of polycations compared with negative groups of DNA.

2. The size of polycation/DNA complexes may vary in different buffers mainly due to their different ionic strengths.

3. In some cases, complex particles may not be spherical. Particles with any extended structures may be ignored with light scattering, which leads to inaccurate results. Therefore, the morphology of complex particles must be checked using transmission electron microscopy or atomic force microscopy, as well as characterizing their size and distribution using light scattering.

4. Cell transfection can be performed in the absence or presence of serum. The effect of serum in the transfection medium will vary depending on the nature of polycations.

5. Cells of high passage number are transfected inefficiently.

References

Ahn, C.-H., Chae, S. Y., Bae, Y. H., and Kim, S. W. (2002). Biodegradable poly(ethylenimine) for plasmid DNA delivery. *J. Control. Release* **80**, 273–282.

Behr, J. P., Kichler, A., and Erbacher, P. (1999). Polyethylenimine: A family of potent polymers for nucleic acid delivery. *In "Nonviral Vectors for Gene Therapy"* (L. Huang, M.-C. Hung, and E. Wagner, eds.), pp. 192–207. Academic Press, New York.

Cherng, J.-Y., Wetering, P. V. D., Talsma, H., Crommelin, D. J. A., and Hennink, W. E. (1996). Effect of size and serum proteins on transfection efficiency of poly((2-dimethylamino)ethyl methacrylate)-plasmid nanoparticles. *Pharm. Res.* **13**, 1038–1042.

Choi, Y. H., Liu, F., Kim, J.-S., Choi, Y. K., Park, J. S., and Kim, S. W. (1998). Polyethylene glycol-grafted poly-L-lysine as polymeric gene carrier. *J. Control. Release* **54**, 39–48.

Haensler, J., and Szoka, F. C., Jr. (1993). Polyamidoamine cascade polymers mediate efficient transfection of cells in culture. *Bioconj. Chem.* **4**, 372–379.

Lee, K. Y., Kwon, I. C., Kim, Y.-H., Jo, W. H., and Jeong, S. Y. (1998). Preparation of chitosan self-aggregates as a gene delivery system. *J. Control. Release* **51**, 213–220.

Wadhwa, M. S., Knoell, D. L., Young, A. P., and Rice, K. G. (1995). Targeted gene delivery with a low molecular weight glycopeptide carrier. *Bioconj. Chem.* **6**, 283–291.

Wu, G. Y., and Wu, C. H. (1987). Receptor-mediated *in vitro* gene transformation by a soluble DNA carrier system. *J. Biol. Chem.* **262**, 4429–4432.

CHAPTER

6

Electroporation of Living Embryos

Takayoshi Inoue, Kristen Correia, and Robb Krumlauf

I. INTRODUCTION

Gene transfer by electroporation (EP) has been widely used to introduce exogenous molecules into both prokaryotic and eukaryotic cells. In the process of EP, transient pores at the cellular membrane are generated by electric shocks, allowing charged macromolecules such as DNAs, RNAs, and proteins to actively penetrate into cells by means of electrophoresis (Fig. 1A). Noncharged molecules can also be incorporated into cells by molecular diffusion. It was Japanese physicians who initially tried to perform gene therapy with EP *in vivo* (Okino and Mohri, 1987; Nishi *et al.*, 1996). However, the main difficulty in applying this approach to living tissues or organisms in the past was that the electric shocks often damaged cells and resulted in substantial cell death. A key breakthrough was the discovery that a rapid series of controlled square wave pulses (Fig. 1B) reduces the level of cell death dramatically. Using this modification, Muramatsu and colleagues (1997) reported the first remarkable results of EP-mediated gene transfer into developing chicken embryos. Based on these observation, chick embryologists around the world are now routinely utilizing this convenient technology perfected in Japan and it is also being applied to many other living tissues and organisms, including mammals (Momose *et al.*, 1999; Akamatsu *et al.*, 1999; Hass *et al.*, 2001; Reviewed in Itasaki *et al.*, 1999; Yasugi and Nakamura, 2000; Shwartz *et al.*, 2001; Inoue and Krumlauf, 2001; Osumi and Inoue, 2001; Takahahi *et al.*, 2002). As EP-mediated gene transfer is relatively nontoxic to cells and/or tissues and it is possible to precisely control the area selected for transfection (Fig. 1A), the method could further provide a powerful tool for human gene therapy. This article summarizes optimal conditions for EP-mediated gene transfer into chicken and mouse embryos, particularly with respect to situations for controlling the area of transfection.

II. PROCEDURES

A. EP-Mediated Gene Transfer in Chicken Embryos at the Early Primitive Streak Stage (HH stage 3+/5−)

Materials and Instrumentation

A variety of methods to efficiently target restricted regions in chicken embryos by EP have been described previously (Muramatsu *et al.*, 1997; Momose *et al.*, 1999; Itasaki *et al.*, 1999; Yasugi and Nakamura, 2000; Schwarts *et al.*, 2001). Above all, Momose *et al.* (1999) reported the microelectroporation method in which they can precisely target a confined region using the various positioning of a needle type and a wire type electrode. Here we introduce a modified protocol to facilitate EP into relatively young chicken embryos (stages HH 3+/5−). The following materials are required for this protocol.

Tape (3M scotch tape 1 in. width, 3M, USA); needle/spatula (#18-568, Miltex Instruments, USA); blunt scissors (#18-1478, Miltex Instruments, USA); transfer pipette (#232-205, Samco, USA); capillary tubes (#GC150F-10, Harvard Apparatus, USA); electrophoresis pipette tips (#23 35 165-6, Eppendorf, Germany); electroporator (CUY21, Bex Co. Ltd., Japan); electronic leads and 0.5-mm-diameter platinum wire electrode (Computech, USA); and forceps (#17-305X, Miltex, USA).

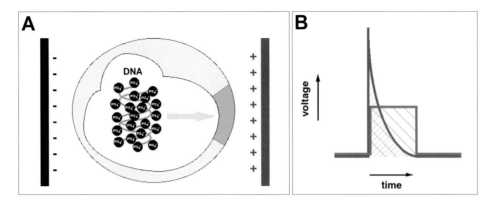

FIGURE 1 (A) The principle of electroporation (EP)-mediated gene transfer. As the DNA molecule includes negatively charged phosphate groups in its backbone (black circles), it naturally charges negative as a whole and will migrate toward the positive electrode (red) in the process of EP. As a result, a restricted area (green) can be targeted in the relevant tissue. (B) For living tissues or embryos, controlled square pulses (red) instead of canonical ones simply decay exponentially (blue) should be applied during EP. Note that the square pulse generates higher amount of energy at relatively lower voltages than the canonical one (the amount of energy is equal to the area with blue or red oblique).

Solutions

1. Plasmid DNA, Qiagen column purified and resuspended in sterile ddH₂O
2. 5% fast green (#EM-4510, VWR)
3. 10X phosphate-buffered saline (PBS): Add 8 g NaCl, 0.2 g KCl, 1.15 g Na₂HPO₄, and 0.2 g KH₂PO₄ to 100 ml of H₂O. Autoclave.

Steps

1. Preparing the DNA. On the day of EP, make 5 μl of solution containing 2 μg/μl plasmid DNA and 0.005% fast green in 1X PBS. It is possible to use two or more different plasmids (see comments later). Centrifuge the DNA solution for 30 s at 13,000 rpm to remove any particulates that might clog the capillary tip. Load 5 μl of the DNA solution into a pulled capillary tube using an electrophoresis pipette tip. Allow the DNA solution to reach the tip of the capillary tube and connect the capillary to a suction pipette.

2. Incubating the eggs. Incubate the eggs on their side until HH stage 3+/5−. This should be approximately 24 h in a humidified incubator at 35°C. Remove the eggs and set a timer for 2 h. Spray the top of the eggs with 70% ethanol and wipe off any debris. Put a piece of tape on the top of each egg, lengthwise. Adhere the tape to the eggshell, making sure that all creases are completely sealed to prevent leaking of egg white. The concave side of the spatula can be used to flatten the tape against the egg.

3. Opening the eggs. Set an egg horizontally on the egg holder. Use the point of the blunt scissors to put a

hole in the top of the egg. Cut a long, 2 × 3-cm oval within the taped area of the egg, using short cuts to prevent cracking, leave the shell attached, and fold back. The embryo should be located in the center of the yolk. If the embryo is not at the center of the yolk, discard the egg. Remove any bubbles and 1–2 ml of thin egg white with a transfer pipette to lower the level of the embryo below the cut edge of the shell. Cover the embryo with a drop of thin egg white using the transfer pipette.

4. Staging the eggs. Place the egg under the microscope. Illuminate the egg from one side to view and stage the embryo (Hamburger and Hamilton 1951). Do **not** inject ink, as this will kill the embryo. Stage 3 embryos are round, whereas stage 4 to 5 embryos are pointed near the posterior end. The streak will be half the embryo length at stage 3 and three-quarters the embryo length at stage 4. At stage 5, the head process is visible anterior to the streak. If the embryo is abnormal in its development (if the shape is not symmetrical or if there is more than one streak), discard the egg. To optimize survival while electroporating the posterior mesoderm, choose embryos that are stage 4+/5−. To electroporate the most anterior mesoderm, such as head mesoderm and anterior somites, choose embryos that are stage 3+/4−.

5. Injecting DNA solution. To make a small hole in the vitelline membrane over the embryo, run the capillary tip over the membrane to create a small fold over the tip. Then lift the tip quickly, cutting a small hole in the fold. Insert the needle tip into the hole in the membrane and inject about 0.25 μl of DNA into the space

between the embryo and the membrane (Fig. 2A). Be careful not to scratch the surface of the embryo with the capillary tip. If the capillary tip becomes clogged, run the tip over the taped edge of the eggshell while expelling the DNA solution. After injection, the DNA should diffuse over the top of the embryo. If the yolk was injected, the DNA will be localized in a small area. If the DNA diffuses past the outside boundary of the embryo, then the DNA was injected on top of the vitelline membrane. In good injections, the primitive streak should be highlighted by the fast green tracer (Fig. 2A inset). Fast green can also highlight scratches in the surface of the embryo made by the capillary tip.

6. Performing electroporation. Set the electroporator to 7 V; 50 ms time length of the pulse (pON); 150 ms time length between pulses (pOFF); and three repetitions of pulses. Position the ends of the upper negative (black) and lower positive (red) wires parallel to one another and 4 mm apart (Fig. 2A). At about 20 mm from the ends, bend both wires up and back to help avoid contacting the eggshell when the electrode is positioned over the embryo. Dip electrode briefly in egg white. Pierce the yolk outside of the area opaca, about 25 mm posterior to the embryo with the forceps. Insert the lower wire and let the yolk settle back in place. Approximately 1 cm of the lower electrode should be contacting the egg white and yolk. Position the upper wire over the streak with the end at the node and the side covering the streak. Approximately 2 mm of the upper electrode should be contacting the surface of the egg white. This situation makes a radial electric field rather than a parallel one, enabling one to target a confined area giving rise to somites. While administering the pulses, small bubbles should form on the upper visible electrode tip. After administering the pulses, carefully remove the electrode and wipe it clean with a paper towel.

7. Reincubation of the egg. Replace the shell and tape up the window using two overlapping pieces of tape. Make sure that all creases are completely sealed to prevent leaking of egg white. Eggs should be returned to the incubator within 2 h of having been removed for maximum viability. After 24 h the electroporated DNA reporters should be expressed in the posterior somites and presomitic mesoderm if electroporated at stage 4+/5– (Figs. 2B–2E) and in the head mesoderm and anterior somites if electroporated at stage 3+/4– (Figs. 2F–2I).

Comments

1. **Promoters.** The expression plasmid pCA (see Section II,B) used in these experiments directs very strong and ubiquitous expression in most tissues of chick embryos. Other plasmids containing promoters

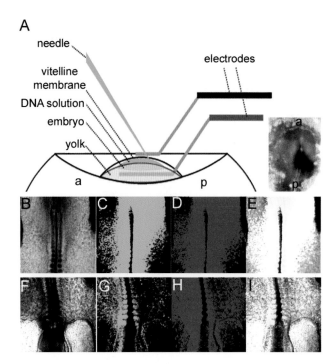

FIGURE 2 (A) Diagram of DNA injection and electroporation. To introduce the DNA solution, the tip of the injection needle is placed between the vitelline membrane and the embryo. To electroporate, the lower, red electrode is positioned within the yolk and the upper, black electrode is placed on top of the vitelline membrane overlying the embryo. The electrodes should be aligned with the anterior (a)/posterior (p) axis of the embryo as shown. (Inset) Stage 3+/4– chick embryo injected with DNA. Anterior is to the top. The DNA solution appears black, the embryo is gray, and the area opaca is white. The DNA solution collects within the streak and also diffuses over the surface of the embryo. (B–E) Stage 10 chick embryo electroporated at stage 4+/5– and incubated *in ovo*. Anterior is to the bottom. The plasmids pCA-dsRed express and pCA-GFP were combined at a concentration of 0.5 and 0.05 µg/µl, respectively. (B) Light microscope image of the posterior end of the embryo; anterior is to the bottom. (C) The CA promoter drives expression of GFP in all electroporated cells. GFP is expressed in somites, the presomitic mesoderm, and the lateral plate mesoderm. The neural tube of this embryo was not electroporated. (D) The same promoter drives expression of dsRED express in the same tissues. (E) Merged image of B, C, and D. Areas where strong GFP and dsRED expression overlap are yellow. (F–I) Stage 10 chick embryo electroporated at stage 3+/4– and incubated *in ovo*. Anterior is to the bottom. The plasmids pCA-dsRed express and cHoxb1 3'pGZ40-GFP were combined at a concentration of 0.25 and 0.75 µg/µl respectively. (F) Light microscope image of the head of the embryo; anterior is to the bottom. (G) The GFP protein driven by the Hoxb1 3' enhancer is expressed strongly in somites and at a lower level in the head mesoderm. Its expression is driven by enhancers within region 3' of chicken Hoxb1, which are linked to the chicken β-globin minimal promoter. (H) The dsRed express protein is driven by a ubiquitous promoter and marks the majority of electroporated cells. It is expressed in the head mesoderm, but not in the neural tube. Most embryos are also electroporated in the neural tube by this method, but this embryo clearly shows that expression in the mesoderm of this embryo is not due to the neural crest. (I) Merged image of F, G, and H. Areas where intense expression of both GFP and dsRed overlap are yellow.

that mediate widespread expression in chick embryos include pCAGGS (Niwa *et al.*, 1991), MiwSV (Tomomura *et al.*, 1990), pBK-CMV (Stratagene, USA), and pCS2 (Turner and Weintraub, 1994). For tissue-specific expression, a choice of unique enhancer sequences can be linked to the minimal β-globin promoter in pGZ40 (Itasaki *et al.*, 1999).

2. **Stage and rate of survival.** The stage of the embryo at the time of EP affects the survival rate. At stage 3+, approximately 50–60% of embryos survive and develop normally. This mainly relates to a decrease in survival as a consequence of windowing of the egg and is not generally an affect of the EP itself. At stage 5–, approximately 80–90% of embryos develop normally.

3. **Multiple plasmids.** Depending on the strength of the promoters, it may be possible to use multiple plasmids. For example, a DNA concentration of 0.05 μg/μl is sufficient to visualize expression of pCA-GFP. Therefore, more DNA (2.45 μg/μl) from a plasmid with a weaker enhancer can be combined with the DNA of a stronger expressing plasmid. Mix both plasmids in the same tube before loading in the capillary tube for reproducibility. In the example shown, GFP is used as a positive control to determine how well the EP has worked (efficiency) and to determine what general area was electroporated.

Pitfalls

1. **Embryo drying.** A sign that the embryo is drying out during incubation is the presence of a hardened area of yolk adhering to the embryo. If the window is sealed well with tape and the embryos are still drying out, rotating the eggs at a 90° angle can help. This results in the uncut shell being positioned over the top of the embryo. Also, transient drying during embryo manipulation can affect the development of the embryo. Be sure to keep a drop of thin egg white on top of the embryo during the procedures.

2. **Positioning the electrode.** Inserting and removing the electrode can be difficult if the window is not large enough or if the embryo is oriented with its rostrocaudal axis along the short axis of the oval window. Keep the microscope on a low power so that the edge of the window can be seen. Rotate the top of the egg toward you so that the edge of the window is lower on the side facing the operator. If the bottom wire contacts the eggshell, pushing down may cause it to bend upward unexpectedly, poking into the embryo or narrowing the distance between the electrodes.

3. **Too much injectant.** If there is a large outpocketing of yolk sac, this is a sign that too much solution is being injected. If the opening of the capillary tip is too large, it will be difficult to inject the proper small volume of DNA solution. Too much fast green in the solution may also be toxic for embryos.

B. EP-Mediated Gene Transfer into Cultured Mouse Embryos

Materials and Instrumentation

For mouse whole embryo culture (WEC), we recommend a rotator bottle system with a continuous gas supply (from BTC engineering, UK, or Type10-0310 from Ikemoto Rika, Japan). The system allows mouse embryos at post-implantation stages (E7.0–E13.0) to develop normally for a few days *in vitro* culture. This time frame covers many of the dynamic events during mouse early organogenesis, permitting analysis of developmental processes over this period with EP (Fig. 3A). Embryo cultures may be initiated from E6.5, but it is hard to target a specific area with EP at this stage. Mouse embryos become more placenta dependent after E10.0 and the placenta does not grow well in the culture system. Hence it is very difficult to keep the *in vitro* embryo culture effective following this stage. To maintain embryos in good condition, the yolk sac and amnion must be opened to directly expose the embryo to culture media (Osumi and Inoue, 2001; Fig. 3L).

To obtain as many embryos as possible for EP, we use the CD-1/ICR mice strain (Charles River, USA). However, other mouse strains, including transgenic and/or mutant mice lines, are equally acceptable for the system. The morning when vaginal plugs are detected is designated as E0.5. Pregnant mice were sacrificed at the developmental stages of interest, as approved by the IACUC Review Committee for Animal Experimentation of the Stowers Institute for Medical Research.

An electroporator that can generate controlled square pulses is essential to EP of mouse embryos. We use CUY-21 (Bex Co. Ltd., Japan), which generates most stable pulses at lower voltages. For mouse experiments, other types of electroporators, such as the BTX-T820 (BTX, USA), are acceptable because relatively higher voltages compared with the chicken experiments are sufficient for efficient EP(see earlier discussion).

For mouse embryos, we used specially designed chamber type electrodes, where the distance between the electrodes is 2.5 cm (Fig. 3B). The electrodes and electric wires are now available commercially from TR Tech/Bex Co. Ltd., Japan. Electrodes made of platinum yield the best results because the metal is resistant to ionization during EP. For DNA injection, microcapillaries (#B100-58-10, Sutter Inc.) are pulled and the solutions are filled by suction with mouth pipette.

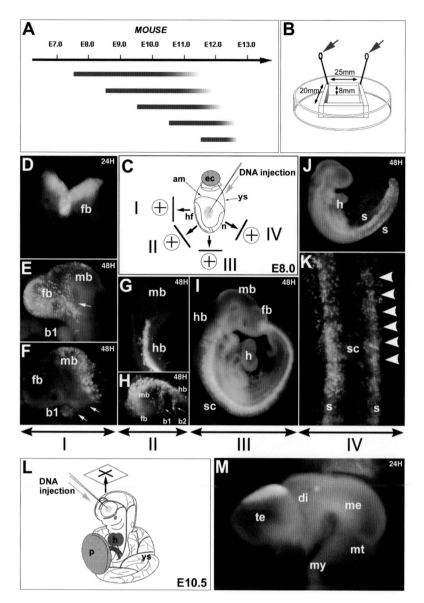

FIGURE 3 (A) Possible time schedules for mouse whole embryo culture after electroporarion (EP). The left edge of a blue bar indicates the start point of culture with EP, and the solid blue area represents the duration in which more than 75% of embryos can grow normally *in vitro*. (B) The electrode chamber designed for mouse embryos at any developmental stages. We made the chamber with silicone blocks on the 100-mm glass dish to be able to set two thin platinum plates (20 × 8 mm each) apart in a 25-mm distance. A lead wire is connected to each electrode and the end is bent to attach to the alligator clips from the electroporator (red arrows). (C) Schematic drawing of a mouse embryo at E8.0 ready for *in vitro* culture. DNA able to direct ubiquitous GFP expression (light blue oval) was microinjected into the amniotic cavity. By rotating the embryo in the electrode chamber filled with Tyrode's solution, various regions (I–IV) facing the positive (+) electrode were targeted by EP. (D–F) Results of EP in targeting region I of C. Embryos cultured *in vitro* for 24 or 48 h (24H/48H) showed a restricted pattern of GFP expression in anterior brain neuroepithelial cells. Neural crest cells emigrated from the forebrain, midbrain, and/or anterior hindbrain were also positive for GFP (white arrows in E and F). (G and H) Results of EP in targeting region II of C. Cells in the hindbrain and neural crest cells migrated out from the hindbrain to branchial arches (white arrows in H) harbored the exogenous GFP expression after 48 h of culture (48H). (I) Results of EP in targeting region III of C. The ventral part of the spinal cord strongly expressed GFP after 48 h of culture (48H). (J) Results of EP in targeting region IV of C. Pairs of posterior somite blocks express the exogenous GFP after 48 h of culture (48H) because cells just posterior to the node give rise to the somitic mesoderm. (K) A higher magnification of J. (L) Schematic drawing of a mouse embryo at E10.5 ready for *in vitro* culture. DNA able to direct GFP expression (light blue circle) was microinjected into the telencephalic vesicle, and the dorsal telencephalon was placed to face the positive (+) electrode in the electrode chamber. (M) After 24 h of culture (24H), strong GFP expression was found only in the dorsal part of the telencephalon. am, amnion; b, branchial arch, di, diencephalon; E, embryonic day; ec, ectoplacental cone; fb, forebrain; h, heart; hb, hindbrain; hf, head fold; mb, midbrain; me, mesencephalon; mt, metencephalon; my, myelencephalon; n, node; p, placenta; s, somite; sc, spinal cord; ys, yolk sac.

All embryo manipulations are performed under the binocular microscope (MZ9₅, Leica), and photographs are taken under a fluorescent microscope (MZ FLIII, Leica) fitted with a filter suitable for GFP.

The procedure described here is partly according to Akamatsu *et al.* (1999), Inoue *et al.* (2001), and Osumi and Inoue (2001).

Solutions

1. *Tyrode's solution, pH 7.4 (TS)*: Perform dissections and EP in TS to keep embryos in the best condition. Five liters of TS contains NaCl, 40.0 g; KCl, 1.0 g; CaCl₂, 1.0 g; MgCl₂·6H₂O, 1.05 g; NaH₂PO₄·2H₂O, 0.285 g; NaHCO₃, 5.0 g; and glucose, 5.0 g. To avoid possible precipitation of salts, we add these reagents in this order and sterilize by filtration. TS stored at 4°C should be prewarmed up to room temperature prior to use.

2. *Culture medium*: Prepare a 1 : 1 mixture of the rat serum and DMEM (#D5546, Sigma) containing L-glutamine and penicillin–streptomycin (100× solutions are available commercially from Gibco-BRL). The method used to collect rat serum was described previously (Hogan *et al.*, 1994).

3. *DNA solutions*: Plasmid DNA solutions with high concentration (~5 μg/μl) are required for mouse EP protocol. The following experiments used the GFP expression vector driven by the cytomegalovirus enhancer and β-actin promoter (pCA-GFP; Inoue *et al.*, 2001). After the Qiagen column and/or CsCl-based DNA purification, dissolve DNA into PBS (see earlier discussion) and add fast green (see earlier discussion) by 0.005% to be able to visualize the solution during the injection/EP processes.

Steps

1. Harvesting mouse embryos. Dissect the mouse embryos in the TS and preincubate them for more than 2 h in the culture system. Mice dissection procedures for WEC were described previously in details (Osumi and Inoue, 2001).

2. Preparations of electrode. Sterilize the electrode chamber briefly with 70% ethanol and then wash with TS three times.

3. Measure resistance. Transfer the embryo(s) from the culture bottle to the chamber and measure the resistance between electrodes. The resistance should be less than or equal to 120 Ω for optimal results. If not, add more TS to reduce the resistance.

4. Injection of DNA. DNA solutions may be injected into the region of interest while manually holding the embryo with forceps. For example, we inject DNA into the amniotic cavity for embryos at the 0–3 somite

stages to direct ubiquitous GFP expression or into the lumen of the neural tube/brain ventricles at E9.0–12.0 to obtain region-specific expression in the nervous system (Figs. 3D–3I and 3M). A mouth pipette can control injections, and the volume of injectant may vary according to the developmental stages, but should be less than 0.5 μl.

5. Positioning embryos. Immediately after injection, center the embryo(s) in the chamber and arrange the embryo(s) so as to face the target region against the positive electrode (Figs. 3C and 3L). Negatively charged DNA molecules will migrate toward the positive electrode during EP. If a stable position cannot be obtained for the embryo(s), a block of agarose gel with a hollowed-out space can be used to hold them.

6. Electroporation. Send electric pulses. While the pON, pOFF, and the number of pulses can be fixed at 50 ms, 950 ms, and five repeats, respectively, the voltage should be changed depending on the embryonic stages. For optimal results we have found that application of 65 V for the 0–3 somite stage mouse embryos, 75 V for E9.5 mouse embryos, and 85 V for E10.5 mouse embryos with the 2.5-cm gap electrode chamber are best. For rat embryos, 1.1 times higher voltages are set than for mouse embryos (also see later for optimization of the EP parameters).

7. Embryo culture. Carefully transfer the embryo(s) into the fresh TS to get rid of any possible contamination of free radicals and/or ionized materials produced around the electrodes by EP. Then the embryo(s) can be put back into the WEC apparatus. The first evidence of exogenous gene expression can normally be detected after 6–8 h of EP, and 24–48 h culture is generally sufficient to see the maximum expression of reporter genes following the EP (Figs. 3D–3K and 3M).

8. Avoiding contamination. The chamber must be washed thoroughly with TS before trying the next round of EP. Repeat steps 3–7 until finishing up all embryos.

Comments

For efficient gene transfer, further optimization of the parameters of EP can be very helpful. The following are critical points in optimization.

i. Generally, a higher voltage is toxic for the cells, but it can yield better results. First try 25 V/cm with five pulses (pON 50 ms/pOFF 950 ms). This is a maximum voltage for any fragile tissues (e.g., early stage embryos).

ii. Keep in mind that the EP energy is proportional to total pulse length and square voltages. Doubling the voltage results in the energy quadrupling, creating heat and damage to cells.

iii. The pulse length can be shortened to increase the voltage and the efficiency as well. For example, while a 2-ms pulse with 300 V generates exactly the same energy as a 50-ms pulse with 60 V, the former condition works much better than the latter condition with respect to the efficiency of gene transfer. It should be noted, however, that too high a voltage damages more cells. Hence this may not be effective for fragile early embryos and/or tissues.

iv. Time length between pulses (pOFF) can affect the EP efficiency. If the electroporator can control this parameter, shorter bursts are better, but it should not be too short to generate a continuous series of pulses.

A disadvantage of WEC is that the method is limited to events between E7.0 and 13.0, as it is difficult for the placenta to develop *in vitro*. However, there are established methods for *in utero* manipulations that could equally be applicable to EP, and several groups have indeed reported elegant results (Fukuchi-Shimogori and Grove, 2001; Saito and Nakatsuji, 2001). The ultrasound backscatter microscope has also been found to be effective at visualizing developing embryos and actually helps in perfecting the transfer genes *in utero* through EP (Takahashi *et al.*, 2002). EP could further be extended to slice cultures of adult tissues or organ culture to investigate questions dealing with events after embryogenesis.

Pitfalls

We use highly concentrated DNA solutions for EP in mouse embryos because we found that solutions less than 1 μg/μl did not yield robust uniform patterns of expression. However, if mosaic expression is useful, then lower DNA concentration, as well as voltage, can be effective. It can be very difficult to load highly concentrated DNA into a capillary, but preheating of the solution for 10 min at 62°C can facilitate the process. Furthermore, the preheating process enhances the efficiency of EP, probably because DNA molecules in the solution stored at 4°C make aggregates, which could influence DNA transportation through cell membranes.

EP generates free radicals and/or ionized materials toxic enough to kill almost all of the cells around the electrodes. Therefore, the electrodes should not touch the embryo and/or tissues during EP and the length between electrodes should be set as distant as possible. It is hard to achieve normal growth of mouse embryo following EP at the somite stage between 4

and 9. The abnormality seen most frequently is a defect in the turning of embryos, so it is possible to examine other events during this period of development.

III. FUTURE PROSPECTIVE

Systems for EP-mediated gene transfer, as described here, are easily set up and EP is indeed an efficient way to evaluate gene function. Most current studies have utilized plasmid DNA. However, dyes, chemical reagents, antibodies, antisense morpholino oligonucleotides, double-strand RNAs, ribozymes, or bacterial artificial chromosomes could be mixed together in any combinations. These might be transferred into specific groups of cells or tissues for which there are not good cell culture models, as any charged macromolecules can be targeted using EP. Several research groups have actually reported excellent results of gene attenuation by dsRNA or morpholino oligonucleotides with EP (Takahashi *et al.*, 2002; Mellitzer *et al.*, 2002; Calegari *et al.*, 2002). In the future, such broad accessibility should drastically change experimental designs in cell and developmental biology. EP will also offer a relatively high-throughput means of assaying or evaluating gene function and regulation, which must be important in light of the flood of information coming from the human genome project.

References

Akamatsu, W., Okano, H. J., Osumi, N., Inoue, T., Nakamura, S., Sakakibara, S., Miura, M., Matsuo, N., Darnell, R. B., and Okano, H. (1999). Mammalian ELAV-like neuronal RNA-binding proteins HuB and HuC promote neuronal development in both the central and the peripheral nervous systems. *Proc. Natl. Acad. Sci. USA* **96**, 9885–9890.

Calegari, F., Haubensak, W., Yang, D., Huttner, W. B., and Buchholz, F. (2002). Tissue-specific RNA interference in postimplantation mouse embryos with endoribonyclease-prepared hort interfering RNA. *Proc. Natl. Acad. Sci. USA* **99**, 14236–14240.

Fukuchi-Shimogori, T., and Grove, E. A. (2001). Neocortex patterning by the secreted signaling molecule FGF8. *Science* **294**, 1071–1074.

Haas, K., Sin, W. C., Javaherian, A., Li, Z., and Cline, H. T. (2001). Single-cell electroporation for gene transfer *in vivo*. *Neuron* **29**, 583–591.

Hamburger, V., and Hamilton, H. L. (1951). A series of normal stages in the development of the chick embryo. *J. Morph.* **88**, 49–92.

Hogan, B., Beddington, R., Costantini, F., and Lacy, E. (1994). Manipulating the Mouse Embryo, 2nd Ed. *Cold Spring Harbor Laboratory*, Cold Spring Harbor, NY.

Inoue, T., and Krumlauf, R. (2001). An impulse to the brain: Using *in vivo* electroporation. *Nature Neurosci.* **4(Suppl.)**, 1156–1158.

Inoue, T., Tanaka, T., Takeichi, M., Chisaka, O., Nakamura, S., and Osumi, N. (2001). Role of cadherins in maintaining the compart-

Somatic Cell Nuclear Transplantation

*Keith H. S. Campbell, Ramiro Alberio, Chris Denning,
and Joon-Hee Lee*

I. INTRODUCTION

The production of live offspring by nuclear transfer using a cultured cell population as nuclear donors was first reported in sheep (Campbell *et al.*, 1996). Since that time, offspring have been reported in a range of species, including sheep (Wilmut *et al.*, 1997), cattle (Cibelli *et al.*, 1998), mice (Wakayama *et al.*, 1998), pigs (Polejaeva *et al.*, 2000), cats (Shin *et al.*, 2002), rabbits (Chesne *et al.*, 2002), horses (Galli *et al.*, 2003), and rats (Zhou *et al.*, 2003), and from a variety of cell types derived from embryos, fetuses, juvenile, and adult animals (for review, see Campbell *et al.*, 2003). The ability to culture large numbers of cells *in vitro*, which can be used as successful nuclear donors, provides unique opportunities for the preservation and genetic modification of defined genomes. Primary cell populations have been used for the production of transgenic animals by gene addition (Schnieke *et al.*, 1997) and also by gene knockout (McCreath *et al.*, 2000; Denning *et al.*, 2001) and knockin strategies (McCreath *et al.*, 2000), producing animals for agricultural, medical, and research applications. Genotype preservation from both live (Wells *et al.*, 1998) and dead (Loi *et al.*, 2001) animals has been demonstrated. Many factors affect the development of embryos reconstructed by nuclear transfer (for review see, Campbell *et al.*, 2003). The methodology described here is based on the production of ovine embryos; with differences in timings, the methods are similar to those used in other species that have been produced by nuclear transfer.

II. MATERIALS AND INSTRUMENTATION

Embryo manipulations are carried out under a Leica MZ12.5 stereomicroscope and under a Leica DMIRBE inverted microscope fitted with differential interference contrast and epifluorescence. Both microscopes are fitted with Linkam M60 warmed stages. A MMO-202ND Narishige three-axis hanging joystick micromanipulator, one IM-5B Narishige injector for oocyte holding, and one Eppendorf Celltram Oil/Vario for cell transfer are fitted to the inverted microscope (Fig. 1A). The oocyte holding system is filled with Fuorinert FC77 (Sigma F-4758). Alternatively, the Narishige IM-9C injector can be used for oocyte holding. This system works properly with air.

Glass pipettes are prepared after pulling borosilicate capillaries (thin wall 1 mm o.d. × 0.8 mm i.d., Intrafil) on a P-97 Sutter Instruments Co. puller. For grinding the pipettes, the EG-400 (Narishige) micropipette grinder is used in conjunction with a microforge MF-830 (Narishige).

Cell fusion is performed with an Eppendorf multiporator fitted with a fusion chamber (electrodes 200 or 500 μm apart; Fig. 1B). For the fusion of somatic cells (15–18 μm) with a MII oocyte, a combination of an AC pulse and DC pulses give a fusion rate of more than 75% on sheep and cattle couplets.

Reagents are from Sigma unless specified. For isolation, culture, and assessment of donor cells, Glasgow MEM (G-5154), 200 mM glutamine (G-7513), 100 mM sodium pyruvate (S-8636), MEM nonessential amino

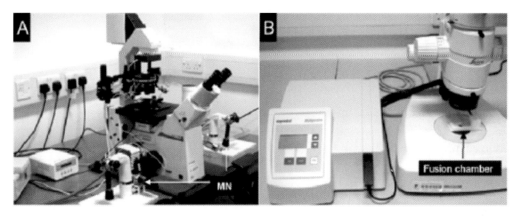

FIGURE 1   (A) Leica microscope fitted with Narishige micromanipulators (MN) for nuclear transplantation. (B) Eppendorf fusion machine with fusion chamber is used under a Leica MZ 12.5 stereomicroscope.

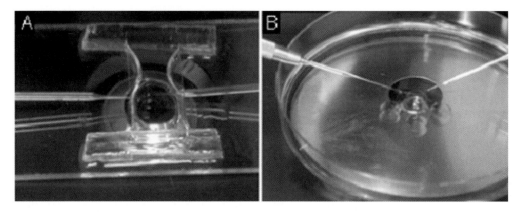

FIGURE 2   A micromanipulation chamber (A) or a plastic petri dish (B) can be used for micromanipulation.

acids (M-7145), trypsin (T-4044), Giemsa solution (G-3032), glycine (G-7126), phosphate-buffered saline (PBS) tablets (P-4417), gelatin (G-1890), trisodium citrate (C-3434), and dimethyl sulfoxide (DMSO) (D-5879) are used. Gentamycin (15750-037) is from Invitrogen/Gibco. Methanol (M/4000/17) and glacial acetic acid (A/0400/PB17) are from Fissons. Fetal calf serum (FCS) (SH-30088) is from HyClone. Vectashield mountant (H-1200) is from Vector Labs. Chamber slides (TKT-210-430L) are from Fisher Scientific. Monoclonal mouse antihuman Ki-67 antigen (clone MIB-1) (M-7240) and FITC-conjugated rabbit antimouse immunoglobulin (F-0261) are from Dako.

III. PROCEDURES

A. Preparation of Tools for Micromanipulation

Tools are prepared differently according to individual preferences relating to the position of the pipettes on the microscope. We will provide the technique used in our laboratory for performing NT with pipettes aligned horizontally to the oocyte (Fig. 2A), although it is also possible to conduct the manipulation at a 45°C angle (Fig. 2B).

1. Holding Pipette

Steps

1. Pull glass capillaries (GC10–100, Intrafil) by hand over a small flame to give a 100- to 150-μm diameter.
2. Mount capillaries with straight end onto the microforge and apply heat until the open tip is almost closed, ensuring an internal diameter of approximately 20 μm.

2. Enucleation Pipette

Steps

1. Pull glass capillaries using the microcapillary puller to give an initial taper, which reduces the diameter of the capillary to slightly greater than the diameter required, with the second taper being almost parallel.

2. Mount a drawn capillary on the microforge. Measure the diameter of the pipette using an eyepiece graticule and break at the required size, between 20 and 25 µm. The break is made by fusing the capillary onto a bead of glass, which is heated using the minimum temperature needed to stick the glass onto the capillary. The power is turned off and the capillary breaks by retraction of the cooled glass bead. Care should be taken to ensure that the capillary is not overheated, as this may cause distortion and thickening of the glass. The tip area must be straight so that it can be properly ground to a point.

3. Mount the pipette in the microforge at a 45° angle and ground it at medium speed ensuring a continuous water flow over the surface where the pipette is being ground.

4. Pipettes with thick walls can be made thin by dipping into 20% hydrofluoric acid for 30–60 s while continuously blowing air through (this prevents acid entering the inside of the pipette). If thin wall capillaries are used, there is no need for hydrofluoric acid dipping.

5. Wash with distilled water to remove acid.

6. Mount the pipette on the microforge with the tip hole in view. Heat the forge to the minimum temperature required to melt the glass bed. Touch the tip of the pipette against the glass bead and pull the tip of the pipette out to a sharp point.

3. Manipulation Chamber

Materials

1. Silicon vacuum grease (Beckman), coverlips, and plastic spacers of 2.5 cm.
2. Siliconized glass slides: Cover clean glass slides with Sigmacote (Sigma SL-2), drain, and let air dry.

Steps

1. Apply a line of wax 2 cm in length along both edges of the upper surface of a siliconized glass slide.
2. Apply silicon vacuum grease to both sides of two plexiglass spacers (2 cm), attach them parallel to the long axis of the slide, and press down firmly.
3. Pipette 60 µl of manipulation medium into the center of the chamber between the spacers. Place a clean coverslip (cleaned in 70% ethanol) on top of the spacers and push down firmly.
4. Turn the chamber upside down and fill with silicon oil until the drop of manipulation medium is enclosed in oil. This seals the ends, preventing evaporation of the manipulation medium.
5. Turn the chamber to the upright position.

B. Preparation of Equipment for Manipulation

Steps

1. Place a prepared manipulation chamber onto the stage of the microscope.
2. Put a small volume of manipulation medium into the holding pipette.
3. Attach the holding pipette to the left-hand side tool holder, ensuring that all air is removed from the hydraulic system.
4. Move the holding pipette into the manipulation chamber.
5. Coat the inside of the injection pipette with sigma-cote by aspirating the solution through the pipette several times. Wash with water and mount onto the pipette holder.
6. Mount the pipette on the right side of the chamber and ensure that the system is free from air bubbles. A slight angle of approximately 5° from horizontal allows the pipette to pick up cells from the bottom of the chamber.

C. Enucleation of Oocytes

Solutions and Equipment

1. Manipulation chamber and microtools: *See Section III,A*
2. MII oocytes
3. *Hyaluronidase stock solution*: Dissolve hyaluronidase in PBS to give 750–1500 IU/mg (H-4272). Dispense into 1-ml (300 IU/ml) aliquots. Store at −20°C.
4. *Cytochalasin B (C-6762) stock solution*: 10 mg/ml in DMSO stored at −20°C
5. *Bisbenzimide (Hoechst 33342) stock solution*: 1 mg/ml in PBS. Prepare 10-µl aliquots and store at −20°C.

Steps

1. Select oocytes surrounded by at last three layers of cumulus cells.

2. Incubate the matured oocytes in a Falcon tube containing 400 µl of hyaluronidase (300 IU/ml) for 2 min at 37°C and then vortex for 4 min.

3. Wash denuded oocytes in modified HEPES-buffered SOF medium (supplemented with 4 mg/ml BSA) and return to maturation medium.

4. At regular intervals, place batches of 10–15 oocytes with extruded polar bodies into mHSOFaac/ 4 mg/ml BSA containing 50 µg/ml bisbenzimide (Hoechst 33342) and incubate for 15 min at 35°C on the heated stage.

5. After incubation, transfer each batch of treated oocytes to the manipulation chamber.

6. Using 20× magnification, pick up and attach a single oocyte to the holding pipette using negative pressure.

7. Change the magnification to 40× DIC and focus on the oocyte held by the pipette.

8. Bring the enucleation pipette into focus. Using the enucleation pipette, rotate the oocyte into a position where the polar body and the area of cytoplasm adjacent to it can be aspirated into the pipette (Fig. 3A).

9. Insert the enucleation pipette through the zona pelucida at a point opposite the holding pipette.

10. Manipulate the enucleation pipette into a position next to the polar body, apply a small amount of negative pressure, and aspirate the polar body and a small amount of cytoplasm from directly beneath it into the pipette (Fig. 3C).

11. Withdraw the pipette from the oocyte and remove the oocyte from the field of view.

12. Turn off the transmitted light source, change to UV illumination (blue light), and examine the aspirated karyoplast (while inside the pipette) for fluorescence using filter block UV-2A. If the metaphase has been removed, it will fluoresce with a blue color; the polar body will also be visible. The metaphase plate fluoresces with a lower intensity than the polar body (Figs. 3B and 3D).

13. Move the enucleated oocyte to the right-hand side of the chamber and discard the aspirated karyoplast form the pipette. If the enucleation was unsuccessful, it is possible to repeat the procedure a second time.

14. Remove completed batches of oocytes from the manipulation chamber and place into a microdrop or dish containing HEPES-SOF (4 mg/ml). Maintain at 37°C until the required number of oocytes have been enucleated.

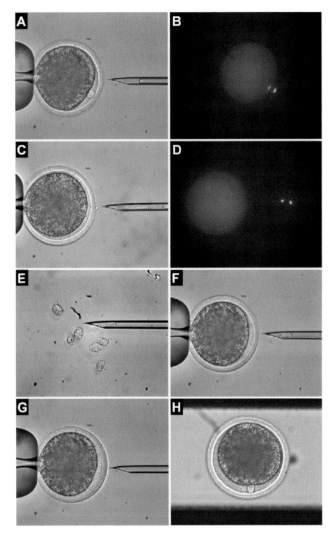

FIGURE 3 Oocyte enucleation and reconstruction of oocyte somatic cell couplets (A) holding the oocyte prior to enucleation (note position of the polar body). (B) Localisation of the polar body (PB) and metaphase plate (MP) by epifluorescence. (C) Enucleated oocyte (D) confirmed by epifluorescence (note PB and MP in the pipette). (E) Somatic cells in suspension. (F) Transfer into enucleated oocyte. (G) Cells after transfer should be in close contact with the oolema. (H) The couplet is positioned between the electrodes for fusion (note perpendicular orientation of the couplet to the electrodes).

D. Embryo Reconstruction

Solution and Equipment

1. Manipulation chamber and microtools
2. Cumulus-free enucleated MII oocytes
3. Serum-starved quiescent donor cells
4. *Calcium-free fusion medium*: 0.3 M manitol with addition 0.1 mM MgSO$_4$ in distilled H$_2$O with an osmolarity of 280 mOsm

Steps

1. Prepare a manipulation chamber, as for the enucleation procedure, containing HEPES-buffered SOF 4 mg/ml BSA.
2. Using a hand-drawn mouth pipette attached to aspirator tube (Sigma; A5177), place the nuclear donor cells into the upper left-hand corner of the chamber and a group of enucleated oocytes into the right hand corner of the chamber.
3. Pick up an enucleated oocyte with the holding pipette.
4. Move the holding pipette to the top.
5. Focus the microscope onto the bottom of the chamber.

6. Move the enucleation pipette to the chamber, maneuver it to the cells, and gently aspirate two to three suitable cells into the pipette (Fig. 3E).

7. Refocus on the enucleated oocyte and move the enucleation pipette until it is in focus.

8. Insert the enucleation pipette through the hole made previously in the zona pelucida. While holding the pipette against the cytoplasm, expel one donor cell into the perivitelline space (Fig. 3F). Ensure contact between the two cells of the couplet (Fig. 3G).

9. Transfer donor cells to groups of 10 oocytes. Upon completion of each batch, wash couplets in washing medium.

10. Place 100 μl of warm (35°C) fusion medium, spanning the electrodes, into the fusion chamber.

11. Pipette the reconstructed oocytes into the fusion medium but outside the electrodes and allow them to settle down for 15–30 s.

12. Using a hand-drawn capillary mouth pipette, place two to three couplets between the electrodes, moving them until the plane of contact between cytoplast and karyoplast is parallel to the electrodes. Precise orientation is necessary for fusion to occur (Fig. 3H).

13. Apply the fusion pulse, which consists of a 5-s AC pulse of 5 V/cm followed by one or two DC pulses of 35 V/cm for 30 μs (for electrodes 200 μm apart).

14. Remove the reconstructed oocytes from the chamber and wash each batch with washing medium.

15. Transfer them into a 50 μl drop of mSOFaaci/4 mg/ml BSA (Holm et al., 1999). Incubate at 39°C for 15–20 min and then examine for cell fusion.

16. Refuse unfused couplets by two DC pulses of 35 V/cm for 30 μs.

17. Incubate fused couplets for 1 h in mSOFaaci and then activate.

18. Activate fused couplets in H-SOF plus 5 μM calcium ionophore (A23187) for 5 min at 35°C and culture subsequently in mSOFaaci with 10 μg/ml of cycloheximide and 7.5 μg/ml cytochalasin B for 5 h.

19. Following activation, wash the reconstructed embryos and culture under oil in 50-μl droplets of mSOFaaci supplemented with 2%(v/v) BME, 1%(v/v) MEM, and 4 mg/ml BSA (fatty acids free) in a humidified atmosphere of 5% CO_2, 5% O_2, and 90% N_2 at 39°C.

20. At day 3 of culture, transfer cleaved embyos into mSOFaaci supplemented with 5% heat-inactivated FBS.

21. At days 7 to 8 transfer embryos surgically into naturally cycling, foster mother ewes, 7 days after estrus.

E. Isolation, Culture, and Preparation of Cells Used as Nuclear Donors

Solutions

1. *Culture medium:* To 500 ml of Glasgow MEM (GMEM) add 56 ml (10%) of FCS, 5.6 ml of L-glutamine, 5.6 ml of sodium pyruvate, and 5.6 ml of nonessential amino acids. Warm to 37°C before use.

2. *Low serum medium:* As in solution 1 but containing 0.5% FCS.

3. *Trypsin solution:* Aliquot and store at −20°C. Thaw and warm to 37°C before use.

4. *Gelatin solution:* Add 1 g of gelatin to 1000 ml Milli-Q water. Autoclave, allow to cool, and filter.

5. *Hypotonic solution:* Add 4 g of trisodium citrate to 1000 ml Milli-Q water. Filter sterilise.

6. *Karyotyping fixative:* Add 3 parts methanol to 1 part acetic acid.

Isolation of a Primary Culture

Steps

1. Add 20 ml of gelatin solution to a T175 flask. Place flask in tissue culture hood until cells have been prepared.

2. Remove the uterus from a 30- to 35-day-old pregnant ewe and immediately remove the fetus.

3. Eviscerate and decapitate the fetus and transfer the carcass to 10 ml PBS supplemented with 100 μg/ml gentamycin. Wash the carcass twice with 10 ml PBS–gentamycin.

4. Place the carcass in a 10-cm tissue culture dish and add 3 ml of trypsin solution. Use scissors to cut cubes of 1 mm. Incubate at 37°C for 10 min.

5. Add 10 ml of culture medium to the dissociating cells and transfer to a 15-ml tube.

6. Dissociate clumps by pipetting.

7. Stand tube for 1 min to allow larger clumps to settle.

8. Aspirate gelatin solution from T175 flask.

9. Avoiding the pelleted cells, transfer the cells in suspension to the gelatinised T175 flask.

10. Resuspend the remaining clumps in the tube in 10 ml culture medium. Repeat steps 6 and 7. Add the cell suspension to the same flask as before.

11. Discard the tube containing the remaining cell clumps.

12. Add culture medium to the T175 flask to a total volume of 50 ml.

13. Add gentamycin to a final concentration of 50 μg/ml and incubate cells for 24 h at 37°C in an atmosphere composed of 5% CO_2, 20% O_2, and 75% N_2.

14. Replace medium with fresh medium containing no antibiotic. Culture cells at 37°C.

15. Record time required for culture to reach confluence. Trypsinise, pellet, and cryopreserve aliquots of 1×10^6 cells in a mixture of 90% growth medium: 10% DMSO.

16. Subsequent passages and culture of cells do not require gelatinized flasks or antibiotics.

F. Preparation of Mitotic Spreads

Steps

1. Culture cells in a T25 flask to subconfluency.

2. Aspirate growth medium and wash once with warm (37°C) PBS.

3. Add 1 ml of trypsin solution. Incubate at 37°C and monitor the culture for detachment of cells.

4. When cells are detached, add 10 ml of culture medium.

5. Decant cells into a 15-ml tube and pellet at 300 g for 5 min.

6. Aspirate medium and resuspend cell pellet in 10 ml fresh growth medium.

7. Add 2 ml of the cell suspension and 8 ml of growth medium to a T25 flask.

8. Incubate for 24 h.

9. Trypsinise cells as described previously and resuspend in 10 ml culture medium.

10. Transfer the cell suspension to a 15-ml tube and pellet the cells at 300 g for 5 min.

11. To resuspend the cells, add 10 ml of hypotonic solution gradually while vortexing.

12. Incubate at room temperature for 15 min to allow cells to swell.

13. Pellet the cells at 300 g for 5 min. Discard supernatant.

14. Resuspend the cells by adding 6 ml of karytyoping fixative dropwise while vortexing.

15. Pellet the cells at 300 g for 5 min. Discard supernatant.

16. Repeat steps 13 and 14 twice more.

17. After the final spin, resuspend the pellet in 0.5 ml karytoping fixative.

18. Prepare spreads by dropping fixed cell suspension onto clean glass slide.

19. Air dry.

20. Stain with 3% Giemsa for 15 min at room temperature.

G. Induction and Assessment of Quiescence in Donor Cells

Solutions

1. *Fixative*: Prepare a 70% : 30% of mixture of methanol : 50 mM glycine. Store at –20°C. (Add 0.38 g glycine to 100 ml Milli-Q water.)

2. *Humidified chamber*: Place a sheet of filter paper into a 10-cm petri dish. Dampen with distilled water.

3. *PBS/1% FCS*: Add 1 ml of FCS to 100 ml sterile PBS

4. *1° antibody working solution*: Dilute Ki-67 antibody (clone MIB-1) 1:150 with PBS/1% FCS.

5. *2° antibody working solution*: Dilute FITC-conjugated rabbit antimouse antibody 1:20 with PBS/1% FCS.

6. Nail varnish.

7. Mountant (Vecta shield) containing 1 μg/ml DAPI.

Steps

1. Trypsinise an 80% confluent monolayer of donor cells and resuspend in growth medium.

2. Replate cells at one-fifteenth the original density in growth medium in 1-well glass chamber slides and continue to culture for 48 h.

3. Wash the monolayer three times with low serum medium and then continue culture in this medium.

4. Exit from the growth cycle; survival in the quiescent state must be established for each cell line. At the time of serum reduction and at 24-h intervals, process one slide for Ki-67 staining as detailed later.

5. Remove culture chamber and rubber gasket from one chamber slide.

6. Rinse in PBS at 4°C and fix in methanol: 50 mM glycine for 20 min at –20°C.

7. Wash in PBS.

8. Block in PBS/1% BSA for 30–45 min at room temperature.

9. Wash in PBS and transfer slide to humidified chamber.

10. Add 500 μl of 1° antibody solution to each slide. Incubate for 1 h at 37°C.

11. Wash three times with PBS/1% FCS.

12. Add 500 μl of 2° antibody solution. Incubate for 1 h at 37°C.

13. Wash three times with PBS/1% FCS.

14. Remove excess liquid. Add two drops of mounting and overlay with a coverslip. Seal edges with nail varnish.

15. Store at 4°C in darkness.

16. Examine using epifluorescence.

17. Record the percentage of Ki-67-positive nuclei. Ki-67 is a nuclear antigen expressed at all active stages of the cell cycle but absent in resting (G0 phase).

IV. PITFALLS

1. Ensure that enucleated oocytes are at MII at the time of enucleation. Oocytes that have activated spontaneously as a result of handling will resume meiosis; this will be visible as a third fluorescent area corresponding to the anaphase/telophase.

2. It is recommended to perform enucleation in early matured oocytes (17–19 h after onset of maturation) to guarantee close contact of the metaphase plate and the polar body. Metaphase plates from matured oocytes tend to move away from the polar body after they have reached the MII stage.

3. Avoid temperature oscillations during manipulation as these can cause spontaneous oocyte activation.

References

Campbell, K. H. S., and Alberio, R. (2003). Reprogramming the genome: role of the cell cycle. *Reprod. Suppl.* **61**, 477–494.

Campbell, K. H. S., McWhir, J., Ritchie, W. A., and Wilmut, I. (1996). Sheep cloned by nuclear transfer from a cultured cell line. *Nature* **380**, 64–66.

Chesne, P., Adenot, P. G., Viglietta, C., Baratte, M., Boulanger, L., and Renard, J. P. (2002). Cloned rabbits produced by nuclear transfer from adult somatic cells. *Nature Biotechnol.* **20**, 366–369.

Cibelli, J. B., Stice, S. L., Golueke, P. J., Kane, J. J., Jerry, J., Blackwell, C., Ponce, d. L. n., and Robl, J. M. (1998). Cloned transgenic calves produced from nonquiescent fetal fibroblasts. *Science* **280**, 1256–1258.

Denning, C., Burl, S., Ainslie, A., Bracken, J., Dinnyes, A., Fletcher, J., King, T., Ritchie, M., Ritchie, W. A., Rollo, M., de Sousa, P., Travers, A., Wilmut, I., and Clark, A. J. (2001). Deletion of the alpha(1,3)galactosyl transferase (GGTA1) gene and the prion protein (PrP) gene in sheep. *Nature Biotechnol.* **19**, 559–562.

Galli, C., Lagutina, I., Crotti, G., Colleoni, S., Turini, P., Ponderato, N., Duchi, R., and Lazzari, G. (2003). Pregnancy: A cloned horse born to its dam twin. *Nature* **424**, 635.

Holm, P., Booth, P. J., Schmidt, M. H., Greve, T., and Callesen, H. (1999). High bovine blastocyst development in a static *in vitro* production system using SOFaa medium supplemented with sodium citrate and myo-inositol with or without serum-proteins. *Theriogenology* **52**, 683–700.

Loi, P., Ptak, G., Barboni, B., Fulka, J., Jr., Cappai, P., and Clinton, M. (2001). Genetic rescue of an endangered mammal by cross-species nuclear transfer using post-mortem somatic cells. *Nature Biotechnol.* **19**, 962–964.

McCreath, K. J., Howcroft, J., Campbell, K. H. S., Colman, A., Schnieke, A. E., and Kind, A. J. (2000). Production of gene-targeted sheep by nuclear transfer from cultured somatic cells. *Nature* **405**, 1066–1069.

Polejaeva, I. A., Chen, S. H., Vaught, T. D., Page, R. L., Mullins, J., Ball, S., Dai, Y., Boone, J., Walker, S., Ayares, D. L., Colman, A., and Campbell, K. H. S. (2000). Cloned pigs produced by nuclear transfer from adult somatic cells. *Nature* **407**, 86–90.

Schnieke, A. E., Kind, A. J., Ritchie, W. A., Mycock, K., Scott, A. R., Ritchie, M., Wilmut, I., Colman, A., and Campbell, K. H. S. (1997). Human factor IX transgenic sheep produced by transfer of nuclei from transfected fetal fibroblasts. *Science* **278**, 2130–2133.

Shin, T., Kraemer, D., Pryor, J., Liu, L., Rugila, J., Howe, L., Buck, S., Murphy, K., Lyons, L., and Westhusin, M. (2002). Cell biology: A cat cloned by nuclear transplantation. *Nature* **415**, 859.

Wakayama, T., Perry, A. C., Zuccotti, M., Johnson, K. R., and Yanagimachi, R. (1998). Full-term development of mice from enucleated oocytes injected with cumulus cell nuclei. *Nature* **394**, 369–374.

Wells, D. N., Misica, P. M., Tervit, H. R., and Vivanco, W. H. (1998). Adult somatic cell nuclear transfer is used to preserve the last surviving cow of the Enderby Island cattle breed. *Reprod. Fertil. Dev.* **10**, 369–378.

Wilmut, I., Schnieke, A. E., McWhir, J., Kind, A. J., and Campbell, K. H. S. (1997) Viable offspring derived from fetal and adult mammalian cells. *Nature* **385**, 810–813.

Zhou, Q., Renard, J. P., Le Friec, G., Brochard, V., Beaujean, N., Cherifi, Y., Fraichard, A., and Cozzi, J. (2003). Generation of fertile cloned rats by regulating oocyte activation. *Science* **302**, 1179.

B

EXPRESSION SYSTEMS

Expression Systems

8

Expression of cDNA in Yeast

Caterina Holz and Christine Lang

I. INTRODUCTION

The ability to express heterologous genes in yeast has become a powerful tool for many biological research techniques. Yeast combines the advantage of eukaryotic posttranslational modifications, such as phosphorylation, glycosylation, disulfide cross-linking, and subcellular targeting, with the expression of soluble proteins in large amounts. The yeast system is also very attractive as both the culture and the protocols for expression are easy to handle. The first species to be employed for the production of heterologous proteins was *Saccharomyces cerevisiae*. The host-vector systems for this organism are now well developed and a wealth of knowledge on its genetics and its physiology has been accumulated. This has led to its frequent use as a model eukaryote to understand the function and properties of many mammalian proteins in recent years (Romanos *et al.*, 1992). Suitable *S. cerevisiae* host strains are available (or can be generated easily) that bear alterations to enhance protein production, such as the elimination of major proteases or the increase of transcription and expression level (Harashima, 1994). Both homologous and heterologous proteins may be produced in the cytoplasm or directed through the secretory pathway. The first approach, the intracellular expression of cDNAs by the yeast *S. cerevisiae*, is the focus of this article.

The most widely used expression vectors are *Escherichia coli*/yeast shuttle plasmids (Bonneaud *et al.*, 1991) that are stabilized mitotically by autonomously replicating sequences (ARS/CEN region, 2-μm origin) or by integration into the yeast genome.

Foreign genes, which are inserted in 2-μm-based plasmids, are maintained in high copy numbers; however, selection is required for growth and protein expression (Rose and Broach, 1990). This article describes the use of such an episomal expression plasmid that is easily introduced into yeast cells by lithium–acetate transformation into competent cells (Gietz *et al.*, 1992) or by electroporation (Becker and Guarente, 1991).

Various strategies are available to induce the expression of the recombinant protein; the specific method chosen depends on the physiological characteristics of the yeast strain used. The use of inducible promoters is advisable when the product itself or its synthesis is deleterious to the cells or, as in cell biology studies, when the impact of the recombinant protein on cellular physiology is to be investigated. This article describes a method based on the expression vector pYEXTHS-BN (Holz *et al.*, 2002), where the transcription of cDNAs is controlled by the Cu^{2+}-induced *CUP1* promoter from the yeast metallothionein gene. This strong promoter is induced rapidly by copper sulfate (0.01 to 1 mM, depending on the copper resistance of the host strain) (Etcheverry, 1990; Ward *et al.*, 1994). The cloned gene is fused to small antigenic epitopes. These epitope tags, the N-terminal His$_6$ tag (Porath, 1992) and the C-terminal StrepII tag (Schmidt *et al.*, 1996), facilitate the subsequent immunological identification and purification of the gene product by two-step affinity chromatography. The intracellular double-tagged protein is released from the cells by disrupting the cell walls by enzymatic, chemical, or mechanical procedures as described here.

II. MATERIALS AND INSTRUMENTATION

The pYEXTHS-BN expression plasmid was constructed in the Protein Structure Factory laboratory (Berlin, Germany). The strain *S. cerevisiae* AH22*ura*3 (MATa, ura3Δ, leu2-23, 112, his4-519, can1) was provided by Berlin University of Technology, Department of Microbiology and Genetics (Germany). NaCl (Cat. No. 1.06404), $NH_4H_2PO_4$ (Cat. No. 1.01126), NH_4Cl (Cat. No. 1.01145), $Na_2HPO_4 \cdot 2H_2O$ (Cat. No. 1.06580), $NaH_2PO_4 \cdot H_2O$ (Cat. No. 1.06346), $MgCl_2 \cdot 6H_2O$ (Cat. No. 1.05833), $CaCl_2 \cdot H_2O$ (Cat. No. 1.02380), KH_2PO_4 (Cat. No. 1.04873), $MgSO_4 \cdot 7H_2O$ (Cat. No. 1.05886), sodium L-glutamate monohydrate (Cat. No. 1.06445), $ZnSO_4 \cdot 7H_2O$ (Cat. No. 1.08883), $FeSO_4 \cdot 7H_2O$ (Cat. No. 1.03965), $CuSO_4 \cdot 5H_2O$ (Cat. No. 1.02790), $MnCl_2 \cdot 4H_2O$ (Cat. No. 1.05927), $Na_2MoO_4 \cdot 2H_2O$ (Cat. No. 1.06521), nicotinic acid (Cat. No. 8.18714), L-histidine HCl (Cat. No. 1.04350), imidazole (Cat. No. 1.047169), isopropanol (Cat. No. 1.09634), 2-mercaptoethanol (Cat. No. 1.15433), glycine (Cat. No. 1.04169), and Tween 20 (Cat. No. 1.00731) are from Merck. *myo*-Inositol (Cat. No. 26310), pyridoxine (Cat. No. 33990), biotin (Cat. No. 15060), sucrose (Cat. No. 35580), dodecylsulfate-Na-salt (SDS) (Cat. No. 20783), Triton X-100 (Cat. No. 37240), and Ponceau S (Cat. No. 33429) are from Serva. Thiamine (Cat. No. T4625), calcium *d*-pantothenate (Cat. No. C8731), bovine serum albumin (BSA) (A-3912), phenylmethylsulfonyl-fluoride (PMSF) (Cat. No. 78830/Fluka), bromphenol blue (Cat. No. B-6131), EDTA sodium salt (Cat. No. E-5134), avidin (Cat. No. A-9275), and desthiobiotin (Cat. No. D-1411) are from Sigma. Tris base (Cat. No. 155004-038) is from Gibco/LifeTechnologies. Concentrated HCl (Cat. No. P074.1), glycerol (Cat. No. 3783.2), methanol (4627.2), milk powder (blotting grade) (Cat. No. T145.1), and glass beads (0.5 mm) (Cat. No. A553.1) are from Roth. Ni-NTA agarose (Cat. No. 30210), penta-His antibody (Cat. No. 34660), and polypropylene columns (1 ml, unpacked columns) (Cat. No. 34924) are from Qiagen. Rabbit antimouse IgG/HRP conjugate (Cat. No. P0260) is from Dako (Denmark), and Strep-Tactin HRP conjugate (Cat. No. 2-1502-001) and Strep-Tactin sepharose (50% suspension) (Cat. No. 2-1201-005) are from IBA GmbH (Göttingen, Germany). The ECL substrate western lightning chemiluminescence reagent plus (Cat. No. NEL 105) is from PerkinElmer Life Science, the polyvenylidene difluoride (PVDF) membrane (Cat. No. IPVH00010) is from Millipore, and the 3 MM chromatography paper (Cat. No. 3030-690) is from Whatman.

The incubator shaker (Innova 4430) is from New Brunswick Scientific. The centrifuge is from Heraeus Instruments (Megafuge 1.0R). The vortex (Vortex-2 genie, Model G-560) is from Scientific Industries Inc., the semidry blotter (Cat. No. 24200) is from H. Hölzel GmbH (Germany), and power supplies are from Bio-Rad (Power Pac 200/300). The imaging system (LAS-1000 luminescent image analyzer, Fujifilm) is from Raytest. The Thermomix Comfort (Cat. No. 5355000.011) is from Eppendorf. The micropipettes are from Gilson.

III. PROCEDURES

A. Copper-Inducible Expression of Proteins (Standard Scale)

The expression plasmid pYEXTHS-BN used here is a pYEXbx (Clontech) derivative, which was constructed to subclone cDNAs carrying a N-terminal *Bam*HI site and C-terminal *Not*I site overhang as polymerase chain reaction fragments (Holz *et al.*, 2002). The expression of cDNAs is induced by adding $CuSO_4$. The strain *S. cerevisiae* AH22*ura*3 (Polakowski *et al.*, 1998) is used as the expression strain. The plasmid DNA containing the cloned cDNA insert is transformed in *S. cerevisiae*, and the selected transformants are transferred to WMVIII medium for protein expression.

Solutions

1. *WM VIII medium*: See Table I
2. *0.5 M $CuSO_4$ stock*: Dissolve 6.2 g in 50 ml distilled water
3. *10× PBS stock*: 100 mM phosphate and 1.5 M NaCl. To make 2 liter, weigh 29.9 g of $Na_2HPO_4 \cdot 2H_2O$, 4.16 g of KH_2PO_4, and 175.32 g of NaCl and adjust to 2 liter with distilled water.
4. *1× PBS*: 8.4 mM Na_2HPO_4, 1.6 mM KH_2PO_4, and 150 mM NaCl. To make 500 ml, take 50 ml of 10× PBS and add distilled water to 500 ml.
5. *1 M Na_2HPO_4 stock*: To make 1 liter, weigh 177.99 g of $Na_2HPO_4 \cdot 2H_2O$ and bring up to 1 liter with distilled water
6. *1 M NaH_2PO_4 stock*: To make 1 liter, weigh 137.99 g of $NaH_2PO_4 \cdot H_2O$ and top up to 1 liter with distilled water
7. *3 M NaCl*: To make 500 ml, weigh 87.66 g of NaCl and adjust to 500 ml with distilled water
8. *100 mM phenylmethylsulfonyl fluoride*: Dissolve 174 mg PMSF in 10 ml isopropanol. Divide the solution into 1-ml aliquots and store at −20°C.

TABLE I Components of WM VIII Protein
Production Medium[a]

Ingredient	Amount
Component I: Autoclaved in 990 ml distilled water	
Sucrose	50 g
$NH_4H_2PO_4$	0.25 g
NH_4Cl	2.8 g
$MgCl_2 \cdot 6H_2O$	0.25 g
$CaCl_2 \cdot 2H_2O$	0.1 g
KH_2PO_4	2.0 g
$MgSO_4 \cdot 7H_2O$	0.55 g
myo-Inositol	0.075 g
Sodium L-glutamate monohydrate	10.0 g
Component II (microelements): Dissolved in 100 ml 10 mM EDTA, pH 8.0, and filter sterilized	
$ZnSO_4 \cdot 7H_2O$	43.7 mg
$FeSO_4 \cdot 7H_2O$	12.5 mg
$CuSO_4 \cdot 5H_2O$	2.5 mg
$MnCl_2 \cdot 4H_2O$	2.5 mg
$Na_2MoO_4 \cdot 2H_2O$	2.5 mg
Component III (vitamin stock solution): Dissolved in 200 ml H_2O, and filter sterilized	
Nicotinic acid (vitamin B)	0.5 g
Pyridoxine (vitamin B_6)	1.25 g
Thiamine (vitamin B_1)	0.5 g
Biotin	0.125 g
Calcium d-pantothenate	2.5 g
Component IV: Filter sterilized in 20 ml distilled water	
L-Histidine HCl	0.4 g

[a] Component I is mixed after autoclaving with 4 ml of component II, 4 ml of component III, and 2 ml of component IV. The pH will be approximately 6.

9. *Lysis buffer*: 100 mM NaPO4, pH 8.0, 300 mM NaC1, and 1% Triton X. To make 100 ml, add 9.3 ml of a 1 M stock solution of Na_2HPO_4, pH 8.0, 680 µl of a 1 M stock solution of NaH_2PO_4, 10 ml of a 3 M stock solution of NaCl, and 1 ml Triton X-100 (100%, Serva). Fill up to 100 ml with distilled water.

10. *1 M Tris–HCl (pH 6.8) stock*: Dissolve 121.1 g Tris base in 800 ml of H_2O. Adjust pH to 6.8 by adding concentrated HCl (about 80 ml), complete to 1 liter with distilled water, and sterilize by autoclaving.

11. *4× SDS PAGE sample buffer*: 0.2 M Tris–HCl, pH 6.8, 8% (w/v) SDS, 40% (v/v) glycerol, 5% (v/v) 2-mercaptoethanol, and 0.4% (w/v) bromphenol blue. To make 50 ml, add 10 ml of a 1 M stock of Tris–HCl, pH 6.8, 4 g of SDS, 20 ml of glycerol, 2.5 ml of β-mercaptoethanol, and 0.2 g bromphenol blue. Top up to 50 ml with distilled water.

Steps

1. Inoculate 20 ml WM VIII medium containing 40 mg/liter histidine in a 100-ml shake flask with a single colony of *S. cerevisiae* AH22*ura*3 bearing the pYEXbx derivative with target CDNA insert as preculture and grow overnight with shaking (200 rpm) at 28°C.

2. Inoculate 100 ml WM VIII medium containing 40 mg/liter histidine in a 500-ml shake flask with 1 ml from the preculture. Grow with shaking (200 rpm) at 28°C for 16–24 h to an OD_{600} of 7–10.

3. Induce the protein expression of the culture by adding 100 µl of a 0.5 M stock of $CuSO_4$ (0.5 mM final concentration) and incubate the culture with shaking for another 3–4 h at 28°C.

4. Collect an aliquot of 7–10 ml of the cell culture (corresponding to OD_{600} = 70) by centrifugation (10 min, 2500 g, 4°C).

5. Resuspend the cell pellet in 5 ml cold PBS. Repeat the spin and discard supernatant.

6. Freeze the washed cells in dry ice-ethanol or liquid nitrogen or store cell pellet at −70°C.

7. Thaw cells at room temperature and resuspend the cells in 300 µl ice-cold lysis buffer and add 3 µl of 100 mM stock solution of PMSF (1 mM final concentration).

8. Add 1 volume of sterile, acid-washed glass beads (0.5 mm) and disrupt the cells by seven cycles of 1 min full-speed vortexing/1 min cooling on ice.

9. Centrifuge the lysate for 15–20 min at 10,000 g at 4°C to remove cellular debris and transfer the supernatant to a fresh tube.

10. Take a 15-µl sample and add 5 µl 4× SDS–PAGE sample buffer, heat the sample at 95°C for 5 min, and store at −20°C for SDS–PAGE and Western blot analysis. *Note:* You cannot detect the recombinant target gene product directly from the total yeast protein after electrophoresis of proteins by staining the gel due to the low proportion of recombinant protein. This requires purifying the recombinant protein from crude extract or detecting the target protein by immunoblotting (Section III,D).

11. Proceed to protocols for purification (Sections III,B and III,C).

B. Affinity Purification of His$_6$-Tagged Proteins via Ni-NTA Agarose

Solutions

1. *1 M imidazole stock*: To make 100 ml, weigh 6.81 g of imidazole and complete to 100 ml with distilled water

2. *Wash buffer*: 50 mM NaPO$_4$, pH 8.0, 300 mM NaC1, and 5 mM imidazole. To prepare 500 ml, add 23.3 ml of a 1 M stock solution of Na_2HPO_4, pH 8.0, 1.7 ml of a 1 M stock solution of NaH_2PO_4, 50 ml of a

3 M stock solution of NaCl, and 2.5 ml of a 1 M stock solution of imidazole. Adjust pH to 8.0 with HCl and adjust volume to 500 ml with distilled water.

3. *Elution buffer*: 50 mM NaPO$_4$, pH 8.0, 300 mM NaCl, and 250 mM imidazole. To prepare 100 ml, add 4.66 ml of a 1 M stock solution of Na$_2$HPO$_4$, pH 8.0, 340 µl of a 1 M stock solution of NaH$_2$PO$_4$, 10 ml of a 3 M stock solution of NaCl, and 25 ml of a 1 M stock solution of imidazole. Adjust pH to 8.0 with HCl and adjust volume to 100 ml with distilled water.

Steps

1. Add 100 µl of a 50% slurry of Ni-NTA resin (100 µl resin has a capacity for 500 µg–1 mg His$_6$-tagged protein) to 300 µl cleared lysate that contains whole cell proteins (Section III,A, step 8) and mix gently by shaking for 60 min at 4°C.

2. Load the lysate Ni-NTA mixture into an unpacked column.

3. Collect the column flow through. Save flow through for SDS–PAGE analysis.

4. Wash three times with 400 µl wash buffer and collect wash fractions. Save wash fractions for SDS–PAGE analysis.

5. Elute the protein three times with 100 µl elution buffer. Collect the elution fractions and analyze by SDS–PAGE.

C. Affinity Purification via StrepTactin

The affinity purification is carried out according to Voss and Skerra (1997) with some modifications.

Solutions

1. *1 M Tris–HCl (pH 8.0) stock*: Dissolve 121.1 g Tris base in 800 ml of H$_2$O. Adjust pH to 8.0 by adding concentrated HCl (about 40 ml), top up to 1 liter with distilled water, and sterilize by autoclaving.

2. *250 mM EDTA stock*: Weigh 9.3 g of EDTA, top up to 100 ml with distilled water, and sterilize by autoclaving

3. *Buffer W*: 100 mM Tris–HCl, pH 8.0, 1 mM EDTA. To make 100 ml, take 10 ml of a 1 M stock solution of Tris–HCl, pH 8.0, and 400 µl of a 250 mM EDTA stock solution. Top up to 100 ml with distilled water.

4. *25 mM Desthiobiotin solution*: 25 mM Desthiobiotin and 100 mM Tris–HCl, pH 8.0. To make 100 ml, add 0.536 g Desthiobiotin, 10 ml of 1 M stock solution of Tris–HCl, pH 8.0, and top up to 100 ml with distilled water.

5. *Buffer E*: 100 mM Tris–HCl pH 8.0, 1 mM EDTA, and 2.5 mM Desthiobiotin. To make 100 ml, add 10 ml of a 1 M stock solution of Tris–HCl, pH 8.0, 400 µl of a 250 mM EDTA stock solution, and 10 ml of a 25 mM

stock solution of Desthiobiotin. Adjust to 100 ml with distilled water.

6. *Avidin solution (2 mg/ml)*: Dissolve 5 mg avidin (65 units) in 2.5 ml buffer W.

Steps

1. Transfer 200 µl of 50% StrepTactin sepharose (IBA GmbH) into an unpacked column and wash twice with 600 µl buffer W.

2. Load 200 µl of the eluate from the Ni-NTA resin (Section III,B step 5) onto the equilibrated column to perform the second step of the two-step affinity chromatography purification or take 200 µl cleared lysate (Section III,A step 8), pretreat this lysate with 1 µl of avidin solution (2 mg/ml) to mask inhibiting biotinylated proteins for 30 min at 4°C, spin down the aggregates formed by centrifugation (15 min, 10,000 g, 4°C), and load the cleared supernatant onto the column.

3. Collect the column flow through. Save flow through for SDS–PAGE analysis.

4. Wash five times with 200 µl buffer W and collect wash fractions. Save wash fractions for SDS–PAGE analysis.

5. Elute protein with 5× 100 µl buffer E. Collect the elution fractions and analyze by SDS–PAGE.

D. Protein Analysis by Western Blot and Immunodetection with Anti-His Antibody or StrepTactin HRP Conjugate (Chemiluminescent Method)

The proteins are separated by 12.5% SDS–PAGE according to Laemmli (1970). Following electrophoresis, proteins in polyacrylamide gel are transferred to polyvinylidene difluoride (PVDF) membranes by semidry electroblotting as described later and characterized by immunodetection (Fig. 1).

Solutions

1. *Transfer buffer*: 25 mM Tris base, 192 mM glycine, and 10% methanol. To make 1 liter, weigh 3.03 g Tris base, 14.4 g glycine, add 100 ml methanol, and adjust to 1 liter with distilled water.

2. *PBST*: 8.4 mM Na$_2$HPO$_4$, 1.6 mM KH$_2$PO$_4$, 150 mM NaCl, and 0.1% (v/v) Tween 20. To make 1 liter, take 100 ml of 10× PBS, 1 ml Tween 20, and top up to 1 liter with distilled water.

3. *Blocking solution*: 2% (w/v) BSA in PBST. To make 100 ml, weigh 2 g of BSA and top up to 100 ml with PBST. Prepare fresh or keep at −20°C.

4. *Secondary antibody dilution buffer*: 5% (w/v) nonfat dried milk (milk powder) in PBST. To make

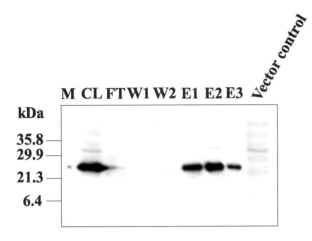

FIGURE 1 Expression and purification of the His$_6$/StrepII-tagged human G-substrate protein (21 kDa) from *S. cerevisiae* as an example. Cleared cell lysate (CL) was prepared and expressed protein purified via Ni-NTA. A clone carrying an empty expression vector was used as the negative control (vector control). The proteins were separated by SDS–PAGE and detected by immunoblotting with StrepTactin HRP conjugate. M, prestained Bio-Rad broad range marker; FT, flow through from Ni-NTA column; W1, wash fraction 1; W2, wash fraction 2; E1–E3, eluates 1–3.

100 ml, weigh 5 g of milk powder and top up to 100 ml with PBST.

Steps

1. Cut eight pieces of Whatman 3 MM paper and a piece of membrane to the same size as the gel. To avoid contamination, always handle the filter paper and membrane with gloves.

2. Wet the PVDF membrane in 100% methanol (or ethanol) for 15 s. Then transfer it to a container of distilled (or Milli-QR) water for 2 min.

3. Equilibrate the membrane for at least 5 min in the transfer buffer.

4. Soak the filter paper in transfer buffer.

5. Immerse the gel in the transfer buffer.

6. Place four sheets of filter paper in the center of the anode electrode plate (positive, usually red), avoiding air bubbles, and place the membrane on top of the filter paper.

7. Place the gel on top of the membrane and place four sheets of filter paper on top of the gel. Air bubbles can be removed by gently rolling a Pasteur pipette over each layer in the sandwich.

8. Place the cathode plate cover (negative, usually black) on the top of the assembled transfer stack.

9. Connect the anode and the cathode lead to their corresponding power supply outputs. Follow the manufacturer's instructions regarding current, voltage, and transfer times. The current density required

is determined by the size of the gel: 1 mA/cm^2 is recommended (1-h transfer).

10. After transferring, mark the orientation of the membrane and the position of bands of the prestained protein standard. You can stain the blot to assess the quality of transfer with Ponceau-S red (reversible stain). Incubate membrane in staining solution (0.5% Ponceau S, 1% acetic acid) with gentle agitation for 2 min. The blot will be destained in distilled water or during the following immunological detection procedure.

11. Incubate the membrane for at least 1 h in blocking solution at room temperature. If using StrepTactin HRP conjugate for detection, subsequently pretreat the membrane in PBST containing 2 µg/ml avidin for 30 min to block biotinylated proteins.

12. Incubate with primary antibody (penta-His antibody, Qiagen) diluted at 1:1000/1:2000 or StrepTactin HRP conjugate diluted at 1:4000 in blocking solution at room temperature for 1 h. *Note:* Do not use dilution buffer containing milk powder for anti-His antibody. This will reduce sensitivity. Do not use milk powder for StrepTactin HRP conjugate either as it contains large amounts of cross-reacting biotin.

13. Wash membrane three times for 10 min each time in PBST at room temperature. When using StrepTactin HRP conjugate, proceed directly to step 16.

14. Incubate with secondary antibody solution for 1 h at room temperature. Either alkaline phosphatase or horseradish peroxidase (HRP)-conjugated antimouse IgG may be used. The rabbit antimouse IgG/HRP conjugate from DAKO yields good results. Dilute according to manufacturer's recommendations. Use 5% nonfat dried milk in PBST for incubation with the secondary antibody when using the chemiluminescent detection method.

15. Wash filter four times for 10 min each time in PBST at room temperature.

16. Perform detection reaction with AP or HRP chemiluminescence reagent and expose to X-ray film or detect by a luminescent imaging system according to the manufacturer's recommendations. Western lightning chemiluminescence reagent plus (PerkinElmer) yields good results.

References

Becker, D. M., and Guarente, L. (1991). High-efficiency transformation of yeast by electroporation. *Methods Enzymol.* **194**, 182–187.

Bonneaud, N., Ozier-Kalogeropoulos, O., Li, G. Y., Labouesse, M., Minvielle-Sebastia, L., and Lacroute, F. (1991). A family of low and high copy replicative, integrative and single-stranded *S. cerevisiae/E. coli* shuttle vectors. *Yeast* **7**, 609–615.

Etcheverry, T. (1990). Induced expression using yeast copper metallothionein promoter. *Methods Enzymol.* **185**, 319–329.

Gietz, D., St Jean, A., Woods, R. A., and Schiestl, R. H. (1992). Improved method for high efficiency transformation of intact yeast cells. *Nucleic Acids Res.* **20**, 1425.

Harashima, S. (1994). Heterologous protein production by yeast host-vector systems. *Bioprocess Technol.* **19**, 137–158.

Holz, C., Hesse, O., Bolotina, N., Stahl, U., and Lang, C. (2002). A micro-scale process for high-throughput expression of cDNAs in the yeast *Saccharomyces cerevisiae*. *Protein Expr Purif.* **25**, 372–378.

Laemmli, U. K. (1970). Cleavage of structural proteins during the assembly of the head of bacteriophage T4. *Nature* **227**, 680–685.

Polakowski, T., Stahl, U., and Lang, C. (1998). Overexpression of a cytosolic hydroxymethylglutaryl-CoA reductase leads to squalene accumulation in yeast. *Appl. Microbiol. Biotechnol.* **49**, 66–71.

Porath, J. (1992). Immobilized metal ion affinity chromatography. *Protein Expr. Purif.* **3**, 263–281.

Romanos, M. A., Scorer, C. A., and Clare, J. J. (1992). Foreign gene expression in yeast: A review. *Yeast.* **8**, 423–488.

Rose, A. B., and Broach, J. R. (1990). Propagation and expression of cloned genes in yeast: 2-micron circle-based vectors. *Methods Enzymol.* **185**, 234–279.

Schmidt, T. G., Koepke, J., Frank, R., and Skerra, A. (1996). Molecular interaction between the Strep-tag affinity peptide and its cognate target, streptavidin. *J. Mol. Biol.* **255**, 753–766.

Voss, S., and Skerra, A. (1997). Mutagenesis of a flexible loop in streptavidin leads to higher affinity for the Strep-tag II peptide and improved performance in recombinant protein purification. *Protein Eng.* **10**, 975–982.

Ward, A. C., Castelli, L. A., Macreadie, I. G., and Azad, A. A. (1994). Vectors for Cu(2+)-inducible production of glutathione S-transferase-fusion proteins for single-step purification from yeast. *Yeast* **10**, 441–449.

9

Semliki Forest Virus Expression System

Maria Ekström, Henrik Garoff, and Helena Andersson

I. INTRODUCTION

The Semliki forest virus (SFV) is a positive-stranded RNA virus belonging to the alphaviruses (Schlesinger and Schlesinger, 1986). The SFV expression system is based on a cDNA copy of the viral genome. The cDNA has been cloned into an SP6-based transcription vector in such a way that exact copies of the alphavirus RNA genome can be transcribed *in vitro* (Liljeström and Garoff, 1991). In the SFV1 vector, genes coding for the viral structural proteins have been replaced with a polylinker region into which heterologous genes can be cloned. The recombinant plasmid then serves as a template for *in vitro* synthesis of recombinant RNA. When introduced into cells, the recombinant RNA self-replicates, as it codes for its own replicase, leading to high synthesis of the heterologous protein while competing out the host protein synthesis.

In the SFVC vector the viral capsid gene is retained and the heterologous gene is cloned directly downstream of the capsid gene by polymerase chain reaction (PCR) (Horton *et al.*, 1989). Due to the presence of a translation-enhancing region in the capsid gene, the expression level reached using the SFVC vector is about 10 times higher than when using the SFV1 vector (Sjöberg *et al.*, 1994).

The recombinant RNA can be either directly introduced into cells by electroporation or packaged *in vitro* by cotransfecting cells with the recombinant RNA and a packaging-deficient helper RNA (SFV/Helper 1 or SFV/Helper 2) encoding the structural proteins of SFV needed for the assembly of new virus particles (Liljeström and Garoff, 1991; Berglund *et al.*, 1993). The virus stock obtained is then used to introduce the recombinant RNA into cells by way of infection. This article describes optimized protocols for the production of heterologous proteins in baby hamster kidney (BHK 21) cells using this system.

II. MATERIALS AND INSTRUMENTATION

Culture medium Glasgow MEM (BHK 21) (Cat. No. 21710), fetal bovine serum (FBS) (Cat. No. 10106), tryptose phosphate broth (Cat. No. 18050), 1 M HEPES (Cat. No. 15630), 200 mM L-glutamine (100×) (Cat. No. 25030), penicillin–streptomycin (Cat. No. 15140), trypsin–EDTA (10×) (Cat. No. 35400), phosphate-buffered saline (PBS) Dulbecco's without Ca^{2+} and Mg^{2+} (Cat. No. 14190), minimum essential medium (MEM) (Cat. No. 21090-022), bovine albumin fraction V solution 7.5% (BSA) (Cat. No. 15260-037), unlabeled methionine (Cat. No. 11086), and PBS Dulbecco's with Ca^{2+} and Mg^{2+} (Cat. No. 14040) are from Invitrogen. Corning/Life Sciences provided 75-cm^2 flasks (Cat. No. 3375), 15-ml centrifuge tubes (Cat. No. 430789), 50-ml centrifuge tubes (Cat. No. 430828), and 35-mm tissue culture plates (Cat. No. 430165). HEPES (Cat. No. 223778) and chymotrypsin (Cat. No. 1418467) are from Roche. Magnesium acetate (MgOAc) (Cat. No. 105819), sodium acetate (NaOAc) (Cat. No. 106268), potassium hydroxide (KOH) (Cat. No. 105033), TitriplexIII, EDTA (Cat. No. 108418), sodium dihydrogen phosphate monohydrate ($NaH_2PO_4 \cdot H_2O$) (Cat. No. 106346), disodium hydrogen phosphate dihydrate ($Na_2HPO_4 \cdot 2H_2O$) (Cat. No. 106580), sodium chloride (NaCl) (Cat. No. 106404), methanol (Cat. No. 106009), sodium hydroxide (NaOH) (Cat. No. 106469), glacial acetic acid (Cat. No. 100066), and isopropanol (Cat. No.

109634) are from Merck. Spermidine (Cat. No. S-2501), dithiothreitol (DTT) (Cat. No. D-9779), phenol : chloroform : isoamylalcohol (Cat. No. P-3803), diethyl pyrocarbonate (Cat. No. D-5758), bromphenol blue (Cat. No. B-6131), gelatin (Cat. No. G-9382), aprotinin (Cat. No. A-6279), phenylmethylsulfonyl fluoride (PMSF) (Cat. No. P-7626), methionine-free MEM (Cat. No. M-3911), and Ficoll 400 (Cat. No. F-4375) are from Sigma-Aldrich. Diguanosine triphosphate sodium (Cat. No. 27-4643-01), SP6 RNA polymerase (Cat. No. E2520Y), ATP (Cat. No. 27-2056-01), CTP (Cat. No. 27-2066-01), GTP (Cat. No. 27-2076-01), and UTP (Cat. No. 27-2086-01) are from Amersham Pharmacia Biotech. SpeI (Cat. No. R6591) and RNasin (Cat. No. N2511) are from Promega. Sea-plaque agarose (Cat. No. 50100) is from FMC Bio Products. Ether (Cat. No. 32203) and boric acid (Cat. No. 31146) are from Riedel-de-Haën. Xylene cyanol FF (Cat. No. 44306 2B) and Nonidet P-40 (NP-40) (Cat. No. 56009 2L) are from BDH Laboratory supplies. Electroporation cuvettes (Cat. No. ECU-104) are from EquiBio. The gene pulser is from Bio-Rad. Tris (Cat. No. 146861) is from Angus buffersbiochemicals. FluorSave Reagent (Cat. No. 345789) is from Calbiochem. Coverslips (18 × 18 mm) (Cat. No. 1.0) are from Menzel-Glaser. [^{35}S]Methionine (Cat. No. AG1094) is from Amersham Biosciences. BHK 21 cells C-13 (Cat. No. CRL 8544) are from American Type Culture Collection.

III. PROCEDURES

A. Preparation of mRNA *in vitro*

Solutions

1. *DEPC-H$_2$O*: To make 1 liter, add 1 ml of diethyl pyrocarbonate to distilled water and mix hard overnight at 37°C with a stirring magnet. Pour into RNase-free bottles and autoclave. Store at room temperature.

2. *3 M NaOAc, pH 6.0*: To make 10 ml, add 2.46 g of NaOAc, adjust pH with glacial acetic acid, and complete the volume with distilled water. Store at room temperature.

3. *1 M HEPES–KOH, pH 7.4, stock solution*: To make 10 ml, add 2.38 g of HEPES to DEPC-H$_2$O, adjust pH to 7.4 with 3 M KOH, and complete the volume to 10 ml with DEPC-H$_2$O. Store at 4°C.

4. *1 M MgOAc stock solution*: To make 10 ml, add 2.14 g of MgOAc to DEPC-H$_2$O. Store at 4°C.

5. *200 mM spermidine–HCl stock solution*: To make 3 ml, add 153 mg of spermidine to DEPC-H$_2$O. Store as aliquots at −20°C.

6. *10x SP6 buffer*: 400 mM HEPES–KOH, pH 7.4, 60 mM MgOAc, 20 mM spermidine–HCl. To make 10 ml, add 4 ml of 1 M HEPES–KOH, pH 7.4, stock solution, 600 μl of 1 M MgOAc stock solution, 1 ml of 200 mM spermidine–HCl stock solution, and complete the volume to 10 ml with DEPC-H$_2$O. Store as aliquots at −20°C.

7. *50 mM DTT*: To make 10 ml, add 77.1 mg of DTT to DEPC-H$_2$O. Store as aliquots at −20°C.

8. *rNTPmix*: 10 mM ATP, 10 mM CTP, 10 mM UTP, 5 mM GTP. To make 200 μl, add 20 μl of 100 mM ATP stock solution, 20 μl of 100 mM CTP stock solution, 20 μl of 100 mM UTP stock solution, 10 μl of 100 mM GTP stock solution, and complete the volume to 200 μl with DEPC-H$_2$O. Store as aliquots at −20°C.

9. *0.5 M EDTA, pH 8.0, stock solution*: To make 10 ml, add 1.86 g of Titriplex III to DEPC-H$_2$O, adjust pH to 8.0 with NaOH, and complete the volume to 10 ml with DEPC-H$_2$O. Store at room temperature.

10. *5x TD*: 20% Ficoll 400, 25 mM EDTA, pH 8.0, 0.05% bromphenol blue, 0.03% xylene cyanol. To make 10 ml, add 2 g of Ficoll 400, 0.5 ml of 0.5 M EDTA pH 8.0 stock solution, 5 mg of bromphenol blue, 3 mg of xylene cyanol, and complete the volume to 10 ml with DEPC-H$_2$O. Store as aliquots at −20°C.

11. *10x TBE*: To make 1 liter, add 60.55 g of Tris base, 30.9 g of boric acid, and 9.3 g of EDTA to distilled water. Store at room temperature.

Steps

1. Linearise 3 μg of plasmid DNA with 3 U SpeI at 37°C for 60 min in a total volume of 50 μl. Verify complete linearisation by gel electrophoresis.

2. Add H$_2$O to a final volume of 100 μl and add 10 μl 3 M sodium acetate, pH 6.0.

3. Add 100 μl phenol : chloroform : isoamylalcohol and mix.

4. Centrifuge at 16,000 g for 1 min at room temperature.

5. Transfer the water phase to a fresh tube.

6. Add 500 μl water saturated ether and mix.

7. Centrifuge at 16,000 g for 20 s at room temperature.

8. Remove the ether phase with a water jet pump.

9. Repeat steps 6–8.

10. Add 250 μl ethanol and incubate for 10 min at −70°C to precipitate the DNA. Centrifuge for 20 min at 12,000 g in a microcentrifuge at 4°C. Wash the pellet with 800 μl 75% ethanol. Centrifuge for 20 min at 12,000 g. Remove ethanol. Dry the pellet in a Speed Vac for 10 min or at room temperature for about 20 min. Resuspend the DNA pellet in 5 μl DEPC-H$_2$O.

11. Mix the following in a 1.5-ml tube at room temperature: 23 μl H$_2$O, 5 μl 10 × SP6 buffer, 5 μl 10 mM

diguanosine triphosphate sodium, 5 µl 50 mM DTT, and 5 µl rNTP mix. Centrifuge at 16,000 g for 10 s at room temperature to remove precipitates.Transfer to a fresh tube: 1.5 µl RNasin (50 units), 5 µl linearised DNA, and 60 U SP6 RNA polymerase

12. Incubate at 37°C for 1 h.

13. Take a 1-µl aliquot into 7 µl DEPC-H₂O, add 2 µl of 5 × TD loading buffer, and run the RNA mix on a 0.7% agarose TBE gel containing 0.02% ethidium bromide to check RNA production.

14. Freeze the rest of the RNA mix on dry ice and then transfer it to –70°C.

B. *In vivo* Packaging of Recombinant RNA into SFV Particles

Solution

BHK 21 medium: 500 ml of BHK medium G-MEM,
25 ml of fetal bovine serum,
50 ml of tryptose phosphate broth,
5 ml of 1 M HEPES,
5 ml of 200 mM glutamine, and
5 ml of 10,000 U/ml penicillin/10,000 µg/ml streptomycin.
Store at 4°C.

Steps

1. Grow BHK 21 cells in a 75-cm² bottle to late log phase in complete BHK medium.
2. Wash cells with 10 ml PBS (without Ca²⁺ and Mg²⁺).
3. Wash cells with 2 ml of 1x trypsin–EDTA (0.5/0.2 mg/ml, respectively) in PBS (without Ca²⁺ and Mg²⁺) and incubate the cells at 37°C for 10 min.
4. Add 2 ml of complete medium and pipette the solution back and forth to detach cells and to remove clumps, add 8 ml of complete BHK medium, and again pipette to remove aggregates.
5. Transfer cells to a 15-ml centrifuge tube and harvest cells by centrifugation for 5 min, 400 g at room temperature.
6. Resuspend cells in 5 ml of PBS (without Ca²⁺ and Mg²⁺).
7. Harvest cells as in step 5.
8. Resuspend cells in 0.8 ml of PBS (without Ca²⁺ and Mg²⁺).
9. Add 20 µl of SFV/Helper 1 or SFV/Helper 2 RNA and 20 µl of SFV recombinant RNA.
10. Transfer the mixture to a 0.4-cm electroporation cuvette.
11. Pulse twice at 850 V/25 µF. The time constant should show 0.4 after each pulse.
12. Transfer cells to 16 ml of complete BHK medium and rinse the cuvette with the same medium to collect all cells. Then transfer the medium to a 75-cm² bottle.
13. Incubate cells for 24 h at 37°C.
14. Clarify the growth medium by centrifugation in a 50-ml centrifuge tube at 1800 g 10 min, 4°C.
15. Aliquot into 1.5-ml Eppendorf tubes and freeze rapidly on dry ice pellets. Store at –70°C.

C. Titer Determination of Helper 1-Packaged Virus by Immunofluorescence

Solutions

1. *Supplemented MEM*: 500 ml of MEM, 5 ml of 1 M HEPES, 5 ml of 200 mM glutamine 5 ml of 10,000 U/ml penicillin/10,000 µg/ml streptomycin, and 14.1 ml of 7.5% BSA. Store at 4°C.
2. *2% PBS–gelatine stock solution*: To make 100 ml, add 2 g of gelatine to PBS. Autoclave. Store at room temperature.
3. *0.2% PBS–gelatine*: To make 100 ml, add 10 ml of 2% PBS–gelatine stock solution and complete the volume to 100 ml with PBS. Store at room temperature.

Steps

1. Grow BHK 21 cells on 18 × 18-mm glasscoverslips to about 80% confluency in complete BHK medium.
2. Pipette 50, 5, and 0.5 µl of recombinant virus stock to Eppendorf tubes and complete the volume to 500 µl with supplemented MEM (dilutions 1 : 10, 1 : 100, 1 : 1000).
3. Remove the medium and wash cells with 2 ml PBS (with Ca²⁺ and Mg²⁺).
4. Add the total virus solution (500 µl) to the cells.
5. Incubate for 60 min at 37°C. Tilt the dishes every 20 min to ensure even distribution of virus particles.
6. Remove virus solution and add 3 ml complete BHK 21 medium and continue incubation for 5 h at 37°C.
7. Rinse coverslips twice with 2 ml PBS (with Ca²⁺ and Mg²⁺). Fix cells in –20°C methanol for 6 min at –20°C.
8. Remove methanol and wash coverslips three times with 2 ml PBS.
9. Add 2 ml of 0.2% PBS–gelatine and incubate for 5 min at room temperature to block nonspecific binding.
10. Replace blocking buffer with the same containing primary antibody and incubate at room temperature for 30 min.
11. Wash cells twice with 2 ml of 0.2% PBS–gelatine and twice with 2 ml of PBS. Incubate for 5 min at

room temperature between each wash (total of 20 min).

12. Dilute secondary antibody in blocking buffer and incubate with cells at room temperature for 30 min.

13. Wash cells twice with 2 ml of 0.2% PBS–gelatine for 5 min and twice with 2 ml of PBS for 5 min.

14. Rinse coverslips in distilled water, dry, and mount in FluorSave reagent.

15. Count positive cells in a fluorescence microscope.

D. Titer Determination of Helper 2-Packaged Virus by Immunofluorescence

Solutions

1. *See solutions in Section III,C*
2. *10 mg/ml chymotrypsin A4 stock solution*: To make 1 ml, add 10 mg of chymotrypsin A4 to PBS with Ca^{2+} and Mg^{2+}. Store as aliquots at –20°C.

Steps

1. Grow BHK 21 cells on 18 × 18-mm glasscoverslips to about 80% confluency in complete BHK medium.

2. Activate virus stock by taking 55.5 µl virus stock solution and add 2.8 µl 10 mg/ml chymotrypsin and incubate for 30 min at room temperature.

3. Inactivate chymotrypsin by adding 29.1 µl aprotinin.

4. Pipette 78.7, 7.9, and 0.8 µl of activated virus to Eppendorf tubes and complete the volume to 500 µl with supplemented MEM (dilutions 1 : 10, 1 : 100, 1 : 1000).

5. Continue from step 3, Section III,C.

E. Metabolic Labeling of Infected Cells

Solutions

1. *Starvation medium*: 100 ml of methionine-free MEM, 1 ml of 200 m*M* glutamine, 1 ml of 1 *M* HEPES, and 1 ml of 10,000 U/ml penicillin/10,000 µg/ml streptomycin. Store at 4°C.

2. *15 mg/ml unlabelled methionine stock solution*: To make 100 ml, add 1.5 g of methionine to distilled water. Sterile filter. Store as aliquots at –20°C.

3. *Chase medium*: 500 ml of MEM, 5 ml of 200 m*M* glutamine, 5 ml of 1 *M* HEPES, and 5 ml of 10,000 U/ml penicillin/10,000 µg/ml streptomycin. Store at 4°C. Add 150 µg/ml of unlabelled methionine just before use from the stock solution.

4. *10% NP40 stock solution*: To make 100 ml, add 10 g of NP-40 to distilled water. Sterile filter. Store at room temperature.

5. *1 M Tris–HCl, pH 7.6, stock solution*: To make 100 ml, add 12.1 g of Tris and adjust pH with HCl. Complete the volume with distilled water. Autoclave. Store at 4°C.

6. *4 M NaCl stock solution*: To make 100 ml, add 23.4 g of NaCl to distilled water. Autoclave. Store at room temperature.

7. *0.2 M EDTA stock solution*: To make 100 ml, add 7.45 g of Titriplex III to distilled water and warm it to dissolve the EDTA. Sterile filter. Store at room temperature.

8. *10 mg/ml PMSF stock solution*: To make 1 ml, add 10 mg of PMSF to isopropanol. Store dark at room temperature. *Note*: Very toxic.

9. *1x lysis buffer*: 1% NP-40, 50 m*M* Tris–HCl, pH 7.6, 150 m*M* NaCl, 2 m*M* EDTA, 10 µg/ml PMSF. To make 100 ml, add 10 ml of 10% NP-40 stock solution, 5 ml of 1 *M* Tris–HCl, pH 7.6 stock solution, 3.75 ml of 4 *M* NaCl stock solution, 1 ml of 0.2 *M* EDTA stock solution, and complete the volume to 100 ml with distilled water. Store at 4°C. Add 10 µg/ml of PMSF just before use from the stock solution.

Steps

1. Grow BHK 21 cells to 80–100% confluency on a 35-mm tissue culture plate in complete BHK medium.

2. Dilute virus in supplemented MEM using a multiplicity of infection of 5.

3. Remove the medium and wash cells with 2 ml of PBS (with Ca^{2+} and Mg^{2+}).

4. Add 500 µl of the virus solution to the cells.

5. Incubate for 60 min at 37°C. Tilt every 20 min.

6. Remove virus solution and add 3 ml of complete BHK 21 medium and continue incubation for 6 h at 37°C.

7. Aspirate growth medium and wash cells twice with 2 ml of PBS (with Ca^{2+} and Mg^{2+}) prewarmed to 37°C.

8. Overlay cells with 2 ml of starvation medium prewarmed to 37°C and incubate plates at 37°C for 30 min.

9. Aspirate medium and replace with 500 µl of the same containing 100 µCi/ml of [^{35}S]methionine and incubate for appropriate pulse time at 37°C.

10. Aspirate pulse medium and wash cells twice with 2 ml of chase medium prewarmed to 37°C.

11. Overlay cells with 2 ml of chase medium and incubate for required chase time at 37°C.

12. Remove medium and wash cells with 2 ml of ice-cold PBS (with Ca^{2+} and Mg^{2+}).

13. Add 300 µl of ice-cold lysis buffer and incubate for 10 min on ice. Resuspend cells and transfer into an Eppendorf tube.

14. Centrifuge for 5 min at 3000 g at 4°C to pellet nuclei. Transfer supernatant to a fresh tube. Store at –70°C.
15. Assay for protein expression by SDS–PAGE and autoradiography.

F. Metabolic Labeling of Transfected Cells

Solutions

See solutions in Section III,E

Steps

See Section III,B, steps 1–8.
9. Add 20 µl of recombinant RNA.
10. Transfer the mixture to a 0.4-cm electroporation cuvette.
11. Pulse twice at 850 V/25 µF. The time constant should show 0.4 after each pulse.
12. Transfer cells to 16 ml of complete BHK medium, rinse the cuvette with the same medium to collect all cells, and transfer 3 ml to a 35-mm tissue culture plate.
13. Incubate cells for 7 h at 37°C.
14. Continue from step 7 Section III,E.

IV. COMMENTS

The cloning of recombinant DNA into the SFVC vector has been done by PCR (Horton *et al.*, 1989). The capsid protein has autoproteolytic activity and will cleave itself from the heterologous protein, provided that the cleavage site is preserved (Melancon and Garoff, 1987). Systematic probing of tolerated amino acid residues round the cleavage site has not been done, but a proline residue in position +2 relative to the cleavage site will inhibit cleavage. Both signal sequence-bearing proteins, as well as myristoylated proteins, have been expressed from the SFVC vector and they are processed normally.

When using SFV/Helper 1 for the production of recombinant virus stocks, a replicative form of the SFV genome may form (Berglund *et al.*, 1993). This probably occurs through recombination promoted by the viral replicase, switching from recombinant to helper RNA in the cotransfected cells (Weiss and Schlesinger, 1991). In nature, SFV can infect humans and care must therefore be taken when using SFV/Helper 1. According to the commission of the European Communities directive 95/30/EC, SFV is classified as a class 2 agent (Official Journal of the European Communities, No. L 155, 6. 7. 1995, pp. 41–42). A corresponding level of containment is recommended for work with SFV (workplace with restricted access, an observation window, and surfaces that are easy to clean).

Therefore, SFV/Helper 2 was designed for use in producing conditionally infectious particles (Berglund *et al.*, 1993). These particles must be activated by chymotrypsin treatment. Using this helper, no replication-proficient virus has been found (Berglund *et al.*, 1993).

SFV has a very broad host range and the recombinant virus stocks can therefore infect many kinds of eukaryotic cells. For transfecting cells using electroporation, the electroporation parameters must be optimized for every cell type. For BHK cells, as many as 10^8 to 10^9 infectious particles can be obtained when cotransfecting 10^7 cells with recombinant and SFV/Helper 2 RNA and incubating for 24 h.

V. PITFALLS

1. Avoid repeated freezing and thawing of the virus stock. This will reduce virus infectivity.
2. For use in electroporation, the cells should not have been passaged more than about 30 times.
3. When amplifying the vector, rearrangments of the DNA may occur. Therefore the vector should be checked thoroughly after amplification by different restriction enzyme digestions.

References

Berglund, P., Sjöberg, M., Garoff, H., Atkins, G. J., Sheahan, B. J., and Liljeström, P. (1993). Semliki Forest virus expression system: Production of conditionally infectious recombinant particles. *Bio/Technology* **11**, 916–920.

Horton, R. M., Hunt, H. D., Ho, S. N., Pullen, J. K., and Pease, L. R. (1989). Engineering hybrid genes without the use of restriction enzymes: Gene splicing by overlap extension. *Gene* **77**, 61–68.

Liljeström, P., and Garoff, H. (1991). A new generation of animal cell expression vectors based on the Semliki Forest virus replicon. *Bio/Technology* **9**, 1356–1361.

Melancon, P., and Garoff, H. (1987). Processing of the Semliki Forest virus structural polyprotein: Role of the capsid protease. *J. Virol.* **61**, 1301–1309.

Schlesinger, S., and Schlesinger, M. J. (eds.) (1986). "The Togaviridae and Flaviviridae." Plenum Press, New York.

Sjöberg, E. M., Suomalainen, M., and Garoff, H. (1994). A significantly improved Semliki Forest virus expression system based on translation enhancer segments from the viral capsid gene. *Bio/Technology* **12**, 1127–1131.

Weiss, B. G., and Schlesinger, S. (1991). Recombination between Sindbis virus RNAs. *J. Virol.* **65**, 4017–4025.

10

Transient Expression of cDNAs in COS-1 Cells: Protein Analysis by Two-Dimensional Gel Electrophoresis

Pavel Gromov, Julio E. Celis, and Peder Madsen

I. INTRODUCTION

Transient and stable expression of exogenous cDNAs in eukaryotic cells provides a powerful approach for studying the corresponding gene products (see also related articles in this volume). Expression of a given protein is achieved by introducing an expression vector, harboring the cDNA of interest downstream an appropriate promoter, into eukaryotic cells. Several techniques are available for introducing DNA into eukaryotic cells (Ravid and Freshney, 1998; Thompson *et al.*, 1999; Schenborn, 2000; Gromov and Celis, 2002): (i) electroporation (Chu *et al.*, 1987; Shigekawa and Dower, 1988; Gehl, 2003), (ii) DEAE-dextran-mediated transfection (Sussman and Milman, 1984; Lopata *et al.*, 1984; Schenborn and Goiffon, 2000a), (iii) calcium phosphate-mediated transfection (Wigler *et al.*, 1978; Chen and Okayama, 1988; Schenborn and Goiffon, 2000b), and (iv) lipofection (Felgner *et al.*, 1987; Whitt *et al.*, 1990; Schenborn and Oler, 2000; Nicolazzi *et al.*, 2003). All four techniques just mentioned can be used for transient transfection, although the method of choice usually depends on target cells, transfection efficiency, cytotoxity, cost, etc. Despite the relative high cost of liposome-mediated transfection, the method enjoys widespread use because of its reproducibility and high transfection rates.

This article describes a protocol for the effective transient transfection of COS-1 cells (Gluzman, 1981) using LipofectAMINE (Gibco, BRL) as a vehicle agent. The technique is carried out essentially according to the manufacturer's recommendation (Gibco, BRL) and is illustrated with the transient overexpression of three small GTP-binding proteins, arf, rab11, and p21-ras, using the pMT21 expression vector (Kaufman *et al.*, 1991). The procedure offers an effective mean to produce considerable amounts of a given protein and, in addition, to faithfully reproduce some posttranslational modifications. In addition, the overexpressed protein can be used for antibody production as well as biochemical studies.

II. MATERIALS AND INSTRUMENTATION

A. Plasmid

Plasmid pMT21 (Kaufman *et al.*, 1991) is used as an expression vector utilizing the adenovirus major late promoter for transcription initiation. The full-length cDNA of interest is subcloned into the pMT21 plasmid according to standard protocols (see, e.g., Sambrook *et al.*, 1990).

B. COS-1 Cells

COS-1 monkey kidney cells (Gluzman, 1981) are grown in complete Dulbecco's modified Eagle's medium (DMEM) containing 10% fetal calf serum (FCS). General procedures for cultured cells are described in details elsewhere (see, e.g., Celis and Celis, 1997; see also article by Mcleady and O'Eonnor in volume 1).

C. Reagents

DMEM (Cat. No. 31966-021) is from GIBCO. Penicillin/streptomycin (Cat. No. A2213) are from Biocrom KG. FCS (Cat. No. 04-001-1A) is from Biological Industries. Tissue culture plates (24-well) (Cat. No. 662 160) are from Greiner. All other reagents and general tissue culture facilities are as described elsewhere (Celis and Celis, 1997).

III. PROCEDURES

The protocol has been slightly modified from the original guidelines provided by Gibco BRL Life Technology.

Solutions

1. *Complete Dulbecco's modified Eagle's medium*: Prepare as described by Celis and Celis (1997). To make 500 ml, mix 445 ml of DMEM medium, 5 ml of 10x stock penicillin/streptomycin, and 50 ml of FCS.
2. *Hank's buffered saline solution*: To make 1 liter, dissolve 0.4 g of KCl, 0.06 g of KH_2PO_4, 0.0621 g of $NaHPO_4 \cdot 2H_2O$, and 8 g of NaCl in 800 ml of distilled water. After dissolving, complete to 1 liter with distilled water.
3. *Lysis solution*: 9.8 M urea, 2% (w/v) Nonidet P-40 (NP-40), 2% ampholytes, pH 7–9, and 100 mM dithiothreitol. Prepare as described by Celis *et al.* in this volume.

Steps

The volumes given are for tranfections carried out in a single well of a 24-well tissue culture plate. For larger tissue culture dishes or multiwell labeling, adjust all amounts accordingly.

1. Seed COS-1 cells in a 24-well tissue culture plate (approximately 3×10^4 cells per well) in 0.3 ml of complete DMEM medium containing 10% of calf serum, 2 mM of glutamine, and antibiotics.
2. Grow the cells at 37°C in a humidified, 5% CO_2 incubator for approximately 24 h or until the cells reach about 80% confluence.
3. For each transfection (single well), prepare the following solutions in sterile plastic conical tubes.
   a. *DNA solution*: Prepare 20 μl of the pMT21 plasmid solution in DMEM medium lacking serum and antibiotics at a concentration of 0.01 mg/ml. Mix gently.
   b. *LipofectAMINE solution*: Add 1.5 μl of the commercial LipofectAMINE reagent stock solution (mix gently prior to use) to 20 μl of DMEM medium lacking serum and antibiotics. Mix gently.
4. To make the transfection mixture, combine the two solutions, mix gently, and incubate at room temperature for 30 min to allow DNA–liposome complexes to form.
5. In the meantime, wash COS-1 cells twice with 0.4 ml of serum-free DMEM medium.
6. For each transfection (single well), add 160 μl of serum-free DMEM medium to the tube containing the transfection mixture. Mix gently and overlay the diluted transfection mixture onto the washed cells.
7. Incubate the cells with DNA–liposome complexes for 8 h at 37°C in a humidified 5%, CO_2 incubator.
8. Following incubation, carefully aspirate the transfection mixture and replace it with 0.3 ml of complete DMEM solution. Incubate the cells for an additional 8–10 h.
9. Label the cells with [^{35}S]methionine for 14 h as described in the article by Celis *et al.* in this volume.
10. Aspirate the labeling medium from the plate and wash cells gently with 1 ml of Hank's buffered saline solution. Repeat the washing step once more. Carefully remove excess Hank's buffered saline solution from the plate. Aspirate as much of the solution as possible using an elongated Pasteur pipette.
11. Resuspend the cells in 40 μl of lysis solution and run two-dimensional (2D) gels as described in the article by Celis *et al.* or by Görg and Weiss in this volume. Alternatively, the cells can be harvested for Northern blotting or prepared for immunostaining.
12. Following 2D gel electrophoresis, dry the gels and subject them to autoradiography.

IV. COMMENTS

In average, about 15–30% transfection efficiency can be achieved as estimated by immunofluorescence. Transcription of the construct starts approximately 10 h after the addition of the transfection mixture to the cells as determined by Northen hybridization. Given the high pMT21 promoter activity, it is possible to harvest the cells for gene activity assays about 24 h after starting transfection. Transfection times may vary depending on the cell type and/or promoter activity.

Representative 2D gel autoradiographs (IEF) of [^{35}S]methionine-labeled proteins from transiently transfected COS-1 cells [ADP-ribosylation factor (Arf), rab11a, and p21-ras] are shown in Fig. 1. Arf is overexpressed as a single polypeptide, whereas the human rab11a and p21-ras constructs yield a set of spots that

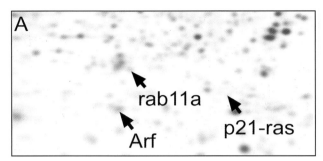

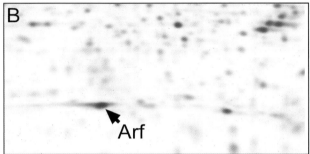

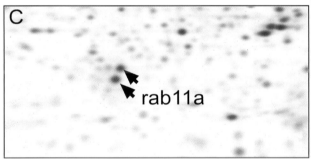

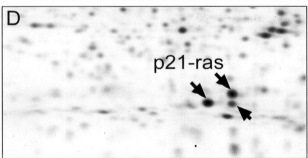

FIGURE 1 Two-dimensional gel (IEF) autoradiographs of [³⁵S]methionine-labeled proteins from COS-1 cells transfected transiently with pMT21 constructs: (A) control, nontransfected cells, (B) ADP-ribosylation factor (Arf), (C) rab11a, and (D) p21-ras.

reflect their posttranslational processing *via* geranylgeranylation and farnesylation, respectively. The COS-1 transfection system seems to reproduce a broad spectrum of posttranslational alterations of proteins and may represent an appropriate model system for the study of many posttranslational modifications.

To obtain significant amounts of the overexpressed protein, the protocol should be scaled up using six-well tissue culture plates. Resuspend the cells from one well in 60 µl of lysis solution, run 2D gels, and detect the overexpressed protein by Coomassie brilliant blue staining. Protein spots can then be electroeluted from the gel and used for immunization.

V. PITFALLS

1. Avoid growing the cells for more than 24 h prior to transfection. Always use the same conditions for cell culturing, as the transfection efficiency is sensitive to cell passage and confluence.

2. Avoid repeated freezing/defrosting of the plasmid DNA stock solution as it may induce one and double strain nicks in plasmid DNA, leading to linearization of supercoiled DNA. Linearized DNA is generally not as efficient a substrate for transient expression as supercoiled DNA. Aliquot the plasmid stock in several tubes (5–10 µl) and store at –70°C.

3. The LipofectAMINE commercial stock solution should be mixed gently before using.

4. Do not add antibacterial agents to transfection media.

References

Celis, J. E., and Celis, A. (1997). General procedures for tissue culture. *In "Cell Biology: A Laboratory Handbook"* (J. E. Celis, N. P. Carter, T. Hunter, K. Simons, D. M. Shotton, J. V. Small, eds.), pp. 5–16. Academic Press, San Diego.

Chen, C., and Okayama, H. (1988). Calcium phosphate-mediated gene transfer: A highly efficient system for stably transforming cells with plasmid DNA. *BioTechniques* **6**, 632–638.

Chu, G., Hayakawa, H., and Berg, P. (1987). Electroporation for the efficient transfection of mammalian cells with DNA. *Nucleic Acid Res.* **15**, 1311–1326.

Felgner, P. L., Gadek, T. R., Holm, M., Roman, R., Chan, H. W., Wenz, M., Northrop, J. P., Ringold, G. M., and Danielson, M. (1987). Lipofection: A highly efficient, lipid-mediated DNA/transfection procedure. *Proc. Natl. Acad. Sci. USA* **84**, 7413–7417.

Gehl J. (2003). Electroporation: Theory and methods, perspectives for drug delivery, gene therapy and research. *Acta Physiol. Scand.* **177**, 437–447.

Gluzman, Y. (1981). SV40-transformed semian cells support the replication of early SV40 mutants. *Cell* **23**, 175–182.

Gromov, P., and Celis, J. E. (2000). Gene transfer and expression in tissue culture cells. *In "Encyclopedia of Life Science"*. Nature Publishing Group, London. http://www.els.net.

Kaufman, R. J., Davies, M. V., Wasley, L. C., and Michnik, D. (1991). Improved vectors for stable expression of foreign genes in mammalian cells by use of the untranslated leader sequence from EMC virus. *Nucleic Acid Res.* **19**, 4485–4490.

Lopata, M. A., Cleveland, D. W., and Sollner-Webb, B. (1984). High-level expression of a chloram-phenicol acetyltransferase gene by DEAE-dextran-mediated DNA transfection coupled with a dimethylsulfoxide or glycerol shock treatment. *Nucleic Acid Res.* **12**, 5707–5717.

Nicolazzi, C., Garinot, M., Mignet, N., Scherman, D., and Bessodes, M. (2003). Cationic lipids for transfection. *Curr. Med. Chem.* **10**, 1263–1277.

Ravid, K., and Freshney, I. (eds.) (1998). "*DNA Transfer to Cultured Cells.*" Wiley-Liss, New York.

Sambrook, J., Fritsh, E. F., and Maniatis, T. (1990). "Molecular Cloning: A Laboratory Manual," 2nd Ed. Cold Spring Harbor Laboratory Press, Cold Spring Harbor, NY.

Schenborn, E. T. (2000). Transfection technologies. *Methods Mol. Biol.* **130**, 91–102.

Schenborn, E. T., and Goiffon, V. (2000a). DEAE-dextran transfection of mammalian cultured cells. *Methods Mol. Biol.* **130**, 147–153.

Schenborn, E. T., and Goiffon, V. (2000b). Calcium phosphate transfection of mammalian cultured cells. *Methods Mol. Biol.* **130**, 135–145.

Schenborn, E. T., and Oler, J. (2000). Liposome-mediated transfection of mammalian cells. *Methods Mol. Biol.* **130**, 155–164.

Shigekawa, K., and Dower, W. J. (1988). Electroporation of eukaryotes and prokaryotes: A general approach to the introduction of macromolecules into cells. *BioTechniques* **6**, 742–751.

Sussman, D. J., and Milman, G. (1984). Short-term, highly-efficiency expression of transfected DNA. *Mol. Cell Biol.* **4**, 1641–1643.

Whitt, M. A., Buonocore, L., Rose, J. K., Ciccarone, V., and Gebeyehu, G. (1990). TranfectACE reagent promotes transient transfection frequencies greater than 90%. *Focus* **13**, 8–12.

Wigler, M., Pellicer, A., Silverstein, S., and Axel, R. (1978). Biochemical transfer of single copy eukaryotic genes using total cellular DNA as donor. *Cell* **14**, 725–731.

11

High-Throughput Purification of Proteins from *Escherichia coli*

Pascal Braun and Joshua LaBaer

I. INTRODUCTION

The preparation of purified protein provides a powerful tool enabling the study of the biophysical characteristics, structure, biochemistry, and functional activity of proteins. With the increased focus on proteomics approaches, there is a growing demand for high-throughput (HT) methods to study protein function. The recent development of protein microarrays and multiwell solution biochemistry enables the study of hundreds or more proteins at a time (Martzen *et al.*, 1999; MacBeath and Schreiber, 2000; Zhu *et al.*, 2001). These advances offer the possibility to help systematically analyze protein–protein interactions, delineate enzyme–substrate networks, and investigate the effects of posttranslational modifications and small molecules on protein interactions or activities (MacBeath, 2002; Zhu *et al.*, 2001).

The use of these new technologies requires access to large numbers of purified proteins. The following protocols describe the expression and purification of proteins from *Escherichia coli* in an automatable 96-well format. The procedure has been shown to give yields in the range of 300 ng to 1 μg per 1 ml bacterial culture for ~50–80% of proteins up to 75 kDa (Braun *et al.*, 2002).

II. MATERIALS AND INSTRUMENTATION

Refrigerated incubator shaker with adjustable temperature (20–37°C) with plate holders (alternatively, two standard incubator shakers: one in the cold room and one at room temperature will also work)

Collection plates (e.g., Corning costar # 3896)

96-deep well blocks 2.2 ml (Marsh Bioproducts, Cat. No. AB0600)

Gas-permeable seals (e.g., Abgene, Cat. No. AB-0718)

96-well filter plate 350 μl, GF/C (Whatman Unifilter, Cat. No. 7700-3301)

Eppendorf Thermoshaker R with 96-well plate holder (S534-41 and S534-46)

(Recommended): A 96-pin inoculator with pin length >2 in. that can reach the bottom of a 96-deep well block

Thermo Finnigen Multidrop 96 DW (MTX Labs, Cat. No. P97115)

Electronic 8-channel pipettes

96-well plate spectrophotometer

Centrifuge with plate holders (best is digital rpm indicator)

SDS–PAGE system that maximizes lane number to handle many samples (e.g., Bio-Rad criterion system—gels with 26 wells)

Aluminum sealing tape (Beckman Reorder# 538619)

III. PROCEDURES

The first step in high-throughput expression and purification of proteins is obtaining the corresponding protein expression constructs. The expression of proteins often requires that the gene sequences of interest be placed (1) under the control of a specific transcriptional promoter (regulated, strong/weak, etc.), (2) near specific translational signals (ribosome binding sites, etc.), (3) in association with a particular selectable marker (ampicillin, tetracycline, etc.) and, in many cases (and nearly all conceivable HT screening scenarios), (4) linked at one or both ends to sequences encoding polypeptide tags to facilitate HT manipulations, such as purification (His$_6$, GST), detection (GFP), or functional readout (Gal4-AD) of the test proteins. The addition of peptide tags requires that the untranslated regions flanking the coding sequences be removed. Finally, the importance of ensuring that the expression clone sequences accurately reflect the natural gene sequence cannot be overemphasized. Even subtle changes in the amino acid sequences of proteins can have profound effects on protein function. The configuration of the clones should be determined in advance for each experiment and all of the clones should have the same configuration.

There is an increasing number of clone collections under construction or already available through both public and private sources that can provide useful starting material for protein expression experiments (Brizuela *et al.*, 2001, 2002; Kikuchi *et al.*, 2003; Reboul *et al.*, 2003; Seki *et al.*, 2002; Strausberg *et al.*, 1999). Some of these clone collections contain full-length cDNAs, requiring that the user extract the coding sequences and place them into the appropriate vectors that support protein expression (Strausberg *et al.*, 1999). Other clone collections have already configured the coding sequences into expression vectors, and their usefulness depends on whether the specific configuration suits the users needs (Martzen *et al.*, 1999). The most flexible approach for protein expression utilizes recombinational cloning vectors, which enable the simultaneous and rapid transfer of the coding sequences into other plasmid vectors—in frame, without mutations (Hartley *et al.*, 2000). These transfers occur in a single-step benchtop reaction that can be automated readily and nearly all vectors can be adapted (with a single fragment blunt ligation) to become recipient vectors. Thus clone collections that employ this approach can support a broad variety of experiments.

For bacterial transformation, several vendors provide competent cells in a 96-well format. The following protocol has been optimized for BL21pLys^S cells.

A. HT Protein Expression

Solutions

1. *TB media*: To prepare 1 liter of Terrific broth, mix 12 g Bacto-tryptone, 24 g yeast extract, and 4 ml glycerol and dissolve in 900 ml ddH$_2$O final volume. In a separate container, prepare a phosphate solution by mixing 2.31 g KH$_2$PO$_4$ with 12.54 g K$_2$HPO$_4$ and bring the final volume to 100 ml. Autoclave both solutions independently. Mix after solutions have cooled down.

2. *Ampicillin*: Dissolve 300 mg ampicillin in 3 ml H$_2$O. Store at $-20°$C. Make fresh weekly.

3. *Chloramphenicol*: Dissolve 340 mg chloramphenicol in 10 ml ethanol, store at $-20°$C.

4. *40% glucose*: Add 200 g L-glucose to 300 ml H$_2$O while stirring. Warm the solution to accelerate sugar dissolution. Once the glucose is in solution, bring the final volume to 500 ml to compensate for any evaporative loss. Filter sterilize and store at room temperature.

5. *1 M IPTG*: Dissolve 2.38 g IPTG in 10 ml H$_2$O. Aliquot and store at $-80°$C.

Steps

1. Aliquot 1 ml media into each well of a 96-deep well block. This is conveniently done using repetitive dispenser such as the Thermo Finnigen Multidrop. Alternatively, a repeat pipetting device (e.g., Eppendorf) can be used.

2. Inoculate cultures with bacteria containing the expression constructs. It is convenient to work with bacteria that are arranged in 96-well footprints. Dipping a sterilized 96-pin inoculation device into 96-well glycerol stocks and then softly pressing on agar will create such a footprint. After the colonies are grown, the same 96-pin device can be used to inoculate 96 O/N cultures simultaneously. Standard inoculation using sterile toothpicks will work, but careful attention must be paid to avoid mixing samples.

3. Grow cultures O/N at 37°C, 375 rpm for 14 h. Some experimentation with growth conditions may be necessary depending on the bacterial strain used. The cells should grow to early plateau, allowing slower growing colonies to catch up. However, avoid late plateau because different samples will reenter growth phase differentially. To avoid day-to-day variation, it is crucial to avoid growing the bacteria longer than 16 h.

4. Prepare a second culture block with 1.5 ml media containing appropriate antibiotics and 2% glucose.

5. After the O/N growth, check OD$_{600}$ of the cultures using a 1:10 dilution. It is advisable to check the OD of the entire plate using a 96-well plate compatible spectrophotometer. However, when many blocks are processed in parallel or if only a standard spectrophotometer is available, sampling at least 24 wells of every plate to ensure even growth is acceptable. The overall variability should be <25%.

Bacteria settle quickly in the 96-well blocks. Whenever transferring bacteria, make sure that the O/N culture block is thoroughly mixed all the time. This can be achieved by maintaining the culture block on a constantly rotating Eppendorf Thermoshaker.

6. Dilute the O/N cultures into the second block to achieve an average final $OD_{600} = 0.1$ Bacterial settling can also cause problems at this step. If possible, pipette all cultures simultaneously or mix the block with the O/N cultures constantly. Otherwise, the samples will get uneven starting densities.

7. Grow at 25°C until average OD_{600} of the culture is ~0.6 (~3.5 h).

8. Induce protein expression (with IPTG, etc.). We use IPTG at 1 mM final concentrations.

9. Express protein for 3 h at 25°C. We have found that cultures at 25°C grow more evenly and tend to have less protein degradation than at higher temperatures. Actual induction temperature and time may need adjustment depending on constructs and conditions.

10. *Optional:* Measure and record the OD_{600} of final culture while maintaining good mixing of the cultures.

11. *Optional:* Take an aliquot for Western blot analysis while maintaining good mixing of the cultures.

12. Harvest the cultures by centrifugation for 10–15 min, 4°C, 5000 g.

13. Decant media.

14. Freeze pellets to –20°C.

B. HT Protein Purification under Denaturing Conditions

Solutions

1. *Guanidine lysis buffer (GLB):* 100 mM NaH$_2$PO$_4$, 10 mM Tris–HCl pH 8.0, 6 M guanidine hydrochloride, and 10 mM β-mercaptoethanol

2. *Ni-NTA matrix:* Ni-NTA can be obtained from Qiagen (Cat. No. 30210). Wash beads several times in GLB. After last wash, resuspend the beads in 10 volumes of GLB, forming a 10% slurry.

3. *Urea wash buffer (UWB):* 100 mM NaH$_2$PO$_4$, 10 mM Tris–HCl, and 8 M urea, adjust to pH 8.0

4. *Elution buffer (EB):* 100 mM NaH$_2$PO$_4$, 10 mM Tris–HCl, 8 M urea, and 500 mM imidazole, adjust to pH 8.0

Steps

1. Thaw the deep well block containing the bacterial pellets for 5–10 min at room temperature.

2. Pipette 100 µl GLB into each well. This step is conveniently done using a repeating device such as a Multidrop.

3. Resuspend pellets by shaking the plate on a Beckman Thermoshaker for 30–60 min at 750 rpm at room temperature. The use of a stationary-inverted 96-

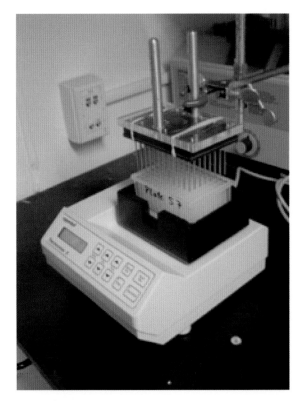

FIGURE 1 A stationary-inverted 96-pin device.

pin device will create enough turbulence to get a better resuspension (Fig. 1). This Swizzler improves lysis under both denaturing and nondenaturing conditions. If the goal is to recover functional protein by renaturation, the shaking speed should be reduced to 300 rpm after 5 min to minimize harmful oxidation of proteins.

4. Prepare two GF/C fiber plates.

 a. Filtration plate (to remove cell debris): Pipette 10 µl GLB into each well of the filtration plate.

 b. Binding plate (to capture the protein): Pipette 5 µl of 10% Ni-NTA slurry equilibrated in GLB buffer (0.5 µl bed volume) into the binding plate. Too much affinity matrix will increase nonspecific binding and lead to more impurities. Add 200 µl lysis buffer. Spin for 1.5 min at 16 g. Seal bottom with aluminum tape. Seal every well thoroughly!

5. Place the filtration plate on top of the binding plate, ensuring that well A1 is above well A1.

6. Transfer lysates from step 3 into filtration plate.

7. Spin the stacked plates at 2000 g, 2 min, 4°C.

8. Seal the top of the binding plate.

9. Rotate plate end over end for 45 min at room temperature.

10. Remove both seals. Remove the bottom seal first.

11. Spin at 50 g for 1.5 min into a waste vessel.

12. Wash the binding plate three times with UWB by adding 250 µl UWB followed by a 50 g spin for 90 s into a waste vessel at room temperature.

13. Place the binding plate on top of a 96-well collection plate.

14. Add 15 µl elution buffer to the binding plate.

15. Incubate at room temperature for 5 min.

16. Spin at 16 g for 1.5 min.

17. Repeat elution (steps 14–16) three more times (60 µl final volume).

18. After the final addition of elution buffer, spin once at 2000 g for 5 min. This hard spin will collect all remaining protein.

C. HT Protein Purification under Nondenaturing Conditions

Solutions

1. *10x phosphate-buffered saline (PBS):* Mix 80 g NaCl, 2.0 g KCl, 14.4 g Na_2PO_4, and 2.4 g KH_2PO_4 and dissolve in 900 ml ddH$_2$O. Adjust pH to 7.2 and bring volume to 1 liter.

2. *1x PBS:* Dilute 100 ml 10x PBS with 900 ml ddH$_2$O

3. *0.5 M EDTA, pH 8.0:* Dissolve 95 g EDTA in 450 ml ddH$_2$O. Start adjusting the pH with NaOH before the powder is dissolved to accelerate the dissolution process. Adjust pH to 8.0 and bring volume to 500 ml.

4. *DNase:* Dissolve 100 mg DNase (Sigma D-4527) in 10 ml H$_2$O. Aliquot in 0.5-ml aliquots and store at −80°C.

5. *Lysozyme:* Dissolve 300 mg lysozyme (enzyme grade) in 6 ml H$_2$O. Aliquot and store at −80°C.

6. *10% Triton X-100:* Slowly pipette 5 ml Triton X-100 in 45 ml H$_2$O. Rotate for a few hours or heat in a water bath until all flakes disappear. Store at room temperature.

7. *40x DNase buffer:* 900 mM $MgCl_2$, and 100 mM $MnCl_2$

8. *Wash buffer 1:* PBS, 2 mM EDTA; and 10% glycerol

9. *Wash buffer 2:* Tris–HCl, 100 mM, pH 8.0; NaCl, 500 mM; EDTA, 2 mM; 0.1% Triton X-100; and 10% glycerol

11. *Glutathione beads (Sigma G4510):* Wash the beads three to six times in wash buffer 1 containing 5 mM dithiothreitol (DTT) (confirm the absence of ethanol smell) and store at 4°C

12. *Lysis buffer:* PBS, 2 mM EDTA; 10% glycerol; 5 mM DTT (*add fresh*), and protease Inhibitors (*add fresh*)

13. *Elution buffer:* Wash buffer 2 + 10 mM reduced glutathione. Verify and readjust the pH to 8.0, if necessary.

14. *Lysozyme mix:* Mix 64 µl lysozyme (50 µg/µl), 320 µl 10% Triton X-100, and 1016 µl lysis buffer. Pipette 8 × 150 µl into one column of a plate. From this reservoir, you can conveniently pipette the mix with an 8-channel pipette in step 4.

15. *DNase mix:* 80 µl DNase (10 mg/ml), 720 µl 40x DNase buffer, and 800 µl lysis buffer. Pipette 8 × 150 µl of this mix into 8 wells of one column of a plate. From this reservoir, one can conveniently pipette the mix with an 8-channel pipette in step 6.

Steps

1. Thaw the deep well block containing the bacterial pellets for 5–10 min at room temperature.

2. Pipette 100 µl lysis buffer into each well. This step is conveniently done using a repeating device such as a Multidrop.

3. Resuspend pellets by shaking the plate on a Beckman Thermoshaker for 30–60 min at 750 rpm, 4°C. The use of a stationary-inverted 96-pin device will create enough turbulence to improve resuspension (Fig. 1). This Swizzler improves lysis under both denaturing and nondenaturing conditions.

4. Add 10 µl lysozyme using an 8-channel pipetting device to transfer the mix from a column of wells of a 96-well plate.

5. Mix for 30 min at 300 rpm, 4°C. The lower speed reduces harmful oxidation of proteins. Preferably use a stationary 96-pin device to increase mixing.

6. Add 10 µl of DNase mix. An 8-channel electronic pipette is adequate.

7. Mix for 15 min at 300 rpm, 4°C.

8. Prepare two GF/C fiber plates.
   a. Filtration plate (to remove cell debris): Pipette 10 µl lysis buffer into each well of the filtration plate to wet the filters.
   b. Binding plate (to capture the protein): Pipette 50 µl of 50% slurry of GSH agarose beads in lysis buffer (25 µl bed volume) into the binding plate.

9. Add 200 µl lysis buffer.

10. Spin for 1.5 min at 16 g.

11. Seal bottom with aluminum tape. Seal every well thoroughly!

12. Place the filtration plate on top of the binding plate, making sure that well A1 is above well A1.

13. Transfer lysates from step 5 into filtration plate.

14. Spin the stacked plates at 2000 g, 2 min, 4°C.

15. Seal the top of the binding plate.

16. Rotate the sealed binding plate end over end for 45 min, 4°C.

17. Remove both seals. Remove the bottom seal first.

18. Spin at 16 *g* for 1.5 min into a waste vessel. Avoid higher speeds, which can dehydrate the matrix and denature the proteins, reducing yields and increasing nonspecific binding.

19. Wash the binding plate three times with wash buffer I by adding 250 µl wash buffer I followed by a 16 *g* spin for 90 s into a waste vessel at 4°C.

20. Wash the binding plate two times with wash buffer II by adding 250 µl wash buffer II followed by a 16 *g* spin for 90 s into a waste vessel at 4°C.

21. Place the binding plate on top of a 96-well collection plate.

22. Add 20 µl elution buffer to the binding plate.

23. Incubate at 4°C for 5 min.

24. Spin at 16 *g* for 1.5 min.

25. Repeat elution (steps 18–20) three more times (80 µl final volume).

26. After the final addition of elution buffer, spin once at 2000 *g* for 5 min. This hard spin will collect all remaining protein.

27. Analyze 10-µl eluate by SDS–PAGE and colloidal Coomassie blue staining

References

Braun, P., Hu, Y., Shen, B., Halleck, A., Koundinya, M., Harlow, E., and LaBaer, J. (2002). Proteome-scale purification of human proteins from bacteria. *Proc. Natl. Acad. Sci. USA* **99**, 2654–2659.

Brizuela, L., Braun, P., and LaBaer, J. (2001). FLEXGene repository: From sequenced genomes to gene repositories for high-throughput functional biology and proteomics. *Mol. Biochem. Parasitol.* **118**, 155–165.

Brizuela, L., Richardson, A., Marsischky, G., and Labaer, J. (2002). The FLEXGene repository: Exploiting the fruits of the genome projects by creating a needed resource to face the challenges of the post-genomic era. *Arch. Med. Res.* **33**, 318.

Hartley, J. L., Temple, G. F., and Brasch, M. A. (2000). DNA cloning using *in vitro* site-specific recombination. *Genome Res.* **10**, 1788–1795.

Kikuchi, S., Satoh, K., Nagata, T., Kawagashira, N., Doi, K., Kishimoto, N., Yazaki, J., Ishikawa, M., Yamada, H., Ooka, H., *et al.* (2003). Collection, mapping, and annotation of over 28,000 cDNA clones from japonica rice. *Science* **301**, 376–379.

MacBeath, G. (2002). Protein microarrays and proteomics. *Nature Genet* **32** (*Suppl.*), 526–532.

MacBeath, G., and Schreiber, S. L. (2000). Printing proteins as microarrays for high-throughput function determination. *Science* **289**, 1760–1763.

Martzen, M. R., McCraith, S. M., Spinelli, S. L., Torres, F. M., Fields, S., Grayhack, E. J., and Phizicky, E. M. (1999). A biochemical genomics approach for identifying genes by the activity of their products. *Science* **286**, 1153–1155.

Reboul, J., Vaglio, P., Rual, J. F., Lamesch, P., Martinez, M., Armstrong, C. M., Li, S., Jacotot, L., Bertin, N., Janky, R., *et al.* (2003). *C. elegans* ORFeome version 1.1: Experimental verification of the genome annotation and resource for proteome-scale protein expression. *Nature Genet* **34**, 35–41.

Seki, M., Narusaka, M., Kamiya, A., Ishida, J., Satou, M., Sakurai, T., Nakajima, M., Enju, A., Akiyama, K., Oono, Y., *et al.* (2002). Functional annotation of a full-length Arabidopsis cDNA collection. *Science* **296**, 141–145.

Strausberg, R. L., Feingold, E. A., Klausner, R. D., and Collins, F. S. (1999). The mammalian gene collection. *Science* **286**, 455–457.

Zhu, H., Bilgin, M., Bangham, R., Hall, D., Casamayor, A., Bertone, P., Lan, N., Jansen, R., Bidlingmaier, S., Houfek, T., *et al.* (2001). Global analysis of protein activities using proteome chips. *Science* **293**, 2101–2105.

PART

C

GENE EXPRESSION PROFILING

Differential Gene Expression

12

Microarrays for Gene Expression Profiling: Fabrication of Oligonucleotide Microarrays, Isolation of RNA, Fluorescent Labeling of cRNA, Hybridization, and Scanning

Mogens Kruhøffer, Nils E. Magnusson, Mads Aaboe, Lars Dyrskjøt, and Torben F. Ørntoft

I. INTRODUCTION

In less than a decade DNA microarray technology has become an indispensable part of functional genomics (Celis *et al.*, 2000). Functional gene expression studies provide insight into changes in the transcription profile resulting from disease, developmental processes, or physiological stimuli. DNA microarrays have especially demonstrated their great potential in human tumour classification and prediction (Thykjaer *et al.*, 2001; Dyrskjot *et al.*, 2003; van de Vijver *et al.*, 2002) but also within drug target discovery, drug action, and toxicity (for an overview and further references, see Heller *et al.*, 1999). Microarrays are not limited to any specific organism but its usefulness increases with increasing availability of sequence information.

A DNA microarray is an orderly arrangement of usually thousands of DNA spots typically less than 150 μm in diameter that provides a medium for hybridisation of labelled nucleic acids extracted from biological samples. The immobilised DNA samples on the array are called the probes and the sample that is labelled and hybridised to the probes is the target (Phimister, 1999). There are several steps in the design and implementation of a DNA microarray experiment and they require specialized robotics,

imaging equipment, and software, which are all available commercially.

Since the first complete microarray protocols were published (Schena *et al.*, 1995; Shalon *et al.*, 1996), a wide variety of designs have developed that are mirrored in the numerous different DNA microarray protocols available. Common for them all are the four essential stages: (1) microarray fabrication, (2) target preparation, (3) hybridisation, and (4) data collection, normalisation, and bioinformatic analysis. In this chapter, protocols are outlined for the fabrication of oligonucleotide microarrays, preparation of amplified cRNA target, and the combination of these in the hybridisation procedure (Fig. 1).

For reviews of different aspects of the technology, please see Eisen and Brown (1999), Kricka (1998), Lemieux *et al.* (1998), Marshall and Hodgson (1998), and Schena and Davis (1999).

II. CHEMICALS

The following chemicals are from Sigma: sodium phosphate monobasic (Cat. No. S-0751), sodium phosphate dibasic (Cat. No. S-0876), Betaine monohydrate (Cat. No. B-2754), EDTA (Cat. No. E-5134), phenol (Cat. No. P-4557), $NaHCO_3$ (Cat. No. S-5761), $ZnCl_2$ (Cat.

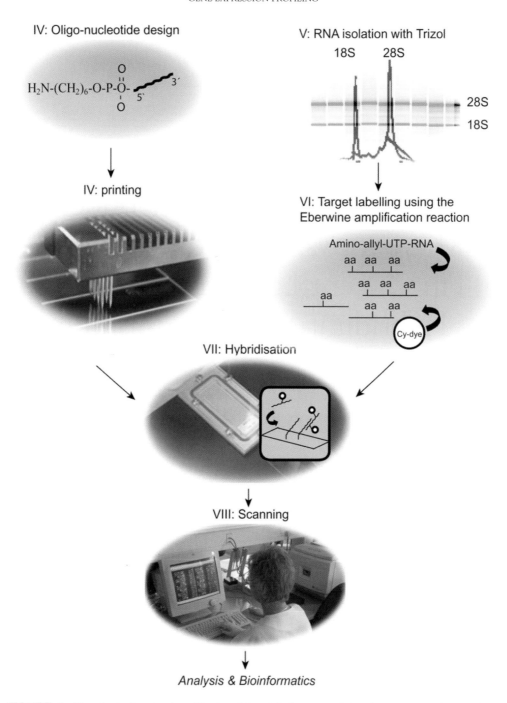

FIGURE 1 Flowchart of protocols outlined in this article for a complete microarray experiment where roman numbers refer to the relevant sections in the text. Upper left corner shows the two main fabrication steps of oligonucleotide microarrays: probe design and printing (Section IV). In the upper right corner, target preparation consisting of RNA isolation and labelling (Sections V and VI). In the middle, the combination of probes and target in the hybridisation reaction (Section VII) (Bottom) Data collection by scanning (Section VIII). Analysis and bioinformatics are beyond the scope of this chapter.

No. Z-4875), 7.5 M ammonium acetate (Cat. No. A-2706), 2-mercaptoethanol (Cat. No. M-3148), dimethyl sulfoxide (DMSO) (Cat. No. D-8779), 5-(3-aminoallyl)-2′-oxyuridine 5′-triphosphate (aa-UTP) 50 mg custom made (1-mg portions, Cat. No. A-5660), and Orange G (Cat. No. O-1625).

The following products are from Calbiochem: NaCl (Cat. No. 567441), Tris base (Cat. No. 648311), Tween 20 (Cat. No. 655204), sodium citrate (Cat. No. 567444), bovine serum albumin (Cat. No. 126575), SDS (Cat. No. 428023), formamide (Cat. No. 69189), and NaOH (Cat. No. 567530).

Glycerol (Cat. No. 818709), chloroform (Cat. No. 1.02445.1000), and isoamylalcohol (Cat. No. 748K 00555679) are from Merck. The tRNA is from Roche (Cat. No. 109495), and SeaKem GTG agarose (Cat. No. 50074) is from BMA products. Fish sperm DNA is from Serva (Cat. No. 18580.01), and ethidium bromide (Cat. No. U532813) is from Amersham Biosciences.

III. PREPARATION OF STOCK SOLUTIONS

1. *1 M Tris*: Add 121.1 g of Tris base in 800 ml of H_2O. Adjust the pH to 9.0 by adding 42 ml of concentrated HCl and bring the volume to 1 liter.

2. *TE (10 mM Tris–Cl, 1 mM EDTA)*: To prepare 1 liter, mix 100 ml of 1 M Tris and 20 ml of 0.5 M EDTA and bring the volume to 1 liter.

3. *10 M NaCl*: Dissolve 399.7 g of NaCl in 800 ml of H_2O and bring the volume to 1 liter.

4. *0.5 M EDTA*: Add 186.1 g of EDTA·2 H_2O to 800 ml of H_2O. Stir on a magnetic stirrer. Adjust the pH to 8.0 with NaOH (~20 g of solid NaOH).

5. *20× SSC*: To make 1 liter, dissolve 175.3 g of NaCl and 88.2 g of sodium citrate in 800 ml of H_2O. Adjust the pH to 7.0 with a 10 M solution of NaOH and bring the volume to 1 liter.

6. *20× SSPE*: To make 1 liter, dissolve 175.3 g of NaCl and 27.6 g of $NaH_2PO_4 \cdot H_2O$ and 7.4 g of EDTA in 800 ml of H_2O. Adjust the pH to 7.4 with a 10 M solution of NaOH and bring the volume to 1 liter.

7. *0.1 M carbonate buffer, pH 9.3*: To prepare 100 ml, dissolve 1.06 g of $NaHCO_3$ in 80 ml of H_2O. Adjust the pH with 10 M NaOH and bring the volume to 100 ml. Store frozen at −20°C in 1-ml aliqouts.

8. *10 M NaOH*: To prepare 1 liter, dissolve 40.0 g of NaOH in 800 ml of H_2O and adjust the volume to 1 liter.

9. *H_2O*: Molecular Biology Water (AccuGene, BMA products Cat. No. 51200), unless otherwise indicated in the text.

IV. FABRICATION OF OLIGONUCLEOTIDE MICROARRAYS

A. Introduction

Oligonucleotide microarrays are usually printed on slides with an activated surface that allows end-point attachment of 5′- or 3′-chemically modified DNA. The common theme is that the glass surface of the microarray slide is coated with active chemical groups, which facilitate the binding of end-modified DNA, typically amino modified. In this way, backbone binding is reduced and most of the immobilized DNA should be available for hybridisation. In the oligonucleotide microarray format, each gene is represented by one or more oligonucleotides that are designed solely on the basis of sequence information. This allows the user to design the probes avoiding cross-hybridisation with repetitive regions or regions that share similarity to other genes (Li and Stormo, 2001; Kane *et al.*, 2000; Hudges *et al.*, 2001). Oligonucleotide arrays have several advantages over cDNA microarrays. Gene-specific oligonucleotides can easily be designed and synthesised so there is no need for verification of bacterial clones and high-throughput production of polymerase chain reaction (PCR) products. In addition, oligonucleotide libraries covering large parts of the transcriptome are available from several companies.

B. Materials and Instrumentation

5′-modified 60-mer oligonucleotides dehydrated in a 96-well format are from DNA-Technology A/S, Denmark. CodeLink microarray slides are from Amersham Biosciences. The VersArray ChipWriter Pro System is from Bio-Rad Laboratories. The Biomek 2000 automated workstation is from Coultier Beckman inc. 384-well plates are from Genetix (Cat. No. x6003). The Array Designer 2.0 software is from Premier Biosoft International. The Hettich Rotina 35 R centrifuge is from Integrated Services TCP Inc. Humid chamber and container with rack are custom made at the Aarhus University workshop. Container and rack for blocking of 25 × 75-mm slides (Cat. No. 115067-0 and Cat. No. 115068-0) are from VWR international.

C. Procedures

1. Oligonucleotide Design

Several programs exist for oligonucleotide design. In Array Designer software, begin with the following criteria:

1. Length = 55–60
2. Distance from 3′ end < 999 bp
3. GC content 35–50%
4. $T_m = 75 \pm 5°C$
5. Hairpin $\Delta G_{max} = 3.0$ Kcal/mol
6. Self dimer $\Delta G_{max} = 6.0$ Kcal/mol
7. Maximum length of repeats ≤ 5 bp
8. Least cross-reactivity (Blast, http://www.ncbi.nlm.nih.gov/blast/)

If no probes are returned, reduce the stringency of one or more parameters in the software.

2. Printing of Microarrays

Solutions

1. *2× printing buffer*: (300 m*M* sodium phosphate and 2 m*M* Betaine, pH 8.5. To prepare 100 ml, mix dissolve 0.41 g sodium phosphate monobasic, 3.785 g sodium phosphate dibasic, and 9 mg Betaine in 90 ml H$_2$O. Adjust pH to 8.5 using 1 *M* NaOH or 1 *M* HCl. Bring final volume to 100 ml with H$_2$O.
2. *Blocking Solution*: 0.1 *M* Tris, 50 m*M* ethanolamine, 0.1% SDS, pH 9.0. To prepare a 1-liter solution, dissolve 12.1 g Trizma base in 900 ml H$_2$O. Stir in 3.05 g ethanolamine and mix thoroughly. Adjust pH to 9.0 using 1 *M* NaOH or 1 *M* HCl. Bring final volume to 1 liter with H$_2$O. Add SDS to blocking solution to a final concentration of 0.1% prior to use.
3. *4× SSC, 0.1% SDS*

Steps

Steps 1–3 can be done manually or automated. We have implemented an automated workstation for all oligonucleotide dilutions and for preparation of source plates for printing.

1. Dissolve the 5′ amino-modified oligonucleotides in H$_2$O to a concentration of 120 μ*M*.
2. To prepare the source plate for printing, transfer 12.5 μl 2× spotting solution into each well of the source plate and then add 12.5 μl of 120 μ*M* oligonucleotide in a predetermined pattern. The final volume is 25 μl with an oligonucleotide concentration of 60 μ*M*.
3. Prepare the microarrayer for printing. This includes equilibrating to 30–35% relative humidity, design of a suitable program for printing, cleaning, and preconditioning of quilted pins. Clean the pins in 20% ethanol and dry them completely. Precondition the quill pins for 30 min to 1 h by printing on ordinary microscope slide with printing solution.
4. Place the slides and the source plate in the microarrayer and run the printing program.
5. Prepare a humid chamber by adding as much solid sodium chloride to H$_2$O as needed to form a 1-cm-deep slurry in the bottom of a plastic container with an airtight lid.
6. For DNA immobilisation, place slides in a humid chamber with approximately 75% relative humidity for 16–24 h at room temperature.
7. After immobilization, residual reactive groups are blocked. Place the slides in a rack (up to 20 slides) and submerge the slides in 300 ml blocking solution prewarmed to 50°C. Agitate gently for 15 min.
8. Discard the blocking solution and rinse the slides twice with H$_2$O at room temperature.
9. Wash the slides with 4× SSC, 0.1% SDS prewarmed to 50°C for 15 min with gentle agitation.
10. Rinse the slides twice by submerging in distilled water. This step removes residual salt and SDS.
11. Dry slides by centrifugation (800 rpm for 3 min).
12. Store the slides dark and desiccated until use.

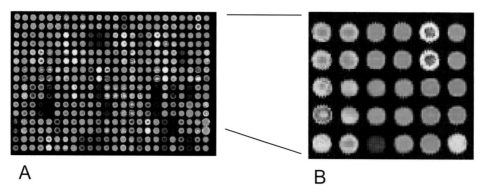

FIGURE 2 Control hybridisation of printed CodeLink slide for verification of spot presence and morphology. (A) Image of a slide with 5′ amine-modified oligonucleotides (60-mer) probes hybridised with a random 25-mer Cy5-labelled oligonucleotide and scanned at 633 nm with laser and PMT settings of 100 and 50%, respectively. (B) Magnified image.

3. Quality Control of Printed Microarrays

During printing of slides, one or more pins may be clogged for a longer or shorter period of time. This can result in abnormal or missing spots on a part of the microarray batch. We recommend performing quality control to evaluate printing errors and probe coupling efficiency for each 20th spotted slide in a batch (Fig. 2).

Solution

5'-Cy5-labelled 25-mer random oligonucleotide from DNA Technology A/S, DK

Steps

1. Dissolve the oligonucleotide in the hybridisation buffer (see Section VII) to a final Concentration of 5 µM
2. Apply 25 µl of the hybridisation mix to the slide as described in Section VII and incubate for 30 min at room temperature.
3. Wash and dry the slide as described in Section VII.
4. Scan the slide with the 633-nm (Cy5) laser at a power setting of 100% and with the PMT setting at 50%.

V. ISOLATION OF TOTAL RNA WITH TRIZOL

A. Introduction

The quality of RNA for microarray experiments must be of high quality. It should be intact and contain no genomic DNA or enzyme inhibiting substances. Several extraction methods have been tested with a wide variety of biological samples. TRIzol has become our method of choice because it gives consistent, reliable results and is considerably less expensive than kit-based products. The outlined protocol is suitable for up to 100 mg of tissue, which should give you an abundant amount of total RNA compared to the 5 µg required per labelling reaction.

All materials should be RNase free and only handled using gloves. Keep a separate set of pipettes solely for RNA work. Glassware should be baked at 180°C for 6 h. The work area can be cleaned using RNase Zap to limit the risk of RNase contamination.

B. Materials and Instrumentation

RNAlater and RNase Zap are from Ambion. Lysing Matrix D (Cat. No. 6913-100) and Q-Bio gene are from Bio101 Systems. The Trizol reagent is from Invitrogen. Bio101 Fastprep, FP120, Savant Instruments Inc.

C. Procedures

Steps

1. After excision of a tissue sample, transfer it as fast as possible to a 1.5-ml microcentrifuge tube. Submerge the tube into a container with liquid nitrogen and then store in a –80°C freezer. Alternatively, add 5 (tissue) volumes of RNAlater and store at 25°C for a couple of days, at 4°C for up to a week, or at –20° or –80°C for long-time storage. Remove RNAlater before continuing with step 2.
2. Transfer the tissue to a lysing matrix placed on ice and add 1 ml TRIzol.
3. Place the matrix in the Fastprep homogeniser. Depending on how tenacious the tissue is, run the homogeniser one to three times for 25 s at 6.0 m/s². Cool the sample on ice between each treatment.
4. Spin the samples at 4°C for 10 min at max speed in microcentrifuge tubes or at 3000 g for 30 min in 15-ml conical tubes.
5. Transfer the homogenate to a microcentrifuge tube.
6. Add 0.2 ml of chloroform/isoamylalcohol (24:1) and shake vigorously for 15 s.
7. Incubate for 5 min on ice and then spin at max speed for 15 min at 4°C.
8. Transfer top (aqueous) layer to a new microcentrifuge tube, being careful not to pick up any of the white interphase layer (the lower phases may be saved for DNA and protein extractions).
9. If double extractions are needed, repeat steps 6–8, except that only 0.1 ml chloroform is added.
10. Precipitate RNA by adding 0.5 ml isopropanol, vortex, and leave on ice for 15 min.
11. Spin at max speed for 30 min at 4°C.
12. Pour off supernatant, add 1 ml cold 70% ethanol, and invert tube to loosen pellet.
13. Spin at max speed for 5 min at 4°C.
14. Pour off supernatant, spin briefly, and remove the remaining ethanol with a pipette.
15. Dry the pellet under a ventilated hood until the RNA is just dry.
16. Dissolve the RNA in 50 µl RNase-free H_2O by vortexing.
17. Measure the optical density at 260 and 280 nm to estimate the RNA concentration and purity. An OD_{260}/OD_{280} ratio of 2.0 indicates that the RNA is pure, but any value down to 1.7 is acceptable. The RNA concentration is calculated by the following equation: Concentration (µg/ml) = OD_{260} × 40 × dilution factor. Verify RNA integrity by running 0.5 µg on a 1% agarose gel. The intensity of the 28S to 18S should have a ratio of approximately 2 : 1. Some species (e.g., drosophila) have 28S and 18S RNA that only move

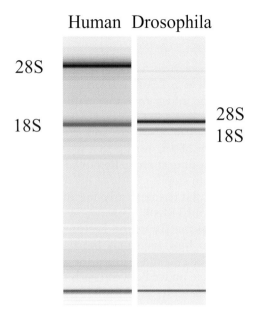

FIGURE 3   Total RNA from human and drosophila separated on a 4% Metaphor agarose gel for 90 min. High-quality RNA will have a 28S to 18S ratio of intensity of approximately 2 : 1.

with slightly different velocities (Fig. 3). In these cases, use a 4% Metaphor agarose gel (FMC Bio Products) and let it run for 90 min.

VI. SYNTHESIS OF LABELLED TARGET USING A MODIFIED EBERWINE AMPLIFICATION PROTOCOL

A. Introduction

We routinely use 5 µg of total RNA per labeling reaction using a modified Eberwine amplification protocol (Eberwine *et al.*, 1992; 't Hoen *et al.*, 2003) but any amount in the range of 200 ng–10 µg of total RNA will work fine. The procedures consist of several steps, starting with the conversion of messenger RNA into first strand cDNA through reverse transcription primed with a T7-tagged oligo(dT). Synthesis of the second-strand cDNA results in a T7-tagged double strand DNA that acts as template in the subsequent *in vitro* transcription (IVT). The incorporation of aminoallyl-UTP during IVT allows the conjugation of reactive-esterified flurophores to the cRNA. There are two clear advantages of this protocol compared to the synthesis of labelled cDNA by incorporation of fluorophore conjugated nucleotides during the reverse transcription: (1) a small amount of RNA down to 200 ng can be used

instead of the otherwise recommended 20–50 µg and (2) there is no incorporation bias during incorporation of aminoallyl nucleotides as opposed to the incorporation of nucleotides coupled to two different sized Cy dyes.

B. Materials and Instrumentation

SuperScript double-stranded cDNA synthesis kit, Invitrogen; oligo(dT)$_{24}$-T7 reverse transcription primer with the sequence 5'-GGCCAGTGAATTGTAATAC GACTCACTATAGGGAGGCGGT$_{24}$), HPLC purified, DNA Technology, DK; gel electrophoresis equipment; PCR machine; MegaScript T7, Ambion; RNeasy kit, Qiagen; Cy dye PostLabelling reactive Dye Pack, Cat. No. RPN-5661, Amersham Biosciences.

C. Procedures

1. Preparation of Double-Stranded DNA

Solutions

1. *Hydrolysis buffer (1 ml):* 1 M NaOH, 2 mM EDTA Prepare by mixing 1 ml of 1 M NaOH and 4 µl of 0.5 M EDTA.
2. *Phenol : chloroform : isoamylalcohol   solution   (25 : 24 : 1)*
3. *7.5 M ammonium acetate*
4. *loading buffer (10 ml):* 0.4 ml of 50 mg/ml Orange G, 5 ml glycerol, and 4.6 ml H$_2$O

Steps

1. Mix total RNA (5 µg) with 100 pmol of oligo(dT)-T7 primer and add H$_2$O to a volume of 9.5 µl. Incubate the tube at 70°C for 10 min and immediately place on ice.

2. Thaw the following kit reagents on ice and then add to the RNA as follows: 2 µl H$_2$O, 4 µl 5× first strand buffer, 2 µl 0.1 M DTT, and 1 µl 10 mM dNTP. Mix gently and incubate at 37°C for 2 min before adding 1.5 µl SuperScript II. Mix gently by stirring with the pipette tip (final volume of 20 µl) and incubate at 42°C for 1 h.

3. Thaw the following reagents on ice and add 90 µl H$_2$O, 30 µl 5× second strand buffer, 3 µl 10 mM dNTP, 1 µl *E.coli* RNase H, 1 µl *E.coli* ligase, and 4 µl *E.coli* DNA polymerase I. Mix gently and incubate at 16°C for 2 h (at this point the reaction may be stored for later handling at −20°C).

4. Purify the cDNA product (total volume of 156.5 µl) by adding 156.5 µl phenol : chloroform : isoamylalcohol solution (25 : 24 : 1), mix carefully, and spin the tube for 10 min at max speed at 4°C.

5. Carefully transfer the water phase to a new tube.

6. Precipitate the cDNA by adding 80 µl of 7.5 M ammonium acetate, 400 µl 96% ethanol (–20°C), and mix carefully. Spin the tube for 30 min at max speed at 4°C.

7. Discard the supernatant and wash the DNA pellet with 200 µl of ice-cold 70% ethanol. Mix carefully and spin the tube for 10 min at max speed at 4°C.

8. Discard the ethanol and let the pellet air dry and then dissolve in 7 µl H_2O.

9. Check the quality of cDNA by gel electrophoresis by adding 3 µl loading buffer to 1 µl of cDNA and load it onto a 1% agarose gel (100 V for 25 min) (Fig. 4).

2. Preparation of Aminoallyl-Labeled cRNA

Place the RNA polymerase mix on ice, it is stored in glycerol and is not frozen at –20°C. Vortex the 10× reaction buffer and the ribonucleotide solutions (ATP, CTP, GTP, and UTP) until they are completely in solution. Once they are thawed, store the ribonucleotides on ice, but keep the reaction buffer at room temperature. Spin all reagents briefly before opening to prevent loss or contamination of material that may be present around the rim of the tube. The spermidine in the 10× reaction buffer can coprecipitate the template DNA if the reaction is assembled on ice. Add the 10× reaction buffer after the water and template DNA are already in the tube.

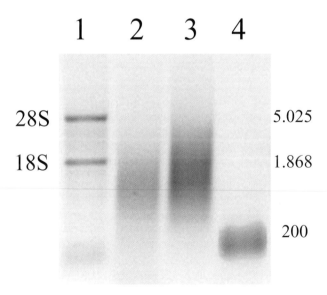

FIGURE 4 Control gel of products from the single steps in the probe preparation procedure. (1) Total RNA; (2) double-stranded cDNA will shown a smear with the highest intensity around 1500 bases; (3) cRNA from IVT should reflect the double-stranded cDNA in terms of size distribution; and (4) fragmented cRNA should be in a narrow range around 100 to 200 bases.

Solution

Prepare a 40 mM aminoallyl-UTP solution (stored at –80°C)

Steps

1. For each reaction (20 µl): 6 µl double-stranded cDNA, 1.1 µl H_2O, and 2 µl 10× reaction buffer. Mix and spin down. Keep the solution at room temperature.

2. Thaw the following reagents on ice and add 2 µl ATP, GTP, CTP (75 mM), 1.0 µl UTP (75 mM), 1.9 µl aa-UTP (40 mM) (the aa-UTP/UTP ratio is 1 : 1), and 2 µl of 10× RNA polymerase mix. Mix thoroughly by gently flicking the tube and then spin tube briefly to collect the reaction mixture at the bottom of the tube. Incubate at 37°C for 6 h or overnight.

3. Check the quality of aa-cRNA by gel electrophoresis: Add 3 µl loading buffer to 1 µl of aa-cRNA and load it onto a 1% agarose gel (100 V for 25 min).

4. Purify the aa-cRNA product using the RNeasy kit as follows: add 350 µl RLT buffer and 3.5 µl 2-mercaptoethanol. Mix and add 250 µl of 96% ethanol.

5. Transfer the sample to an RNeasy spin column and spin at max speed for 15 s.

6. Collect the run through and transfer it to the spin column again. Spin at max speed for 15 s.

7. Add 500 µl RPE and spin at max speed for 15 s.

8. Repeat step 7.

9. Transfer the spin column to a new tube and spin at max speed for 2 min to dry the filter in the column.

10. Elute the cRNA product with 50 µl H_2O. Place the column in a heating block (65°C) for 5 min.

11. Spin the column at the lowest possible speed for 1 min and continue at max speed for 1 min.

12. Repeat step 11.

13. Measure the absorbance at 260 and 280 nm (1 µl sample + 59 µl H_2O) and calculate the RNA concentration (1 OD = 40 µg/ml).

14. Aliquots (5 µg) of RNA may be stored at –80°C.

3. cRNA Conjugation with Cy Dyes

As the dyes are susceptible to photobleaching, all steps involving the labelled sample should be performed with as little light exposure as possible.

Solutions

1. *0.1 M carbonate buffer, pH 9.3*: To prepare 100 ml, dissolve 0.84 g of $NaHCO_3$ in 80 ml of H_2O. Adjust the pH with 10 M NaOH and bring the volume to 100 ml. Store frozen at –20°C in 1-ml aliquouts.

2. *1 M Tris (pH 9)*

3. *50 mM ZnCl₂*: To prepare 100 ml, dissolve 0.68 g of $ZnCl_2$ in 100 ml in H_2O. Store at room temperature.

4. *7.5 M ammonium acetate*
5. *70 and 96% ethanol*
6. *DMSO*

Steps

1. Speed-Vac 5 µg aa-cRNA to dryness. Resuspend in 4.5 µl 0.1 M carbonate buffer.
2. Add 9 µl DMSO to the Cy-dye vial and mix until the dye is completely dissolved. Add 4.5 µl of this mixture to the aa-cRNA solution.
3. Incubate at room temperature for 2.5 h.
4. To quench the remaining free Cy dyes, add 9 µl 1 *M* Tris (from −20°C) to the tube. Incubate at room temperature for 15 min.
5. Fragment the labelled RNA by adding 4.5 µl of 50 m*M* ZnCl$_2$ to the tube. It is important to mix the ZnCl$_2$ solution carefully before use. Incubate at 60°C for 30 min.
6. Precipitate the labelled RNA by first adding 2.25 µl of 7.5 M ammonium acetate to the tube. Then add 56.25 µl of 96% ethanol (4°C). Mix carefully.
7. Incubate at −20°C for 30 min or overnight.
8. Spin for 30 min at max speed (4°C).
9. Wash pellet with 70% ethanol (−20°C) and spin for 10 min at max speed (4°C).
10. Repeat step 9.
11. Remove the supernatant and dry the pellet for 5 min. This can be done in a Speed-Vac without heating.
12. Resolve the pellet in 60 µl H$_2$O and measure the absorbance at 260, 280, 550, and 650 nm.
13. Calculate the frequency of incorporation (FOI), which is the number of dye molecules per 1000 nucleotides. The optimal incorporation is 20–50 FOI.

FOI (dyes/1000 nucleotides): pmol dye × 340 per ng of cRNA

ng cRNA = A_{260} × 40 × volume (µl)

pmol dye = **Cy3:** A_{550} × volume (µl)/0.15

Cy5: A_{650} × volume (µl)/0.25

VII. HYBRIDISATION

A. Introduction

The protocol outlined here is optimised for hybridisation of cRNA target to 60-mer oligonucleotide probes on CodeLink-activated slides. Other systems may benefit from different temperatures and hybridisation times and a different composition of the hybridisation buffer.

B. Materials and Instrumentation

Hybridisation chambers were custom made in John Kreitler's laboratory at Washington University School of Medicine (St. Louis, MO). LifterSlips are from Erie Scientific Company (Cat. No. 221x25-2-4635). The hybridisation oven is from Promega. The DNA 120 Speed-Vac is from Thermo Savant. The Plactronic rectangular hot plate (Cat. No. 6156100) is from J.P Selecta s.a.

C. Procedures

Solutions

1. The volume of hybridisation mix for one hybridisation (using LifterSlips) is 20 µl hybridisation buffer. To prepare 100 µl of hybridisation mix, combine 30 µl formamide [final concentration 30% formamide], 25 µl 20× SSC [final concentration 5× SSC], 1 µl 10% SDS [final concentration 0.1% SDS], 5 µl fish sperm DNA (0.66 mg/ml) [final concentration 0.033 mg/ml FS DNA], 5 µl tRNA (0.66 mg/ml tRNA) [final concentration 0.033 mg/ml tRNA], 10 µl 0.5% BSA [final concentration 0.05% BSA], and 24 µl H$_2$O. *Note: Formamide, SSC, and SDS should be filtered through a 0.22-µm syringe filter.*

2. *Nonstringent wash buffer (400 ml):* Mix 120 ml 20× SSPE, 0.2 ml 10% Tween 20, and 279.8 ml H$_2$O

3. *Stringent wash buffer (400 ml):* Mix 1.2 ml 20× SSPE and 398.8 ml H$_2$O

Steps

1. Combine 1 µg of each labelled cRNA (both Cy3 and Cy5, 1 : 1 ratio) and dry the combined target sample in a Speed-Vac for 30 min until the volume is only a few microliters (~2 µl). Avoid drying the sample completely.

2. Resuspend the sample in 20 µl hybridisation mix. Vortex slightly and incubate for 10 min at room temperature.

3. Place the slide and the LifterSlip on the heating plate (42°C) for a few minutes (array side up!). The LifterSlip should be cleaned with 96% ethanol and dried before use.

4. Denature the sample at 95°C for 2 min and let cool to 42°C in the prewarmed hybridisation oven until use.

5. Spin the sample for 5 min at 12,000 *g* and apply it onto the LifterSlip, avoiding air bubbles (Fig. 5A) and attach the slide (array side down!) to the liquid (Fig. 5B)—the LifterSlip should stick to the slide. Place the slide with the LifterSlip in the hybridisation chamber. To avoid drying out, add 70 µl 3× SSC to the reservoirs in the hybridisation chamber (Fig. 5C) and

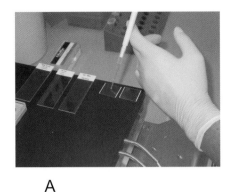 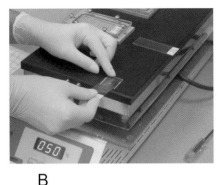 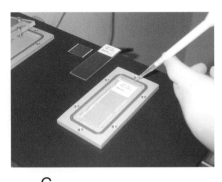

A B C

FIGURE 5 Overview of the hybridisation procedure. (A) Applying the target sample to the LifterSlip. (B) Attaching the LifterSlip to the slide. (C) Humidifying the hybridisation chamber. See text for details.

A B

FIGURE 6 Overview of washing procedure. (A) Removing the LifterSlip from the slide. (B) Washing the slides. Slides have been placed in the rack.

make sure that the hybridisation chamber is closed tightly. Place it in the hybridisation oven at 42°C and incubate for 16 h. *These steps should be performed in a clean environment, as dust particles trapped within the hybridisation area will show on the scanning and hamper the results.*

Steps

1. Take out the hybridisation chamber with the slide from the hybridisation oven.
2. Remove the LifterSlip carefully with *a pair of tweezers* (Fig. 6A). Do not touch the array area. Try to remove it *vertically*, not horizontally.
3. Quickly transfer the slide into the slide rack and immediately submerge the slide rack into the nonstringent wash buffer (Fig. 6B). It is essential that the time used for these first steps (2–3) be minimised to ensure that the array area does not dry out.
4. Wash the slide for 30 s with gentle agitation in the nonstringent wash buffer.

5. Transfer the slide rack to the stringent wash buffer and wash for 10 s with gentle agitation.
6. Dry the slide by centrifugation (800 rpm for 3 min).

VIII. SCANNING

A. Introduction

Many different laser scanner and CCD camera imaging systems are available. This section outlines a few principles in the laser scanning imaging process, but it is recommended to read the manufactures instructions carefully for a detailed description.

B. Materials and Instrumentation

ScanArray 4000 laser scanner (Perkin Elmer)

Steps

After the slides have been dried they are ready to be scanned. For optimal fluorescence it is advisable to scan the slides immediately after they have been dried. If the slides cannot be scanned at this time, store them dark and desiccated.

1. Turn on the laser scanner and allow the 543-nm (Cy3) and 633-nm (Cy5) lasers to warm up. Set the desired parameters on the scanner, e.g., laser power and PMT gain and resolution. We recommend fixing the laser power at 100% and automatically adjust PMT settings. Otherwise, we recommend the following settings: 100% laser power and 55% PMT for the Cy5 fluorophore and 100% power and 70% PMT for the Cy3 fluorophore.

2. Select the area to be scanned. The (X, Y) values defining the position of the grid on the slide are set in the printing procedure.

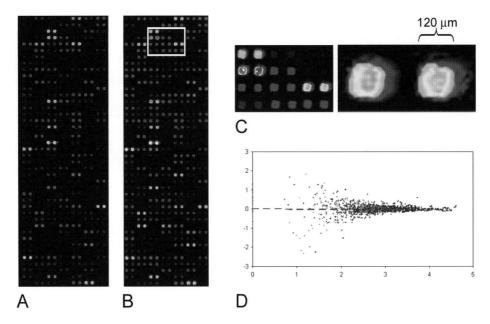

FIGURE 7   Two colour hybridisation. Scanned image of a Cy3- (A) and Cy5-labelled (B) cRNA hybridised overnight at 42°C. The same cRNA was labelled with the two dyes in order to measure the degree of variation between the two channels. (B) Enlarged view of spots. (C) MA plot of data generated from the two images. Data were normalised using a Lowess nonlinear normalisation and background corrected. The dotted line illustrates the regression line through the points. The correlation coefficient is 0.96 and indicates a low variation in a self versus self-hybridisation experiment.

3. Clean the backside of the slide to be scanned with a soft paper towel wetted with 96% ethanol. Allow the ethanol to evaporate.

4. Load the microarray (face up) into the scanner and scan the slide.

5. Save the Cy3 and Cy5 data images in an appropriate format for data analysis (Fig. 7).

IX. PITFALLS

1. Irregular spot morphology: Check the printing pins. Make sure that the temperature and humidity are kept constant when printing the slides.

2. The spotting solution should not contain aliphatic amino groups (e.g., Tris buffer) or nucleophiles (e.g., ethanolamine, lysine, and free ammonium), as these substances will compete for the active groups on the slide and inhibit the covalent coupling of the DNA to the slide. Detergents (e.g., Tween 20, SDS) should not be included in the printing solution, as they will decrease the binding efficiency on the CodeLink slides.

3. Uneven or high background after blocking: Make sure not to extend the time of blocking. Always use clean and freshly prepared solutions. Always clean glassware thoroughly after use.

4. Uneven or high background after hybridisation: Make sure that the target sample is washed properly after precipitation and does not contain free dye molecules. Be sure to use clean washing solutions and lab ware. Even trace amounts of dyes in lab ware may lead to increased background. Check that the labelling efficiency is in the desired range (20–50 dye molecules per 1000 nucleotides). Be sure the target has not dried out during incubation. If needed the sample volume may be increased.

5. Low signals after hybridisation: Check the labelling efficiency. Scan the slide again with increased laser power or PMT.

6. The spots are overlapping: Check that the relative humidity is below 35% during spotting.

References

Celis, J. E., Kruhøffer, M., Gromova, I., Frederiksen, C., Østergaard, M., Thykjaer, T., Gromov, P., Yu, Y., Pálsdóttir, H., Magnusson, N., and Ørntoft, T. F. (2000). Gene expression profiling: Monitoring transcription and translation products using DNA microarrays and proteomics. *FEBS Let.* **480**, 1–15.

Dyrskjot, L., Thykjaer, T., Kruhøffer, M., Jensen, J. L., Marcussen, N., Hamilton-Dutoit, S., Wolf, H., and Orntoft, T. F. (2003). Identifying distinct classes of bladder carcinoma using microarrays. *Nature Genet.* **33**, 90–96.

Eberwine, J., Spencer, C., Miyashiro, K., Mackler, S., and Finnell, R. (1992). Complementary DNA synthesis in situ: Methods and applications. *Methods Enzymol.* **216**, 80–100.

Eisen, M. B., and Brown, P. O. (1999). DNA arrays for analysis of gene expression. *Methods Enzymol.* **303**, 179–205.

Hughes, T. R., Mao, M., Jones, A. R., Burchard, J., Marton, M. J., Shannon, K. W., Lefkowitz, S. M., Ziman, M., Schelter, J. M., Meyer, M. R., Kobayashi, S., Davis, C., Dai, H., He, Y. D., Stephaniants, S. B., Cavet, G., Walker, W. L., West, A., Coffey, E., Shoemaker, D. D., Stoughton, R., Blanchard, A. P., Friend, S. H., and Linsley, P. S. (2001). Expression profiling using microarrays fabricated by an ink-jet oligonucleotide synthesizer. *Nature Biotechnol.* **19**, 342 – 347.

Kane, M. D., Jatkoe, T. A., Stumpf, C. R., Lu, J., Thomas, J. D., and Madore, S. J. (2000). Assessment of the sensitivity and specificity of oligonucleotide (50mer) microarrays. *Nucleic Acids Res.* **28**, 4552–4557.

Kricka, L. (1998). Revolution on a square centimeter. *Nature Biotechnol.* **16**, 513.

Lemieux, B., Aharoni, A., and Schena, M. (1998). Overview of DNA chip technology. *Mol. Breed.* **4**, 277–289.

Li, F., and Stormo, G. D. (2001). Selection of optimal oligos for gene expression arrays, *Bioinformatics* **17**, 1067–1076.

Marshall, A., and Hodgson, J. (1998). DNA chips: An array of possibilities. *Nature Biotechnol.* **16**, 27–31.

Phimster, B. (1999). *Nature Genet.* **21** (Suppl.), 1.

Schena, M., and Davis, R. W. (1999). Genes, genomes and chips. *In* "DNA Microarrays: A Practical Approach" (M. Schena, ed.). Oxford Univ. Press, Oxford.

Schena, M., Heller, R. A., Theriault, T. P., Konrad, K., Lachenmeier, E., and Davis, R. W. (1998). Microarrays: Biotechnology's discovery platform for functional genomics. *Trends Biotechnol.* **16**, 301–306.

Schena, M., Shalon, D., Davis, R. W., and Brown, P. O. (1995). Quantitative monitoring of gene expression patterns with a complementary DNA microarray. *Science* **20**, 270, 467–470.

Shalon, D., Smith, S. J., and Brown, P. O. (1996). A DNA microarray system for analyzing complex DNA samples using two-color fluorescent probe hybridization. *Genome Res.* **6**, 639–645.

't Hoen, P. A., de Kort, F., van Ommen, G. J., and den Dunnen, J.T (2003). Fluorescent labelling of cRNA for microarray applications. *Nucleic Acids Res.* **31**, e20.

Thykjaer, T., Workman, C., Kruhøffer, M., Demtroder, K., Wolf, H., Andersen, L. D., Frederiksen, C. M., Knudsen, S., and Orntoft, T. F. (2001). Identification of gene expression patterns in superficial and invasive human bladder cancer. *Cancer Res.* **61**, 2492–2499.

13

ArrayExpress: A Public Repository for Microarray Data

Helen Parkinson, Susanna-Assunta Sansone, Ugis Sarkans, Philippe Rocca-Serra, and Alvis Brazma

I. INTRODUCTION

ArrayExpress is a public repository for microarray data developed and maintained at the European Bioinformatics Institute. ArrayExpress has three major goals: (1) to serve as an archive for microarray data associated with scientific publications and other research, (2) to provide easy access to microarray data in a standard format for the research community, and (3) to facilitate the sharing of microarray designs and experimental protocols. ArrayExpress has been online since 2002 (Brazma *et al.*, 2003). ArrayExpress contains data from over 3400 hybridisations from 10 different organisms, including human and all major model organisms (as of January 2004). ArrayExpress accepts data from all microarray platforms and supports the following experiment types: gene expression experiments, CHG, and chromatin IP.

ArrayExpress supports two standards developed by the Microarray Gene Expression Data (MGED) Society (http://www.mged.org): minimum information about a microarray experiment (MIAME) (Brazma *et al.*, 2001), a content standard, and microarray expression data mark-up language (MAGE-ML), a format standard (Spellman *et al.*, 2002). MIAME specifies the minimum information that has to be provided about a microarray experiment to enable its unambiguous interpretation and potential reproducibility. According to MIAME conventions, a microarray *experiment* is defined as a set of related hybridisations (often related to a publication). To satisfy MIAME requirements, the description of the overall experiment structure-sample, hybridisation, array design, and protocol(s) used in the experiment, alongside raw and normalised data files-should be provided (Fig. 1). MIAME compliant data are loaded into ArrayExpress in MAGE-ML format and the ArrayExpress implementation is based on the MAGE-OM.

Since December 2002, many of the major scientific journals, including *The Lancet*, the *EMBO Journal* and journals from the *Nature* group require MIAME compliant data submissions to public repositories such as ArrayExpress and GEO (Edgar *et al.*, 2002). A third public repository, CIBEX, is currently under development at DDBJ, Japan. As a result of journal requirements, both the number of submissions and the amount of data in ArrayExpress are growing rapidly.

Data in ArrayExpress can be accessed via the Web interface at http://www.ebi.ac.uk/arrayexpress. Users requiring programmatic access should contact arrayexpress@ebi.ac.uk for advice.

ArrayExpress is organised into three sections: *arrays*, *protocols*, and *experiments*. Each experiment, array, and protocol in ArrayExpress is given a unique accession number (see Section III for format information). ArrayExpress provides a password-protected log-in service for private submissions—this option can be used for anonymous reviewing of microarray experiment-based publications or for submitters to access their data before release. When data are released, this facility is withdrawn.

This article describes how to submit and access data in ArrayExpress.

II. DATA SUBMISSION

Data can be submitted to ArrayExpress either online using the microarray experiment submission and annotation tool MIAMExpress (http://www.ebi.

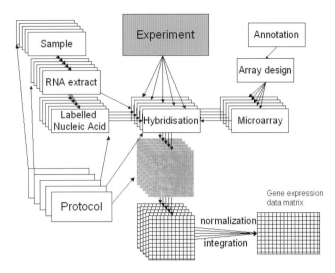

FIGURE 1 A schematic diagram of MIAME requirements showing relationships among the sections, including experiment structure, sample(s) description, hybridisation(s), array design(s), and protocols used.

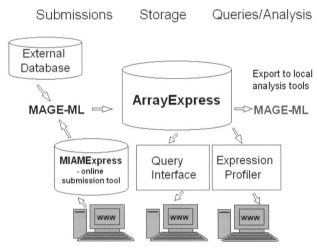

FIGURE 2 A schematic diagram of the ArrayExpress components, including data submission routes, data query, and data analysis.

ac.uk/miamexpress) or as a MAGE-ML-formatted file (Fig. 2). The second route is discussed briefly at the end of this section. The user opens an account, enters the experimental annotation, links the data files (normally stored on the user's local PC or file server accessible to the browser), and completes data submission. A typical submission of an experiment consisting of about 20 hybridisations will usually take 1 h (assuming that the submitter has read the basic instructions provided in the online help page). This section describes submission via MIAMExpress in detail.

To start a submission, users are required to create an account and provide contact details. They receive an email with their log-in details and can log-in again at any time whilst the submission is incomplete. A schematic diagram of a MIAMExpress submission is shown in Fig. 3. We suggest that users submit in the order: *arrays*, *protocols*, and *experiment(s)*. This is because the *arrays* and *protocols* are used within an *experiment* submission and observing this order speeds up the submission process. Users can try out MIAMExpress by clicking on the "Try me" option.

MIAMExpress is based on the MIAME checklist (Ball *et al.*, 2002) and the required fields appear as a series of text boxes, each of which has links to contextual help.

Whenever possible MIAMExpress uses drop-down lists of controlled vocabularies, helping the user select the appropriate term for the required field, and free text boxes are provided to add new terms. MIAMExpress is a generic tool, which is suitable for any species or domain, and is suitable for single and dual channel experiments and gene expression and CGH data submissions. MIAMExpress has been developed further for the toxicogenomics community (see Section II,C,6) and an Arabidopsis version is under development; both these versions have added domain-specific information and use specific vocabularies.

In MIAMExpress, as in MIAME, an *experiment* is defined as a group of related hybridizations. A completed submission consists of three parts: *arrays*, *protocols*, and *experiments*. The same *protocols* and *arrays* can be reused in many experiments, thereby making subsequent submissions quicker, especially if commercial arrays are used. MIAMExpress allows users to log back in to a submission at any time until it is completed. Once a submission is complete, data are exported from the MIAMExpress database (as MAGE-ML) and curated by the ArrayExpress curation team.

A. Arrays

Users need to submit details of the design of the array only if they are using a self-made array. The manufacturers normally provide details of commercial arrays. Users can check if their arrays are already present in the database using the "Browse existing arrays" link. If a particular commercial array is not present, the user should contact the ArrayExpress curation team (arrayexpress@ebi.ac.uk). ArrayExpress contains x commercial array designs from suppliers such as Agilent, Affymetrix, and Incyte and x user-supplied array designs, including many from academic facilities (figures from January 2004).

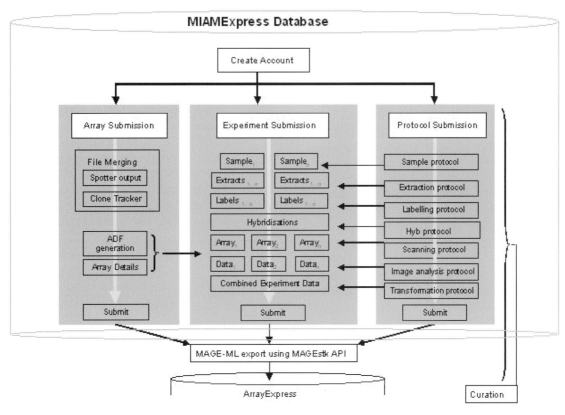

FIGURE 3 A schematic representation of the MIAMExpress submission process showing the branching tree structure of submissions.

Array designs are submitted in Array Design File (ADF) format; a simple tab delimited format, which is generated by merging a spotter output file and a clone tracker file. The ADF provides information layout (from the spotter output file) and the biological annotation for each spot (from the clone tracker file) in a single file. To allow users to access up-to-date annotation for their arrays, the ADF converter tool has been developed. It consists of a web form through which users submit a series of database identifiers, which describe the sequences on the array. These are submitted to EnsMart (Kasprzyk *et al.*, 2004), the output of the query is current sequence annotation (gene names, chromosomal location, and GO terms) in the form of database identifiers preformatted as an ADF-compatible file. Full help on the ADF is provided (http://www.ebi.ac.uk/miamexpress/help/adf/ADF_Help.htm) and the ArrayExpress curation team provides assistance in helping users generate ADF files for their arrays. MIAMExpress also allows submission of MIAME required information: species represented on the array, the surface/substrate type, a text description of the array, and finally the protocol used to manufacture the array. On

submission of the array, users can use it immediately in an experiment submission and the ArrayExpress curation team will perform curation.

B. Protocols

All protocols (except those that relate to array design) are entered from a single protocol submission page. The different protocol types are sample growth, sample treatment, sample pooling, nucleic acid extraction, labelling, hybridisation, scanning, and data transformation (normalization). Different protocol types have different (MIAME) required parameters, e.g., when describing the labelling process, users must select which nucleic acid was labelled from a dropdown list of defined terms. Unchanged protocols can be reused in subsequent submissions.

C. Experiments

MIAMExpress employs a top-down annotation approach starting from the general information about the experiment, describing each sample and the extract(s) prepared from it, to a hybridisation(s) process [linking labelled extract(s) to arrays], and the files (Fig. 3). Thus the user generates a branching tree structure as they enter information.

1. General Information

At the beginning of a submission, users are required to provide general information about the experiment, including a brief description and selecting an experiment design type and experimental factors from drop-down lists. For example, in an experiment where a sample is treated with a compound, the design type is "compound treatment" and factors could be the name of the compound. In cases where the supplied terms are not useful, users are encouraged to submit their own via the "other" text box present on all drop-downs. Information on publications and quality control methods can be provided. Users are asked to indicate the date for the experiment to be made public. Finally, in this section the total number of samples used in the experiment is provided and these are created for annotation. Samples can be added or deleted at any time during the submission process simply by returning to this experiment page and using the "Create" or "Delete" option.

2. Samples

Each sample is given a default name; we suggest that this is changed to one meaningful to the user. Species, sample type, and disease state information is supplied in boxes and the user decides which information is relevant for their experiment. This information can be copied across all samples for ease of submission. MIAME requires that sample-related protocols should be supplied and these are linked on this page. These can also be copied and pasted across multiple samples to limit the time needed for submission. It is also possible to navigate between samples by using the "Next" or "Previous" link; alternatively, these can be navigated by the list provided on the experiment page.

3. Extracts and Labelled Extracts

Each sample has one or more extracts—in single channel experiments usually there will be only one per sample. Users proceed from defining and describing a sample to creating a relationship between samples and extracts (Screenshot 1). Users define how many extracts they need and a drop-down list appears. Relevant samples are selected from this list. Pooling is supported as multiple samples can be linked to a single extract. For example, several organs from different mice may need to be pooled to provide sufficient nucleic acid for labelling. A "View/Edit" button opens

the page for each extract. All extracts require an extraction protocol to be linked and where pooling is performed the pooling protocol can also be linked from this page. Protocol information can be propagated to all subsequent extracts, meaning users need only define this for the first sample. Users should complete all the extract annotation before proceeding to providing information about labelling.

Users are required to create labelled extracts for each extract. For a two-channel experiment with a dye swap, there will typically be two labelled extracts per extract: cy3 and cy5. For a single-channel system (like Affymetrix) the labelled extract-to-extract relationship is one to one. Once the labelled extracts are created, the labelling protocol is linked and this information can be propagated to all subsequent labelled extracts. The relationships among the samples, extracts, and labelled extracts are shown in Fig. 3.

4. Hybridizations

Hybridizations are created after the samples, extracts, and labeled extracts are annotated completely. If there are unannotated objects in the sample extract-labeled extract hierarchy (Fig. 3), an error page appears with hyperlinks to each object so they can be completed.

Users must select the number of hybridizations they want to create. Hybridizations can be added or deleted using the "Create" or "Delete" option. Each hybridization can be linked to the appropriate label by selecting them from a drop-down list. Array designs are selected from a list of the users own and publicly available designs.

5. Data Files

Data files are linked to hybridizations by uploading them from a local machine via a browse button. There is normally one raw and one normalized per hybridization. At least one data file per hybridization must be provided. MIAMExpress accepts data files in meta row, meta column data format, and block, row, and column data format (e.g., GenePix), which covers the majority of scanners. Full information and guidance on formatting cDNA and Affymetrix data files can be found in the contextual help (http://www.ebi.ac.uk/miamexpress/help/data_files.html). Relevant protocols must also be provided at this point (hybridization, scanning, normalization) and, as before, can be copied across all hybridizations after the first one has been annotated. Finally, a summary data file—the "Final Gene Expression Matrix" (FGEM)—may be provided. Typically this file will be generated from some transformation of the normalized data files and may be averaged across replicates. For example, for a four-point time course experiment (0, 1, 2, 3), with two replicates per time point, the following data files may be submitted: eight raw data files (two per time point, one per hybridization), eight normalized files (two per time point, one per hybridization), and a FGEM file. Each column represents a single time point, where values have been averaged across the two replicates. Users can specify which hybridizations contributed to each column in the FGEM by providing the hybridization identifiers in the header row. Multiple quantitation types can also be provided in the same way. Rows in the FGEM typically correspond to genes and are matched with gene information provided in the ADF. On completion of submission, the experiment submission is locked; it can be released by curators, for example, to add additional information. The submission is exported from MIAMExpress in MAGE-ML format, an accession number is assigned, and it is loaded into ArrayExpress. All submissions are held confidential until the release date specified by the user.

6. Tox-MIAMExpress

Tox-MIAMExpress (http://www.ebi.ac.uk/tox-miamexpress/) is a specific version of the annotation and submission tool that links microarray data with biological end points (e.g., clinical observations, clinical chemistry, and histopathology evaluation) (Mattes *et al.*, 2004). While maintaining the general functionality of MIAMExpress, Tox-MIAMExpress provides the users with additional fields to describe toxicogenomics or pharmacogenomics specific details, such as treatment with a compound. Tox-MIAMExpress also allows submission of additional detail at the sample level. Users can provide the usual information on species and detail complex treatments such as compound/drug, dose, and delivery method. Details of blood/urine analysis commonly needed in toxicogenomics or pharmacogenomics experiments are provided via drop-down lists of terms and units. Pathological findings related to samples are entered onto a downloadable Excel spreadsheet (http://www.ebi.ac.uk/tox-miamexpress/pics/Pathology-Observation.xls). Protocols are linked to these sample treatments, in addition to the usual protocols required by MIAMExpress. Users provide information, including sample biometrics (e.g., body weight, organ weight), the target organ, and select nominal measurements, to describe any pathological changes of the organ (morphologies). This is then uploaded to Tox-MIAMExpress; the information is stored in the database and exported on completion of submission.

D. Direct MAGE-ML Submissions

Preparing MAGE-ML files is usually not a practical option for small laboratories. This direct submission route is used by laboratories with their own local microarray database systems or LIMS. In this case, a MAGE-ML export pipeline can be established from the source database directly to ArrayExpress. As of January 2004, MAGE-ML pipelines have been established from a number of local databases, including the Stanford Microarray Database (SMD) (Gollub 2003), TIGR microarray database (Saeed 2003), and German Genome Resource Centre (RZPD). Several microarray informatics tools, such as GeneSpring 6.1 (http://www.silicongenetics.com/), J-Express (http://www.ii.uib.no/~bjarted/jexpress/), and BioConductor (http://www.ebi.ac.uk/~ele/ext/submitter.html) are currently implementing MAGE-ML export and import. These developments will simplify data submission for many submitters and illustrate the utility of MAGE-ML as a data exchange format. Development of the pipeline from the SMD is described in an article available from the ArrayExpress Web site, which describes mapping a database to the MAGE-OM and implementing export using open source tools. If you wish to establish such a pipeline, please check the documentation on the ArrayExpress Web site (http://www.ebi.ac.uk/~ele/ext/submitter.html) or contact ArrayExpress at arraysubs@ebi.ac.uk for detailed advice.

III. ACCESSING DATA IN ARRAYEXPRESS

Querying data in ArrayExpress requires a web browser. Accession numbers in ArrayExpress are of the form E-XXXX-N, A-XXXX-N, or P-XXXX-N for experiments, array designs, and protocols respectively, where XXXX denotes an abbreviation of data source or data processing software and N is a number. For instance, an experiment submitted via a pipeline from the Stanford Microarray Database has accession numbers E-SMDB-N. All experiments submitted via MIAMExpress have accession numbers E-MEXP-N.

All routes to access and retrieve data from ArrayExpress consist of two basic steps.

1. Select the submission, i.e., experiment, array, or protocol, of interest by browsing or querying the database.
2. Explore the required information from the chosen submission and retrieve data.

Browsing or querying returns a brief description of the selected experiment, array, or protocol, which we will refer to as a *top-level* description. The top-level description consists of a brief summary of the submission and provides hyperlinks, which the users follow to explore data in detail or to perform specific tasks, such as retrieving raw or normalised data.

A. Retrieving the Top-Level Description

There are four ways to retrieve top-level descriptions.

1. Click on "View" experiments, arrays, or protocols.
2. Click on "Browse database."
3. Click on "Query database."
4. Click on "Log-in" (to access for submitters to non-public password-protected data).

The first two options are used to get an overview of the database. Query and log-in are described in detail in the following sections.

1. Query Database: Accessing Publicly Available Data

Clicking on "Query database" brings up the standard query interface page, which is organised into three separate sections: experiments, arrays, and protocols. The user is logged into the database as a "guest" by default and can only query public datasets. Experiments, arrays, or protocols can be retrieved quickly by their respective accession numbers or by their attributes, such as species, laboratory, or array type used in the experiment. For example, to retrieve all experiments from a single species *C. elegans*, select *Caenorhabiditis elegans* from the drop-down menu in the "organism" field of experiment menu and click on "query." This will retrieve all experiments that used samples of this species.

Combined queries are possible (for instance, by species and array type). If no query fields are completed, the query will retrieve all experiments, arrays, or protocols, respectively. Queries are case insensitive and wild cards are permitted. A query can be revised by going back to the query page (using the browser back button) and revising the selection criteria. For example, to retrieve all experiments involving human samples, you can select *Homo sapiens* from the drop-down list in the organism field and query for experiments. The user then can go back to the main query page and additionally choose *Affymetrix* from the array field to retrieve only experiments annotated as *Homo sapiens* and using Affymetrix arrays. If the query does not retrieve any results, this means that query criteria are too stringent and that there are no submissions in

the database satisfying this particular combination of parameters. In this case the user should check the selected query parameters and relax them or use the browse options to explore data at a higher level.

2. Log-in: Accessing Password-Protected Data

Clicking on "Log-in" brings up the log-in page, where the user has to type in a user name and password. Passwords are only provided to the data submitter and authorised persons (with the permission of the submitter), such as journal editors and reviewers. Passwords are used to protect unreleased datasets until the specified release date. Once logged in, the query interface is identical to the query page described in Section IIIA.

B. Exploring the Experiment and Retrieving Data

Once an experiment, array, or protocol is selected, users can retrieve further details. For instance, experiments may contain raw and normalised data consisting of different quantitation types. The interface allows the user to select and export data from specific quantitation types from raw and/or normalised data. Users can also compare array designs or view protocols. This section describes data retrieval and exploration of experiments and protocols in detail.

Top-level information for experiments provides a brief free text description supplied by the submitter and an autogenerated experiment description, which indicates the number of samples and arrays used, hybridizations, and data files produced, as well as information on sample attributes, e.g., species. Hyperlinks help the user navigate to complete information on authors, laboratory, and type of experiment. By clicking on "Retrieve data," users retrieve a summary page of data associated with the particular experiment. The information is organized in two or more columns.

Column 1: Lists available data files and their type (raw or normalised). Raw data files are displayed first.
Column 2: List of quantitation types, with links to descriptions.
Column 3: List of sequence annotation, such as "Database accessions" and "Gene names" (only for arrays submitted after July 2003).

Any (or all) of the items in the columns can be exported to a tab-delimited data file. By clicking on "Export data," users can export the selected information and view them in tab-delimited format by clicking "View data matrix." The data matrix will appear in a new browser window and can be saved locally for future use. Alternatively, selected data can be imported directly into Expression Profiler for analysis online (Vilo *et al.*, 2003).

Users can export data from several hybridisations/ normalisations in a single file if the same array has been used, if data are about the same set of features (spots), reporters, or composite sequences (genes), and the same quantitation types are provided. In other cases, where, for example, only a subset of genes have expression values, data are exported as multiple files as the file will have different numbers of rows and/or columns.

C. Arrays

Users can query array designs by name, provider, and species. Details of the array layout and annotation are provided either as a spreadsheet (the ADF used to submit the array design through MIAMExpress) or as a simple tab-delimited file, both containing active hyperlinks to respective databases used for annotation. Both formats contain a single line for every feature (spot) on the array, except for Affymetrix arrays, where each line corresponds to a probe set due to the large number of rows in Affymetrix raw data files. Users can request complete Affymetrix data files by emailing arrayexpress@ebi.ac.uk and stating the accession number.

D. Protocols

When querying the protocols, users retrieve a list, including basic information (name, type, description) and protocol parameters.

IV. FUTURE PLANS

ArrayExpress repository can currently be queried by a number of parameters, including species, array design provider, and experiment type, as described earlier, returning a set of experiments, array designs, or protocols. However, querying the repository, users cannot retrieve, for example, all experiments where expression of a certain gene has been tested or the expression values of genes from a certain GO category and for samples treated under certain conditions.

To enable such granular queries, a data warehouse is under development. Users will be able to query parameters, including gene names, gene properties, ranges of expression values, and sample properties. The resulting query will return lists of genes, samples, or matrices of expression values (across many experiments). The database schema is very different from

that of the ArrayExpress repository: whereas the repository has to be able to store all experimental details (MIAME and beyond) and import all well-formed MAGE-ML submissions, the main issue of the data warehouse is how well it can support queries of many different types.

The data warehouse will only contain a subset of the good quality experiments deposited in the database. Data from the repository to the data warehouse cannot be transferred automatically. It will require curation and selection of specific experimental properties important from the data mining perspective. For example, in the repository there is no distinction made among experiments on which measurements are the best estimates of gene expression levels and which specific sample properties (apart from experimental factors) are important from the data mining perspective.

A unique interface will allow users to query the repository and the data warehouse. Links from the data warehouse to the repository will be implemented to allow users to navigate from gene expression data to the experiment context.

Acknowledgments

The ArrayExpress design and software development is coordinated by Ugis Sarkans, with a major contribution from Gonzalo Garcia Lara. Mohammad Shojatalab coordinates MIAMExpress development, with major contributions from Niran Abeygunawardena and Sergio Contrino. The has been largely developed by the ArrayExpress curation team: Helen Parkinson, Ele Holloway, and Gaurab Mukherjee and Philippe Rocca-Serra and Susanna-Assunta Sansone. Susanna-Assunta Sansone developed the specification for Tox-MIAMExpress and implementation has been largely the work of Sergio Contrino. Helen Parkinson has coordinated the data curation and construction of MAGE-ML pipelines to ArrayExpress, with major contributions from Philippe Rocca-Serra, Ele Holloway, and the software development team. Anjan Sharma has developed a data management tool allowing curators to manage the growing number of data submissions. Ahmet Oezcimen is the ArrayExpress database administrator. Jaak Vilo, Misha Kapushesky, and Patrick Kemmeren developed expression Profiler. ArrayExpress is largely funded by the TEMBLOR grant from the European Commission, with contributions from International Life Sciences Institute, Health and Environmental Safety Institute (ILSI-HESI) and from CAGE grant from the European Commission. Incyte Genomics provided the initial funding. We also thank the MGED community, particularly Catherine

A. Ball, Gavin Sherlock, Pall Spellman, John Quacken-bush, and Chris Stoeckert.

References

Ball, C. A., Sherlock, G., *et al.* (2002). Standards for microarray data. *Science* **298**(5593), 539.

Brazma, A., Hingamp, P., *et al.* (2001). Minimum information about a microarray experiment (MIAME)-toward standards for microarray data. *Nature Genet.* **29**(4), 365–371.

Brazma, A., Parkinson, H., *et al.* (2003). ArrayExpress: A public repository for microarray gene expression data at the EBI. *Nucleic Acids Res.* **31**(1), 68–71.

Edgar, R., Domrachev, M., and Lash, A. E. (2002) *Nucleic Acids Res.* **30**(1), 207–210.

Gollub, J., Ball, C. A., Binkley, G., Demeter, J., Finkelstein, D. B., Hebert, J. M., Hernandez-Boussard, T., Jin, H., Kaloper, M., Matese, J. C., Schroeder, M., Brown, P. O., Botstein, D., and Sherlock, G. (2003). The Stanford Microarray Database: Data access and quality assessment tools. *Nucleic Acids Res.* **31**(1), 94–96.

Kasprzyk, A., Keefe, D., Smedley, D., London, D., Spooner, W., Melsopp, C., Hammond, M., Rocca-Serra, P., Cox, T., and Birney, E., (2004). EnsMart: A generic system for fast and flexible access to biological data. *Genome Res.* **14**, 160–169.

Mattes, W. B., Sansone, S.-A., Bushel, P. R., and Waters, M. D., (2004). Database development in toxicogenomics: Issues and efforts. *Environ, Health Perspect, Toxicogenom.* **doi:10.1289/txg.6697**.

Saeed, A. I., Sharov, V., White, J., Li, J., Liang, W., Bhagabati, N., Braisted, J., Klapa, M., Currier, T., Thiagarajan, M., Sturn, A., Snuffin, M., Rezantsev, A., Popov, D., Ryltsov, A., Kostukovich, E., Borisovsky, I., Liu, Z., Vinsavich, A., Trush, V., and Quacken-bush (2003). TM4: A free, open-source system for microarray data management and analysis. *Biotechniques* **34**(2), 374–378.

Spellman, P. T., Miller, M., *et al.* (2002). Design and implementation of microarray gene expression markup language (MAGE-ML). *Genome Biol.* **3**(9).

Vilo, J., Kapushesky, M., Kemmeren, P., Sarkans, U., and Brazma, A. (2003). "*Expression Profiler*." Springer-Verlag, New York.

14

Serial Analysis of Gene Expression (SAGE): Detailed Protocol for Generating SAGE Catalogs of Mammalian Cell Transcriptomes

Sergey V. Anisimov, Kirill V. Tarasov, and Kenneth R. Boheler

I. INTRODUCTION

Serial analysis of gene expression (SAGE) is a sequencing-based technique that permits the simultaneous evaluation of thousands of transcripts in a single assay. SAGE relies on two major principles: (1) short DNA sequences are sufficient for identifying individual gene products and (2) concatenation (linking together) of short DNA sequences or tags increases the efficiency of identifying expressed mRNAs in a sequence-based assay. The transcript profile generated by SAGE currently relies on 14 to 21 base nucleotide sequences (SAGE tags) for gene identification (Velculescu, 1995; Saha, 2002). The technique generates large numbers of short tags, originating from the last (most 3′) unique location of a restriction enzyme recognition site in a cDNA generated from a single transcript. When the tags are sequenced, the technique can theoretically identify up to 4^{10} (1,048,576) unique transcripts; however, the newer LongSAGE method can distinguish 4^{17} different tags, a number sufficient to be virtually unique even within the whole genome.

II. MATERIALS AND INSTRUMENTATION

Kits: Messagemaker kit (Cat. No. 10298-016) and cDNA synthesis kit (Cat. No. 18267-013) are from Invit-

rogen. Deoxynucleotide triphosphates (dNTP) (Cat. No. US77100) are from Amersham.

Enzymes: T4 DNA Ligase (1 and 5 U/µl; Cat. Nos. 15224-017 and 15224-041) is from Invitrogen. Restriction enzymes *NlaIII* (Cat. No. 125S), *BsmFI* (Cat. No. 572S), *SphI* (Cat. No. 182S), and T4 polynucleotide kinase (PNK, Cat. No. 201S) with corresponding buffers are from New England Biolabs. DNA polymerase I Klenow fragment (Cat. No. M2201) is from Promega. AmpliTaq Gold polymerase (Cat. No. 4311814) is from Applied Biosystems.

DNA molecular weight markers: 25 bp (Cat. No. 10597-011), 100 bp (Cat. No. 15628-050), and 1 kb (Cat. No. 15615-016) are from Invitrogen.

Reagents: Dimethyl sulphoxide (DMSO, Cat. No. D2650), ammonium sulfate [(NH₄)₂SO₄, Cat. No. A5132], magnesium chloride (MgCl₂, Cat. No. M8266), potassium phosphate, monobasic (KH₂PO₄, Cat. No. P5379), potassium phosphate, dibasic (K₂HPO₄, Cat. No. P2222), sodium perchlorate (NaClO₄, Cat. No. S1513), and β-mercaptoethanol (Cat. No. M7522) are from Sigma. Agarose (Cat. No. 15510-027) is from Invitrogen, ATP (Cat. No. 27-1006.01) is from Amersham, and NaCl (Cat. No. 7581) is from Mallinckrodt.

Electrophoresis reagents: Polyacrylamide solutions (acrylamide : bisacrylamide, 19 : 1 (Cat. No. 161-0144) and 37.5:1 (Cat. No. 161-0148)), *N,N,N′,N′*-tetramethylenediamine (TEMED, Cat. No. 161-0800), and ammonium persulfate (Cat. No. 161-0700) are from Bio-Rad. TBE polyacrylamide precast gels (20%, Cat. No. EC6315) are from Novex.

Solutions: Ethylenediaminetetraacetic acid, disodium (Na₂EDTA, 0.5 M, pH 8.0, Cat. No. 3400-1003, Digene), phenol–chlorophorm–isoamyl alcohol mix (PC8, 25:24:1, pH 8.0, Cat. No. 0883, Amresco), isopropanol (Cat. No. 9084-01, J.T. Baker Inc.), and absolute ethanol (Cat. No. 201096, Warner-Graham Co.) are from commercial vendors. Tris/acetate/EDTA (TAE, 50×, Cat. No. 330-008-161), Tris–HCl (1 M, pH 7.5; Cat. No. 351-006-100; 1 M, pH 8.0 Cat. No. 351-007-100), sodium acetate (C₂H₃NaO₂, 3 M, pH 5.2, Cat. No. 351-035-060), NaCl (5 M, Cat. No. 351-036-100), and EDTA (0.25 M, pH 7.0, Cat. No. 118-090-060) are from Quality Biological. Diethylpyrocarbonate (DEPC, 0.1%)-treated water (Cat. 395-000), Tris/borate/EDTA buffer (TBE, 10×, pH 8.3, Cat. No. 336-000), and SOC media (Cat. No. 396-110) are from BioSource.

Other materials: Tryptone (Cat. No. 0123-17), yeast extract (Cat. No. 0127-01-7) and Bacto–agar (Cat. No. 0140-01) are from Difco. ElectroMAX DH10B cells (Cat. No. 18290-015), pZErO-1 (Cat. No. 2500-01), and zeocin (Cat. No. R250-01) are from Invitrogen. SYBR green I (Cat. No. S-7567) is from Molecular Probes, and glycogen (Cat. No. 901 393) is from Boehringer. Dynabeads M-280 (Cat. No. 112.05) are from Dynal.

Oligo-DNA linkers and primers are synthesized by Invitrogen. *Linker 1A* (gel purified): TTTGGA TTTGCTGGTGCAGTACAACTAGGCTTAATA GGGACATG; *linker 1B* (gel purified): TCCCTAT TAAGCCTAGTTGTACTGCACCAGCAAATCC [amino mod. C7]; *linker 2A* (gel purified): TTTCTG CTCGAATTCAAGCTTCTAACGATGTACGGGG ACATG; *linker 2B* (gel purified): TCCCCGTACA TCGTTAGAAGCTTGAATTCGAGCAG[amino mod. C7]; *primer 1*: GGATTTGCTGGTGCAGTACA; *primer 2*: CTGCTCGAATTCAAGCTTCT; *5′-biotinylated oligo(dT)* (gel purified): [biotin]T₁₈; *M13 forward*: GTAAAACGACGGCCAGT; *M13 reverse*: GGAAA CAGCTATGACCATG.

Equipment: GeneAmp PCR systems 9600 and 9700 are from Perkin-Elmer. For polyacrylamide electrophoresis, use a protean II electrophoresis unit (Bio-Rad) or similar. For agarose electrophoresis, use Hoefer HE33 (Pharmacia) or similar. For precast polyacrylamide gels, use XCell II (Novex). Power supply: Power PAC 3000 (Bio-Rad) or similar. Electroporation is performed on Gene-Pulser II/Pulse Controller II system (Bio-Rad) in 0.2-cm cuvettes (Bio-Rad, Cat. No. 16086). The magnetic particle concentrator is of MPC-E type from Dynal; RT6000 centrifuge and H1000B rotor are from Sorvall.

Required standard laboratory equipment: Millipore H₂O system, bacterial incubator, orbital shaker, autoclave, imaging system with UV light, sequencer, and microfuge.

Plastics: SpinX microcentrifuge tubes (Cat. No. 8160) are from Costar, Hot Start 50 tubes (Cat. No. 6002) are from Molecular Bio Products, Nunc sterile 96-well plates with lids, U-shape (Cat. No. 81-6668-03) and SealPlates, sterile (Cat. No. 05-6125-12) are from PGC Scientific. Thermo-Fast 96-well detection plate Cat. No. AB-1100) and adhesive sealing sheets (Cat. No. AB-0558) are from Marsh Bioproducts. MicroAmp reaction tubes (Cat. No. 801-0580) and MicroAmp caps (Cat. No. 801-0535) are from Applied Biosystems.

Software: SAGE sequence analysis: SAGE 2000 V4.13 or better is recommended (available at http://www.sagenet.org); Database download: CuteFTP (GlobalSCAPE, Inc.), DOS to UNIX Text Converter (Streamline Solutions), WinRAR for Windows (any manufacturer); MS Access and MS Excel (Microsoft).

III. PROCEDURES

A. SAGE Protocol

This protocol is modified from those of Velculescu *et al.* (1995), Kenzelmann and Muhlemann (1999), and Anisimov *et al.* (2002). The whole protocol takes about 5–6 days, but this time will vary depending on the experience of the investigator.

Solutions

1. *2× binding and washing buffer (B + W):* 10 mM Tris–HCl (pH 7.5), 1 mM EDTA, 2.0 M NaCl. For 100 ml final, add 1 ml of 1 M Tris–HCl (pH 7.5), 1 ml 0.1 M EDTA, and 11.69 g of NaCl, adjust the volume with sterile H₂O, and store at room temperature.

2. *LoTE:* 3 mM Tris–HCl (pH 7.5), 0.2 mM EDTA (pH 7.5). For 50 ml final, add 150 μl 1 M stock solution of Tris–HCl and 100 μl 0.1 M stock solution of EDTA, adjust the volume with sterile H₂O, and store at 4°C

3. *10× PCR Buffer:* 166 mM (NH₄)₂SO₄, 670 mM Tris, pH 8.8, 67 mM MgCl₂, 100 mM β-mercaptoethanol. For 50 ml final, add 1.1 g of ammonium sulfate, 16.75 ml of Tris 2 M stock solution, 0.32 g of magnesium chloride, and 390 μl of β-mercaptoethanol, adjust the volume with sterile H₂O, distribute into 0.5-ml aliquots, and store at −20°C.

4. *Low salt LB agar plates with zeocin:* 1% tryptone, 0.5% yeast extract, 0.5% NaCl, 1.5% bacto-agar, 50 μg/ml zeocin. For 500 ml final, add 5 g of tryptone, 2.5 g of yeast extract, and 2.5 g of NaCl to 450 ml of dH₂O, stir, adjust pH of the solution to 7.5 with NaOH, add 7.5 g of Bacto-agar, and adjust the volume to 500 ml. Autoclave, cool to 55°C, add 250 μl of zeocin

100 µg/µl stock, and pour plates. Seal with Parafilm and foil and store at 4° for up to 1 month in the dark.

5. *TB medium base:* Add 12 *g* of tryptone, 24 *g* of yeast extract, 4 ml of glycerol to 900 ml dH$_2$O final volume and autoclave

6. *Phosphate buffer:* Add 2.3 *g* of KH$_2$PO$_4$ and 12.5 *g* of K$_2$HPO$_4$ to 100 ml dH$_2$O final volume and filter sterilize

7. *TB medium/zeocin:* For 1 liter, add 900 ml of TB medium base, 100 ml of phosphate buffer, and add 500 µl of zeocin (100 µg/µl). Store in dark room at 4° for up to 2 weeks.

8. *12% polyacrylamide solution:* For a 20 × 16 × 0.1-cm gel, add 29.4 ml of dH$_2$O, 13.1 ml of 40% PAAG mix (acrylamide : bisacrylamide, 19 : 1), 875 µl of 50× TAE, 438 µl of 10% ammonium persulfate, and 37.5 µl of TEMED

9. *8% polyacrylamide solution:* For a 20 × 16 × 0.1-cm gel, add 33.75 ml of dH$_2$O, 8.75 ml of 40% PAAG mix (Acrylamide : bisacrylamide, 37.5 : 1), 875 µl of 50× TAE, 438 µl of 10% ammonium persulfate, and 37.5 µl of TEMED

Preliminary and Routine Procedures

All incubations are performed in a water bath.

1. *Kinasing (phosphorylation) reaction for linkers:* In each of two Eppendorf tubes labeled "linker 1" and "linker 2," add 6 µl of LoTE, 2 µl of 10× kinase buffer, 2 µl of 10 m*M* ATP, and 1 µl of PNK (10 U/µl). Add 9 µl of linker 1B (350 ng/µl) to "linker 1" and 9 µl of linker 2B (350 ng/µl) to "linker 2". Incubate both tubes at 37°C for 30 min and then at 65°C for 10 min. Add 9 µl of linker 1A to "linker 1" and 9 µl of linker 2A to "linker 2." Incubate at 95°C for 2 min, transfer to 65°C for 10 min, then to 37°C for 10 min, and finally to room temperature for 20 min. Kinased linkers are stable at –20°C for up to a year.

2. *Control reaction for linkers:* Prepare four Eppendorf tubes marked as "1+," "1–," "2+," and "2–." Add 20 µl of LoTE, 8 µl of 5× ligation buffer, and 1 µl of kinased linker (linker 1 to tubes marked "1" and linker 2 to tubes marked "2") to each tube. Incubate all four tubes at 50°C for 2 min and then at room temperature for 15 min. Add 2 µl of T4 ligase (1 U/µl) to tubes marked "+" and 2 µl of LoTE to tubes marked "–" and incubate at 16°C for 2 h. Run 1 µl of each sample on 20% precast TBE polyacrylamide gel at 120 V for 3 h and stain with SYBR green I (20 µl/200 ml of 1× TBE buffer) for 15 min while shaking. Observe linkers on the gel under UV light. Only linkers dimerized with >70% efficiency (based on band fluorescence intensity) are acceptable.

3. *Prepare pZeRO-1:* To 4 µl of dH$_2$O, add 1 µl of restriction enzyme (RE) buffer 2, 2.5 µl of pZeRO-1 plasmid (1 µg/µl), 2.5 µl of *SphI* (5 U/µl), and incubate at 37°C for 25 min. Add 90 µl of TE (final concentration is 25 ng/µl) and heat inactivate enzyme at 68°C for 15 min. Quench on ice and use immediately. Linearized pZeRO-1 plasmid can be stored at –20°C, but its cloning efficiency decreases with time.

4. *Phenol–chloroform extraction:* Bring sample volume to 200 µl with LoTE and add an equal volume (200 µl) of PC8 and vortex. Separate the aqueous phase, which contains the DNA, from the organic phase by centrifugation in a microfuge at 7800 *g* for 2 min at room temperature. Transfer the aqueous phase into a fresh Eppendorf tube, while ensuring that no organic solution is included. Reextract if necessary.

5. *Ethanol (EtOH) precipitation:* DNA in aqueous solution can be precipitated by EtOH in the presence of sodium or ammonium acetate. Because of the small amount of DNA and small fragment sizes, glycogen is added as a career. For a 200-µl sample volume in LoTE, add 3 µl of glycogen (20 mg/ml), one-half volume 10 *M* ammonium acetate, and 2.2 volumes ice-cold (–20°C) 100% ethanol and vortex. Pellet the precipitated DNA by centrifugation in a microfuge at 15,300 *g* for 10 min. Remove the ethanol with care and always wash with 75% EtOH to remove any traces of salt. Centrifuge at 7800 *g* for 10 min, discard supernatant, dry the pellet, and resuspend in LoTE. In certain cases, an increased volume of ethanol or 10 min dry ice/ethanol incubation is required on the precipitation step, as indicated in the text.

6. *Isopropanol precipitation:* Bring sample volume to 450 µl with LoTE, add 3 µl of glycogen, 150 µl of 2 *M* sodium perchlorate and 330 µl of isopropanol. Vortex and spin at 15,300 *g* for 10 min at room temperature. Aspirate supernatant, dry the pellet, and resuspend in LoTE.

Steps

1. Prepare poly(A)$^+$ (mRNA) from total RNA with the Messagemaker kit according to the manufacturer's instructions (see Comment 1 and Pitfalls 1 and 2). A total quantity of 0.5–1.0 mg of total RNA gives the best mRNA yield and is convenient for handling.

2. Prepare cDNA from 5 µg of mRNA using a cDNA synthesis kit following the manufacturer's instructions. A 5′-biotinylated oligo(dT)$_{18}$ (500 ng/µl) is, however, suggested for first strand synthesis (see Comment 2). After the second strand reaction is terminated by the addition of 25 µl of 0.25 *M* Na$_2$EDTA (pH 7.5), extract cDNA sample with PC8, EtOH precipitate, and resuspend the pellet in 20 µl of LoTE.

3. To half the cDNA sample (10 µl), add 163 µl of LoTE, 2 µl of BSA (100×), and 5 µl of *Nla*III (anchoring enzyme, 10 U/µl) (see Pitfall 3). Incubate at 37°C for

1 h, extract with PC8, EtOH precipitate, and resuspend in 20 μl of LoTE. Use the second half of the cDNA sample as a backup.

4. Add 100 μl of resuspended M-280 streptavidin magnetic beads (Dynabeads) to two Eppendorf tubes marked "1" and "2". Use the magnetic particle concentrator to immobilize the beads and discard the storage buffer. Wash beads once with 200 μl 1×B+W buffer and discard the buffer. Add 100 μl of 2×B+W buffer, resuspend, and add 90 μl of dH₂O and 10 μl of cDNA digestion products to each Eppendorf tube. Incubate at room temperature in suspension for 15 min. Wash beads three times with 200 μl 1×B+W and then once with 200 μl LoTE, removing the wash each time after immobilizing the beads using the magnetic particle concentrator. Proceed immediately to step 5.

5. Resuspend the Dynabeads in 25 μl of LoTE and 8 μl of 5× ligase buffer. Add 5 μl of kinased linker 1 to the tube marked "1" and 5 μl of kinased linker 2 to the tube marked "2". Incubate both tubes at 50°C for 2 min and then at room temperature for 15 min. Add 2 μl of T4 ligase (high concentration, 5 U/μl) to each tube and incubate at 16°C for 2 h. Afterward, wash the beads four times with 200 μl 1×B+W and then two times with 1× RE buffer 4. Proceed immediately to step 6.

6. Resuspend the contents of tubes "1" and "2" in 84 μl of LoTE and then add 10 μl of 10× RE buffer 4, 2 μl of 100× BSA, and 2 μl of BsmFI (2 U/μl, tagging enzyme). Incubate at 65°C for 1 h, mixing intermittently, immobilize Dynabeads using magnetic particle concentrator, and collect supernatants. Extract samples with PC8, EtOH precipitate, and resuspend pellets in 10-μl volumes of LoTE.

7. To each of these tagging enzyme digestion products, add 32.4 μl of dH₂O, 5 μl of 10× second strand buffer (from cDNA synthesis kit), 1 μl of 100× BSA, 1 μl of dNTPs (25 mM), and 0.6 μl of Klenow (5 U/μl) to fill in the 5' DNA overhangs. Incubate at 37°C for 30 min, add 150 μl of LoTE, extract samples with PC8, and precipitate with a high concentration of EtOH [add 3 μl of glycogen, 100 μl of 10 M ammonium acetate, and 900 μl of ice cold (−20°C) 100% EtOH]. Vortex and pellet the precipitated DNA by centrifugation in a microfuge at 15,300 g for 10 min; remove the EtOH with care and wash with 75% EtOH. Centrifuge at 7800 g for 10 min, discard supernatant, and dry the pellets. Resuspend pellets in 6 μl of LoTE.

8. For ditag ligations, mix 2 μl of blunt-ended samples from the previous step, add 1.2 μl of 5× ligase buffer, and 0.8 μl of high concentration T4 ligase (5 U/μl). For negative control reactions, mix 2 μl volumes of blunt-ended samples from the previous step, 1.2 μl of 5× Ligase buffer, and 0.8 μl of dH₂O. Incubate at 16°C

overnight, followed by the addition of 14 μl of LoTE. Proceed directly to PCR step.

9. Take 1-μl aliquots to make serial LoTE dilutions: 1/10, 1/25, 1/50, and 1/100 for ditag ligation reactions, 1/10 and 1/25 for negative control reactions, and one LoTE sample as an additional negative control. Prepare a PCR mix using 31.0 μl of sterile water, 1 μl of selected ligation reaction or control dilution, 5 μl of 10×PCR Gold buffer, 3 μl of MgCl₂ (25 mM), 3 μl of DMSO, 4 μl of dNTPs (25 mM), 1 μl each of primers (350 ng/μl each), and 1.0 μl of AmpliTaq Gold polymerase (5 U/μl). Use one strip (8 tubes) of MicroAmp reaction tubes for PCR. Cycling conditions (optimized for Gene Amp PCR system 9700) are as follows: 10 min at 95°C, followed by 28 cycles (30 s at 95°C, 1 min at 55°C, 1 min at 72°C), final extension at 72°C for 5 min, and a hold at 4°C.

10. Mix 20 μl of each PCR product with 4 μl of 6× loading buffer. Load the samples and 100-bp DNA ladder (one lane) to the wells of a 10-well 20% precast Novex TBE polyacrylamide gel. Use 1× TBE as a running buffer. Set voltage to 30 V until the entire sample has entered the gel and then increase the voltage to 120 V. Terminate the reaction when the bromphenol blue stain in the loading buffer migrates to the bottom of the gel. Stain gel with SYBR green I (20 μl/200 ml of 1× TBE buffer) for 15 min while shaking. Visualize by UV. Ligated ditags are located at ~102 bp. Another bright band is present at ~80 bp (linker–linker ligation), and the remaining bands are considerably less bright. The negative control reaction should lack the 102-bp band, and PCR with LoTE as a template (negative control) should not produce any bands (Fig. 1). The optimal dilution of the ditag ligation reaction products for large-scale PCR is considered based on 102-bp band brightness to background ratio and highest dilution ratio. Usually, 1/50–1/100 dilutions produce best results with comparatively low background and a large number of possible PCR reactions. If little or no ditags are visualized, the backup volume of blunt-ended cDNA tags (2 μl for each of the linkers) could be used to repeat steps 8–10 (without negative control); alternatively, backup cDNA could be used to repeat the protocol starting from step 3.

11. Once the conditions have been optimized, perform a large-scale PCR with a fresh LoTE dilution of the selected ditag ligation product as a template. Perform the reaction as described in step 10, except use a Thermo-Fast 96-well detection plate. Cover plate with adhesive sealing sheets and spin for 3 min at 600 g on a RT6000 centrifuge using a H1000B rotor (Sorvall) and adapters for 96-well plates. Run about 110–120 50-μl reactions in the GeneAmp 9700.

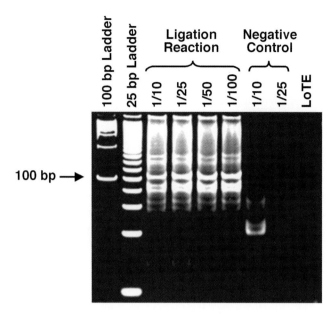

FIGURE 1 PCR amplification of ditags. PCR products of the serial LoTE dilution of ditag ligation reactions (1/10 to 1/100), negative control reaction (1/10 to 1/25), and LoTE-template reaction resolved on a 20% TBE polyacrylamide gel (Novex). Ditag ligation reaction products are represented by a major 102-bp band, whereas other bands represent products of linker–linker ligation (~80 bp) and unspecific products.

12. Combine the PCR products into 12 Eppendorf tubes (~450 μl in each), extract with PC8, EtOH precipitate (using 1 ml of absolute ethanol), and resuspend the pellets in 18 μl of LoTE (288 μl total). Add 57.6 μl of 6× loading buffer and load the entire volume into 8 wells of a 12% polyacrylamide gel (acrylamide : bisacrylamide, 19 : 1). Load a 100-bp DNA ladder to both end wells of the gel. Use 1× TAE as a running buffer. Electrophorese at 30 V until all the samples have fully entered the gel and then increase the voltage to 130 V. Terminate the reaction when bromphenol blue in the loading buffer migrates to the bottom of the gel. Stain gel with SYBR green I (20 μl/200 ml of 1× TAE buffer) for 15 min while shaking. Pierce the bottoms of eight 0.5-ml test tubes with a needle or surgical blade to form a hole with a diameter of about 0.8–1.0 mm and place them in 1.5-ml Eppendorf tubes. Visualize the gel by UV, and using a surgical blade, cut off the 102-bp bands and put the DNA into individual pierced tubes. Spin tubes at 15,300 g for 2 min at room temperature and then discard 0.5-ml tubes. Add 300 μl of LoTE to each Eppendorf and vortex. Incubate at 65°C for 15 min, mixing intermittently. Transfer the contents of each tube to individual SpinX tubes and spin at 15,300 g for 2 min at room temperature. Discard SpinX cartridges with a solid phase, EtOH precipitate

the samples (aqueous phase) using 940 μl of absolute ethanol, and resuspend pellets in 14.8 μl volumes of LoTE (118.5 μl total).

13. Combine the entire contents into a one Eppendorf tube, add 15 μl of 10× RE buffer 4, 1.5 μl of 100× BSA, and 15 μl of *Nla*III (10 U/μl). Incubate at 37°C for 1 h and 15 min, extract sample with PC8, and add 50 μl of LoTE to bring the sample volume to 200 μl. EtOH precipitate on dry ice/ethanol bath [add 3 μl of glycogen (20 mg/ml), 67 μl of 10 M ammonium acetate and 733 μl of ice-cold (–20°C) 100% EtOH] and vortex. Place the sample in a dry ice/100% EtOH bath for 10 min. Pellet the precipitated DNA by centrifugation in a microfuge at 15,300 g for 15 min at 4°C. Remove the EtOH with care and wash with 75% EtOH. Centrifuge at 7800 g for 10 min at room temperature, discard supernatant, dry the pellet, and resuspend in 32 μl of LoTE (see Comments 3 and 4).

14. Add 6.4 μl of 6× loading buffer and load onto two wells of 12% polyacrylamide gel (acrylamide: bisacrylamide, 19:1). Two lanes are reserved for 100- and 25-bp DNA ladders. Run and stain the gel as described in step 12. Using a surgical blade, cut 24 to 26-bp bands from the gel (Fig.2) and place in individual bottom-pierced 0.5-ml tubes. Spin tubes at 15,300 g for 2 min at room temperature and then discard 0.5-ml tubes. Add 300 μl of LoTE to each Eppendorf and vortex. Incubate at 37°C for 15 min, mixing intermittently, transfer the contents of each Eppendorf tube to individual SpinX tubes, and spin at 15,300 g for 2 min at room temperature. Discard SpinX cartridges with a solid phase and distribute the aqueous contents to three Eppendorf tubes. EtOH precipitate on dry ice/ethanol bath (as described in step 13). Resuspend and combine the pellets in 7 μl of LoTE.

15. To the pooled purified ditags (7 μl total), add 2 μl of 5× ligation buffer and 1 μl of high concentration T4 ligase (5 U/μl). Ligate ditags to form concatemers by incubation at 16°C for 3 h (see Comment 5), add 90 μl of LoTE, and continue to incubate at 60°C for 5 min. Extract the sample with PC8, EtOH precipitate, and resuspend the pellet in 10 μl of LoTE.

16. Incubate the concatemers at 60°C for 5 min, add 190 μl of LoTE, extract with PC8, EtOH precipitate, and resuspend the pellet in 10 μl of dH₂O. Incubate at 60°C for 5 min and immediately load onto one well of an 8% polyacrylamide gel (acrylamide:bisacrylamide, 37.5:1) (see Comment 6). Employ the 100-bp DNA ladder and 1-kb DNA ladder as markers. Run and stain the gel as described in step 12. Using a surgical blade, cut and separate concatemers of 600–1100 ("light" cluster) and 1100–2500 bp ("heavy" cluster) (Fig.3) and place each into two labeled bottom-pierced 0.5-ml tubes. Spin at 15,300 g for 2 min at room

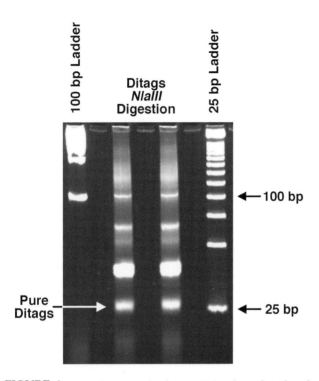

FIGURE 2 Pure ditags resolved on a 12% polyacrylamide gel. Bands represent (starting from the top) undigested linker–ditag–linker structures, products of incomplete anchoring enzyme digestion (one linker–ditag structures), pool of linkers, and pure ditags.

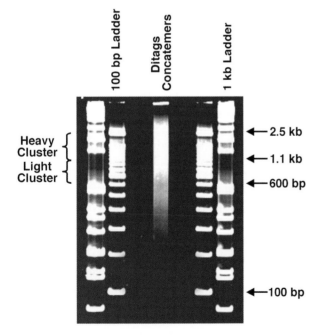

FIGURE 3 Ditag concatemers resolved on an 8% polyacrylamide gel using the "hot" variation of the original protocol (see Comment 6). Recommended "heavy" and "light" concatemer cluster sizes are 1100–2500 and 600–1100 bp, respectively.

temperature and discard 0.5-ml tubes. Add 300 μl of LoTE to each Eppendorf and vortex. Incubate at 65°C for 15 min, mixing intermittently. Transfer the contents of each Eppendorf tube to two SpinX tubes and spin at 15,300 g for 2 min at room temperature. Discard SpinX cartridges and transfer the aqueous phase from two SpinX tubes to one Eppendorf tube. EtOH precipitate using 933 μl of absolute ethanol and resuspend pellets in 5 μl total volume of LoTE for each "light" and "heavy" cluster of the sample.

17. To 5 μl of purified concatemers, add 2 μl of 5× ligase buffer, 1 μl of pZeRO-1 cut with *Sph*I (25 ng/μl), and 1 μl of T4 ligase (1 U/μl). Incubate at 16°C overnight, add 191 μl of LoTE to the samples, extract with PC8, EtOH precipitate, wash pellets three times with 70% ethanol, and resuspend pellets in 6 μl of LoTE. Use self-ligated pZeRO-1 vector as a negative control.

18. Electroporate bacteria with concatenated DNA by adding 2 μl of each sample to 40 μl of freshly thawed ElectroMAX DH10B cells on ice. Mix the contents gently by rotating a pipette tip in a test tube, rather than by trituration. On ice, transfer the mixture to the bottom of 0.2-cm electroporation cuvettes (avoiding bubbles) and electroporate: Voltage setting, 2.5 kV; capacitor, 25 μF; and resistance, 200 Ω. After the electric pulse, immediately transfer the cells to a 15-ml bacterial tube with 1 ml of prewarmed SOC media (no zeocin). Place in a bacterial shaker and incubate at 37°C for 40 min at 200 rpm.

19. Plate 80-μl aliquots of the cell/SOC solution onto 10-cm petri dishes with low salt LB agar plates supplemented with zeocin (50 ng/μl) (see Comment 7 and Pitfall 4). Incubate plates for about 20 h at 37°C.

20. Identify ~15–20 individual colonies from each "light" and "heavy" fraction to run a PCR control to determine cloning efficiency and the presence of concatemers in the clones (see Pitfall 5). For this purpose, prepare a 25-μl PCR reaction containing 18 μl of dH₂O, 2.5 μl of custom-made 10× PCR buffer (see *Solutions*), 1.5 μl of DMSO, 1.5 μl of dNTPs (25 m*M*), 0.5 μl of each M13 forward and reverse primer (both 350 ng/μl), and 0.1 μl of *Taq* polymerase (0.5 U), inoculated with a single bacterial colony picked with a sterile toothpick. Cycling conditions are: 2 min at 95°C, followed by 25 cycles (30 s at 95°C, 1 min 30 s at 56°C, 1 min 30 s at 70°C), final extension 70°C for 5 min, and hold at 4°C.

21. To the 5-μl aliquots of control PCR reactions, add 1 μl of 6× loading buffer and load onto an 1.2% agarose gel containing 1 μg/ml ethidium bromide and a 100-bp DNA ladder. Use 0.5× TBE as the running buffer, and electrophorese at 120 V. Estimate cloning

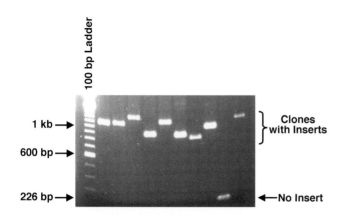

FIGURE 4 PCR control of cloning efficiency. A minor portion of the clones may lack inserts (PCR product will be of 226 bp size), whereas the majority should contain inserts with a size that varies as a function of concatemer length.

```
>FC16a06.seq                                    →
GATACACTACTATAGGGCGAATTGGGCCCTCTAGATGCATGTCTGTTTCT
GTTTATGTATTATCATGCAGACGGAAGGGCTTACTTAGCTCATGCCTGCT
CACGGCTCCACTTGGCCATGCAGGCCACACAGATCTCATTCTCATGGTTG
CTGAGAAGCCATAGCGGTCCATGGTGCTCCCACTATTGCAGAGGCCCATG
ATCCTGTGCTGTGGGGCTTGCCCATGATGGCAATTTCTGGACTCAGGGCA
TGTCTCAAACCCAATCAACATCAACATGTGGGCAAAGCCTTCCAAAGTCC
ATGTGTGGGTCTGGGGGCCATTGGGCATGTCTGACTTCCAGGGAGACCCA
GTCATGGTCTGGACGCGGGCTCCGGAGTTCATGGGAGACCCGCTTTTATT
ACTGTCCATGCTGCACTCCTGAGAACTGGGCACCCATGCTGCTGTAAAAC
TCACGGAATCCATGCTCCCTGCCCTAGTCTGTTCTGGCATGGCTGCCCTC
CCAACGGTGGTGCATGCTTACACGTCTCTGCCGCCTTCCCCATGCGCTGG
TTCCTCCTCAGCAACCATGTCCCTATTAAGGGCCTTCATCCCATGCCTAC
TAACCAGCAGGGATGGTCCATGGTGGCTCACTATGGAAGTCAGACATGTA
AGCTGTGAGCAACCAGCTCAGCATGGACGCGGAGCGTAACTAAGAGACAT
GGGTACAGGATAAAGCCGCGGCGCATGGGCAAGCCCCAGGAGTGGTGAGA
CATGTTGATTTTTTTGTTTTTTGCCTGCATGTTGATCCCCATCAT
```

FIGURE 5 Raw data file of the SAGE sequencing project. Correct concatemer structure with anchoring enzyme (*Nla*III) spacer (in bold) between 24- and 26-bp ditags. Arrow indicates the starting point of the insert.

efficiency by calculating the ratio of clones with inserts to the total number of clones amplified with PCR. Background clones without inserts will appear on the gel as 226-bp bands (Fig.4).

22. For those clones with inserts, add 180 μl of LoTE to the remaining PCR volume (20 μl), extract with PC8, add 250 μl of LoTE, and precipitate samples with isopropanol. Resuspend each pellet in 10 μl of LoTE.

23. For test sequencing, use forward and/or reverse M13 primers to ensure appropriate ditag formation. If inserts of the test clones demonstrate correct concatemer structure (Fig.5) and the cloning efficiency is satisfactory, the SAGE library should be sequenced until a sufficient number of tags (i.e., 40,000) can be

identified. We recommend using a commercial vendor for sequencing (e.g., Agencourt Bioscience Corp.).

24. Many commercial sequencing services require libraries to be submitted as frozen glycerol stocks. For this, inoculate the wells of a 96-well plate containing 100 μl of TB/zeocin media with bacterial colonies using sterile toothpicks. We recommend that 1 well in each plate (H12) be filled with sterile LoTE to serve as a negative control for the sequencing reactions. Incubate plates for about 20 h at 37°C.

25. Copy the "master plates" by transferring 15 μl of the bacterial cultures to the corresponding wells of another set of 96-well plates containing 100 μl of TB/zeocin media. Stop further bacterial growth in the "master plates" by adding 40 μl of sterile 50% glycerol to the wells. Seal plates with adhesive SealPlate film and store at –80°C until the project is completed successfully.

26. Incubate the "copy plates" for about 18 h at 37°C. Stop the bacterial growth in the "copy plates" by adding 60 μl of sterile 50% glycerol to the wells. Seal plates with adhesive SealPlate film, freeze at –80°C, and ship for sequencing on dry ice (see Comment 8).

B. SAGE Analysis

This analysis protocol is based on the original SAGE 2000 software (Kinzler, 2000) and a separate SAGE analysis software protocol is available online from Invitrogen (I-SAGE kit; http://www.invitrogen.com/downloads/sagesoftware_man.pdf; Invitrogen Corp., 2001).

Steps

1. Database Creation

The aim is to create a complete, up-to-date data set to match with the SAGE tags generated in your library.

1. Download a current release of a GenBank database (e.g., rodent, primate) from the National Center of Biotechnology Information (NCBI) Web site using ftp software (e.g., CuteFTP) from ftp.ncbi.nlm.nih.gov/Genbank/.

2. Extract GenBank sequence files from the archives using WinRAR or comparable software.

3. Convert the GenBank sequence file format from UNIX to DOS using Streamline Solutions or other appropriate software. Raw sequence files should be marked with the suffix *.seq.

4. Open SAGE 2000 software. Go to "Database, Create" to generate a new database [suggested name should include species identification (e.g., Mm, mouse; Hs, human), nonredundant (NR) and/or EST affiliation, and numeric index of the dataset release].

Individual databases are recommended for NR and EST data sets. "Add Files" to the database using GenBank under "File Format" and in the cDNA mode under "Method."

2. SAGE Project Creation and Analysis

The aim is to create and analyze a gene expression profile based on SAGE library sequence data (SAGE catalog).

5. To create a new project, hit "Project/New Project" and use the following settings: anchoring enzyme, *Nla*III; tag length, 10 bp; and ditag length, 24 bp. Copy raw sequence files (see Comment 9) to a folder, where a new project has been created.

6. Extract SAGE tags from the sequences with the "Project/Add Tags" option. Reset "Stop position" constant to 2000 bp (this also depends on the sequencer type used for large-scale sequencing, producing reliable readings of various length). For preliminary analysis (preprocessing), we recommend performing the analysis on each individually selected file (i.e., select file, and hit "Analyze" until all the files have been analyzed). Record the output (i.e., number of "Good Tags" and "Duplicate Dimers") on a spreadsheet.

7. Once the preliminary analysis is complete, use spreadsheets to remove files associated with background clones (no SAGE tags detected) and clones that produced a minor number of "Good Tags" and/or "Duplicate Dimers" (≤ 4 tags or ≤ 2 duplicate dimers; tags derived from such clones may contain sequencing errors).

8. Create a new project as in step 5 and reanalyze sequence files that passed preprocessing criteria only, but this time use the "Auto Analyze" option.

9. We recommend that $(A)_{10}$ be excluded from the list of tags (accessible via "Analyze/Report/Exclude Tags Manager" path: hit "Add", type "1" and hit "OK"). An $(A)_{10}$ SAGE tag is generally uninformative, gives multiple matches to databases entries, and may originate from an anchoring enzyme recognition site located immediately before the poly(A) tail.

10. We recommend that the expression profile of your sequence data be created using the "Analyze/ Report" path. For primary analysis, "Minimum tag count for report" should be reset to 1 (via "Settings/Minimum tag count for report" path) and "Options/Exclude Tags" checked. Using additional features, the expression profile (identified tags and absolute tag count) could be saved in MS Access database format (by having "Options/Save Output as MS Access File" option checked), and tags could be linked to selected SAGE database [by opening a selected database (see steps 1–4) and having "Options/Link to

Database" option checked]. The latter permits direct matching of tags to the GenBank database. Also, more than one SAGE catalog can be analyzed simultaneously using the "All open projects" option.

11. The analysis of SAGE catalogs by linking them sequentially to NR and EST GenBank databases is recommended. To supplement GenBank data with UniGene data (which has a different structure) we recommend direct one-by-one online analysis via SAGEmap applet (http://www.ncbi.nlm.nih.gov/ SAGE/) due to the notable presence of false positives in the "downloadable" "SAGEmap_tag_ug" data sets (available for human, mouse, rat, and a few other species at ftp://ftp.ncbi.nih.gov/pub/sage/map/).

12. For the comparison of multiple SAGE catalogs, we recommend MS Excel and MS Access statistical functions, not the SAGE 2000 function (accessible via "Analyze/Compare" path, with selected normalization value). MS Access files can be easily reformatted to MS Excel.

3. Outside Comparisons

The aim is to include publicly available SAGE data in the analysis. We encourage investigators to submit their SAGE data to the Gene Expression Omnibus (GEO) database.

13. Interlibrary comparisons among SAGE data are available via public databases, most importantly Entrez ProbeSet (http://www.ncbi.nlm.nih.gov/ SAGE/index.cgi?cmd=libsearch). Individual SAGE catalogs of interest (use keywords such as "mouse," "cancer," and "brain") can be accessed and downloaded (using "Full table view" option). These are then converted from plain text to MS Excel table, tag abundance normalized to tags per million (tpm) value, and comparisons performed in convenient MS Excel or MS Access formats.

IV. COMMENTS

1. This protocol is optimized for "standard" SAGE using the *Nla*III/*Bsm*FI enzyme pair. Other anchoring and restriction enzymes could be used to generate longer tags (LongSAGE; Saha *et al.*, 2002) or to generate SAGE tags from genes lacking the *Nla*III recognition site (*Sau*3A is suggested and is recognized by SAGEmap; Lash *et al.*, 2000). Certain modifications of the SAGE protocol are also available for SAGE applications with a limited amount of starting material (microSAGE, Datson *et al.*, 1999; miniSAGE, Ye *et al.*, 2000; SAR-SAGE, Vilain *et al.*, 2003).

2. For cDNA synthesis, we recommend using α^{32}P dCTP labeling only to test the protocol and skip it

for the generation of most SAGE libraries. If used to calculate cDNA yield, isotope labeling allows tracing of the cDNA up to the tagging enzyme digestion step.

3. A modification of the ditag purification step, based on the employment of 5′-biotinylated linkers for PCR, followed by the magnetic separation of anchoring enzyme digestion products, has also been suggested (Powell, 1998). Although this modification shortens the ditag purification stage, it does not always result in good yields of ditags.

4. A few modifications of the original SAGE protocol are suggested to improve the yield of anchoring enzyme digestion of ditags (Angelastro et al., 2000; Damgaard Nielsen et al., 2003). We find that complete NlaIII digestion is readily ensured by increasing the amount of enzyme, the overall reaction volume, and the digestion time (150 U, 150 µl, 1 h and 15 min, recommended).

5. Although concatenation time can vary from 45 min to overnight, which depends on the volume of ditags purified, we find that a 3-h concatenation produces good results.

6. A modification of the original SAGE protocol allows better resolution of concatemers in polyacrylamide gels, improving the overall yield of SAGE libraries (Kenzelmann and Muhlemann, 1999): 65°C, 15 min, chill on ice, 10 min; and then load onto the gel. We recommend using the "hot" variation of the same modification: 60°C, 5 min; PC8 extract; EtOH precipitate, 60°C, 5 min; and load immediately.

7. A X-Gal/IPTG blue/white selection can be employed to distinguish between bacterial colonies containing or lacking inserts (Angelastro et al., 2002) This technique is most useful when the number of bacteria without inserts is high.

8. With pZeRO-1, we recommend the M13 forward primer for sequencing. For clones originated from the "heavy" cluster of concatemers, an additional M13 reverse-primed sequence could be beneficial, yielding more tags from long inserts, but it leads to rapid accumulation of duplicate dimers due to the partial overlapping of sequences.

9. We recommend that each individual clone be designated by a nomenclature that contains a project script, plate number, and clone position in a 96-well plate. For example, file "EX15A08.seq" indicates association with experiment X, plate #15, well A8. All *.seq files generated in this manner represent individual SAGE library raw sequence files.

10. Other useful SAGE-related Web links:

SAGE home page: http://www.sagenet.org/

SAGEmap home page: http://www.ncbi.nlm.nih.gov/SAGE/

UniLib: http://www.ncbi.nlm.nih.gov/UniLib/index.cgi

SAGE Genie: http://cgap.nci.nih.gov/SAGE

GEO: http://www.ncbi.nlm.nih.gov/geo/

V. PITFALLS

1. The quality of the mRNA and the purity of oligo-DNA linkers and primers are of a great importance. We recommend the technique of Chirgwin et al. (1979) for the preparation of total RNA. Test linker kinasing efficiency before proceeding with the protocol.

2. Preciseness and RNase- or DNase-free conditions are important at most steps. Use DEPC-treated glassware and plasticware on the early steps of the work and aerosol barrier pipette tips throughout the whole experiment.

3. The commonly used anchoring enzyme NlaIII has a half-life of only about 3 months at −20°C; therefore, use only fresh batches of enzyme for each SAGE project.

4. Zeocin is light sensitive and comparatively unstable. TB/zeocin media can be kept in the dark for up to 2 weeks. Low salt LB agar/zeocin plates should be protected from light immediately after pouring.

5. Cloning efficiency (ratio of clones with inserts to the total number of clones) is the major concern of researchers employing the SAGE method. Reasons for the appearance of background clones with the pZeRO-1/zeocin system are not completely clear. Suggested strategies rely on specific concatemer cluster sizes to achieve an acceptable range of cloning efficiency (i.e., >80%). While small concatemers routinely have high cloning efficiency, the number of SAGE tags extracted from each single clone is low. Extremely long concatemers (>2500 bp) have a tendency to circularize, preventing effective cloning. The suggested concatemer cluster sizes (600–1100 bp for "light" and 1100–2500 for "heavy") seem to have the best tag yield/clones sequenced ratio, allowing direct large-scale sequencing. In other cases, an additional PCR step or X-Gal/IPTG selection could be required to select clones with insert before sequencing.

References

Angelastro, J.M., Klimaschewski, L.P., and Vitolo, O.V. (2000). Improved NlaIII digestion of PAGE-purified 102 bp ditags by addition of a single purification step in both the SAGE and microSAGE protocols. Nucleic Acids Res. 15, p.E62.

Angelastro, J.M., Ryu, E.J., Torocsik, B., Fiske, B.K., and Greene, L.A. (2002). Blue-white selection step enhances the yield of SAGE concatemers. Biotechniques 32, 484–486.

Anisimov, S.V., Tarasov, K.V., Tweedie, D., Stern, M.D., Wobus, A.M., and Boheler, K.R. (2002). SAGE identification of gene transcripts with profiles unique to pluripotent mouse R1 embryonic stem cells. *Genomics* **79**, 169–176.

Chirgwin, J.M., Przybyla, A.E., MacDonald, R.J., and Rutter, W.J. (1979). Isolation of biologically active ribonucleic acid from sources enriched in ribonuclease. *Biochemistry* **18**, 5294–5299.

Damgaard Nielsen, M., Millichip, M., and Josefsen, K. (2003). High-performance liquid chromatography purification of 26-bp serial analysis of gene expression ditags results in higher yields, longer concatemers, and substantial time savings. *Anal. Biochemi.* **313**, 128–132.

Datson, N.A., van der Perk-de Jong, J., van den Berg, M.P., de Kloet, E.R., and Vreugdenhil, E. (1999). MicroSAGE: A modified procedure for serial analysis of gene expression in limited amounts of tissue. *Nucleic Acids Res.* **27**, 1300–1307.

Kenzelmann, M., and Muhlemann, K. (1999). Substantially enhanced cloning efficiency of SAGE (serial analysis of gene expression) by adding a heating step to the original protocol. *Nucleic Acids Res.* **27**, 917–918.

Lash, A.E., Tolstoshev, C.M., Wagner, L., Schuler, G.D., Strausberg, R.L., Riggins, G.J., and Altschul, S.F. (2000). SAGEmap: A public gene expression resource. *Genome Res.* **10**, 1051–1060.

Powell, J. (1998). Enhanced concatemer cloning-a modification to the SAGE (serial analysis of gene expression) technique. *Nucleic Acids Res.* **26**, 3445–3446.

Saha, S., Sparks, A.B., Rago, C., Akmaev, V., Wang, C.J., Vogelstein, B., Kinzler, K.W., and Velculescu, V.E. (2002). Using the transcriptome to annotate the genome. *Nature Biotechnol.* **20**, 508–512.

Ye, S.Q., Zhang, L.Q., Zheng, F., Virgil, D., and Kwiterovich, P.O. (2000). miniSAGE: Gene expression profiling using serial analysis of gene expression from 1 microg total RNA. *Anal. Biochem.* **287**, 144–152.

Velculescu, V.E., Zhang, L., Vogelstein, B., and Kinzler, K.W. (1995). Serial analysis of gene expression. *Science* **270**, 484–487.

Vilain, C., Libert, F., Venet, D., Costagliola, S., and Vassart, G. (2003). Small amplified RNA-SAGE: an alternative approach to study transcriptome from limiting amount of mRNA. *Nucleic Acids Res.* **31**, E24.

15

Representational Difference Analysis: A Methodology to Study Differential Gene Expression

Marcus Frohme and Jörg D. Hoheisel

I. INTRODUCTION

Representational difference analysis (RDA) was originally developed for the identification of differences between two complex genomes (Lisitsyn *et al.*, 1993). Adapted slightly, cDNA-RDA was established by Hubank and Schatz in 1994. It aims at the detection of differentially expressed genes, comparing two sources of RNA, e.g., tumour and normal tissue, by means of a polymerase chain reaction (PCR)-coupled subtractive and kinetic enrichment procedure (Fig. 1). First, the two complex mRNA populations are reverse transcribed into cDNA. An initial digest with a restriction enzyme is followed by the addition of PCR adapters and subsequent amplification, thereby reducing the complexity of each sample. The resulting mixtures of gene fragments are named "representations" (Fig. 2). Differences in transcript levels in the tissues analysed are imaged in the representations by relative differences in the abundance of the respective cDNA fragments. In a subtractive hybridisation reaction with a strong kinetic component, one representation (#2)—named "driver"—serves as a competitor during the reannealing of the other (#1), which is called "tester" (Fig. 1). Due to an earlier change of adapters, only double-stranded tester molecules and therefore predominantly fragments unique to or overrepresented in the tester are exponentially amplified. The product is named "difference product" and may need further enrichment, acting as a tester in a subsequent round of RDA with an increased ratio of driver to tester, for example. Swapping the role of the two representations reverses the analysis mode, then identifying transcripts that are more abundant in representation #2 compared to #1.

cDNA-RDA has been proven to be a powerful tool for the identification of differentially expressed genes in many screens in both eukaryotic and prokaryotic systems. RDA is less prone to produce false positives than differential display and is easier to perform. The possibility of modifying the driver's composition, e.g., by spiking with certain fragments, offers an option of suppressing unwanted products. In comparison to microarray techniques, RDA has the advantage that there is no preselection in the set of analysed genes. Thus, unknown genes and unexpected splice variants are found. Moreover, difference products may serve as highly informative probe molecules on DNA arrays or be used as hybridisation samples. The protocols described here are similar to the originals of Hubank and Schatz (1994) but contain modifications found to be useful.

II. MATERIALS AND INSTRUMENTATION

Because the technique is very prone to pick up DNA contaminations, always use allocated (aliquots of) chemicals, buffers, plasticware, and so on and avoid common stocks. Take all precautions for RNA work but remember that DNA contaminations are a problem too that does not disappear by autoclaving. The preparation of most chemicals and buffers follows standard protocols using chemicals from various suppliers. However, always use high-quality grade products (p.a.

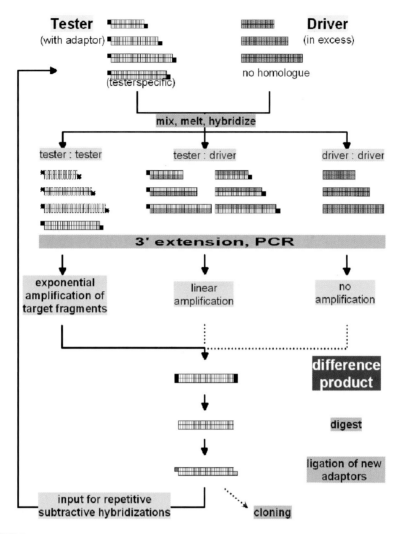

FIGURE 1 **Schematic representation of the RDA procedure**. Two cDNA populations are cut with a restriction enzyme. Subsequently, adapter molecules are added to the ends. The material is then PCR amplified to produce representations. The adapter of the tester sample is removed and replaced by a new molecule of different sequence (not shown). Tester and driver are mixed, denatured, and reannealed. During this process, the majority of tester molecules form duplexes with fitting driver molecules. Only fragments overrepresented in the tester or even lacking a homologue in the driver population will be amplified exponentially. The resulting difference product is either processed further or acts as a tester in another round of RDA.

or molecular biology grade) and prepare aliquots from freshly opened supplies.

Diethylpyrocarbonate (DEPC) is from Roth, Germany (Cat. No. K028.2), is diluted 1 : 1000 in water and autoclaved twice before being added to buffers. Low melting point (LMP) agarose is from Serva (Cat. No. 11408). The 50 × TAE stock solution for electrophoresis buffer consists of 2 M Tris–acetate (pH 8.2) and 50 mM EDTA. Various types of agarose are used for analytical gels as well as standard molecular weight markers with bands between 0.1 and 1 kb (100-bp ladders, pUC19/SauIIIa digest, etc.); never use these markers on preparative tester gels, however. The ethidium bromide (Roth, Germany, Cat. No. 2218.1) stock solution (10 mg/ml) for gel staining is diluted 1 : 10,000 prior to use. Phenol/chloroform/isoamylal-

cohol (25 : 24 : 1) is equilibrated with Tris to a pH of 7.5 to 8.0 (Roth, Germany, Cat. No. UN2821); glycogen (1 µg/µl for molecular biology) is from Ambion (Cat. No. 9510); alternatively, linear acrylamide (Cat. No. 9520) may be used in precipitation.

Other chemicals are ethanol and 2-propanol (iso-propanol); 1 M HCl for cleaning the gel chamber; 3 M sodium acetate (pH 5.3) (adjust pH with acetic acid, but use aliquots when checking the pH); 10 M ammonium acetate (sterile filter only, do not autoclave); 1 M Tris–HCl (pH 8.8) (8.5); 0.5 M EDTA stock (pH 8); TE: 10 mM Tris–HCl (pH 8.5); 1 mM EDTA; T1: 1 mM Tris–HCl (pH 8.5); T10: 10 mM Tris–HCl (pH 8.5); T50: 50 mM Tris–HCl (pH 8.8); 1 M MgCl$_2$; 1 M ammonium sulfate; 5x PCR buffer: 335 mM Tris–HCl (pH 8.8), 20 mM MgCl$_2$, 80 mM ammonium sulfate, 166 µg/ml

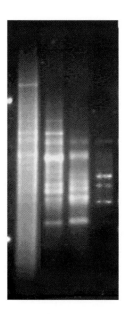

FIGURE 2   Difference products on an agarose gel. The four lanes of a 2.5% TAE agarose gel are shown. (Left to right) A tester representation and the three difference products resulting from iterative RDA are presented. The reduction in complexity is apparent. The white dots on the left margin indicate fragment sizes of 955 and 258 bp, respectively.

bovine serum albumm (BSA) (New England Biolabs, Cat. No. B9001S); nucleotides: dATP, dCTP, dGTP, and dTTP, each at 100 mM (Fermentas, Cat. No. R0181); dNTP mix containing 4 mM of each nucleotide; mineral oil (filtered) from Sigma (Cat. No. M 8662); alternatively, chill-out liquid wax from MJ Research is used; 10x ligase buffer: use buffer supplied with enzyme from New England Biolabs—if not available, prepare 500 mM Tris–HCl (pH 7.8), 100 mM MgCl$_2$, 100 mM dithiothreitot (DTT), 10 mM ATP—store frozen, mix vigorously before use; 300 mM EPPS (N-[2-hydroxyethyl]-piperazine-N'-[3-propanesulfonic acid] (pH 8.0) from Sigma (Cat. No. E-9502); 3x EE: 30 mM EPPS (pH 8.0), 3 mM EDTA; 5 M NaCl; yeast tRNA (Roche, Cat. No. 109495): set up as stock solution of 10 mg/ml; 0.5xTE/tRNA: mix tRNA stock with 1 volume TE; Qiaquick PCR purification kit (Cat. No. 28104); Qiaquick buffer QG (Cat. No. 19063); DpnII: 10 U/µl from New England Biolabs (Cat. No. R0543L); T4 DNA ligase: 400 U/µl from New England Biolabs (Cat. No. M0202S) (if using ligase from another supplier, check the unit definition); mung bean nuclease: 10 U/µl from New England Biolabs (Cat. No. M0250S); Taq polymerase (5 U/µl) is purchased from different suppliers.

All oligonucleotides (final concentration 1 mg/ml) are purchased as HPLC-purified material.

R-Bam-12:   d(GATCCTCGGTGA)

R-Bam-24:   d(AGCACTCTCCAGCCTCTCACCGAG)

J-Bgl-12:   d(GATCTGTTCATG)

J-Bgl-24:   d(ACCGACGTCGACTATCCATGAACA)

N-Bgl-12:   d(GATCTTCCCTCG)

N-Bgl-24:   d(AGGCAACTGTGCTATCCGA
               GGGAA)

Reagents for fluorescence measurements include TNE buffer: 100 mM Tris–HCl (pH 7.4), 10 mM EDTA, 2 M NaCl; Hoechst-dye 33258 (BisBenzimide H33258; Sigma, Cat. No. B 2883); DNA standard: calf thymus DNA from Sigma (Cat. No. D 4764).

Plasticware should be resistant to phenol and chloroform and amicable to centrifugation at high speed. PCR plasticware is from ABgene: 96-well PCR plates with disposable covers; 8-well PCR strips with additional lid strips; and individual PCR tubes. Also 1.5 and 2-ml reaction tubes are required, as well as larger reaction vessels of up to 30 ml volume. For all pipetting steps, use filter tips in order to avoid contamination.

All technical equipment necessary is conventional and standard in molecular biology laboratories. However, it should be set aside for RDA experiments in order to avoid contamination. Required are the following: a minigel chamber and a power supply to run agarose gels at about 100 V; a set of pipettes; a small centrifuge for 2-ml reaction tubes; a centrifuge for spinning 30-ml tubes at 12,000 g; a UV transilluminator with camera; a vortex shaker; a timer; incubation blocks to keep reactions at precise temperatures between 37 and 100°C, preferably with a mixer function (e.g., Eppendorff Thermomixer comfort); a water bath or other equipment to keep reaction vessels at 14°C; and a refrigerator or cold room for storage. In principle, all incubation steps can be done in a thermocycler. The PCR machine should have a hot-lid function and hold 96-well plates (e.g., MJ Research PTC 200). For fluorescence measurements, we use a DyNAquant 200 device (Hoefer Scientific).

III. PROCEDURES

A. RNA Extraction and cDNA Synthesis

For RNA preparation, we recommend using a protocol that is well established for the particular biological system. Generally, our experiences with phenol/guanidinium-thiocyanate protocols (Trizol reagent from Invitrogen, Cat. No. 15596-026) are better than with column-based methods. In some experiments, we performed DNase treatments of total RNA preparations. However, we cannot generally recommend this because RNA degradation was observed several times. If required, however, use RQi RNase free DNase from Promega (Cat. No. M6101) with the

RNasin ribonuclease inhibitor (Cat. No. N2111) according to the supplier's recommendations. For mRNA isolation, magnetic beads are highly recommended (Dynal Dynabeads mRNA purification kit, Cat. No. 610.06). An estimation of the mRNA amount may be done by adsorption measurements. Make sure that the sample can be recovered from the cuvette.

Synthesis of double-stranded cDNA is preferably done following the protocol provided by Invitrogen with the Superscript double-stranded cDNA synthesis kit (Cat. No. 11917-010). Primer is $dT_{(15)}$ or $dT_{(13)}$ with an anchor of two random bases at the 5' terminus. Radioactive label, used frequently for quantification and quality control, should only be added to an aliquot rather than to the whole cDNA preparation. The final yield of double-stranded cDNA should be similar to the initial amount of mRNA.

B. Generation of Representations; Preparation of Driver and Tester

1. For a digest, use 1–1.5 µg cDNA of good quality. If less than 1 µg is cleaved, the eventual number of cycling reactions may have to be increased.

2. Mix 10 µl 10x *Dpn*II buffer, 5 µl *Dpn*II, cDNA, and water to a total of 100 µl, incubate at 37°C for at least 2 h, inactivate the enzyme at 65°C for 10 min, and add 500 µl PB buffer of the Qiaquick kit.

3. Place a Qiaquick spin column in a collection tube (without lid), apply sample to column, and centrifuge 60 s to bind the cDNA. Discard the flow through and put the column back into the tube. To wash, add 750 µl of the supplied PE buffer, incubate 5 min, and spin for about 60 s. Discard flow through, put the column back into the tube, and spin for 1 min to remove residual buffer. Place column into fresh tube to elute the cDNA. Add 42 µl EB buffer to the centre of the membrane, incubate for 1 min, and spin for 1 min. Continue with the next step immediately.

4. Mix 8 µl *R-Bam-24*, 4 µl *R-Bam-12*, 6 µl 10x ligase buffer, and 39 µl of the Qiaquick column eluate. Place sample in a heating block of 50°C and let the block cool down for more than 1 h in a cold room until the block temperature is below 14°C. Place tube on ice and add 3 µl T4 DNA ligase for an overnight incubation at 14°C. Subsequently, dilute the ligation by the addition of 140 µl TE.

5. Prepare a PCR reaction to check whether digest and ligation worked. Mix 69 µl water, 20 µl 5x PCR buffer, 8.5 µl dNTP, 1 µl *R-Bam-24*, and 1 µl ligation product. Incubate at 72°C for 3 min to melt off the 12-mer oligomer and add 0.5 µl *Taq* polymerase. Incubate for 5 min at 72°C to fill in the ends. Cycle 20 times: 1 min at 95°C and 3 min at 72°C. Finally, keep the sample at 72°C for 10 min. Check 5 µl on a 2.5% gel. A DNA smear between 0.1 and 1 Kb should be visible.

It is highly recommended to perform a negative and a positive control. Both representations should have a similar yield in quality and quantity. If not, and especially if the input amount of cDNA is low, the cycle numbers have to be adjusted. Perform the PCR with additional cycles and remove 5 µl every second or third cycle during the 72°C incubation, starting at cycle 17. Check the products on a gel. With an increase in cycle number the faint DNA smear will grow in intensity, finally reaching a plateau. The plateau is characterised by a smear toward the slot or a shift of the product to smaller sizes—going along with a decrease in quality. Pick an optimal number of cycles for each representation, avoiding the plateau phase, but with a good yield. Both representations used in an RDA reaction must look similar on a gel.

6. Perform PCR for the production of more material, following the protocol established earlier. For a driver preparation, perform identical 100-µl reactions in all wells of a 96-well microtiter plate. Eight reactions are sufficient for preparing a tester, as less material is needed. Mix enzyme, reaction buffer, and DNA first before distributing the mix into the wells. Always include a negative control without DNA. For most analyses, reciprocal experiments are recommended: each cDNA serves as both driver and tester. In this case, a total of one 96-well plate worth of DNA is sufficient. After PCR, check a representative number of products on a 2.5% gel.

7. Combine the 96 driver reactions (or the 8 tester reactions) in one tube, add an equal volume of phenol/chloroform, and vortex vigorously. Spin briefly for phase separation and transfer the supernatant into a new tube. Warming the sample slightly may clear the phases if cloudy. Repeat the extraction, add one volume of chloroform, and vortex again. Spin and recover the supernatant. Do avoid the interphase. If necessary, repeat the extraction. Add 0.1 volume sodium acetate, 1 volume cold isopropanol, mix, and precipitate for 20 min at −20°C; longer incubation may result in salt precipitation. Spin for 15–30 min at high speed (>10,000 g). Remove the supernatant carefully and wash pellet with 70% ethanol. Spin again a few minutes, remove the ethanol, and dry the pellet (preferably at 37°C and not in a vacuum). Resuspend a driver in 200 µl T10 and a tester in 20 µl. Determine the concentration with a fluorometer.

8. ***For the preparation of a driver***, mix 300 µg DNA, 140 µl *Dpn*II buffer, 75 µl *Dpn*II, and water to a final volume of 1400 µl. Incubate at 37°C for 4 h and stop the reaction at 65°C for 15 min. Extract twice with phenol/chloroform and once with chloroform and precipitate with cold isopropanol as described earlier.

Resuspend the DNA in 200 µl T10 and determine its concentration. Dilute the digested driver to about 600 ng/µl and determine the concentration again before diluting the driver to precisely 500 ng/µl.

9. *For the preparation of a tester*, mix 10 µg DNA, 10 µl *Dpn*II buffer, 5 µl *Dpn*II, and water to a final volume of 100 µl. Incubate at 37°C for 2 h, incubate at 65°C for 15 min, and add 500 µl PB buffer. Alternatively, take 10 µg *Dpn*II-digested driver (= 20 µl) and add 100 µl PB buffer. Purify the DNA with a Qiaquick spin column as described in Section III, B, **step 3**. Elute the DNA finally with 50 µl EB buffer.

10. Clean one minigel chamber for each tester; do not run different testers in one gel. Wash the chamber with detergent and rinse with water extensively; incubate all tools (combs, spacer, etc.) for 30 min in 1 *M* HCl and wash with plenty of water.

11. Load 10 µg tester DNA on a 1.3% LMP-agarose gel in TAE buffer without any other DNA (no marker!). Use only a small amount of loading buffer; even better, add glycerol (up to 10%) only and run bromphenol blue in a separate lane. Electrophoresis is at 100 V for a short time. The bromphenol blue should have migrated about 2 cm. Stain the gel in a fresh ethidium bromide solution (use once only). Check on a clean long wavelength UV transilluminator at low (preparative) intensity. There should be a smear of DNA and, directly below, a band from the adapter molecules. Cut out the smear portion (up to the slot), transfer with a clean scalpel to a tube, of which you know the weight, and determine the weight of the agarose plug.

12. For DNA extraction, use Qiaex spin colums with the Qiaex protocol. Melt the agarose plug at 50 to 55°C in three gel volumes QG buffer for 10 min. The buffer has a pH indicator. If the mixture is orange or purple, add sodium acetate (about 10 µl) until the colour turns yellow. Add one gel volume of isopropanol and mix. Place a Qiaquick spin column in a collection tube (without lid), apply the sample to the column, and centrifuge for 60 s to bind DNA. Proceed as in Section III, B, **step 3**. Elute the DNA with 30 µl EB buffer. Check the concentration fluorometrically.

13. Mix 1 µg purified tester DNA, 2 µl *J-Bgl-12*, 1 µl *J-Bgl-24*, 3 µl 10x ligase buffer, and water to a total reaction volume of 30 µl. Place in a heating block of 50°C and let it cool more than 1 h until the block temperature is below 14°C. Place the reaction on ice, add 1.5 µl T4 DNA ligase, and incubate overnight at 14°C. Dilute to approximately 10 ng/µl by addition of 70 µl T10 and finally incubate the tester at 50°C for 10 min.

14. Check the tester by performing PCR. Mix 69 µl water, 20 µl 5x PCR buffer, 8.5 µl dNTPs, 1 µl *J-Bgl-24*, and 1 µl tester. Incubate at 72°C for 3 min, add 0.5 µl *Taq* polymerase, and incubate at 72°C for 5 min. Cycle

25 times between 95°C for 1 min and 70°C for 3 min. Subsequently, incubate at 72°C for 10 min, cool down, and check 5 µl on a 2.5% agarose gel. The amplified tester should visually resemble the representation.

C. Subtractive Hybridisation and Generation of Difference Products

1. For the *first subtractive hybridisation*, mix 80 µl digested driver (40 µg) with 40 µl tester (400 ng) (ratio 100 : 1) and incubate for 5 min at 95°C for denaturation. Extract once with 120 µl phenol/chloroform/isoamylalcohol and once with 120 µl chloroform. Add 30 µl ammonium acetate and 380 µl cold ethanol, mix, and precipitate at –70°C for 10 min. Spin at high speed for 15 min, remove supernatant carefully, wash with 70% ethanol, spin again, remove supernatant, and dry the pellet. Resuspend the DNA very thoroughly in 4 µl 3xEE by pipetting and vortexing or, preferably, by incubation in a heat mixer at 37°C for 10 min. Spin briefly, transfer into a new tube, and overlay with two drops of mineral oil (30–60 µl). Denature for 5 min at 98–100°C and mix immediately with 1 µl 5 *M* NaCl; penetration of the hot mineral oil with the pipette tip may be difficult. Place the reaction immediately at 67°C and incubate for 20 h.

2. Add 8 µl 0.5 × TE/tRNA and mix. Remove 8 µl mineral oil with same tip. Add 25 µl TE, mix thoroughly, and again remove some mineral oil using the very same tip. Add 362 µl TE, vortex, and place on ice for immediate use or store frozen.

3. Set up eight 100-µl PCR reactions for each experiment (and control). Mix 60 µl water, 20 µl 5x PCR buffer, 8.5 µl dNTPs, and 10 µl diluted hybridisation mix. Incubate at 72°C for 3 min, add 0.5 µl *Taq* polymerase, incubate at 72°C for 5 min, and add 1 µl *J-Bgl-24*. Pause the cycler at 72°C for all additions. Cycle 10 times between 95°C for 1 min and 70°C for 3 min. Then incubate at 72°C for 10 min and cool the sample.

4. Combine the eight reactions in a bigger tube and add 4 ml PB buffer. Place one Qiaquick spin column in a collection tube (without lid), apply 750 µl of sample, and centrifuge about 60 s. Discard the flow through and put the column back into the empty tube. Repeat this until the whole sample is loaded onto the column. Add 750 µl PE buffer, incubate for 5 min, and spin for about 60 s. Discard the flow through, put the column back into the tube, and spin for 1 min to remove residual buffer. Place column into a fresh tube. To elute DNA, add 42 µl T1, incubate for 1 min, and spin for 1 min.

5. Mix 14 µl water, 4 µl 10x mung bean nuclease buffer, 2 µl mung bean nuclease, 20 µl of the eluted DNA, and incubate at 30°C for 35 min. Stop the reaction by the addition of 160 µl T50 and an incubation at 98°C for 5 min. Chill on ice.

6. Set up a second PCR (two 100-μl reactions) on ice. Mix 60 μl water, 20 μl 5x PCR buffer, 8.5 μl dNTP, and 1 μl *J-Bgl-24*. On ice, add 10 μl of the DNA digested with mung bean nuclease. Incubate at 95°C for 1 min, cool to 80°C, add 0.5 μl *Taq* polymerase, and cycle 18 times between 95°C for 1 min and 70°C for 3 min. Incubate at 72°C for 10 min and cool on ice. Combine the two reactions and check 5 μl on a 2.5% gel. The crude product may still look similar to the representations but usually first discrete bands become visible.

7. Add 1 ml PB buffer and place the sample on a Qiaquick spin column in a collection tube. Apply 500 μl to column and centrifuge about 60 s. Discard flow through and put column back into the tube. Repeat the last two steps. Add 750 μl PE buffer, incubate for 5 min, and spin for about 60 s. Discard flow through, place column back into collection tube, and spin for 1 min to remove residual buffer. Place column into fresh tube (without lid). Add 85 μl buffer T1, incubate for 1 min, and spin for 1 min to elute DNA. Check DNA concentration fluorometrically. The product is *difference product 1 (DP1)*.

8. To remove the adaptors, mix 10 μl *DpnII* buffer, 3 μl *DpnII*, 5 μg DP1, and water up to 100 μl reaction volume. Incubate at 37°C for 2 h. Stop reaction at 65°C for 15 min, add 500 μl PB buffer, and continue as described in Section III, B, **step 3**.

9. For ligation of new adaptors, mix 6 μl 10x ligase buffer, 8 μl *N-Bgl-24*, 4 μl *N-Bgl-12*, 200 ng DNA, and water to a total volume of 60 μl. Place in a heating block and continue as described in Section III, B, **step 4**. Dilute ligation reaction to approximately 1.25 ng/μl by the addition of 100 μl T10. Stop reaction at 50°C for 10 min.

10. For a test PCR, set up a 100-μl reaction as described in Section III, B, **step 5**, but with *N-Bgl-24* and 25 cycles. The amplified product should look similar to DP1.

11. For the *second subtractive hybridisation*, mix 80 μl digested driver (40 μg) with 40 μl diluted DP1 with *N-Bgl* adapter (50 ng) (ratio 800 : 1) and continue as described in Section III, C, **steps 1 and 2**.

12. For the first PCR after the second subtractive hybridisation, use *N-Bgl-24* primer and follow the protocol described in Section III, C, **steps 3 and 4**.

13. For mung bean nuclease digest, follow the protocol described in Section III, C, **step 6**.

14. For the second PCR, again use *N-Bgl-24* primer and follow the protocol described in Section III, C, **step 6**. On the gel, there should be less smear and more distinct bands.

15. To clean the PCR product, follow the protocol of Section III, C, **step 7**, resulting in *difference product 2 (DP2)*.

16. To change the adapters, follow protocols in Section III, C, **step 8** and in Section III, B, **step 3**.

17. For ligation of new adapters (*J-Bgl*), follow the protocol of Section III, C, **step 9** using oligomers *J-Bgl-12* and *J-Bgl-24*. After an overnight ligation, dilute the reaction to 1 ng/μl by the addition of 140 μl TE and incubate at 50°C for 10 min. Dilute 2 μl of this with 778 μl T10 and add 20 μl (20 μg) yeast tRNA. The final DNA concentration is 2.5 pg/μl.

18. For a test PCR, follow the protocol in Section III, B, **step 14** using 1 μl (2.5 pg) DP2 with *J-Bgl* cassette as template. In a separate reaction, use 1 μl of digested DP2 as negative control. The amplified product should look similar to DP2.

19. For a *third subtractive hybridisation*, mix 80 μl digested driver (40 μg) with 40 μl diluted DP2 with *J-Bgl* cassette (100 pg) (ratio 400,000 : 1) and continue as in Section III, C, **steps 1 and 2**.

20. Continue as in Section III, C, **steps 3 to 6**. For the second PCR, use 18 cycles instead of 29. If individual bands with no or little background smear appear, your *difference product 3 (DP3)* is probably fine. If you want further enrichment, you may continue with a fourth RDA cycle using *N-Bgl* primers and a tester-to-driver ratio of 1 : 8,000,000 and/or competition with selected sequences. Start cloning the DNA fragments in the difference product either directly or after further purification.

D. Cloning and Analysis of Difference Products

Preferably the difference products are directly cloned into TA-cloning vectors (making use of the A overhang produced by the *Taq* polymerase), such as pGEM-T-Easy of Promega (Cat. No. A1360). A representative number of clones, determined according to the complexity of the band pattern on the gel, but at least eight times the number of visible bands), may be analysed by sequencing, restriction pattern analysis, hybridisation assays, and other procedures. Isolation and cloning of individual bands from the gel are also possible, but must be followed by analysis of several clones, as frequently there is more than one fragment. The latter process also selects against very rare fragments, which are present but insufficient in mass to be visible as a band.

RDA and arraying techniques are powerful tools for the identification of differential gene expression. Combining them can either produce microarrays that contain exclusively highly relevant probe molecules, focusing on the relevant information that can be gathered from hybridisation experiments, or simplify the analysis of global difference products. For the former, individual difference products of the second or third iterative RDA cycle are best suited, as they should represent differentially transcribed genes mainly and thus be very informative. Several libraries that result from different difference analyses may be combined in one array. Initial arrays can be used for hybridising pools made of individual clones in order to identify repeatedly cloned fragments and thus reduce the redundancy. Validation of the RDA process is possible by a comparative hybridisation of the initially used cDNA or the driver and tester representations on such arrays (Chang *et al.*, 1998; Welford *et al.*, 1998; Frohme *et al.*, 2000). Only fragments that originate from the respective difference products should give a signal in such an experiment. Furthermore RDA libraries may complement specific cDNA libraries used for arrays.

Another option is to use labelled difference products as targets for hybridisation on arrays. Because differential sequences are enriched already, differences in signal intensities will be more pronounced and therefore simpler to be detected, probably without the need for differential hybridisations using different dyes. Thus *in silico* analyses may become superfluous (Geng *et al.*, 1998; Kim *et al.*, 2001). More quantitative RDA-based analyses in connection with microarrays techniques were presented by Andersson and colleagues (2002, 2003).

IV. COMMENTS, PITFALLS, AND ALTERNATIVE PROTOCOLS

1. By increasing the initial cycle number, we could apply RDA to a few nanograms of original material, although with limited success. Also, this process may cause bias and will result in more false-positive fragments.

2. The starting material for RDA has to be selected carefully, not only because of the high demands in purity and quality, but also with respect of the experimental design. The result of a crude comparison of some tumour vs normal tissue may reflect the immune response rather than oncogenesis, for example.

3. Supplementing the driver with unwanted fragments, material from certain tissues or rDNA, for example, is a means to eliminate those in the difference products (Gress *et al.*, 1997, Wada *et al.*, 1997, Ushijima *et al.*, 1997; Iwama *et al.*, 1998, Bowler *et al.*, 1999).

4. Frequently, DNA concentrations are estimated by comparing visually on an agarose gel the DNA of unknown concentration to a DNA of well-defined concentration. Because the representations and difference products are a smear with more or less distinct banding, this method is not very precise. Using adsorption measurement instead is obscured by the unused nucleotides and primers in the reaction mixture. By far the best results are obtained with fluorescence measurements.

5. Various publications suggest the use of other driver-to-tester ratios and longer hybridisation (Vician *et al.*, 1997; Pastorian *et al.*, 2000; Birkenmeyer *et al.*, 2003). This may be helpful when the standard protocol fails. However, we found that alterations in ratios by a factor of up to fivefold did not remarkably alter the resulting difference products.

6. For most purposes, *Dpn*II is an appropriate restriction enzyme. Digestion of the PCR products may even be done directly in the amplification reaction after the addition of magnesium chloride (Vician *et al.*, 1997). *Dpn*II could be replaced by *Sau*3A1 when a cDNA library cloned in recombinant bacteria is used as source material rather than RNA. A different representation of the cDNA or a bias toward longer fragments could be achieved with different enzymes (Edman *et al.*, 1997; Morris *et al.*, 1998; Pastorian *et al.*, 2000; Felske, 2002). Note that the primer/adapter sequences have to be adapted then.

7. Several protocols describe omission of the digestion of single-stranded DNA with the mung bean nuclease (e.g., Welford *et al.*, 1998). Instead it was suggested to dilute the first PCR product (Pastorian *et al.*, 2000).

8. Extended use of nucleases (mung bean nuclease and exonuclease III) was reported (Kuvbachieva and Goffinet, 2002) to remove single-stranded DNA and unwanted (driver) duplexes prior to PCR. For this, thiodeoyxnucleotides were used to protect the tester hybrids from degradation.

9. Several authors describe improvements of RDA by alternative primers/adapters.
   a. It is reported that linear amplification of driver:tester heteroduplexes can be avoided by use of different primer/adaptors for generation of the driver and tester representations (Felske, 2002). Another protocol also uses different primers as well as a randomised amplification of the driver cDNA (Herblot *et al.*, 1997).
   b. Additional primer pairs avoid repeated use of the original adapter/primer pair, thereby reducing the risk of carryover (Vician *et al.*, 1997; Pastorian *et al.*, 2000; Birkenmeyer *et al.*, 2003).
   c. More selective PCR primers have been suggested (Birkenmeyer *et al.*, 2003) to reduce the complexity of the representation. Nested primers (Michiels *et al.*, 1998) might have a similar effect.

10. Spin columns used for the purification of PCR products may be replaced by Amicon or Centricon units (Vician *et al.*, 1997) or by adapting RDA to solid-phase procedures (Odeberg *et al.*, 2000). The latter makes use of representations generated with biotinylated primers. These can be removed after restriction cleavage with streptavidin-coated paramagnetic beads. In another protocol (Chu and Paul, 1998), driver–tester or driver–driver hybrids are removed by magnetic beads after subtractive hybridisation.

11. An initial melt-depletion procedure has been discussed (Hubank and Schatz, 1994). This process could make the resulting representations a better substrate for RDA. This modification uses denaturation followed by a short hybridization to let the more abundant DNA species reanneal before generation of the representations by PCR.

12. Changes in the relative composition of substrates during the RDA procedure were suspected to result in inefficient subtraction (Birkenmeyer *et al.*, 2003). Using a driver, that has gone through a prior RDA subtraction procedure against itself should account for this problem.

13. In one publication (Luo *et al.*, 1999), a conversion of tester–driver hybrids to driver only by another mung bean nuclease digest was suggested. This "differential subtraction chain" was reported to enhance RDA considerably.

14. The "phenol emulsion reassociation technique" was applied to improve the reassociation kinetics of the driver during subtractive hybridisation (Becker *et al.*, 2001).

References

Andersson, T., Borang, S., Unneberg, P., Wirta, V., Thelin, A., Lundeberg, J., and Odeberg, J. (2003). Shotgun sequencing and microarray analysis of RDA transcripts. *Gene* **310**, 39–47.

Andersson, T., Unneberg, P., Nilsson, P., Odeberg, J., Quackenbush, J., and Lundeberg, J. (2002). Monitoring of representational difference analysis subtraction procedures by global microarrays. *Biotechniques* **32**, 1348–1358.

Becker, P., Hufnagle, W., Peters, G., and Herrmann, M. (2001). Detection of differential gene expression in biofilm-forming versus planktonic populations of *Staphylococcus aureus* using micro-representational-difference analysis. *Appl. Environ. Microbiol.* **67**, 2958–2965.

Birkenmeyer, L. G., Leary, T. P., Muerhoff, A. S., Dawson, G. J., Mushahwar, I. K., and Desai, S. M. (2003). Selectively primed adaptive driver RDA (SPAD-RDA): An improved method for subtractive hybridisation. *J. Med. Virol.* **71**, 150-159.

Bowler, L. D., Hubank, M., and Spratt, B. G. (1999). Representational difference analysis of cDNA for the detection of differential gene expression in bacteria: Development using a model of iron-regulated gene expression in *Neisseria meningitidis*. *Microbiology* **145**, 3529–3537.

Chang, D. D., Park, N. H., Denny, C. T., Nelson, S. F., and Pe, M. (1998). Characterization of transformation related genes in oral cancer cells. *Oncogene* **16**, 1921–1930.

Chu, C. C., and Paul, W. E. (1998). Expressed genes in interleukin-4 treated B cells identified by cDNA representational difference analysis. *Mol. Immunol.* **35**, 487–502.

Edman, C. F., Prigent, S. A., Schipper, A., and Feramisco, J. R. (1997). Identification of ErbB3-stimulated genes using modified representational difference analysis. *Biochem. J.* **323**, 113–118.

Felske, A. (2002). Streamlined representational difference analysis for comprehensive studies of numerous genomes. *J. Microbiol. Methods* **50**, 305–311.

Frohme, M., Scharm, B., Delius, H., Knecht, R., and Hoheisel, J. D. (2000). Use of representational difference analysis and cDNA arrays for transcriptional profiling of tumor tissue. *Ann. N. Y. Acad. Sci.* **910**, 85–105.

Geng, M., Wallrapp, C., Müller-Pillasch, F., Frohme, M., Hoheisel, J. D., and Gress, T. M. (1998). Isolation of differentially expressed genes by combining representational difference analysis (RDA) and cDNA library arrays. *Biotechniques* **25**, 434–438.

Gress, T. M., Wallrapp, C., Frohme, M., Müller-Pillasch, F., Lacher, U., Friess, H., Buchler, M., Adler, G., and Hoheisel, J. D. (1997). Identification of genes with specific expression in pancreatic cancer by cDNA representational difference analysis. *Genes Chrom. Cancer* **19**, 97–103.

Herblot, S., Vekris, A., Rouzaut, A., Najeme, F., de Miguel, C., Bezian, J. H., and Bonnet, J. (1997). Selection of down-regulated sequences along the monocytic differentiation of leukemic HL60 cells. *FEBS Lett.* **414**, 146–152.

Hubank, M., and Schatz, D. G. (1994). Abstract Identifying differences in mRNA expression by representational difference analysis of cDNA. *Nucleic Acids Res.* **22**, 5640–5648.

Iwama, A., Zhang, P., Darlington, G. J., McKercher, S. R., Maki, R., and Tenen, D. G. (1998). Use of RDA analysis of knockout mice to identify myeloid genes regulated *in vivo* by PU.1 and C/EBPalpha. *Nucleic Acids Res.* **26**, 3034–3043.

Kim, S., Zeller, K., Dang, C. V., Sandgren, E. P., and Lee, L. A. (2001). A strategy to identify differentially expressed genes using representational difference analysis and cDNA arrays. *Anal. Biochem.* **288**, 141–148.

Kuvbachieva, A. A., and Goffinet, A. M. (2002). A modification of representational difference analysis, with application to the cloning of a candidate in the reelin signalling pathway. *BMC Mol. Biol.* **3**, 6.

Lisitsyn, N., Lisitsyn, N., and Wigler, M. (1993). Cloning the differences between two complex genomes. *Science* **259**, 946–951.

Luo, J. H., Puc, J. A., Slosberg, E. D., Yao, Y., Bruce, J. N., Wright, T. C., Becich, M. J., and Parsons, R. (1999). Differential subtraction chain, a method for identifying differences in genomic DNA and mRNA. *Nucleic Acids Res.* **27**, e24.

Michiels, L., Van Leuven, F., van den Oord, J. J., De Wolf-Peeters, C., and Delabie, J. (1998). Representational difference analysis using minute quantities of DNA. *Nucleic Acids Res.* **26**, 3608–3610.

Morris, M. E., Viswanathan, N., Kuhlman, S., Davis, F. C., and Weitz, C. J. (1998). A screen for genes induced in the suprachiasmatic nucleus by light. *Science* **279**, 1544–1547.

Odeberg, J., Wood, T., Blucher, A., Rafter, J., Norstedt, G., and Lundeberg, J. (2000). A cDNA RDA protocol using solid-phase technology suited for analysis in small tissue samples. *Biomol. Eng.* **17**, 1–9.

Pastorian, K., Hawel, L., and Byus, C. V. (2000). Optimization of cDNA representational difference analysis for the identification of differentially expressed mRNAs. *Anal. Biochem.* **283**, 89–98.

Vician, L., Basconcillo, R., and Herschman, H. R. (1997). Abstract Identification of genes preferentially induced by nerve growth factor versus epidermal growth factor in PC12 pheochromocytoma cells by means of representational difference analysis. *J. Neurosci. Res.* **50**, 32–43.

Wada, J., Kumar, A., Ota, K., Wallner, E., Batlle, D. C., and Kanwar, Y. S. (1997). Representational difference analysis of cDNA of genes expressed in embryonic kidney. *Kidney Int.* **51**, 1629–1638.

Welford, S. M., Gregg, J., Chen, E., Garrison, D., Sorensen, P. H., Denny, C. T., and Nelson, S. F. (1998). Detection of differentially expressed genes in primary tumor tissues using representational differences analysis coupled to microarray hybridisation. *Nucleic Acids Res.* **26**, 3059–3065.

16

Single Cell Gene Expression Profiling: Multiplexed Expression Fluorescence *in situ* Hybridization: Application to the Analysis of Cultured Cells

Jeffrey M. Levsky, Steven A. Braut, and Robert H. Singer

I. INTRODUCTION

Most current methods of measuring gene expression rely on averaging many cellular responses or artificial amplification steps to reach a detectable threshold of signal. In contradistinction, *in situ* assays circumvent these procedures to yield direct single cell expression information. Fluorescence *in situ* hybridization (FISH) is the gold standard for localization of nucleic acids (Fauth and Speicher, 2001; van der Ploeg, 2000). The introduction of amino-allyl-modified bases (Langer *et al.*, 1981) allowed the chemical synthesis of multiply labeled fluorescent oligomer hybridization probes (Femino *et al.*, 1998; Kislauskis *et al.*, 1993). This in turn allowed the application of multicolor/multispectral FISH (Nederlof *et al.*, 1990) to visualization of multiple RNA species simultaneously (Levsky *et al.*, 2002). With the introduction of visualization to gene expression assays we began to understand the complexity of behavior at the cell level, allowing reinvestigation of assumed consistencies of cell populations with single cell resolution (Elowitz *et al.*, 2002; Levsky *et al.*, 2002; Levsky and Singer, 2003).

II. MATERIALS AND INSTRUMENTATION

Oligomer probes are designed with OLIGO software from Molecular Biology Insights and synthesized on an Applied Biosystems automated DNA/RNA synthesizer (Model 394). Solid-support synthesis columns are from Applied Biosystems (dA, Cat. No. 400949; dC, Cat. No. 400950; dmf-dG, Cat. No. 401184; T, Cat. No. 400952). Phosphoramidites (dA, Cat. No. 10-1000-10; dC, Cat. No. 10-1010-10; dmf-dG, Cat. No. 10-1029-10; dT, Cat. No. 10-1030-10) and an amino-allyl-modified base for attachment of ester-conjugated fluorophores (C6-dT, Cat. No. 10-1039-05) are obtained from Glen Research. Oligonucleotide purification cartridges (OPC, Cat. No. 400771) and 2 M triethylamine acetate (TEAA, Cat. No. 400613) are from Applied Biosystems, and the anhydrous acetonitrile (Cat. No. 40-4050-50) is from Glen Research. Trifluoroacetic acid (TFA, Cat. No. BP618-500) is from Fisher, and triethylamine (TEA, Cat. No. T-0886) is from Sigma.

Fluorophores are from Amersham (Cy3, Cat. No. PA23001; Cy3.5, Cat. No. PA23501; Cy5, Cat. No. PA25001) and Molecular Probes (Oregon green 488,

Cat. No. O-6147; Alexa Fluor 488, Cat. No. A-10235 or A-20191). Sodium canbonate for labeling buffer (Cat. No. BP357-1), 25-ml pipettes for make-shift size-exclusion chromatography columns (Cat. No. 13-674-41E), and Pasteur pipettes (Cat. No. 13-678-20D) for filling the columns are from Fisher. Sephadex G-50 resin (Cat. No. 55897-100G) for purification is from Sigma. Columns are fashioned by removing the cotton from the top of one of these pipettes and using a portion of it to plug up the tip. Secure the "column" vertically to a ring stand and cap the tip with a 1.5-ml Eppendorf tube to prevent liquid loss. The vacuum concentrator system used is from Savant (Speed-Vac), and the UV spectrophotometer used for measurements of probe concentrations and labeling efficiencies is from Beckman (DU640).

Glass coverslips (Cat. No. 12-542B), glass slides (Cat. No. 12-518-103), 12 N HCl (Cat. No. A144-212), gelatin (Cat. No. G8-500), Parafilm (Cat. No. 13-374-12), forceps for coverslip manipulation (Cat. No. 08-953-E), and magnesium chloride (Cat. No. BP214-500) are from Fisher. The 20% paraformaldehyde for preparation of fixative (PFA, Cat. No. 15713) and coplin jars for washes (Cat. No. 72242-01) are from Electron Microscopy Sciences. 10× phosphate-buffered saline (PBS, Cat. No. 1 666 789), 20× sodium chloride/sodium citrate (SSC, Cat. No. 1 666 681), purified bovine serum albumin (BSA, Cat. No. 711 454), and *Escherichia coli* tRNA (Cat. No. 109 541) are from Roche. Triton X-100 (Cat. No. T-9284), formamide (Cat. No. F-4761), sheared salmon sperm DNA (ssDNA, Cat. No. D-7656), 4',6-diamidino-2-phenylindole (DAPI, Cat. No. D-9564), and diethylpyrocarbonate (DEPC, Cat. No. D-5758) are from Sigma. The glass plates used for hybridization (Cat. No. 165-1824) are from Bio-Rad. The proLong Gold antifade reagent for mounting slides (Cat. No. P36934) is from Molecular Probes.

Upright fluorescence microscopes from Olympus are used to image multiple spectral signatures from the FISH specimens (Models AX70 and BX51) with a piezoelectric translator from Physik Instrumente (Cat. No. PZ54 E) to generate three-dimensional image stacks. Alternatively, microscopes featuring an internal harmonic drive may be used (e.g., BX61 from Olympus). Illumination is provided by a 100-W mercury arc lamp. Microscopes are outfitted with Olympus PlanApo 60x, 1.4 NA objectives and Chroma HiQ band-pass filters to separate fluorescence signals. Although other methods have been introduced to discern multiple fluorescence signals from chromosomes (Schrock *et al.*, 1996), they have not been applied successfully to detection of mRNA transcription sites. In Levsky (2002), we performed color coding of transcripts using the following filters from Chroma: DAPI

(Cat. No. 31000), FITC (Cat. No. 41001), Cy3 (Cat. No. SP-102v1), Cy3.5 (Cat. No. SP-103v1), and Cy5 (Cat. No. 41008). High-resolution, low-noise fluorescence images are captured using charge-coupled device (CCD) cameras from Roper Scientific [Models CH-350(502) and CoolSNAP-HQ]. Acquisition and data manipulation are performed using IPLab software from Scanalytics. To ease data processing and avoid manual manipulations that introduce bias, we code our own filtering and data analysis software in the JAVA programming language using the Java development kit and advanced imaging library from Sun Microsystems.

III. PROCEDURES

A. Preparation of Fluorescent Oligomer Hybridization Probes

This procedure is according to Kislauskis (1993).

Solutions

1. *Diethylpyrocarbonate-treated distilled water (DDW)*: To make 1 liter, add 0.5 ml DEPC to 1 liter of distilled water. Shake or stir until DEPC is well distributed and then autoclave. Prepare enough of this to use in all other solutions.

2. *Labeling buffer (0.1 M Na_2CO_3, pH 9.0)*: To make 100 ml, weigh 1.06 g Na_2CO_3 and complete to 100 ml with DDW. Adjust pH to 9.0 by adding 10 N NaOH and store at 4°C.

3. *2 M TEAB stock*: To make 500 ml, take 138.3 ml (101 g) TEA and fill to 400 ml with DDW. Use dry ice to bubble in CO_2 until pH is below 8.0. Complete to 500 ml with DDW and store at 4°C. *Note: TEA is extremely hazardous so take care when handling. Use glassware instead of plasticware when measuring and transporting.*

4. *Filtration column running buffer (10 mM TEAB)*: To make 1 liter, take 5 ml of 2 M TEAB stock solution and complete to 1 liter with DDW. Store at 4°C.

5. *Filtration column running matrix*: To make approximately 200 ml, pour 200 ml of 10 mM TEAB into an Erlenmeyer flask. Add approximately 50 g of Sephadex G-50 and swirl to absorb the liquid. Suspension will settle. Store at 4°C. Prior to use, apply vacuum pressure to the flask to degas the suspension for at least 2 h before pouring matrix.

Steps

1. Having selected a gene of interest, choose four to five regions for probe fabrication, each 50 bases in length. Adjust search parameters within the OLIGO

software to receive best possible sequences for gene detection. Several considerations for probe design are:

    i. Spanning different areas of the mRNA increases chances detection; intronic regions should be avoided.

    ii. 50% GC content (or close to this) is optimal.

    iii. Highly stable hairpins should be avoided.

    iv. There must be enough well-spaced residues for substitution of modified bases. This depends on the modifier used. We space five modified thymidine residues at eight or more bases apart.

    v. The sequences must not cross-react significantly with other mRNAs. Use BLAST to test this (see http://ncbi.nih.gov/blast).

2. Prepare the reversed antisense sequence of each designed oligonucleotide, substituting the modified bases appropriately.

3. Synthesize the oligonucleotides according to synthesizer specifications at a 0.2 μM scale, specifying *TRITYL-ON*. Deprotect the crude products in a 65°C water bath for 1 h.

4. Aliquot the crude product into 200- to 300-µl portions and set one aside for immediate purification. Vacuum dry the remaining aliquots and then resuspend each pellet in 1.0 ml 10 mM TEAB plus 5 µl TEA. Store these at −80°C for future use. *Note: As the aliquots dry, the solutions become increasingly acidic and may cause detritylation of the oligonucleotides. To avoid this, add a drop of TEA to each tube periodically while drying them.*

5. Purify the remaining aliquot using the OPC according to recommended procedures (Applied Biosystems). Vacuum dry the final pure product and then resuspend in 100 µl DDW. Determine concentration of product using OD measurements at 260 nm.

6. Prepare a probe mixture with equal amounts of each oligonucleotide to obtain a final amount of 20 µg—either 4 or 5 µg of each oligonucleotide depending on how many were synthesized. Vacuum dry this mixture.

7. Resuspend the pellet in 10 µl labeling buffer and add it to the reaction vial containing approximately 1.0 mg of dye–ester conjugate. Alternatively, oligonucleotides can be labeled according to the manufacturer's specifications (Amersham or Molecular Probes). Vortex and leave at room temperature in the dark overnight.

8. Assemble a size-exclusion chromatography column by transferring 10 mM TEAB via a glass Pasteur pipette into the prepared 25-ml pipette/column until the liquid level is about a third of the way up. Add the G-50 suspension in the same manner and, as the matrix settles, remove the 1.5-ml tube "cap" to allow the matrix to settle above the cotton-plug stop, while permitting liquid to pass through. After the

matrix has filled the pipette, pack it down with a continuous flow of TEAB for 10–15 min. This can be accomplished most easily using a siphoning system attached to the column.

9. Once the matrix has packed, remove the siphoning attachment and allow the buffer to run down to the level of the matrix, taking care not to let it run below. Add the 10 µl volume of probe/dye mixture to the column and wash the reaction vial with an additional 200–300 µl of fresh TEAB. Add the wash to the column and allow it to begin to run down into the matrix. When the dye product has been absorbed into the matrix, refill the column with buffer and reattach the siphoning system to provide continuous liquid flow.

10. As the labeled probe mixture runs down the column it will separate into two bands. The first, faster band will contain the desired pure product. Collect column eluates in 1.0-ml fractions to include this first band. Vacuum dry these fractions. Resuspend the selected fractions in DDW to achieve a total volume of 100 µl.

11. Measure OD of the final sample to determine final concentration and labeling efficiency for the product according to specifications of the dye manufacturers (Amersham, Molecular Probes). A final concentration of 40 ng/µl would indicate that all 20 µg of oligonucleotide initially labeled has been collected.

12. Labeled probe can be stored at 4°C or at −20°C for longer term storage.

B. Preparation of Cell Samples

Solutions

1. *Coverslips in 0.5% gelatin*: To make 200 ml, sterilize a box of coverslips by boiling in 0.1 N HCl for 20 min. Rinse and wash the coverslips in DDW several times. Weigh 1.0 g of gelatin and complete to 200 ml DDW. Stir and warm to dissolve completely. Transfer sterilized coverslips to gelatin solution and autoclave for 20 min. Store at 4°C.

2. *$10\times$ PBS stock*: To make 500 ml of DEPC-treated $10\times$ PBS, take 500 ml $10\times$ PBS and add 250 µl DEPC. Stir or shake to dissolve; autoclave.

3. *1 M MgCl$_2$ stock*: To make 100 ml, weigh 20.3 g MgCl$_2$ and complete to 100 ml with DDW.

4. *Washing solution (PBSM)*: To make 1 liter, take 100 ml $10\times$ PBS stock, add 5 ml 1 M MgCl2 stock, and complete to 1 liter with DDW.

5. *Extractant (PBST)*: To make 1 liter, take 100 ml $10\times$ PBS stock, add 5 ml Triton X-100, and complete to 1 liter with DDW. Stir gently to allow Triton to dissolve completely. *Note: This extractant has been used successfully to remove cytoplasm in cultured DLD-1 cells. The strength of the extractant must be optimized for each cell*

type to obtain optimal reduction of cytoplasmic background without damaging nuclei or loss of cells.

6. *Fixative (4% PFA)*: To make 50 ml, take one 10-ml vial of 20% paraformaldehyde stock, add 5 ml 10× PBS stock, and complete to 50 ml with DDW. Store at 4°C.

Steps

1. Grow cells under standard conditions and seed onto gelatinized coverslips in a petri dish. Cells are grown to empirically determined confluence such that they are sparse enough to facilitate automated separation of nuclei during image processing and dense enough to have significant amounts for analysis.

2. Any treatment steps, such as serum starvation and stimulation, can be performed at this point before fixation.

3. Wash the cells briefly with ice-cold PBSM.

4. Extract the cells for 60 s in PBST at room temperature.

5. Wash the cells twice briefly with ice-cold PBSM.

6. Fix the cells with the PFA fixative solution for 20 min at room temperature.

7. Wash the cells again twice briefly with ice-cold PBSM.

8. Fixed coverslips may be stored at 4°C in PBSM until use. *Note: Further extraction and background reduction can be obtained for some cell types by storage in 70% ethanol at 4°C. In some cases this can cause cells to detach from cover slips.*

C. Hybridization

This procedure is modified from Femino (1998) and Levsky (2002).

Solutions

1. *Washing solution (PBSM)*: To make 1 liter, take 100 ml 10× PBS stock, add 5 ml 1 M MgCl2 stock, and complete to 1 liter with DDW.

2. *Pre/posthybridization wash (50% formamide/2× SSC)*: To make 500 ml, take 250 ml formamide, add 50 ml 20× SSC stock, and complete to 500 ml with DDW.

3. *Probe competitor solution (ssDNA/tRNA)*: To make 100 μl of 10 mg/ml total concentration competitor, take 50 μl of 10 mg/ml sheared salmon sperm DNA and add 50 μl 10 mg/ml *E. coli* tRNA (prepared from solid by adding 10 mg to 1.0 ml DDW). Store at −20°C.

4. *Hybridization buffer*: To make 100 μl, take 60 μl DDW and add 20 μl BSA and 20 μl 20× SSC stock. Prepare fresh and hold on ice. *Note: This volume is sufficient for 10 hybridization reactions (10 coverslips).*

5. *Low-salt wash solution (2× SSC)*: To make 500 ml, take 50 ml 20× SSC stock and complete to 500 ml with DDW.

6. *Lower-salt wash solution (1855C)*: To make 500 ml, take 25 ml 20× SSC stock and complete to 500 ml with DDW.

7. *Nuclear stain solution (DAPI)*: To make 1 liter, take 100 ml 10× PBS stock, add 50 μl 10 mg/ml DAPI stock (prepared from solid by adding 10 mg to 1.0 ml DDW), and complete to 1 liter with DDW. Shake or mix to dissolve DAPI completely and store at 4°C.

8. *Mounting medium*: Use the ProLong Gold antifade reagent according to manufacturer's specifications (Molecular Probes). About 25 μl of medium is needed per coverslip.

Steps

1. Test hybridization before color coding and multiple transcript detection. We start by using two bright dyes (Cy3 and Cy5) to show transcription sites. After this, assign each gene an arbitrary color code using combinations of dyes and test singly. Only after results are reproducible is multiplex detection performed.

2. Using forceps, place fixed coverslips vertically in a coplin jar, keeping note of which side has the cells on it. Rehydrate and wash the cells in PBSM for 10 min. *Note: All washes are at room temperature unless otherwise noted.*

3. Equilibrate the cells in prehybridization solution for 10 min.

4. Aliquot probe mixtures for gene(s) to be detected into tubes for each different combination of targets to be assayed. *Note: As a starting concentration, combine 20 ng of each probe for the 20 μl total final reaction volume. Optimal concentrations of the different probe mixtures are determined empirically by balancing the resultant colors detected upon imaging transcription sites.*

5. Add competitor solution to the probe mixture(s) in 100-fold excess. Vacuum dry this mixture, taking care not to overdry.

6. Resuspend the dry pellet in 10 μl formamide and place the tubes on a heating block at 85°C for 5–10 min and then place immediately on ice.

7. Add 10 μl of hybridization buffer to each tube, giving a final reaction volume of 20 μl.

8. Wrap a glass/plastic plate with Parafilm to allow enough working space for the amount of reactions you have. Dot each 20-μl reaction volume onto the plate, far enough apart such that coverslips can be placed over each volume without overlap.

9. Remove cover slips from prehybridization solution and blot off excess liquid. Place each coverslip—*cell side down*—on the hybridization mix already dotted onto the plate.

10. Wrap another layer of Parafilm over the plate and coverslips to seal the reactions. Press around the edges with a pen or similar instrument.

11. Incubate the plate at 37°C for 3 h, along with a sufficient amount of prehybridization solution to wash the coverslips twice after hybridization.

12. Remove the top layer of Parafilm and carefully lift the lower layer so that the coverslips can be easily removed without excessive manipulation. Place the coverslips back into coplin jars with the prewarmed wash, keeping track of the cell side, and incubate for 20 min at 37°C. Change and repeat this wash for another 20 min.

13. Remove ProLong Gold antifade reagent from −20°C and allow it to reach room temperature.

14. Change the solution with 2× SSC and incubate at room temperature for 10 min.

15. Change the solution with 1855C and incubate at room temperature for 10 min.

16. Change the solution with PBSM and incubate at room temperature for 10 min.

17. Counterstain nuclei by changing the solution with the prepared DAPI and incubating at room temperature for 1 min and then washing with PBSM.

18. Change the PBSM and keep at room temperature until ready to mount.

19. Mount each coverslip (cell side down) onto glass slides, using freshly prepared antifade mounting medium. Blot off excess liquid and store at −20°C.

D. Microscopy and Image Analysis

These procedures are from Levsky (2002).

Steps

1. Image stacks are acquired with high index oil immersion on a fluorescence microscope outfitted for optical sectioning. We use a step size of 0.5 μm to generate image volumes, as transcription site signals are bright and do not require more finely spaced planes on our setup. For future processing steps and for detecting less bright signals, closer optical sections may be needed. We use the 60× objective and additional magnification (when necessary) to yield digital images of roughly 100-nm per pixel resolution. The total magnification should be adjusted to yield similar resolution given the physical size of elements on the CCD camera used. High resolution enables morphometric processing of the signals.

2. Image volumes from different fluorescent channels are normalized by contrast enhancement to ensure interpretation is independent of relative intensity. This can be performed with commercial software, such as IPLab (Scanalytics), for the entire three-dimensional image stacks at one time to ensure that the sample is analyzed evenly. The "black value" for the enhancement should be set to the approximate extranuclear noise level for the sample, which can vary markedly. The "white value" should be slightly above the intensity for the center of the brightest signals, namely nuclear sites of transcription.

3. Digital signal enhancement can be approached by two methods: direct analysis of three-dimensional images or splitting the image into two-dimensional slices, slice-by-slice processing, and, finally, collation of data into a three-dimensional representation. Both approaches require similar filtering algorithms, but currently available implementations generally require decomposition into slices as they can only process two-dimensional images. Either way, the basic method of signal enhancement is simple convolution filtering using a kernel that approximates the size of the target signal. This implies that the kernel should be adjusted to approximate the size of empirically observed sites of transcription, as determined by the magnification used in image acquisition. The designed kernel should include surround penalty to decrease the chances of false-positive detection of larger areas of fluorescence noise (intrinsic or extrinsic to the sample). The center, or positively scored part of the kernel, should be large enough to ignore specular noise and camera defects, which can appear as highly intense single pixels.

4. Positive detection of sites of transcription depends on empiric selection of a threshold. If contrast enhancement (step 1) was performed correctly, this should allow a single color level to be used to distinguish between background levels and transcription site color codes in all fluorescence channels. This procedure may be performed using a segmentation algorithm for each color combination used for detection in the image, such that one singles out sites of each identity one at a time. Finding this tedious, we prefer to detect all suprathreshold signals and determine the color-coded identity at once by coding a simple algorithm. This procedure involves scanning the image pixel by pixel for suprathreshold signal, recording each putative signal, and marking off contiguous regions surrounding the signal such that they are not scored more than once. Location of the signal and intensity in all color bands (both point wise and with surrounding area) are recorded. The intensity values are compared with the threshold and identity of the site is assigned.

5. For visualization purposes, a pseudo-colored, flattened two-dimensional rendering is prepared. For a background, we prefer to use the middle Z slice of the nuclear counterstain image. Transcription site locations and identities (now arbitrarily pseudo-colored

and depicted with an artificial marker in the image) are shown. We have added a number adjoining the site to mark the Z section from which the center of the site was detected. This is necessary, as filtered and threshold-corrected data contain more than three colors of images and cannot be depicted unambiguously in red–green–blue color systems.

6. Nuclear bounds are generated by binarization and simple flood fill of the nuclear counterstain image. Binarization requires a single threshold to be chosen to distinguish between intranuclear and extranuclear; this can usually be done given appropriate exposure of the counterstain. Flood-fill algorithms will only work with discretely separated nuclei and must be modified significantly to interpret overlapping signals. Nuclei for which the flood-fill-defined area extends to the edge of the image plane should be ruled out for further analysis as their contents are incompletely imaged.

7. Joining the results of steps 4 and 5 now yields data of single cell gene expression profiles—a set of transcription sites for each nucleus analyzed. Each transcription site detected in a field is placed onto the flood-fill map and assigned a nucleus. Sets of nuclear data are exported for further statistical study.

IV. PITFALLS

1. The overlap of fluorophore colors should be considered carefully when designing a bar-coding scheme. Consideration of the strength of fluorophores, the separation between emission spectra, excitation characteristics of the lamp, and the filter sets to be used to discern signals is critical.

2. When assembling the G-50 column and loading sample do not let the liquid level run below the matrix. This will create cracks and bubbles, potentially disrupting complete band separation and adding to contamination of product with free dye.

3. Poorly labeled probes (<40%) can fail to detect transcription sites. To increase labeling efficiency, multiple serial labelings and purifications can be performed.

4. Probe mixtures that have a suspiciously high level of labeling (>80%) may contain free dye, which will increase background. Multiple purifications by G-50 column can be used to remedy this.

5. When placing the coverslips down onto the Parafilm-coated plates, care should be exercised to avoid bubbles occurring, thereby preventing total contact of the probe with the coverslip. Also take care not to touch or move the coverslips excessively once they are placed onto the Parafilm, as this may contribute to cell detachment and damage.

6. Some cell types have high inherent autofluorescence obscuring nuclear signals. Careful processing with adequate extraction can remedy this at times. Additional processing steps may be necessary for recalcitrant noise problems.

7. Transcript color codes in which the colors are balanced inadequately may "decay" such that the observed signal is misinterpreted as a different color code containing a subset of the original code. This is especially problematic under conditions of low transcript abundance and with less intense fluorophores (such as FITC derivatives). Color codes must be tuned carefully before multiplex detection.

References

Elowitz, M. B., Levine, A. J., Siggia, E. D., and Swain, P. S. (2002). Stochastic gene expression in a single cell. *Science* **297**, 1183–1186.

Fauth, C., and Speicher, M. R. (2001). Classifying by colors: FISH-based genome analysis. *Cytogenet Cell Genet* **93**, 1–10.

Femino, A. M., Fay, F. S., Fogarty, K., and Singer, R. H. (1998). Visualization of single RNA transcripts in situ. *Science* **280**, 585–590.

Kislauskis, E. H., Li, Z., Singer, R. H., and Taneja, K. L. (1993). Isoform-specific 3'-untranslated sequences sort alpha-cardiac and beta-cytoplasmic actin messenger RNAs to different cytoplasmic compartments. *J. Cell Biol.* **123**, 165–172.

Langer, P. R., Waldrop, A. A., and Ward, D. C. (1981). Enzymatic synthesis of biotin-labeled polynucleotides: Novel nucleic acid affinity probes. *Proc. Natl. Acad. Sci. USA* **78**, 6633–6637.

Levsky, J. M., Shenoy, S. M., Pezo, R. C., and Singer, R. H. (2002). Single-cell gene expression profiling. *Science* **297**, 836–840.

Levsky, J. M., and Singer, R. H. (2003). Gene expression and the myth of the average cell. *Trends Cell Biol.* **13**, 4–6.

Nederlof, P. M., van der Flier, S., Wiegant, J., Raap, A. K., Tanke, H. J., Ploem, J. S., and van der Ploeg, M. (1990). Multiple fluorescence in situ hybridization. *Cytometry* **11**, 126–131.

Schrock, E., du Manoir, S., Veldman, T., Schoell, B., Wienberg, J., Ferguson-Smith, M. A., Ning, Y., Ledbetter, D. H., Bar-Am, I., Soenksen, D., Garini, Y., and Ried, T. (1996). Multicolor spectral karyotyping of human chromosomes. *Science* **273**, 494–497.

van der Ploeg, M. (2000). Cytochemical nucleic acid research during the twentieth century. *Eur. J. Histochem.* **44**, 7–42.

PART

D

PROTEINS

Protein Determination and Analysis

17

Protein Determination

Martin Guttenberger

I. INTRODUCTION

The protein content of tissues or samples can serve a number of purposes: It can be a research topic of its own (e.g., in nutritional studies; Hoffmann *et al.*, 2002), a loading control in gel electrophoresis (Ünlü *et al.*, 1997), or a reference quantity in biochemical (e.g., yields in protein purification) or physiological (e.g., specific activities of enzyme preparations; Guttenberger *et al.*, 1994) investigations. In addition, with the advent of proteomics, there is an increasing need for protein quantitation in complex sample buffers containing detergents and urea as potentially interfering compounds (Ünlü *et al.*, 1997). In any case, care should be taken to obtain correct results. This article focuses on three techniques and outlines the specific pros and cons.

II. MATERIALS AND INSTRUMENTATION

The following reagents are from the indicated suppliers. All other reagents are of analytical grade (Merck):

A. Lowry Assay

From Lowry *et al.* (1951): Folin–Ciocalteu phenol reagent (Merck, Cat. No. 1.09001). A detergent-compatible modification of the Lowry assay is available as a kit (Bio-Rad 500-0116).

B. Bradford Assay

From Bradford (1976): Coomassie brilliant blue G-250 (Serva Blue G, Serva, Cat. No. 35050). The reagent for this assay is available commercially from Bio-Rad (Cat. No. 500-0006).

C. Neuhoff Assay (Dot-Blot Assay)

From Guttenberger *et al.* (1991) and Neuhoff *et al.* (1979): Ammonium sulfate for biochemical purposes (Merck, Cat. No. 1.01211), benzoxanthene yellow (Hoechst 2495, Merck Biosciences, Cat. No. 382057, available upon request), cellulose acetate membranes (Sartorius, Cat. No. SM 11200), glycine, and SDS (Serva, Cat. Nos. 23390 and 20763, respectively). Commercially available ammonium sulfate frequently contains substantial amounts of undefined UV-absorbing and fluorescing substances. These lead to more or less yellowish solutions. Use only colourless solutions to avoid possible interference in fluorometry.

Solutions are prepared from bidistilled water. Bovine serum albumin (BSA, fraction V, Roche, Cat. No. 735086) is used as a standard protein. Ninety-six-well, flat-bottomed polystyrene microtiter plates (Greiner, Cat. No. 655101) are used for the photometric tests.

III. PROCEDURES

With respect to convenience and speed, microplate reader assays are described where appropriate. These assays can be read easily in conventional instruments by employing microcuvettes or by scaling up the volumes (fivefold).

The composition of the sample (extraction) buffer requires thought with respect to the avoidance of arti-factual alterations of the protein and to the compatibility with the intended experimental procedures. The former requires strict control of adverse enzyme activities (especially proteases and phenol oxidases) and, in the case of plant tissues, of interactions with secondary metabolites. A convenient, semiquantitative assay for proteolytic activities allowing for the screening of suitable inhibitors was described by Gallagher *et al.* (1986). There is some uncertainty as to which assay gives the most reliable results in combination with extracts from plant tissues rich in phenolic substances. The influence of such substances can never be predicted. It is therefore imperative to minimize interaction of these substances with protein in the course of sample preparation. For a more detailed discussion of this problem, see Guttenberger *et al.* (1994).

A frequent source of ambiguity is the use of the term "soluble protein." Soluble as opposed to membrane-bound proteins stay in solution during centrifugation for 1 h at 105,000 *g* (Hjelmeland and Chrambach, 1984).

All assays described in this article quantitate protein relative to a standard protein. The choice of the standard protein can markedly influence the result. This requires special attention for proteins with a high content of certain amino acids (e.g., aromatic, acidic, or basic amino acids). For most accurate results, choose a standard protein with similar amino acid composition or, if not available, compare different assays and standard proteins. Alternatively, employ a modified Lowry procedure that allows for absolute quantitation of protein (Raghupathi and Diwan, 1994).

The most efficient way to prepare an exact dilution series of the standard protein employs a handheld dispenser (e.g., Eppendorf multipette). Typically a six-point series is pipetted according to Table I. In any case, avoid a concentration gradient of the sample buffer. Usually samples and standards may be kept at −20°C for a couple of weeks. For longer storage intervals, keep at −80°C.

A. Lowry Assay

See Lowry *et al.* (1951).

Solutions

Note: For samples low in protein (0.02 mg ml^{-1} or less), prepare reagents A and B at double strength.

1. *Reagent A*: 2% (w/v) sodium carbonate (Na_2CO_3) in 0.10 *N* NaOH. To make 1 litre of reagent A (5000

TABLE I   Pipetting Scheme for Preparation of a Standard Dilution Series[a]

Concentration	Blank	0.2×	0.4×	0.6×	0.8×	1.0×
Water	5	4	3	2	1	0
Standard protein (2×)	0	1	2	3	4	5
Buffer (2×)	5	5	5	5	5	5

[a] To prepare 1 ml of each concentration, 1 volume corresponds to 0.1 ml.

determinations), dissolve 20 g Na_2CO_3 in 1 litre 0.10 *M* NaOH. Keep at room temperature in tightly closed screw-cap plastic bottles.

2. *Reagent B*: 0.5% $CuSO_4 \cdot 5H_2O$ in 1% sodium or potassium tartrate. To make 20 ml of reagent B, dissolve 0.1 g $CuSO_4 \cdot 5H_2O$ in 20 ml 1% tartrate (0.2 g sodium or potassium tartrate dissolved in 20 ml water). Keep at room temperature.

3. *Reagent C (alkaline copper solution)*: Mix 25 ml of reagent A and 0.5 ml of reagent B. Prepare fresh each day.

4. *Reagent D (Folin–Ciocalteu phenol reagent)*: Dilute with an equal volume of water just prior to use

Steps

1. Place 40 μl of sample (protein concentration 0.02–1 mg ml^{-1}) or blank into cavities of a microplate or into appropriate test tubes.

2. Add 200 μl of reagent C and mix. Allow to stand for at least 10 min.

3. Add 20 μl of reagent D and mix *immediately*. Allow to stand for 30 min or longer.

4. Read the samples in a microplate reader or any other photometer at 750 nm.

Modifications

1. The sample volume may be raised to 140 μl when samples are low in protein (0.02 mg ml^{-1} or less). In this case, employ double-strength reagent C.

2. If samples have been dissolved in 0.5 *M* NaOH (recommended for resolubilization of acid precipitates), omit NaOH from reagent A.

B. Bradford Assay

See Bradford (1976).

Solutions

1. *Protein reagent stock solution*: 0.05% (w/v) Coomassie brilliant blue G-250, 23.8% (v/v) ethanol, 42.5% (w/v) phosphoric acid. To make 200 ml of stock solution (5000 determinations), dissolve 0.1 g Serva

blue G in 50 ml 95% ethanol (denatured ethanol works as well), add 100 ml 85% phosphoric acid, and make up to 200 ml by adding water. The stock solution is available commercially (Bio-Rad). Keep at 4°C. The reagent contains phosphoric acid and ethanol or methanol. Handle with due care (especially when employing a dispenser)!

2. *Protein reagent*: Prepare from the stock solution by diluting in water (1 : 5). Filter immediately prior to use.

Steps

1. Place 4 µl of sample (protein concentration 0.1– 1 mg ml^{-1}) or blank into cavities of a microplate or into appropriate test tubes.
2. Add 200 µl of protein reagent and mix. Allow to stand for at least 5 min.
3. Read the samples within 1 h in a microplate reader or any other photometer at 595 nm.

Modifications

1. For improved linearity and sensitivity, compute the ratio of the absorbances, 590 nm over 450 nm (Zor and Selinger, 1996).
2. Microassay: For diluted samples (less than 0.1 mg ml^{-1}), proceed as follows: Employ 200 µl of sample and add 50 µl of protein reagent stock.

C. Dot-Blot Assay

See Guttenberger *et al.* (1991). Do not change the chemistry of the membranes. Nitrocellulose will dissolve in the staining solution; PVDF membranes develop a strong background.

Solutions

1. *Benzoxanthene stock*: To prepare the stock solution add 1 ml of water to 0.5 g of the fluorescent dye (as supplied, weighing not necessary); keep at −20°C. The toxicity of benzoxanthene is not thoroughly studied, it might be mutagenic!

2. *Destaining solution*: Methanol/acetic acid (90/10, v/v). To make 1 litre, mix 100 ml acetic acid and 900 ml methanol.

3. *Staining solution*: To obtain 100 ml, dilute 80 µl benzoxanthene stock in 100 ml destaining solution. Be sure to pour the destaining solution onto the stock solution to prevent the latter from clotting. Keep staining and destaining solutions in tightly closed screw-cap bottles at 4°C in the dark. They are stable for months and can be used repeatedly. Take due care in handling the highly volatile solutions containing methanol!

4. *SDS stock*: To make 30 ml of 10% (w/v) SDS stock solution, dissolve 3 g SDS in approximately 20 ml of water, stir, and make up to 30 ml (allow some time for settling of foam). Keep at room temperature; it is stable for at least 1 year.

5. *Elution buffer*: 0.25 *M* glycine–sulfuric acid buffer (pH 3.6) and 0.02% (w/v) SDS. To prepare 1 litre, dissolve 18.8 g glycine in approximately 900 ml water and add 15 ml of 0.5 *M* sulfuric acid. Slight deviations from pH 3.6 are tolerable. Add 2 ml SDS stock and make up to 1 litre. Keep at room temperature; it is stable for months.

The following solutions are not needed for the standard protocol.

6. *Washing solution A*: Saturated ammonium sulfate, adjust to pH 7.0 with Tris. To make 1 litre, stir ammonium sulfate in warm water (do not heat excessively). Let the solution cool to room temperature overnight and titrate to pH 7.0 with a concentrated (approximately 2 *M*) solution of Tris (usually approximately 1 ml is required). Keep at room temperature. As ammonium sulfate tends to produce lumps in the storage bottle it might be easier to weigh the entire bottle, add some water, remove the resulting slurry, and weigh the empty bottle again. To produce a saturated solution (53.1%, w/v), dissolve 760 g ammonium sulfate in 1 litre water.

7. *Washing solution B*: Methanol/acetic acid/water (50/10/40, v/v). To make 1 litre, mix 100 ml acetic acid and 500 ml methanol; make up to 1 litre. Keep at 4°C.

8. *Drying solution*: 1-Butanol/methanol/acetic acid (60/30/10, v/v). To make 0.1 litre, mix 10 ml acetic acid, 30 ml methanol, and 60 ml butanol. Keep at 4°C; use up to six times.

Steps

The dot-blot assay is a versatile tool; its different modifications enable one to cope with almost every potentially interfering substance. In the following description the steps for all modifications are included.

1. Preparation of filter sheets (cellulose acetate membrane). *Handle the sheets with clean forceps and scissors, do not touch!* Cut one corner to aid in orientation during processing of the sheet. Mark the points of sample application (see later). Mount the membrane in such a way that the points of sample application are not supported (otherwise a loss of protein due to absorption through the membrane may be encountered). There are two different ways to achieve these requirements.

a. For routine assays it is recommended to mount the sheets in a special dot-blot apparatus (Fig. 1). Mark dot areas by piercing the sheets through small holes in the upper part of the device.

A

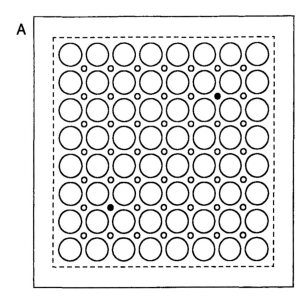

B

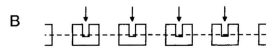

FIGURE 1 Dot-blot apparatus. (A) Top view. (B) Section along the diagonal. The apparatus has not been drawn to scale. Dashed lines indicate the position of the cellulose acetate membrane. Large circles correspond to the application points, small ones to the holes that are used for piercing the membrane (arrows in B), and solid small ones to the position of the pins that hold together the apparatus.

b. For occasional assays, mark the application points by impressing a grid (approximately 1-cm edge length) onto the filter surface (use a blunt blade and a clean support, preferably a glass plate covering a sheet of graph paper). Mount the sheets on a wire grating (preferably made from stainless steel, fixation by means of adhesive tape is recommended; cut off the taped areas prior to staining).

2. Apply samples (0.01–10 mg ml^{-1}) to the membrane sheets in aliquots of 2 µl (piston pipettes are highly recommended; well-rinsed capillary pipettes may be used instead). Leave to dry for a couple of minutes. Dilute samples may be assayed by applying samples repeatedly (let the sample dry prior to the next application).

3. Perform heat fixation. *Note: This step is imperative for samples containing SDS whereas it might prove deleterious to samples lacking SDS!* Bake the dot-blot membranes on a clean glass plate for 10 min at 120°C (oven or heating plate).

4. Remove interfering substances. *Note: This step is optional! Its use depends on the presence of potentially interfering substances (mainly carrier ampholytes, but also peptides and the buffer PIPES).* Remove interfering

substances prior to protein staining by vigorous shaking in washing solution A (3 × 5 min), followed by gentle agitation in washing solution B (3 × 2 min).

5. Stain and destain. Perform staining (10 min) and destaining (5, 5, and 15 min) in closed trays (polyethylene food boxes work very well) on a laboratory shaker at ambient temperature. For the last destaining bath, employ fresh destaining solution; discard the first destaining bath. The incubation times given here represent the minimal time intervals needed. As long as the vessels are closed tightly, each of these steps may be delayed according to convenience (in case of the last destaining bath, rinse in fresh destaining solution before proceeding).

6. Dry the stained membrane sheets. To facilitate cutting dot areas from the sheets, the following drying step is recommended. Shake the membranes in drying solution for exactly 2 min, mount them between two clamps[1] (Fig. 2), and leave them to dry in a fume hood. The dried sheets may be stored in the dark for later analysis.

7. Elute. Prior to elution, cut the dots from the membrane sheet. Perform elution (45 min in 2 ml of elution buffer) in glass scintillation vials on a laboratory shaker at ambient temperature (bright illumination should be avoided). Dried sheets have to be rewetted in destaining solution prior to immersion in elution buffer. It is recommended to dispense the destaining solution (25 µl) and the elution buffer with appropriate repetitive devices (e.g., Eppendorf multipette and Brand dispensette, respectively).

8. Take readings in a fluorometer (e.g., Luminescence Spectrometer LS 50B; Perkin-Elmer; Beaconsfield, UK) at 425 (excitation) and 475 (emission) nm.

Modification

Skip elution and take readings directly from the wet membrane sheets (step 6) with a video documentation system (e.g., DIANA, Raytest GmbH, Straubenhardt, Germany; Hoffmann *et al.*, 2002). Depending on the choice of filters, there might be considerable deviation from linearity.

IV. COMMENTS

With the exception of protein solutions, most stock solutions have a long shelf life. Discard any stock solu-

---

[1] Test for chemical resistance prior to first use: The edges of the clamp can be protected by a piece of silicon tubing cut open along one side.

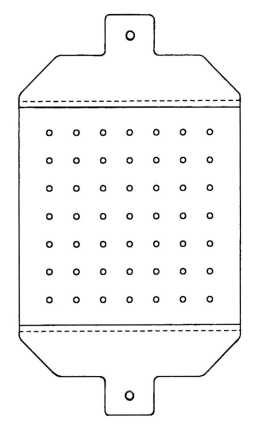

FIGURE 2 Membrane mounted for drying. Be sure to mount the drying membranes between two clamps of sufficient size to prevent distortion by uneven shrinkage. The weight of the lower clamp should keep the membrane spread evenly.

tion that changed its original appearance (e.g., got cloudy or discoloured).

Calculate standard curves according to the method of least squares. Appropriate algorithms are provided with scientific calculators and most spreadsheet programs for personal computers. It is better to compute standard curves employing single readings instead of means. Be aware of the basic assumptions made in regression analysis. For additional reading on the statistics of standard curves, compare Sokal and Rohlf (1995).

A. Lowry Assay

Pros: The Lowry assay exhibits the best accuracy with regard to absolute protein concentrations due to the chemical reaction with polypeptides. It is also useful for the quantitation of oligopeptides. This contrasts with the other two methods, which, as dye-binding assays, exhibit more variation depending on the different reactivity of the given proteins (standards as well as samples).

Cons: High sensitivity to potentially interfering substances; least shelf life of the reagents employed.

Recommendation: Employ where absolute protein contents are of interest.

B. Bradford Assay

Pros: The assay is widespread because of its ease of performance (only one stable reagent is needed, low sensitivity to potentially interfering substances, unsurpassed rapidity), its sensitivity, and its low cost.

Cons: High blank values, requires dual-wavelength readings for linearity, and possibly rather high deviations from absolute protein values (depending on the choice of standard protein).

Recommendation: Employ where relative protein contents are sufficient (in most cases such as electrophoresis) and where the assay shows no interference by sample constituents (compare Bradford, 1976).

C. Dot-Blot Assay

Pros: The dot-blot assay combines high sensitivity, an extended range of linearity (20 ng to 20 µg), and high tolerance to potentially interfering substances. The sample is not used up during assay. Hence, it may be reprobed[2] (Fig. 3) for immunological tests or detection of glycoproteins (Neuhoff *et al.*, 1981).

Cons: More demanding and time-consuming than the other assays and rather expensive (chemicals and instrumentation).

Recommendation: Employ where (1) the other assays show interference, especially with complex sample buffers used in one-dimensional[3] and two-dimensional[4] electrophoresis; (2) the amount of sample is limited and/or reprobing of the dotted samples is desirable; or (3) the mere detection of protein in aliquots, e.g., from column chromatography, is needed (spot 0.2–2 µl onto membrane, process according to

---

[2] Sheets containing single dot areas can be marked conveniently by cutting the edges (Fig. 3, Neuhoff *et al.*, 1979).

[3] Sample buffer according to Laemmli (1970): 62.5 mM Tris–HCl (pH 6.8), 2% (w/v) SDS, 10% (v/v) glycerol, 5% (v/v) 2-mercaptoethanol, and 0.001% (w/v) bromphenol blue. Range of the assay: 0.04 to 10 mg ml^{-1}, i.e., 80 ng to 20 µg in the test.

[4] Sample (lysis) buffer according to O'Farrell (1975): 9.5 M urea, 2% (w/v) Nonidet P-40, 5% (v/v) 2-mercaptoethanol, and 2% (w/v) carrier ampholytes. Standards are prepared by a stepwise dilution of the BSA stock solution in a modified sample buffer lacking carrier ampholytes. These are added from a doubly concentrated stock solution (4%, w/v) in sample buffer. Range of the assay: 0.02 to 8 mg ml^{-1}, i.e., 40 ng to 16 µg in the test.

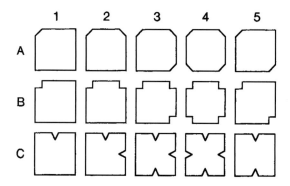

FIGURE 3 Useful incision patterns employed for marking membrane sheets prior to reprobing. Additional patterns may be generated by combination.

standard protocol, prevent evaporation by covering the destained membrane with a thin glass plate, view under UV light).

V. PITFALLS

1. Solutions containing protein exhibit an altered surface tension. Avoid foaming and pipette slowly and steadily.

2. Extraction or precipitation steps to eliminate interfering substances should be carefully controlled for complete recovery of protein (Lowry *et al.*, 1951). The more demanding dot-blot assay frequently is a good alternative because of a considerable gain of convenience and accuracy with respect to a simplified sample preparation.

3. Omission of known interfering buffer components from just those samples that are intended for protein determination is strongly discouraged as the solubility of proteins might be influenced (carrier ampholytes, e.g., enhance solubilization of membrane proteins in two-dimensional electrophoresis sample buffer; for references, see Guttenberger *et al.*, 1991).

4. In the case of photometers/fluorometers operating with filters (usually microplate readers), the correct wavelength may not be available. Instead, a similar wavelength may be employed [Lowry assay: 530–800 nm, Bradford assay: 540–620 nm, dot-blot assay: 366–450 nm (excitation), 450–520 nm (emission)]. In the case of fluorometry, allow for a sufficient wavelength interval between excitation and emission (consult the operating instructions of your instrument). Be aware that considerable deviations from the standard wavelengths will be at the expense of linearity and sensitivity.

5. In microplates it is important to achieve uniform menisci: Prick air bubbles with a thin wire and mix the plates on a gyratory shaker.

6. Analysis of dilute samples by application of larger sample volumes also increases the amount of potentially interfering substances. Include appropriate controls.

A. Lowry Assay

1. Many reagents used commonly in protein extraction interfere with this assay. The main groups of interfering substances are reductants (e.g., sulfhydryl compounds such as mercaptoethanol, reducing sugars such as glucose), chelating agents (e.g., EDTA), amine derivatives (many common buffering substances such as Tris), and detergents (e.g., Triton, SDS). A detailed list of interfering substances, along with remedies and tolerable limits, is provided by Petersen (1979).

2. Reagent D is not stable at a basic pH. Immediate mixing after the addition of reagent D is imperative. In microplates the use of a small plastic spatula is convenient for this purpose (change or rinse between samples).

3. The colour reaction takes about 80 min to come to completion. Prior to this, reading of samples over an extended period of time will give rise to experimental error (more than 20%; Kirazov *et al.*, 1993). Keep the reading interval to a minimum. Alternatively, both incubation steps can be cut to 3 min by raising the incubation temperature to 37°C (Shakir *et al.*, 1994). As the time to reach thermal equilibration will depend on the experimental setup, a test run in comparison to the original method is recommended.

B. Bradford Assay

1. The commonly used standard protein BSA is highly reactive in this dye-binding assay. As a consequence the protein content of the samples is underestimated. This systematic error does not matter in comparative analyses but brings about wrong absolute values. Bovine γ-globulin is a preferable standard.

2. The standard curves are not strictly linear in the original version of the assay. If the necessary equipment for the recommended dual-wavelength ratio is not available, do not extend the range of standard concentrations beyond one order of magnitude or do not calculate standard curves by means of linear regression.

3. Samples containing detergents (1% will interfere) must be diluted (if possible) or precipitated (compare Section V.2) prior to analysis.

4. The protein–dye complex is insoluble and will precipitate with time (Marshall and Williams, 1992). For highest accuracy, take readings within an interval between 5 and 20 min after addition of the reagent. With crude extracts (e.g., from mycelia of certain fungi), this interval may be considerably shorter—too short to take meaningful readings. In this case, alter the way of sample preparation or use another assay.

5. Plastic and glassware (especially quartz glass) tend to bind dye. Remove the resulting blue colour by one of the following procedures: (1) Rinse with glassware detergent (avoid strongly alkaline detergents with cuvettes; rinse thoroughly to remove detergent again), (2) rinse with ethanol or methanol, or (3) soak in 0.1 M HCl (takes several hours).

C. Dot-Blot Assay

1. Generally, it is imperative to prevent the membrane sheets from drying during one of the transfer steps (residual acetic acid will destroy the filter matrix).

2. In case of highly variable results, inspection of the stained filters (last destaining bath or dried) under UV illumination may be helpful: Background staining resulting from improper handling of the membranes will be visible (do not use UV-irradiated membranes for quantitative analyses).

3. After the washing procedure, thorough rinsing in washing solution B is imperative. Ammonium sulfate accumulating in the staining solution will interfere with the assay.

4. Although the dot-blot assay is extremely insensitive to potentially interfering substances, it is advisable to include appropriate controls (at least blank buffer and buffer plus standard).

5. In the case of buffers containing detergent plus carrier ampholytes, the storage conditions and the number of freeze–thaw cycles may prove important. Use fresh solutions or run appropriate controls.

6. If membrane sheets turn transparent upon drying, they have not been equilibrated properly in the drying solution (keep in time: 2 min) or the drying solution has been diluted by accumulation of destaining solution (do not reuse the drying solution too often).

References

Bradford, M. M. (1976). A rapid and sensitive method for the quantitation of microgram quantities of protein utilizing the principle of protein-dye binding. *Anal. Biochem.* **72**, 248–254.

Gallagher, S. R., Carroll, E. J., Jr., and Leonard, R. T. (1986). A sensitive diffusion plate assay for screening inhibitors of protease activity in plant cell fractions. *Plant Physiol.* **81**, 869–874.

Guttenberger, M., Neuhoff, V., and Hampp, R. (1991). A dot-blot assay for quantitation of nanogram amounts of protein in the presence of carrier ampholytes and other possibly interfering substances. *Anal. Biochem.* **196**, 99–103.

Guttenberger, M., Schaeffer, C., and Hampp, R. (1994). Kinetic and electrophoretic characterization of NADP dependent dehydrogenases from root tissues of Norway spruce (*Picea abies* [L.] Karst.) employing a rapid one-step extraction procedure. *Trees* **8**, 191–197.

Hjelmeland, L. M., and Chrambach, A. (1984). Solubilization of functional membrane proteins. *Methods Enzymol.* **104**, 305–318.

Hoffmann, E. M., Muetzel, S., and Becker, K. (2002). A modified dot-blot method of protein determination applied in the tannin-protein precipitation assay to facilitate the evaluation of tannin activity in animal feeds. *Br. J. Nutr.* **87**, 421–426.

Kirazov, L. P., Venkov, L. G., and Kirazov, E. P. (1993). Comparison of the Lowry and the Bradford protein assays as applied for protein estimation of membrane-containing fractions. *Anal. Biochem.* **208**, 44–48.

Laemmli, U. K. (1970). Cleavage of structural proteins during the assembly of the head of bacteriophage T4. *Nature* **227**, 680–685.

Lowry, O. H., Rosebrough, N. J., Farr, A. L., and Randall, R. J. (1951). Protein measurement with the Folin phenol reagent. *J. Biol. Chem.* **193**, 265–275.

Marshall, T., and Williams, K. M. (1992). Coomassie blue protein dye-binding assays measure formation of an insoluble protein-dye complex. *Anal. Biochem.* **204**, 107–109.

Neuhoff, V., Ewers, E., and Huether, G. (1981). Spot analysis for glycoprotein determination in the nanogram range. *Hoppe-Seyler's Z. Physiol. Chem.* **362**, 1427–1434.

Neuhoff, V., Philipp, K., Zimmer, H.-G., and Mesecke, S. (1979). A simple, versatile, sensitive and volume-independent method for quantitative protein determination which is independent of other external influences. *Hoppe-Seyler's Z. Physiol. Chem.* **360**, 1657–1670.

O'Farrell, P. H. (1975). High resolution two-dimensional electrophoresis of proteins. *J. Biol. Chem.* **250**, 4007–4021.

Peterson, G. L. (1979). Review of the Folin phenol protein quantitation method of Lowry, Rosebrough, Farr and Randall. *Anal. Biochem.* **100**, 201–220.

Raghupathi, R. N., and Diwan, A. M. (1994). A protocol for protein estimation that gives a nearly constant color yield with simple proteins and nullifies the effects of four known interfering agents: Microestimation of peptide groups. *Anal. Biochem.* **219**, 356–359.

Shakir, F. K., Audilet, D., Drake, A. J., III, and Shakir, K. M. M. (1994). A rapid protein determination by modification ot the Lowry procedure. *Anal. Biochem.* **216**, 232–233.

Sokal, R. R., and Rohlf, F. J. (1995). "Biometry." Freeman, New York.

Ünlü, M., Morgan, M. E., and Minden, J. S. (1997). Difference gel electrophoresis: A single gel method for detecting changes in protein extracts. *Electrophoresis* **18**, 2071–2077.

Zor, T., and Selinger, Z. (1996). Linearization of the Bradford protein assay increases its sensitivity: Theoretical and experimental studies. *Anal. Biochem.* **236**, 302–308.

18

Phosphopeptide Mapping: A Basic Protocol

Jill Meisenhelder and Peter van der Geer

I. INTRODUCTION

Peptide mapping is a technique in which a radioactively labeled protein is digested with a sequence specific protease. The resulting peptides are separated in two dimensions on a thin-layer cellulose (TLC) plate by electrophoresis and chromatography. The peptides are visualized by autoradiography, giving rise to a peptide map.

Peptide maps of ^{35}S-labeled proteins are used most often to find out whether two polypeptides are related. Peptide maps of ^{32}P-labeled proteins are used to obtain information about the phosphorylation of the protein under investigation. Proteins can be labeled *in vivo* by incubating cells in the presence of [^{32}P]orthophosphate or by incubating them *in vitro* with an appropriate protein kinase in the presence of [γ-^{32}P]ATP. Proteins are usually separated from other contaminating proteins by SDS–PAGE and then subjected to phosphopeptide mapping or phosphoamino acid analysis.

II. MATERIALS AND INSTRUMENTATION

HTLE 7000 electrophoresis system (CBS Scientific, Del Mar, CA)

pH 1.9 electrophoresis buffer: 50 ml formic acid (88%, w/v), 156 ml glacial acetic acid, and 1794 ml deionized water

pH 3.5 electrophoresis buffer: 100 ml glacial acetic acid, 10 ml pyridine, and 1890 ml deionized water

pH 4.72 electrophoresis buffer: 100 ml n-butanol, 50 ml pyridine, 50 ml glacial acetic acid, and 1800 ml deionized water

pH 8.9 buffer: 20 g $(NH_4)_2CO_3$ and 2000 ml deionized water

Regular chromatography buffer: 785 ml n-butanol, 607 ml pyridine, 122 ml glacial acetic acid, and 486 ml deionized water

Phospho-chromatography buffer: 750 ml n-butanol, 500 ml pyridine, 150 ml glacial acetic acid, and 600 ml deionized water

Isobutyric acid buffer: 1250 ml isobutyric acid, 38 ml n-butanol, 96 ml pyridine, 58 ml glacial acetic acid, and 558 ml deionized water

Phosphoamino acid stocks: 1 mg/ml each in deionized water is stable for years at −20°C.

50 mM NH_4HCO_3 pH 7.3–7.6. Make up fresh; lower the pH by bubbling CO_2 through it if necessary. The pH of this buffer will drift overnight toward pH 8.0.

1.5-ml microfuge tubes with plastic pestles can be obtained from Kimble Kontes (Vineland, NJ). These pestles fit nicely into Sarstedt screw-cap microcentrifuge tubes.

RNase A is dissolved in deionized water at 1 mg/ml, boiled for 5 min, and stored at −20°C.

TPCK-treated trypsin (Worthington Lakewood, NJ) can be dissolved at 1 mg/ml in 1 mM HCl and is stable at −70°C for years.

III. PROCEDURES

A. Phosphopeptide Mapping

1. Separate the ^{32}P-labeled protein of interest from other contaminants by resolving the sample by

SDS–PAGE. Dry the gel onto Whatman 3MM paper, mark the paper backing around the gel with radioactive or fluorescent ink, and expose the gel to X-ray film. Line the gel up with the film using the markings on the paper backing and autorad, localize the protein of interest, and cut the protein band out of the gel with a clean, single edge razor or a surgical blade. Remove the paper backing from the gel slices by scraping gently with a razor blade.

2. Extract the protein from the gel by grinding the gel into small fragments. Place the gel slice(s) in a 1.7-ml screw-cap tube and hydrate briefly in 500 μl 50 mM NH$_4$HCO$_3$, pH 7.3–7.6. Grind the gel to small pieces using a fitted plastic, disposable pestle. Add 500 μl more NH$_4$HCO$_3$, 10 μl 10% SDS, and 10 μl βME, vortex, boil for 5 min, and extract for at least 4 h on an agitator at room temperature.

3. Spin down the gel bits by centrifugation in a microfuge for 5 min at 2000 rpm, transfer the supernatant to a new microfuge tube, and store at 4°C. This supernatant represents volume X(μl). Add (1300-X) μl more NH$_4$HCO$_3$ to the gel bits, vortex, and extract again for at least 4 h on an agitator at room temperature. Spin down the gel bits and combine this supernatant with the first extract.

4. Clear the (combined) extract by centrifugation. Spin 15 min at 15,000 rpm in a microfuge at room temperature. Transfer the supernatant to a new tube, leaving the final 20 μl behind to avoid transfer of particulate material. Repeat this step one or two more times. It is important that the final extract is free of any particulate materials (gel and paper bits).

5. Concentrate the protein by TCA precipitation. Add 20 μl RNase A (1 mg/ml) to the protein extract, mix, and incubate 20 min on ice. Add 250 μl ice-cold 100% TCA, mix, and incubate 1 h on ice. Spin 15 min at 15,000 rpm in a microfuge at 4°C and remove the supernatant. Add 0.5 ml 100% cold ethanol to the pellet, invert the tube, and spin 10 min at 15,000 rpm in a microfuge and remove the supernatant. Spin again briefly, remove residual ethanol, and briefly air dry the pellet.

6. To avoid the formation of oxidation-state isomers, oxidize the protein to completion by incubation in performic acid. Performic acid is formed by incubating 9 parts 98% formic acid with 1 part H$_2$O$_2$ for 30–60 min at room temperature and then cool on ice. Resuspend the TCA pellet in 50 μl cold performic acid, incubate for 1 h on ice, add 400 μl deionized water, freeze, and lyophilize.

7. In order to analyze the different phosphorylation sites, digest the protein with a sequence-specific protease. We routinely use trypsin because it works well on denatured protein and its specificity is well charac-

terized. Resuspend the oxidized protein in 50 μl 50 mM NH$_4$HCO$_3$, pH 8.0. Add 10 μl 1 mg/ml TPCK-treated trypsin, vortex, and incubate for 4–16 h at 37°C. Add a second aliquot of trypsin, vortex, and incubate again for 4–16 h at 37°C.

8. Now subject the sample to several rounds of lyophilization to remove the ammonium bicarbonate. Add 400 μl deionized water to the sample, mix, and lyophilize. Repeat this procedure two to three times and then spin the final rinse for 5 min at 15,000 rpm in a microfuge and transfer the supernatant to a new tube and lyophilize.

The peptides are now ready for application onto a 20 × 20 TLC plate. Electrophoresis will be used for separation in the first dimension. Three different buffer systems are commonly used for electrophoresis (pH 1.9, pH 4.72, and pH 8.9). All three buffer systems should be tried to determine which will best separate the tryptic phosphopeptides of a particular protein.

9. Dissolve the final pellet in 5–10 μl of the electrophoresis buffer to be used; use deionized water instead of pH 8.9 buffer. Spot the peptide mix using a gel-loading tip fitted to an adjustable micropipette. Keep the sample on as small an area on the plate as possible by spotting the samples 0.3–0.5 μl at a time, drying the sample between spottings. Spot the sample 3 cm from the bottom of the plate and 5 cm from the left side for electrophoresis at pH 1.9 or 4.72 or 10 cm from the left hand side for electrophoresis at pH 8.9 (see Fig. 1a). Mark origins on the plate with a blunt, soft pencil. We like to spot 1 μl of marker dye mixture 2 cm from the top of the plate above the sample origin (dye origin first dimension, Fig. 1a). The mobilities of marker dyes can be used as standards when comparing different maps.

In our laboratories we use the HTLE 7000 electrophoresis system. This system should be connected to a power supply, cooling water, and an air line with a pressure regulator and should be set up according to the manufacturer's directions (Fig. 2).

10. Wet the TLC plate with electrophoresis buffer immediately before placing it onto the electrophoresis apparatus as described in Fig. 1. The plate should be damp with no puddles present. Shut off the air on the HTLE 7000, remove the securing pins, the neoprene pad, the Teflon insulator, and the upper polyethylene protector sheet and fold the electrophoresis wicks back over the buffer tanks. Wipe excess buffer from the upper and lower polyethylene protector sheets and place the TLC plate onto the lower polyethylene sheet. Fold the electrophoresis wicks so that they overlap ~1 cm onto the TLC plate. Reassemble the apparatus, insert the securing pins, and adjust the air pressure to

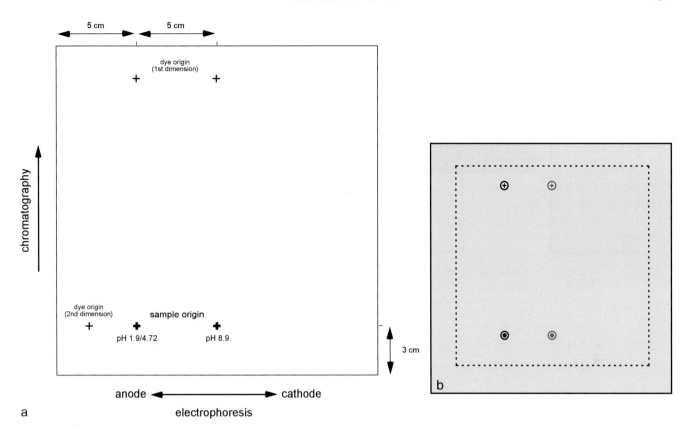

FIGURE 1 Applying the peptide mixture onto a TLC plate. (a) Phosphopeptide mixtures are usually separated by electrophoresis in the horizontal dimension and chromatography in the vertical dimension. The mixture is spotted 3 cm from the bottom of the plate and 5 cm from the left side for electrophoresis at pH 1.9 or pH 4.72 or in the center of the plate for electrophoresis at pH 8.9 (with the anode on the left and the cathode on the right after placing the plate on the HTLE 7000). We usually mark the sample and dye origin on the TLC plate with a soft blunt pencil. (b) Plates are wetted with electrophoresis buffer using a blotter composed of two layers of Whatman 3MM paper. Separate blotters are used for different electrophoresis buffers and each blotter can be used many times. A blotter contains two 1.5-cm-diameter holes that correspond to the sample and dye origin. The blotter is soaked in electrophoresis buffer and excess buffer is removed by blotting with a piece of 3MM paper. The blotter is then placed on the TLC plate so that the sample and dye origins are in the center of the two holes. The blotter is pushed onto the plate with the palm of the hand so that buffer is transferred from the blotter onto the plate. The edge of the holes is pushed onto the plate so that buffer moves from the blotter toward the center of the holes. This will concentrate the sample and dye on their origins. When the plate is completely wet, the blotter is removed and the plate is placed on the HTLE 7000. The electrophoresis apparatus is reassembled (Fig. 2) and electrophoresis is started.

10 lbs/in.^{2} Turn on the cooling water and switch on the high voltage power supply. We normally run maps for 20–30 min at 1000 V. When the run is finished, disassemble the apparatus and air dry the plate.

Running times and origins can be adjusted for individual proteins. We recommend increasing the running time rather than the voltage to get better separation of the peptides on the map.

11. Separate peptides in the vertical dimension by ascending chromatography. A plastic tank (57 × 23 × 57 cm) is available from CBS Scientific (Del Mar, CA 92014). These tanks can hold up to eight TLC plates.

Tanks need to be equilibrated with chromatography buffer for at least 24 h before the run. Three different types of chromatography buffer are commonly used in our laboratories (regular chromatography, phospho-chromatography, and isobutyric acid buffers). We recommend trying the phospho-chromatography buffer first.

We like to spot 1.0 μl of marker dye in the right or left margin of the plate at the same level as the sample (dye origin second dimension, Fig. 1a). Place all plates in the tank at the same time, leaning them at the same angle; replace the lid and do not open the tank

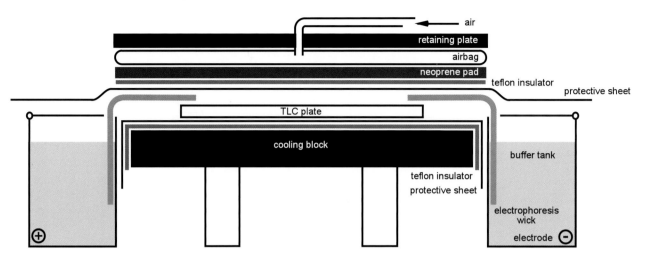

FIGURE 2   Setting up the HTLE 7000. Fill the buffer tanks with 600 ml freshly prepared electrophoresis buffer. Cover the cooling plate and fitted Teflon insulator with a polyethylene protector sheet that can be tucked between the cooling plate and the buffer tanks. Insert the wet electrophoresis wicks (14 × 20-cm double-layer Whatman 3 MM paper) into the buffer tanks and fold them over the cooling plate. Cover the cooling plate and electrophoresis wicks with the top polyethylene protector sheet, which should extend over the buffer tanks. Add the Teflon protector sheet, the neoprene pad, and close the cover. Insert the two pins to secure the cover before turning up the air pressure to 10 lbs/in.[2] Immediately before starting electrophoresis, wet a plate as described in Fig. 1. Then turn off the air pressure, take out the pins, open the apparatus, and remove the neoprene pad, the Teflon protector, and the polyethylene sheet. Fold the electrophoresis wicks backward over the buffer tanks. Dry the bottom and top protector sheets with a Kimwipe, place the TLC plate on the apparatus, and fold the electrophoresis wicks over the cooling plate so that they overlap ~1 cm with the TLC plate. Place the polyethylene protector sheet on top of the TLC plate and reassemble the apparatus. Adjust the air pressure to 10 lbs/in.[2], turn on the cooling water, and start electrophoresis.

again while chromatography is in progress. The front advances more slowly as it climbs higher on the plate. Let chromatography proceed until the buffer front reaches 1–2 cm from the top of the plate. This can take from 8 to 16 h depending on the lot of cellulose plates, the ambient temperature, and the age of the buffer. At this point remove all plates from the tank and let them air dry. Mark the plates with fluorescent or radioactive ink and expose to X-ray film in the presence of an intensifier screen or to a PhosphorImager screen.

B. Phosphoamino Acid Analysis

1. To analyze the phosphoamino acid content of a phosphoprotein, the protein can be isolated exactly as described for phosphopeptide mapping (Section III,A, steps 1–6). Transfer a fraction of the sample resuspended in performic acid (step 6) to a screw-cap microfuge tube and lyophilize. If the entire sample is to be used for phosphoamino acid analysis, use the ethanol-washed TCA precipitate (step 5).

2. Hydrolyze the protein to liberate the individual phosphoamino acids. Dissolve the sample in 50 μl 6 M HCl and incubate for 1 h at 110°C.

3. Remove the hydrochloric acid by lyophilization.

4. Dissolve the amino acid mixture in 5–10 μl pH 1.9 buffer containing unlabeled phosphoamino acids as standards (70 μg/ml of each) by vortexing vigorously. Spin for 5 min at maximal speed in a microfuge to pellet particulate matter.

5. Spot the samples on a TLC plate as described for phosphopeptide mapping (step 9). Four samples can be analyzed simultaneously on a single TLC plate (Fig. 3a).

6. After all samples are spotted, wet the plate with pH 1.9 buffer using a blotter containing five holes that correspond to the four sample origins and a dye origin, as described for phosphopeptide mapping (step 10, Fig. 3b). Place the wetted plate immediately on the HTLE 7000 containing pH 1.9 buffer in the tanks and reassemble the apparatus. Resolve the samples in the first dimension by electrophoresis for 20 min at 1.5 kV (Fig. 3a, lower left) and air dry the plates.

7. Before running electrophoresis in the second dimension, wet the plate with pH 3.5 buffer containing ~0.1 mM EDTA using three rectangular blotters 10, 6, and 4 cm wide. Place blotters on the plate so that a ~1.5-cm strip of the plate containing the sample separated in the first dimension remains uncovered (Fig.

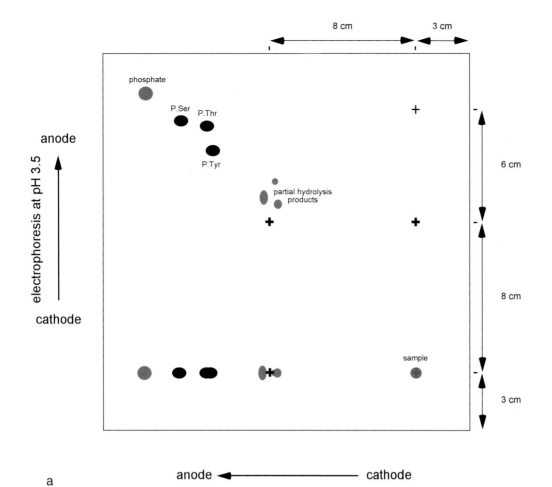

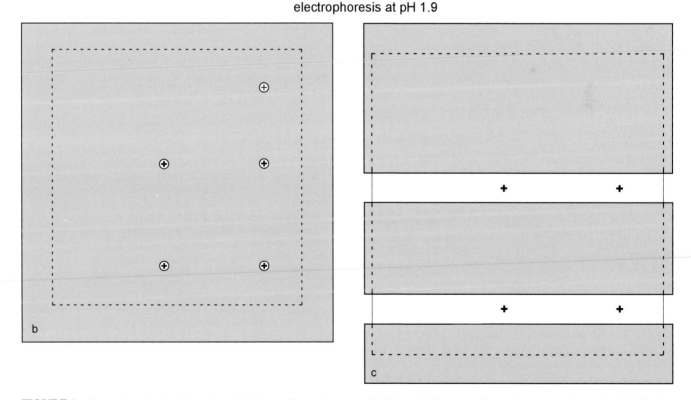

FIGURE 3 **Separation of phosphoamino acids in two dimensions on a TLC plate.** (a) Four samples can be separated on a single TLC plate at the same time. The four sample origins and an origin for marker dye are marked on the TLC plate with a blunt, soft pencil. Samples are first resolved in the horizontal dimension by electrophoresis at pH 1.9 (lower left quadrant). Subsequently the samples are resolved in the vertical dimension by electrophoresis at pH 3.5 (upper left quadrant). (b) After spotting the samples and marker dye, the plate is wetted with a blotter containing five holes, as shown. This will concentrate the samples and dye at their origins. (c) After electrophoresis in the first dimension, the plate is air dried. The plate is wetted with pH 3.5 buffer using three rectangular strips of Whatman 3M paper as blotters. Buffer will move from each strip as a horizontal line converging with the buffer from the adjacent strips, thereby concentrating the sample on this midpoint. When the plate is completely wet it is turned 90° counterclockwise and placed on the HTLE 7000 apparatus.

3c). By pressing the blotters onto the plate the buffer will migrate from the blotters and concentrate the samples on a line midway between the blotter edges (Fig. 3c). Turn the plate 90° counterclockwise to place it on the HTLE 7000. Reassemble the apparatus, now containing pH 3.5 buffer in the tanks, and carry out electrophoresis for 16 min at 1.3 kV (Fig. 3a, upper left). After separation in the second dimension, disassemble the apparatus and dry the plate.

8. To visualize the standard phosphoamino acids, spray the plate with 0.25% ninhydrin in acetone. Develop the ninhydrin stain by placing the plate for 5–15 min at 65°C.

9. Mark the plate around the edges with radioactive or fluorescent ink and expose it to X-ray film in the presence of an intensifier screen. After developing, align the X-ray film with the TLC plate using the radioactive or fluorescent markings. Phosphoamino acids are identified by comparing the mobility of the radioactive phosphoamino acids seen on the film with the mobility of the ninhydrin-stained phosphoamino acid standards.

IV. COMMENTS

Peptide maps provide information about the phosphorylation status of a protein. Only radioactively labeled peptides are visualized by autoradiography. In principle, every spot on the map represents a phosphorylation site; there are, however, exceptions (Boyle *et al.*, 1991; Meisenhelder *et al.*, 1999). Occasionally, two phosphorylation sites are close together and consequently are contained within a single peptide. If this is the case, a single spot may represent two phosphorylation sites. It is important to realize that single and doubly phosphorylated forms of a particular peptide will be resolved from each other because adding a phosphate group will add a negative charge to the peptide. It is also possible that as a consequence of partial digestion, multiple peptides/spots are generated that represent a single phosphorylation site.

We have described the basic protocol in which proteins are isolated from an SDS–PAGE gel before digestion with proteases. An alternative protocol exists in which proteins are transferred from the gel to a membrane. Membrane-bound proteins are then digested or hydrolyzed (Luo *et al.*, 1990, 1991). Peptides and phosphoamino acids detach from the membrane and are analyzed as decribed here. A possible problem with this protocol is that not all peptides detach from the membrane, resulting in loss of information.

If the ultimate goal is to identify phosphorylation sites, then spots/peptides on the phosphopeptide map have to be matched to the amino acid sequence of the protein of interest. This is done best by making a list of possible peptides that can be generated by the protease used to cleave the protein. Now peptides can be eliminated from this list by gathering more information about the phosphopeptides of interest (Gould and Hunter, 1988; van der Geer and Hunter, 1990). The peptides can be isolated from the TLC plate and used to answer additional questions. What is the phosphoamino acid content of the peptide? Does it contain cleavage sites for other proteases? Which residue within the peptide is phosphorylated? Double-labeling experiments can be used to find out whether the peptide contains cysteine or methionine. Information obtained in these experiments can be used to eliminate candidate peptides from the list of possibilities until only one or two are left. The protein can also be analyzed using the peptide mapping program available at www.genestream.org; in this way some candidate peptides can be eliminated because their predicted mobility is substantially different from actual mobility on the map of the peptide in question. As verification of the identity of a phosphorylation site, the mutant protein in which the phosphorylation site has been changed to a nonphosphorylatable amino acid can be subjected to phosphopeptide mapping to confirm that the phosphorylation site of interest/spot(s) on the map has been eliminated.

REFERENCES

Boyle, W. J., van der Geer, P., and Hunter, T. (1991). Phosphopeptide mapping and phosphoamino acid analysis by two-dimensional separation on thin-layer cellulose plates. *Methods Enzymol.* **201**, 110–148.

Gould, K. L., and Hunter, T. (1988). Platelet-derived growth factor induces multisite phosphorylation of pp60c-src and increases its protein-tyrosine kinase activity. *Mol. Cell. Biol.* **8**, 3345–3356.

Luo, K., Hurley, T. R., and Sefton, B. M. (1990). Transfer of proteins to membranes facilitates both cyanogen bromide cleavage and two-dimensional proteolytic mapping. *Oncogene* **5**, 921–923.

Luo, K., Hurley, T. R., and Sefton, B. M. (1991). Cyanogen bromide cleavage and proteolytic peptide mapping of proteins immobilized to membranes. *Methods Enzymol.* 149–152.

Meisenhelder, J., Hunter, T., and van der Geer, P. (1999). Phosphopeptide mapping and identification of phosphorylation sites. *In* "Current Protocols in Protein Science", pp. 13.19.11–13.19.28. Wiley, New York.

van der Geer, P., and Hunter, T. (1990). Identification of tyrosine 706 in the kinase insert as the major colony-stimulated factor 1 (CSF-1)-stimulated autophosphorylation site in the CSF-1 receptor in a murine macrophage cell line. *Mol. Cell. Biol.* **10**, 2991–3002.

Coupling of Fluorescent Tags to Proteins

Markus Grubinger and Mario Gimona

I. INTRODUCTION

Fluorescence microscopy in conjunction with fluorescent probes offers a powerful approach for gathering information about the localization of molecules in fixed and living cells. Fluorescent probes can be conjugated to proteins, nucleic acids, lipid analogs, antibodies, and ions. Microinjection of these conjugates, together with low light video or digital microscopy, can be used to study the organization and dynamics of specific molecules in live cells. The most widespread application for fluorescently labeled probes is that of secondary antibodies for immunofluorescence. Hence, most protocols for protein–dye conjugation are optimized for this special class of proteins. However, a large number of cytoskeletal proteins have been labeled successfully, particularly using thiol-reactive iodoacetamides and maleimides. Among them are actin, ADF, α-actinin, caldemon, calponin, desmin, kinesin, myosin, spectrin, tropomyosin, troponin, and tubulin. Likewise, the technique is useful for labeling extracellular matrix components such as fibronectin and collagen.

As a general rule, the attachment of a fluorescent label to a protein requires, in most cases, a compromise between optimal labeling efficiency and maximal preservation of protein function. This is explained easily by the pH optima at which some of the labeling reactions work best and the isoelectric points and solubility requirements of the protein of choice, which may often interfere. Because protein function should be regarded as the principal goal, lower labeling efficiencies should be accepted.

Overviews of the most commonly used commercial fluorescent dyes and their characteristics are given by Wang *et al.* (1982), DeBiasio *et al.* (1987), and Mason (1993). The *"Molecular Probes Handbook of Fluorescent Probes and Research Products"* by Haugland (2002) is also very informative.

Here is a short list of commonly used fluorescent dyes to label proteins through their thiol or amine groups:

1. **Thiol-reactive probes**: *Iodoacetamides*, e.g., iodoacetamide tetramethyl rhodamine (IAMTR), and *maleimides*, e.g., tetramethylrhodamine-5-maleimide.
2. **Amine-reactive probes**: *Isothiocyanates*, e.g., fluorescein isothiocyanate (FITC), *sulfonyl chlorides*, e.g., lissamine-rhodamine, and s*uccinimidyl esters*, e.g., 5′-carboxytetramethylrhodamine single isomer (5′-TAMRA-SE), 5′-carboxyfluorescein, (5′-FAM-SE), and Alexa 488 carboxylic acid succinimidyl ester dilithium salt.

All amine reactive probes will conjugate with aliphatic nonprotonated amines of proteins (or other amine-containing molecules). Consequently, the reaction has to take place in a slightly basic buffer to deprotonate the amine groups. For isothiocyanates and sulfonyl chlorides, 100–200 mM Na-bicarbonate at pH 9.0 or 50 mM sodium borate buffer at pH 8.5 are the most widely used conditions. Succinimidyl esters can be used at a pH range of 7.5–8.5; 100–200 mM Na-bicarbonate at pH 8.3 (no correction of pH needed) is ideal for this class of reagents. Alternatively, 50 mM phosphate buffer at pH 7.5 can be used. Avoid the use of amine-containing buffers such as Tris or glycine when coupling to reactive amine groups!

II. MATERIALS

Chemicals

NaCl	(M_r 58.44)	Merck, Cat. No.	1.06404
NaH$_2$PO$_4$	(M_r 156.01)	Merck, Cat. No.	1.06345
NaHCO$_3$	(M_r 84.01)	Merck, Cat. No.	1.06329
Na$_2$CO$_3$	(M_r 286.14)	Merck, Cat. No.	6391
Hydroxylamine	(M_r 69.49)	Sigma,	H-9876
Imidazole	(M_r 68.08)	Merck, Cat. No.	1.04716
MgCl$_2$	(M_r 203.30)	Merck, Cat. No.	5833
Sucrose	(M_r 342.39)	Fluka, Cat. No.	84-100

Prepacked Amersham Pharmacia PD-10 column (G-25) (Amersham Pharmacia Biotech, Cat. No. 17-0851-01; 30 prepacked columns per unit)

Column holders (any style; see Fig. 1A for a typical example)

Protein (1–5 mg/ml) in 100 mM Na-Bicarbonate, pH 8.3, or 10 mM phosphate buffer, pH 7.5–8.0

III. PROCEDURES

A. Coupling with Iodoacetamides

Solutions

1. *5′-IATMR (cysteine reactive):* Dissolve in dimethylformamide (DMF); final concentration 1 mg/ml

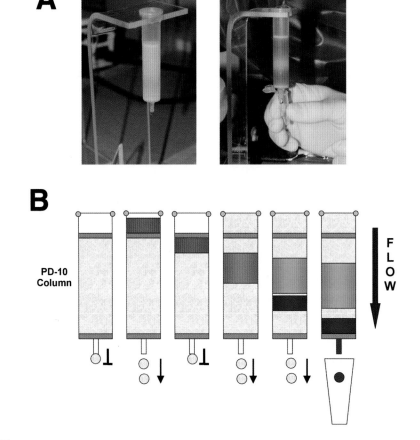

FIGURE 1 (A) Examples for the separation of labeled protein from free uncoupled dye on a PD-10 column. (Left) Alexa 488 maleimide-coupled calponin. (Right) Alexa 568 succinimidyl ester-coupled α-actinin. (B) Events of protein–dye separation using a PD-10 column (see Section VI,A).

2. *10× phosphate buffer 100 mM (NaH₂PO₄), pH 8.0:* Add 1.56 g to 100 ml water. Adjust pH with phosphoric acid

3. *Carbonate buffer 1 M (NaHCO₃/Na₂CO₃), pH 9.0:* Add 4.2 g NaHCO₃ and 14.3 g Na₂CO₃ to 100 ml distilled water

4. *NaCl stock 5 M (NaCl):* Add 29.2 g to 100 ml distilled water

Steps

1. Dialyze the protein solution against 10 m*M* phosphate buffer, pH 8.0, overnight and measure volume.

2. Add 1/4 volume of 1 *M* carbonate buffer, pH 9.0, to the antibody or protein solution in a small glass vial embedded in ice in a medium-sized plastic or styrofoam beaker and add a small magnetic bar.

3. Stir gently on a magnetic stirrer (vigorous stirring will cause the protein to denature!).

4. Add the fluorophore in two to four aliquots in 10-min intervals (use 20–40 mg of dye/mg of antibody or protein). Let the coupling reaction proceed for a total of 60 min.

5. Load the coupled protein on a PD-10 column (1 × 10 cm) equilibrated in PBS or EDC buffer.

6. Collect the faster migrating band (dark red/violet) in a fresh sterile tube.

7. (Optional) Measure the coupling efficiency (F/P value; fluorophore to protein).

Estimation of Fluorophore/Protein Ratio

Molar concentrations of dye and protein are calculated and the ratio of these values is the molar dye/protein ratio. As an example, the molar extinction coefficient for 5'-IATR at 549 nm is 97,000 M^{-1} cm^{-1}. The ratio calculation is corrected for the absorbance at 280 nm (approximately 5% of the absorbance at 549 nm). A ratio of 2 : 1–5 : 1 is ideal.

B. Coupling with Succinimidyl Esters (SE)

Solutions

1. *200 mM Na-Bicarbonate, pH 8.3:* To make 100 ml, add 3.02 g to 100 ml distilled water

2. *1.5 M Hydroxylamine, pH 8.5:* Add 1.5 g to 5 ml of distilled water and adjust the pH to 8.5 with NaOH. Dilute 1 : 1 with distilled water prior to use.

3. *PBS, pH 7.4:* To make 1 litre, add 0.36 g NaH₂PO₄, 1.1 g Na₂HPO₄, and 9 g NaCl to 100 ml distilled water and adjust pH to 7.4 with HCl

4. *EDC buffer, pH 7.0:* 10 m*M* imidazole, 30 m*M* NaCl, 2 m*M* MgCl₂ (optional), and 4 m*M* NaN₃. To make 1 litre, add (from stocks) 10 ml imidazole (1 *M*), 6

ml NaCl (5 *M*), 2 ml MgCl₂ (1 *M*), and 1 ml NaN₃ (1 *M*).

5. 5'-Carboxytramethylrhodamine succinimidyl ester, single isomer or 5'-carboxyfluorescein succinimidyl ester, single isomer at 5 mg/ml in DMF.

Steps

1. Dialyze the antibody or protein solution against 100 m*M* Na-bicarbonate, pH 8.3, overnight and measure volume and concentration.

2. Place the protein (0.75–1.0 ml) solution in a small glass vial or Eppendorf tube embedded in ice in a medium-sized plastic or styrofoam beaker and add a small magnetic bar.

3. Stir gently on a magnetic stirrer (vigorous stirring will cause the protein to denature!).

4. Add 10–50 µl of the fluorophore (corresponds to about 0.1–0.25 mg of amine reactive dye). About one-third to one-fourth of the dye will conjugate to the protein.

5. Stop the reaction by adding 100 µl of freshly prepared 1.5 *M* hydroxylamine solution and incubate for 1 h at 4°C on ice.

6. Separate the conjugate from the unreacted fluorochrome and the hydroxylamine by gel filtration. Load the coupled protein on a PD-10 column equilibrated in PBS, pH 7.4, or EDC buffer.

7. Collect the faster migrating band in a fresh sterile tube (see Fig. 1).

IV. COMMENTS

1. The concentration of the protein solution should not be lower than 1 mg/ml (2–5 mg/ml is ideal).

2. Basic pH (carbonate buffer, pH 9.0) is necessary for obtaining satisfying coupling efficiency. Lower pH, however, may be necessary to stabilize the protein of interest but results in reduced coupling efficiency.

3. 5'-IATMR rhodamine is a specially purified rhodamine 5'-isomer. It should be dissolved in dimethylformamide (avoid dimethyl sulfoxide).

4. Add NaN₃ to the coupled proteins and store in small aliquots at −70°C.

5. Tris or other amine-containing buffer must be avoided.

V. PITFALLS

1. Aggregates, as well as under- and overlabelled proteins, are sources for artefacts.

2. 5'-IATMR, like other rhodamines, is extremely light sensitive and susceptible to repeated temperature changes. Aliquots should therefore be protected from light and stored at −70° C (for prolonged storage).

VI. PRACTICAL NOTES

A. Separation of Labeled Proteins from Uncoupled Free Dye by Gel Filtration Using a PD-10 Column

The scheme in Fig. 1B highlights the sequence of events during separation.

After the sealed bottom has been removed by a firm cut at the first rim (using a strong blade or a pair of scissors), the column is equilibrated with 5–7 volumes of buffer. *Note:* Due to the presence of a sinter layer atop the column bed, the columns will not run dry.

1. Equilibrate the column about 10 min before coupling is finished and let the flow come to a halt
2. Quickly add the protein–label mixture. *Note:* Due to the top sinter the gel bed cannot be damaged and the probe can be applied in a single flush. Avoid applying the sample dropwise as this will cause uneven penetration of the probe.
3. Let the probe enter the gel and wait until the flow has come to a halt.
4. Fill the top reservoir with buffer and let the separation proceed.
5. In the last one-fourth of the column a three-phase separation should become visible featuring a bright zone separating the labeled protein (dark, fast migrating band) from the free dye (light, slower migrating band).
6. Collect the dark band containing only labeled protein. The dilution factor is about 1.5, so the entire peak fraction should fit into a single 1.5-ml Eppendorf tube.

B. Storage of Aliquots of Purified Conjugates by Liquid N₂ Infusion

As most fluorescently labeled protein probes will be used for microinjection, the storage of probes in small aliquots is required. In order to avoid hundreds of small tubes populating your freezer, the following storage method is recommended.

Add 1/10 volume of 0.6 g/ml sucrose (dissolved in the desired buffer) to the labeled protein and mix gently. Place a liquid N_2-resistant plastic beaker in a styrofoam box and fill the box and the beaker with liquid N_2. Pipette *single* drops of about 30–60 µl (using a 200-µl Gilson pipette and a yellow tip) directly onto the N_2 surface of the beaker. Avoid multiple drops on the surface as they will fuse and generate larger aggregates. The drop will float for about 10–20 s until the entire liquid is frozen and then sink rapidly to the bottom. Continue until the entire probe is at the bottom of the beaker. Prepare a labelled, prechilled 15-ml polycarbonate tube and gently maneuver the frozen drops into the tube and store at −70°C. Individual drops can be removed with a pair of tweezers and processed for microinjection.

References

Brantzaeg, P. (1973). Conjugates of immunoglobulin G with different fluorochromes. I. Characterization by anionic exchange chromatography. *Scand. J. Immunol. Suppl.* **2**, 273–290.

DeBiasio, R., Bright, G. R., Ernst, L. A., Waggoner, A. S., and Taylor, D. L. (1987). Five-parameter fluorescence imaging: Wound healing of living Swiss 3T3 cells. *J. Cell Biol.* **105**, 1613–1622.

Haugland, R. P. (2002). Handbook of Fluorescent Probes and Research Chemicals, 9th Ed. Molecular Probes Inc.

Mason, W. T. (1993). Fluorescent and Luminescent Probes for Biological Activity: A Practical Guide to Technology for Quantitative Real-Time Analysis. Academic Press, New York.

Wang, K., Feramisco, J. R., and Ash, J. F. (1982). Fluorescent localization of contractile proteins in tissue culture cells. *Methods Enzymol.* **85**, 514–562.

Radioiodination of Proteins and Peptides

Martin Béhé, Martin Gotthardt, and Thomas M. Behr

I. INTRODUCTION

Radioiodination of biomolecules was established in 1948, when ^{131}I was the first radioisotope of iodine to be used for the labeling of a, at that time, polyclonal antikidney serum, performed and described by Pressmen and Keighley. The technical production of iodine-131 was pioneered in 1938 by Seaborg and Livingood. A variety of other iodine isotopes exist (Table I), from which ^{123}I, ^{125}I, and ^{131}I are the most widely used for the labeling of biomolecules for *in vitro* (i.e., radioimmunoassay) and *in vivo* (i.e., pharmacokinetic and metabolism) applications. [123]I, ^{125}I, and ^{131}I are γ emitters that can technically be detected easily, whereas [124]I emits positrons, which annihilate with an electron to produce two photons with an energy of 512 keV each. Due to this property, it can be used for positron emission tomography (PET). All these nuclides can be detected directly *in vitro* or *ex vivo* without a scintillation cocktail, whereas for *in vivo* imaging, the most suitable nuclides are [124]I for PET and [123]I (as well as, under certain conditions, ^{131}I) for single photon emission computer tomography (SPECT). [125]I is not suitable for *in vivo* imaging, due to its low energy of 35 keV, which is absorbed within a very short path length (a few millimeters, at best). However, this makes it very useful, together with the long half-life of 60 days for *ex vivo* (microautoradiography) and *in vitro* studies. [131]I decays by emitting photons with high energies (284, 364, and 637 keV) and electrons (0.6 and 0.8 MeV), the latter of which may cause substantial radiation exposure to tissues, rendering the isotope less optimal for its use in experiments. The main value of this isotope is its clinical application as a therapeutic nuclide in various therapies, including benign and malignant thyroid disorders or radioimmunotherapy.

All isotopes are available as NaI in neutral or basic solution, whereas ^{123}I, ^{125}I, and ^{131}I are commercially produced isotopes; [124]I can be delivered only by specialized cyclotron facilities. Because iodide (I^-) is a nonreactive form, it must be activated by an oxidizing agent to a reactive cationic species (I^+), which allows a spontaneous electrophilic substitution on aromatic rings with a good leaving group such as H^+ in p-kresol (tyrosine) or 4-methylimidazol (histidine) (Fig. 1). The iodination place is pH driven. The tyrosine moiety is labeled mainly at a pH around 7.5, whereas at a pH around 8.5, mainly the histidine is labeled, at a much lower yield though.

The tyrosine moiety can be labeled twice (Fig. 1). The second step, yielding di-iodinated tyrosyl moieties, occurs faster than the monoiodination reaction. Due to the stoichiometry of the reactants, in proteins, usually monoiodotyrosine residues prevail, whereas in peptides, di-iodinated tyrosyl moieties prevail at sufficiently high specific activities (Table II).

The possibility of rapid enzymatic deiodination of the mono- or di-iodinated tyrosine is a disadvantage. Two different kinds of iodination are known: (a) direct iodination as described earlier or (b) indirect iodination with the Bolton and Hunter reagents or similar via free amino groups.

II. MATERIALS AND INSTRUMENTS

Na[125]I (Cat. No. IMS30) and [125]I Bolton and Hunter reagent (Cat. No. IM5861 or IM5862) are from

TABLE I   Relevant Isotopes of Iodine

	$t_{1/2}$	γ energy (keV)	β energy (MeV)
^{123}I	13.1 h	159 (97.7%)	(Auger/conversion e$^-$)
^{124}I	4.17 days	603, 1691	β^+ 2.1
^{125}I	60.1 days	35 (100%)	(Auger/conversion e$^-$)
^{126}I	13.0 days	389, 688	β^+ 1.1; β^- 0.9, 1.3
^{127}I	Stable		
^{129}I	1.59×10^7 year	40	β^- 0.2
^{131}I	8.04 days	364 (83%)	β^- 0.6, 0.8
		637 (6.7%)	
		284 (6.9%)	

TABLE II   Iodine Molecules per mCi (or per MBq), as well as Antibody Molecules per mg of Protein

	$t_{1/2}$	N (mol)/mCi	N (molecules)/MBq
^{123}I	13.1 h	4.2×10^{-12}	1.1×10^{-13}
^{124}I	4.2 days	3.3×10^{-11}	9.1×10^{-13}
^{125}I	60.1 days	4.7×10^{-10}	1.3×10^{-11}
^{131}I	8.04 days	6.2×10^{-11}	1.7×10^{-12}

	N (mol)/mg	
IgG	6.7×10^{-9}	
(Fab')$_2$	1.0×10^{-8}	
Fab'	2.0×10^{-8}	

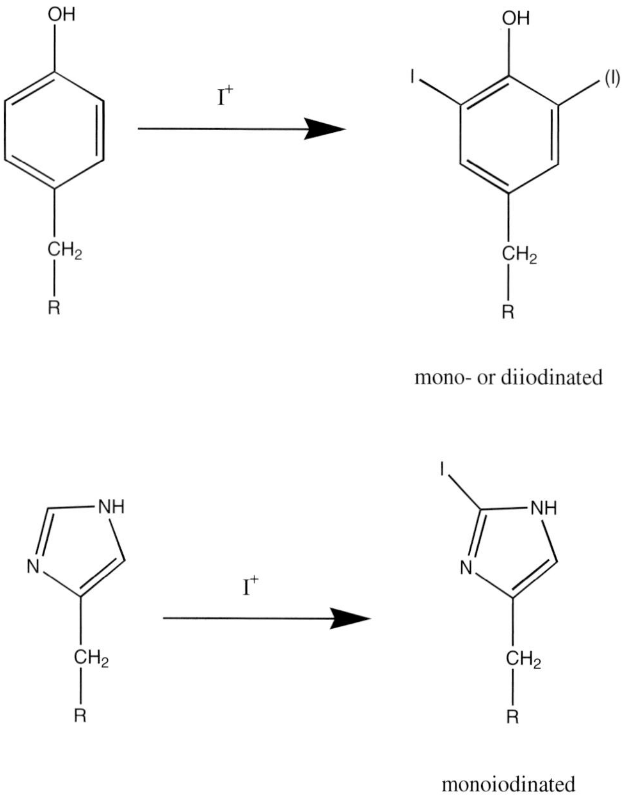

FIGURE 1 Schematic representation of the radioiodination procedure.

Amersham Bioscience. IM5862 is the di-iodo derivative used to achieve a higher specific activity. Na^{131}I (I-131-S) and Na^{123}I (I-123-S) are delivered by MDS Nordion. All radioisotopes of iodine are purchased in no carrier-added form. Iodo-Gen (T0656), chloramine-T (C9887), sodium metabisulfite (S1516), 3-(4-hydroxyphenyl) propionic acid, N-hydroxysuccinimide ester (H1256, Bolton–Hunter reagent), sodium acetate (S5636), chloroform (C5312), dichlormethane (32,026-

0), phosphate-buffered saline (PBS) solution (P5119), sodium dihydrogen phosphate (S9638), disodium hydrogen phosphate (S9390), 1 M hydrochloric acid solution (H9892), boric acid (B0394), 1 M sodium hydroxide solution (S2770), 0.9% saline (8776), trifluoroacetic acid (T8506), HPLC water (27,073-3), and HPLC acetonitrile (57472-4) are from Sigma-Aldrich.

Tyr^3-octreotide and Exendin-3 are synthesized by Bachem. Minigastrin (G0267) is from Sigma-Aldrich.

Rituximab is a mouse–human chimeric anti-CD20 antibody (clone IDEC-C2B8, Mabthera), which is obtained commercially from Roche Pharma (via a pharmacy).

We use 1.5-ml tubes from Eppendorf (0030 120.086). The IB-F silical gel thin-layer strips (4463-02) are from Baker.

Radioactivity is measured with a Cobra II quantum automatic gamma counter from Perkin Elmer. HPLC is performed on a 535 pump and 545 UV/vis analyzer from Biotek with online flow radioactive analyzer 500TR from Perkin Elmer. PD-10 columns (17-0851-01) are from Amersham Bioscience. The HPLC protein column Bio-Silect SEC 250-5 (125-0476) is from Bio-Rad. The HPLC column is a CC 250/4.6 Nucleosil 120-5 C-18 (721712.46) from Marcherey-Nagel. The Speed-Vac Savant (SPD 101 B) is from Thermo Life Science connected with a MD4C vacuum pump (69 62 92) from Vacubrand.

III. PROCEDURES

A. Iodogen Radioiodination of an Antibody

Solutions

1. *Iodogene solution:* Dissolve 2.5 mg iodogene in 10 ml of chloroform or dichloromethane.

2. *0.05 M phosphate buffer, pH 7.4:* Prepare by dissolving 2.87 g Na_2HPO_4 and 0.66 g NaH_2PO_4 in 500 ml water (pH control must give a pH value of 7.4)

3. *0.1 M NaOH:* Dilute 1 M sodium hydroxide solution (S2770) 10 times

Steps

1. Add 200 μl of Iodogen solution (corresponding to 50 μg of Iodogen) to a 1.5-ml tube. Evaporate the chloroform or dichloromethane under gentle heating in a water bath (40–50°C) with constant and homogeneous rotation, plating the Iodogen homogeneously onto the inner surface of the vial. Batches of 20–30 vials can be produced simultaneously and stored for several months at –20°C.

2. Buffer 67 μg of the antibody Rituximab (1 mg of protein per 555–740 MBq (15–20 mCi) of ^{131}I, for molar ratios and other iodine isotopes, see Tables I and II) in 100 μl 0.05 M phosphate buffer, pH 7.4, and put into the Iodogen vial.

3. Add 37 MBq of ^{131}I in 5 μL 0.1 M NaOH to the reaction vial.

4. Stop the reaction after incubating for 30 min at room temperature by removing the reaction solution from the Iodogen vial to another 1.5-ml tube.

5. For quality control and purification, see special section.

B. Chloramine-T Radioiodination of Tyr³-Octreotide

Solutions

1. *0.05 M sodium acetate buffer, pH 4.2:* Dissolve 410 mg of sodium acetate in 100 ml water. Adjust the pH to 4.2 by adding 1 M HCl.

2. *Tyr³-octreotide solution:* Dissolve 5.3 mg tyr³-octreotide in 10 ml 0.05 M sodium acetate buffer, pH 4.2

3. *0.05 M phosphate buffer:* Prepare 0.05 M phosphate buffer, pH 7.4, by dissolving 2.87 g Na_2HPO_4 and 0.66 g NaH_2PO_4 in 500 ml water (pH control must give a pH value of 7.4)

4. *Phosphate buffered saline (PBS)*

5. *Sodium meta-bisulfite:* Dissolve 20 mg sodium meta-bisulfite in 100 ml PBS

6. *Na¹²⁵I solution:* Available commercially

Steps

1. Add 3 μl of a 0.51 M tyr³-octreotide solution in 0.05 M sodium acetate buffer, pH 4.2, to 20 μl of 0.05 M phosphate buffer, pH 7.4, into a 1.5-ml reaction tube.

2. Add 48 MBq of Na¹²⁵I solution in 2.8 μL 10^{-5} M NaOH.

3. Start the reaction by 1.6 μg chloramine-T in 20 μl 0.05 M phosphate buffer, pH 7.4.

4. Stop the reaction after incubating 1 min at room temperature by adding 20 μg sodium meta-bisulfite in 100 μl PBS.

5. For quality control and purification, see special section.

C. Production of ¹²⁵I-Labeled Bolton–Hunter Reagent

Radioiodination by the Bolton–Hunter procedure was performed essentially according to the original description (13).

Solutions

1. *PBS:* Available commercially

2. *Bolton–Hunter reagent solution:* Available commercially

3. *Chloramine-T solution:* Dissolve 2 mg of chloramine-T in 400 μl 0.25 M PBS, pH 7.4

4. *Sodium meta-bisulphite solution:* Solve 1.2 mg sodium meta-bisulphite in 600 μl 0.05 M PBS, pH 7.4

Steps

1. Add 37 MBq of Na¹²⁵I solution in 2.2 μl 10^{-5} M NaOH to 10 μl of Bolton–Hunter reagent solution in a 1.5-ml vial.

2. Add 40 μl of the chloramine-T solution to the reaction vial to start the reaction.

3. Stop the reaction after 10 s by the addition of the sodium meta-bisulphite solution.

4. After the addition of 200 μl dimethylformamide, extract the radioiodinated Bolton–Hunter reagent with two 500-μl portions of benzene.

D. Conjugation with the Bolton–Hunter Reagent

Solution

0.1 M sodium borate, pH 8.5: Dissolve 618 mg boric acid in 90 ml water. Adjust the pH with 1 M NaOH to pH 8.5 and make up volume to 100 ml with pure water.

Steps

1. Evaporate the solvent of the ¹²⁵I-labeled Bolton–Hunter reagent (from Section IIIC or commercially bought) in a hood under a gentle stream of dry nitrogen at room temperature.

2. Add 250 ng of minigastrin in 200 μl ice-cold borate buffer to the Bolton–Hunter tube.

3. Vortex the mixture for 2 h on ice.

4. Stop the reaction by removing the reaction solution from the reaction tube.

5. For quality control and purification, see special section.

E. Radioiodination of Exendin-3 on the Histidine Moiety

Solutions

1. *Iodogene solution:* Dissolve 2.5 mg in 10 ml of chloroform or dichloromethane.

2. *0.05 M Tris buffer, pH 8.5:* Dissolve 0.606 g of tris base (M_r 121.14) in approximately 90 ml of pure water. Titrate to pH 8.7 with 1 M HCl. Make up volume to 100 ml with pure water.

Steps

1. Add 200 µl of Iodogen solution (corresponding to 50 µg of Iodogen) to a 1.5-ml tube. Evaporate the chloroform or dichloromethane under gentle heating in a water bath (40–50°C) and constant and homogeneous rotation, plating the Iodogen homogeneously onto the inner surface of the vial. Batches of 20–30 vials can be produced simultaneously and stored for several months at −20°C.

2. Buffer 10 µg of exendin-3 in 100 µl 0.05 M tris buffer, pH 8.5, and put into the iodogen vial.

3. Add 37 MBq of ^{125}I in 2.2 µl 0.01 M NaOH to the reaction vial.

4. Stop the reaction after incubating 60 min at room temperature by removing the reaction solution from the Iodogen vial to another Eppendorf tube.

5. For quality control and purification, see special section.

F. Quality Control of Radioiodinated Antibody with Thin-Layer Chromatography

Solution

0.9% saline: Available commercially

Steps

1. Cover the floor of the chromatography tank with 0.9% saline to a depth of 5 mm.

2. Spot a droplet (approximately 1–5 µl) of the final product (radioiodinated antibody or protein) onto an ITLC silicagel IB-F flexible strip with a length of 10 cm at a distance of approximately 1 cm from the bottom.

3. After 20–30 s, place the strip into the chromatography tank.

4. Allow the strip to develop until the solvent front migrates to approximately 1 cm from the top of the

strip. At this point, remove the strip from the tank and allow to dry.

5. Cut the strip in half.

6. Count its lower half, containing the radiolabeled protein, as well as its upper half, containing the unincorporated radionuclide ("free iodine"), in a γ counter.

7. Determine the amount of incorporated radioiodine as follows: Incorporated = 100% × (counts or activity of lower strip half)/(total counts or activity).

G. Quality Control and Purification of Radioiodinated Antibody with Low-Pressure PD-10 Coloumn (Fig. 2)

Solution

0.1 M phosphate buffer, pH 7.4: Dissolve 12 g of NaH_2PO_4 in approximately 900 ml of pure water. Titrate to pH 7.42 at the laboratory temperature of 20°C with monovalent strong base or acid as needed. Make up volume to 1000 ml with pure water.

Steps

1. Preequilibrate the PD-10 column with 40 ml phosphate buffer, pH 7.4

2. Apply 370 kBq (for purification: 37 MBq) of the final radiolabeled product onto the PD-10 column

3. Elute the PD-10 column with 20 ml of phosphate buffer, pH 7.4, collecting 0.5-ml fractions.

4. Count the activity in each fraction.

5. Purification: Combine the three samples of the first peak with the highest activity.

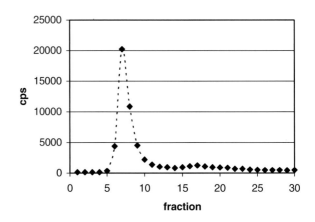

FIGURE 2 Typical elution profile of a radioiodinated monoclonal antibody (complete IgG) on a PD-10 column (Sephadex G-25 pre-equilibrated in a 15 × 50-mm polypropylene tube). The labeled IgG elutes in fractions 6–9, unbound iodine in fractions 15–22 (fraction size: 0.5 ml).

H. Quality Control and Purification of Radioiodinated Antibody with High-Pressure Liquid Chromatography (HPLC)

Solution

0.1 M phosphate buffer, pH 7.4: Dissolve 12 g of NaH_2PO_4 in approximately 900 ml of pure water. Titrate to pH 7.42 at the laboratory temperature of 20°C with monovalent strong base or acid as needed. Make up volume to 1000 ml with pure water.

Steps

1. Preequilibrate the size-exclusion HPLC column with 40 ml phosphate buffer, pH 7.4
2. Apply 37 kBq (for purification: 37 MBq or larger activities) of the final radiolabeled product onto the HPLC column.
3. Elute the substances with phosphate buffer, pH 7.4.
4. Collect the samples with a sample collector (purification or quality control) and measure with an online radioactivity detector (quality control).
5. Purification: Combine the three samples of the first peak with the highest activity.

I. Quality Control and Purification of Radioiodinated Peptides with HPLC

Solutions

1. *0.1% TFA:* Add 1 ml trifluoroacetic acid in 1 liter HPLC water.
2. *Acetonitrile HPLC:* Available commercially

Steps

1. First preequilibrate the C-18 HPLC column by starting an empty run (see Table III).
2. Apply 37 kBq (for purification: 37 MBq) of the final radiolabeled product onto the HPLC column.
3. Elute the substances with the gradient (Table III).

TABLE III   Gradient for HPLC Quality Control and Purification for Radiolabelled Peptides

	Flow ml/min	0.1% TFA (%)	Acetonitrile (%)
0 min	1	95	5
5 min	1	95	5
20 min	1	45	55
25 min	1	45	55
40 min	1	0	100
45 min	1	0	100

4. Collect samples with a sample collector (purification or quality control) and measure on a γ counter or measure with an online radioactivity detector (quality control).
5. The acetonitrile can be evaporated on a Speed-Vac at a temperature of 60°C.
6. Purification: The radioionated peptide peak can be used for further experiments.

IV. PITFALLS

1. The iodination procedure must be performed within a hood because elementary iodine can be produced during synthesis, which is volatile.
2. The protein or peptide must be in the reaction vial before adding the radioiodine to avoid the formation of elementary iodine.
3. There must not be any additive proteins or substances with tyrosine-like structures in the solutions, as they would be radioiodinated too.
4. The solution with the iodinated protein or peptide should be adjusted to a protein content of about 2% with albumin after iodination to avoid unspecific deposition on the equipment.

V. COMMENTS

The described procedures can be changed easily between the isotopes and can be applied to proteins and their fragments or peptides without too much difficulty. The quality control part must be modified for any particular substance and can be applied to a purification procedure. The protocols can be transferred between the different iodine isotopes without major problems.

Immunoreactivity testing must be performed for antibodies and their fragments.

For the described Iodogen or chloramine-T radioiodination procedure, incorporation yields were between 60 and 90%, regardless of the antibody isotype or molecular form (IgG or fragments). After purification, the percentage of unbound radionuclide fell below 2% in all cases. Also, tyrosyl residues containing peptides were labeled successfully with yields between 60 and 85% using these procedures. With immunoconjugates, their molecular integrity was maintained at over 90% by using this Iodogen methodology, whereas chloramine-T iodination of $F(ab')_2$ fragments led to an up to 50% degradation to monovalent Fab' fragments. With peptides, oxidation of methionine residues to their respective sulfoxide derivatives was only 15–30% with Iodogen, which stands in contrast to over 90% with the chloramine-T methodology. Finally, the tested immunoreactivities were above 85% in all cases (regardless of whether Iodogen or chloramine-T methodology was used).

Much higher variability in labeling yields was observed while using the Bolton–Hunter procedure (with incorporation yields between 20 and 55%). However, in accordance to the lack of contact to oxidants or reducing agents by using this procedure, no methionine oxidation was observed.

Sample Fractionation for Proteomics

21

Free-Flow Electrophoresis

Peter J. A. Weber, Gerhard Weber, and Christoph Eckerskorn

I. INTRODUCTION

Without prefractionation and enrichment, none of the existing techniques for the analysis of proteomes, such as two-dimensional electrophoresis (2-DE), or chromatographic methods, such as reversed-phase high-performance liquid chromatography (RP-HPLC), will be able to cope with the enormous complexity of biological samples and the extremely wide dynamic range of the protein concentrations. For example, this means that *low abundant proteins* are very likely to be hidden by highly expressed species.

Free-flow *isoelectric focusing* (Hannig, 1961; Krivankova and Bocek, 1998) of *protein mixtures* (Bernardo *et al.*, 2000; Hoffmann *et al.*, 2001; Maida *et al.*, 2000; Weber and Bocek, 1996; Weber and Bocek, 1998) is one of the methods that fulfil the prerequisites to meet these *prefractionation* demands, i.e., to increase the amount of low abundant proteins and to dramatically reduce the complexity of protein mixtures. This is based on (1) the continuous operation principle, which allows the processing of large sample amounts; (2) the absence of any kind of gel or matrix that increases the recoveries of the proteins and makes this method highly compatible to virtually all kinds of follow-up analyses; and (3) the gentle procedure, which allows the fractionation of active enzymes and even protein complexes.

The power and resolution of free-flow isoelectric focusing are illustrated with the analysis of pig serum using a Pro Team *free-flow electrophoresis* (FFE) instrument. Detailed information about the instrument can be found in http://www.tecan.com. Please follow the "Proteomics" and "Fractionation" links.

II. MATERIALS AND INSTRUMENTATION

Pro Team HPMC (hydroxypropylmethylcellulose, Cat. No. 5170709) and Pro Team glycerol (Cat. No. 5170708), as well as the Pro Team FFE reagent basic kit (Cat. No. B132001) containing Prolyte 1, Prolyte 2, Prolyte 3, SPADNS (sulfanilic acid azochromotrop), and coloured pI markers, are from Tecan (Grödig, Austria). 1 M NaOH (Cat. No. 35256), 1 M H$_2$SO$_4$ (Cat. No. 35276), and petroleum benzene (Cat. No. 32248) are from Riedel-de Haën (Seelze, Germany). Urea (analytical grade, Cat. No. 24524) is from Serva (Heidelberg, Germany). Isopropanol (Cat. No. 9866.4) is from Karl Roth (Karlsruhe, Germany).

The Pro Team FFE apparatus is from Tecan (Grödig, Austria). It is equipped with seven 0.64-mm (i.d.) media tubes, one 1.42-mm (i.d.) counterflow tube, one 0.51-mm (i.d.) sample tube, a 0.4-mm spacer, and 0.65-mm filter paper strips. The water cooler IC006 (P ≥ 350 W) is from Huber (Offenburg, Germany).

III. PROCEDURES

A. Preparation and Running of a Denaturing Free-Flow Isoelectric Focusing Experiment

Solutions

1. *Anode stabilisation media (inlet 1):* 14.5% (w/w) glycerol, 0.12% (w/w) HPMC, 100 mM H$_2$SO$_4$, and 8 M urea. To make 60 ml (~68.9 g), add 6.0 g of 1 M H$_2$SO$_4$, 14.0 g distilled water, 10.0 g 0.8% (w/w) HPMC, 10.0 g glycerol, and 28.9 g urea. Mix thor-

157

oughly by stirring. The solution should not be heated. Actual consumption: ~10 g/h.

2. *Separation media 1* (*inlet 2*): 14.5% (w/w) glycerol, 0.12% (w/w) HPMC, 11.6% (w/w) Prolyte 1, and 8 M urea. To make 60 ml (~68.9 g), add 8.0 g Prolyte 1, 12.0 g distilled water, 10.0 g 0.8% (w/w) HPMC, 10.0 g glycerol, and 28.9 g urea. Mix thoroughly by stirring. The solution should not be heated. Actual consumption: ~10 g/h.

3. *Separation media 2* (*inlet 3 + 4*): 14.5% (w/w) glycerol, 0.12% (w/w) HPMC, 19.4% (w/w) Prolyte 2, and 8 M urea. To make 120 ml (~137.8 g), add 26.7 g Prolyte 2, 13.3 g distilled water, 20.0 g 0.8% (w/w) HPMC, 20.0 g glycerol, and 57.8 g urea. Mix thoroughly by stirring. The solution should not be heated. Actual consumption: ~20 g/h.

4. *Separation media 3* (*inlet 5*): 14.5% (w/w) glycerol, 0.12% (w/w) HPMC, 14.5% (w/w) Prolyte 3, and 8 M urea. To make 60 ml (~68.9 g), add 10.0 g Prolyte 3, 10.0 g distilled water, 10.0 g 0.8% (w/w) HPMC, 10.0 g glycerol, and 28.9 g urea. Mix thoroughly by stirring. The solution should not be heated. Actual consumption: ~10 g/h.

5. *Cathode stabilisation media* (*inlet 6 + 7*): 14.5% (w/w) glycerol, 0.12% (w/w) HPMC, 100 mM NaOH, and 8 M urea. To make 120 ml (~137.8 g), add 12.0 g 1 M NaOH, 28.0 g distilled water, 20.0 g 0.8% (w/w) HPMC, 20.0 g glycerol, and 57.8 g urea. Mix thoroughly by stirring. The solution should not be heated. Actual consumption: ~20 g/h.

6. *Counterflow media* (*inlet 8*): 14.5% (w/w) glycerol, 0.12% (w/w) HPMC, and 8 M urea. To make 288 ml (~330.5 g), add 96.0 g distilled water, 48.0 g 0.8% (w/w) HPMC, 48.0 g glycerol, and 138.5 g urea. Mix thoroughly by stirring. The solution should not be heated. Actual consumption: ~46 g/h.

7. *Electrolyte anode circuit* (*+ve*): 100 mM H_2SO_4. To make 400 g, add 40.0 g 1 M H_2SO_4 and 360.0 g of distilled water. Mix thoroughly by stirring. Actual consumption: none, amount lasts for one working day.

8. *Electrolyte cathode circuit* (*–ve*): 100 mM NaOH. To make 400 g, add 40.0 g 1 M NaOH and 360.0 g of distilled water. Mix thoroughly by stirring. Actual consumption: none, amount lasts for one working day.

1. Disassembly, Cleaning, Reassembly, and Filling of the Instrument

Steps

1. Switch on the cooler first and set it to 10°C.

2. Attach the fractionation plate to the separation chamber front part via the magnetic holder and move the separation chamber to the vertical position.

3. Reduce the force of all separation chamber clamps (turn counterclockwise), open the clamps pairwise from the outside to the inside, and open the separation chamber by carefully pulling the front part.

4. Put the electrode membranes into a 1 : 1 mixture of glycerol : isopropanol and put the paper strips in distilled water.

5. Make sure that the sample tube is connected to the middle sample inlet and that the two other sample inlets are closed.

6. Clean the interior surfaces of the separation chamber with lint-free paper towels in the following order: distilled water—isopropanol—petroleum benzene—isopropanol—high-purity water. Use a separate paper towel for each cleaning operation.

7. Place the wet 0.4-mm spacer on the front plate of the separation chamber with even distance to the electrode seals. Take care that the separation media inlets do not get covered.

8. Place membranes on electrodes starting from the top to the bottom. The smooth side of the membrane should face towards the electrode seal. The membrane must not protrude over the electrode seal. Subsequently, place the paper strips congruently on the membranes in the same fashion.

9. Quickly move the separation chamber front part towards the back part, ensuring that the separation chamber front part is parallel with the back part. Then, close the middle pair of clamps simultaneously. Afterwards, close adjacent clamps pairwise. Finally, increase the clamping force pairwise starting from the centre by opening the pair of clamps, turning the clamps clockwise, and closing the clamps (the water drops underneath the spacer should get displaced, the membranes and filters should not be visible).

10. Place the fractionation plate on the fraction collector housing.

11. To fill the instrument with water, tilt the separation chamber 45° and turn the bubble trap in the filling position.

12. Place all media and counterflow tubes in a bottle with distilled water. Place the sample tube in an empty 2-ml reaction tube without tightening the screw.

13. Open the three-way stopcock on the counterflow tube in all directions and open the Luer lock closure on the cock. Place the upper counterflow tube outlet in the fractionation tray and the three-way stopcock in the bottle grid.

14. Close the media pump tube cassettes, start the pump (direction "IN") with a delivery rate of 50 rpm, and wait until the separation chamber is filled halfway. Reverse pumping direction and empty the separation chamber until the inlet areas of the spacer are reached and all air bubbles are gone. Reverse the pumping

direction again and fill the chamber bubble free until the counterflow reservoir is nearly full. Then reduce the pump speed to 20 rpm and fill the remaining part of the counterflow reservoir.

15. Fill the counterflow tube, including the bubble trap at maximum speed, reduce the flow rate to 20 rpm, and connect the Luer lock closure of the three-way stopcock. Finally, close the remaining opening of the three-way stopcock. The fractionation tubes will start dripping.

16. Tilt the chamber to the horizontal position and tap the fractionation plate on the fractionation collector housing. If a fractionation tube fails to deliver, connect a syringe to the corresponding fractionation tube and suck until liquid is coming out of the tube.

17. Dry the outside of the separation chamber and control for any kind of leakage.

18. Start the sample pump with 4 rpm (direction "IN") and tighten the adjusting screw until the tube starts to deliver. Subsequently, change the pumping direction until no air remains in the tube. Then stop the sample pump.

2. Quality Control and Calibration of the Pumps

Steps

1. To check the flow profile and the proper assembly of the instrument, dilute 500 µl of Pro Team SPADNS with 50 ml of distilled water. Then switch off the media pump, place the media tubes of inlets 2, 4, and 6 in the SPADNS solution, and leave the remaining media tubes of inlet 1, 3, 5, and 7 as well as the counterflow tube of inlet 8 in distilled water.

2. Run the media pump with 40 rpm (direction "IN") for at least 6 min: In the separation chamber, red and colourless stripes should appear that flow in parallel, with identical width, and sharp boundaries (see Fig. 1). The equivalent pattern should appear in a 96-well plate. If this is not the case, try to readjust the clamps and make sure that the sample pump screw is closed. Other reasons might be surface contamination of the separation chamber; incorrectly installed spacer, seals, paper strips, or membranes; and partially clogged or leaking media tubes.

3. To continue, switch off the media pump, return the media tubes of inlets 2, 4, and 6 to distilled water, and run the media pump at 40 rpm (direction "IN") to rinse the separation chamber for at least 6 min.

4. To calibrate the sample pump, fill a 2-ml reaction tube with approximately 1.5 ml of distilled water, weigh it accurately to a milligram, and immerse the sample tube into it.

5. Call up the dialogue "Calibrate sample pump" and follow the instructions.

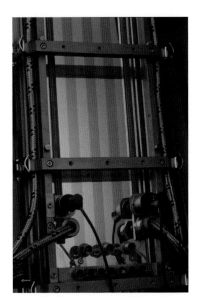

FIGURE 1 Stripe test to check the flow profile and the proper assembly of the Pro Team FFE instrument.

6. Afterwards, weigh the reaction tube again and save the calibration.

7. To calibrate the media pump, fill an appropriate bottle with approximately 200 ml of distilled water, weigh it accurately to 10 mg, and immerse the media tubes (inlets 1–7 without counterflow tube 8) in the calibration solution.

8. Call up the dialogue "Calibrate media pump" and follow the instructions.

9. Afterwards, weigh the bottle again and save the calibration.

10. To fill the separation chamber with separation media, make sure that the media pump is switched off.

11. Immerse the liquid circuit tubes in the appropriate media (+ve: anode circuit, −ve: cathode circuit), close the safety cover of the electrode circuit bottles and the media pump, and turn on the electrode pump (flowing air bubbles in the electrode ducts indicate the proper function of the pump).

12. Immerse the media tubes into the corresponding separation media and the counterflow tube into the counterflow media.

13. Turn on the media pump with 20 rpm for 3 s (direction "OUT"). Tap the media bottles against the bottle grid. Reverse the direction of the media pump and rinse the separation chamber with 20 rpm for 15 min. Then adjust the rate of delivery to ~57 ml/h.

14. Set the voltage to 1200 V and the current to 50 mA, switch on the high voltage, and wait approximately 15 min until the current reached a stable minimum.

FIGURE 2 Performance test to check the functionality of the Pro Team FFE instrument. Red and the six yellow markers are separated according to their different pI values and mimic the actual separation of the proteins.

15. To check the performance of the instrument (Fig. 2), dilute the Pro Team pI markers 1 : 10 with separation media 2 and apply them via the middle sample inlet at a sample flow rate of 0.5 ml/h. As soon as coloured drops are visible at the fractionation tubes, wait 10 more minutes. Then collect the separated pI marker fractions in a 96-well plate by placing it on the drawer and moving it under the fractionation tubing outlets. You can avoid cross-contaminations during the collection by tapping the fractionation plate on the fractionation collector housing just before introducing and removing the drawer. The width of the red and the six yellow pI markers should be no more than 3–4 wells, otherwise the separation media have not been prepared properly or the separation chamber has not been assembled properly.

3. Fractionation of the Protein Sample by Free-Flow Isoelectric Focusing

Steps

1. Prepare the sample solution in accordance with the information mentioned in Section IV.

2. To avoid carryover when applying a new sample, make sure that the separation chamber and the sample tube get rinsed for 30 min (media pump direction "IN", ~57 ml/h; sample pump direction "OUT", 2 ml/h), that the voltage is on during this time, and that the current has reached its stable minimum.

3. Stop the sample pump, snip the end of the sample tube to create a tiny bubble at its end, immerse the tube into the sample vial, and turn on the sample pump (direction "IN") with a high flow rate until the bubble reaches the sample inlet. Then reduce the flow rate to ~1 ml/h.

4. As soon as the sample reaches the fractionation manifold (indicated by the red Pro Team SPADNS that you added to the sample), start collecting the sample fractions in a 96-well plate by placing it on the drawer and moving it under the fractionation tubing outlets. As soon as the red dye leaves the separation chamber you can stop the sample collection.

4. Shutting Down the Instrument

Steps

1. For active rinsing of the instrument you have to inactivate the instrument first, i.e., make sure that the high voltage is switched off as well as the media pump and the electrode pump.

2. Move the separation chamber to its horizontal position, place the counterflow tube, as well as the media tubes, in a bottle containing at least 1 liter of distilled water, and place the sample tube in an empty 2-ml reaction tube.

3. Operate the media pump for 30 min with 50 rpm (direction "IN") and the sample pump for 30 min with 4 rpm (direction "OUT").

4. To rinse the electrodes, remove the two "longer" electrode tubes signed with "+" or "−" from the electrode media and operate the electrode pump until the electrode ducts are almost free of electrode media. Then switch off the electrode pump. Remove the bottles with electrode media, place all four electrode tubes in a vessel containing approximately 500 ml of distilled water, and operate the electrode pump for approximately 10 min. Remove the two "longer" electrode tubes signed with "+" or "−" from the distilled water and operate the electrode pump until the electrode ducts are almost free of water. Finally, switch off electrode pump and remove the water vessel.

5. For passive rinsing of the instrument, exchange the 96-well plate drawer for a rinse tray with approximately 2 liters of distilled water. While operating the media pump (direction "IN"), dunk the fractionation plate into the rinse tray and stop the media pump. Place the counterflow and media tubes onto the bottle grid and remove the water vessel. Turn the bubble trap in the draining position. Tilt the chamber 45° upwards. Open every pair of clamps and relax clamping force for three complete revolutions (turn counterclockwise) and close it again. Remove the media pump safety cover and release the tube cassettes by pushing on the lower right side. Release the adjusting screw of the sample pump and place the sample tube on the bottle

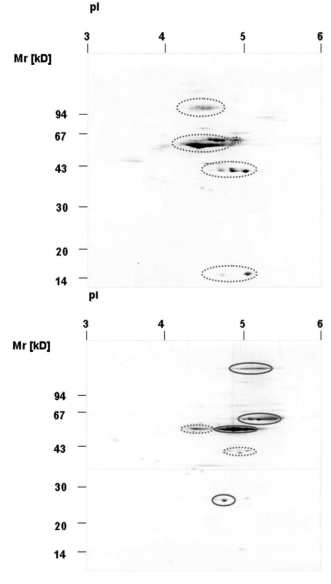

FIGURE 3 Silver-stained 2-DE analysis of two individual fractions of FFE–IEF-separated pig serum. Top: fraction 31; bottom: fraction 33. Coloured ellipses indicate the minor overlap of the two protein patterns.

grid. Finally, switch off the FFE and the cooler. If the FFE is used within the next 24 h it remains in this state. If the FFE is not used for a longer period of time, open the chamber on the next day. Rinse the spacer and filter paper with distilled water and store them dry. Store the membranes in glycerol : isopropanol (1 : 1).

B. Gel Examples

Figure 3 shows representative two-dimensional gels of individual FFE fractions after separation by IEF-

FFE. For more information, the reader is encouraged to visit Tecan's Web site at http://www.tecan.com.

IV. PITFALLS

1. Use only high-purity water as well as chemicals of the highest grade available.

2. Never use acetone, powders (gloves!), silicon (latex gloves!), oil, grease, or adhesive in proximity of the separation chamber.

3. Air bubbles must not enter the separation chamber, as this would disturb the laminar flow profile and the resolution of the fractionation.

4. Make sure to prepare all solutions very carefully and to mix them thoroughly by stirring. The temperature of media should not exceed the working temperature by more than 15°C.

5. For optimal results, prepare the media freshly every day.

6. Prepare all solutions through weighing, as the high viscosity of glycerol and HPMC prevents accurate volumetric measurements.

7. Before fractionating a real sample, make sure to run two quality control experiments (see Figs. 1 and 2) to check the proper assembly and the functionality of the instrument.

8. The chemical and physical properties (density, conductivity, and viscosity) of the sample and the separation media should be similar, i.e., it is useful to dilute a concentrated sample with separation media.

9. To visualize the actual position of the colourless samples within the separation chamber, mix it with 1% of Pro Team SPADNS.

10. The total salt concentration in the sample should be less than 50 mM.

11. Turbid protein samples are an indication that protein precipitation has already occurred or that insoluble cell components are still present. You have to clarify turbid sample solutions by filtration or centrifugation before use.

12. You can add nonionic detergents such as CHAPS, octylglucoside, Triton X-100, and Triton X-114 (0.1–1%), as well as reducing agents such as dithiothreitol (up to 50 mM) or uncharged phosphines, to the samples to increase solubility.

13. Precipitation of proteins will appear as white lines in the separation chamber. You can tolerate precipitation as long as no immobile "islands" are formed. Otherwise you have to reduce the sample delivery rate or the concentration of the sample.

References

Bernardo, K., Krut, O., Wiegmann, K., Kreder, D., Micheli, M., Schafer, R., Sickman, A., Schmidt, W. E., Schroder, J. M., Meyer, H. E., Sandhoff, K., and Kroenke, M. (2000). Purification and characterization of a magnesium-dependent neutral sphingomyelinase from bovine brain. *J. Biol. Chem.* **275**, 7641–7647.

Hannig, K. (1961). Die trägerfreie kontinuierliche Elektrophorese und ihre Anwendung. *Fresenius Zeitschrift Anal. Chem.* **181**, 244–274.

Hoffmann, P., Ji, H., Moritz, R. L., Connolly, L. M., Frecklington, D. F., Layton, M. J., Eddes, J. S., and Simpson, R. J. (2001). Continuous free-flow electrophoresis separation of cytosolic proteins from the human colon carcinoma cell line LIM 1215: A non two-dimensional gel electrophoresis-based proteome analysis strategy. *Proteomics* **1**, 807–818.

Krivankova, L., and Bocek, P. (1998). Continuous free-flow electrophoresis. *Electrophoresis* **19**, 1064–1074.

Maida, R., Krieger, J., Gebauer, T., Lange, U., and Ziegelberger, G. (2000). Three pheromone-binding proteins in olfactory sensilla of the two silkmoth species *Antheraea polyphemus* and *Antheraea pernyi*. *Eur. J. Biochem.* **267**, 2899–2908.

Weber, G., and Bocek, P. (1996). Optimized continuous flow electrophoresis. *Electrophoresis* **17**, 1906–1910.

Weber, G., and Bocek, P. (1998). Recent developments in preparative free flow isoelectric focusing. *Electrophoresis* **19**, 1649–1653.

Gel Electrophoresis

Gel-Based Proteomics: High-Resolution Two-Dimensional Gel Electrophoresis of Proteins. Isoelectric Focusing and Nonequilibrium pH Gradient Electrophoresis

Julio E. Celis, Signe Trentemølle, and Pavel Gromov

I. INTRODUCTION

The sequencing of the human and other important genomes is only the beginning of the quest to understand the functionality of cells, tissues, and organs in both health and disease. Together with advances in bioinformatics, this development has paved the way to the revolution in biology and medicine that we are experiencing today. We are rapidly moving from the study of single molecules to the analysis of complex biological systems, and the current explosion of emerging technologies within proteomics and functional genomics (see other articles in this volume) promises to elicit major advances in medicine in the near future. In particular, proteomic technologies are expected to play a key role in the study and treatment of diseases as they provide invaluable resources to define and characterize regulatory and functional networks, investigate the precise molecular defect in diseased tissues and biological fluids, and for developing specific reagents to precisely pinpoint a particular disease or stage of a disease. For drug discovery, proteomics assist with powerful tools for identifying new clinically relevant drug targets and provide functional insight for drug development.

High-resolution two-dimensional (2D) polyacrylamide gel electrophoresis (PAGE), often referred as

gel-based proteomics, multidimensional chromatography, and protein biochips in combination with mass spectrometry (McDonald and Yates, 2002; Yip and Lomas, 2002; Wu and McCoss, 2002 and references therein), are among the proteomic tools that are available for biomarker and drug target discovery. Considerable work is currently underway to explore applications of nongel-based proteomics in various areas of biology as this technology has much to offer.

2D PAGE is often considered the method of choice to separate complex protein mixtures present in cells, tissues, and fluids (Cash and Kroll, 2003; Ong and Pandey, 2001; Celis and Gromov, 1999). The technique separates proteins in terms of both their isoelectric points (pI) and molecular weights (M_r), and it is essentially a stepwise separation tool that combines isoelectric focusing and SDS–polyacrylamide gel electrophoresis. Using the current 2D PAGE technologies it is possible to (i) separate complex protein mixtures into their individual polypeptide components, (ii) compare the protein expression profiles of sample pairs (normal versus transformed cells, cells at different stages of growth or differentiation, etc.), and (iii) choose a condition of interest, e.g., the addition of a cytokine or a drug to a given cell type or tissue, and allow the cell or tissue to reveal the global protein behavioral response under conditions in which all of the detected proteins can be analyzed, both qualitatively and quantitatively in relation to each other (Celis and Olsen, 1994). Protein profiles can be scanned and quantitated to search for protein differences (changes in the levels of preexisting proteins, induction of new products, coregulated polypeptides), and interesting targets or molecular signatures can be identified using additional proteomic technologies, such as mass spectrometry and Western blotting. Furthermore, by carrying out studies in a systematic manner, it is possible to store the information in comprehensive 2D PAGE databases that record how genes are regulated in health and disease (see, e.g., http://proteomics.cancer.dk; Celis *et al.*, 1995, 1998; Gromov *et al.*, 2002).

Today, 2D PAGE can be carried out using two separation modes in the first dimension: (i) conventional isoelectic focusing (IEF) gels and (ii) immobilised pH gradient (IPG) gels (see article by Görg and Weiss in this volume). In the first case, the pH gradient is generated and maintained by special amphoteric compounds, carrier ampholytes, that migrate and stack according to their pI when an electric field is applied. In contrast, IPGs are an integral part of the polyacrylamide matrix, a fact that is achieved by copolymerization of several nonamphoteric buffering species with various pK values (Immobilines), within the fibres of a gel (Bellqwist, 1982). This article presents protocols to resolve proteins based on conventional carrier ampholytes.

Dulbecco's modified Eagle medium (DMEM, Cat. No. 31966-021) is from GIBCO. Penicillin/streptomycin (Cat. No. A2213) are from Biocrom KG. Fetal calf serum (FCS, Cat. No. 04-001-1A) is from Biological Industries. The [^{35}S]methionine (Cat. No. SJ 204) and Amplify fluorographic reagent (Cat. No. NAMP100) are from Amersham. The 96-well plates (Cat. No. 655 180) and 50-ml culture flasks (Cat. No. 690 160) are from Greiner.

First-dimension glass tubes (14 cm in length and 2 mm inside diameter) are from Euro-GLAS. Prior to use they are washed with a solution containing 60 ml of alcohol and 40 ml of HCl. The glass tubes should be immersed in this solution for at least 30 min. Afterward they are washed thoroughly with glass-distilled water. Spacers are cut from 1-mm-thick polystyrene plates (Metzoplast SB/Hk). First- and second-dimension chambers, as well as the rack to hold the first dimension tubes, are homemade. X-ray films (X-OMAT UV; 18 × 24 cm, Cat. No. 524 9792) are from Kodak. The scalpels (Paragon No. 11) are from Paragon, and the long (Cat. No. V2A 1415 LL-10) and short (Cat. No. V2A 1406 LL-7) needles are from Acufirm. The aspiration pump (recirculates water) for drying gels is from Holm & Halby (HETO, Cat No. SUE 300Q).

Power supplies are from Pharmacia Biotech (EPS 500/400) or similar. The orbital shaker (Red Rotor PR70) is from Pharmacia.

II. MATERIALS AND INSTRUMENTATION

Ampholines are from Pharmacia Biotech (pH 3.5–10, Cat. No. 80-1125-87; pH 7–9, Cat. No. 80-1125-94, and pH 8–9.5, Cat. No. 80-1125-95; the pH 8–9.5 ampholyte can be replaced by SERVALYT 9-11, Cat. No. 42909) and Serva (pH 5–7, Cat. No. 42905. Acrylamide (Cat. No. 161-0100), *N,N*'-methylenebisacrylamide (Cat. No. 161-0200), *N,N,N*'*N*'-tetramethylenediamine (TEMED, Cat. No. 161-0800), agarose (Cat. No. 162-0100), and ammonium persulfate (Cat. No. 161-0700) are from Bio-Rad. Dithiothreitol (DTT, Cat. No. D-0632) and bromphenol blue (Cat. No. B-6131) are from Sigma. Glycine (Cat. No. 808822) and urea (Cat. No. 821527) are from ICN Biomedical. Tris base (Cat. No.6483111) and Nonidet P-40 (NP-40, Cat. No. 492015) are from Calbiochem. SDS (Cat. No. 20763) is from Serva. Acrylamide (Cat. No. 10674) from Serva has also been used for the second dimension with essentially the same results. Filter-Count is from Packard (Cat. No. 6013149).

III. PROCEDURES

A. Sample Preparation

1. Labeling of Cultured Cells with [^{35}S]Methionine Solutions

1. *Complete Dulbecco's modified Eagle's medium*: To make 500 ml, mix 445 ml of DMEM medium, 5 ml of 10× stock penicillin/streptomycin, and 50 ml of FCS.
2. *Methionine-free solution*: Supplement MEM lacking methionine with antibiotics (100 U/ml penicillin– 100 µg/ml streptomycin) and 2% dialyzed (against 0.9% NaCl) fetal calf serum (FCS). Dispense in 1-ml aliquots and keep at −80°C.
3. *[^{35}S]Methionine (SJ 204, Amersham)*: Aliquot in 100-µCi portions in sterile 1-ml cryotubes. Keep at −20°C. Freeze dry just before use.
4. *Labeling medium*: Add 0.1 ml of MEM lacking methionine to each ampoule containing 100 µCi of [^{35}S]methionine.

Steps

1. Grow the cells in complete DMEM and seed in a 96-well microtiter plate. Leave in a 37°C humidified, 5% CO_2 incubator until they reach the desired density (3000–4000 cells per well).
2. Freeze dry the [^{35}S]methionine and resuspend in labeling medium at a concentration of 1 mCi/ml. For one well, one needs 100 µCi of [^{35}S]methionine in 0.1 ml of labeling medium.
3. Remove the medium from the well with the aid of a sterile, drawn-out (under a flame) Pasteur pipette. Wash once with labeling medium. Add the labelling medium containing the radioactivity.
4. Wrap the plate in Saran wrap and place in the 37°C humidified, 5% CO_2 incubator for 16 h or shorter period (if necessary).
5. At the end of the labeling period, remove the medium with the aid of a drawn-out Pasteur pipette. Keep the medium if you want to analyze secreted or externalised proteins. Place the 96-well plate at an angle to facilitate removal of the liquid. Dispose of the radioactive medium according to the regulations enforced in your laboratory.
6. Resuspend the cells in 0.1 ml of O'Farrell's lysis solution. Pipette up and down (avoid foaming). Keep at −20°C until use.
7. Apply about 10^6 cpm to the first-dimension gels (IEF and NEPHGE) as described in Section III,B steps 6 and 7.

2. Labeling of Tissue Samples with [^{35}S]Methionine

1. Place the tissue sample on ice immediately after dissection and transport to the laboratory as fast as possible.
2. Remove clots and contaminating tissue with the aid of a scalpel. Rinse the piece two to three times in Hank's solution.
3. Mince the tissue sample in small specimens (about 1 mm^3) with the aid of a scalpel.
4. Place a tissue specimen in a 10-ml sterile plastic conical tube containing 0.2 ml of MEM lacking cold methionine and containing 2% dialyzed FCS and 200 µCi of [^{35}S]methionine and incubate for 16 h at 37°C in a humidified 5% CO_2 incubator.
5. Following incubation, carefully aspirate the medium and resuspend the tissue specimen in 0.2–0.3 ml of lysis solution. Homogenize using a small glass homogenizer.
6. Apply 20–50 µl to the first-dimension gels (IEF and/or NEPHGE) as described in Section III,B steps 6 and 7.

3. Preparation of Cell Extracts from Cultured Cells for Silver Staining

1. Plate cells in 50-ml culture flasks and grow until they reach 80% confluence.
2. Wash the monolayer three times with Hank's buffered saline. Carefully aspirate the fluid with the aid of an extended Pasteur pipette.
3. Add 0.6 ml of lysis solution. Rock at room temperature for a couple of min.
4. Aspirate and keep the extract at −20°C until use. Usually 20–30 µl of the sample can be applied to the first-dimension gel.

4. Preparation of Tissue Extracts for Silver Staining

1. Place the tissue sample on ice immediately after dissection and proceed further as described in Section III,A,2, steps 1–3.
2. Place four to six small tissue pieces in a glass homogenizer and add 1–2 ml of lysis solution. Homogenize at room temperature until the suspension clears up.
3. Keep at −20 °C until use. Usually 20–30 µl of the sample can be applied to the first-dimension gel.

B. Preparation and Running of First-Dimension Gels (IEF, NEPHGE)

This procedure is modified from those of O'Farrell (1975), O'Farrell et al. (1977), and Bravo (1984).

Solutions

1. *Lysis solution:* 9.8 M urea, 2% (w/v) NP-40, 2% ampholytes, pH 7–9, and 100 mM DTT. To make 50 ml, add 29.42 g of urea, 10 ml of a 10% stock solution of NP-40, 1 ml of ampholytes, pH 7–9, and 0.771 g of DTT. After dissolving, complete to 50 ml with distilled water. The solution should not be heated. Aliquot in 2-ml portions and keep at −20°C.

2. *Overlay solution:* 8 M urea, 1% ampholytes, pH 7–9, 5% (w/v) NP-40, and 100 mM DTT. To make 25 ml, add 12.012 g of urea, 0.25 ml of ampholytes, pH 7–9, 12.5 ml of a 10% stock solution of NP-40, and 0.386 g of DTT. After dissolving, complete to 25 ml with distilled water. The solution should not be heated. Aliquot in 2-ml portions and keep at −20°C.

3. *Equilibration solution:* 0.06 M Tris–HCl, pH 6.8, 2% SDS, 100 mM DTT, and 10% glycerol. To make 250 ml, add 15 ml of a 1 M stock solution of Tris–HCl, pH 6.8, 50 ml of a 10% stock solution of SDS, 3.857 g of DTT, and 28.73 ml of glycerol (87% concentration). After dissolving, complete to 250 ml with distilled water. Store at room temperature.

4. *Acrylamide solution:* 28.38% (w/v) acrylamide and 1.62% (w/v) *N,N'*-methylenebisacrylamide. To make 100 ml, add 28.38 g of acrylamide and 1.62 g of bisacrylamide. After dissolving, complete to 100 ml with distilled water. Filter if necessary. Store at 4°C and use within 3 to 4 weeks.

5. *NP-40:* 10% (w/v) NP-40 in H_2O. To make 100 ml, weigh 10 g of NP-40 and complete to 100 ml with distilled water. Dissolve carefully. Store at room temperature.

6. *Agarose solution:* 0.06 M Tris–HCl, pH 6.8, 2% SDS, 100 mM DTT, 10% glycerol, 1% agarose, and 0.002% bromphenol blue. To make 250 ml, add 15 ml of a 1 M stock solution of Tris–HCl, pH 6.8, 50 ml of a 10% stock solution of SDS, 3.857 g of DTT, 28.73 ml of glycerol (87% concentration), 2.5 g of agarose, and 2.5 ml of a 0.2% stock solution of bromphenol blue. Add distilled water and heat in a microwave oven. Complete to 250 ml with distilled water and aliquot in 20-ml portions while the solution is still warm. Keep at 4°C.

7. *1 M NaOH stock:* Weigh 4 g of NaOH and complete to 100 ml with distilled water. Keep at 4°C for no more than 2 weeks.

8. *1 M H_3PO_4 stock:* To make 100 ml, take 6.74 ml of H_3PO_4 (87%) and complete to 100 ml with distilled water. Keep at room temperature.

9. *20 mM NaOH:* To make 500 ml, take 10 ml of 1 M NaOH and complete to 500 ml with distilled water. Prepare fresh.

10. *10 mM H_3PO_4:* To make 500 ml, take 5 ml of 1 M H_3PO_4 and complete to 500 ml with distilled water. Prepare fresh.

Steps

1. Mark the glass tubes with a line (use Easy Marker from Engraver or a diamond-tipped pencil) 12.5 cm from the bottom (Fig. 1A, f). Seal the bottom end of the tube by wrapping with Parafilm and place it standing up in a rack (Fig. 1B).

2a. To make 12 first-dimensional IEF gels, use 4.12 g urea; 0.975 ml of acrylamide solution; 1.5 ml of 10% NP-40, 1.5 ml of H_2O; 0.30 ml of carrier ampholytes, pH range 5–7; 0.10 ml of carrier ampholytes, pH range 3.5–10; 15 µl of 10% ammonium persulfate; and 10 µl of TEMED.

2b. To make 12 first-dimensional NEPHGE gels, use 4.12 g urea; 0.975 ml of acrylamide solution; 1.5 ml of 10% NP-40; 1.69 ml of H_2O; 0.170 ml of carrier ampholytes, pH range 7–9; 0.020 ml of carrier ampholytes, pH range 9–11; 15 µl of 10% ammonium persulfate; and 10.5 µl of TEMED.

2c. Mix the urea, H_2O, acrylamide, NP-40, and ampholytes (kept at –20°C in 1-ml aliquots) in a tube

containing a vacuum outlet (Fig. 1A, g). Swirl the solution gently until the urea is dissolved. The solution should not be heated. Add ammonium persulfate and TEMED, mix gently, and degas using a vacuum pump. Use a clean rubber stopper to control the vacuum.

3. Pour the solution into a 55-mm culture dish. Aspirate the liquid with a 10-ml syringe and add to the thin glass tubes (Fig. 1A, f) using a long needle (Fig. 1A, d). Insert the tip of the needle into the bottom of the tube and slowly fill to the mark to avoid air bubbles while moving up the needle (Fig. 1B).

4. Overlay the gel mix with 10 µl of glass-distilled water and leave to polymerise for 45 min. In the meantime, fill the lower chamber of the first dimension (Fig. 1C) with 250 ml of 10 mM H_3PO_4 (+; IEF gels) or 250 ml of 20 mM NaOH (–; NEPHGE gels).

5. Take the tubes from the rack and remove the Parafilm using a scalpel. Remove excess liquid from the upper part of the tube by shaking and dry using a thin strip of Whatman 3MM paper (Fig. 1A, c). Insert the tubes into the chamber, which holds up to 12 tubes (Fig. 1C). Tap the bottom of the tubes to remove trapped air bubbles.

6. Prerun IEF gels before adding the sample. First add 10 µl of lysis solution and then 10 µl of overlay solution. Use a Gilson microman pipette to apply the solutions. Fill the tubes as well as the upper chamger (–) with 20 mM NaOH. Prerun gels at room temperature for 15 min at 200 V, 30 min at 300 V, and 60 min at 400 V. After prerunning, disconnect the power supply and discard the upper and bottom solutions. Remove the tubes and wash the top of the gels with distilled water. Dry with a thin strip of Whatmann 3MM paper and apply the sample (up to 50 µl in lysis solution). Add 10 µl of overlay solution and fill the tubes with 20 mM NaOH. Fill the upper chamber with 20 mM NaOH (–) and the bottom one with 20 mM H_3PO_4. Run for 19 h at 400 V at room temperature.

7. NEPHGE gels are not prerun. Add the sample in lysis solution (up to 50 µl). Cover with 10 µl of overlay solution. Add 250 ml of 20 mM NaOH to the bottom champer (–) and fill the tubes and the upper chamber with 10 mM H_3PO_4 (+). Run the gels for 4.5 h at 400 V at room temperature.

8. Before stopping the run, add 3.5 ml of equilibration solution to 35-mm tissue culture dishes marked with the gel number in both the bottom part and the lid. Turn off the power supply and take the gels out with the aid of a syringe (Fig. 1A, h) filled with glass-distilled water. First, use a short needle (Fig. 1A, e) to loosen the gel at both ends of the tube. Then extrude the gel with the aid of pressure applied by a 20-ml syringe (Fig. 1D). Collect the gel in a sieve and place in the 35-mm culture dish containing the equilibration

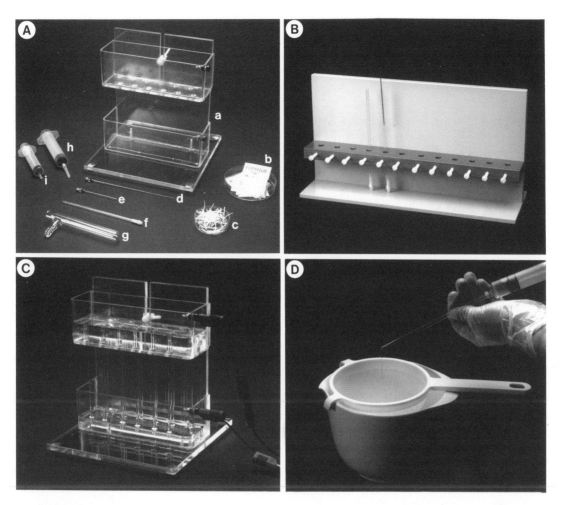

FIGURE 1 (A) First-dimension chamber and accessories for first-dimension. (a) First-dimension chamber, (b) Parafilm, (c) paper strips, (d) long needle for filling the tubes, (e) short needle for extruding the gel, (f) first-dimension tubes, (g) vacuum tube, (h) syringe connecting to a piece of rubber tubing, and (i) syringe. (B) Filling first-dimensional glass tubes with gel solution. (C) First-dimension chamber filled with tubes. (D) Extruding the first-dimension gel.

solution. Leave 2–5 min at room temperature and store at –20°C until use. Samples can be stored for at least 2 months under these conditions.

C. Second Dimension: SDS–Polyacrylamide (15%) Gel Electrophoresis

This procedure is performed essentially according to Laemmli (1970).

Solutions

1. *Solution A:* To make 500 ml, add 150 g of acrylamide and 0.75 g of bisacrylamide. After dissolving, complete to 500 ml with distilled water. Filter if necessary. Aliquot 100-ml portions and store at 4°C.

2. *Solution B:* To make 1 liter of 1.5 M Tris–HCl, pH 8.7, add 181.6 g of Tris base and titrate with HCl. Complete to 1 liter with distilled water. Aliquot 200-ml portions and store at 4°C.

3. *Solution C:* To make 1 liter of 1 M Tris–HCl, pH 6.8, add 121.1 g of Tris base and titrate with HCl. Complete to 1 liter with distilled water. Aliquot in 200-ml portions and store at 4°C.

4. *Solution D:* To make 100 ml, add 10 g of acrylamide and 0.5 g of bisacrylamide. Complete to 100 ml with distilled water. Filter if necessary. Aliquot in 200-ml portions and store at 4°C.

5. *10% SDS:* To make 1 liter, weigh 100 g of SDS and complete to 1 liter with distilled water. Filter if necessary. Store at room temperature.

6. *10% ammonium persulfate:* To make 10 ml, weigh 1 g of ammonium persulfate and complete to 10 ml with distilled water. This solution should be prepared just before use.

7. *Electrode buffer:* To make 1 liter of a 5× solution, add 30.3 g of Tris base, 144 g of glycine, and add 50 ml of 10% SDS solution. Complete to 1 liter with distilled water. Store at room temperature.

Steps

1. Store clean glass plates in dust-free boxes. One of the plates is 16.5 cm wide and 20 cm high and has a notch 2 cm deep and 13 cm wide. Cover the edges of the plate with a thin line of Vaseline. Use a plastic 10-ml syringe filled with vaseline and fitted with a Gilson tip (Fig. 2A). Place 1-mm-thick polystyrene spacers at the edges of the plate and cover with Vaseline (Fig. 2A). Place a small piece of paper without lines containing the gel number (written with pencil) at the corner of the plate (Fig. 2A).

2. Assemble the rectangular glass place 16.5 cm wide and 20 cm high together with the notched plate and spacers. Make sure that the vertical spacers are in contact with the horizontal one at the bottom. Hold the assembled plates together with the aid of fold-back clamps. Mark a line 2.5 cm from the top of the notched plate.

3. To make six 15% separation gels, mix the following solutions in a 250-ml filter flask containing a magnetic stirrer: solution A (acrylamide : bisacrylamide, 30 : 0.15), 75.0 ml; 10% SDS, 1.5 ml; solution B (1.5 M Tris–HCl, pH 8.8), 37.5 ml; H_2O, 35.22 ml; 10% ammonium persulfate, 750 µl; and TEMED, 30 µl.

4. Add ammonium persulfate and TEMED to the separation gel solution just before degassing using a vacuum pump. Pour the solution into the assembled plates until the marked line and overlay with distilled water. Leave the gels to polymerise for approximately 1 h.

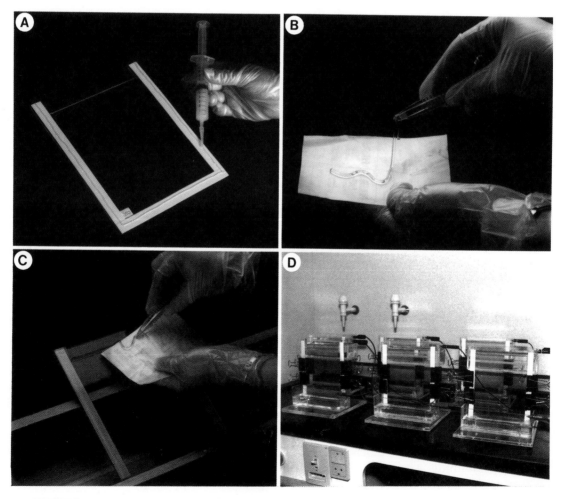

FIGURE 2 (A) Covering the spacers with Vaseline. (B) First-dimension gel prior to application to second-dimension gel. (C) Application of first-dimension gel to second-dimension gel. (D) Second-dimension chambers.

5. Remove excess liquid and dry the top of the gel with a strip of Whatman 3MM paper (2 × 9 cm).

6. To make 5% stacking solution for six gels, mix the following solutions: solution D (acrylamide : bisacrylamide, 10 : 0.5), 15.0 ml; 10% SDS, 0.3 ml; solution C (1.0 M Tris–HCl, pH 6.8), 3.6 ml; H_2O, 10.8 ml; 10% ammonium persulfate, 0.24 ml; and TEMED, 12 μl.

7. After degassing, add the stacking gel solution and insert a polystyrene spacer a few millimetres into the assembled plates. Keep in place with the aid of a fold-out clamp. Leave the gels to polymerise for approximately 1 h.

8. When the gel has polymerised, remove the top spacer and clean the top of the gel with a strip of Whatman 3MM paper. Remove the clamps as well as the horizontal spacer at the bottom with the aid of a spatula. Clean the space between the two glass plates using a thin spatula covered with tissue paper.

9. Lay the gels at an angle in order to facilitate application of the first dimension (Fig. 2C). Take out the culture dishes containing the first-dimension gels from the freezer 20 min before application. Once they are defrosted, melt the agarose solution in a microwave oven and immediately cover the top of the stacking gel with a small amount of agarose to fill the space left by the spacer.

10. Collect the first-dimension gel into a sieve (Fig. 1D) and place it on a piece of Parafilm (Fig. 2B). Place the gel carefully on top of the second dimension with the aid of plastic tweezers (Fig. 2C). Do not stretch the gel. Cover the gels with 2–3 ml of melted agarose. Eliminate air bubbles by pushing them out with the same pipette.

11. Clamp the gel plates to the electrophoresis chambers, which have been prefilled with 1× electrode buffer. Fill the upper chamber with enough electrode buffer to cover the agarose. Remove air bubbles at the bottom of the gel with electrode buffer using a 10-ml syringe joined to a bent needle.

12. Connect the electrodes (upper, –; lower, +) to the power supply. Run the gels at 10 mA for 4 h and at 3 mA overnight at room temperature (until the tracking dye has reached 1 cm from the bottom) (Fig. 2D). At the end of the run turn off the power supply, disassemble the plates, and remove the stacking gel with the aid of a scalpel. Process the separation gel for autoradiography, fluorography (see later), or for staining with either silver (see article by Gromova and Celis in this volume) or Coomassie brilliant blue. Gels can be used directly for blotting (see article by Celis *et al.* in Volume 1).

Figure 3 shows several representative autoradiographs and silver-stained gels of normal (Fig. 3A) and malignant tissues (Figs. 3B–3D) run under the condi-

tions (IEF and NEPHGE) described in this article. For more information, the reader is encouraged to visit the group's Web site at http://proteomics.cancer.dk.

D. Other Procedures

1. Fluorography

This protocol is essentially from Amersham. The procedure increases the detection efficiency 1000-fold for ^3H and 15-fold for ^{35}S.

Solutions

1. *Fixation solution:* To make 1 liter, add 450 ml of methanol and 75 ml of acetic acid. Complete to 1 liter with distilled water.
2. *Amplify fluorographic reagent:* Available commercially from Amersham (Cat. No. NAMP100).

Steps

1. Place the gel in a rectangular glass pie dish (24 × 19 cm) and fix for 60 min at room temperature in fixative solution. Shake while fixing.
2. Place gel in 120 ml of amplifying fluorographic reagent and agitate for 30 min.
3. Dry gels.
4. Sensitize X-ray films by preflashing in the dark and expose the gels at –80°C in cassettes.

2. Quantitation of [^{35}S]Methionine-Labeled Protein Spots Excised from Two-Dimensional Gels

Steps

1. Localize protein spots with the aid of the X-ray film (Fig. 3). Before exposing, make four crosses at the corner of the gels using radioactive ink. Excise the proteins from dry gels using a scalpel.
2. Place gel pieces in counting vials containing 4 ml of Filter-Count, leave for 1 h, and count for 5 min in a scintillation counter.

IV. COMMENTS

Using the protocols described in this article, it is possible to resolve proteins having apparent molecular masses between 8.5 and 230 kDa and pI values from 4 to 12 (see also http://proteomics.cancer.dk). Autoradiographs can be quantitated using phosphorimaging autoradiography (BioRad; Amersham; Fuji). Several softwares for the analysing of 2D protein images are available. These include PDQUEST (BioRad), Z3 (Compugen), Phoretix (Nonlinear Dynamics), GelFox (Imaxia), ProteinMine (Scimagix), and Melanie, as well as several others. By carrying out

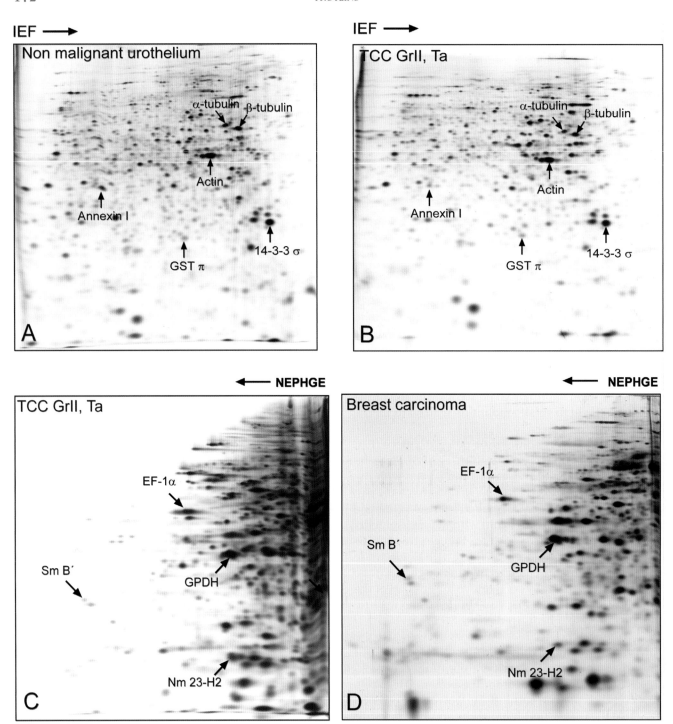

FIGURE 3 Two-dimensional patterns of whole protein extracts from human normal urothelium (A), a transitional cell carcinoma (B and C), and a breast tumor (D). Proteins were separated by 2D PAGE—IEF (A and B) and 2D PAGE—NEPHGE (C and D) and were visualized by autoradiography (A–C) and silver staining (D). The identity of a few proteins is indicated for reference.

Transitional Cell Carcinomas-IEF database

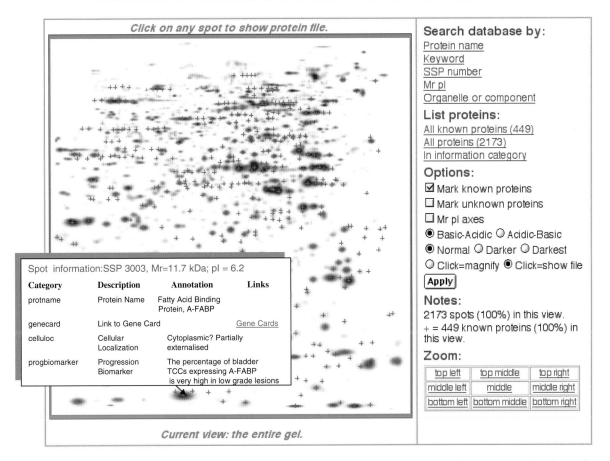

FIGURE 4 Master synthetic image of human bladder TCC proteins separated by IEF 2D PAGE as depicted in the Internet (http://proteomics.cancer.dk). Proteins flagged with a cross correspond to known proteins. By clicking on any spot, it is possible to open a file that contains protein information available in the database as well as links to other related Web sites. Only part of the file for A-FABP is shown.

systematic studies it is possible to store protein information in comprehensive 2D PAGE databases that record how genes are regulated in health and disease (see, e.g., http://proteomics.cancer.dk) (Fig. 4). As these databases achieve critical mass of data, they will become valuables sources of information for expediting the identification of signaling pathways and components that are affected in diseases (Celis *et al.*, 1991a,b, 1998; Gromov *et al.*, 2002).

V. PITFALLS

1. Do not heat the sample when dissolving in lysis solution or when defrosting.

2. For optimal results, titrate the ampholytes. Mix in various proportions, maintaining the final percentage fixed, and run first and second dimensions. Select the combination that gives the best separation (well-focused spots, no streaking, etc.). Store them at $-20°C$ in 1- or 2-ml aliquots. Ampholytes can be stored for many years at $-20°C$.

3. Make sure that all glassware is washed properly. Rinse with glass-distilled water before drying.

4. The pH of the $1.5\,M$ Tris–HCl, pH 8.8, solution should be checked carefully and regularly.

5. Use sterile, disposable pipettes to dispense the solutions. Do not allow other colleagues to use your solutions.

References

Anderson, N. L., and Anderson, N. G. (1978). Analytical technique for cell fractions. XXII. Two-dimensional analysis of serum and tissue proteins: Multiple gradient-slab gel electronporesis. *Anal. Biochem.* **85**, 341–354.

Bravo, R. (1984). Two-dimensional gel electrophoresis: A guide for the beginner. *In* "Two-Dimensional Gel Electrophoresis of Proteins: Methods and Applications" (J. E. Celis *et al.*, eds.), pp. 3–36. Academic Press, New York.

Celis, J. E. (ed.) (1996). Electrophoresis in cell biology. *Electrophoresis* **11**.

Celis, J. E., and Bravo, R. (eds.) (1984). "Two-Dimensional Gel Electrophoresis of Proteins: Methods and Applications." Academic Press, New York.

Celis, J. E., Gromov, P., Østergaard, M., Madsen, P., Honoré, B., Dejgaard, K., Olsen, E. Vorum, H., Kristensen, D. B., Gromova, I., Haunsø, A., Van Damme, J., Puype, M., Vandekerckhove, J., and Rasmussen, H. H. (1996). *FEBS Lett.* **398**, 129–134.

Celis, J. E., Leffers, H., Rasmussen, H. H., Madsen, P., Honore, B., Gesser, B., Dejgaard, K., Olsen, E., Ratz, G. P., Lauridsen, J. B., *et al.* (1991a). The master two-dimensional gel database of human AMA cell proteins: Towards linking protein and genome sequence and mapping information (update 1991). *Electrophoresis* **12**, 765–801.

Celis, J. E., Østergaard, M., Jensen, N. A., Gromova, I., Rasmussen, H. H., and Gromov, P. (1998). Human and mouse proteomic databases: Novel resources in the protein universe. *FEBS Lett.* **430**, 64–72.

Celis, J. E., Rasmussen, H. H., Leffers, H., Madsen, P., Honore, B., Gesser, B., Dejgaard, K., and Vandekerckhove, J. (1991b). Human cellular protein patterns and their link to genome DNA sequence data: Usefulness of two-dimensional gel electrophoresis and microsequencing. *FASEB J.* **5**, 2200–2208.

Celis, J. E., Rasmussen, H. H., Olsen, E., Gromov, P., Madsen, P., Leffers, H., Honoré, B., Dejgaard, K., Vorum, H., Kristensen, D.

B., Østergaard, M., Haunsø, A., Jensen, N. A., Celis, A., Basse, B., Lauridsen, J. B., Ratz, G. P., Andersen, A. H., Walbum, E., Kjærgaard, I., Andersen, I., Puype, M., Van Damme, J., and Vandekerckhove, J. (1995). The human keratinocyte two-dimensional gel protein database (update 1995): Mapping components of signal transduction pathways. *Electrophoresis* **12**, 2217–2184.

Gromov. P., Østergaard, M., Gromova, I., and Celis, J. E. (2002). Human proteomic databases: A powerful resource for functional genomics in health and disease. *Prog. Biophys. Mol. Biol.* **80**, 3–22.

Klose, J. (1975). Protein mapping by combined isoelectric focusing and electrophoresis of mouse tissues: A novel approach to testing for induced point mutations in mammals. *Humangenetik* **26**, 231–243.

Laemmli, U. K. (1970). Cleavage of structural proteins during the assembly of the head of bacteriophage T4. *Nature (London)* **227**, 680–685.

Laskey, R. A., and Mills, A. D. (1975). Quantitative film detection of ^{3}H and ^{14}C in polyacrylamide gels by fluorography. *Eur. J. Biochem.* **56**, 335–341.

McDonald, W. H., and Yates, J. R., 3rd (2002). Shotgun proteomics and biomarker discovery. *Dis. Mark.* **18**, 99–105.

O'Farrell, P. H. (1975). High resolution two-dimensional electrophoresis of proteins. *J. Biol. Chem.* **250**, 4007–4021.

O'Farrell, P. Z., Goodman, H. M., and O'Farrell, P. H. (1977). High resolution two dimensional electrophoresis of basic as well as acidic proteins. *Cell* **12**, 1133–1142.

Ong, S. E., and Pandey, A. (2001) An evaluation of the use of two-dimensional gel electrophoresis in proteomics. *Biomol. Eng.* **18**, 195–205.

Wu, C. C., and McCoss, M. J. (2002). Shotgun proteomics: Tools for the analysis of complex biological systems. *Curr. Opin. Mol. Ther.* **4**, 242–250.

Yip, T. T., and Lomas, L. (2002). SELDI ProteinChip array in onco-proteomic research. *Cancer Res. Treatment* **1**, 273–274.

23

High-Resolution Two-Dimensional Electrophoresis with Immobilized pH Gradients for Proteome Analysis

Angelika Görg and Walter Weiss

I. INTRODUCTION

Although promising progress has been made in the development of alternative protein separation techniques for proteome analysis, such as multidimensional chromatography–tandem mass spectrometry, there is still no generally applicable method that can replace two-dimensional gel electrophoresis (2D PAGE) in its ability to simultaneously separate and display thousands of proteins from complex biological samples. 2D PAGE separates proteins according to two independent parameters, i.e., isoelectric point (pI) in the first dimension and molecular mass (M_r) in the second dimension, by coupling isoelectric focusing (IEF) and sodium dodecyl sulfate polyacrylamide gel electrophoresis (SDS–PAGE). In comparison with the classical O'Farrell method (1975) of 2D PAGE based on the use of carrier ampholyte-generated pH gradients, 2D PAGE using immobilized pH gradients (IPG) in the first dimension (IPG-Dalt) (Görg *et al.*, 1988) has proved to be extremely flexible with respect to the requirements of proteome analysis.

For proteome analysis it is essential to generate reproducible, high-resolution protein separations. Using the classical 2D PAGE approach of O'Farrell, it is, however, often difficult to obtain reproducible results even within a single laboratory, let alone between different laboratories. The problem of limited reproducibility is largely due to the synthetic carrier ampholytes (CA) used to generate the pH gradient required for IEF, for reasons such as batch-to-batch

variability of CAs, pH gradient instability over time, cathodic drift etc. with resultant loss of alkaline proteins. These problems have been largely overcome by the development of immobilized pH gradients (IPG) (Bjellqvist *et al.*, 1982) based on the use of the bifunctional Immobiline reagents, a series of 10 chemically well-defined acrylamide derivatives with the general structure CH_2=CH–CO–NH–R, where R contains either a carboxyl or an amino group. These form a series of buffers with different pK values between pK 1 and >12. Because the reactive end is copolymerized with the acrylamide matrix, extremely stable pH gradients are generated, allowing true steady-state IEF with increased reproducibility, as has been demonstrated in several interlaboratory comparisons (Corbett *et al.*, 1994; Blomberg *et al.*, 1995).

More than a decade ago, a basic protocol of IPG-Dalt was described by Görg *et al.* (1988), summarizing the critical parameters inherent to isoelectric focusing with IPGs and a number of experimental conditions that were not part of the classical 2D electrophoresis repertoire with CAs. In principle, this protocol is still valid today: The first dimension of IPG-Dalt, IEF, is performed in individual 3-mm-wide IPG gel strips cast on GelBond PAGfilm (either ready-made Immobiline DryStrips or laboratory made). Samples are applied either by cup loading or by in-gel rehydration. After IEF, the IPG strips are equilibrated with SDS buffer in the presence of urea, glycerol, dithiothreitol (DTT), and iodoacetamide and are applied onto horizontal or vertical SDS gels in the second dimension.

II. MATERIALS AND INSTRUMENTATION

Immobiline chemicals (Cat. Nos. 80125570–80125575), Pharmalyte pH 3–10 (Cat. No. 17-0456-01), Immobiline DryStrips pH 4–7, 3–10 and 3–10NL (Cat. Nos. 17-1233-01, 17-1234-01, and 17-1235-01), narrow-range Immobiline DryStrips covering the pH range between pH 3.5 and pH 6.7 (Cat. Nos. 17-6001-83–17-6001-87), ExcelGel SDS and buffer strips (Cat. No. 17-1236-01 and Cat. No. 80-1129-42), IEF electrode strips (Cat. No. 18-1004-40), CHAPS (Cat. No. 17-1314-01), acrylamide (Cat. No. 17-1302-01), N, N'-methylenebisacrylamide (Cat. No. 17-1304-01), TEMED (Cat. No. 17-1312-01), GelBond PAGfilm (Cat. No. 80-1129-37), and DryStrip cover fluid (Cat. No. 17-1335-01) are from Amersham Biosciences. Sodium dodecyl sulfate (SDS) (Cat. No. 20763), ammonium persulfate (Cat. No. 13375), glycine (Cat. No. 23390), kerosene (Cat. No. 26940), and ion exchanger (Serdolite MB-1, Cat. No. 40701) are from Serva. Trizma base (Cat. No. T-1503), DTT (Cat. No. D-0632), acrylamido buffers (pK 1.0, 10.3, and 12.0; Cat. No. 84885), thiourea (Cat. No. T8656), and iodoacetamide (Cat. No. I-6125) are from Sigma-Aldrich. Glycerol (Cat. No. 4093), urea (Cat. No. 84879), and a serine protease inhibitor (Pefabloc, Cat. No. 24839) are from VWR. Dimethylchlorosilane (Cat. No. 40140) is from Fluka. Water is deionized using the Milli-Q system of Millipore.

Apparatus for isoelectric focusing and horizontal SDS electrophoresis (Multiphor II, Cat. No. 18-1018-06), multiple vertical SDS electrophoresis (Ettan DALT Vertical System, Cat. No. 80-6466-46), thermostatic circulator (Multitemp III, Cat. No. 18-1102-78), power supply (Multidrive XL, Cat. No. 18-1013-68), IPGphor (Cat. No. 80-6469-88), IPGphor strip holders (Cat. No. 80-6416-68), cup-loading strip holders (Cat. No. 80-6459-43), IPG DryStrip reswelling tray (Cat. No. 80-6465-32), IPG DryStrip kit (Cat. No. 18-1004-30), and gradient gel kit (including glass plates with a 0.5-mm-thick U frame and gradient maker, Cat. No. 80-1013-74) are from Amersham Biosciences. Prior to use the glass plates are washed with a mild detergent, rinsed with deionized water, and air dried. If new glass plates are used, pipette 1–2 ml of dimethylchlorsilane (diluted 1 + 9 in trichlorethane) on the glass plate that bears the U frame and distribute it evenly with a fuzz-free filter paper. Let it dry for a few minutes, rinse again with water, and let it air dry. Repeat this procedure occasionally in order to prevent the gels from sticking to the glass plates.

III. PROCEDURES

A. Preparation of First-Dimensional IPG Gel Strips

Solutions

1. *Acrylamide/bisacrylamide solution (30% T, 3% C)*: 29.1% (w/v) acrylamide, 0.9% (w/v) N, N'-methylenebisacrylamide. To make 100 ml of the solution, dissolve 29.1 g of acrylamide and 0.9 g of bisacrylamide in deionized water and fill up to 100 ml. Add 1 g of Serdolite MB-1, stir for 10 min, and filter. This solution can be stored for 1 week at 4°C; however, for optimum results it is advisable to prepare it freshly the day you use it.

2. *Ammonium persulfate solution*: 40% (w/v) in deionized water. To prepare 1 ml of the solution, dissolve 0.4 g of ammonium persulfate in 1 ml of deionized water. This solution should be prepared freshly just before use.

3. *Solutions for casting immobiline gels*: To prepare 15 ml each of acidic and basic solutions, mix chemicals and reagent solutions as described in **Table I**. A huge selection of recipes for many types of narrow or broad pH gradient have been calculated (Righetti, 1990). Table I describes several of our favourite pH gradients for 2D electrophoresis.

Steps

IPG slab gels for 180-mm separation distance (250 × 180 × 0.5 mm³) are cast on GelBond PAGfilm. After polymerization, the IPG gels are washed extensively with deionized water, air dried, and stored frozen until used.

1. To assemble the polymerisation cassette, wet the plain glass plate (size 260 × 200 mm²) with a few drops of water. Place the Gelbond PAGfilm, hydrophilic side upward, on the wetted surface of the plain glass plate. The GelBond PAGfilm should overlap the upper edge of the glass plate for 1–2 mm to facilitate filling of the cassette. Expel excess water with a roller. Place the glass plate that bears the U frame (0.5 mm thick) on top of the Gelbond PAGfilm and clamp the cassette together (**Fig. 1A**). Put it in the refrigerator for 30 min.

2. To cast the IPG gel, pipette 12.0 ml of the acidic, dense solution into the mixing chamber of the gradient mixer. The outlet and connecting line between the mixing chamber and reservoir have to be closed! Add 7.5 µl of TEMED and 12 µl of ammonium persulfate and mix. Open the connecting line between the chambers for a second to release any air bubbles.

TABLE I   Recipes for Casting Immobiline Gels with pH Gradients[a]

Linear pH gradient	pH 4–7		pH 4–9		pH 6–12		pH 4–12		pH 3–12	
	Acidic solution (pH 4)	Basic solution (pH 7)	Acidic solution (pH 4)	Basic solution (pH 9)	Acidic solution (pH 6)	Basic solution (pH 12)	Acidic solution (pH 4)	Basic solution (pH 12)	Acidic solution (pH 3)	Basic solution (pH 12)
Immobiline pK 1.0	—	—	—	—	—	—	—	—	1287 µl	—
Immobiline pK 3.6	578 l	302 µl	829 µl	147 µl	1367 µl	—	950 µl	—	306 µl	—
Immobiline pK 4.6	110 µl	738 µl	235 µl	424 µl	—	—	352 µl	74 µl	414 µl	—
Immobiline pK 6.2	450 µl	151 µl	232 µl	360 µl	188 µl	251 µl	319 µl	206 µl	558 µl	336 µl
Immobiline pK 7.0	—	269 µl	22 µl	296 µl	323 µl	125 µl	294 µl	103 µl	496 µl	168 µl
Immobiline pK 8.5	—	—	250 µl	71 µl	365 µl	84 µl	48 µl	522 µl	112 µl	699 µl
Immobiline pK 9.3	—	876 µl	221 µl	663 µl	497 µl	32 µl	52 µl	219 µl	84 µl	157 µl
Immobiline pK 10.0	—	—	—	—	50 µl	485 µl	41 µl	325 µl	25 µl	342 µl
Immobiline pK > 13	—	—	—	—	—	345 µl	—	531 µl	—	258 µl
Acrylamide/ bisacrylamide	2.0 ml	2.0 ml	2.0 ml	2.0 ml	2.25 ml	2.25 ml	2.5 ml	2.5 ml	2.25 ml	2.25 ml
deionized water	8.9 ml	10.7 ml	8.3 ml	11.1 ml	7.0 ml	10.5 ml	7.4 ml	10.5 ml	6.45 ml	10.8 ml
Glycerol (100%)	3.75 g	—	3.75 g	—	3.75 g	—	3.75 g	—	3.75 g	—
TEMED (100%)	10.0 µl	10.0 µl	10.0 µl	10.0 µl	10.0 µl	10.0 µl	10.0 µl	10.0 µl	10.0 µl	10.0 µl
Persulfate (40%)	15.0 µl	15.0 µl	15.0 µl	15.0 µl	15.0 µl	15.0 µl	15.0 µl	15.0 µl	15.0 µl	15.0 µl
Final volume	15.0 ml	15.0 ml	15.0 ml	15.0 ml	15.0 ml	15.0 ml	15.0 ml	15.0 ml	15.0 ml	15.0 ml

[a] For effective polymerization, acidic and basic solutions are adjusted to pH 7 with 4 N sodium hydroxide and 4 N acetic acid, respectively, before adding the poymerization catalysts (TEMED and ammonium persulfate).

3. Pipette 12.0 ml of the basic, light solution into the reservoir of the gradient mixer. Add 7.5 µl of TEMED and 12 µl of ammonium persulfate and mix with a spatula.

4. Switch on the magnetic stirrer at a reproducible and rapid rate; however, avoid excessive vortex. Remove the polymerisation cassette from the refrigerator and put it underneath the outlet of the gradient mixer. Open the valve connecting the chambers and, immediately afterwards, the pinchcock on the outlet tubing so that the gradient mixture is applied centrally into the cassette from a height of about 5 cm just by gravity flow. Formation of the gradient is completed in 2–3 min (**Fig. 1B**).

5. Keep the mold at room temperature for 15 min to allow adequate leveling of the density gradient. Then polymerize the gel for 1 h at 50°C in a heating cabinet.

6. After polymerization, wash the IPG gel for 1 h with 10-min changes of deionized water (500 ml each) in a glass tray on a rocking platform. Then equilibrate the gel in 2% (w/v) glycerol for 30 min and dry it overnight at room temperature, using a fan, in a dust-free cabinet. Afterward, protect the surface of the dry gel with a sheet of plastic film. The dried IPG gel can be stored in a sealed plastic bag at –20°C for at least several months without loss of function. Dried IPG

gels in several pH ranges are also available commercially (Immobiline DryPlate).

7. For IEF in individual IPG gel strips, cut the dried IPG gels or the ready-made Immobiline DryPlates into 3-mm-wide strips with the help of a paper cutter (**Fig. 1C**). Alternatively, ready-cut IPG strips (Immobiline DryStrip) can also be used.

B. Running of First-Dimensional IPG Strips

The first dimension of IPG-Dalt, isoelectric focusing, is performed in individual 3-mm-wide IPG gel strips cast on GelBond PAGfilm. Instead of laboratory-made gels, ready-made gels (Immobiline DryStrip) can be used. Samples are applied either by in-gel rehydration or by cup loading after IPG strip rehydration (*see* **Table II and Fig. 2**). IPG-IEF can be simplified by using an integrated system, the IPGphor. This instrument features strip holders that provide rehydration of individual IPG strips with or without sample, optional separate sample cup loading, and subsequent IEF, all without handling the strip after it is placed in a ceramic strip holder.

Solutions

1. *Lysis solution*: 9.5 M urea, 2% (w/v) CHAPS, 2% (v/v) Pharmalyte pH 3–10, 1% (w/v) DTT. To prepare

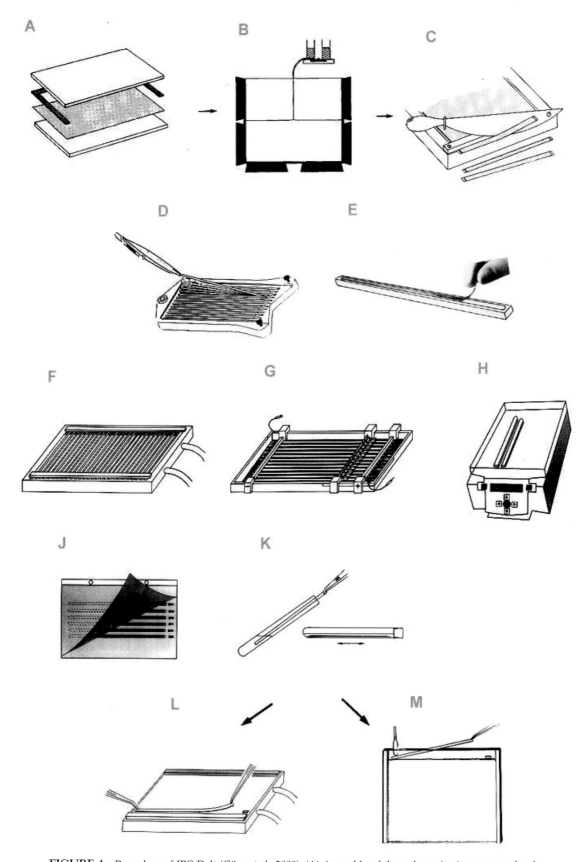

FIGURE 1 Procedure of IPG-Dalt (Görg *et al.*, 2000). (**A**) Assembly of the polymerisation cassette for the preparation of IPG and SDS gels on plastic backings (glass plates, GelBond PAGfilm and 0.5-mm-thick U frame). (**B**) Casting of IPG gels. (**C**) Cutting of washed and dried IPG slab gels (or IPG DryPlates) into individual IPG strips. (**D**) Rehydration of individual IPG strips in the IPG DryStrip reswelling tray. (**E**) Rehydration of individual IPG strips in the IPGphor strip holder. (**F**) IEF of individual IPG gel strips directly on the cooling plate of the IEF chamber. (**G**) IEF of individual IPG gel strips in the IPG DryStrip kit. (**H**) IEF of individual IPG gel strips on the IPGphor. (**J**) Storage of IPG strips after IEF. (**K**) Equilibration of IPG strips prior to SDS–PAGE. (**L**) Transfer of the equilibrated IPG strip onto onto the surface of a ready-made horizontal SDS gel along the cathodic buffer strip. (**M**) Loading of the equilibrated IPG strip onto onto the surface of a vertical SDS gel.

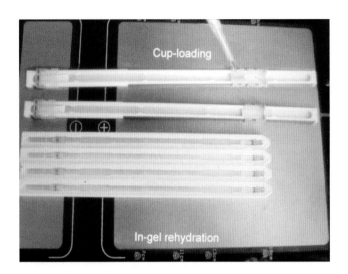

FIGURE 2 IPGphor: Sample application by in-gel rehydration and by cup loading.

TABLE II Sample Application

In-gel rehydration	Wide pH range IPGs between pH 3 and pH 12 (analytical and micropreparative runs)
Cup loading at the anode	Narrow pH range IPGs at the basic extreme (e.g., IPG 9–12)
Cup loading at the cathode	Narrow pH range IPGs at the acidic extreme (e.g., IPG 2.5–5)

50 ml of lysis solution, dissolve 30.0 g of urea in deionized water and make up to 50 ml. Add 0.5 g of Serdolite MB-1, stir for 10 min, and filter. Add 1.0 g of CHAPS, 0.5 g DTT, 1.0 ml of Pharmalyte pH 3–10, and 50 mg of Pefabloc to 48 ml of the urea solution.

2. *IPG strip rehydration solution*: $8\,M$ urea, 0.5% (w/v) CHAPS, 15 mM DTT, 0.5% (v/v) Pharmalyte pH 3–10. To prepare 50 ml of the solution, dissolve 25.0 g of urea in deionized water and complete to 50 ml. Add 0.5 g of Serdolite MB-1, stir for 10 min, and filter. To 48 ml of this solution add 0.25 g of CHAPS, 0.25 ml Pharmalyte pH 3–10 (40% w/v), and 100 mg of DTT and complete to 50 ml with deionized water.

Notes

a. For solubilization of the more hydrophobic proteins, *thiourea/urea lysis solution* [2 M thiourea, 5–7 M urea, 2–4% (w/v) CHAPS, and/or sulfobetaine detergents, 1% DTT, 2% (v/v) carrier ampholytes] and *rehydration solution* consisting of a mixture of urea/thiourea [6 M urea, 2 M thiourea, 1% (w/v) CHAPS,

15 mM DTT, 0.5% (v/v) Pharmalyte pH 3–10] are recommended (Rabilloud *et al.*, 1997).

b. Lysis solution and IPG strip rehydration solution should always be prepared freshly. Alternatively, small aliquots (1 ml) can be stored at −80°C. Lysis and rehydration solution thawn once should not be refrozen again. Never heat urea solutions above 37°C! Otherwise, protein carbamylation may occur.

c. It is important that the urea solution is deionized with an ion exchanger prior to adding the other chemicals because urea in aqueous solution exists in equilibrium with ammonium cyanate, which can react with the NH_3^+ of protein side chains (e.g., lysine) and introduce charge artifacts (= additional spots) on the 2D gel.

1. Rehydration of IPG DryStrips

Prior to IEF, the dried IPG strips must be rehydrated to their original thickness of 0.5 mm. IPG DryStrips are rehydrated either with sample already dissolved in rehydration buffer ("sample in-gel rehydration") or with rehydration buffer only, followed by sample application by "cup loading." Sample in-gel rehydration is generally not recommended for samples containing very high molecular weight, very alkaline, and/or very hydrophobic proteins, as these are taken up into the gel with difficulties only. In these cases, cup loading is preferred.

a. Rehydration of IPG DryStrips Using the IPG DryStrip Reswelling Tray

1. For *sample in-gel* rehydration (Rabilloud *et al.*, 1994), directly solubilize a cell lysate or tissue sample (1–10 mg protein/ml) in an appropriate quantity of rehydration buffer. For 180-mm-long and 3-mm-wide IPG dry strips, pipette 350 µl of this solution into the grooves of the IPG DryStrip reswelling tray (**Fig. 1D**). For longer or shorter IPG strips, the rehydration volume has to be adjusted accordingly.

2. Remove the protective covers from the surface of the IPG DryStrips and apply the IPG strips, gel side down, into the grooves without trapping air bubbles. Then cover the IPG strip, which must still be moveable and not stick to the tray, with silicone oil (which prevents drying out during reswelling) or IPG cover fluid and rehydrate the strips overnight at approximately 20°C. Higher temperatures (>37°C) hold the risk of protein carbamylation, whereas lower temperatures (<10°C) should be avoided to prevent urea crystallization on the IPG gel.

3. For *cup loading*, rehydrate the IPG dry strips overnight in rehydration buffer in the reswelling tray as described earlier, however without sample.

b. Sample In-Gel Rehydration of IPG DryStrips Using IPGphor Strip Holders

1. Solubilize proteins in sample solubilization buffer and dilute the extract (protein concentration ≈10 mg/ml) with IPG strip rehydration solution (dilution: 1 + 1 for micropreparative runs, and 1 + 19 for analytical runs, respectively) to a final volume of 350–400 µl for 180-mm-long IPG strips.

2. Apply the required number of IPGphor strip holders (**Fig. 1E**) onto the cooling plate/electrode contact area of the IPGphor.

3. Pipette 350 µl of sample-containing rehydration solution (for 180-mm-long IPG strips) into the strip holder base. For shorter (e.g., 110 mm) IPG strips in shorter strip holders, use correspondingly less liquid.

4. Peel off the protective cover sheets from the IPG strip and slowly lower the IPG strip (gel side down) onto the rehydration solution. Avoid trapping air bubbles. The IPG strip must still be moveable and not stick to the tray. Cover the IPG strips with 1–2 ml of cover fluid and apply the plastic cover. Pressure blocks on the underside of the cover assure that the IPG strip keeps in good contact with the electrodes as the gel swells.

2. IEF on the Multiphor Flat-Bed Electrophoresis Apparatus

1. After IPG DryStrip rehydration, rinse the rehydrated IPG strips with distilled water for a few seconds and then blot them between two sheets of moist filter paper to remove excess rehydration buffer in order to avoid urea crystallization on the gel surface, which is held responsible for prolonged IEF and "empty" vertical lanes in the 2D pattern.

2. Apply up to 40 IPG strips (*rehydrated with or without sample solution*) side by side and 1–2 mm apart onto the surface of the kerosene-wetted cooling plate (**Fig. 1F**) of the horizontal flat-bed electrophoresis apparatus (e.g., Multiphor, Amersham Biosciences). It is of utmost importance that the acidic ends of the IPG gel strips face towards the anode.

3. Soak electrode paper strips (cut from 1-mm-thick filter paper, e.g., MN 440, Machery & Nagel) with deionized water, blot against filter paper to remove excess liquid, and place them on top of the aligned IPG gel strips at the cathodic and anodic ends. If samples have already been applied by sample in-gel rehydration, proceed as described in step 5).

4. In case of sample application by cup loading, apply the samples (20 µl; protein concentration 5–10 mg/ml) into silicon rubber frames (size: 2 × 5 mm²) or special sample cups placed at either the anodic or the cathodic end of the IPG strips. (see **Fig. 1F**). For analytical purposes, load 50–100 µg of protein onto a single, 180-mm-long IPG gel strip. For micropreparative purposes, up to several milligrams of protein may be applied with the help of a special strip tray, the IPG DryStrip kit (**Fig. 1G**) (*see* **Notes**).

5. Position the electrodes and press them down on top of the IEF electrode paper strips. Place the lid on the electrofocusing chamber, connect the cables to the power supply, and start IEF. For maximal sample entry, the initial voltage should be limited to 150 V for 60 min and then increased progressively until 3500 V is attained. Current and power settings should be limited to 0.05 mA and 0.2 W per IPG strip, respectively. The optimum focusing temperature is 20°C (Görg *et al.*, 1991). The time required for the run depends on several factors, including the type of sample, the amount of protein applied, the length of the IPG strips, and the pH gradient being used. Some typical running conditions are given in **Table III** (Görg *et al.*, 2000).

6. After termination of IEF, the IPG strips can be used immediately for the second dimension. Alternatively, strips can be stored between two sheets of plastic film at –70°C for several months.

Notes

a. When basic IPG gradients are used for the first dimension (e.g., IPG 6–10), horizontal streaking can often be observed at the basic end of 2D protein profiles. This problem may be resolved by applying an extra electrode strip soaked in 20 m*M* DTT on the surface of the IPG strip alongside the cathodic electrode strip (Görg *et al.*, 1995). This has the advantage that the DTT within the gel, which migrates towards the anode during IEF, is replenished by the DTT released from the strip at the cathode. Alternative approaches are (i) to use the noncharged reducing agent, tributyl phosphine, which does not migrate during IEF (Herbert, 1999), (ii) to substitute DTT in the rehydration buffer of the IPG strip by a disulfide such as hydroxyethyl disulfide (Olsson *et al.*, 2002), or (iii) to apply high voltages (8000 V) for short running times (Wildgruber *et al.*, 2002).

b. After sample entry, the filter papers beneath the anode and cathode should be replaced with fresh ones. This is because salt contaminants have quickly moved through the gel and have now collected in the electrode papers. For high salt and/or protein concentrations, it is recommended to change the filter paper strips several times. In case of IEF with very alkaline, narrow range IPGs, such as IPG 10–12, this procedure should be repeated once an hour.

c. For micropreparative IEF or for running very alkaline IPGs exceeding pH 10, a special strip tray (IPG DryStrip kit) is recommended that has a frame that fits

TABLE III   Running Conditions Using the Multiphor

Gel length	180 mm
Temperature	20°C
Current max.	0.05 mA per IPG strip
Power max.	0.2 W per IPG strip
Voltage max.	3500 V

I. Analytical IEF
  Initial IEF
    Cup loading (20–50 μl)
      150 V, 1 h
      300 V, 1–3 h
      600 V, 1 h
    In-gel rehydration (350 μl)
      150 V, 1 h
      300 V, 1–3 h

  IEF to the steady state at 3500 V

1–1.5 pH units			4 pH units		7 pH units		
e.g., IPG 5–6	24 h		IPG 4–8	10 h	IPG 3–10 L	6 h	
e.g., IPG 4–5.5	20 h		IPG 6–10	10 h	IPG 3–10 NL	6 h	

3 pH units			5–6 pH units		8–9 pH units		
IPG 4–7	12 h		IPG 4–9	8 h	IPG 3–12	6 h	
IPG 6–9	12 h		IPG 6–12	8 h	IPG 4–12	8 h	

II. Extended Separation Distances (240 mm)
  IEF to the steady state at 3500 V
      IPG 3–12   8 h
      IPG 4–12   12 h
      IPG 5–6   40 h

III. Micropreparative IEF
  Initial IEF
    Cup loading (100 μl)
      50 V, 12–16 h
      300 V, 1 h
    In-gel rehydration (350 μl)
      50 V, 12–16 h
      300 V, 1 h
  IEF to the steady state at 3500 V
    Focusing time of analytical IEF plus approximately 50%

[a] From Görg et al. (2000).

on the cooling plate of the Multiphor (**Fig. 1G**). This tray is equipped with a corrugated plastic plate that contains grooves allowing easy alignment of the IPG strips. In addition, the tray is fitted with bars carrying the electrodes and a bar with sample cups allowing application of samples at any desired point on the gel surface. The advantages are that the cups can handle a larger quantity of sample solution (100 μl can be applied at a time, but it is possible to apply a total of up to 200 μl portion by portion, onto a single IPG gel), and the frame allows one to cover the IPG strips with a layer of silicone oil that protects the gel from the effects of the atmosphere during IEF. In case of very basic pH gradients exceeding pH 10 (e.g., IPG 3–12, 4–12, 6–12, 9–12, 10–12) or narrow-range ("zoom" gels)

pH gradients with 1.0 or 1.5 pH units with extended running time (>24 h), the IPG strips *must* be covered by a layer of silicone oil.

d. In case of cup loading, it is not recommended to apply proteins at (or proximate to) the pH area that corresponds with their pI in order to avoid that they are uncharged and poorly soluble and thus prone to precipitation at the sample application site.

e. The IEF run should be performed at 20°C, as at lower temperatures there is a risk of urea crystallization and at higher temperatures carbamylation might occur. Precise temperature control is also important because it has been found that alterations in the relative positions of some proteins on the final 2D patterns may happen (Görg, 1991).

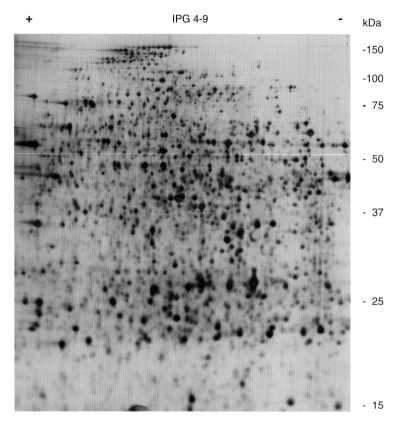

FIGURE 3 IPG-Dalt of mouse liver proteins. First dimension: IEF with IPG 4–9 run on the IPGphor. Separation distance: 24 cm. Sample application by cup loading near the anode. Second dimension: vertical SDS–PAGE, 13% T. Silver stain.

3. IEF on the IPGphor

Using the IPGphor, sample in-gel rehydration and IEF can be carried out automatically according to the programmed settings, preferably overnight. Alternatively, the IPGphor can also be used with a cup-loading procedure, which allows the application of quantities up to 100 µl. The instrument can accommodate up to 12 individual strip holders and incorporates Peltier cooling with precise temperature control and a programmable 8000-V power supply. The IPGphor saves about a day's worth of work by combining sample application and rehydration, as well as by starting the run at preprogrammed times, and by running the IEF at rather high voltages (Fig. 3).

a. IEF following In-Gel Rehydration of Sample

1. Apply the required number of strip holders onto the cooling plate/electrode contact area of the IPGphor. Pipette the appropriate amount of sample-containing rehydration solution into the strip holders, lower the IPG dry strips gel side down into the rehydration solution, and overlay with cover fluid (**Fig. 1E**).

2. Program the instrument (**Fig. 1H**) (desired rehydration time, volthours, voltage gradient). Apply low voltage (30–50 V) during IPG strip rehydration for improved entry of high M_r proteins into the polyacrylamide gel matrix (Görg *et al.*, 1999, 2000).

3. After the IPG gel strips have been rehydrated (which requires 6 h at least, but typically overnight), IEF starts according to the programmed parameters (*see* **Table IV**).

4. After completion of IEF, store those IPG gel strips that are not used immediately for a second-dimension run between two sheets of plastic film at −70°C (**Fig. 1J**).

b. IEF after Sample Application by Cup Loading

When protein separation is performed in basic pH ranges, much better separations are obtained by applying sample *via* cup loading on separately rehydrated IPG strips than by sample in-gel rehydration. Sample cup loading is accomplished with a special cup-loading ("universal") IPGphor strip holder.

TABLE IV   IPGphor Running Conditions[a]

Gel length	180 mm			
Temperature	20°C			
Current max.	0.05 mA per IPG strip			
Voltage max.	8000 V			
I. Analytical IEF				
Reswelling	30 V, 12–16 h			
Initial IEF:	200 V, 1 h			
	500 V, 1 h			
	1000 V, 1 h			
IEF to the steady state:	Gradient from 1000 to 8000 V within 30 min			
	8000 V to the steady state, depending on the pH			
	used used			

1–1.5 pH units		4 pH units		7 pH units	
e.g., IPG 5–6	8 h	IPG 4–8 4 h		IPG 3–10 L	3 h
e.g., IPG 4–5.5	8 h			IPG 3–10 NL	3 h
3 pH units		5–6 pH units		8–9 pH units	
IPG 4–7	4 h	IPG 4–9 4 h		IPG 3–12	3 h
				IPG 4–12	3 h

II. Micropreparative IEF	
Reswelling	30 V, 12–16 h
IEF to the steady state	Focusing time of analytical IEF
	+ additional 50% (approximate)

[a] From Görg et al. (2000).

1. Rehydrate IPG dry strips with rehydration buffer, without sample, in an IPG DryStrip reswelling tray (**Fig. 1D**) as described earlier for the Multiphor instrument.

2. Apply the required number of the cup-loading IPGphor strip holders onto the cooling plate/electrode contact area of the IPGphor instrument (**Fig. 1H**), and apply the rehydrated IPG gel strips into the cup-loading strip holders, gel side upwards and acidic ends facing towards the anode.

3. Moisten two filter paper electrode pads (size: 4 × 10 mm²) with deionzed water, remove excess liquid by blotting with a filter paper, and apply the moistened filter paper pads at the anodic and cathodic ends of the IPG strip between the IPG gel and the electrodes. This is particularly important when analyzing samples containing high salt amounts.

4. Position the movable electrodes above the electrode filter paper pads and gently press the electrodes on the filter papers. Filter papers should be replaced after 2–3 h.

5. Position the movable sample cup near either the anode or the cathode and gently press the sample cup onto the surface of the IPG gel strip. Overlay the IPG strip with 5 ml of cover fluid (do not use silicone oil or kerosene!). If cover fluid leaks into the sample cup, rearrange the cup and remove the cover fluid from the cup with the help of tissue paper. Check again for leakage before pipetting the sample (20–100 µl) into the cup.

6. Program the instrument (desired volt hours, voltage gradient, temperature, etc.) and run IEF according to the settings recommended in **Table IV**.

Note

When IPG strips with separation distances <11 cm are used, voltage should be limited to 5000 V.

C. Second Dimension: Horizontal SDS Gel Electrophoresis

Solutions

1. *Tris–HCl buffer*: 1.5 M Tris–HCl, pH 8.8, 0.4% (w/v) SDS. To make 250 ml, dissolve 45.5 g of Trizma base and 1 g of SDS in about 200 ml of deionized water. Adjust the pH of the solution with 4 N HCl and fill up to 250 ml with deionized water. Add 25 mg of sodium azide and filter. The buffer can be stored at 4°C up to 2 weeks.

2. *Equilibration buffer*: 6 M urea, 30% (w/v) glycerol, 2% (w/v) SDS in 0.05 M Tris–HCl buffer, pH 8.8. To make 500 ml, add 180 g of urea, 150 g of glycerol, 10 g of SDS, and 16.7 ml of gel buffer. Dissolve in deionized water and fill up to 500 ml. The buffer can be stored at room temperature up to 2 weeks.

3. *Bromphenol blue solution*: 0.25% (w/v) of bromphenol blue in stacking gel buffer. To make 10 ml, dissolve 25 mg of bromphenol blue in 10 ml of gel buffer. Store at 4°C.

Steps

For the second dimension, an SDS pore gradient gel (0.5 mm thick on GelBond PAGfilm) is applied on the cooling plate of the horizontal electrophoresis unit. Then the equilibrated IPG gel strip is simply placed gel side down onto the surface of the SDS gel alongside the cathodic electrode buffer strip without any embedding procedure. Horizontal setups are particularly suited for the use of ready-made gels on film supports. This section describes the procedure for ready-made ExcelGels. For casting and running of laboratory-made horizontal SDS gels, *see* Görg and Weiss (2000).

1. Equilibration of IPG Gel Strips

The IPG gel strips are equilibrated twice, 15 min each (*see* **Table V**). During the second equilibration, 260 mM iodoacetamide is added to the equilibration buffer in order to remove excess DTT (responsible for the "point streaking" in silver-stained patterns) and to alkylate sulfhydryl groups for subsequent MALDI-MS.

1. Dissolve 100 mg of DTT in 10 ml of equilibration buffer (= equilibration buffer I). Take out the focused IPG gel strips from the freezer and place them into individual test tubes (200 mm long, 20 mm i.d.). Add 10 ml of equilibration buffer I and 50 µl of the bromphenol blue solution. Seal the test tubes with Parafilm, rock them for 15 min on a shaker (**Fig. 1K**), and then pour off the equilibration buffer.

TABLE V   **Equilibration of IPG Strips**

Reagent	Purpose		
50 mM Tris–HCl, pH 8.8			
+2% SDS	Improved protein transfer onto SDS gel		
+6 M urea	Improved protein transfer onto SDS gel		
+30% glycerol	Improved protein transfer onto SDS gel		
+1% DTT	Complete reduction of disulphide bonds		
+4.8% iodoacetamide	Removal of excess DTT (point streaking) Alkylation of SH groups for MALDI-MS		
		DTT	Iodoacetamide
First step 15 min		+	−
Second step 15 min		−	+

2. Dissolve 480 mg of iodoacetamide in 10 ml of equilibration buffer (= equilibration buffer II). Add equilibration buffer II and 50 µl of bromphenol blue solution to the test tube as just described and equilibrate for another 15 min on a rocker.

3. After the second equilibration, place the IPG gel strip on a piece of moist filter paper to remove excess equilibration buffer. The strip should be turned up at one edge for a few minutes to drain off excess equilibration buffer.

2. Horizontal SDS–PAGE with Ready-Made Gels

Ready-made SDS gels (ExcelGel, $250 \times 200 \times 0.5\ mm^3$, on plastic backing; Amersham Biosciences) in combination with polyacrylamide buffer strips are used.

1. Equilibrate the IPG gel strips as described earlier (**Table V**).

2. While the strips are being equilibrated, begin the assembly of the SDS ExcelGel for the second dimension: Remove the ExcelGel from its foil package. Pipette 2–3 ml of kerosene on the cooling plate of the horizontal electrophoresis unit (15°C). Remove the protective cover from the top of the ExcelGel and place the gel on the cooling plate, with cut-off edge toward the anode. Avoid trapping air bubbles between the gel and the cooling block.

3. Peel back the protective foil of the cathodic SDS buffer strip. Wet your gloves with a few drops of deionized water and place the buffer strip on the cathodic end of the gel. Avoid trapping air bubbles between gel surface and buffer strip.

4. Repeat this procedure with the anodic buffer strip.

5. Place the equilibrated IPG gel strips gel side down on the surface of the ExcelGel, 2–3 mm apart from the cathodic buffer strip (**Fig. 1L**).

6. Press gently on top of the IPG gel strips with forceps to remove any trapped air bubbles.

7. Align the electrodes with the buffer strips and lower the electrode holder carefully onto the buffer strips.

8. Start SDS–PAGE at 100 V for about 60 min with a limit of 20 mA. When the bromphenol blue tracking dye has moved 4–5 mm from the IPG gel strip, interrupt the run, remove the IPG gel strip, and move the cathodic buffer strip forward so that it just covers the former contact area of the IPG gel strip. Readjust the electrodes and continue with electrophoresis at 800 V and 35 mA for about 180 min until the bromphenol blue dye front has reached the anodic buffer strip.

9. Proceed with protein staining with Coomassie blue, silver nitrate, or fluorescent dyes or with blotting.

D. Second Dimension: Multiple Vertical SDS–Polyacrylamide Gel Electrophoresis

First-dimension IEF and the equilibration step are performed as described previously, no matter whether the second dimension is run horizontally or vertically. After equilibration, the IPG gel strip is placed on top of the vertical SDS gel, with or without embedding in agarose.

Vertical SDS–PAGE is performed as described by Görg & Weiss (2000). A stacking gel is usually not necessary. For multiple runs, the Ettan DALT II (Amersham Biosciences) vertical electrophoresis apparatus is recommended because this system allows a large batch of SDS slab gels (up to 12) to be run under identical conditions. Ready-made gels on plastic backing are also available (Amersham Biosciences).

Solutions

1. *Gel buffer*: 1.5 M Tris–HCl, pH 8.6, 0.4% (w/v) SDS (Laemmli, 1970).

To make 500 ml, dissolve 90.85 g of Trizma base and 2.0 g of SDS in about 400 ml of deionized water. Adjust to pH 8.6 with 4 N HCl and fill up to 500 ml with deionized water. Add 50 mg of sodium azide and filter. The buffer can be stored at 4°C up to 2 weeks.

2. *Acrylamide/bisacrylamide solution (30.8% T, 2.6% C)*: 30% (w/v) acrylamide, 0.8% (w/v) methylenebisacrylamide in deionized water. To make 1000 ml, dissolve 300.0 g of acrylamide and 8.0 g of methylenebisacrylamide in deionized water and fill up to 1000 ml. Add 1 g of Serdolit MB-1, stir for 10 min, and filter. The solution can be stored up to 2 weeks in a refrigerator.

3. *Ammonium persulfate solution*: 10% (w/v) of ammonium persulfate in deionized water. To prepare 10 ml of the solution, dissolve 1.0 g of ammonium persulfate in 10 ml of deionized water. This solution should be prepared freshly just before use.

4. *Displacing solution*: 50% (v/v) glycerol in deionized water, 0.01% (w/v) bromphenol blue. To make 500 ml, mix 250 ml of glycerol (100%) with 250 ml of deionized water, add 50 mg of bromphenol blue, and stir for a few minutes.

5. *Overlay buffer*: Buffer-saturated 2-butanol. To make 30 ml, mix 20 ml of gel buffer with 30 ml of 2-butanol, wait for a few minutes, and pipette off the butanol layer.

6. *Electrode buffer*: To make 5 liter of electrode buffer stock solution, dissolve 58.0 g of Trizma base, 299.6 g of glycine, and 19.9 g of SDS in deionized water and complete to 5.0 liter.

7. *Equilibration buffer*: 6 M urea, 30% (w/v) glycerol, 2% (w/v) SDS in 0.05 M Tris–HCl buffer, pH 8.6. To

make 500 ml, add 180 g of urea, 150 g of glycerol, 10 g of SDS, and 16.7 ml of gel buffer. Dissolve in deionized water and fill up to 500 ml. The buffer can be stored at room temperature up to 2 weeks.

8. *Agarose solution*: Suspend 0.5% (w/v) agarose in electrode buffer and melt it in a boiling water bath or in a microwave oven.

1. Casting of Vertical SDS Gels

Steps

1. The polymerisation cassettes (200×250 mm^2) are made in the shape of books consisting of two glass plates connected by a hinge strip with two 1.0-mm-thick spacers in between them.

2. Stack 14 cassettes vertically into the gel-casting box with the hinge strips to the right, interspersed with plastic sheets (e.g., 0.05-mm-thick polyester sheets).

3. Put the front plate of the casting box in place and screw on the nuts (hand tight).

4. Connect a polyethylene tube (*i.d.* 5 mm) to a funnel held in a ring stand at a level of about 30 cm above the top of the casting box. Place the other end of the tube in the grommet in the casting box side chamber.

5. Fill the side chamber with 100 ml of displacing solution.

6. Immediately before gel casting, add TEMED and ammonium persulfate solutions to the gel solution (**Table VI**). To cast the gels, pour the gel solution (about 830 ml) into the funnel. Avoid introduction of any air bubbles into the tube!

7. When pouring is complete, remove the tube from the side chamber grommet so that the level of the displacing solution in the side chamber falls.

8. Very carefully pipette about 1 ml of overlay buffer onto the top of each gel in order to obtain a smooth, flat gel top surface.

TABLE VI   Recipes for Casting Vertical SDS Gels

	7.5% T 2.6% C	10% T 2.6% C	12.5% T 2.6% C	15% T 2.6% C
Acrylamide/ bisacrylamide (30.8%T, 2.6%C)	202 ml	270 ml	337 ml	404 ml
gel buffer	208 ml	208 ml	208 ml	208 ml
Glycerol (100%)	41 g	41 g	41 g	41 g
Deionized water	383 ml	315 ml	248 ml	181 ml
TEMED (100%)	42 µl	42 µl	42 µl	42 µl
Ammonium persulfate (10%)	6.0 ml	6.0 ml	6.0 ml	6.0 ml
Final volume	830 ml	830 ml	830 ml	830 ml

9. Allow the gels to polymerize for at least 3 h (better: overnight) at approximately 20°C.

10. Remove the front of the casting box and carefully unload the gel cassettes from the box using a razor blade to separate the cassettes. Remove the polyester sheets that had been placed between the individual cassettes.

11. Wash each cassette with water to remove any acrylamide adhered to the outer surface and drain excess liquid off the top surface. Discard unsatisfactory gels; in general the gels opposite to the front and rear plate of the gel casting box (due to the uneven thickness of these gels).

12. Gels not needed at the moment can be wrapped in plastic wrap and stored in a refrigerator (4°C) for 1–2 days.

2. Multiple Vertical SDS–PAGE Using Ettan Dalt II

Steps

1. Pour 1875 ml of electrode buffer stock solution and 5625 ml of deionized water in the lower electrophoresis buffer tank and turn on cooling (20°C).

2. Support the SDS gel cassettes in a vertical position to facilitate application of the first-dimension IPG gel strips.

3. Equilibrate the IPG gel strip as described earlier for horizontal SDS gels. Immerse it in SDS electrode buffer for a few seconds to facilitate insertion of the IPG strip between the two glass plates of the gel cassette.

4. Place the IPG gel strip on top of an SDS gel and overlay it with 2 ml of hot agarose solution (75°C). Carefully press the IPG strip with a spatula onto the surface of the SDS gel to achieve complete contact (**Fig. 1M**). If it is desired to coelectrophorese M_r marker proteins, soak a piece of filter paper (2×4 mm^2) with 5 µl of SDS marker proteins dissolved in electrophoresis buffer, let it dry (!), and apply it to the left or right of the IPG strip. Allow the agarose to solidify for at least 5 min and then place the slab gel into the electrophoresis apparatus (*see* later). Repeat this procedure for the remaining IPG strips. *Note*: Embedding in agarose is not absolutely necessary, but it ensures better contact between the IPG gel strip and the top of the SDS gel.

5. Wet the gel cassettes by dipping then into electrode buffer for a few seconds and then insert them in the electrophoresis apparatus. Pour 1250 ml of electrophoresis buffer stock solution and 1250 ml of deionized water in the upper electrophoresis tank, mix, and start SDS electrophoresis.

6. Run the SDS–PAGE gels with 50 mA (100 V maximum setting) for about 2 h. Then continue with 175 mA (200 V maximum setting) for about 16 h. *Note*: In contrast to the procedure of horizontal SDS–PAGE it is not necessary to remove the IPG gel strips from the surface of the vertical SDS gel once the proteins have migrated out of the IPG gel strip.

7. Terminate the run when the bromphenol blue tracking dye has migrated off the lower end of the gel.

8. Open the cassettes carefully with a plastic spatula. Use the spatula to separate the agarose overlay from the polyacrylamide gel. Peel the gel from the glass plate carefully, lifting it by the lower edge, and place it in a box of stain solution or transfer buffer, respectively. Then continue with protein staining or blotting.

IV. COMMENTS

Since the early 1990s, IPG-Dalt has constantly been refined to meet the requirements of proteome analysis. In particular, (i) the development of basic IPGs up to pH 12 has facilitated the analysis of very alkaline proteins (Görg *et al.*, 1997, 1998, 1999; Wildgruber *et al.* 2002); (ii) the introduction of overlapping narrow-range IPGs to stretch the first dimension permits increased resolution (Δ pI = 0.01/cm) (Görg *et al.*, 1988), as well as higher loading capacity for the analysis of less abundant proteins (Wildgruber *et al.*, 2000); and (iii) the availability of ready-made IPG strips and integrated devices such as the IPGphor (Islam *et al.*, 1998) have paved the way towards automation (Görg *et al.*, 1999, 2000) (Fig. 4).

Using the protocols described in this article, it is possible to resolve proteins having apparent molecular masses between 10 and 200 kDa and pI values from 2.5 to 12. The protein-loading capacity for micropreparative runs is up to several milligrams of protein if narrow pH range IPGs (0.5–1 pH units wide) are used.

V. PITFALLS

1. Blot the rehydrated IPG gel strips to remove excess rehydration solution; as urea crystallization on the surface of the IPG gel strips might occur, which will disturb IEF patterns or prolong focusing time to reach the steady state.

2. Make sure that the orientation of the IPG gel strips on the cooling block of the IEF chamber is correct (acidic end towards anode). Check the temperature of the cooling block (20°C).

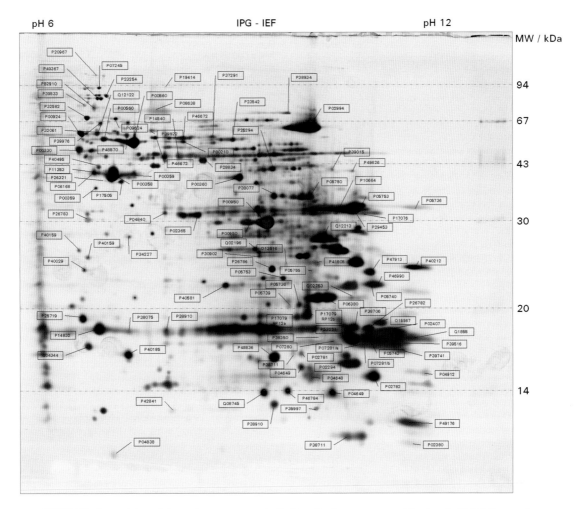

FIGURE 4 Web-based 2D reference map of alkaline *Saccharomyces cerevisiae* (strain FY1679) proteins showing 106 mapped and identified spots annotated by SWISS-PROT accession numbers. First dimension: IPG 6–12. Second dimension: SDS–PAGE, 15% T. Spot identification by MALDI-TOF-MS (Ettan z² MALDI-TOF, Amersham Biosciences) (with permission from Wildgruber *et al.*, 2002) (http://www.wzw.tum.de/proteomik).

3. The sample solution should not be too concentrated to avoid protein precipitation at the sample application point. If you are in doubt, dilute the sample with lysis solution and apply a larger volume (>20 μl) instead.

4. For improved sample entry, start IEF at low voltage (150 V for 30 min, followed by 300 V for 60 min). Extend these times when the sample contains high amounts of salts or when large sample amounts (>50 μl) are applied *via* cup loading.

5. Sample in-gel rehydration: Use IPGphor and apply low voltage (30–50 V) during IPG strip rehydration for improved entry of high M_r proteins into the polyacrylamide gel matrix.

6. When analyzing samples containing high salt amounts, apply moistened filter paper pads at the anodic and cathodic ends of the IPG strip between the IPG gel surface and the electrodes and replace these filter papers with fresh ones after several hours.

7. Remove the IPG gel strip from the surface of the *horizontal* SDS–PAGE gel as soon as the bromphenol blue dye front has migrated 4–5 mm off the IPG gel strip. Then move the cathodic electrode wick (or buffer strip) forward so that it overlaps the area the IPG gel strip once covered.

References

Bjellqvist, B., Ek, K., Righetti, P. G., Gianazza, E., Görg, A., Westermeier, R., and Postel, W. (1982). Isoelectric focusing in immobilized pH gradients: Principle, methodology and some applications. *J. Biochem. Biophys. Methods* **6**, 317–339.

Blomberg, A., Blomberg, L., Norbeck, J., Fey, S., Mose Larsen, P., Larsen, M., Roepstorff, P., Degand, H., Boutry, M., Posch, A., and Görg, A. (1995). Interlaboratory reproducibility of yeast protein patterns analyzed by immobilized pH gradient two-dimensional gel electrophoresis. *Electrophoresis* **16**, 1935–1945.

Bossi, A., Righetti, P. G., Vecchio, G., and Severinsen, S. (1994). Focusing of alkaline proteases in pH 10–12 immobilized gradients. *Electrophoresis* **15**, 1535–1540.

Corbett, J. M., Dunn, M. J., Posch, A., and Gorg, A. (1994). Positional reproducibility of protein spots in two-dimensional polyacrylamide gel electrophoresis using immobilised pH gradient isoelectric focusing in the first dimension: An interlaboratory comparison. *Electrophoresis* **15**, 1205–1211.

Görg, A., Boguth, G., Obermaier, C., Posch, A., and Weiss, W. (1995). Two-dimensional polyacrylamide gel electrophoresis with immobilized pH gradients in the first dimension (IPG-Dalt): The state of the art and the controversy of vertical versus horizontal systems. *Electrophoresis* **16**, 1079–1086.

Görg, A., Boguth, G., Obermaier, C., and Weiss, W. (1998). Two-dimensional electrophoresis of proteins in an immobilized pH 4–12 gradient. *Electrophoresis* **19**, 1516–1519.

Görg, A., Obermaier, C., Boguth, G., Csordas, A., Diaz, J. J., and Madjar, J. J. (1997). Very alkaline immobilized pH gradients for two-dimensional electrophoresis of ribosomal and nuclear proteins. *Electrophoresis* **18**, 328–337.

Görg, A., Obermaier, C., Boguth, G., Harder, A., Scheibe, B., Wildgruber, R., and Weiss, W. (2000). The current state of two-dimensional electrophoresis with immobilized pH gradients. *Electrophoresis* **21**, 1037–1053.

Görg, A., Obermaier, C., Boguth, G., and Weiss, W. (1999). Recent developments in 2-D gel electrophoresis with immobilized pH gradients: Wide pH gradients up to pH 12, longer separation distances and simplified procedures. *Electrophoresis* **20**, 712–717.

Görg, A., Postel, W., Friedrich, C., Kuick, R., Strahler, J. R., and Hanash, S. M. (1991). *Electrophoresis* **12**, 653–658.

Görg, A., Postel, W., and Günther, S. (1988). The current state of two-dimensional electrophoresis with immobilized pH gradients. *Electrophoresis* **9**, 531–546.

Görg, A., and Weiss. (2000). 2D electrophoresis with immobilized pH gradients. *In* "Proteome Research: Two Dimensional Electrophoresis and Identification Methods" (T. Rabilloud, ed.), pp. 57–106. Springer, Berlin.

Herbert, B. (1999). Advances in protein solubilisation for two-dimensional electrophoresis. *Electrophoresis* **20**, 660–663.

Islam, R., Ko, C., and Landers, T. (1998). A new approach to rapid immobilised pH gradient IEF for 2-D electrophoresis. *Sci. Tools* **3**, 14–15.

Laemmli, U. K. (1970). Cleavage of structural proteins during the assembly of the head of bacteriophage T4. *Nature* **227**, 680–685.

O' Farrell, P. H. (1975). High resolution two-dimensional electrophoresis of proteins. *J. Biol. Chem.* **250**, 4007–4021.

O'Farrell, P. Z., Goodman, H. M., and O'Farrell, P. H. (1977). High resolution two-dimensional electrophoresis of basic as well as acidic proteins. *Cell* **12**, 1133–1142.

Olsson, I., Larsson, K., Palmgren, R., and Bjellqvist, B. (2002). Organic disulfides as a means to generate streak-free two-dimensional maps with narrow range IPG strips as first dimension. *Proteomics* **2**, 1630–1632.

Rabilloud, T., Adessi, C., Giraudel, A., and Lunardi, J. (1997). Improvement of the solubilization of proteins in two-dimensional electrophoresis with immobilized pH gradients. *Electrophoresis* **18**, 307–316.

Rabilloud, T., Valette, C., and Lawrence, J. J. (1994). Sample application by in-gel rehydration improves the resolution of two-dimensional electrophoresis with immobilized pH-gradients in the first dimension. *Electrophoresis* **15**, 1552–1558.

Righetti, P. G. (1990). "Immobilized pH Gradients: Theory and Methodology." *Elsevier*, Amsterdam.

Wildgruber, R., Harder, A., Obermaier, C., Boguth, G., Weiss, W., Larsen, P. M., Fey, S., and Görg, A. (2000). Towards higher resolution: Two-dimensional electrophoresis of *Saccharomyces cerevisiae* proteins using overlapping narrow immobilized pH gradients. *Electrophoresis* **21**, 2610–2616.

Wildgruber, R., Reil, G., Drews, O., Parlar, H., and Görg, A. (2002). Web-based two-dimensional database of *Saccharomyces cerevisiae* proteins using immobilized pH gradients from pH 6 to pH 12 and matrix-assisted laser desorption/ionization-time of flight mass spectrometry. *Proteomics* **2**, 727–732.

CHAPTER

24

Two-Dimensional Difference Gel Electrophoresis: Application for the Analysis of Differential Protein Expression in Multiple Biological Samples

John F. Timms

I. INTRODUCTION

Fluorescence two-dimensional difference gel electrophoresis (2D-DIGE) is a recently developed 2D gel-based proteomics technique that provides a sensitive, rapid, and quantitative analysis of differential protein expression between biological samples. Developed by Minden and co-workers (Unlu *et al.*, 1997), the technique utilizes charge- and mass-matched chemical derivatives of spectrally distinct fluorescent cyanine dyes (Cy3 and Cy5), which are used to covalently label lysine residues in different samples prior to mixing and separation on the same 2D gel. Resolved, labelled proteins are then detected at appropriate excitation and emission wavelengths using a multiwavelength fluorescence detection device and the signals compared. As well as reducing the number of gels that need to be run, differential labelling and mixing mean that samples are subjected to the same handling procedures and microenvironments during 2D separation, raising the confidence with which protein changes can be detected and quantified. Because fluorescence detection also provides a superior linear dynamic range of detection and sensitivity to other methods (Patton, 2000), this technology is suited to the analysis of biological samples with their large dynamic ranges of protein abundance.

The 2D-DIGE methodology is now commercialized as the Ettan DIGE proteomics system (GE

Healthcare), with a third dye (Cy2) introduced, allowing simultaneous analysis of three samples on a single gel. In expression profiling experiments, one dye is used to label an internal standard to be run on all gels against pairs of test samples labelled with the other two dyes. This allows the direct comparison of ratios of expression across multiple samples and gels, improving the ability to distinguish biological variation from gel-to-gel variation. Because the labelling strategy employed is compatible with downstream identification of gel spots by mass spectrometry (MS) (Tonge *et al.*, 2001; Gharbi *et al.*, 2002), 2D-DIGE is of particular use as a reproducible, high-throughput proteomic technology. This article describes the necessary materials and instrumentation, experimental design, and work flow for the preparation and labelling of samples for 2D-DIGE analysis. Image capture, analysis, and spot picking for MS identification are also described.

II. MATERIALS AND INSTRUMENTATION

The CyDye DIGE fluors *N*-hydroxysuccinimidyl (NHS) esters of Cy2 (Prod. Code RPK0272), Cy3 (Prod. No. RPK0273), and Cy5 (Prod. No. RPK0275) are from GE Healthcare. NHS-Cy3 and NHS-Cy5 are also synthesized in-house following the original published protocol (Unlu *et al.*, 1997), with modifications (unpublished data). Anhydrous 99.8% *N,N*-dimethylformamide (DMF, Cat No. 22,705-6) is from Aldrich, urea (Cat. No. U-0631), dithiothreitol (DTT, Cat. No. D-9163), thiourea (Cat. No. T-8656), and L-lysine (Cat. No. L-5626) are from Sigma. 1.5 *M* Tris solution, pH 8.8 (Prod. No. 20-79000-10), 1 *M* Tris solution, pH 6.8 (Prod. No. 20-7901-10), and 10× Tris–glycine SDS electrophoresis buffer (Prod. No. 20-64) are from Severn Biotech Ltd. Phosphate-buffered saline (PBS, Cat No. 14190-094) is from GIBCO. Nonidet P-40 (NP-40, Prod. No. 56009D2L), bromphenol blue (Prod. No. 200173J), methanol (Prod. No. 10158BG), and glacial acetic acid (Prod. No. 27013BV) are from VWR. 3-[(3-Cholamidopropyl)-dimethylammonio]-1-propanesulfonate (CHAPS, Cat No. B2006) and sodium dodecyl sulphate (SDS, Cat. No. B2008) are from Melford Laboratories Ltd. Coomassie protein assay reagent (Prod. No. 1856210) is from Pierce, S&S weighing papers (Cat. No. Z13411-2) are from Aldrich, and low-gelling temperature agarose (Cat. No. 05075) is from Fluka.

Ampholines (pH 3.5–10, Cat. No. 80-1125-87), Pharmalyte (pH 3–10, Cat. No. 17-0456-01), Immobiline DryStrip reswelling tray (Cat. No. 80-6465-32), Immobiline DryStrip IEF gels (IPG strips, Cat. No. 17-6002-45), Multiphor II electrophoresis unit (Cat. No. 18-1018-06), Ettan DALT low fluorescence glass plates (27 × 22 cm, Cat. No. 80-6475-58), Plus One Repel Silane (Cat. No. 17-1332-01), Plus One Bind Silane (Cat. No. 17-1330-01), reference markers (Cat. No. 18-1143-34), Ettan DALT 12-gel caster, separation unit, and power supply (Cat. Nos. 80-6467-22 and 80-6466-27), Typhoon 9400 imager (Cat. No. 60-0038-54), Ettan Spot Picker (Cat. No. 18-1145-28), and DeCyder differential analysis software (Cat. No. 56-3202-70) are from GE Healthcare. SYPRO Ruby protein gel stain (Cat. No. S-12000) is from Molecular Probes, and colloidal Coomassie brilliant blue G-250 tablets (CBB, Cat No. K26283182) are from Merck. The Immobilon-P polyvinylidene fluoride transfer membrane (PVDF, Cat. No. IPVH00010) is from Millipore.

III. PROCEDURES

A. Experimental Design

The following steps are guidelines for the design of 2D-DIGE expression profiling experiments. Several experiments are outlined to illustrate how samples can be fluorescently labelled and mixed for 2D gel separation and image analysis so that statistically meaningful data can be acquired. The throughput of the technique is, however, dependent upon the 2D gel running, image capture, and analysis capabilities of the laboratory. We routinely run 12-gel experiments providing accurate differential expression data for 24 samples, including an internal standard run on all gels for accurate spot matching and quantitation. The same gels are then poststained and proteins of interest are picked for MS identification. 2D-DIGE is applicable for the analysis of total cell lysates and complex protein mixtures from cultured cells, whole tissues, sorted or fractionated cells, whole organisms (*E. coli, S. pombe, C. elegans*, etc.), cellular subfractions (membrane, nuclear, cytoplasmic, etc.), or affinity-purified protein fractions.

Steps

1. Three spectrally distinct fluorescent CyDyes (Cy2, Cy3, and Cy5) can be used for differential labelling of protein samples. In the simplest expression profiling experiment, two individual samples are labelled with two different dyes, mixed, and resolved on a single 2D gel (see Section III,B). Because the same protein isoforms in each sample will comigrate, one can accurately measure differential expression as the ratio of the fluorescence intensities of comigrating spots. Thus, the problem of gel-to-gel variation is avoided. This type of single gel experiment is useful where only limited sample quantities are available, e.g., laser capture dissection-procured normal and cancer cells (Zhou *et al.*, 2002).

2. To obtain statistically meaningful expression changes, at least triplicate samples should be labelled and analysed on separate gels. These may be biological replicates, e.g., three separately grown cell cultures or tissue from three individual animals, or may be experimental replicates, where three aliquots of the same sample are compared across different gels. Differential expression can then be taken as an average fold change (e.g., the average spot intensity ratio between differentially labelled spots matching across all three gels) with statistical confidence provided by applying a *t* test. Of note we have found considerable interanimal variation in liver lysates from mice (unpublished data), and it is therefore advisable to analyse samples from at least five individual animals for each treatment or condition to provide statistically meaningful data.

3. 2D-DIGE analysis is further improved by running a Cy-labelled internal standard on all gels against pairs of test samples, labelled with the other two CyDyes. This increases the ability to distinguish biological variation from gel-to-gel variation by increasing the confidence with which spots can be matched across gels and by allowing the direct comparison of expression ratios across samples. An equal

pool of all samples (including biological replicates) is best employed as an internal standard as it will contain proteins present in all samples. It is created simply by mixing and labelling equal amounts of protein from every sample and should provide sufficient material for the number of gels to be run. Equal amounts of standard and test samples are then resolved on each gel.

4. In an experiment comparing 100 μg of protein from cell A and cell B, lysates from triplicate cultures are prepared and protein concentrations determined (six samples). For simplicity, samples are adjusted to the same protein concentration by adding lysis buffer. Then 100 μg of each is labelled with Cy3 or Cy5 as shown in Table I. A pool consisting of a mixture of 50 μg of each of all the replicate samples (300 μg total) is labelled with Cy2. Following labelling, the samples are mixed appropriately for separation on three 2D gels as shown in Table I. This scheme controls for dye bias, although labelling combinations are interchangeable so long as each gel is loaded with samples labelled with distinct dyes. This experiment generates nine images for matching, cross-comparison, and statistical analysis in the biological variance analysis (BVA) module of DeCyder software.

5. For complex comparisons we recommend running 12 gels at once. This allows imaging within a day and fits with our downstream laboratory work flow. Although more gels can be run in an individual experiment, consistency may be compromised by running gels at different times or on different electrophoresis units, or may be impractical depending on man power, resources, and automation. Still, it is possible to compare 24 different samples in a single 12-gel 2D-DIGE experiment. For example, our laboratory was able to analyse lysates from two cell lines subjected to growth factor stimulation for four different time periods (8 conditions) using triplicate cultures (Gharbi et al., 2002). The 24 lysates generated were labelled with Cy3 or Cy5 and run in pairs against the internal standard (a pool of all samples) labelled with Cy2. This generated 36 images for DeCyder BVA analysis, with an image acquisition time of ~8 h using a Typhoon 9400 imager.

B. Preparation of CyDye-Labelled Samples for 2D Electrophoresis

The following procedure can be applied for the preparation and labelling of multiple samples for 2D-DIGE comparative protein expression analysis. The procedure is based on the Ettan DIGE System (GE Healthcare), with some modifications. The protocol is designed to generate differentially labelled samples that are compatible with all systems for 2D gel electrophoretic separation. The principles and applications of 2D gel electrophoresis are discussed in more detail elsewhere in this volume. For brevity, the sample preparation is outlined for 24-cm pH 3–10 NL-immobilised pH gradient (IPG) gels. Accordingly, final volumes, protein loads, and IPG buffers may differ depending on the size and pH range of the first-dimension gels.

Solutions

1. *CyDyes (NHS-Cy2, -Cy3, -Cy5)*: From lyophilized powder (stored at −20°C), reconstitute to 1 mM stock by dissolving in the appropriate volume of anhydrous DMF. Keep stock solutions in the dark at −20°C; they are stable for up to 4 months.

2. *Lysis buffer*: 8 M urea, 2 M thiourea, 4% (w/v) CHAPS, 0.5% NP-40 (w/v), 10 mM Tris–HCl, pH 8.3. To make 100 ml, dissolve 48 g of urea and 15.2 g of thiourea in 50 ml of distilled H$_2$O. Add 4 g CHAPS, 0.5 g NP-40, and 0.67 ml of 1.5 M Tris, pH 8.8, solution. This should give a final pH of 8.3. Make up to final volume, aliquot, and store at −20°C. Do not heat.

3. *40% (w/v) CHAPS*: To make 50 ml, dissolve 20 g CHAPS in distilled H$_2$O and complete to 50 ml. Store at room temperature.

4. *10% (w/v) NP-40*: To make 50 ml, dilute 5 g of 100% NP-40 in distilled H$_2$O and complete to 50 ml. Store at room temperature.

5. *L-lysine solution*: 10 mM L-lysine in H$_2$O. Dissolve 9.1 mg in 5 ml distilled H$_2$O. Aliquot and store at −20°C.

6. *DTT solution*: 1.3 M DTT in H$_2$O. To make 10 ml, dissolve 2 g DTT in distilled H$_2$O and complete to 10 ml. Aliquot and store at −20°C. Do not heat.

7. *Ampholines/Pharmalyte mix*: Mix equal volumes of ampholines (pH 3.5–10) and Pharmalyte (pH 3–10). Store at 4°C. These broad pH range IPG buffers can be replaced with narrow range buffers depending on the first-dimension pH range.

8. *Bromphenol blue*: 0.2% (w/v) bromphenol blue in H$_2$O. To make 10 ml, weigh 20 mg bromphenol blue and complete to 10 ml with distilled H$_2$O. Filter and store at room temperature.

TABLE I   Example of Differential Labelling, Mixing, and Loading for Statistical Comparison of Protein Expression in Two Cell Lines Using 2D-DIGE

	Cy3	Cy5	Cy2
Gel 1	100 μg A, replicate 1	100 μg B, replicate 1	100 μg pool
Gel 2	100 μg B, replicate 2	100 μg A, replicate 2	100 μg pool
Gel 3	100 μg A, replicate 3	100 μg B, replicate 3	100 μg pool

Steps

1. Wash cultured or fractionated cells, whole organisms, or tissues in PBS or, if possible, a low-salt buffer that does not compromise cellular integrity. Salts should be kept to a minimum so drain well. Subcellular or affinity-purified fractions should be prepared at high protein concentration (>2.5 mg/ml) in a low-salt buffer (<10 mM) or dialysed against a low-salt buffer.

2. Lyse cells in lysis buffer using appropriate physical disruption (sonnication, grinding, homogenisation, repeated passage through a 25-gauge needle). Do not let samples heat up. A volume of buffer should be used to give a final protein concentration of at least 1 mg/ml. For cellular fractions in a known volume of low-salt buffer and at >2.5 mg/ml, add urea, thiourea, 10% NP-40, 40% CHAPS, and 1.5 M Tris, pH 8.8, solution to give final concentrations of 8 M urea, 2 M thiourea, 4% (w/v) CHAPS, 0.5% (w/v) NP-40, and 10 mM Tris (same as lysis buffer). Rotate on a wheel at room temperature until reagents have dissolved. Because the volume is increased substantially upon reagent addition, amounts should be calculated for 2.5 times the original sample volume. Thus, for a 1-ml sample, add 1.2 g urea, 0.38 g thiourea, 250 µl 40% CHAPS, 125 µl 10% NP-40, 16.67 µl 1.5 M Tris, pH 8.8, and make to 2.5 ml with lysis buffer. Use weighing papers to avoid static during weighing. The final pH should be ~8.3, the optimum for NHS-CyDye labelling.

3. Determine protein concentrations using Pierce Coomassie protein assay reagent according to the manufacturer's instructions, using BSA in lysis buffer to generate a standard curve. It is recommended that at least four replicate assays are performed for each sample for accurate protein determination. Dilute concentrated samples with lysis buffer if necessary. For ease, samples should be adjusted to the same protein concentration at this point using lysis buffer.

4. Aliquot desired amount of sample into tubes for CyDye labelling. Typically we label 100 µg of protein in triplicate using a random combination of Cy3 and Cy5 across the sample set (See Experimental Design). Mix equal amounts of protein from each sample to create an internal standard. This is labelled with Cy2 and should provide enough material for the number of gels to be run (100 µg/gel).

5. Label samples by the addition of 4 pmol of the appropriate CyDye per microgram of protein (400 pmol/100 µg, equivalent to 4 µM for a 1-mg/ml sample). CyDye stocks can be diluted with anhydrous DMF to avoid pipetting submicroliter volumes. Incu-

bate samples on ice in the dark for 30 min. Note that protein lysates are viscous so ensure thorough mixing at all steps to avoid non-uniform labelling.

6. Quench reactions by adding a 20-fold molar excess of L-lysine. For 400 pmol CyDye, add 0.8 µl of 10 mM L-lysine solution. Incubate on ice in the dark for 10 min.

7. Mix Cy3- and Cy5-labelled samples appropriately and add a 100-µg aliquot of the Cy2-labelled pool (to give 300 µg total protein). Note that the final volume should be less than that required for reswelling of IPG strips (450 µl for 24-cm strips). This reswelling volume dictates the practical lower limit for sample protein concentrations.

8. Reduce samples by adding 1.3 M DTT to a final concentration of 65 mM (22 µl). Add carrier ampholines/Pharmalyte mix to a final concentration of 2% (v/v) (9 µl) and 1 µl of 0.2% bromphenol blue. Adjust volume to 450 µl with lysis buffer. Spin samples briefly.

9. Rehydrate Immobiline DryStrip pH 3–10 NL gels with samples overnight in the dark at room temperature in a reswelling tray according to the manufacturer's instructions (passive rehydration method). Other methods of sample loading (cup loading, rehydration under voltage) can also be applied depending on user preference.

10. Perform 2D electrophoresis following guidelines for the type of system employed, but see Section III,C for recommended modifications.

Comments

1. Primary amines and reducing agents should be avoided as they interfere with CyDye labelling. These include carrier ampholines/Pharmalytes and DTT, which are therefore added after labelling but prior to 2D-PAGE.

2. It is often necessary to use protease, kinase, and phosphatase inhibitors for the preparation of lysates and cellular fractions. We have found that aprotinin (17 µg/ml), pepstatin A (1 µg/ml), leupeptin (1 µg/ml), EDTA (1 mM), okadaic acid (1 µM), fenvalerate (5 µM), BpVphen (5 µM), and sodium orthovanadate (2 mM), at the final concentrations shown, do not interfere with CyDye labelling.

3. The quantity of CyDye used for labelling is limiting in the reaction and only ~3% of protein molecules are labelled on an average of one lysine residue (minimal labelling). This minimal labelling approach maintains sample solubility and prevents heterogeneous labelling that would lead to vertical spot trains. However, because 436 Da (Cy2), 467 Da (Cy3), or 465 Da (Cy5) is added to the 3% of labelled molecules, a slight shift in migration is observed between

CyDye and poststained images (Gharbi *et al.*, 2002). This is more noticeable in the lower molecular weight range and necessitates poststaining with a general protein stain to attain accurate picking of the majority (97%) of unlabelled protein (see Section III,D).

C. Preparation of 2D Gels, Imaging, and Image Analysis

Isoelectric focusing and second-dimension poly-acrylamide gel electrophoresis of CyDye-labelled samples can be performed on any system following the manufacturer's instructions. However, inclusion of the following steps is recommended for high sensitivity, reproducibility, accuracy in the determination of differential expression, and precise excision of protein features for MS identification. The steps are detailed for use with the Multiphor II IEF and Ettan DALT 12 PAGE separation systems for 24 × 20-cm 2D gels, but are generally applicable to other systems. All gel preparation and running steps should ideally be performed in a dedicated clean room to avoid contamination with particulates and nonsample proteins, such as skin and hair keratins. Image analysis and statistical analysis can be performed using various 2D gel analysis softwares (e.g., Melanie, Phoretix, ImageMaster), although DeCyder software is tailored specifically for use with the 2D-DIGE system and is relatively simple to use. Instructions for analysis using DeCyder software are found in the DeCyder software user manual.

Solutions

1. *Bind saline solution*: For twelve 24 × 20-cm plates, mix 16 µl of Plus One bind saline, 400 µl glacial acetic acid, 16 ml ethanol, and 3.6 ml distilled H_2O.

2. *Equilibration buffer*: 6 M urea, 30% (v/v) glycerol, 50 mM Tris–HCl, pH 6.8, 2% (w/v) SDS. To make 200 ml, dissolve 72 g urea in 100 ml distilled H_2O. Add 60 ml of 100% glycerol, 10 ml of 1 M Tris, pH 6.8, solution, and 4 g SDS. Dissolve all powders and adjust volume to 200 ml with distilled H_2O. Aliquot and store at –20°C.

3. *Agarose overlay*: 0.5% (w/v) low-melting point agarose in 1× SDS electrophoresis buffer. To make 200 ml, melt 1 g of agarose in 200 ml of 1× SDS electrophoresis buffer in a microwave on low heat. Add bromphenol blue solution to give a pale blue colour.

Steps

1. Prior to gel casting, treat low-fluorescence glass plates for gel bonding by applying 1.5 ml of fresh bind saline solution per plate and wiping over one surface

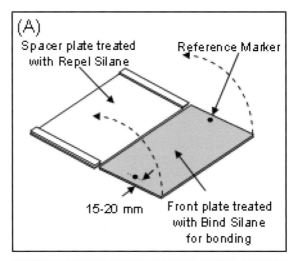

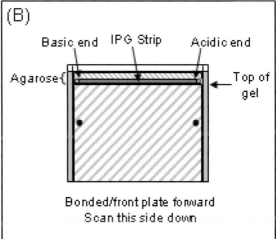

FIGURE 1 (A) Treatment of plates for bonding and reference marker positioning. (B) Casting and loading of second-dimension gels. Based on the Ettan DALT 24-cm strip format.

with a lint-free tissue. Leave plates to dry for a minimum of 1 h. Note that only one plate in each set should be treated; treat the smaller, nonspacer "front plate" if using Ettan DALT 24-cm gel plates (Fig. 1A). Bonding allows easier handling of gels during scanning, protein staining, storage, and, importantly, automated spot excision.

2. Treat the inner surface of clean and dry "spacer plates" with Repel Silane to ensure easy separation after running (Fig. 1A). Apply PlusOne Repel Silane solution to a lint-free tissue and wipe over the surface. Leave to dry for 10 min. Use in a well-ventilated area. Remove excess Repel Saline by wiping with a clean tissue, rinse with ethanol, then with distilled H_2O, and dry with a tissue.

3. Stick two reference markers to the bonded surface of the plates. These should be placed half-way

down the plates and 15–20 mm in from each edge (Fig. 1A). These markers are used as references for determining cutting coordinates for automated spot picking using the Ettan Spot Picker.

4. Assemble plates in casting chamber and cast gels according to the manufacturer's guidelines.

5. Perform IEF in the dark according to the manufacturer's guidelines.

6. Equilibrate strips for 15 min in equilibration buffer containing 65 mM DTT (reduction) and then 15 min in the same buffer containing 240 mM iodoacetamide. (alkylation).

7. Rinse strips in 1× SDS electrophoresis buffer and place onto the top of second-dimension gels in melted 0.5% agarose overlay, with the basic end of the strip towards the left hand side when the bonded plate is facing forward (Fig. 1B).

8. Run second-dimension gels until the dye front has completely run off to avoid fluorescence signals from bromphenol blue and free dye. For the Ettan DALT *twelve* system, this can be achieved by running 12% gels for 16 h at 2.2 W per gel.

9. Images are best acquired directly after the 2D run by scanning gels between their glass plates using a Typhoon 9400 imager or similar device. Ensure that both outer plate surfaces are clean and dry before scanning and that the bonded plate is the lower plate on the scanner bed. If the strip is placed correctly (Fig. 1B), the resulting image will not need to be rotated and give a consensus image with the acidic end to the left. Alternatively, gels can be scanned after fixing with the gel facing up from the bonded plate in the scanner, giving the same orientation of the image.

10. Perform an initial low-resolution scan (1000 μm) for one gel on the Cy2, Cy3, and Cy5 channels with the photomultiplier tube (PMT) voltages set low (e.g., 500 V). The optimal excitation/emission wavelengths for fluorescence detection using the Typhoon 9400 are 488/520 nm for Cy2, 532/580 nm for Cy3, and 633/680 nm for Cy5, although other instrumentation may vary slightly. An image is then built up by the scanner for each channel and is converted to grey-scale pixel values.

11. Using ImageQuant software for the Typhoon 9400, establish maximum pixel values in various user-defined, spot-rich regions of each image and adjust the PMT voltages for a second low-resolution scan to give similar maximum pixel values (within 10%) on each channel and without saturating the signal from the most intense peaks (i.e., <90,000 pixels). As a guide, increasing the PMT voltage by 30 V roughly doubles the pixel value. Repeat scans may be required until values are within 10% for the three channels. PMT voltages can be increased further to enhance the detec-tion of low-intensity features, whilst producing tolerable saturation of only a few of the most abundant protein features.

12. Once set for the first gel, use the same PMT voltages for the whole set of gels scanning at 100-μm resolution. A 24 × 20-cm gel image takes ~10 min to acquire per channel and two gels are scanned simultaneously. Images are generated as .gel files, the same format as .tif files.

13. Crop overlayed images in ImageQuant and import into DeCyder Batch Analysis software for subsequent BVA analysis according to the DeCyder software user manual. Differential expression can also be detected visually using Adobe Photoshop by overlaying coloured images (made in the Channel Mixer) and merging using the "Multiply" option in "Layers" (Fig. 2D).

Comments

1. Low-fluorescence glass plates should be used to reduce background.

2. Bind Saline is extremely resistant to removal, and cleaned plates previously treated are still likely to bind acrylamide with subsequent use. For this reason, dedicated treated plates are marked with a diamond pen and reused in the same orientation for subsequent experiments. Bind and Repel saline should be reapplied for subsequent runs.

3. Plates with bonded gels are best cleaned by scraping with a sturdy straight-edged decorator's scraper in warm water with detergent. It is important to remove all gel material, as this produces a fluorescent signal in the Cy3 channel when dried.

4. CyDye-labelled, gel-separated proteins can also be visualized on membranes following electroblotting. The blotted PVDF membrane is scanned using the Typhoon Imager immediately after transfer, wet and face down under a low-fluorescence glass plate. These membranes are subsequently used for immunoblotting with specific antibodies, and the immunoblot signal is aligned directly with the CyDye signal. This alignment can be used for spot identification, validation of MS, or to identify post translationally modified proteins such as phosphoproteins. Note that gels must not be bonded and the plates used must never have been treated with bind saline.

D. Poststaining and Spot Excision

Bonded gels must be poststained to allow accurate picking (see earlier discussion). We have found that both Sypro Ruby and colloidal Coomassie brilliant blue (CBB) protein stains can be used in conjunction with CyDye labelling (Fig. 2). These general protein

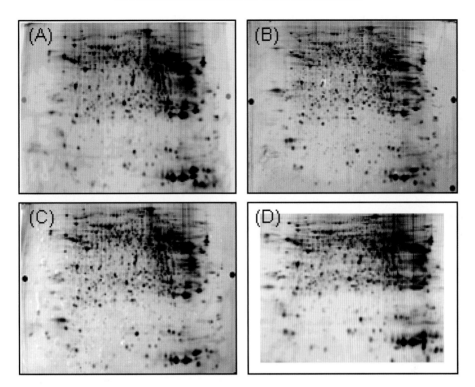

FIGURE 2   (A) Cy5 fluorescence image of 100 mg of mouse liver homogenate separated on a 24-cm pH 3–10 NL IPG strip and 12% PAGE gel. (B) SyproRuby poststained gel of 300 mg of mixed CyDye-labelled liver homogenate. (C) Colloidal CBB poststained gel of 300 mg of mixed CyDye-labelled liver homogenate. This is the same gel as shown in A. (D) Adobe Photoshop-generated Cy3/Cy5 coloured overlay of WT (red) and mutant (blue) mouse liver lysates showing differential expression.

staining methods are sensitive down to the low nanogram level and are reported to be compatible with downstream mass spectrometric identification of proteins (Scheler *et al.*, 1998; Berggren *et al.*, 2000; Lopez *et al.*, 2000; Gharbi *et al.*, 2002). MS-compatible silver staining (Shevchenko *et al.*, 1996) is not recommended for bonded gels due to its insensitivity and variability from gel to gel.

Solutions

1. *Fixing solution*: 35% (v/v) methanol, 7.5% (v/v) acetic acid in distilled H_2O

2. *Colloidal CBB fixing solution*: For colloidal Coomassie brilliant blue staining, fix gels in 35% (v/v) ethanol, 2% (v/v) phosphoric acid in distilled H_2O

3. *Colloidal CBB staining solution*: 34% (v/v) methanol, 17% (w/v) ammonium sulphate, 3% (v/v) phosphoric acid in distilled H_2O

Steps

1. All gel staining steps should be performed in a dedicated clean room to avoid contamination.

2. After CyDye fluorescence scanning, remove spacer plate and immerse gels in fixing solution and incubate overnight with gentle shaking. Fixed and bonded gels can now be stored for many months at 4°C by sealing in plastic bags with 50 ml of 1% (v/v) acetic acid. The CyDye fluorescence signal is also detectable after several months of storage.

3. For poststaining with Sypro Ruby protein stain (Berggren *et al.*, 2000), wash fixed gels for 10 min in distilled H_2O and incubate in Sypro Ruby stain for at least 3 h on a shaking platform in the dark. Pour off the stain and wash the gel in distilled H_2O or destain [10% (v/v) methanol, 6% (v/v) acetic acid] for three times 10 min. Drain and dry the outer surface of the glass plate and scan gel-side up in a Typhoon 9400 imager at the appropriate excitation/emission wavelength for the Sypro Ruby protein stain.

4. The colloidal CCB G-250 staining method has been modified from that of Neuhoff *et al.* (1988). Fix gels in colloidal CBB fixing solution for at least 3 h on a shaking platform. Wash three times for 30 min with distilled H_2O and incubate in CCB staining solution for 1 h. Add one crushed CCB tablet (25 mg) per 50 ml of staining solution (0.5 g/liter) and leave to stain for 2–3 days. No destaining step is required to visualise proteins. Stained gels can be imaged on a densitometer or

on the Typhoon scanner using the red laser and no emission filters.

5. Align poststained and CyDye gel images to identify spots of interest for cutting. Alignment and spot identification can be carried out by comparing images by eye or using Adobe Photoshop to overlay images. A shift in molecular weight between poststained and CyDye images should be apparent due to the increased mass of the dye-labelled fraction of proteins (Gharbi *et al.*, 2002).

6. For automated spot picking, input and process poststained images in DeCyder software and create a pick list for the spots of interest by comparing with the results of the BVA analysis. To facilitate sample tracking and later data matching with MS results, the poststained image can be imported and matched within the current experimental BVA work space. This means that any spot picked according to the poststained image will have the same master spot number as in the BVA quantitative analysis. Define the positions of the two reference markers in DeCyder (left then right) and export the pick list coordinate file (.txt) to the spot picker controller.

7. Excise chosen spots from the poststained gel. In the case of visible colloidal CBB-stained gels, this can be done manually with a glass Pasteur pipette or gel-plug cutting pipette or on a robotic picker incorporating a "click-n-pick" format, such as the Ettan Spot Picker. The gel is best submerged under 1–2 mm of distilled water, and picking performed in a dedicated clean room.

8. For automated picking using an Ettan Spot Picker, open the imported pick list and align the instrument with the reference markers according to the manufacturer's instructions. Pick and collect spots in 96-well plates in 200 µl of water, drain, and store at −20°C prior to MS analysis. Sample preparation and protein identification by mass spectrometry are detailed elsewhere.

Comment

Harsh fixatives (e.g., >35% methanol) should not be used on bonded gels as they cause overshrinkage and cracking of gels.

References

Berggren, K., Chernokalskaya, E., Steinberg, T. H., Kemper, C., Lopez, M. F., Diwu, Z., Haugland, R. P., and Patton, W. F. (2000). Background-free, high sensitivity staining of proteins in one- and two-dimensional sodium dodecyl sulfate-polyacrylamide gels using a luminescent ruthenium complex. *Electrophoresis* **21**(12), 2509–2521.

Gharbi, S., Gaffney, P., Yang, A., Zvelebil, M. J., Cramer, R., Waterfield, M. D., and Timms, J. F. (2002). Evaluation of two-dimensional differential gel electrophoresis for proteomic expression analysis of a model breast cancer cell system. *Mol. Cell Proteomics* **1**(2), 91–98.

Lopez, M. F., Berggren, K., Chernokalskaya, E., Lazarev, A., Robinson, M., and Patton, W. F. (2000). A comparison of silver stain and SYPRO ruby protein gel stain with respect to protein detection in two-dimensional gels and identification by peptide mass profiling. *Electrophoresis* **21**(17), 3673–3683.

Neuhoff, V., Arold, N., Taube, D., and Ehrhardt, W. (1988). Improved staining of proteins in polyacrylamide gels including isoelectric focusing gels with clear background at nanogram sensitivity using Coomassie brilliant blue G-250 and R-250. *Electrophoresis* **9**(6), 255–262.

Patton, W. F. (2000). A thousand points of light: The application of fluorescence detection technologies to two-dimensional gel electrophoresis and proteomics. *Electrophoresis* **21**(6), 1123–1144.

Scheler, C., Lamer, S., Pan, Z., Li, X. P., Salnikow, J., and Jungblut, P. (1998). Peptide mass fingerprint sequence coverage from differently stained proteins on two-dimensional electrophoresis patterns by matrix assisted laser desorption/ionization-mass spectrometry (MALDI-MS). *Electrophoresis* **19**(6), 918–927.

Shevchenko, A., Wilm, M., Vorm, O., and Mann, M. (1996). Mass spectrometric sequencing of proteins silver-stained polyacrylamide gels. *Anal. Chem.* **68**(5), 850–858.

Tonge, R., Shaw, J., Middleton, B., Rowlinson, R., Rayner, S., Young, J., Pognan, F., Hawkins, E., Currie, I., and Davison, M. (2001). Validation and development of fluorescence two-dimensional differential gel electrophoresis proteomics technology. *Proteomics* **1**(3), 377–396.

Unlu, M., Morgan, M. E., and Minden, J. S. (1997). Difference gel electrophoresis: A single gel method for detecting changes in protein extracts. *Electrophoresis* **18**(11), 2071–2077.

Zhou, G., Li, H., DeCamp, D., Chen, S., Shu, H., Gong, Y., Flaig, M., Gillespie, J. W., Hu, N., Taylor, P. R., *et al.* (2002). 2D differential in-gel electrophoresis for the identification of esophageal scans cell cancer-specific protein markers. *Mol. Cell Proteomics* **1**(2), 117–124.

Affinity Electrophoresis for Studies of Biospecific Interactions: High-Resolution Two-Dimensional Affinity Electrophoresis for Separation of Hapten-Specific Polyclonal Antibodies into Monoclonal Antibodies in Murine Blood Plasma

Kazuyuki Nakamura, Masanori Fujimoto, Yasuhiro Kuramitsu, and Kazusuke Takeo

I. INTRODUCTION

Affinity electrophoresis (AEP) was developed as a novel technique for separation of biomolecules by biospecific interactions with their ligands in an electric field (Nakamura, 1959). The techniques of rocket immunoelectrophoresis (Laurell, 1966) and crossed immunoelectrophoresis (Svenson and Axelsen, 1972) are based on the same principle. AEP has been applied not only for separation of a tiny amount of those biomolecules, but also for determination of dissociation constants (K_d) of those interactions (Takeo and Nakamura, 1972; Bog-Hansen, 1973; Horejsi and Kocourek, 1974; Caron *et al.*, 1975; Takeo and Kabat, 1978; Nakamura *et al.*, 1980; Shimura and Kasai, 1982) at different pH (Ek *et al.*, 1980; Mimura *et al.*, 1992) and temperatures for calculation of thermodynamic parameters (Tanaka *et al.*, 1986; Kashiwagi *et al.*, 1991).

Two-dimensional affinity electrophroesis (2DAEP), which was newly developed by a combination of isoelectric focusing (IEF) with AEP (Takeo *et al.*, 1983), has been used for studies of the immune response *in vivo*.

Hapten-specific polyclonal IgG antibodies, which were produced by immunization of rabbits and mice with the hapten of dinitrophenyl (DNP)- or fluorescein isothiocyanate (FITC)-conjugated protein carriers (Takeo *et al.*, 1992; Nakamura *et al.*, 1993), were separated into a large number of groups of IgG spots by 2DAEP, and each of the groups showed an identical affinity to the hapten but a different isoelectric point (pI) as in the case of monoclonal antibodies specific for the hapten. Diversification, affinity maturation, and subclass switching of the hapten-specific antibodies *in vivo* were evidenced in the course of immunization of a single mouse (Nakamura and Takeo, 1998). This article describes the procedures for 2DAEP used for separation of anti-DNP antibodies in murine blood plasma.

II. MATERIALS AND INSTRUMENTATIONS

Acrylamide (Cat. No. 00809-85), N, N'-methylenebisacrylamide (Bis) (Cat. No. 22402-02), and

N, N, N', N'-tetramethylethylenediamine (TEMED) (Cat. No. 33401-72) are from Nacalai tesque (Kyoto, Japan). Carrier ampholytes, Parmalite pH 4–6.5 (Cat. No. 17-0452-01), Pharmalite pH 6.5–9 (Cat. No. 17-0454-01), and Pharmalite pH 5–8 (Cat. No. 17-0453-01) are from Amersham Biosciences (Little Chalfront, Buckinghamshire, UK). Chicken albumin (Cat. No. A-5503), dinitrofluorobenzene (DNFB)(Cat. No. D1529), dinitrophenyl glycine (Cat. No. 9504), β-alanine (Cat. No. A7752), L-lysine (Cat. No. L5501), human γ-globulins (HGG) (Cat. No. G4386), Tris (Cat. No. T6066), and 4-chloro-1-naphthol (Cat. No. C8890) are from Sigma (St.Louis, Mo). Sodium hydroxide (Cat. No. 28-2940), glycine (Cat. No. 12-1210-5), glycerol (Cat. No. 12-1120-5), urea (Cat. No. 32-0280-5), ammonium peroxodisulphate (APS) (Cat. No. 01-4910-2), potassium hydroxide (Cat. No. 24-4670-5), sucrose (Cat. No. 28-0010-5), methylene blue trihydrate (Cat. No. 19-3200-2), riboflavin (Cat. No. R4500), hydrochloric acid (Cat. No. 13-1700-5), sodium chloride(Cat. No. 28-2270-5), Tween 20 (Cat. No. 30-5450-5), and methanol (Cat. No. 19-2410-8) are from Sigma-Aldrich-Japan(Tokyo, Japan). L-Glutamic acid (Cat. No. 074-00505), hydrogen peroxide (Cat. No. 086-07445), and acetic acid (Cat. No. 012-00245) are from Wako (Osaka, Japan). POD-conjugated rabbit IgG fraction to goat IgG Fc (Cat. No. 55360) and goat antihuman γ-globulin antisera (Cat. No. 55074) are from ICN Pharmaceuticals, Inc. (Cappel Products). Other chemicals are of the highest available purity obtained from various sources.

A. Preparation of Antihapten Antisera

A BALB/c mouse (female, 2 months old) was immunized by an intraperitoneal injection of 50 μg of an antigen, hapten-conjugated protein carrier of DNP-conjugated chicken serum albumin (DNP-CSA) in an emulsion with an equal volume of Freund's complete adjuvant as the primary immunization. The second immunization was performed by an intraperitoneal injection of 50 μg of the antigen 2 weeks after the primary immunization and a boosting was made by an intraperitoneal injection of 100 μg of the antigen weeks after the second immunization. In the course of immunization, 150 μl of whole blood was taken by a puncture of the veniplex lining of an eyeball of a deeply anesthetized mouse every week. The blood was allowed to stand at 37°C for 2 h, and the blood clot was removed by centrifugation at 1000 *g* for 5 min to yield 70 μl of clear antisera. Antisera were stored at 4°C by adding 0.1% sodium azide and were submitted to 2DAEP.

B. Preparation of Water-Soluble DNP-Conjugated Noncross-Linked Acrylamide–Allylamine Copolymer (DNP-PA) as an Affinity Ligand

Acrylamide monomer (40 g) and 4 g of allylamine were dissolved with 400 ml of distilled water in a 500-ml Erlenmeyer flask, followed by the addition of 0.8 ml of TEMED and 50 ml of 0.8% freshly prepared ammonium peroxodisulphate solution with gentle mixing. A small volume of distilled water was overlaid on top of the solution to shut off oxygen and was allow to stand overnight at 30°C. The solution was then extensively dialyzed against tap water for 3 days and distilled water with four changes a day for 2 days. Sodium bicarbonate (4.5 g) was added to the dialyzed solution, and 4.8 g of dinitrofluorobenzene in 15 ml of acetone was mixed by dropwise addition. The mixture was gently stirred overnight in the dark at 25°C using an evaporator without sucking.

The reaction mixture was then dialyzed against tap water for 5 days and distilled water with four changes a day for 2 days. The dialyzed solution was concentrated *in vacuo* until the DNP concentration reached 5–10 m*M* for storage at 4°C in the dark.

The concentration of DNP residue in DNP-conjugated noncross-linked acrylamide–allyamine copolymer (Fig. 1) was calculated from the absorbance at 360 nm in 0.1 *M* sodium hydroxide, using DNP-glycine as a standard (molar absorption at 360 nm was 17530 liter/mol/cm).

III. PROCEDURES

A. Preparation and Running of First-Dimension Gels (IEF)

Solutions

1. *Working solutions for capillary IEF polyacrylamide gels*: To make 6.0 ml of IEF gel solution, prepare the working solutions as shown in Table I
2. *Running buffer solution for anode*: To make 1 liter of 40 m*M* L-glutamic acid solution, dissolve 5.95 g of L-glutamic acid in distilled water and complete to 1 liter
3. *Running buffer solution for cathode*: To make 1 liter of 1 *M* NaOH, dissolve 40.0 g sodium hydroxide in distilled water and complete to 1 liter
4. *Sample solution*: To make 0.1 ml sample solution, mix 0.027 ml of antisera or purified antibodies with 0.055 ml of phosphate-buffered saline, pH 7.4 (PBS), and add 0.03 g of sucrose

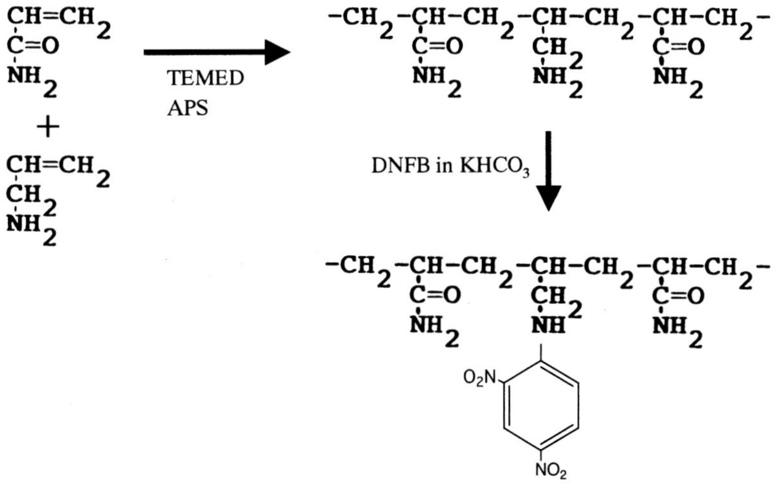

FIGURE 1 Preparation of water-soluble dinitrophenyl-conjugated noncross-linked acrylamide–allyamine copolymer (DNP-PA). TEMED, N, N, N', N'-tetramethylethylenediamine; APS, ammonium peroxodisulphate; DNFB, dinitrofluorobenze.

TABLE I Preparation of Working Solutions for Capillary IEF Gels

Stock solutions and reagent	Volume and weight
Acrylamide solution	0.6 ml
Acrylamide 48.5 g	
Bis 1.5 g	
Dissolved in distilled water to 100 ml	
Carrier ampholyte	
Pharmalite	
pH 4–6.5	0.0094 ml
pH 6.5–9	0.0094 ml
pH 5–8	0.0094 ml
Glycerol solution	0.5 ml
Glycerol 40.0 g	
Dissolved in 100 ml distilled water	
L-Lysine solution	0.8 ml
L-Lysine 1.58 g	
Dissolved in distilled water to 100 ml	
Distilled water	0.98 ml
Urea	2.162 g
APS solution[a]	0.143 ml
Ammonium peroxodisulphate 0.21 g	
dissolved in distilled water to 10 ml	
TEMED	6 µl
Total volume	6.0 ml

[a] APS solution is prepared freshly before mixing with the deaerated acrylamide gel solution.

Steps

1. Mark the glass capillary (1.2 mm in inner diameter and 8.5 cm in height, Cat. No. 2-000-100, Drummond Scientific Co., Broomall, PA) at 7.5 cm from the bottom.

2. Prepare the gel solution by mixing 0.6 ml of the acrylamide solution, 0.094 ml of Pharmalite pH 4–6.5, 0.094 ml of Pharmalite pH 6.5–9, 0.188 ml of Pharmalite 5–8, 1.5 ml of 40% glycerol solution, 0.8 ml of L-lysine solution with 0.98 ml distilled water, and add 2.162 g of urea to yield 5.867 ml of gel solution. Deaerate the gel solution and mix with 0.143 ml of APS solution and 6 µl of TEMED well just before preparation of gels.

3. Pour two-thirds of the gel solution into a plastic cylindrical chamber (Fig. 2: 1.2 cm in inner diameter and 8.0 cm in height) and immerse the glass capillaries in the solution, avoiding the formation of air bubbles in the capillaries, which are tied with a rubber band.

4. Pour the residual solution into the chamber until the meniscus in the capillaries reaches the marker 7.5 cm from the bottom.

5. Stand the chamber at 25°C for 10 min and find a new surface of top of polyacrylamide gel, which is formed 0.6 mm beneath the original meniscus.

6. Remove the rubber band from the capillaries and seal the top of capillaries in the chamber with Parafilm tightly to avoid drying up during storage at 4°C. Use the capillary gels within 4 weeks.

7. Take the capillary from the chamber and wipe up gel crumbs clinging to the outer surface of the capillary.

8. Pour 10 µl of the sample solution onto the top of the capillary by a microsyringe and overlay the running buffer of 40 mM L-glutamic acid (for anode), filling up the capillary.

9. Insert the capillary into a hole of a cylindrical rubber connector and set the connector to the upper buffer reservoir (Fig. 2) tightly. After setting six capillaries and a thermometer with the rubber connector to the upper buffer reservoir (for anode) and avoid air bubbles on the bottom of gels with a small volume of the running buffer of 1 M NaOH for cathode.

10. Place the upper reservoir on the lower buffer reservoir (for cathode), which is filled with the running buffer of 1 M NaOH, and place the cooling device in the middle of the apparatus (Fig. 2, right).

11. Run the electrophoresis using a constant voltage power supply (Cat. No. 2197-010, Pharmacia LKB, Germany) with stepwise elevation of the voltage from 250 V for 15 min to 500 V for 15 min, 1000 V for 15 min, and 2000 V for 2 h and keep the temperature at 15°C using the cooling device.

12. Detach the capillaries from the apparatus and subject to second-dimension gels (AEP).

B. Preparation and Running Second-Dimension Gels (AEP)

Solutions

1. *Working solutions for AEP polyacrylamide gels*: To make 8.0 ml of the AEP gel solution, prepare the

TABLE II   Preparation of Working Solutions for AEP Separating Gels[a]

Stock solution	Volume of working solution for a gel	
	Control gel	Affinity gel
Acrylamide solution	1.0 ml	1.0 ml
Acrylamide 40.0 g		
Bis 1.067 g		
Dissolved in distilled water to 100 ml		
Gel buffer solution	2.0 ml	2.0 ml
1 M KOH 48.0 ml		
Acetic acid 17.2 ml		
TEMED 4.0 ml		
Dissolved in distilled water to 100 ml		
DNP-PA solution		
DNP-PA (at any concentration)	—	2.73 ml
Distilled water	2.73 ml	—
Sucrose solution	2.0 ml	2.0 ml
Sucrose 40.0 g		
Dissolved in 100 ml distilled water		
APS solution[b]	0.27 ml	0.27 ml
Ammonium peroxodisulphate 0.21 g		
Dissolved in distilled water to 10 ml		
Total volume	8.0 ml	8.0 ml

[a] Polyacrylamide gels are 5.13% in T, 2.59% in C, pH 4.3.
[b] APS solution is prepared freshly before mixing with the deaerated acrylamide gel solution.

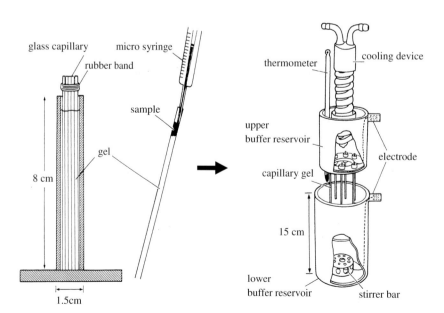

FIGURE 2   Preparation of capillary gels and apparatus for first-dimension isoelectric focusing (IEF).

TABLE III   Preparation of Working Solutions for AEP Stacking Gel[a]

Stock solution	Volume of solution for four gels
Acrylamide solution	1.5 ml
Acrylamide	10.0 g
Bis	2.5 g
Dissolved in distilled water to 100 ml	
Gel buffer solution	1.5 ml
1 M KOH	48.0 ml
Acetic acid	2.87 ml
TEMED	0.46 ml
Dissolved in distilled water to 100 ml	
Sucrose solution	1.5 ml
Sucrose	40.0 g
Dissolved in 100 ml distilled water	
Distilled water	0.75 ml
Riboflavin solution[b]	0.75 ml
Riboflavin	4.0 mg
Dissolved in 100 ml distilled water	
Total volume	6.0 ml

[a] Polyacrylamide gels are 3.13% in T, 20.0% in C, pH 6.7.

[b] The solution should be prepared freshly or kept at 4°C in a brown bottle to use within a few weeks before mixing with the deaerated gel solution.

TABLE IV   Preparation of Stacking Solution for AEP Gel

Stock solution	Volume of solution for four gels
Buffer solution	3.0 ml
1 M KOH	48.0 ml
Acetic acid	2.87 ml
Dissolved in distilled water to 100 ml	
Sucrose solution	1.5 ml
Sucrose	40.0 g
Dissolved in 100 ml distilled water	
Distilled water	3.7 ml
Methylene blue solution	3.8 ml
Methylene blue	50 mg
Dissolved in 100 ml distilled water	
Total volume	12.0 ml

working solutions as shown in Tables II and III for separating gel and stacking gel in Reisfeld's buffer system, respectively.

2. *Running buffer solution(Reisfeld, 1962)*: To make 500 ml of running buffer, pH 4.5, mix 100 ml of the stock solution (dissolve 31.2 g of β-alanine and 8.0 ml acetic acid with distilled water to make 1 liter) and 400 ml of distilled water. Check the pH before use.

3. *Stacking solution (Reisfeld, 1962)*: To make 12 ml of the stacking solution, prepare the working solutions as shown in Table IV. Check the pH before use.

Steps

All steps from 1 to 10 must be finished before the completion of the first-dimension IEF.

1. Mark the slab gel cassette (inner space; 100 mm in height, 85 mm in width, 1 mm in thickness, Cat. Nos. SE-400 and SE-401, Marysol, Tokyo, Japan) at 7.5 cm from the bottom and seal the bottom and both sides of the cassette.

2. Stand the cassette at 37°C in a dry incubator until gel preparation.

3. Prepare the AEP separating gel solution by mixing 1.0 ml of acrylamide solution, 2.0 ml of buffer solution, 2.73 ml of DNP-PA solution (substitute with 2.73 ml of distilled water for control gel solution), and

2.0 ml of 40% (w/v) sucrose solution to yield 7.73 ml of gel solution as shown in Table II.

4. Deaerate the gel solution and mix with 0.27 ml of APS solution well just before preparation of the gel.

5. Pour the gel solution into the cassette until the meniscus reaches the marker 7.5 cm from the bottom.

6. Overlay the DNP-PA solution of the same concentration as in the gel (substitute with distilled water for control gel) on top of the gel solution carefully with a syringe to avoid disturbances to the gel solution meniscus.

7. Stand the cassette at 37°C in a dry incubator to complete gelification within 30–40 min and find a new surface of gel beneath the solution.

8. Discard the residual solution on top of the gel and wash with 1 ml of the stacking gel solution, which has been prepared as in Table III, to remove residual DNP-PA on top of the gel. Shade the part cassette for the separating gel by wrapping with aluminum foil.

9. Place a comb in the cassette to prepare the stacking gel in 1 cm height on the top of the separating gel and pour the stacking gel solution, avoiding the formation of air bubbles.

10. Illuminate the stacking gel solution with a fluorescent lamp to complete gelification the stacking gel in 20–30 min.

11. Remove the comb from the cassette and find the stacking gel as shown in Fig. 3 (left). The narrow wells in both sides of the stacking gel are for marker proteins, if necessary.

12. Remove the seal of the bottom of the gel cassette and place two sets of the gel cassette onto the apparatus for AEP as shown in Fig. 3 (right), to fixing the cassette with clips.

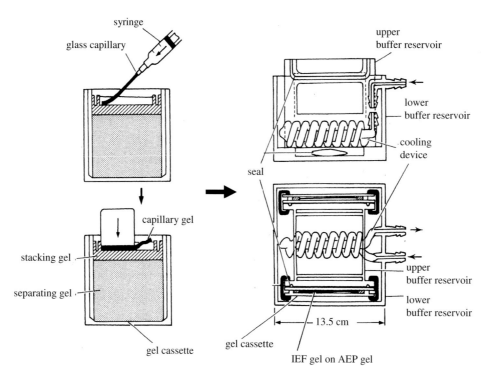

FIGURE 3 Preparation of affinity gels and apparatus for second-dimension affinity electrophoresis (AEP). The IEF capillary gel is put on top of the stacking gel of AEP tightly with a metal plate as indicated by the fine arrow (left). Water at the desired temperature circulates in the glass cooling device, which is connected to a thermostat apparatus.

13. Fill the lower buffer reservoir with the running buffer solution carefully, avoiding the formation of air bubbles at the bottom of the gel.

14. Fill the middle well of the stacking gel with the stacking, solution which has been prepared as described in Table IV.

15. Place the IEF capillary gel by pushing out from the glass capillary with an expeller of a syringe filled with the stacking solution as shown in Fig. 3 (left) and tightly fit on the top of the stacking gel.

16. Fill the upper buffer reservoir with the running buffer and run the electrophoresis at 50 V for 15 min and 80 V for 2.5 h until the methylene blue reaches the bottom of the separating gel.

17. Using a water circulating cooling device (Cat. No. LKB 2219-001, Pharmacia LKB, Germany) keep the temperature during electrophoresis at the desired temperature.

C. Immunoblotting

Solution

Blotting buffer solution: To make 4 liter of electrode buffer solution of 50 mM glycine, pH 2.5, mix 200 ml of 1 M glycine solution (dissolve 75.07 g of glycine with distilled water to 1 liter) with 100 ml of 1 M HCl solution and add distilled water to 4 liters.

Steps

Steps 1–4 must be completed before completion of the second electrophoresis of AEP.

1. Immerse two sheets of nitrocellulose (NC) membrane (Millipore, pore size 0.22 μm, 100 × 75 mm), four sheets of Whatman 3 MM filter paper (Cat. No. 3030917) in 100 × 80 mm, and two pieces of Scotch-Brite 3 M nylon sponge pad in 100 × 85 mm, into 1 liter of the blotting buffer solution and deaerate extensively with suction by an aspirator for 1 h.

2. Pour the deaerate blotting buffer into the buffer reservoir of the blotting apparatus with carbon electrodes (Fig. 4, left).

3. Immerse the cassette holder into 3 liters of the blotting buffer solution in a tray (32 × 23 × 6 cm).

4. Put a sheet of the Whatman 3 MM filter paper on a window of the cassette holder and lay the NC membrane on the filter paper to avoid leaving air bubbles between the sheets.

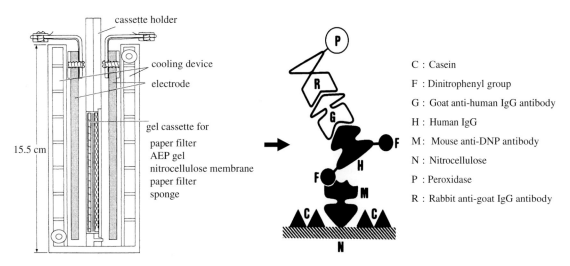

FIGURE 4 Blotting apparatus and a scheme for immunostaining of mouse anti-DNP antibodies on the NC membrane.

C : Casein
F : Dinitrophenyl group
G : Goat anti-human IgG antibody
H : Human IgG
M : Mouse anti-DNP antibody
N : Nitrocellulose
P : Peroxidase
R : Rabbit anti-goat IgG antibody

5. Remove the AEP gel cassette from the upper buffer reservoir (Fig. 3) immediately after the completion of AEP and leave the gel on the either side of the cassette plate.

6. Make a small hole with a metal punch (2 mm in diameter) at the corner of the bottom of the AEP gel on the anodic side of IEF.

7. Release the AEP gel from the cassette plate carefully using a rectangular metal spatular of (85 × 85 mm) and lay on the NC membrane carefully to avoid leaving air bubbles.

8. Overlay another sheet of the Whatman 3 MM filter paper on the AEP gel carefully and put on the sheet of nylon sponge pad.

9. Close the cassette holder and place the holder in the middle of the electrodes of the blotting apparatus as shown in Fig. 4.

10. Fill the buffer reservoir with the blotting buffer, if necessary, and run the electrophoresis at 5 V for 2 h.

11. Keep the temperature at 6°C using the cooling device.

12. After completion of blotting, remove the NC membrane from the AEP gel and immerse the NC membrane into Tris-buffered saline, pH 7.5 (TBS), to wash out the blotting buffer overnight at 4°C.

13. Start steps for immunostaining of mouse anti-DNP antibodies blotted on the NC membrane as shown in Table V.

14. Carry out all of the procedures for immunostaining in Tupperware (10 × 14 × 3 cm) at room temperature.

15. Take photographs of the patterns of the anti-DNP antibodies spots on the NC membrane as shown in Fig. 5 soon after immunostaining and store the NC membrane by wrapping with aluminum foil at 4°C if necessary.

IV. COMMENTS

Dissociation constants for interactions between each group of anti-DNP antibodies in mouse antisera and the DNP group can be determined using gels containing a series of different concentrations of DNP-PA for second-dimension AEP. The migration distance of an anti-DNP antibody is decreased by increasing the concentration of DNP-PA in the gel, and the dissociation constant (K_d) can be calculated by the equation $1/r = (1/R_o)[1 + (c/K_d)]$, where r is the relative migration distance of the antibody in the presence of DNP-PA, R_o is the relative migration distance of the antibody in the absence of DNP-PA, and c is the concentration of DNP-PA. To obtain reproducible results, the voltage of the electric field and the temperature of the gels should be kept constant during AEP.

TABLE V   Procedure for Immunostaining of Anti-DNP Antibodies Blotted on NC Membrane[a]

No.	Step	Solution	Volume (ml) per NC membrane	Incubation time
1.	Washing	TBS[b]	100	Overnight
2.	Blocking	Skimmed milk[c]	20	1 h
3.	Washing	TBS	50	30 s
4.	Reaction with DNP–conjugate	DNP-conjugated HGG in 1% skimmed milk[d]	10	2 h
5.	Washing (repeat four times)	Tween-TBS[e]	50	10 min
6.	Reaction with anti-HGG antibody	Goat anti-HGG antisera in 1% skimmed milk[f]	10	2 h
7.	Washing (repeat four times)	Tween-TBS	50	10 min
8.	Reaction with POD-conjugated antibody	POD-conjugated rabbit antigoat IgG antibodies in 1% skimmed milk[g]	10	2 h
9.	Washing (repeat four times)	Tween-TBS	50	10 min
10.	Visualization	POD substrate solution[h]	20	5–15 min
11.	Washing (repeat four times)	Distilled water	50	5 min

[a] Carry out all steps of the procedure at room temperature with gentle shaking.

[b] Tris-buffered saline: dissolve 2 M Tris 50 ml, 1 M HCl 81 ml, and NaCl 43.5 g in distilled water to 5 liters.

[c] Dissolve 5 g of skimmed milk in 100 ml of TBS.

[d] Prepare DNP-conjugated human γ-globulin (HGG) (0.1 mg/ml) in 1% skimmed milk of TBS.

[e] Dissolve 50 mg of Tween 20 in 100 ml of TBS.

[f] Dissolve 0.08 ml of goat anti-HGG antisera in 1% skimmed milk of TBS.

[g] Dissolve 0.02 ml of peroxidase (POD)-conjugated rabbit antigoat IgG antibody solution in 1% skimmed milk of TBS.

[h] Dissolve 0.06 g of 4-chloro-1-naphthol in 20 ml of cold methanol and mix with 100 ml of TBS and 0.06 ml of 30% of hydrogen peroxide just before visualization.

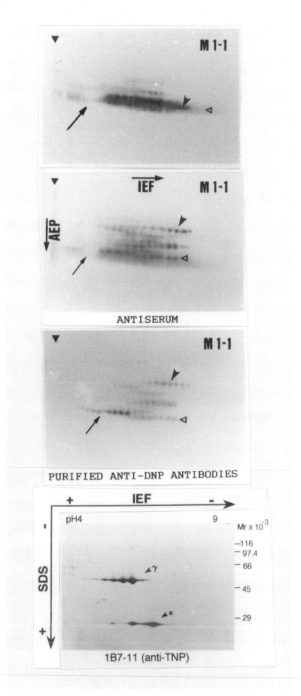

FIGURE 5 Separations of anti-DNP antibodies in mouse antisera by 2DAEP and of a mouse monoclonal antibody specific to the trinitrophenyl (TNP) group by 2DE. The 2DAEP patterns of anti-DNP antibodies in the absence of affinity ligand (a) and the presence of 0.1 mM DNP-PA (b) are shown. (c) The 2DAEP patterns of 2 µg of anti-DNP antibodies, which were purified from mouse antisera by affinity chromatography with DNP-conjugated lysine Sepharose in the presence of 0.1 mM DNP-PA. Arrows indicate the position of serum albumin which interferes with the blotting of anti-DNP antibodies on the NC membrane. Arrowheads indicate a group of IgG spots with an identical affinity to the ligand of DNP but different isoelectric points (pI), as in the case of the monoclonal antibody. (d) The 2DE patterns of the monoclonal antibody (IB7-11) specific to TNP separated by a combination of IEF and SDS–polyacrylamide gel electrophoresis. The γ and κ indicate the IgG heavy chain and light chain, respectively. Chains show the molecular heterogeneity with different pI values.

References

Bog-Hansen, T. C. (1973). *Anal. Biochem.* **56**, 480–488.

Caron, M., *et al.* (1975). *J. Chromatogr.* **103**, 160–165.

Ek, K., and Righetti, P. J. (1980). *Electrophoresis* **1**, 137–140.

Horejsi, V., and Kocourek, J. (1974). *Biochim. Biophys. Acta* **336**, 338–343.

Kashiwagi, S., *et al.* (1991). *Electrophoresis* **12**, 420–424.

Laurell, C.-B. (1966). *Anal. Biochem.* **15**, 45–52.

Mimura, Y., *et al.* (1992). *J. Chromatogr.* **597**, 345–350.

Nakamura, K., *et al.* (1980). *J. Chromatogr.* **192**, 351–362.

Nakamura, K., *et al.* (1993). *Electrophoresis* **14**, 81–87.

Nakamura, K., and Takeo, K. (1998). *J. Chromatogr. B.* **715**, 125–136.

Nakamura, S. (1966). *"Cross Electrophoresis."* Igaku Shoin, Tokyo and Elsevier, Amsterdam.

Nakamura, S., *et al.* (1959). *Nature (London)* **184**, 638–639.

Reisfeld, R. A., *et al.* (1962). *Nature* **195**, 281–283.

Shimura, K., and Kasai, K. (1982). *J. Biochem.* **92**, 1615–1622.

Svenson, P. J., and Axelson, N. H. (1972). *J. Immunol. Methods* **1**, 169–172.

Takeo, K. (1987). Affinity electrophoresis. *In "Advances in Electrophoresis"* (A. Chrambach, M.J. Dunn, and B. J. Radola, eds.), pp. 229–279. VCH, Weinheim.

Takeo, K., and Kabat, E. A. (1978). *J. Immunol.* **121**, 2305–2310.

Takeo, K., and Nakamura, S. (1972). *Arch. Biochem. Biophys.* **153**, 1–7.

Takeo, K., *et al.* (1983). *In "Electrophoresis 82"* (D. Stathakos, ed.), pp. 277–283. Walter de Gruyter, Berlin.

Takeo, K., *et al.* (1992). *J. Chromatogr.* **597**, 365–376.

Tanaka, T., *et al.* (1986). *Electrophoresis* **7**, 204–209.

26

Image Analysis and Quantitation

Patricia M. Palagi, Daniel Walther, Gérard Bouchet, Sonja Voordijk, and Ron D. Appel

I. INTRODUCTION

Many proteomics studies involve comparisons of two-dimensional electrophoresis (**2-DE**) **gels** to identify protein expression changes between different samples. They need efficient **image analysis** software to automatically analyse gel images and extract pertinent biological data. The major steps in such analyses include detection of protein spots in the gels, finding corresponding spots among gels, computation of **protein expression** modifications, and statistical interpretation.

Currently, at least 10 different commercial software packages for the analysis of 2-DE gel images are available commercially. Some of them are descendents of the first generation of tools and software to analyse 2-DE gels, such as Gellab (Lemkin and Lipkin, 1981), THYCO (Anderson *et al.*, 1981), and Melanie (Appel *et al.*, 1997). Although each of these new systems has its own philosophy and approaches, most of them provide the basic operations and functionalities necessary to carry out a complete gel study. The objective of this article is to describe the main steps in a 2-DE gel analysis necessary to find out differently expressed proteins as performed with Melanie software version 4.

II. MATERIALS AND INSTRUMETATION

A. Software

The Melanie software (version 4) is developed at the Swiss Institute of Bioinformatics. It is commercialised under the name ImageMaster 2D Platinum by Amersham Biosciences in collaboration with GeneBio.[1] A demonstration version of the program and support documentation are freely available from GeneBio's Web site (www.genebio.com). Melanie Viewer, a reduced version of this software, can be freely downloaded from the ExPASy server (http://www.expasy.org).

To use the on-line manual and to access remote databases over the Internet with Melanie, Internet Explorer (Microsoft Corporation), Netscape Navigator (Netscape Communications Corporation), or Mozilla (The Mozilla Foundation) have to be installed on the computer.

B. Image Capture

Gel images may be produced with a large variety of image capture devices, ranging from flatbed document scanners, camera systems, densitometers, phosphor imagers, or fluorescence scanners. The default output format for most imaging equipment, and definitely the most appropriate one for further analysis by 2D software, is Tag Image File Format (TIFF, Aldus Corporation). This is the recommended format for use with Melanie, although the software can read some other file types.

The scanning resolution of the gel image is very important, as it influences the amount of visible details in the image. A low resolution corresponds to a large pixel size or a small number of pixels (or dots) per inch. When the image resolution is too low, the automatic spot detection becomes more difficult. However,

---

[1] Melanie 4 is currently also available from Bruker Daltonics integrated with their PROTEINEER spII spot picking robot.

when the scan resolution gets too high, the image file becomes very large, and this can slow down the gel analysis. A resolution of 100–200 µM (or about 250–150 dpi) is a good compromise.

The range of possible grey levels (intensity values) in a picture varies according to the image depth (number of bites used to represent a pixel). Images scanned with a higher image depth contain more information. In the case of an 8-bit image, for instance, one pixel has 256 possible grey values (0 to 255). A 16-bit image (65536 grey levels) may reveal subtler but often significant changes; however, it requires more memory. An image depth of at least 12 bits is judicious.

C. Computer Requirements

The Melanie software runs on any of the current Windows operating systems, i.e., 98/ME/NT/2000/XP.

The minimum recommended virtual memory is 256 MB, which is enough to open and process a large number of gels.

Melanie functions properly with a colour resolution of 8 bits (256 colours). However, to use its 3D View module, the colour resolution should be set to 24 bits (16.7 million grey values). It is also recommended to use a screen resolution of at least 1024 × 768 pixels.

III. PROCEDURES

A. Opening Gels and Setting up a Workspace

The investigation of six gels is explained step by step hereafter to describe the usual reasoning when carrying on a whole analysis of 2-DE gels. The examples and images shown in this article were generated with gels from a study of aortic smooth muscle cells from newborn (4-day-old) and aged (18-month-old) rats (Cremona et al., 1995). Each population has three gels and are called henceforth newborn and aged.

To start a new work session, import the gel images with the import function and, if necessary, choose a reduction factor. It is highly recommended to setup a workspace as soon as gels have been imported. It facilitates to organise the gel experiments and to work in a personalized environment. The workspace holds information on the relationships between gels such as their organisation into populations (classes) and their reference gel (the gel chosen to make the connection with the other gels). The workspace allows organising gels into projects that reflect the structure and design of the experimental studies facilitating the subsequent work.

   i. Click on the Melanie *Workspace* icon in the toolbar to display the Melanie *Workspace* window and create a new workspace.

   ii. Inside the workspace, create a new project with the name *"Aortic smooth muscle cells."* Select the newborn gels and add a new class to the project. The selected gels will immediately be allocated to the new class.

   iii. Repeat the same procedure for aged gels.

   iv. Choose the best representative gel among the six gels and set it as the reference gel.

   v. Position the mouse cursor on the class names and right click to open the classes and make these settings active.

B. Viewing and Manipulating Gels

The following usual operation is to adjust the image contrast to improve its visualisation, i.e., to visually better differentiate the spots from the background. This kind of operation is often indicated because of differences between the images and the screen display depths. To adjust the gels contrast perform the following steps.

   i. Select one or several gels and draw a region in these gels to get a preview of the contrast mapping modifications.

   ii. Choose *View → Adjust Contrast → Current* from the menu.

   iii. Select the gel for which you would like to display the histogram function by choosing it from the *Image* list.

   iv. In the *Gel Display Settings* window, change the minimum and maximum grey levels by displacing the slider borders.

   v. Click on OK to apply the visual changes to all selected gels.

In the case where many gels are opened simultaneously, their visible parts may become too small. Stacking gels, by displaying one gel on top of the others, thus creating a pile of gels, then becomes a good alternative to display and compare gels. To stack two or more gels, select them and drag them onto one of the display cells. The concept of stacking gels is very helpful to visually discover differences among gels and to add annotations. An example of stacked gels is given in Fig. 1.

A *Transparent* mode can also be used to visualise any similarities or differences between two gels by using a colour overlay of red and cyan. When the pixel colours

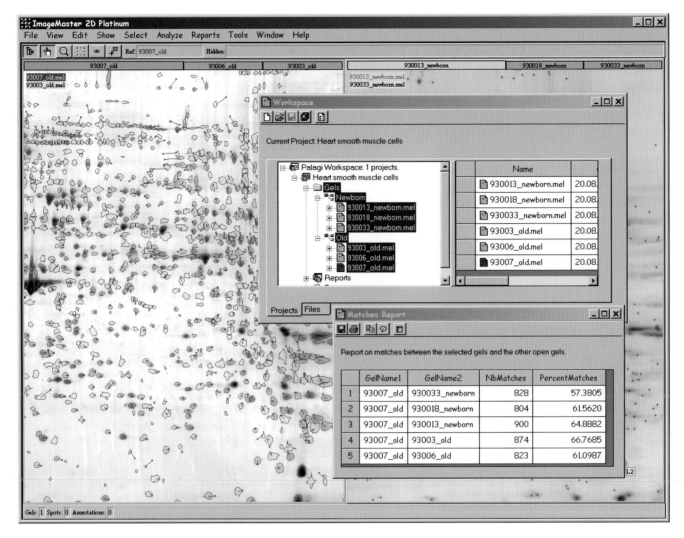

FIGURE 1 Spots are delimited by outlined shapes. Cell display at the left shows three gels in stacked mode. The paired spots in matched gels are linked by blue pair vectors. The Workspace window is displayed over the gels as well as a *Matches Report*.

of the two superimposed gels are added: overlapping spots appear as shades of grey, red and blue spots are present only in one of the gels, halos of red or blue around dark spots indicate that the protein is over- or underexpressed. This means that the fewer colours seen in the transparent mode, the more similar the gels are. To compare gels in the transparent mode:

i. Stack two or more gels.
ii. Select the stack reference.
iii. Choose *View → Stack → Show Transparent* from the menu.

Other useful functions are usually available at gel image analysis software. In case gels were scanned at the wrong orientation, they can be corrected with a rotate function. Cropping gels allows creating new

gels with only defined regions. Scaling gels is particularly useful in the case of very large images where a reduction in size may significantly decrease time and memory required for the analysis. A calibration procedure can be very useful to compensate for image differences caused by variations in experimental conditions (e.g., protein loading, staining) and scanning properties (e.g., image depth).

C. Detecting and Quantifying Spots

The elementary component of a gel is the spot, a shape that can be detected automatically by a spot detection algorithm or adjusted manually by the user. It delimits a more or less tiny region in the gel where a protein or a mixture of proteins is present. Each spot in a gel has an associated spot ID (a unique sequential number) automatically assigned to it when it is created. Moreover, a spot is automatically quantified, i.e., its optical density, area, and volume are computed. The spot detection algorithm of the Melanie software is optimised to identify a maximum number of proteins, while minimizing the number of artifacts detected. Three parameters have to be set to locate the spots automatically.

i. Smooth: it fixes the number of times the image is filtered before detecting spots, using a smooth-by-diffusion algorithm. The smooth parameter has to be optimised to detect all real spots and split as much as possible any overlapping spots.

ii. Min area: it eliminates spots that have an area smaller than a given threshold, eliminating dust particles that consist of a few very dark pixels (artifacts or noise).

iii. Saliency: it is based on the spot curvature and indicates how prominent a spot is. Real spots generally have high saliency values, whereas artifacts and background noise have small saliencies. Although the saliency is a very efficient quantity for filtering spots, it is also very dependent on the gels (e.g., gel resolution and image depth).

To detect spots automatically:

i. Select the gels from which spots will be detected.
ii. Draw a gel region with the *Region* tool on one or more selected gels in zones with representative spots.
iii. Choose *Edit → Spots → Auto-Detect* from the menu.
iv. The *Detect Spots* window appears on the screen, and spots in the active regions in selected gels are detected with the default parameters (Fig. 2).
v. Adjust the detection parameters if necessary. The default parameters are optimised to typical SDS–PAGE gels; however, refining the saliency parameter may be indicated. Each change in one of the spot detection parameters is immediately reflected in the selected region helping to choose the parameters. Using the cursor information window is very helpful to find optimal values for the spot filtering.
vi. Click on OK to detect all spots in the selected gels using the parameter values having been set. The spot shapes will be displayed over the gels.

Although it should be avoided to edit spots, they can be created, modified, merged, or deleted manually.

Once spots have been detected, the amount of protein present in each spot is computed automatically. Among the measured quantitative protein values, the most often used in analyses is the relative volume (%Vol). It is a normalized value that remains relatively independent of uninteresting variations between gels, particularly caused by varying experimental conditions. This measure takes into account variations due to protein loading and staining by considering the total volume over all the spots in the gel.

D. Annotating Spots and Pixels

Individual pixels and spots in a gel image may be indicated by annotations. In Melanie, annotations are active elements in the gel analyses and they can have several different purposes, e.g., be used to calibrate, align, and match gels or to be employed to mark spots with their proper information such as protein name, accession number, and so on. Annotations may also be used to mark spots with common characteristics, thus creating subsets. Annotations also offer the possibility to link and associate gel objects to external query engines or data sources of any format (text, html, spreadsheet, multimedia, 2-DE database entry) located locally or on the Internet.

An annotation is defined by its position on the gel (*X* and *Y* coordinates) and its set of labels. Each label belongs to a predefined or user-defined category. Among the available predefined annotation categories, some will be important for the explanations given in this chapter.

i. Ac: This category is provided to hold the accession number (AC) of the protein taken from a user-selected database, e.g., Swiss 2D-PAGE (Hoogland *et al.*, 2000) or Swiss-Prot (Boeckmann *et al.*, 2003), and can be the link to the remote database query engine of Melanie. When such a link is defined, a double click on a label of this category displays the corresponding protein entry in the chosen database with the selected browser software.

ii. Landmark: This is used to mark pixels or spots in the gels as reference points, for the operations of gel alignment or matching, and for the calculation of corresponding locations between gels. Two annotations are considered to refer to the same point in different gels when they hold identical labels.

To create an annotation:

i. Activate the *Annotation* tool.
ii. Double click on the pixel or spot in the gel where the annotation should be located.
iii. In the pop-up window, enter the name of a new category or choose one of the existing categories by clicking on its name.
iv. When a new category is created, the *Create Category* window will appear.

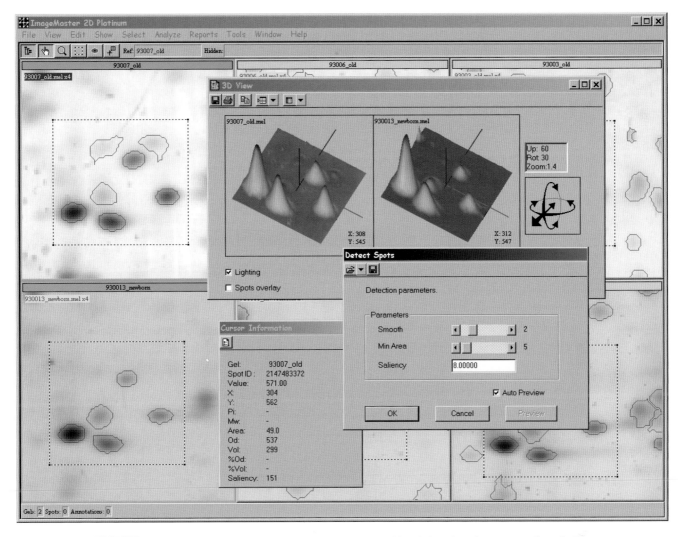

FIGURE 2 Three-dimensional view window details spot profiles of the selected regions on the gels. The cursor information window gives spot value information. Selected regions on the gels are updated in real time when adjusting spot detection parameters in the detect spots window.

v. Type the desired label in the next dialog box and click OK.

vi. The annotation is created and its label is displayed on the gel.

The predefined category *Set:* is used to mark spots with common properties by indicating that they belong to a set. The labels in such a category do not contain specific information. They only display the name of the set to which they belong.

To create a set:

i. Select one or many spots.

ii. Choose *Edit → Annotations → Add Labels* from the menu.

iii. In the pop-up window, click on the category called *Set:* and add a key word, which will be the name

of the set. For instance, to mark spots that were found to be differently expressed and should be exported to a spot excision robot, the final category name might indicate *Set:ToPick*.

iv. A label containing the name of the set (e.g., *ToPick*) will be attached to the selected spots.

Figure 3 shows annotations of categories *Landmark*, *Set:Old*, *Set:Verified*, *Set:ToPick*, and *Comments*.

E. Matching Gels

After spots have been detected, and annotations of category *Landmark* have been possibly added to the gels, the next essential step is to match gel images, i.e., find the corresponding protein spots in different gels.

FIGURE 3 Annotations of numerous categories such as *Landmark, Set:ToPick, Set:Old,* and *Set:Verified* are displayed on the gels. Group 6506 has its histogram highlighted in the *Classes+Groups Histogram* window as well as in the *Classes Report* window.

A gel-matching algorithm compares two gel images to find *Pairs* of related spots, i.e., spots describing the same protein in both gels.

Matching two gels in Melanie means finding all the pairs between spots of the two gels. Matching several gels means picking out a *Reference gel* and then successively matching each gel with the reference gel. In this way, spots in all gels may be compared with spots in the reference gel.

All spots in selected gels that are paired with a given spot in the reference gel form a *Group*. A spot group is the basic element for analysing spot variations across gels and for producing reports and histograms, as well as for performing statistical and clustering analysis. Moreover, when several gels have been matched to a given reference gel, this reference gel provides a unique numbering scheme for spots across all gels. Indeed, each paired spot in a gel image may be associated to the corresponding Spot ID in the reference gel. The Spot ID in the reference gel is then called the *Group ID*.

To match two or more gels automatically:

i. Select the gels to be matched (including the match reference).
ii. Choose *Edit → Pairs → Auto-Match Gels* from the menu.
iii. Set the reference gel for matching in the pop-up dialog box, i.e., specify to what image the other gels should be matched and click *OK*.
iv. All selected gels are matched with the chosen reference gel.

When matching is completed, Melanie gives the total number of pairs found. In the example given in Fig. 1, gel *93007_old* is the reference gel and about 4200 pairs were found among this gel and the other 5. There are 840 pairs in average per gel as it can be seen on the *Matches Report* window.

F. Analysing and Reporting Data

There are numerous ways of finding variations in protein expression among gels with Melanie but only one of them is detailed in this article.

When populations of gels are known, e.g., when comparing gels of newborn-rat tissues against aged-rat tissue samples, the analysis is based on this classification. Consider that the classes are defined as described previously in Section IIIA. The next step is to find out which are the characteristic spots of each class, i.e., proteins that are expressed differently. The class spot values may be summarized by statistical and overlapping descriptors, such as mean, standard deviation, gap, ratio, and normalization. To investigate groups of spots according to these descriptors one possible way is to

  i. Select the gels and then select the groups to be studied with *Select → Groups → All.*
 ii. Choose *Analyze → Classes Report.*
iii. In the pop-up list, select the %Vol value type to be displayed.
 iv. Accept the default statistics (mean 100% and mean-squared deviation 100%) in the subsequent dialog box.
  v. Change the *Displayed value* at the top of the *Classes Report* window for *Ratio*, rank the report in descending order (by clicking on the column headers), and select the rows showing a ratio from the highest value until 2.
 vi. Create a new report by choosing the *Report from selection* option in the *Classes Report* window.
vii. In the new window report, change the *Displayed value* for *Gap*. Rank the gap values in descending order. Select all rows from the Gap report and create a *Classes + Groups Histograms.*
viii. Use the created reports and the *Select on Gels* function on the reports to verify the pertinence of the given results. Use the green arrows on the report menu to select rows one by one.
 ix. In the Gap report, create an annotation of category "Set" with name *Verified* and type Boolean. When results are reliable, select the field *Verified* in the Gap report (Fig. 3).
  x. When finished, select all rows in the Gap report, refine the selection using the column "Verified"

with value 1, and reselect the spots on the gels and on the displayed reports with the function "Select on gels + reports."

Melanie also proposes statistical tests to help investigate the significance of the resulting spot groups: two-sample *t* test, Mann–Whitney *U* test, and Kolmogorov–Smirnov test. The principle of those tests is to calculate the probability of observing data sampled from populations with different means by chance or by fact. In order to get the statistical test results:

  i. Choose Analyze → Statistical Tests.
 ii. In the pop-up list, select the %Vol value type.
iii. Choose one or more of the statistical tests among the two-sample *t* test, Mann–Whitney *U* test, and Kolmogorov–Smirnov test to be displayed.
 iv. Sort, for example, the *t* test values in the report in descending order by clicking on the column header on the top of the *Statistical Tests Report* window.
  v. Reselect the first 30 groups that have the highest two-sample *t* test value to concentrate the analysis on the most significant spot differences between classes.
 vi. Click the *Classes + Groups Histograms* button at the top of the *Statistical Tests Report* window and then on each histogram to check up the obtained results.
vii. To obtain another view of the results, click the *Classes + Groups Histograms* button at the top of the *Classes Histograms* window and then on each histogram to check up the obtained results.
viii. Mark the resulting spots with labels from the "Set:" category and name *t* test.

Based on the explained procedures, groups composed of spots whose quantification values are unusual may be located. The detected variations can result from protein expression changes among gels or can be due to an inadequate detection or matching operation. Therefore, this analysis is not only useful for investigating extracted data, but also for controlling them.

Among the other Melanie functions to analyse gels, factor analysis and heuristic clustering are two options to check when gels correspond to experimentally known populations. These functions do not rely on any class attribution already set up; they blindly classify gels according to a global similarity and identify the characteristic spots of each population.

G. Integrating Data

Importing and exporting data to and from 2-DE analysis software are fundamental procedures. To make data produced available for processing by other applications or to import information coming from external sources, Melanie uses the common XML format. The main interest of the XML format, besides being used directly by this software, is that external applications can easily extract necessary data.

Melanie exports all gel-related data into a single XML format file, which may include all available information on a set of gels, together with spots (shape, quantification, aligned coordinates), annotations, and pair information.

Spot coordinates can also be exported in XML format to spot excision robots. On the other hand, once the spots have been identified, by mass spectrometry analysis or Edman degradation, for example, the accession number of the identified proteins, as well as other identification data, may be imported from an XML file to annotate the gels

IV. COMMENTS

Working with many gels at a time may be a tiring task, especially when the images are of bad quality. Melanie 4.0 tools for controlling and automating gel analyses may make repetitive tasks less tedious.

The *History* guarantees a better control over the gel analysis study; the operations that have been performed during the current and preceding work sessions can be checked. The History operator consists of a list of actions that have been carried out on the open gels, the parameters, and the selection criteria used at the different steps.

To display a history window

i. Select the gels for which you would like to display the History.
ii. Choose *Edit* → *History* → *Show* from the menu.
iii. To insert a marker in the history, choose *Edit* → *History* → *Insert Marker* from the menu.
iv. To clear the list of actions, choose *Edit* → *History* → *Clear*.

The History function is directly related to the Script function. The *Script* operation enables the automation of parts of the analysis process. A script is a sequence of instructions that is carried out automatically by Melanie when it is run. It is a kind of program that may be encoded by the user without any programming knowledge, just by cutting and pasting some desired actions from the History to the Script. The easiest way to create a script is to

i. Carry out the desired sequence of operations on a set of gels.

ii. Display the *History* of these gels.
iii. Copy the required actions to a new script by selecting the actions and pressing the *New Script* icon in the History toolbar.

A *Script* window will then be displayed on the screen. It contains a list of action descriptors in the Script navigator (actions list), as well as a toolbar at the top of the window. The toolbar contains icons that correspond to standard functionalities, such as copy, paste, save, and print, which create and run the scripts.

In addition to *History* and *Script* operators, the Undo operator corrects mistakes and helps to better control the gel analysis processing. Any earlier state of an analysis may be selected in the action descriptor list of the Undo/Redo operator, which is a particular sequence of actions that can be cancelled any time. Through this multiple undo function, whole parts of the gel analysis may be recovered, avoiding errors.

V. PITFALLS

1. Spot editing: Quantitative protein data, especially the spot volume, are highly dependent on an optimal and reproducible definition of the spot borders and a correct splitting of partially overlapping spots. To guarantee reproducibility of quantitative work, it is highly recommended to create spots by using the automatic spot detection algorithm and to avoid manual editing as much as possible. At most, spots should be separated where necessary.

2. Be critical when matching gels: When the gels are very distorted or different, automatic matching may fail. The choice of landmarks or pairs is very important to obtain good matching results. During matching with Melanie version 4.0, landmarks essentially correct global deformations of gels. Therefore, it is recommended not to put landmarks on spots in locally distorted regions because this can worsen the matching results around such regions.

References

Anderson, N. L., Taylor, J., Scandora, A. E., Coulter, B. P., and Anderson, N. G. (1981). The TYCHO system for computer analysis of two-dimensional gel electrophoresis patterns. *Clin Chem.* **27**(11),1807–1820.

Appel, R. D., Hochstrasser, D. F., Roch, C., Funk, M., Muller, A. F., and Pellegrini, C. (1988). Automatic classification of two-dimensional gel electrophoresis pictures by heuristic clustering analysis: A step toward machine learning. *Electrophoresis* **9**,136–142.

Appel, R. D., Palagi, P. M., Walther, D., Vargas, J. R., Sanchez, J. C., Ravier, F., Pasquali, C., and Hochstrasser, D. F. (1997). MelanieII: A third generation software package for analysis of two-dimensional electrophoresis images. I. Features and user interface. *Electrophoresis* **18**,2724–2734.

Boeckmann, B., Bairoch, A., Apweiler, R., Blatter, M.-C., Estreicher, A., Gasteiger, E., Martin, M. J., Michoud, K., O'Donovan, C., Phan, I., Pilbout, S., and Schneider, M. (2003). The SWISS-PROT protein knowledgebase and its supplement TrEMBL in 2003. *Nucleic Acids Res.* **31**,365–370.

Cremona, O., Muda, M., Appel, R. D., Frutiger, S., Hughes, G. J., Hochstrasser, D. F., Geinoz, A., and Gabbiani, G. (1995). Differential protein expression in aortic smooth muscle cells cultured from newborn and aged rats. *Exp. Cell Res.* **217**(2),280–287.

Hoogland, C., Sanchez, J.-C., Tonella, L., Binz, P.-A., Bairoch, A., Hochstrasser, D. F., and Appel, R. D. (2000). The 1999 SWISS-2DPAGE database update. *Nucleic Acids Res.* **28**,286–288.

Lemkin, P. F., and Lipkin, L. E. (1981). GELLAB: A computer system for 2D gel electrophoresis analysis. I. Segmentation of spots and system preliminaries. *Comput. Biomed. Res.* **14**(3),272–297.

Detection of Proteins in Gels

27

Protein Detection in Gels by Silver Staining: A Procedure Compatible with Mass Spectrometry

Irina Gromova and Julio E. Celis

I. INTRODUCTION

Silver staining is one of the procedures, in addition to Coomassie blue, R and G types, (Neuhoff *et al.*, 1988) and fluorescent dyes (Steinberg *et al.*, 1996; Patton, 2002; see also article by Patton in this volume), that are available for detecting proteins separated by gel electrophoresis. Switzer *et al.*, (1979) introduced silver staining in 1979, a technique that today provides a very sensitive tool for protein visualization with a detection level down to the 0.3- to 10-ng level.

The basic mechanisms underlying silver staining of proteins in gels are relatively well understood. Basically, protein detection depends on the binding of silver ions to the amino acid side chains, primary the sulfhydril and carboxyil groups of proteins (Switzer *et al.*, 1979; Oakley *et al.*, 1980; Merril *et al.*, 1981, 1986), followed by reduction to free metallic silver (Rabilloud, 1990, 1999). The protein bands are visualized as spots where the reduction occurs and, as a result, the image of protein distribution within the gel is based on the difference in oxidation–reduction potential between the gel area occupied by the proteins and the free adjacent sites. A number of alterations in the silver-staining procedure can shift the oxidation–reduction equilibrium in a way that gel-separated proteins will be visualized as either positively or negatively stained bands (Merril *et al.*, 1986).

Silver staining protocols can be divided into two general categories: (1) silver amine or alkaline methods and (2) silver nitrate or acidic methods (Merril, 1990). In general, the detection level using the various procedures is determined by how quickly the protein bands develop in relationship to the background (e.g., signal-to-noise ratio). The silver amine or alkaline methods usually have lower background and, as a result, are most sensitive but require extended procedures. Acidic protocols, however, are faster but slightly less sensitive. A comparative analysis of the sensitivity of a number of silver staining procedures has been published (Sorensen *et al.*, 2002; Mortz *et al.*, 2001). Clearly, each protocol has different advantages regarding timing, sensitivity, cost, and compatibility with other analytical methods, especially mass spectrometry (MS), a tool that is being used in combination with gel electrophoresis or chromatographic methods for rapid protein identification (see various articles in this volume). Until recently, most of the silver staining protocols used glutaraldehyde-based sensitizers in the fixing and sensitization step, thus introducing chemical modifications into proteins. The utilization of those chemicals causes the cross-linking of two lysine residues within protein chains, which affects MS analysis by hampering trypsin digestion and highly reduces protein extraction from the gel (Rabilloud, 1990).

Several modifications of the silver nitrate staining procedure have been developed for visualizing proteins that can be subsequently digested, recovered from the gel, and subjected to MS analysis (Shevchenko *et al.*, 1996; Yan *et al.*, 2000). This article describes a procedure that is slightly modified from these.

II. MATERIALS AND INSTRUMENTATION

Ultrapure water (>18 MΩ/cm resistance) for preparation of all buffers as well as during the washing steps is recommended. Use high-quality laboratory reagents that can be purchased from any chemical company.

III. PROCEDURE

To achieve the best results, i.e., high sensitivity and low background, it is very important to follow closely the incubation time throughout all steps as given in the protocol.

Solutions

1. *Fixation solution*: 50% ethanol (or methanol), 12% acetic acid, 0.05% formalin. To make 1 liter, add 120 ml of glacial acetic acid to 500 ml of 96% ethanol and 500 µl of 35% formaldehyde (commercial formalin is 35% formaldehyde). Complete to final volume with deionized water. Store at room temperature.

2. *Washing solution*: 20% ethanol (or methanol). To make 1 liter, add 200 ml 96% ethanol to 800 ml of deionized water. Store at room temperature.

3. *Sensitizing solution:* 0.02% (w/v) sodium thiosulfate ($Na_2S_2O_3$). To make 1 liter, add 200 mg of sodium thiosulfate anhydrate to a small volume of deionized water, mix well, and bring to the final volume of 1 liter.

4. *Staining solution*: 0.2% (w/v) silver nitrate ($AgNO_3$), 0.076% formalin. Prepare fresh. To make 1 liter, add 2 g of $AgNO_3$ to a small amount of deionized water. Add 760 µl of 35% formaldehyde. Dissolve and bring to final volume with deionized water. Precool the solution at 4°C before using.

5. *Developing solution*: 6% (w/v) sodium carbonate (Na_2CO_3), 0.0004% (w/v) sodium thiosulfate ($Na_2S_2O_3$), 0.05% formalin. To make 1 liter, add 60 g Na_2CO_3 to a small amount of deionized water and dissolve. Add 4 mg of sodium thiosulfate anhydrate to a small volume of deionized water and dissolve. Mix both solutions, add 500 µl of 35% formaldehyde, and bring to the final volume with water. Store at room temperature.

6. *Terminating solution*: 12% acetic acid. To make 1 liter, add 120 ml of glacial acetic acid to 500 ml of deionized water. Mix well and bring to the final volume with water. Store at room temperature.

7. *Drying solution*: 20% ethanol. To make 1 liter, add 200 ml of ethanol to 800 ml of deionized water. Mix well. Store at room temperature.

Steps

Use powder-free rubber gloves throughout the procedure. Wash the gloves with water during the staining procedure. The gel fixation and washing procedure can be carried in a staining try (polypropylene trays are recommended), but make sure that these are only used for silver staining. The size of the container has to be big enough to perform free movement of the gel during the shaking. For each step use sufficient volumes of the solutions to fully immerse the gels. Close the plastic trays or place the trays on top of each other to protect the gels from dust. Perform all steps at room temperature, one gel per tray, placed on a shaker at a very gentle speed. Do not touch the gel with bare hands or metal objects during handling. Plastic or Teflon bars (or ordinary glass pipettes) can be used to handle the gel. The staining procedure can be performed on any type of rotary shakers.

1. After electrophoresis, remove the gel from the cassette and place into a tray containing the appropriate volume of fixing solution. Soak the gel in this solution for approximately 2 h. Fixation will restrict protein movement from the gel matrix and will remove interfering ions and detergent from the gel. Fixation can also be done overnight. It may improve the sensitivity of the staining and decrease the background.

2. Discard the fixative solution and wash the gel in 20% ethanol for 20 min. Change the solution three times to remove remaining detergent ions as well as fixation acid from the gel.

Note: We recommend using 20% ethanol solution instead of deionized water to prevent swallowing of the gel. The size of the gel can be restored by incubating the gel in 20% ethanol for 20 min if water is used during the washing step

3. Discard the ethanol solution and add enough volume of the sensitizing solution. Incubate for 2 min with gentle rotation. It will increase the sensitivity and the contrast of the staining.

4. Discard the sensitizing solution and wash the gel twice, 1 min each time, in deionized water. Discard the water.

5. Add the cold silver staining solution and shake for 20 min to allow the silver ions to bind to proteins. *Note*: Do not pour the staining solution directly on the gel as it may result in unequal background. Add the solution to the corner of the tray.

6. After staining is complete, pour off the staining solution and rinse the gel with a large volume of deionized water for 20–60 s to remove excess unbound silver ions. Repeat the washing once more. *Note*:

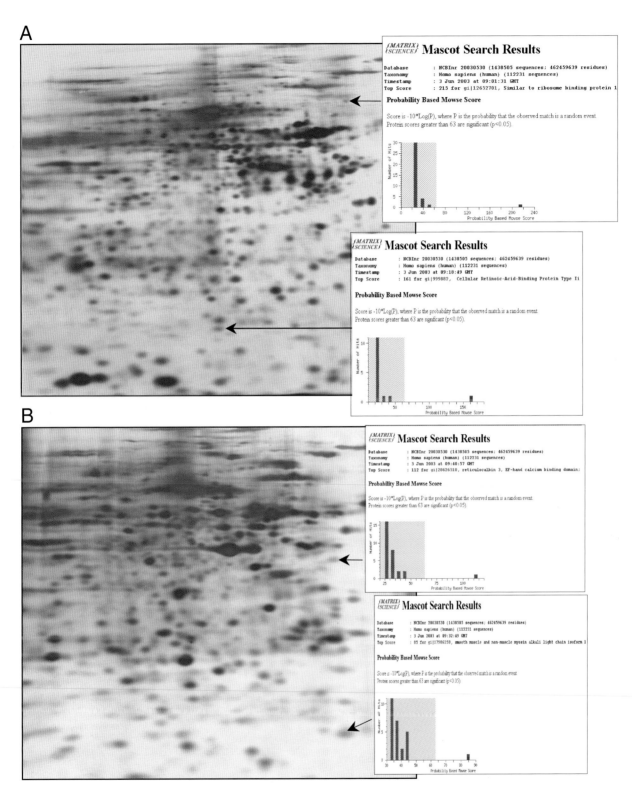

FIGURE 1   Gel image of normal human colon, location 7 (A), and human breast tumor biopsy (B) separated by two-dimensional gel IEF electrophoresis. Protein spots labeled on the gel images were identified by matrix-assisted laser desorption/ionization time-of-flight MS analysis. MS analysis of proteins resolved by gel electrophoresis utilizes extraction of the protein spot from the stained gel, followed by trypsin digestion, measurement, and database analysis (see also other articles in this volume).

Washing the gel for more than 1 min will remove silver ions from the gel, resulting in decreased sensitivity.

7. Rinse the gel shortly with the developing solution. Discard the solution.

8. Add new portion of the developing solution and develop the protein image by incubating the gel in 300 ml of developing solution for 2–5 min. The reaction can be stopped as soon as the desired intensity of the bands is reached.

9. Stop the reduction reaction by adding 50 ml of terminating solution directly to the gel while still immersed in the developing solution. Gently agitate the gel for 10 min. Development is stopped as soon as "bubbling" is over.

10. Moist gels can be kept in 12% acetic acid at 4°C in sealed plastic bags or placed in the drying solution for 2 h prior to vacuum drying.

Figure 1 shows silver-stained gels of whole protein extracts from a normal colon tissue biopsy and a breast tumor biopsy separated by two-dimensional gel electrophoresis as described by Celis *et al.*, in this volume. Since the sensitivity of silver staining is in the same range as modern mass spectrometry, it makes this staining one of the most attractive techniques for protein visualization before MS analysis. Protein bands of various intensity can be excised from the gels and identified by MS analysis (see various articles in this volume).

IV. COMMENTS

When choosing the silver staining protocol it is necessary to remember that not all proteins are stained equally by this technique. Thus, several classes of highly negative charged proteins, including proteoglycans and mucins, which contain high levels of sulfated sugar residues, and some very acidic proteins are detected poorly by silver staining (Goldberg *et al.*, 1997). Note that the linear dynamic range of the stain is restricted to approximately the 10-fold range, thus hampering the use of this method for quantitative protein expression analysis.

We replaced methanol for ethanol in all fixative solutions because of methanol toxicity. However, the use of ethanol in combination with acetic acid can result in the formation of ethyl acetate, which may interfere with protein identification by mass spectrometry. Several silver staining kits that offer improved sensitivity and that are compatibility with subsequent mass spectrometric analysis are available

commercially, including: Silver Stain PlusOne, Amersham Pharmacia Biotech, Amersham, UK; and SilverQuest silver staining kit, Invitrogen, USA.

V. PITFALLS

1. To increase the sensitivity of the staining, use extended washing after fixation to remove all residual acid. This extra washing will reduce the background during development.

2. Development of the gel for a long period of time can decrease the yield of the peptides for subsequent mass spectrometric analysis. This is due to the fact that mainly unstained peptides from the inner part of the gel are eluted to the solutions following "in gel" tryptic digestion of the proteins.

3. Negative staining can be observed when an excess of protein is applied.

4. In some cases, artificial bands with a molecular mass of around 50–70 kDa, as well as streaking or yellow background, can be observed due to the presence of a high concentration of reducing agents such as 2-mercaptoethanol or dithiothreitol in the sample buffer.

5. It has been reported that the recovery of peptides from the gel for MS analysis could be increased by destaining of the silver-stained protein bands immediately after the staining procedure (Gharahdaghi *et al.*, 1999).

References

Gharahdaghi, F., Weinberg, C. R., Meagher, D. A., Imai, B. S., and Mische, S. M. (1999). Mass spectrometric identification of proteins from silver-stained polyacrylamide gel: A method for the removal of silver ions to enhance sensitivity. *Electrophoresis* **20**, 601–605.

Goldberg, H. A., and Warner, K. J. (1997). The staining of acidic proteins on polyacrylamide gels: Enhanced sensitivity and stability of "Stains-all" staining in combination with silver nitrate. *Anal. Biochem.* **251**, 227–233.

Merril, C. R. (1990). Silver staining of proteins and DNA. *Nature* **343**, 779–780.

Merril, C. R., Dunau, M. L., and Goldman, D. (1981). A rapid sensitive silver stain for polypeptides in polyacrylamide gels. *Anal. Biochem.* **110**, 201–207.

Merril, C. R., and Pratt, M. E. (1986). A silver stain for the rapid quantitative detection of proteins or nucleic acids on membranes or thin layer plates. *Anal. Biochem.* **156**, 96–110.

Mortz, E., Krogh, T. N., Vorum, H., and Gorg, A. (2001). Improved silver staining protocols for high sensitivity protein identification using matrix-assisted laser desorption/ionization-time of flight analysis. *Proteomics* **1**, 1359–1363.

Neuhoff, V., Arold, N., Taube, D., and Ehrhardt, W. (1988). Improved staining of proteins in polyacrylamide gels including isoelectric focusing gels with clear background at nanogram sensitivity using Coomassie brilliant blue G-250 and R-250. *Electrophoresis* **9**, 255–262.

Oakley, B. R., Kirsch, D. R., and Morris, N. R. (1980). A simplified ultrasensitive silver stain for detecting proteins in polyacrylamide gels. *Anal. Biochem.* **105**, 361–363.

Patton, W. F. (2002). Detection technologies in proteome analysis. *J. Chromatogr. B Anal. Technol. Biomed. Life Sci.* **771**, 3–31.

Rabilloud, T. (1990). Mechanisms of protein silver staining in polyacrylamide gels: A 10-year synthesis. *Electrophoresis* **10**, 785–794.

Rabilloud, T. (1999). Silver staining of 2-D electrophoresis gels. *Methods Mol Biol.* **112**, 297–305.

Shevchenko, A., Wilm, M., Vorm, O., and Mann, M. (1996). Mass spectrometric sequencing of proteins silver-stained polyacrylamide gels. *Anal. Chem.* **68**, 850–858.

Sorensen, B. K, Hojrup, P., Ostergard, E., Jorgensen, C. S., Enghild, J., Ryder, L. R., and Houen, G. (2002). Silver staining of proteins on electroblotting membranes and intensification of silver staining of proteins separated by polyacrylamide gel electrophoresis. *Anal. Biochem.* **304**, 33–41.

Steinberg, T. H., Jones, L. J., Haugland, R. P., and Singer, V. L. (1996). SYPRO orange and SYPRO red protein gel stains: One-step fluorescent staining of denaturing gels for detection of nanogram levels of protein. *Anal. Biochem.* **239**, 223–237.

Switzer, R. C., 3rd, Merril, C. R., and Shifrin, S. (1979). A highly sensitive silver stain for detecting proteins and peptides in polyacrylamide gels. *Anal. Biochem.* **98**, 231–237.

Yan, J. X., Wait, R., Berkelman, T., Harry, R. A., Westbrook, J. A., Wheeler, C. H., and Dunn, M. J. (2000). A modified silver staining protocol for visualization of proteins compatible with matrix-assisted laser desorption/ionization and electrospray ionization-mass spectrometry. *Electrophoresis* **17**, 3666–3672.

28

Fluorescence Detection of Proteins in Gels Using SYPRO Dyes

Wayne F. Patton

I. INTRODUCTION

Operationally, fluorescent, noncovalent staining methods using SYPRO dyes resemble traditional, colorimetric staining procedures such as colloidal Coomassie blue dye staining. After electrophoresis, gels are incubated in a stain solution and proteins are visualized based upon differential dye binding to protein bands relative to the polyacrylamide gel matrix. Because proteins are not covalently modified with dye molecules and staining is performed postelectrophoretically, no alteration in the migration of proteins during electrophoresis occurs. These fluorescence-based staining methods are also highly compatible with downstream microchemical methods, such as Edman-based protein sequencing and peptide mass profiling by matrix-assisted laser desorption time-of-flight mass spectrometry (MALDI-TOF MS).

SYPRO Orange dye, SYPRO Red dye, and SYPRO Tangerine dye bind noncovalently to proteins in gels through interaction with SDS micelles (Steinberg *et al.*, 1996a,b, 1997, 2000). These fluorophores are virtually nonfluorescent in aqueous solution, but they fluoresce in nonpolar solvents or when associated with SDS–protein complexes. Protein quantitation with fluorophores of this type is generally more reliable than that achieved with fluorophores that label primary amines alone (Patton, 2000). The dyes offer detection sensitivities comparable to those of colloidal Coomassie blue staining methods (Patton, 2000).

Because the staining properties of SYPRO Orange and SYPRO Red dyes are similar, the two protocols describing them have been grouped together in this article. SYPRO Orange protein gel stain is slightly

brighter, whereas SYPRO Red protein gel stain has somewhat lower background fluorescence in gels. These dyes offer high sensitivity and rapid staining of SDS–polyacrylamide gels using a simple, one-step, 30- to 60-min staining procedure, with no destaining required (Steinberg *et al.*, 1996a,b, 1997, 2000). Staining exhibits low protein-to-protein variability, high selectivity for proteins, and a broad linear detection range extending over three orders of magnitude. The 4- to 10-ng detection sensitivity of these SYPRO dyes is as high as rapid silver staining and colloidal Coomassie blue staining methods (Steinberg *et al.*, 1996a,b, 1997, 2000; Patton, 2000, 2002). Both dyes are efficiently excited by UV and by visible illumination. Thus, stained gels can be viewed and photographed with a standard laboratory UV transilluminator, CCD camera, or any of a variety of laser scanners, in conjunction with the proper filters. SYPRO Orange stain (excitation/emission: ~470/569 nm) is preferable for argon ion or second-harmonic generation (SHG) laser-based instruments, and SYPRO Red stain (excitation/emission: ~547/631 nm) is preferable for green He-Ne or Nd : YAG lasers.

It is possible to stain SDS–PAGE gels during electrophoresis using the SYPRO stains in the running buffer; however, detection sensitivity in this case is four- to eight-fold poorer (Steinberg *et al.*, 1996b). After electrophoresis, gels are briefly washed prior to visualizing proteins. In contrast to many silver staining and reverse staining methods, SYPRO dyes do not stain nucleic acids or bacterial lipopolysaccharides to a significant extent (Steinberg *et al.*, 1996a,b, 1997, 2000). Gels may be completely destained in 30% aqueous methanol (Steinberg *et al.*, 1996b). SYPRO Orange and Red protein gel stains are not suitable for staining proteins on blotting membrane or in

225

isoelectric focusing (IEF) gels and show reduced sensitivity when staining proteins in two-dimensional gels. Both SYPRO Orange and SYPRO Red stain require acetic acid, making them less suitable for applications involving electroblotting, electroelution, or measuring enzyme activity. While acetic acid may be omitted from the SYPRO Orange and SYPRO Red staining solutions, and proteins then recovered from gels, this yields substantially reduced sensitivity of detection and increased protein-to-protein variability in staining (Steinberg et al., 1996b, 2000).

SYPRO Tangerine protein gel stain is a versatile stain for detecting proteins separated by SDS–polyacrylamide gel electrophoresis (Steinberg et al., 2000). Like SYPRO Orange and SYPRO Red stains, it offers high sensitivity, a rapid and simple staining procedure, low protein-to-protein variability, high selectivity for proteins, and a broad linear range of detection. Staining is performed in a nonfixative solution that permits subsequent electroblotting, electroelution, or detection of enzyme activity. Proteins stained without fixation can be used for further analysis by zymography (in-gel enzyme activity assay), provided SDS does not inactivate the protein of interest. Stained proteins can also be eluted easily from gels and used for further analysis. The stain is fully compatible with Edman-based sequencing and mass spectrometry (Steinberg et al., 2000). In addition, staining does not interfere with the transfer of proteins to blotting membranes, allowing visualization of proteins *before* proceeding with Western blotting or other blotting applications. Small regions of a gel or even individual bands can be excised before blotting. This enables one to use much smaller amounts of transfer membrane and immunodetection reagents. After transfer to membranes, proteins can be visualized using SYPRO Ruby protein blot stain. If protein fixation is preferred, the dye can be used with 7% acetic acid or 12.5% trichloroacetic acid (Steinberg et al., 2000). In this case, however, one should expect slightly higher background staining than with SYPRO Orange and SYPRO Red stains. SYPRO Tangerine protein gel stain is not suitable for staining proteins in IEF gels and shows only moderate sensitivity when staining proteins on two-dimensional gels.

SYPRO Ruby protein gel stain differs from the other SYPRO dyes that all bind through intercalation into SDS micelles (Berggren et al., 2000, 2002; Patton, 2000, 2002). Instead, this stain binds to proteins by a mechanism that is quite similar to Coomassie blue stain, via direct electrostatic interaction with basic amino acid residues (Patton, 2000). SYPRO Ruby dye readily stains glycoproteins, lipoproteins, calcium-binding proteins, fibrillar proteins, and other difficult-to-stain proteins. The dye is used in a simple staining procedure and is ideal for high-throughput gel staining. The stain is as sensitive as the best silver staining methods available and superior to them in terms of ease of use, linear dynamic range, and compatibility with downstream microchemical characterization techniques (Lopez et al., 2001; Nishihara and Champion, 2002; Gerner et al., 2002). SYPRO Ruby protein gel stain is an ultrasensitive dye for detecting proteins separated by SDS–polyacrylamide gels or two-dimensional gels (Berggren et al., 2000, 2002). The background fluorescence is low and the linear dynamic range of the stain extends over three orders of magnitude and shows low protein-to-protein variation. The stain is more sensitive than colloidal Coomassie blue dye and SYPRO Orange, Red, or Tangerine dyes and is comparable in sensitivity to the best available silver stains (Patton, 2000, 2002). The stain is ready to use and gels cannot overstain. Staining protocols are simple, although optimal staining incubation requires about 4 h, slower than times required for SYPRO Orange, SYPRO Red, and SYPRO Tangerine stains. Staining times are not critical though and staining can be performed overnight. SYPRO Ruby protein gel stain will not stain extraneous nucleic acids or lipopolysaccharides and is compatible with further downstream microchemical processing. SYPRO Ruby protein gel stain can be used with many types of gels, including two-dimensional gels, Tris–glycine SDS gels, Tris–tricine precast SDS gels, isoelectric focusing gels, and nondenaturing gels. SYPRO Ruby stain is also compatible with gels adhering to plastic backings, although the inherent blue fluorescence of the plastic must be removed with an appropriate emission filter. The stain does not interfere with subsequent analysis of proteins by Edman-based sequencing or mass spectrometry and the stain is especially well suited for peptide mass profiling using MALDI-TOF mass spectrometry (Berggren et al., 2000, 2001, 2002; Lopez et al., 2001). Stained gels can be visualized with a 300-nm UV transilluminator, various laser scanners, or other blue light-emitting sources. The dye maximally emits at about 610 nm. SYPRO Ruby protein gel stain has exceptional photostability, allowing long exposure times for maximum sensitivity.

This article presents protocols for staining proteins in SDS–polyacrylamide and two-dimensional gels using the different SYPRO dyes. The following protocols describe several steps following the preparation and running of SDS–polyacrylamide gels or two-dimensional gels. Some issues regarding further processing of proteins in the gels are also included. In general, gel electrophoresis should be performed according to standard procedures (Laemmli, 1970;

TABLE I

Protein stain	Excitation/ emission	Major applications	Features
SYPRO Ruby protein gel stain	280 nm, 450 nm /610 nm	Two-dimensional gels, IEF gels, one-dimensional (1D) SDS–PAGE	Highest sensitivity (1–2 ng/band; comparable to best silver staining methods), linear quantitation range over 3 orders of magnitude, compatible with fluorescence-based phosphoprotein and glycoprotein detection (multiplexed proteomics technology)
SYPRO Orange protein gel stain	300 nm, 470 nm /570 nm	1D SDS–PAGE	Good sensitivity (4–10 ng/band; comparable to colloidal Coomassie blue stain), linear quantitation range over 3 orders of magnitude
SYPRO Red protein gel stain	300 nm, 550 nm /630 nm	1D SDS–PAGE	Good sensitivity (4–10 ng/band; comparable to colloidal Coomassie blue stain), linear quantitation range over 3 orders of magnitude
SYPRO Tangerine protein gel stain	300 nm, 490 nm /640 nm	1D SDS–PAGE, staining before blotting, zymography, electroelution	Good sensitivity (4–10 ng/band; comparable to colloidal Coomassie blue stain), linear quantitation range over 3 orders of magnitude, requires no alcohol or acid fixatives, ideal for protein elution from gels

O'Farrell, 1975). There are several important considerations to take into account when choosing effective and appropriate stains for an application. Table I outlines key features of the stains discussed in this article. Important notes regarding the protocols are included at the end of the article.

II. PROTOCOLS USING SYPRO ORANGE AND SYPRO RED PROTEIN GEL STAINS

A. Materials and Reagents

SYPRO Orange protein gel stain (Molecular Probes, Inc., Cat. No. S-6650, S-6651) and SYPRO Red protein gel stain (Molecular Probes, Inc., Cat. No. S-6653, S-6654) are provided as 5000× concentrated solutions in dimethyl sulfoxide (DMSO), either as a single vial containing 500 µl of stock solution or as a set of 10 vials, each containing 50 µl of stock solution. In each case, enough reagent is supplied to prepare a total of 2.5 liter of working stain solution, which is sufficient to stain ~50 polyacrylamide minigels. Before opening the vial, warm it to room temperature to avoid water condensation and subsequent precipitation. After thawing completely, briefly centrifuge the vial in a microcentrifuge to deposit the DMSO solution at the bottom of the vial. If particles of dye are present, redissolve them by briefly sonicating the tube or vortexing it vigorously after warming. Staining should be performed in plastic dishes (preferably polypropylene) (*see* Section V,B).

B. Gel Electrophoresis

1. SDS–PAGE

Prepare and run SDS–polyacrylamide gels according to standard protocols (Laemmli, 1970). Originally it was recommended to use 0.05% SDS in the running buffer instead of the usual 0.1% SDS. However, this recommendation was based upon separating molecular weight markers and would not be advisable for biological specimens containing hydrophobic proteins. The use of standard, 0.1% SDS in the running buffer decreases overall detection sensitivity only slightly and improves the resolution of protein bands in gels.

2. Two-Dimensional Gels

Neither SYPRO Orange nor SYPRO Red protein gel stain is recommended for high sensitivity detection of proteins in two-dimensional gels (See Table I). High-quality silver stains and SYPRO Ruby Protein Gel Stain offer better detection sensitivity.

C. Staining Protocol

1. Clean and thoroughly rinse the staining dishes before use. Residual detergent in staining dishes will compromise the detection of proteins.

2. Prepare the staining solution by diluting the stock SYPRO reagent 1 : 5000 in 7.5% (v/v) acetic acid and mixing vigorously. For 50 µl of stock, dilute into 250 ml acetic acid solution (*see* Section V).

3. Pour the staining solution into a small, clean plastic dish. For one or two standard-size minigels, use ~50 ml to 100 ml of staining solution; for larger gels, use 500 to 750 ml) (*see* Section V).

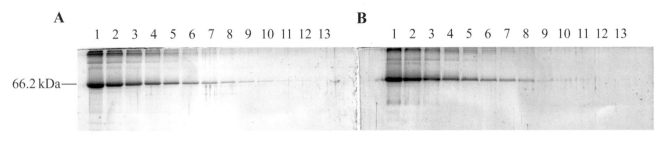

FIGURE 1 Comparison of detection sensitivity and brightness of SYPRO Orange protein gel stain and SYPRO Ruby protein gel stain using a laser-based gel-imaging system. Solution-quantified bovine serum albumin standard (Cat. No. P7656, Sigma Chemical Company, Saint Louis, MO) was applied to lanes of 10% SDS–polyacrylamide gels (1000–0.25 ng/lane) and separated by electrophoresis using standard procedures (Laemmli, 1970). Gels were then stained using either SYPRO Ruby protein gel stain or SYPRO Orange protein gel stain as described in the text. Subsequently, gels were imaged using a Fuji FLA 3000 laser-based gel scanner. Gels were scanned using the 473-nm second-harmonic generation (SHG) laser and 580-nm long-pass filter. (A) Gel stained with SYPRO Ruby protein gel stain and imaged using a laser-based gel scanner. (B) Gel stained with SYPRO Orange protein gel stain and imaged using a laser-based gel scanner. SYPRO Orange protein gel stain was capable of detecting 2 ng of bovine serum albumin while SYPRO Ruby protein gel stain was capable of detecting 0.5 ng of the protein. Figure courtesy of Ms. Courtenay Hart, Molecular Probes, Inc.

4. Place the gel into the staining solution. Protect the gel and staining solution from light at all times by covering the container with a lid or with aluminum foil.

5. Gently agitate the gel in stain solution at room temperature (50 rpm on an orbital shaker). Staining times range from 10 to 60 min, depending on the thickness and percentage of polyacrylamide in the gel. For 1-mm-thick 15% polyacrylamide gels, the signal is typically optimal at 40 to 60 min of staining. For standard SDS–PAGE minigels incubate for 40 to 60 min. Large gels sometimes require a preincubation in 7.5% acetic acid for 15 min prior to staining to reduce background fluorescence due to excess SDS in the gel.

6. Rinse briefly (<1 min) with 7.5% acetic acid to remove excess stain from the gel surface and to avoid accumulation of fluorescent dye on the surface of the transilluminator or gel scanner.

D. Viewing, Photographing, and Storing the Gel

View the stained gel on a standard 300-nm UV transilluminator (*see* Section V,E). Gels may be left in staining solution overnight without losing sensitivity. However, photographs should be taken as soon as possible after staining.

Gels may be photographed with a Polaroid or CCD camera. Use Polaroid 667 black-and-white print film and the SYPRO protein gel stain photographic filter (Molecular Probes, Inc., Cat. No. S-6656). Exposure times vary with the intensity of the illumination source; for an f stop of 4.5, use 2 to 5 s for SYPRO Orange stain and 3 to 8 s for SYPRO Red stain. CCD cameras and laser scanners provide high sensitivity detection; contact the manufacturer to determine the optimal filter sets (*see* Section V).

E. Destaining Gels

To destain gels, incubate overnight in 0.1% Tween 20. Alternatively, if thorough destaining is desired, incubate for prolonged periods in several changes of 7.5% acetic acid or 30% methanol.

III. PROTOCOL FOR SYPRO TANGERINE PROTEIN GEL STAIN

A. Materials and Reagents

SYPRO Tangerine protein gel stain (Molecular Probes, Inc., Cat. No. S-12010) is provided in a 500-μl unit size, as a 5000× concentrated solution in DMSO. One 500-μl unit size prepares a total of 2.5 liter of working stain solution, which is sufficient to stain ~50 polyacrylamide minigels. Before opening the vial, warm it to room temperature to avoid water condensation and subsequent precipitation. After thawing completely, briefly centrifuge the vial in a microcentrifuge to deposit the DMSO solution at the bottom of the vial. If particles of dye are present, redissolve them by briefly sonicating the tube or vortexing it vigorously after warming.

A

B

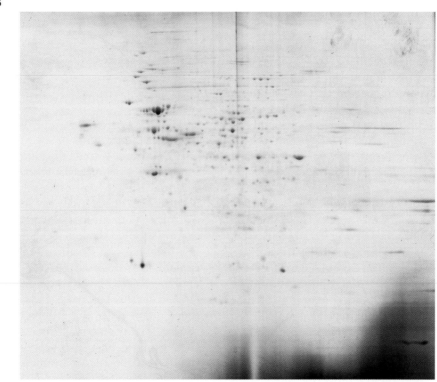

FIGURE 2 Comparison of the detection sensitivity of SYPRO Orange protein gel stain and SYPRO Ruby protein gel stain using a laser-based gel-imaging system. Jurkat cell extract (250 μg) was separated on a tube gel as a first dimension followed by 12.5% SDS–PAGE. Two-dimensional gels were stained with either SYPRO Ruby dye or SYPRO Orange dye overnight. Gels were destained for 2–3 h before imaging. (A) Gel stained with SYPRO Ruby protein gel stain. (B) Gel stained with SYPRO Orange protein gel stain. Figure courtesy of Dr. Birte Schulenberg.

1. Buffers

SYPRO Tangerine staining solution is prepared by diluting the stock reagent 1 : 5000 in one of a variety of buffers as described in the protocol. Staining should be performed in plastic dishes (preferably polypropylene) (*see* Section V,A).

2. SDS–PAGE

Prepare and run SDS–polyacrylamide gels according to standard protocols (Laemmli, 1970). The use of standard, 0.1% SDS in the running buffer is recommended to ensure complete solubilization of hydrophobic proteins.

3. 2-D Gels

SYPRO Tangerine gel stain is not recommended for high sensitivity detection of proteins in two-dimensional gels (See Table I). High-quality silver stains and SYPRO Ruby protein gel stain offer better detection sensitivity.

B. Staining Procedure

1. Clean and *thoroughly rinse* the staining dishes before use. Residual detergent in staining dishes will compromise the detection of proteins.

2. Prepare the staining solution by diluting the stock SYPRO Tangerine reagent 1 : 5000 in an appropriate buffer and mixing vigorously.

2a. If the proteins are to be used for electroelution, electroblotting, or zymography, dilute the stock solution into 50 mM phosphate, 150 mM NaCl, pH 7.0. For 50 µl of stock, dilute into 250 ml buffer. *Note: If no fixative is used before or during staining, some diffusion of the protein bands may occur, especially for smaller proteins.*

2b. Alternatively, use one of a wide range of buffers that are compatible with the stain. These include formate, pH 4.0; citrate, pH 4.5; acetate, pH 5.0; MES, pH 6.0; imidazole, pH 7.0; HEPES, pH 7.5; Tris acetate, pH 8.0; Tris–HCl, pH 8.5; Tris borate, 20 mM EDTA, pH 9.0; and bicarbonate, pH 10.0. Buffers should be prepared as 50–100 mM solutions containing 150 mM NaCl. The stock dye solution may also be diluted directly into 150 mM NaCl. For 50 µl of stock, dilute into 250 ml buffer.

2c. For fixative staining to minimize diffusion of the proteins, dilute the SYPRO Tangerine stock solution in 7.5% (v/v) acetic acid. For low percentage gels and for very small proteins, 10% acetic acid will result in better retention of the protein in the gel without compromising sensitivity (Steinberg *et al.*, 2000). However, note that acetic acid and other fixatives will

interfere with the transfer of proteins to blotting membranes.

3. Pour the staining solution into a small plastic dish. For one or two standard-size minigels, use ~50 to 100 ml of staining solution; for larger gels, use 500 to 750 ml (*see* Section V,A).

4. Place the gel into the staining solution. Protect the gel and staining solution from light at all times by covering the container with a lid or with aluminum foil.

5. Gently agitate the gel in stain solution at room temperature (50 rpm on an orbital shaker). The staining times range from 10 to 60 min, depending on the thickness and percentage of polyacrylamide in the gel. For 1-mm–thick 15% polyacrylamide gels, the signal is typically optimal at 30 to 60 min of staining. For standard SDS–PAGE minigels prepared with 0.1% SDS in the running buffer, incubate for 30 to 60 min. Large gels, including large two-dimensional gels, sometimes require a preincubation in 7.5% acetic acid for 30 min prior to staining to reduce background fluorescence due to excess SDS in the gel. It is important to note, however, that acetic acid interferes with transfer to blots.

6. Use the following step if the proteins are to be transferred to a blot. After staining, incubate the gel in Western blotting buffer containing 0.1% SDS. The SDS is not absolutely required, but it helps in the transfer of some proteins to the blot.

C. Viewing, Photographing, and Storing the Gel

Detect the proteins in the stained gel using a standard 300-nm UV transilluminator, a blue-light transilluminator, 473-nm SHG laser, 488-nm argon ion, or 473-nm He–Ne laser-based imaging system (*see* Section V,E). Gels may be left in staining solution overnight without loss of sensitivity. However, photographs should be taken as soon as possible after staining. Gels may be documented by a Polaroid or CCD camera. Use Polaroid 667 black-and-white print film and the SYPRO protein gel stain photographic filter (Molecular Probes, Inc., Cat. No. S-6656). Exposure times vary with the intensity of the illumination source; for an f stop of 4.5, use a 2- to 5-s exposure time. CCD cameras and laser scanners provide high sensitivity; contact the manufacturer to determine the optimal filter sets (*see* Section V,F).

D. Destaining Gels

SYPRO Tangerine stain is readily destained by incubation in 7% acetic acid or 30% methanol.

E. Notes for SYPRO Tangerine protein gel stain

The SDS front at the bottom of the gel stains very heavily with SYPRO Tangerine stain. Unless the proteins that interest you are comigrating with the SDS front, it will be advantageous to run the SDS front off the gel. Colored stains and marker dyes, as well as commercially prestained protein markers, may interfere with SYPRO Tangerine dye staining and quench fluorescence.

IV. PROTOCOL FOR SYPRO RUBY PROTEIN GEL STAIN

A. Materials and Reagents

SYPRO Ruby protein gel stain (Molecular Probes, Inc., Cat. No. S-12000, S-12001, S-21900) is provided ready to use in 200 ml volume (will stain ~4 minigels), 1 liter volume (~20 minigels or 2–3 large-format gels), or 5 liter volume (~100 minigels or 10–15 large-format gels). Staining should be performed in plastic dishes (preferably polypropylene) (*see* Section V,B).

B. Protocol

Prepare and run SDS–PAGE or two-dimensional PAGE according to standard protocols (Laemmli, 1970; O'Farrell, 1975). Perform staining with SYPRO Ruby due using continuous, gentle agitation (e.g., on an orbital shaker at 50 rpm).

1. Clean and *thoroughly rinse* the staining dishes before use. Residual detergent in staining dishes will compromise the detection of proteins.

2. A range of fixatives have been validated for use with SYPRO Ruby protein gel stain, including 40% ethanol/10% acetic acid, 10% ethanol/7% acetic, acid 25% ethanol/12.5% trichloroacetic acid, and 10% ethanol/3% phosphoric acid. Harsher fixatives, such as 40% ethanol/10% acetic acid, are recommended as they retain proteins in gels better. Fix the gel in a plastic dish for 30 min. This step improves the sensitivity of the stain in two-dimensional gels, but is optional for one-dimensional SDS–PAGE gels.

3. Pour the staining solution into a small, clean plastic dish. For one or two standard-size minigels, use ~50 to 100 ml of staining solution; for larger gels, use 500 to 750 ml (*see* Section V,A).

4. Place the gel into the staining solution. Protect the gel and staining solution from light at all times by covering the container with a lid or with aluminum foil.

5. Gently agitate the gel in stain solution at room temperature (50 rpm on an orbital shaker). The staining times range from 90 min to 4 h, depending on the thickness and percentage of polyacrylamide in the gel. Specific staining can be seen in as little as 30 min. However, a minimum of 4 h of staining is required for maximum sensitivity and linearity. For convenience, gels may be left in the dye solution overnight or longer without overstaining.

6. *Optional.* After staining, rinse the gel in deionized water for 30–60 min to decrease background fluorescence. To better decrease background fluorescence the gel can be washed in a mixture of 10% methanol and 7% acetic acid for 30 min instead of water. The gel may be monitored periodically using UV illumination to determine the level of background fluorescence.

C. Viewing, Photographing, and Storing the Gel

View the stained gel on a standard 300-nm UV or a blue-light transilluminator (*see* Section V,E). Gels may also be visualized using various laser scanners: 473-nm (SHG) laser, 488-nm argon ion laser, and 532-nm (YAG) laser. Alternatively, use a xenon arc lamp, blue fluorescent light, or blue light-emitting diode (LED) source. Gels may be left in staining solution overnight without losing sensitivity. However, images should be acquired as soon as possible after staining. Gels may be imaged by a Polaroid or CCD camera. Use Polaroid 667 black-and-white print film and the SYPRO protein gel stain photographic filter (Molecular Probes, Inc., Cat. No. S-6656). Exposure times vary with the intensity of the illumination source; for an f stop of 4.5, try 1 s. CCD cameras and laser scanners provide high sensitivity; contact the manufacturer to determine the optimal filter sets (*see* Section V,F). To dry the stained gel for permanent storage, incubate the gel in a solution of 2% glycerol for 30 min. Dry the stained gel using a gel dryer. Note that proteins present at very low levels may no longer be detectable after gel drying.

V. NOTES

A. Staining

Minimal staining volumes for typical gel sizes:

50 ml for 8 cm × 10 cm × 0.75 mm gels (minigels)
330 ml for 16 cm × 20 cm × 1 mm gels
500 ml for 20 cm × 20 cm × 1 mm gels
or ~10 times the volume of the gel for other gel sizes

B. Staining Containers

Polypropylene dishes, such as Rubbermaid Servin' Savers, are the optimal containers for staining because the high-density plastic adsorbs only a minimal amount of the dye. Clean and rinse the staining containers well before use, as detergent will interfere with staining. Some rinse the containers with ethanol before use. For small gels, circular staining dishes provide the best fluid dynamics on orbital shakers, resulting in less dye aggregation and better staining. For large-format two-dimensional gels, polyvinyl chloride photographic staining trays, such as Photoquip Cesco-Lite 8 × 10-in. photographic trays (Genomic Solutions, Ann Arbor, MI), also work well. Another convenient staining option for large-format two-dimensional gels uses the Clearview three-drawer organizer (Sterilite, Cat. No. 1790, Townsend, MA). This polypropylene box provides a convenient format for staining three gels per unit and is available at many department stores. Glass dishes are not recommended as they have a tendency to bind dye.

C. Fixing Gels

For low percentage gels and for very small proteins, 10% acetic acid solution will result in better retention of the protein in the gel without compromising sensitivity. Acetic acid will interfere with transfer of the proteins to a blot. Therefore, for applications in which blotting will follow electrophoresis, use SYPRO Tangerine protein gel stain, which does not require acetic acid fixation.

D. Storing Gels

Always store gels in the dark to prevent photobleaching. When gels are stored in the staining solution, the signal decreases somewhat after several days; however, depending on the amount of protein in bands of interest, gels may retain a usable signal for many weeks. Gels may be dried between sheets of cellophane (BioRad Laboratories), although there is sometimes a slight decrease in sensitivity.

E. Viewing Gels

Viewing the Gel with UV Transillumination

Place the gel directly on the transilluminator; do not use plastic wraps or plastic backing. It is important to clean the surface of the transilluminator after each use with deionized water and a soft cloth (such as cheesecloth), as fluorescent dyes, such as SYPRO stains, will accumulate on the glass surface and cause a high background fluorescence.

The polyester backing on some precast gels is highly fluorescent. For maximum sensitivity using a UV transilluminator, the gel should be placed polyacrylamide side down and an emission filter, such as the SYPRO protein gel stain photographic filter (S-6656), used to screen out the blue fluorescence of the plastic. The use of a blue-light transilluminator or laser scanner will reduce the amount of fluorescence from the plastic backing so that the gel may be placed polyester side down.

Noticeable photobleaching can occur after several minutes of exposure to ultraviolet light. If a gel becomes photobleached, it can be restained by simply returning it to the staining solution.

F. Imaging Gels

Because of the low fluorescence, it is possible to take advantage of the integrating capability of photographic or CCD cameras and use long exposure times to increase the sensitivity, often making bands visible that are not visible to the eye. Images are best obtained by digitizing at about 1024 × 1024 pixels resolution with 12- or 16-bit gray scale levels per pixel. Contact the camera manufacturer for recommendations on filter sets. A CCD camera-based image analysis system can gather quantitative information that will allow comparison of fluorescence intensities between different bands or spots.

Acknowledgments

I gratefully acknowledge the many scientists who have contributed to the SYPRO dye development program over the years. These include Tom Steinberg, Kiera Berggren, Birte Schulenberg, Richard Haugland, Vicki Singer, Courtenay Hart, Brad Arnold, Nick Smith, Mary Nunally, and Laurie Jones. SYPRO is a registered trademark of Molecular Probes, Inc.

References

Berggren, K., Chernokalskaya, E., Lopez, M., Beechem, J., and Patton, W. (2001). Comparison of three different fluorescent visualization strategies for detection of *Escherichia coli* ATP synthase subunits after SDS-polyacrylamide gel electrophoresis. *Proteomics* **1**, 54–65.

Berggren, K., Chernokalskaya, E., Steinberg, T., Kemper, C., Lopez, M., Diwu, Z., Haugland, R., and Patton, W. (2000). Background-free, high-sensitivity staining of proteins in one- and two-dimensional sodium dodecyl sulfate-polyacrylamide gels using a luminescent ruthenium complex. *Electrophoresis* **21**, 2509–2521.

Berggren, K., Schulenberg, B., Lopez, M., Steinberg, T., Bogdanova, A., Smejkal, G., Wang, A., and Patton, W. (2002). An improved formulation of SYPRO Ruby protein gel stain: Comparison with the original formulation and with a ruthenium II tris (bathophenanthroline disulfonate) formulation. *Proteomics* **2**, 486–498.

Gerner, C., Vejda, S., Gelbmann, D., Bayer, E., Gotzmann, J., Schulte-Hermann, R., and Mikulits, W. (2002). Concomitant determination of absolute values of cellular protein amounts, synthesis rates, and turnover rates by quantitative proteome profiling. *Mol. Cell Proteomics* **1**, 528–537.

Laemmli, U. (1970). Cleavage of structural proteins during the assembly of the head of bacteriophage T4. *Nature* **227**, 680–685.

Lopez, M., Berggren, K., Chernokalskaya, E., Lazarev, A., Robinson, M., and Patton, W. (2000). A comparison of silver stain and SYPRO Ruby protein gel stain with respect to protein detection in two-dimensional gels and identification by peptide mass profiling. *Electrophoresis* **21**, 3673–3683.

Nishihara, J., and Champion, K. (2002). Quantitative evaluation of proteins in one- and two-dimensional polyacrylamide gels using a fluorescent stain. *Electrophoresis* **23**, 2203–2215.

O'Farrell, P. (1975). High resolution two-dimensional electrophoresis of proteins. *J. Biol. Chem.* **250**, 4007–4021.

Patton, W. (2000). A thousand points of light; the application of fluorescence detection technologies to two-dimensional gel electrophoresis and proteomics. *Electrophoresis* **21**, 1123–1144.

Patton, W. (2002). Detection technologies in proteome analysis. *J. Chromatogr. B Biomed. Appl.* **771**, 3–31.

Patton, W., and Beechem, J. (2002). Rainbow's end: The quest for multiplexed fluorescence quantitative analysis in proteomics. *Curr. Opin. Chem. Biol.* **6**, 63–69.

Steinberg, T., Haugland, R., Singer V., and Jones, L. (1996a). Applications of SYPRO Orange and SYPRO Red protein gel stains. *Anal. Biochem.* **239**, 238–245.

Steinberg, T., Jones, L., Haugland, R., and Singer, V. (1996b). SYPRO Orange and SYPRO Red protein gel stains: One-step fluorescent staining of denaturing gels for detection of nanogram levels of protein. *Anal. Biochem.* **239**, 223–237.

Steinberg, T., Lauber, W., Berggren, K., Kemper, C., Yue, S., and Patton, W. (2000). Fluorescence detection of proteins in SDS-polyacrylamide gels using environmentally benign, non-fixative, saline solution. *Electrophoresis* **21**, 497–508.

Steinberg, T., Martin, K., Berggren, K., Kemper, C., Jones, L., Diwu, Z., Haugland, R., and Patton, W. (2001). Rapid and simple single nanogram detection of glycoproteins in polyacrylamide gels and on electroblots. *Proteomics* **1**, 841–855.

Steinberg, T., White, H., and Singer, V. (1997). Optimal filter combinations for photographing SYPRO Orange or SYPRO Red dye-stained gels. *Anal. Biochem.* **248**, 168–172.

29

Autoradiography and Fluorography: Film-Based Techniques for Imaging Radioactivity in Flat Samples

Eric Quéméneur

I. INTRODUCTION

Autoradiography (ARG) is the photography-derived technique used to visualize the distribution of a compound labelled radioactively with either a β or γ emitting isotope in a biochemical sample. It has become a fundamental tool, particularly since slab gel electrophoresis has established as an inevitable techniques for separating and analysing complex mixtures of biomolecules. Any other flat samples such as blotting and dotting membranes, thin-layer chromatography plates, or microscopy slides are suitable. It is worth comparing ARG with numeric methods (phosphorimagers) that have popularised in many laboratories for 10 years. Indeed, films appear superior in terms of traceability and resolution. The resolution of a film depends on the size of metal grain in the photographic emulsion. A simple calculation shows that, assuming a grain diameter of 0.2 μm, a film displays a resolution of about 127,000 dpi compared to 600 dpi for a common laser printer. This resolution allows a possible magnification of up to 500-fold until the resolution of human eye (about 100 μm) is surpassed. In other terms, a $10 \times 10\text{-cm}^2$ autoradiographic picture is equivalent to a 25 million pixel digital image. Furthermore, an autoradiogram is exactly the same size as the sample, making it easy the precise location of a "hot spot" in a complex pattern.

Basically, autoradiographic techniques divide into two modes. The first one, direct ARG, is the direct exposure of film by β particles or γ rays emitted by the sample. In the second mode, the film is sensitized indirectly by the secondary light generated upon excitation of a "fluor" (fluorography) or a "phosphor" (indirect ARG with intensifying screen) by the radioactive emission. Autoradiography and fluorography look simple in appearence, but some fundamental notions should be known and kept in mind for obtaining sensitive, resolutive, and reproducible results. Among them, the notion that sensitivity and resolution are antagonistic concepts that cannot be matched simultaneously. Because of their path lengths, lower energy isotopes such as ^{3}H, ^{35}S, or ^{14}C have better resolution than higher ones such as ^{32}P or ^{125}I. This detection is not sensitive and indirect methods based on the emission of UV/blue photon promote sensitivity but decrease resolution (Fig. 1). Other important notions are discussed in Section IV.

II. MATERIALS AND INSTRUMENTATION

For catalog numbers of reagent, products, and equipment, see Table I. Dark room equipped with appropriate lighting systems: the red light could be a 7.5- to 15-W bulb covered with a red Kodak GBX-2 filter.

Labelling the corners of the sample to assist the accurate alignment of the sample to its autoradioradiogram is useful. Use either a nib and radioactive ink prepared by mixing 100 μl of Indian ink with a few microliters of the diluted radioisotope solution used in the experiment. Alternatively, a pen containing a

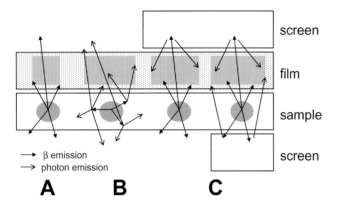

FIGURE 1 The underlying principle of the major detection methods and the way to assemble the various components in the cassette correctly. (A) Direct ARG; (B) fluorography; and (C) indirect ARG with either one or two intensifying screens. Note the variation in image size (resolution) depending on the nature and path of radiations.

phosphorescent ink or luminescent stickers can prove more convenient (e.g., Glogos II Autorad Markers, Stratagene).

Microscale standards (Amersham Biosciences) are convenient calibration strips for relative quantitation in all mode of detection. They are available for ^{3}H, ^{14}C, and ^{125}I with activity scales ranging from 3.74 to 4048 Bq/mg, from 3.7 to 31,890 Bq/g, and from 44 to 23,900 Bq/mg, respectively. For films and intensifying screens, see Table I for the appropriate choice.

II. PROCEDURES

Three procedures are reported here. The choice depends practically on the nature of the isotope (energy of emitted radiations) and on the yield of radi-

TABLE I Catalog Numbers of Reagents, Products, and Equipments Used

Darkroom safelight kit (Kodak, Cat. # 852 1429)
GBX-2 filter, 10 × 12 in. (Kodak, Cat. # 141 6940)
Pen with phosphorescent ink (Soquelec Cat. # R20105)
Glogos II Autorad Markers, 100 luminescent stickers (Stratagene, Cat. # 420 201)
Microscale standards
  ^{3}H microscales: 10 strips 0.111–4.07 kBq/mg + 10 strips 3.7–592 Bq/mg (Amersham Cat. # RPA510)
  ^{14}C microscales: 20 strips 1.15–32.7 kBq/g (Amersham Cat. # RPA504L)
  ^{14}C microscales: 20 strips 3.7–3700 Bq/g (Amersham Cat. # RPA511L)
  ^{125}I microscales: 20 strips 0.046–23.7 kBq/mg (Amersham Cat. # RPA523L)
Films
  Hyperfilm MP, 20.3 × 25.4 cm, 75 sheets/box (Amersham, Cat. # RPN 1678K)
  BioMax MS-2, 20.3 × 25.4 cm, 50 sheets/box (Kodak, Cat. # 837 7616)
  BioMax MR-2, 20.3 × 25.4 cm, 50 sheets/box (Kodak, Cat. # 895 2855)
  X-Omat AR-2, 20.3 × 25.4 cm, 50 sheets/box (Kodak, Cat. # 165 1579)
  Cronex 10T, 10 × 12 in., 100 sheets/box (Agfa Medical, Cat. # LF5E3)
Intensifying screens
  Hyperscreen, 20.3 × 25.4 cm (Amersham, Cat. # RPN 1669)
  BioMax MS Screen, 20.3 × 25.4 cm (Kodak, Cat. # 851 8706)
  TransScreen HE, 20.3 × 25.4 cm (Kodak, Cat. # 856 3959)
  TransScreen LE, 20.3 × 25.4 cm (Kodak, Cat. # 162 2034)
  Optex HighPlus (MCI Optonix, Cat. # 6102)
Cassettes
  BioMax Cassette, 20.3 × 25.4 cm (Kodak Cat. # 820 9140)
  Hypercassette, 20.3 × 25.4 cm (Amersham, Cat. # RPN 11649)

En^{3}Hance spray for fluorography, 2 oz. (NEN, Cat. # NEF 9700)
Enligthning rapid autoradiography enhancer, 1 liter (NEN)
Amplify fluorographic reagent, 1 liter (Amersham, Cat. # NAMP 100)
PPO (2,5-diphenyloxazole) scintillation grade, Sigma-Aldrich, Product # D210404
Cellulose paper Grade 3 MM Chr (thickness 0.34 mm), 20 × 25 cm, 100 sheets/box (Whatman, Cat. # 3030-866)
Cellulose paper grade 1 Chr (thickness 0.18 mm), 25 × 25 cm, 100 sheets/box (Whatman, Cat. # 3001-878)
Mineral oil, "Nujol" (Sigma-Aldrich, Cat. # M1180)
Orange-filtered preflash unit Sensitize (Amersham, Cat. # RPN 2051)
Kodak GBX developer, concentrate to make 3.8 liter (Kodak, Cat. # 190 0943)
Kodak GBX fixer, concentrate to make 3.8 liter (Kodak, Cat. # 190 2485)
Metal hanger for 20.3 × 25.4-cm films (Kodak, Cat. # 150 2764)

olabeling (amount of activity expected in a spot). A compromise has to be found between sensitivity and resolution. Table III might help in making the choice and in estimating the necessary exposure time.

A. Direct Autoradiography

This is the method of choice for high resolution whatever the isotope used, although it is rarely suitable for ^{3}H, ^{14}C, or ^{35}S under normal labelling conditions *in vivo*.

Steps

1. Protect the cassette from radioactivity by placing a Saran wrap or plastic sheet under the sample.

2. Put the sample in the cassette. Preferably, the sample should be dry. If not, cover it with Saran wrap in order to avoid sticking of the sample to the film. However, for a sample containing ^{3}H, it is important to remove any barrier such as Saran wrap or a cellophane sheet that may quench the emitted radiations.

3. If necessary, label distinctly two corners of the sample using the radioactive ink. This will help in the future superimposition of the gel and the film. Avoid moisture or glove powder on the surface of the sample.

4. In the dark room, place the film in close contact with the sample. When using single-coated films as recommended for ^{3}H (Fig. 2), the sensitive face of the film should face the sample (Fig. 1).

5. Close the cassette carefully and let it stand at room temperature for the necessary exposure time. Refer to Table II for an estimate of this exposure time. An overnight exposure would be sufficient for the detection of most ^{32}P-containing samples, whereas a 2- to 3-day exposure might be necessary for ^{35}S.

6. Remove the film from the cassette in the darkroom and process it according to the manufacturer's instructions or as described in Section III,E.

TABLE II   Choosing the Right Combination of Films and Products

Priority	Method	Film	Screen
Speed, sensitivity	Indirect ARG *(high energy isotopes)*	Hyperfilm™ MP[a] BioMax™ MS[b]	Hyperscreen™[a] BioMax™ MS Screen[b]
	Fluorography *(low energy isotopes)*	X-Omat™ AR (XAR)[b] Super RX[c]	TranScreen™ HE[b] *(high energy isotopes)*
	Direct ARG *(high energy isotopes)*	Cronex 10T[d]	TranScreen LE[b] *(low energy isotopes)* Optex HighPlus[e]
Resolution	Direct ARG	BioMax™ MR[b] Hyperfilm™ 3H[a] Hyperfilm™ βmax[a]	

[a] Amersham;
[b] Eastman Kodak;
[c] Fuji;
[d] Agfa;
[e] MCI Optonix.

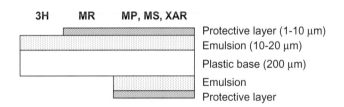

FIGURE 2   The structure of some commercially available films. The film types are ordered by increasing sensitivity but decreasing resolution. See Table II for film names and manufacturers.

TABLE III   Choosing the Optimal Detection Method

Isotope	Radiation	E_{max} (MeV)	Path max in air	Minimal activity detected in 24 h in a 0.1-cm^2 band (in Bq)		
				Direct ARG	Fluorography	Indirect ARG
^{3}H	β$^-$	0.018	6 mm	15,000	15	inadequate
^{14}C, ^{35}S	β$^-$	0.156, 0.167	25 cm	10	400	inadequate
^{32}P	β$^-$	1.710	8 m	1	Inadequate	0.1
^{125}I	γ	0.035	>10 m	2.5	Inadequate	0.2
	X	0.027				
	Auger e$^-$	0.030				

B. Fluorography

Fluorography is particularly recommended for low-energy isotopes, when sensitivity is required and loss of resolution is acceptable. In pratice, it might be the only way to visualise ^{3}H-labeled compounds.

Solutions

Commercial reagents such as En^{3}Hance (NEN), Enligthning (NEN), or Amplify (Amersham) are efficient and convenient, but expensive.

Steps

1. Stain and destain the sample as usual, avoid TCA for fixing proteins and BET for staining nucleic acids because this would subsequently quench the fluorescence.
2. Work in a fume hood. Soak the sample in the scintillant reagent with gentle agitation for 1 h. The tray should contain enough reagent so that the sample is free floating and does not stick to the walls.
3. Discard the reagent and move to the 4°C cold room. Add slowly 500 ml cold 10% glycerol along the walls of the tray and shake slowly for 1 h. The low temperature prevents the fluor from diffusing out of the gel. The gel will turn white opaque after this step.
4. Put the sample on the top of a thick Whatman paper, cover it with a Saran wrap, and dry the ensemble under vacuum for 2 h at 60°C.
5. Let the sample reach the ambient temperature before putting it in the autoradiography cassette. In the dark room, prepare the cassette in the following order: gel/Saran/film. Exposure should be performed at −70/−80°C. For film processing, see Section III,E.

For seldom use of fluorography if the commercial reagent is not available in the laboratory, or if cost is a limiting factor, an efficient home-made reagent can be made: 20% 2,5-diphenyloxazole (PPO) in undiluted acetic acid (Skinner, 1983). The following procedure works well with most types of samples.

1. Soak sample in undiluted acetic acid for 5 min.
2. Impregnate with the reagent for 1.5 h with gentle agitation.
3. Soak in water for 30 min.
4. Dry and expose to film at −70°C.

C. Indirect Autoradiography with Intensifying Screens

The use of intensifying screens is not efficient for low-energy isotopes such as ^{3}H, ^{14}C, ^{33}P, or ^{35}S because the path of their β emission is too short to cross the film and reach the screen.

Steps

Steps 1–3 are the same as in Section III,A.

4. Place the film over the sample.
5. Place the screen over the film so that the matt white side (the side where the fluorescent material stands) faces the film. Note that the screen must have been kept in the dark for several hours before being introduced in the cassette in order to avoid a glow effect that would increase background on the film.
6. Close the cassette carefully and put it at −70/−80°C for the necessary exposure time (see Table III for estimation). By default, a 24-h exposure might be fine.

For transparent samples such as polyacrylamide gels, a second screen could be placed under the sample (white face up) to further increase sensitivity. Blots can be rendered translucent by impregnation with mineral oil (Rust, 1987). Kodak has introduced the TranScreen to solve the problem of the film attenuating the β particle before it reaches the screen. In our experience, with ^{3}H, the result has about twofold higher sensitivity compared to the conventional screen.

D. Film Preflashing

The preflashing of films increases their sensitivity for small amounts of radioactivity and expands the range for linear response. It is used only with photon-based detection (Sections III,B and III,C) and has no effect for direct autoradiography or for large amounts of radioactivity in short exposures. Preflashed films are only valid for a couple of days and must be stored at −70/−80°C if not used immediately.

Steps

1. In the dark room, install the orange-filtered flash, e.g., Sensitize (Amersham), at 50 cm above the film area.
2. Cut six film strips (just large enough to be processed) and cover them with a sheet of Whatman paper No. 1. Keep one of the strips in a closed cassette.
3. Make five consecutive flashes of 1 ms each while removing a strip between each in order to obtain a set of strips ranging from 1- to 5-ms exposure times.
4. Process the six strips and cut them to the format of a spectrophotometer cuvette holder (usually 1 × 4 cm).

5. Read absorbances at 545 nm on a spectrophotomer, plot the values of exposure time versus absorbance, and select the exposure time that gives an A_{545} in the range of 0.1–0.2.

6. In the dark room, select the film to be used for fluorography or with intensifying screens. Preflash it for the selected exposure time. The preflashed side of the film should face the sample when preparing the cassette (Fig. 1).

E. Processing of Films

Most laboratories dealing with a large number of autoradiograms have an automated processor. If this is the case, for optimal results with your sample, make sure that chemicals are fresh and the machine is regularly cleaned up. This article provides a method for the manual processing of films to those who are not equipped with such a machine or seldom use autoradiographic methods. Furthermore, some films are not protected by an antiscratch layer in order to maximize sensitivity to low-energy isotopes, e.g., Hyperfilms ^{3}H and βmax. They must be processed manually. The present method should work well in most cases, but it is obviously preferable to refer to the manufacturer's specific instructions.

Solutions

If only a single film or few films should be processed, prepare three trays or deep tanks for multiple processings containing the following solutions.

1. *Developer*: Kodak GBX developer
2. *Rinser*: Kodak GBX indicator stop or a large volume of tap water
3. *Fixer*: Kodak GBX fixer

Minimizing volumes is an environment-friendly attitude, but good results are obtained with generous volumes of fresh reagents. Collect carefully used solutions 1 and 3 in a dedicated can, which should then be evacuated as a toxic chemical waste.

Steps

1. In the dark room, open the cassette and remove the film. Gloves should not be damp or leave powder on the film. If the cassette comes from the –70°C freezer, let it reach the ambient temperature before opening.

2. Handle the film with suitable pliers or attach it to a metal hanger.

3. Immerse quickly in developer. Dislodge air bubbles by tapping the film against the tank wall. Then do not agitate during development, which will occur within 2–5 min depending on the temperature of the solution.

4. Remove film from the developer and transfer it to the rinser. Wash with continuous, moderate agitation for 30 s.

5. Place the film in the fixer for 10 min. Moderate agitation is recommended. The background should become uniformly transparent.

6. Wash in a large tank filled with running water for 5 min.

7. Hang the film on a line with a suitable clip attached to one corner and let it dry at room temperature in a dust-free area.

IV. COMMENTS

Whenever possible, blotting samples separated previously by electrophoresis from the gel matrix onto the surface of a nitrocellulose membrane should be performed. This has a minor impact on resolution but it improves tremendously the detection of all types of isotopes, particularly that of low-energy isotopes (Quéméneur, 1995). Films should be handled carefully when unpacking to avoid electrostatic artifacts that would print on the film.

V. PITFALLS

When working with high-energy isotopes such as ^{32}P or ^{125}I, do not stack cassettes during exposure because the path of the emission is sufficient to generate phantom images on the neighbouring film. In the same idea, do not store films too close to the working area. Film storage is extremely important for reproducible high-quality results. Ideal storage conditions are 10–20°C and 30–50% hygrometry. Avoid piling up film packages and use an upright position to diminish physical strains on the surface of the films.

The quality of the dark room and its equipment is critical. The inactinic safelight should be at a minimal distance of 1 m. The performance of the red filter decrease with time and it is reasonable to change it every 5 years.

References

Laskey, R. A. (1993). Efficient detection of biomolecules by autoradiography, fluorography or chemilumiscence, Review booklet #23, 2nd Ed.—Amersham Biosciences.

Laskey, R. A. (2002). Radioisotope detection using X-ray film. *In* *"Radioisotopes in Biology"* (R. J. Slater, ed.), 2nd Ed.", pp. 63–83. Oxford Univ. Press, Oxford.

Perng, G., Rulli, R. D., Wilson, D. L., and Perry, G. W. (1988). A comparison of fluorographic methods for the detection of 35S-labeled proteins in polyacrylamide gels. *Anal. Biochem.* **175**, 387–392.

Quéméneur, E., and Simonnet, F. (1995). Increased sensitivity of autoradiography and fluorography by membrane blotting. *Biotechniques* **18**, 100–103.

Rust, S., Kunke, H., and Assman, G. (1987). Mineral oil enhances the autoradiographic detection of ^{32}P-labeled nucleic acids bound to nitrocellulose membranes. *Anal. Biochem.* **163**, 196–199.

Skinner, M. K., and Griswold, M. D. (1983). Fluorographic detection of radioactivity in polyacrylamide gels with 2,5-diphenyloxazole in acetic acid and its comparison with existing procedures. *Biochem. J.* **209**, 281–284.

Gel Profiling of Posttranslationally Modified Proteins

30

Two-Dimensional Gel Profiling of Posttranslationally Modified Proteins by *in vivo* Isotope Labeling

Pavel Gromov and Julio E. Celis

I. INTRODUCTION

The repertoire of posttranslational modifications, which some cellular proteins may undergo after synthesis, falls into two main categories: chemical modifications and processing. Chemical modification involves the linkage of a chemical group to the terminal amino acid or carboxyl groups of the backbone or to reactive groups in the side chains of internal residues. Protein processing involves the proteolytic removal of the polypeptide segments from the premature protein chain. In some cases, both types of protein modifications are closely coupled. Such posttranslational modifications may alter the activity, life span, interactions, and/or cellular localization of proteins, depending on the nature of the modification(s). The most common chemical modifications include acetylation, methylation, phosphorylation, glycosylation, lipid-mediated modifications, and ADP-ribosylation, as well as several others (Mumby, 2001; Fu *et al.*, 2002; Spiro, 2002; Corda *et al.*, 2002; Sinensky, 2000; Cohen, 2000; Casey, 1995 and references therein). New multiplexing tools suitable for general protein detection, including posttranslational variants, have been employed in proteome analysis over the years and are reviewed by Patton (2002).

Each modification causes changes in the molecular weight and often in the charge of a protein, a fact that makes two-dimensional (2D) PAGE well suited for the detection of many posttranslational modifications in combination with mass spectrometry and/or Western immunoblotting if the epitope structure of the modified molecule is not altered by the modification (see article by Celis *et al.* in Volume 2). Several mass spectrometric approaches have been developed for the identification and analysis of various posttranslational modifications and have been reviewed by Abersold and Mann (2003) and by Sickmann *et al.* (2002).

In many cases, however, the identification and analysis of posttranslational modifications can be achieved by *in vivo* or *in vitro* labeling of the proteins with an appropriated isotope-labeled metabolite. Once the radiolabeled ligand is covalently attached to the protein, it can be readily detected on a gel or in a blot using autoradiography. This article describes protocols for 2D gel mapping of posttranslationally modified proteins as applied to the analysis of phosphorylated, glycosylated, and lipidated proteins (palmytoylated, myristoylated, farnesylated, and geranylgeranylated) from human keratinocytes and transformed human amnion cells (AMA).

II. MATERIALS AND INSTRUMENTATION

A. Transformed Human Amnion Cells

AMA cells are grown in complete Dulbecco's modified Eagle's medium (DMEM) containing 10% fetal calf serum (FCS). General procedures for culturing these cells are described in details elsewhere (Celis and Celis, 1997; see also article by Micleady and O'Commo in Volume 1).

B. Noncultured Human Keratinocytes

Noncultured unfractionated human epidermal keratinocytes are prepared from normal skin epidermis as described by Celis *et al.* (1995).

C. Reagents

DMEM (Cat. No. 31966-021) is from GIBCO. Penicillin/streptomycin (Cat. No. A2213) is from Biocrom KG. Dulbecco's modified Eagle's phosphate-deficient medium (Cat. No. 16-423-49) is from ICN. FCS (Cat. No. 04-001-1A) is obtained from Biological Industries. Tissue culture plates (24-well) (Cat. No. 662 160) are from Greiner. All other reagents and general tissue culture facilities are as described elsewhere (Celis and Celis, 1997). 2,5-Diphenyloxazole (PPO, Cat. No. D-4630) is from Sigma, and dimethylether (Cat. No. 823277) is from Merck. [^{35}S]Methionine (Cat. No. SJ 204), [^{32}P]orthophosphate (Cat. No. PBS 13A), [^{3}H]mannose (Cat. No. TRK364), [^{3}H]palmitic acid (Cat. No. TRK909), [^{3}H]myristic acid (Cat. No. TRK907), [^{3}H]farnesyl pyrophosphate (Cat. No. TRK917), and [^{3}H]geranylgeranyl pyrophosphate (Cat. No. TRK918) are from Amersham. Lovastatin is from Merck.

III. PROCEDURES

The protocols for metabolic labeling of posttranslationally modified proteins are illustrated using noncultured unfractionated human epidermal keratinocytes and AMA cells, but can be applied to other cultured cell lines. The volumes given in the following procedure are for labeling in a single well of a 24-well plastic culture plate. For larger tissue culture dishes or multiwell labeling, adjust all amounts accordingly.

A. Phosphorylation

Eukaryotic protein phosphorylation is a reversible covalent addition of phosphate to a protein molecule by means of formation of an ester bond between the phosphoryl group and serine, threonine, or tyrosine residues. Protein phosphorylation is one of the most common posttranslational modifications and is of crucial importance for many regulatory processes. Net phosphorylation is regulated by a complex cascade of protein kinases and phosphatases that catalyze phosphorylation and dephosphorylation reactions, respectively (Cohen, 2000 and references therein). While

mass spectrometric methods have taken a leading role in the identification of phosphoproteins, protein radiolabeling with ^{32}P inorganic phosphate is still an effective and inexpensive method for the detection of the ^{32}P subproteome using 2D gels (Mason *et al.*, 1998). This article presents a protocol for the detection of phosphorylated proteins that is based on the metabolic incorporation of [^{32}P]orthophosphate into cultured AMA cells and noncultured human keratinocytes, followed by 2D PAGE and autoradiography.

Solutions

1. *Complete Dulbecco's modified Eagle's medium*: To make 500 ml, mix 445 ml of DMEM medium, 5 ml of 10× stock penicillin/streptomycin, and 50 ml of FCS.

2. *Complete DMEM phosphate-free medium*: Prepare as described in the article by Celis and Celis (1997) using commercial DMEM phosphate-deficient medium.

3. *Hank's buffered saline solution*: To make 1 liter, dissolve 0.4 g of KCl, 0.06 g of KH_2PO_4, 0.0621 g of $NaHPO_4 \cdot 2H_2O$, and 8 g of NaCl in 800 ml of distilled water. After dissolving, complete to 1 liter with distilled water.

4. *Lysis solution*: 9.8 M urea, 2% (w/v) Nonidet P-40 (NP-40), 2% ampholytes, pH 7–9, and 100 mM dithiothreitol, (DTT). Prepare as described in the article by Celis *et al.* in this volume.

5. *[^{32}P]orthophosphate-labeling solution*: To prepare 0.25 ml, add 50 µl of commercial aqueous [^{32}P]orthophosphate solution (10 mCi/ml; HCl free) to 200 µl of complete DMEM phosphate-free medium.

Steps

1. Seed AMA cells in a 24-well tissue culture plate (approximately 3×10^4 cells per well) in 0.3 ml of complete DMEM medium containing 10% calf serum, 2 mM glutamine, and antibiotics.

2. Incubate the cells at 37°C in a humidified, 5% CO_2 incubator for approximately 24 h, or until the cells reach about 50% confluence.

3. Prior to labeling, aspirate the medium and wash twice with phosphate-free medium. Add 0.3 ml of phosphate-free medium and incubate the cells for 1 h.

4. Remove the phosphate-free medium and overlay with 0.25 ml of [^{32}P]orthophosphate-labeling solution (2 mCi/ml).

5. Incubate the cells at 37°C in a humidified, 5% CO_2 incubator for 8 h, or a shorter period if necessary.

6. Carefully remove the medium containing [^{32}P]orthophosphate from the plate and gently wash the cells twice with 1 ml of Hank's buffered

saline solution. Dispose of radioactive solutions according to the safety procedures enforced in your laboratory.

7. Repeat the washing twice more.

8. Carefully remove excess Hank's buffered saline solution from the plate using an elongated Pasteur pipette.

9. Resuspend the cells in 50 µl of lysis buffer and run 2D gels as described by Celis *et al.* and by Görg and Weiss in this volume.

10. Dry the gels and subject to phosphorimaging or to X-ray film autoradiography at –70°C using an amplifying screen.

Comments

Representative autoradiographs (IEF gel) of phosphoproteins from human keratinocytes and AMA cells labeled metabolically with [^{32}P]orthophosphate are shown in Fig. 1. To facilitate the identification of phosphorylated proteins, we recommend adding small amounts of [^{35}S]methionine-labeled proteins from the same cell type to the ^{32}P-labeled protein mixture. 2D gels can then be autoradiographed using two films placed on top of each other. The first film, which is placed in direct contact with the dried gel, visualizes both ^{35}S and ^{32}P isotopes, whereas the second one reveals only ^{32}P.

Identification of ^{32}P-labeled proteins can also be done in combination with 2D gel blot immunodetection. In this case, following 2D gel electrophoresis, the proteins are electroblotted to the nitrocellulose membrane and the blot is probed with the antibody of interest, e.g., against phosphotyrosine, prior or after autoradiography (see article by Celis *et al.* in Volume 1).

B. Glycosylation

Protein glycosylation is perhaps one of the most abundant and structurally diverse types of posttranslational modification. Formation of the amino acid–sugar bond is a critical event in the biosynthesis of glycoproteins and leads to diverse biological functions (Spiro, 2002 and references therein). Glycoproteins can be detected by autoradiography after metabolic incorporation of ^{3}H or ^{14}C sugars into cultured cells or tissues (Chandra *et al.*, 1998). The procedure for revealing glycosylated proteins described here is based on the metabolic incorporation of [^{3}H]mannose into cultured AMA cells, followed by 2D gel electrophoresis.

Solutions

1. *Complete Dulbecco's modified Eagle's medium*: Prepare as described in Section III,A.

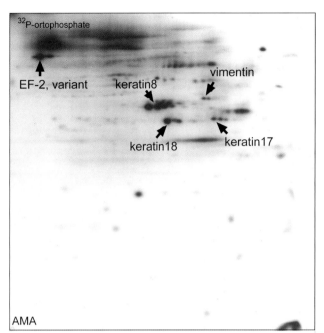

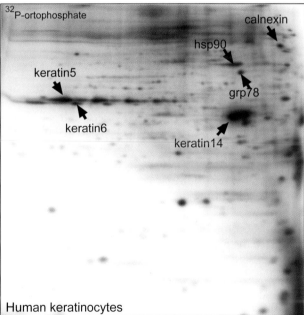

FIGURE 1 Two-dimensional (IEF) autoradiographs of whole protein extracts from AMA cells (top) and human keratinocytes (bottom) labeled with [^{32}P]orthophosphate. Several phosphoproteins are indicated as references.

2. *Hank's buffered saline solution*: Prepare as described in Section III,A.

3. *[^{3}H]Mannose–labeling solution*: 200 µCi/ml. Evaporate 50 µCi of commercial [^{3}H]mannose solution using a Speed–Vac centrifuge or by directing a gentle stream of nitrogen gas onto the surface of the solution (during this operation the temperature of

the solution should not exceed 30°C). Resuspend in 0.25 ml of complete DMEM medium.

4. *Lysis solution*: 9.8 M urea, 2% (w/v) NP-40, 2% ampholytes, pH 7–9, and 100 mM DTT. Prepare as described in Section III,A.

5. *Amplifying solution*: 7% of 2,5-diphenoloxazole (PPO) in dimethylether. To make 100 ml, weigh 7 g of PPO and complete to 100 ml with dimethylether. Store at –20°C in a hermetic glass vessel.

Steps

1. Plate and grow AMA cells in complete DMEM medium as described in steps 1 and 2 in Section III,A.

2. Aspirate the medium using an elongated Pasteur pipette and replace it with 0.25 ml of the [³H]mannose-labeling solution.

3. Incubate the cells at 37°C in a humidified, 5% CO_2 incubator for 2 h.

4. Aspirate the labeling solution and wash the cells twice with Hank's buffered saline solution.

5. Carefully aspirate the excess of Hank's buffered saline solution using an elongated Pasteur pipette.

6. Resuspend the cells in 50 µl of lysis solution and run 2D gels as described by Celis *et al.* or by Görg and Weiss in this volume.

7. Following 2D gel electrophoresis, transfer the proteins onto a nitrocellulose membrane by electroblotting (see article by Celis *et al.* in Volume 1).

8. Dry the nitrocellulose blot overnight at room temperature.

9. Pour 100 ml of amplifying solution in a rectangular glass container.

10. Immerse the nitrocellulose blot into the amplifying solution for 1 s.

11. Place the nitrocellulose blot on the filter paper with the protein-bearing side facing upward and dry for 30 min.

12. Expose the dried nitrocellulose blot to an X-ray film for 1–7 days at –80°C.

Comments

A representative 2D gel (IEF) fluorograph of AMA cell proteins labeled with [³H]mannose is shown in Fig. 2. Following exposure to an X-ray film, the membrane can be stained with amido black to aid in the identification of polypeptide spots. ³H-labeled polypeptides can also be revealed by fluorography of dried gels stained with silver and saturated with the amplifying solution (Amersham; see also article by Celis *et al.* in this volume). However, ³H fluorography from dried gels requires longer exposure times as compared to fluorography using nitrocellulose blots. To facilitate

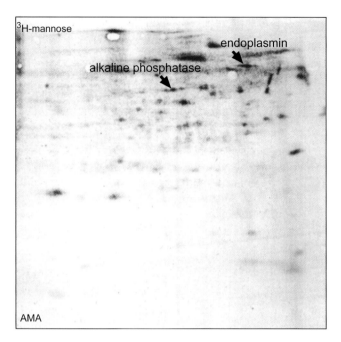

FIGURE 2 Two-dimensional (IEF) fluorograph of whole protein extracts from AMA cells labeled with [³H]mannose. Several glycoproteins are indicated as references.

protein identification, we recommend adding a concentrated, unlabeled AMA protein extract to the ³H-labeled protein sample prior to 2D gel electrophoresis.

C. Palmitoylation and Myristoylation

Protein lipidation involves co- or posttranslational modification by specific lipids. For most lipid-modified proteins, attached lipids appear to direct or enhance the interaction with both membrane lipids and other proteins, resulting in their specific membrane localization (Casey, 1995). Lipid modified proteins are classified based on the identity of the attached lipid. Palmitoylation and myristoylation are the result of the cotranslational addition of the saturated 16-carbon fatty acid palmitate or 14-carbon fatty acid myristate, respectively, to a glycine residue at the N or C terminus of the protein. The procedure for revealing lipidated proteins described here is based on the metabolic incorporation of either [³H]palmitate or [³H]myristate into cultured AMA cells, followed by 2D gel electrophoresis.

Solutions

1. *Complete Dulbecco's modified Eagle's medium*: Prepare as described in Section III,A.

2. *Hank's buffered saline solution*: Prepare as described in Section III,A.

3. *[³H]Palmitic acid–labeling solution*: 200 µCi/ml. Evaporate 50 µCi of commercial [³H]palmitic acid solution (supplied in ethanol) using a Speed–Vac centrifuge or by directing a gentle stream of nitrogen gas onto the surface of the solution (during this operation the temperature of the solution should not exceed 30°C). Resuspend the label in 0.25 ml of complete DMEM medium.

4. *[³H]Myristic acid–labeling solution*: Prepare as described earlier for [³H]palmitic acid solution

5. *Lysis solution*: 9.8 *M* urea, 2% (w/v) NP-40, 2% ampholytes, pH 7–9, and 100 m*M* DTT. Prepare as described in Section III,A.

6. *Amplifying solution*: 7% of PPO in dimethylether. Prepare as described in Section III,B.

Steps

Grow, label, and handle AMA cells as described in Section III,B.

Comments

Representative 2D gel blot fluorographs, with IEF in the first dimension, of [³H]palmitoylated (Fig. 3A) and [³H]myristoylated (Fig. 3B) proteins from cultured AMA cells are shown in Fig 3. Spots indicated with arrowheads are labeled with both fatty acids.

D. Isoprenylation: Farnesylation and Geranylgeranylation

Posttranslational modifications of protein with isoprenoids play important roles in targeting a number of proteins to the plasma membrane, as well as in protein–protein interactions, and membrane-associated protein traffic (Sinensky, 2000 and references therein). Protein isoprenylation consists in the covalent attachment of 15-carbon isoprenoid farnesyl or 20-carbon isoprenoid geranylgeranyl via a stable thioether bond to a cysteine residue located in the C-terminal "CAAX," "CC," or "CXC" boxes (Clarke, 1992; Cox and Der, 1992). The method for detecting isoprenylated proteins is based on the specific inhibition of endogenous mevalonate (prenoid precursor) synthesis by lovastatin, followed by metabolic labeling of isoprenylated proteins *in vivo* with either [³H]farnesyl- or [³H]geranylgeranyl pyrophosphate. The protocol described here follows closely those of Danesi *et al.* (1995) and Gromov *et al.* (1996).

Solutions

1. *Complete Dulbecco's modified Eagle's medium*: Prepare as described in Section III,A.

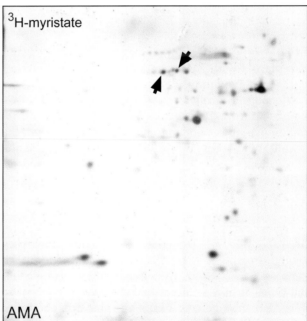

FIGURE 3 Two-dimensional (IEF) fluorographs of whole protein extracts from AMA cells labeled with [³H]palmitic acid and [³H]myristic acid. Proteins indicated with closed arrowheads incorporated both lipids.

2. *Hank's buffered saline solution*: Prepare as described in Section III,A.

3. *10 mM lovastatin*

4. *[³H]Farnesyl-PP labeling solution*: 50 µCi/ml. Evaporate 12.5 µCi of commercial [³H]farnesyl-PP solution using a Speed–Vac centrifuge or by directing

a gentle stream of nitrogen gas onto the surface of the solution (during this operation the temperature of the solution should not exceed 30°C). Resuspend in 0.25 ml of complete DMEM medium and add 0.5 ml of 10 mM lovastatin.

5. *[³H]Geranylgeranyl-PP labeling solution*: Prepare as described previously for [³H]farnesyl-PP

6. *Lysis solution*: 9.8 M urea, 2% (w/v) NP-40, 2% ampholytes, pH 7–9, and 100 mM DTT. Prepare as described in Section III,A.

7. *Amplifying solution*: 7% of PPO in dimethylether. Prepare as described in Section III,B.

Steps

1. Plate and grow AMA cells in complete DMEM medium until they reach about 80% confluence as desribed in steps 1 and 2 of Section III,A.

2. Aspirate the medium from the plate and replace it with 0.25 ml of fresh, complete DMEM medium. Add 1.25 µl of 10 mM lovastatin (final concentration is 50 µM).

3. Incubate the cells at 37°C in a humidified, 5% CO_2 incubator for 6 h.

4. Following incubation, remove the medium and replace it with 50 µCi/ml of either [³H]farnesyl- or [³H]geranylgeranyl-PP labeling solutions.

5. Incubate the cells at 37°C in humidified, 5% CO_2 incubator for 16 h.

6. Remove the labeling medium and rinse the cells twice with 0.5 ml of Hank's buffered saline solution.

7. Proceed as described in Section III,B, steps 5–12.

Comments

Lovastatin induces growth inhibition and apoptosis, especially at high concentration and after prolonged treatment (Perez-Sala and Mollinedo, 1994; Patterson *et al.*, 1994). Therefore, when labeling other cell types, it may be necessary to lower its concentration. Also, the efficiency of labeling may differ between cell types due to possible differences in the uptake of the isoprenoids.

Some isoprenylated proteins can be modified by both farnesyl and geranylgeranyl (Gromov *et al.*, 1996). Representative ³H fluorographs of AMA proteins labeled with [³H]farnesyl- or [³H]geranylgeranyl-PP are shown in Figs. 4A and 4B, respectively.

IV. PITFALLS

1. Use as short a labeling time as possible in order to reduce the effect of secondary labeling.

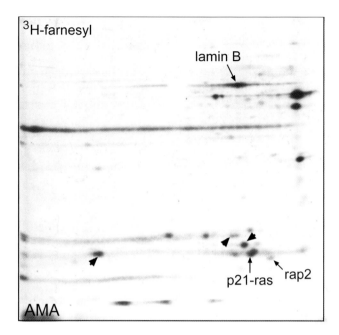

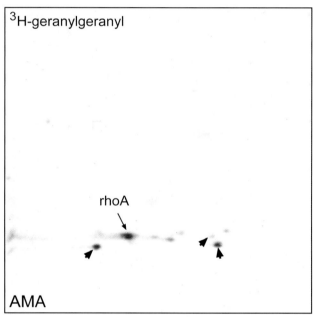

FIGURE 4 Two-dimensional (IEF) fluorographs of whole protein extracts proteins from AMA cells labeled with [³H]farnesyl-PP (top) and [³H]geranylgeranyl-PP (bottom). Several prenylated proteins are indicated with arrows. Proteins labeled with both isoprenoids are indicated with closed arrowheads.

2. Proteins that carry the same posttranslational modification may require different labeling times due to differences in protein metabolism.

3. Do not immerse the nitrocellulose filter into the amplifying solution longer than 1 s as the dimethylether destroys the membrane.

References

Aebersold, R., and Mann, M. (2003). Mass spectrometry-based proteomics. *Nature* **422**, 198–207.

Casey, P. J. (1995). Protein lipidation in cell signaling. *Science* **268**, 221–225.

Celis, A., and Celis, J. E. (1997). General procedures for tissue culture. *In "Cell Biology: A Laboratory Handbook"* (J. E. Celis, N. Carter, T. Hunter, D. Shotton, K. Simons, J. V. Small, eds.), Vol. 1, pp 5–16. Academic Press, San Diego.

Celis, J. E., Rasmussen, H. H., Gromov, P., Olsen, E., Madsen, P., Leffers, H., Honoré, B., Dejgaard, K., Vorum, H., Kristenesen, D. B., Øsregaard, M., Haunsø, A., Jensen, N. A., Celis, A., Basse, B., Lauridsen, J. B., Ratz, G. P., Andersen, A. H., Walbum, E., Kjærgaard, I., Andersen, I., Puype, M., Damme, J. V., and Vandekerckhove, J. (1995). The human keratinocyte two-dimensional gel protein database (update 1995): Mapping components of signal transdauction pathways. *Electrophoresis* **16**, 2177–2240.

Chandra, N., Spiro, M., and Spiro, J. (1998). Identification of a glycoprotein from rat liver mitochondrial inner membrane and demonstration of its origin in the endoplasmic reticulum. *J. Biol. Chem.* **273**, 19715–19721.

Clarke, S. (1992). Protein isoprenylation and methylation at carboxyterminal cystein residues. *Annu. Rev. Biochem.* **61**, 355–386.

Cohen, P. (2000). The regulation of protein function by multisite phosphorylation: A 25 year update. *Trends Biochem. Sci.* **25**, 596–601.

Cox, A. D., and Der, C. J. (1992). Protein prenylation: More than just glue? *Curr. Opin. Cell Biol.* **4**, 1008–1016.

Danesi, R., Mc Lellan, C. A., and Myers, C. E. (1995). Specific labeling of isoprenylated proteins: Application to study inhibitors of the post-translational farnesylation and geranylgeranylation. *Biochem. Biophys. Res. Commun.* **206**, 637–643.

Fu, M., Wang, C., Wang, J., Zafonte, B. T., Lisanti, M. P., and Pestell, R. G. (2002). Acetylation in hormone signaling and the cell cycle. *Cytokine Growth Factor Rev.* **13**, 259–276.

Gromov, P., and Celis, J. E. (1996). Identification of isoprenyl modified proteins metabolically labeled with [³H]farnesyl- and [³H]geranylgeranyl-pyrophosphate. *Electrophoresis* **17**, 1728–1733.

Mason, G. G., Murray, R. Z., Pappin, D., and Rivett, A. J. (1998). Phosphorylation of the ATPase subunits of the 26S proteasome. *FEBS Lett.* **430**, 269–274.

Mumby, M. (2002). A new role for protein methylation: Switching partners the phosphatase ball. *Sci. STKE* **79**, PE1.

Patterson, S. D., Grossman, J. S., D'Andrea, P., and Latter, G. I. (1994). Reduced numatrin/B23/nucleosphosmin labeling in apoptotic jurcat T lymphoblasts. *J. Biol. Chem.* **270**, 9429–9436.

Patton, W. F. (2002). Detection technologies in proteome analysis. *J. Chromatogr. B* **771**, 3–31.

Perez-Sala, D., and Mollinedo, F. (1994). Inhibition of isoprenoid biosynthesis induces apoptosis in human promyelocytic HL-60 cells. *Biochem. Biophys. Res. Commun.* **199**, 1209–1215.

Sickmann, A., Mreyen, M., and Meyer, H. E. (2002). Identification of modified proteins by mass spectrometry. *IUBMB Life* **54**, 51–57.

Sinensky, M. (2000). Recent advances in the study of prenylated proteins. *Biochim. Biophys. Acta* **1484**, 93–106.

Spiro, R. G. (2002). Protein glycosylation: Nature, distribution, enzymatic formation, and disease implications of glycopeptide bonds. *Glycobiology* **12**, 43R–56R.

Protein/Protein and Protein/Small Molecule Interactions

31

Immunoprecipitation of Proteins under Nondenaturing Conditions

Jiri Lukas, Jiri Bartek, and Klaus Hansen

I. INTRODUCTION

Immunoprecipitation of native proteins has proven to be a powerful and widely used approach in addressing questions related to the nature of a single protein or protein complexes existing under different biological conditions. A number of characteristics can be revealed using this method: (1) what is the relative molecular weight of the protein under study, (2) does it contain any posttranslational modifications, such as phosphorylation, acetylation, and glycosylation, (3) is the protein part of a larger multiprotein complex, (4) does it interact with nucleic acids or other ligands, and (5) does the level of the protein change upon growth factor stimulation, during progression through the different phases of the cell cycle, or during the transition between active proliferation and differentiation?

Combined with the recent improvements of protein microsequencing techniques and mass spectrometry, immunoprecipitation also gives the researcher an option to obtain sequence information from unknown proteins identified through coimmunoprecipitation and thereby collect data on multiprotein complexes. Very recently, several papers have described high-throughput analysis of multiprotein complexes in yeast using the mass spectrometry approach in a technique called high-throughput mass spectrometric protein complex identification (HMS-PCI) (Ho *et al.*, 2002) and tandem-affinity purification (TAP) and mass spectrometry (Puig *et al.*, 2001; Gavin *et al.*, 2002). The future will surely bring this kind of screening technique into focus, as it can, compared to conventional two-hybrid screens (which detect only those binary interactions not influenced by posttranslational

modifications), detect interactions that require protein modifications.

A critical prerequisite for successful analysis of immunoprecipitated native proteins is the quality of the primary antigen-specific antibodies. For the most straightforward interpretation of results, such a reagent should form specific immunocomplexes with the antigen in its native form without dissociating other associated proteins. In this context, one obvious possibility is to express the protein under investigation as an epitope-tagged protein from an exogenous promoter or even from an endogenous promoter (when possible), which gives the chance to immunoprecipitate the protein using specific antibodies against the tag epitope, such as Myc, Flag, and His. These tags can even be combined. Various strategies in how to obtain such reagents are described elsewhere (Harlow and Lane, 1988; Erica Golemis, 2002). This article shares experiences gained with immunoprecipitation under native conditions that have been obtained during studies of various proteins involved in cell cycle regulation (Lukas *et al.*, 1995; Hansen *et al.*, 2001). We provide a detailed description of an optimized immunoprecipitation protocol that could serve as a basis for isolating native proteins and protein complexes from mammalian cells.

II. MATERIALS AND INSTRUMENTATION

The following chemicals are from Sigma-Aldrich (see catalogue numbers in parentheses). HEPES (H-7523), NaCl (S-3014), EDTA (E-5134), EGTA (E-4378), glycerol (G-5516), Tween 20 (P-1379), Triton X-100 (X-

100), IGEPAL CA-630 (I7771), sodium dodecyl sulfate, SDS (L-4509), dithiothreitol, DTT (D-9779), β-glycerophosphate (G-6251), sodium fluoride, NaF (S-1504), sodium orthovanadate, Na$_3$-VO$_4$ (S-6508), leupeptine (hemisulfate salt, L-2884), aprotinin (A-1153), phenylmethylsulfonyl fluoride, PMSF (P-7626), bromphenol blue (B-7021), and myelin basic protein, MBP (M1891). Histone H1 is from Roche (223549).

Protein G–Sepharose 4 fast flow (17-0618-01) and protein A–Sepharose 4 fast flow (17-0974-01) are from Pharmacia Biotech.

Safe-Lock 1.5-ml polypropylene tubes are from Eppendorf (0030 123.328), and 15-ml conical polypropylene tubes are from Corning (430791). Cell scrapers (179693), 92-mm tissue culture dishes (150350), and 80-cm^2 tissue culture flasks (153732) are from Nunc. Protein concentration is measured by the Bio-Rad protein assay kit from Bio-Rad (500-0006).

III. PROCEDURES

Stock Solutions

1. *1 M HEPES, pH 7.5*: To make 1 liter, dissolve 238.3 g of HEPES in 800 ml of distilled water, adjust the pH to 7.5 with 10 N NaOH, and complete the volume to 1 liter. Sterilize by autoclaving; store at room temperature.

2. *1 M Tris, pH 7.5*: To make 1 liter, dissolve 121.1 g of Trizma base in 800 ml of distilled water (room temperature). Adjust the pH to 7.5 by concentrated HCl and fill up to 1 liter with distilled water. Autoclave and store at room temperature.

3. *5 M NaCl*: Disolve 292.2 g of NaCl in 800 ml of distilled water. Adjust the volume to 1 liter, autoclave, and store at room temperature.

4. *0.5 M EDTA*: To make 100 ml, add 18.6 g EDTA to 80 ml of distilled water. Stir vigorously on a magnetic stirrer. Adjust the pH to 8.0 with concentrated HCl and let the powder dissolve completely (the disodium salt of EDTA will not go into solution until the pH of the solution reaches approximately 8.0 by the addition of HCl). Adjust the final volume to 100 ml, autoclave, and store at room temperature.

5. *0.5 M EGTA*: To make 100 ml, add 19.0 g of EGTA to 80 ml distilled water. Stir and adjust the pH to 8.0 with HCl (see EDTA preparation). Adjust the volume to 100 ml, autoclave, and store at room temperature.

6. *1 M DTT (100× stock)*: Dissolve 5 g of DTT in 32 ml of distilled water. Sterilize by filtration (do not autoclave!), dispense into 1-ml aliquots, and store at −20°C.

7. *1 M β-glycerophosphate (100× stock)*: To make 100 ml, dissolve 21.6 g of β-glycerophosphate in 80 ml of distilled water. Adjust the volume 100 ml, autoclave, and store at room temperature.

8. *0.5 M NaF (500× stock)*: To make 100 ml, dissolve 2.1 g of NaF in a total amount of 100 ml of distilled water. Autoclave and store a room temperature.

9. *0.1 M Na$_3$VO$_4$ (100× stock)*: Dissolve 200 mg of Na$_3$VO$_4$ in 10.8 ml of distilled water, sterilize by filtration, dispense in multiple (100 μl) aliquots, and freeze at −20°C. Once recovered from the freezer, use the batch instantly and only once. Do not refreeze repeatedly.

10. *10 mg/ml leupeptin (1000× stock)*: Dissolve 25 mg of leupeptin in 2.5 ml of precooled distilled water, dispense into 0.5-ml aliquots, and freeze at −20°C.

11. *2 mg/ml aprotinin (1000× stock)*: Dissolve 10 mg of aprotinin in 5 ml of distilled water, dispense into 0.5-ml aliquots, and freeze at −20°C. Do not refreeze repeatedly; once recovered, store the batch for a maximum of 1–2 weeks at 4°C.

12. *0.1 M PMSF (1000× stock)*: To make 10 ml, dissolve 174 mg of PMSF in pure isopropanol (store at 4°C) (PMSF is highly unstable in aqueous solutions).

13. *2× Laemmli SDS sample buffer*: 100 mM Tris–HCl, pH 6.8, 200 mM DTT, 4% SDS, 20% glucerol, and approximately 0.2% bromphenol blue. To make 10 ml, mix 1 ml of Tris–HCl, pH 6.8, 2 ml of 1 M DTT, 4 ml of 10% SDS, 2 ml glycerol, and 1 ml of distilled water. Add traces of bromphenol blue powder to obtain the desired blue color and mix well. Divide into 0.5- to 1-ml aliquots and freeze at −20°C.

Buffers

A number of different lysis buffers have been described in the literature to effectively extract native proteins from mammalian cells. The following sections offer three different protein extraction buffers that have been used repeatedly and successfully in cell cycle studies for the evaluation of protein–protein interactions and for functional assays such as measuring *in vitro* kinase activity of the immunoprecipitated protein complexes.

1. *Lysis buffer 1 (Matsushime et al., 1994)*: 50 mM HEPES, pH 7.5, 150 mM NaCl, 1 mM EDTA, 2.5 mM EGTA, 10% (v/v) glycerol, and 0.1% Tween 20. To make 1 liter of 1× basic stock solution, add 50 ml of 1 M HEPES, pH 7.5, 30 ml of 5 M NaCl, 2 ml 0.5 M EDTA, 5 ml of 0.5 M EGTA, 100 ml glycerol, and 1 ml Tween 20 into 812 ml of distilled water. Stir well on a magnetic stirrer and store at 4°C. Immediately prior to use, add DTT (1 : 1000 from a 1 M stock); phosphatase inhibitors: NaF (1 : 500 from a 0.5 M stock),

β-glycerophosphate (1 : 100 from a 1 M stock), and Na$_3$VO$_4$ (1 : 1000 from 0.1 M stock); and protease inhibitors: leupeptin (1 : 1000 from a 10-mg/ml stock), aprotinin (1 : 1000 from a 2-mg/ml stock), and PMSF (1 : 1000 from a 0.1 M stock). Keep on ice throughout the whole procedure.

2. *Lysis buffer 2 (Jenkins and Xiong, 1996)*: 50 mM Tris–HCl, pH 7.5, 150 mM NaCl, and 0.5% (v/v) IGEPAL CA-630. To make 1 liter of 1× basic stock, add 50 ml of 1 M Tris–HCl, pH 7.5, 30 ml of 5 M NaCl, and 5 ml IGEPAL CA-630 into 915 ml of distilled water. Store at 4°C. Immediately prior to use, add DTT, phosphatase inhibitors, and protease inhibitors as described for buffer 1.

3. *Lysis buffer 3 (Pagano et al., 1993)*: 50 mM Tris, pH 7.5, 250 mM NaCl, 5 mM EDTA, and 0.1% (v/v) Triton X-100. To make 1 liter of 1× basic stock, add 50 ml of 1 M Tris–HCl, pH 7.5, 50 ml of 5 M NaCl, 10 ml of 0.5 M EDTA, and 1 ml Triton X-100 into 889 ml of distilled water. Store at 4°C. Immediately prior to use, add DTT, phophastase inhibitors, and protease inhibitors as described for buffer 1.

4. *Kinase assay buffer*: 20 mM HEPES, pH 7.2, 1 mM DTT, 10 mM MgCl$_2$, 10 mM MnCl$_2$, 2.5 mM EGTA, 1 mM NaF, 0.2 mM sodium orthovanadate, 2.5 μg/ml leupeptin, and 2 μg/ml aprotinin; should be prepared fresh from stock solutions upon use.

A. Cell Lysis

For all three lysis buffers, highly effective extraction of native proteins can be achieved by the following protocol.

Steps

1. Wash the cell monolayer twice with ice-cold phosphate-buffered saline (PBS) using 10 ml of PBS per washing step for a surface corresponding to a 92-mm diameter tissue culture dish or to a 80-cm^2 tissue culture flask.

2. Add 2.5 ml of ice-cold PBS and dislodge the cells with a cell scraper. Transfer the cell suspension into a prechilled 15-ml polypropylene test tube and repeat the same procedure with another 2.5 ml in order to recover the cells quantitatively. Spin for 5 min in a precooled (4°C) centrifuge (1000 g), discard the supernatant, wash the pellet briefly with 5 ml of cold PBS, and spin again. The cell pellet is now ready for lysis or, for many assays, can be frozen quickly by dipping the tube into liquid nitrogen and stored at −80°C until use.

3. Lyse the cells by adding 3–5 pellet volumes of ice-cold lysis buffer, vortex vigorously (4°C) for 10 s, and keep on ice for an additional 30 min. Throughout this period, resuspend the cells by brief vortexing every 5–10 min (4°C).

In the case of buffer 1, efficient lysis has been reported (Matsushime *et al.*, 1994) that involves resuspension of the cell pellet in lysis buffer and subsequent brief sonication on ice. We have successfully reproduced this procedure in our laboratory using a Branson sonifier 250 (2 × 10-s pulses at output level position 6).

In several cases it can be advantageous to avoid the scraping of cells into PBS and to perform a more instant lysis procedure based on adding lysis buffer directly to the cell monolayer (200 μl per surface corresponding to a 92-mm diameter tissue culture dish; see previous discussion for further specifications) that has been washed previously three times with ice-cold PBS. Distribute the lysis buffer on the cell monolayer (after draining off PBS) and collect the lysate with a cell scraper. Transfer the cell lysate into a prechilled 1.5-ml Eppendorf tube. Incubate the lysate on ice for an additional 30 min with occasional brief vortexing (4°C) in order to obtain an efficient protein extraction.

4. Centrifuge the protein extract in a microfuge cooled down to 4°C for 15 min at 20,000 g to pellet cell debris. Transfer the cleared extract to a clean test tube prechilled on ice and measure the total amount of extracted protein (in our laboratory, we use the Bio-Rad protein assay system and exactly follow the manufacturer´s protocol with the protein standards supplied with the kit).

B. Preclearing

Preequilibrate the protein A(G)–Sepharose beads supplied by the manufacturer by three successive rounds of resuspension in 10× bead volume of lysis buffer and gentle pelleting by brief (10 s) spinning in a microfuge. To eliminate nonspecific contaminants that can potentially associate with the beads, mix the cell lysate [up to 2 mg of total protein in total volume of 1 ml per tube with 50 μl of preeqiulibrated protein A(G)–Sepharose (in a 50% slurry)] and rotate in the cold room for 30 min. Pellet the beads by a 5-min centrifugation in a precooled microfuge (20,000 g) and carefully transfer the supernatant into a clean prechilled Eppendorf tube, leaving the beads behind. The protein extract is now ready for immunoprecipitation with specific antibodies. To improve the preclearing step, one can use general nonspecific control IgG (precoupled to proteinA/G-Sepharose) from the same species as the specific antibody used for the final immunoprecipitation. It is an advantage to use chemically cross-linked control IgG (Harlow and Lane, 1988).

C. Immunoprecipitation

Steps

1. To presaturate protein A(G)–Sepharose beads with antibodies, aliquot 10–20 μl of beads (50% slurry), preequilibrated with the chosen lysis buffer, into Eppendorf tubes containing 0.5 ml of lysis buffer. Add the desired antibodies in saturating amounts. The amount of antibody varies significantly depending on the titer and source of a particular batch and should be determined beforehand. As a rough guide, we recommend starting with 1 μg of purified immunoglobulin, 1–2 μl of mouse ascites, 100–200 μl of hybridoma supernatant, or 2–4 μl of crude rabbit antiserum per 10 μl of beads (note that protein A–Sepharose is particularly suitable for all antibodies of rabbit origin and for mouse IgG2 subclasses. For other subclasses of mouse immunoglobulins, use protein G–Sepharose in order to achieve high-affinity binding). Rotate slowly for 1 h in the cold room and then wash the beads three times in 2 ml of lysis buffer [to pellet the beads between the washing steps, centrifuge briefly in a cooled microfuge (5000 g)].

2. Add the protein extract to a 10- to 20-μl aliquot (50% slurry) of beads precoated with the desired antibody and adjust the volume with the lysis buffer to 0.5–1 ml. The total amount of protein input in each sample depends very much on the type of assay and the relative abundance of the protein under study. Thus for sensitive functional assays, such as measuring *in vitro* kinase activity, as little as 50 μg of total protein could be sufficient, but the usual input ranges between 200 μg and 2 mg of total extracted proteins. Close the tubes and rotate end over end in the cold room for 90 min up to 2–3 h (it is not recommended to immunoprecipitate overnight, as this will increase the risk of proteolysis and dephosphorylation of proteins, even in the presence of diverse inhibitors).

3. Pellet the beads for 10 s in a cooled microfuge (5000 g) and wash four times by resuspending the beads in 1 ml of of lysis buffer. Gently invert the tubes several times between each washing step.

4. After the last wash, aspirate the lysis buffer above the beads carefully (we recommend using a blunt-ended 25-gauge needle connected to a vacuum pump). For kinase reactions, continue to step 5, for other immunoprecipitations, continue to step 6.

5. For kinase reactions, wash the beads twice in 1 ml of kinase assay buffer and remove excess liquid above the beads. Start the kinase reaction by adding kinase assay buffer including the appropriate protein substrate (from 1–2 μg per reaction) and ATP. The amount of nonlabeled ATP to be added to the reaction depends on the kinase under study but should normally be in the range of 15–200 μM (final concentration). Furthermore, it is convenient to include [□^{32}P]-ATP (1–10 μCi per reaction; >3000 Ci/mmol) in order to be able to quantitate the incorporation of phosphate in the target substrate by exposure on X-ray film or a PhosphorImager screen. The kinase reaction should take place at 30°C for 10–30 min (should be optimized for the kinase under investigation). Reactions are terminated by the addition of 2× LSB containing 5 mM EDTA and heating for 5 min at 95°C. In cases where a phosphospecific antibody has been developed, it is possible to omit the isotope and instead perform immunoblotting after electrophoresis.

6. For one-dimensional SDS-PAGE analysis, resuspend the beads in 30 μl of 2× Laemmli SDS sample buffer and heat the samples on a heating block (95°C) for 4–5 min. Finally, centrifuge the tubes in a microfuge for 1 min at room temperature and load the samples directly on the gel by use of thin gel-loading tips.

In case the size of the protein of interest is close to 50 or 25 kDa, which also corresponds to the size of the heavy and light chains of immunoglobulins used for the immunoprecipitation, it can be an advantage to use chemically cross-linked antibodies as described by Harlow and Lane (1988). Because the cross-linking of antibodies preferentially takes place between the heavy chain and protein A/G, it is recommended to elute the immunoprecipitated proteins using Laemmli sample buffer without DTT or other reducing agent and, after heating at 95°C for 4–5 min, transfer the eluted material to a clean tube, avoiding beads, and thereafter add a similar volume of Laemmli sample buffer containing the appropiate amount of DTT or other reducing agent and then heat to 95°C again in order to reduce disulfide bonds in the immunoprecipitated proteins. The sample is now ready to load on the gel as described earlier. By using this two-step elution procedure, you can almost completely avoid disturbing signals from heavy and light chains upon Western blotting or staining of the gel.

Figure 1 shows an example of an immunoprecipitation of the cyclin-dependent kinase 4 (Cdk4) from primary mouse fibroblasts in complex with its associated subunits: D-type cyclins and Cdk inhibitors. Figure 2A shows an example of *in vitro* kinase assays using specific antibodies against three different kinase complexes in immunoprecipitation: Erk1 and Erk2 combined, cyclin D/Cdk4(6), and cyclin E/Cdk2. Three different substrates were used in the *in vitro* kinase reaction, which also included [□^{32}P]-ATP. The right-hand side of Fig. 2A shows the effect of the

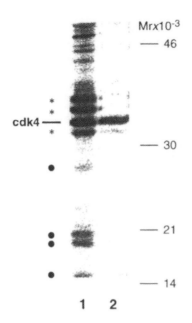

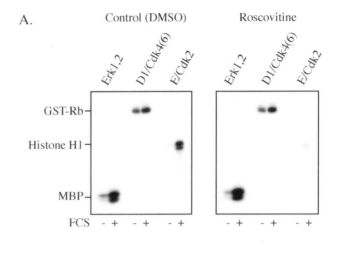

A.

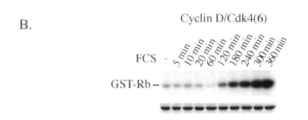

B.

FIGURE 1 Primary mouse embryonic fibroblasts were labeled with [35S]methionine, lysed in buffer 1, and immunoprecipitated with rabbit polyclonal antibody against the Cdk2 C-terminal peptide (kindly provided by Dr. M. Pagano). Lane 1 shows a labeled Cdk4 protein coprecipitated along with the associated subunits: D-type cyclins (marked by asterisks) and Cdk inhibitors (marked by closed circles). The parallel reaction in lane 2 was subsequently heated and incubated with SDS in order to dissociate specific protein–protein interactions and was reimmunoprecipitated with the same antibody.

Cdk2-specific inhibitor roscovitine when included in the final kinase reaction. Figure 2B shows a time course of activation of cyclin D/Cdk4(6) after the release of serum-starved T98G cells (a human glioblastoma cell line). It is obvious that the activity of the complex increases with time after release as cells progress through G1 and approach the G1/S phase transition. The bottom part of Fig. 2B indicates that the loading on the gel was equal for all samples by staining for the substrate GST-Rb.

FIGURE 2 Human glioblastoma cells (T98G) were starved (0.1% FCS) for 48 h before readdition of serum (10%). After 5 min a lysate was prepared for measuring Erk1 and Erk2 activation; at 6 h a lysate was made to measure cyclin D/Cdk4(6) activation; and at 12 h a lysate was made in order to measure cyclin E/Cdk2 activation. The following antibodies were used: Santa Cruz antibodies against Erk1 (Santa-Cruz SC-093) and Erk2 (Santa-Cruz SC-154) were mixed; DCS 6 mouse monoclonal antibody to cyclin D1 was used to immuno-precipitate cyclin D1/Cdk4(6) complexes; and the mouse mono-clonal antibody to cyclin E, HE172, was used to immunoprecipitate cyclin E/Cdk2 complexes. In the *in vitro* kinase reactions, myelin basic protein (MBP) was used as a substrate for Erks, whereas the GST-Rb fusion protein was used as a substrate for Cdk4(6) and histone H1 as a substrate for Cdk2. The kinase reactions in A were performed in the absence (DMSO control) or presence of the Cdk2 inhibitor roscovitine (5 μM final concentration).

References

Gavin, A. C., *et al.* (2002). Functional organization of the yeast pro-teome by systematic analysis of protein complexes. *Nature* **415**, 141–147.

Golemis, E. (2002). "Protein–Protein Interactions." Cold Spring Harbor Laboratory Press, Cold Spring Harbor, NY.

Hansen, K., Farkas, T., Lukas, J., Holm, K., Rönnstrand, L., and Bartek, J. (2001). Phosphorylation-dependent and -independent functions of p130 cooperate to evoke a sustained G1 block. *EMBO J.* **20**, 422–432.

Harlow, E., and Lane, D. (1988). "Antibodies: A Laboratory Manual." Cold Spring Harbor Laboratory Press, Cold Spring Harbor, NY.

Ho, Y., *et al.* (2002). Systematic identification of protein complexes in *Saccharomyces cerevisiae* by mass spectrometry. *Nature* **415**, 180–183.

Jenkins, C. W., and Xiong, Y. (1996). Immunoprecipitation and immunoblotting in cell cycle studies. *In* "Cell Cycle Materials and Methods" (M. Pagano, ed.), pp. 250–264. Springer Lab Manual.

Lukas, J., Bartkova, J., Rohde, M., Strauss, M., and Bartek, J. (1995). Cyclin D1 is dispensable for G1 control in retinoblastoma gene-deficient cell independently of cdk4 activity. *Mol. Cell. Biol.* **15**, 2600–2611.

Matsushime, H., Quelle, D. E., Shurtleff, S. A., Shibuya, M., and Sherr, C. J. (1994). D-type cyclin-dependent kinase activity in mamalian cells. *Mol. Cell. Biol.* **14**, 2066–2076.

Pagano, M., Pepperkok, R., Lukas, J., Baldin, V., Ansorge, W., Bartek, J., and Draetta, G. (1993). Regulation of the cell cycle by the cdk2 protein kinase in cultured human fibroblasts. *J. Cell Biol.* **121**, 101–111.

Puig, O., Caspary, F., Rigaut, G., Rutz, B., Bouveret, E., Bragado-Nilsson, E., Wilm, M., and Seraphin, B. (2001). The tandem affinity purification (TAP) method: A general procedure of protein complex purification. *Methods* **24**(3), 218–229.

Xiong, Y., Hannon, G., Zhang, H., Casso, D., Kobayashi, R., and Beach, D. (1993). p21 is a universal inhibitor of the cyclin kinases. *Nature* **366**, 701–704.

Nondenaturing Polyacrylamide Gel Electrophoresis as a Method for Studying Protein Interactions: Applications in the Analysis of Mitochondrial Oxidative Phosphorylation Complexes

Joél Smet, Bart Devreese, Jozef Van Beeumen, and Rudy N. A. Van Coster

I. INTRODUCTION

Under native PAGE conditions, polypeptides retain their higher order structure, enzymatic activity, and interaction with other polypeptides. The migration of proteins depends on many factors, including size, shape, and native charge. The resolution of nondenaturing electrophoresis is generally not as high as that of SDS–PAGE, but the technique is useful when the native structure or enzymatic activity of a protein must be preserved following electrophoresis. One straight-forward approach to native electrophoresis is to omit the sodiumdodecylsulfate (SDS) and the reducing agent dithiothreitol (DTT) from the standard Laemmli SDS protocol (Amersham Bioscience Web site: http://www4.amershambioscience.com). The separation of water-soluble proteins using the Laemmli system can simply be done by replacing SDS with Triton X. The published methods can be devided into two classes: (i) relying on the own charge of the protein to determine the anodic or cathodic migration at a given pH and (ii) using charged, mild detergents to induce a charge shift so that all proteins binding the detergent migrate in the same direction. The methods for native electrophoresis of membrane proteins, however, suffer from many drawbacks. Any new method designed to separate them must include a charge shift method. Two components were introduced for that purpose: Coomassie blue G for the separation of multiprotein complexes ("blue native PAGE") and taurodeoxycholate for the separation of lower molecular mass proteins ("native PAGE") (Schägger and von Jagow, 1991).

The blue native PAGE technique is particularly useful for the characterization of mitochondrial oxidative phosphorylation (OXPHOS) enzymes. Combining this separation technique with histochemical staining makes it possible to quantify mitochondrial enzymes *in situ* (Zerbetto *et al.*, 1997). It allows to evaluate the enzymatic activities of the complexes I, II, IV, and V in heart and skeletal muscle, liver and cultured skin fibroblasts and to detect deficiencies. Also, the amount of protein in the complexes I, II, III, IV, and V can be evaluated using silver or Coomassie staining. Often the complexes are even visible in the gel without additional staining. When the background is high, immunoblotting has to be used to visualise the amount of complex proteins (Van Coster *et al.*, 2001).

Combining blue native PAGE with SDS–PAGE reveals a two-dimensional pattern showing the individual subunits of the five OXPHOS multienzyme complexes. The implementation of mass spectrometric techniques, e.g., mass fingerprinting and mass spectrometric sequence analysis, allows the unambiguous identification of the individual subunits (Devreese *et al.*, 2002).

II. MATERIALS AND INSTRUMENTATION

Tris–HCl (Cat. No. T1378), sucrose (Cat. No. S0389), glycerol (Cat. No. G6279), phenylmethylsulfonyl fluoride (PMSF, Cat. No. S0389), dimethyl sulfoxide (DMSO, Cat. No. D5879), Tricine (Cat. No. T0377), aminocaproic acid (Cat. No. A7824), bis–Tris (Cat. No. 9754), dodecyl β-D-maltoside (Cat. No. D4641), nitro blue tetrazolium (NBT, Cat. No. N6876), phenazinemethosulfate (PMS, Cat. No. P9625), succinic acid (Cat. No. S2378), 3,3'-diaminobenzidine (DAB, Cat. No. D8001), catalase (Cat. No. C9322), cytochrome c (Cat. No. C7752), magnesium sulfate (Cat. No. M7506), lead(II) nitrate (Cat. No. L7281), adenosinetriphosphate (ATP, Cat. No. A5394), Tween 20 (Cat. No. P1379), and the Gelbond film (Cat. No. E0389) are from Sigma. Serva Blue G (Cat. No. 35050) is from Serva. Acrylamide (Cat. No. 161-0101), *N,N'*-methylenebisacrylamide (bisacrylamide, Cat. No. 161-0201), ammonium persulfate (APS, Cat. No. 161-0700), *N,N,N',N*-tetramethylenediamine (TEMED, Cat. No. 161-0800), the gel-staining Coomassie Bio-Safe solution (Cat. No. 161-0787), the blotting grade blocker (Cat. No. 170-6404), the gel-drying solution (Cat. No.161-0752), and the transparent membranes (Cat. No. 165-0963) are from Bio-Rad. SDS (Cat. No. 44215) is from BDH Laboratories. Nicotinamide adenine dinucleotide, reduced form (NADH, Cat. No. 107735), is from Roche. The silver stain kit GelCode SilverSNAP (Cat. No. 24602) and the Bradford reagent for protein determination (Cat. No. ZZ23238) are from Pierce. Whatman paper (Cat. No. 3030672) is from VWR. PVDF immunoblot membranes (Cat. RPN2020F), ECL films (Cat. No. RPN 3103K), the secondary IgG HRP-linked whole AB to mouse (Cat. No. NA931), and the ECL plus kit (Cat. No. RPN2132) are from Amersham Biosciences. The different primary antibodies against OXPHOS subunits are from Molecular Probes (http://www.probes.com). The homogeniser Model L42 is from Schwaben Präsizion Nordlingen. The glass/glass dual tube tissue homogenisers of different volumes are from Kontes, distributed by Helma Benelux. The air-driven ultra-centrifuge airfuge (Cat. No. 340401) and the microcentrifuge tubes (Cat. No. 344718) are from Analis. Minigels (gel size 8.3 × 7.3 cm) are run on a Mini-Protean three electrophoresis cell (Cat. No. 165-3302). All parts, such as glass plates, combs, casting stand, spacers, electrode assembly, tank lid with power cables, and power supply (Model 200/2.0), are from

Bio-Rad. The Mini Trans-Blot cell (Cat. No. 170-3930) and the Hydrotech gel-drying system (Model 583, Cat. No. 165-1745) equipped with a Hydrotech vacuum pump (Cat. No. 165-1782) are from Bio-Rad. The hot plate (Präzitherm, type PZ 28-2 T), the Inolab pH level 1 pH meter, and the orbital shaker (a GFL model 3006) are from VWR. Pipettes (P1000, P100, and P25) are from Hamilton. The Sharp JX-330 scanner is from Amersham Biosciences.

III. PROCEDURES

A. Isolation of Mitochondria from Tissues and Cultured Fibroblasts

This procedure is performed according to Scholte *et al.* (1992). In order to isolate a sufficient amount of mitochondria, at least 100 mg (wet weight) of tissue (muscle, heart, liver) is needed. Starting from cultured fibroblasts, after harvesting, a pellet volume of 100 μl (approximately three T75 Falcon flasks) is needed.

B. Solubilization of Mitochondria

Solutions

1. *Protease inhibitor solution*: 1 mM PMSF in DMSO. Prepare a 25-ml solution by dissolving 4.3 mg PMSF into 25 ml DMSO, aliquot in 1-ml portions, and store at −20°C.

2. *Solution A*: 750 mM aminocaproic acid, 50 mM Bis–Tris, and 20 μM PMSF. To make 25 ml, dissolve 2.46 g aminocaproic acid and 0.26 g Bis–Tris and add 20 μl of 1 mM PMSF in DMSO solution; adjust pH to 7.4 and adjust volume to 25 ml. Divide into single-use aliquots of 1 ml and store at −20°C. Samples can be stored under these conditions up to 3 months.

3. *Solution B*: Laurylmaltoside 10%. Dissolve 0.5 g in 5 ml distilled water, divide into single-use aliquots of 150 μl, and store at −20°C. Samples can be stored under these conditions maximally for 3 months.

Steps

1. Place the mitochondrial pellets (stored at −80°C) on ice.
2. Add a mixture of 96 μl solution A and 12 μl solution B equivalent to a mitochondrial pellet resulting from approximately. 100 mg of tissue or a fibroblast pellet of 100 μl.
3. Dissolve the pellet by pipetting up and down and by vortexing the tube strongly.
4. Pipette the solubilized material into a microcentrifuge tube (max 200 μl/tube).
5. Centrifugate in the airfuge at 100,000 g for 15 min.

6. Pipette the clear supernatant containing the oxidative phosphorylation enzyme complexes into an Eppendorf tube; avoid pipetting material from the pellet, which is discarded.

7. Use 5 μl of the solution to determine the protein content according to the Bradford (1970) method.

8. Store at –80°C if the supernatant cannot be submitted to electrophoresis within 2 h after preparation.

C. First-Dimension BN-Polyacrylamide Gel Electrophoresis

This is a procedure according to Schägger (1995).

Solutions

1. *Cathodal buffer (colorless)*: 50 mM tricine, 15 mM bis–Tris. To make 1 liter, dissolve 9.0 g tricine and 3.1 g bis–Tris in distilled water, adjust pH to 7.0 and complete to 1 liter with distilled water. Store for a maximum of 2 months at 4°C.

2. *Cathodal buffer (colored)*: Colorless buffer + 0.02% Serva Blue G. Add 0.2 g Serva Blue G to 1 liter of colorless cathodal buffer.

3. *Anodal buffer (5× concentrated)*: 250 mM bis–Tris. To make 1 liter, dissolve 52 g bis–Tris in distilled water, adjust pH to 7.0, and complete to 1 liter. Dilute prior to use 1/5 with distilled water. Store for 2 months maximum at 4°C.

4. *Acrylamide/bisacrylamide mixture (49.5% T, 3% C)*: To make a 250-ml solution, add 3.13 g bisacrylamide to 250 ml of the 40% solution from the supplier. Devide into 10-ml aliquots, which are stored at –20°C for 3 months maximum.

5. *Gel buffer (3× concentrated)*: 1.5 M aminocaproic acid and 150 mM bis–Tris. To make 100 ml, dissolve 19.7 g aminocaproic acid and 3.1 g bis–Tris, adjust pH to 7.0, and complete to 100 ml with distilled water. Divide into 10-ml aliquots, which are stored at –20°C for 3 months maximum.

6. *10% ammonium persulfate solution*: To make 10 ml, dissolve 1 g APS in 10 ml distilled water and divide into 1-ml aliquots, which are stored at –20°C for 3 months maximum. Use within 24 h.

7. *Sample loading buffer*: 5% Serva Blue G and 750 mM aminocaproic acid. Prepare a solution of 1 ml by dissolving 50 mg Serva Blue G and 98 mg aminocaproic acid in a final volume of 1 ml distilled water. Divide into single-use 100 μl aliquots and store them for 3 months maximum.

Steps

1. Assemble the Mini-Protean gel apparatus according to the manual of the supplier. Gel dimensions are $8.0 \times 7.3 \times 0.1$ cm.

TABLE I

Reagents	5% T	13% T	Stacking gel 4% T
A–B mixture (49.5% T–3% C)	0.8 ml	1.7 ml	0.4 ml
Gel buffer (3×)	2.1 ml	1.8 ml	1.5 ml
Glycerol	—	0.9 ml	—
AD	3.4 ml	1.0 ml	2.6 ml
Total volume	**6.3 ml**	**5.4 ml**	**4.5 ml**
APS (10%)	35 μl	18 μl	36 μl
TEMED	3.5 μl	1.8 μl	3.6 μl

2. Prepare a 5% and 13% T gel solution in order to make a 5–13% resolving gel gradient (see Table I).

3. Finally add the APS and TEMED to the 5 and 13% solutions and vortex both solutions.

4. Pipette 2× 100 μl of the 5% solution into the 13% solution and vortex.

5. Pipette 2× 100 μl between the glass plates at the left and right side of both gels.

6. Repeat steps 4 and 5 until you reach a level of approximately 1 cm under the comb.

7. Overlay with water and leave to polymerize. After a minimum of 1 h, place at 4°C overnight.

8. Prepare the 4% stacking gel solution.

9. Remove the water using Whatman paper and pour the solution on top of the resolving gel. Insert the comb (e.g., 10-well comb) and let polymerize for 45 min.

10. Poor the cold (<10°C) colored cathodal buffer between the two glass plates at a level 0.5 cm below the bottom of the sample wells.

11. Add 5 μl of sample loading buffer to 100 μl of the solubilized mitochondrial sample, which is kept on ice, and vortex.

12. Load the calculated volume (depending of the amount of protein) of the sample into the sample wells (20–100 μg of protein/lane). Two lanes are needed if you want to evaluate the activities by activity staining (see Section III,F).

13. Completely fill the space between the glass plates with colored cathodal buffer.

14. Place the gel assembly into the anodal buffer tank and poor the anodal buffer into the tank up to a level above the bottom end of the glass plates.

15. Cool the apparatus to 4°C and connect the electrodes to the power supply.

16. Run the gels at 75 V for 1 h.

17. Replace the colored cathodal buffer by the colorless buffer without rinsing the cathodal buffer chamber and run the gels at 200 V until the Serva Blue tracking dye has run off the gel completely

(approximately 3 h). The gels can be used immediately for protein staining (see Section III,D), immunoblotting (see Section III,E), activity staining (see Section III,F), or can be stored for several months at −80°C.

D. Protein Staining

1. Coomassie Staining

Steps

1. Fix the gels in a 50/40/10 mixture of distilled water/methanol/acetic acid by shaking on an orbital shaker for 30 min.
2. Stain the gels by transferring them into the Bio-Safe Coomassie staining solution and place them on the shaker for 60 min.
3. Destain the gels by transferring them in a tray filled with distilled water and put on the shaker for 60 min, refresh, and continue destaining on the shaker overnight.
4. Put the gels in a drying solution for 30 min.
5. The gels then can be dried on Whatman paper or between two transparent membranes according to the manufacturer's protocols. Normally a 2-h drying time at 80°C is sufficient (Fig. 1).

2. Silver Staining

Steps

1. Fix the gels in 50 ml of a 50/40/10 mixture of distilled water/methanol/acetic acid by shaking on an orbital shaker for 30 min.
2. Wash the gels twice for 5 min with 50 ml 10% ethanol solution and twice for 5 min with distilled water.
3. Stain the gels with 50 ml of a 50/1 mixture of the staining/enhancing solution for 30 min.
4. Rinse with distilled water for 1 min.
5. Transfer the gels into 50 ml of a 50/1 mixture of developer/enhancer solution for 2 to 5 min.
6. When the bands are clearly visible, stop the development by placing the gels in 50 ml of a 5% acetic acid solution.
7. Dry the gels on Whatman paper or between two transparent membranes according to the manufacturer's protocols. Normally a 2-h drying time at 80°C is sufficient (Fig. 1).

E. Immunoblotting

Solutions

1. *Towbin blotting buffer* (Towbin *et al.*, 1979): 10× concentrate (0.25 *M* Tris and 1.92 *M* glycine, pH 8.5).

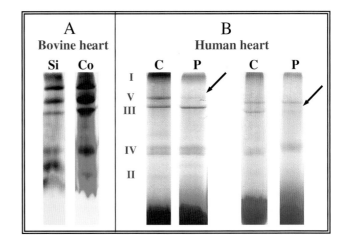

FIGURE 1   (Left) Complex proteins are visualized using protein staining (silver and Coomassie). (Right) Complex proteins are visible prior to any protein staining. Arrows mark complexes with low amounts of protein.

To make 1 liter (10×), dissolve 30.3 g Tris and 144 g glycine in distilled water, complete to 1 liter.

2. *Phosphate-saline buffer*: 10× concentrate (0.8 *M* Na_2HPO_4, 0.2 *M* NaH_2PO_4, 1 *M* NaCl). To make 250 ml (10×), dissolve 28.4 g Na_2HPO_4, 6.9 g NaH_2PO_4, and 14.5 g NaCl, complete to 250 ml with distilled water.

3. *Washing buffer (0.1% Tween 20)*: Dilute 100 ml 10× phosphate-saline buffer with 900 ml distilled water and add 1 ml Tween 20, adjust pH to 7.5.

4. *Blocking reagent*: 5% blocking reagent in washing buffer. To make 100 ml solution, dissolve 5 g blocking reagent in a final volume of 100 ml washing buffer.

5. *Primary antibody dilution*: Make a dilution of the primary OXPHOS antibodies into the blocking solution. Minimally, 20 ml is needed for incubation of the two membranes. When using the ECLplus system, a concentration of 1 μg/20 ml primary antibody is, in most cases, sufficient to detect even very low amounts of protein.

6. *Secondary antibody dilution*: Make a 1/2000 dilution of the secondary HRP antibody: 10 μl of antibody solution in 20 ml blocking solution.

7. *ECLplus detection reagent*: Place reagents A and B at room temperature for 1 h. Just prior to the detection procedure, mix reagents A and B in a 40 : 1 ratio. For two membranes, 4 ml of solution A and 100 μl of solution B are needed. Keep away from light.

Steps

1. Cut the PVDF membranes and Whatman papers (two per gel) in a size such that they completely cover the gel.

2. Dip the PVDF membranes briefly into 100% methanol and incubate them, together with the Whatman papers and the sponges, for 30 min in the Towbin blotting buffer.

3. Fill the blotting cassette with sponge–paper–gel–membrane–paper–sponge (in this order) and close the cassette. Make sure no air bubbles are trapped between the gel and the membrane.

4. Place the cassette into the holder, with the membrane facing the cathode. Insert the ice box and fill the blotting chamber with 600 ml Towbin blotting buffer.

5. Connect the electrodes and run for 1 h at 100 V.

6. Remove the membranes from the cassette, dip them briefly into methanol, and place them in 20 ml washing buffer for 30 min.

7. Incubate the membranes in 20 ml blocking solution for at least 1 h on the orbital shaker.

8. Remove the blocking solution, add 20 ml of the primary antibody solution, and incubate on the orbital shaker for a minimum of 1 h.

9. Remove the primary antibody solution and wash the membranes at least three times for 5 min with 30 ml of washing buffer on the shaker.

10. Add 20 ml of the 1/2000 dilution of the secondary HRP antibody and incubate using the orbital shaker for at least 1 h.

11. Remove the secondary antibody solution and wash the membranes at least three times for 5 min with 30 ml of washing buffer.

12. Place the membrane, protein side facing up, in a tray and add the detection reagent on the whole surface of the membrane using a pipette. Let react for 5 min. Take the membranes with a forceps, let excess reagent drip off, and fix the membranes between two sheets of transparent membrane in a film cassette.

13. Place an ECL hyperfilm on the membrane for several time intervals in order to obtain the best signal-to-noise result on the film (Fig. 2).

14. After detection, wash the membranes three times for 5 min with 30 ml of washing buffer on the shaker; these can then be stored in a sealed plastic bag for several months at –20°C.

F. In-Gel Activity Staining

This is the procedure according to Zerbetto *et al.* (1997), with certain modifications.

Staining Solutions

1. *Complex I (10×)*: 2 mM Tris–HCl, pH 7.4, 0.1 mg/ml NADH, and 2.5 mg/ml NBT. To make

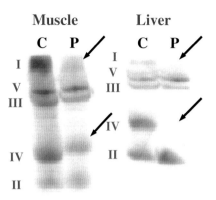

FIGURE 2 Immunoblotting of OXPHOS complexes using one specific antibody to one subunit of each complex and ECLplus detection. Arrows mark low CRM of the complex in patient lanes (P) as compared to controls (C).

25 ml of a 10× solution, dissolve 6.0 mg Tris–HCl, 2.5 mg NADH, and 62.5 mg NBT in distilled water. Adjust pH to 7.4, complete to 25 ml, and divide into 3-ml single-use aliquots, which are kept at –80°C.

2. *Complex II*: 4.5 mM EDTA, 10 mM KCN, 0.2 mM PMS, 84 mM succinic acid, and 10 mM NBT in a 1.5 mM phosphate buffer, pH 7.4. Prepare 0.1 M phosphate buffer, pH 7.4, by mixing a 100-ml solution of 1.36 g KH$_2$PO$_4$ and a 100-ml solution of 1.74 g K$_2$HPO$_4$. Prepare a 1.5 mM buffer solution by adding 98.5 ml distilled water to 1.5 ml 0.1 M buffer. To make 25 ml solution, dissolve 33 mg EDTA, 16 mg KCN, 1.5 mg PMS, 0.56 g succinic acid, and 200 mg NBT in the 0.1 M phosphate buffer, adjust pH to 7.4, complete to 25 ml, and divide into 3-ml single-use aliquots, which are kept at –80°C.

3. *Complex IV*: 5 mg DAB dissolved in 9 ml phosphate buffer (0.05 M, pH 7.4), 1 ml catalase (20 µg/ml), 10 mg cytochrome c, and 750 mg sucrose. Prepare a 0.05 M phosphate buffer by adding 13.5 ml distilled water to 13.5 ml 0.1 M phosphate buffer (see complex II). Dissolve 2 mg catalase in 100 ml distilled water. To make 30 ml solution, dissolve 15 mg DAB, 30 mg cytochrome c, and 2.25 g sucrose, add 3 ml of the catalase solution, complete to 27 ml with 0.05 M phosphate buffer, adjust pH to 7.4, and divide into 5-ml single-use aliquots, which are kept at –80°C.

4. *Complex V*: 35 mM Tris–HCl, 270 mM glycine, 14 mM MgSO$_4$, 0.2% Pb(NO$_3$)$_2$, and 8 mM ATP, pH 7.8. To make 25 ml solution, dissolve 0.11 g Tris–HCl, 0.51 g glycine, 42 mg MgSO$_4$, 50 mg Pb(NO$_3$)$_2$, and 0.11 g ATP, adjust pH to 7.8, complete to 25 ml, adjust pH to 7.4, and divide into 3-ml single-use aliquots, which are kept at –80°C.

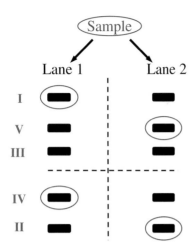

FIGURE 3 Schematic drawing of the five OXPHOS complexes and cutting of the gel.

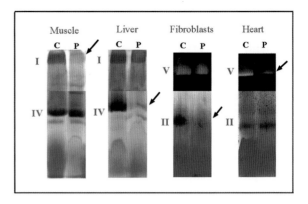

FIGURE 4 Activity staining of OXPHOS complexes in different tissues. Arrows mark deficient complexes in patients (P) in comparison with controls (C).

Steps

1. Thaw the staining solutions. Prepare a 1/10 dilution of the 10× complex I staining solution by adding 9 ml distilled water to 1 ml of the (10×) solution. Pour approximately 1 ml of each staining solution for one gel into the four staining trays (one tray/complex) (dimensions: e.g., 8 × 15 cm).

2. The gels were kept at −80°C, first thaw the gel quickly by immersing the plate in a water bath. Divide the gel as shown in Fig. 3. Using a sharp scalpel knife, divide the gels into four parts (dotted lines). The upper part of the first lane is used for complex I staining and the lower part for complex IV staining. The upper part of the second lane is used for complex V staining and the lower part for complex II staining.

3. Place the gels into the appropriate staining solutions. Cover with the rest of the staining solution and place on a hot plate at 37°C.

4. Keep the gels humid by adding 1 ml of an isotonic saline solution (0.87 g NaCl per 100 ml distilled water) at regular time intervals.

5. The bands of complexes I and II reach their maximum intensity after 3–4 h and the bands of complexes IV and V after 4–6 h.

6. Thereafter, fix the gels for 15 min in 50 ml of a mixture of 50% methanol and 10% acetic acid, except gels stained for complex V, which are rinsed with distilled water.

7. Place the gels between two Gelbond films and scan in transmission mode using a JX-330 scanner. To evaluate the white complex V band, scan the gel in the reflection mode using a black background. Staining for complex I results in a dark blue/purple band, and staining of Complex V gives a white band, both visible in the upper part of the gel. In the lower part of the gel, staining for complex IV results in a brown band and complex II staining results in a dark purple band. (Fig. 4).

8. The gels then can be dried on Whatman paper or between two transparent membranes according to the manufacturers protocols. Normally 2 h of drying time at 80°C is sufficient.

References

Bradford, M. M. (1970). *Anal. Biochem.* **74**, 248–254.

Devreese, B., Vanrobaeys, F., Smet, J., Van Beeumen, J., and Van Coster, R. (2002) Mass spectrometric identification of mitochondrial oxidative phosphorylation subunits separated by two-dimensional blue-native polyacrylamide gel electrophoresis. *Electrophoresis* **23**(15), 2525–2533.

Schägger, H. (1995). Quantification of oxidative phosphorylation enzymes after blue native electrophoresis and two-dimensional resolution: Normal complex I protein amounts in Parkinson's disease conflict with reduced catalytic activities. *Electrophoresis* **16**(5), 763–770.

Schägger, H., and von Jagow, G. (1991). Blue native electrophoresis for isolation of membrane protein complexes in enzymatically active form. *Anal. Biochem.* **199**(2), 223–231.

Scholte, H. R., Ross, J. D., Blow, W., Boonman, A. M., van Diggelen, O. P., Hall, C. L., Huijmans, J. G., Luyt-Houwen, I. E., Kleijer, W. J., de Klerk, J. B., *et al.* (1992). Assessment of deficiencies of fatty acyl-CoA dehydrogenases in fibroblasts, muscle and liver. *J. Inherit. Metab. Dis.* **15**(3), 347–352.

Towbin, H., *et al.* (1979). *Proc. Natl. Acad. Sci. USA* **76**, 4350.

Van Coster, R., Smet, J., George, E., De Meirleir, L., Seneca, S., Van Hove, J., Sebire, G., Verhelst, H., De Bleecker, J., Van Vlem, B., Verloo, P., and Leroy, J. (2001). Blue native polyacrylamide gel electrophoresis: A powerful tool in diagnosis of oxidative phosphorylation defects. *Pediatr. Res.* **50**(5), 658–665.

Zerbetto, E., Vergani, L., and Dabbeni-Sala, F. (1997). Quantification of muscle mitochondrial oxidative phosphorylation enzymes via histochemical staining of blue native polyacrylamide gels. *Electrophoresis* **18**(11), 2059–2064.

33

Affinity Purification with Natural Immobilized Ligands

Nisha Philip and Timothy A. Haystead

I. INTRODUCTION

This article discusses affinity chromatography-based techniques to study protein interaction between other proteins or small ligands. The first procedure involves the use of toxin microcystin LR (MC-LR) conjugated to a biotin or Sepharose matrix to biochemically examine phosphatase–protein interactions. MC linked to biotin has the advantage of using mild conditions to elute bound proteins and also the holoenzyme components remain intact. Using this technique we identified various PP-1-binding proteins, including regulatory subunits from different tissues (Campos *et al.*, 1996).

The second method, termed proteome mining, is employed to discover protein targets of drugs. A natural ligand, immobilized in an orientation that favors binding with its protein interactors, is used to capture a subproteome. We have, in the past, used the ligand ATP to isolate the purine-binding proteome, which includes various protein and nonprotein kinases, dehydrogenases, heat shock proteins, and other ATP using enzymes (Haystead *et al.*, 1993, 1994; Davies *et al.*, 1994; Shellman *et al.*, 1999). This process first involves the application of saturating amounts of tissue or cell extract to the matrix. Once the ATP-binding proteome is captured, various drugs are applied to the column. If the drug can compete with the natural ligand, the protein is eluted. Proteins are resolved by 1DE and identified by mass spectrometry. In our laboratory, this technique was used successfully to identify targets of antimalarials in the human purine-binding proteome (Graves *et al.*, 2002).

This method can be modified to identify protein–protein interactions in the purine-binding proteome. Once the ATP-binding proteome is captured on the resin, specific inhibitors to the protein of interest can be used to elute it and its interactors. The proteins are resolved by 1DE and identified by mass spectrometry.

II. MATERIALS AND INSTRUMENTATION

Microcystin-LR is from Alexis biochemicals (Cat. No. 350-012-M001). Ethanol is from Aaper alcohol, and dimethyl sulfoxide (DMSO) is from Mallinckrodt (Cat. No. 5507-04). Cysteamine hydrochloride (Cat. No. M6500), sodium hydroxide (Cat. No. 93065), acetonitrile (Cat. No. A998), glacial acetic acid (Cat. No. A6283), sodium bicarbonate (Cat. No. S6014), Tris-hydrochloric acid (Cat. No. T6066), EDTA (Cat. No. E5134), EGTA (Cat. No. E4378), dithiothreitol (Cat. No. D9779), Triton X-100 (Cat. No. T9284), sodium chloride (Cat. No. S9625), and magnesium chloride (Cat. No. M0250) are from Sigma. Trifluroacetic acid (Cat. No. 9470-01) is from J. T. Baker. The protease inhibitors leupeptin (Cat. No. 1017101), phenylmethylsulfonyl fluoride (PMSF, Cat. No. 0837091), aprotonin (Cat. No. 0236624), and pepstatin (Cat. No. 0253286) are from Roche. EZ-Link sulpho-NHS-biotin (Cat. No. 21217) and avidin Sepharose (Cat. No. 20347) are from Pierce. NP-40 (Cat. No. 492015) is from Calbiochem.

III. PROCEDURES

A. Synthesis of Microcystin Biotin

This procedure is modified from that of Campos (1996).

Solutions

1. *5 M NaOH stock*: Weigh 20 g NaOH and complete to 100 ml with Millipore water. Store at room temperature
2. *0.1 M NaHCO₃*: Add 8.14 g NaHCO₃ to 100 ml Millipore water
3. *Buffer A*: 25 mM Tris, pH 7.5, 1.5 mM EGTA, 4 mM EDTA, 1 mM DTT plus protease inhibitors (4 µg/ml leupeptin, 0.1 mM PMSF, 0.1 µg/ml aprotonin, and 0.1 µg/ml pepstatin)
4. *Buffer B*: Buffer A with 0.5% Triton X-100
5. *Buffer C*: Buffer A with 0.6 M NaCl and 0.5% Triton X-100

Steps

1. Dissolve 3 mg microcystin-LR in 90 µl ethanol and mix with 150 µl water, 200 µl DMSO, 100 µl cysteamine hydrochloride solution, and 67 µl 5 M NaOH (add NaOH last).
2. All components need to be degassed with nitrogen before mixing.
3. Incubate reaction at 50°C for 60 min.
4. Cool solution on ice.
5. Mix in an equal volume of glacial acetic acid (617 µl).
6. Dilute sample to 6 ml with 0.1% TFA.
7. Reduce pH to 1.5 by dropwise addition of TFA (18–20 µl).
8. Apply to C18 Sep Pak resin that has already been equilibrated in 0.1% TFA.
9. Wash with 0.1% TFA in 10% acetonitrile.
10. Elute with 10 ml of 0.1% TFA in 100% acetonitrile.
11. Dry by evaporating the solution at 45°C in a Speed-Vac.
12. Dissolve MC-AET in 200 µl methanol. Make up volume to 900 µl by adding 0.1 M NaHCO₃ (pH 8.3).
13. Dissolve 50 mg NHS-LC-biotin in DMSO.
14. Set up reaction MC-AET dissolved in methanol and NHS-LC-biotin dissolved in DMSO and rotate for 1 h at 37°C.
15. Apply to C18 Sep Pak resin equilibrated in 0.1% TFA.
16. Wash with 0.1% TFA in 10% acetonitrile.
17. Elute with 10 ml of 0.1% TFA in 100% acetonitrile.
18. Dry by evaporating at 45°C in a Speed-Vac.
19. Dissolve in 200 µl ethanol and store at −20°C.

B. Affinity Purification of PP-1 and PP-2A Holoenzymes

Steps

1. Powder frozen tissue (muscle, bladder, or liver) in liquid nitrogen.
2. Weigh 30 g wet weight of tissue and homogenize in 10 ml buffer A. Centrifuge at 15,000 g for 90 min at 4°C.
3. Keep the supernatant aside. Homogenize the pellet in 25 ml buffer B.
4. Recentrifuge at 15,000 g for 10 min at 4°C.
5. Decant supernatant off and homogenize pellet in buffer C. Incubate the homogenate for 30 min at 4°C and add equal volume of buffer C.
6. Centrifuge at 15,000 g for 30 min.
7. Dialyze the supernatant against 20 volumes of buffer A for 6 h.
8. Both cytosolic and particulate fractions are used for further analysis.
9. Preclear the endogenous biotinylated proteins in the lysate by passing over avidin–Sepharose (2 × 5-cm column) equilibrated in buffer A.
10. Wash avidin–Sepharose with 10 volumes of buffer A.
11. Mix the precleared lysate with 10 µM MC-biotin and apply to the washed avidin–Sepharose.
12. Wash the column extensively with buffer A containing 1 M NaCl followed by buffer A.
13. Elute bound proteins with buffer A containing 1 mM biotin.
14. Analyze the column fractions by SDS–PAGE and silver staining (Fig. 1).

C. Proteome Mining

This technique has been modified from Haystead (1993) and Graves (2002).

Solutions

1. *Lysis buffer*: 25 mM Tris–HCl, pH 7.5, 60 mM MgCl2, 60 mM NaCl, 1 mM DTT, 0.2% NP-40, and protease inhibitors (4 µg/ml leupeptin, 0.1 mM PMSF, 0.1 µg/ml aprotonin, and 0.1 µg/ml pepstatin).
2. *Buffer A*: 25 mM Tris–HCl, pH 7.5, 60 mM MgCl2, 60 mM NaCl, 1 mM DTT
3. *Buffer B*: Buffer A + 0.5 M NaCl

Steps

1. Homogenize 10 g of tissue in 40 ml of lysis buffer until mixture turns frothy.
2. Spin lysate in ultracentrifuge at 100,000 g for 1 h at 4°C.
3. Filter the supernatant over glass wool and also through a 0.2-µm filter.
4. Wash ATP–Sepharose resin in 50 ml low salt buffer to yield a bed volume of 1 ml.
5. Combine the supernatant with the ATP–Sepharose and rotate for 30 min at room temperature.

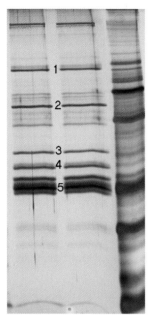

Proteins identified
1. Histone deacetylase
2. PP2A regulatory subunit
3. Actin
4. PP1 catalytic subunit
5. PP2A catalytic subunit

FIGURE 1 Microcystin–biotin eluate of rabbit muscle lysate.

6. Wash with 50 ml of buffer B followed by 50 ml buffer A.
7. Combine resin with 100 μM drug (e.g., staurosporine) in buffer A at pH 7.5 and rotate for 15 min at room temperature.
8. Collect the eluate and resolve by SDS–PAGE and silver staining.
9. Identify eluted proteins by mass spectrometry (Fig. 2).

IV. COMMENTS

To identify binding partners of proteins bound to the ATP resin, the column is subjected to milder wash conditions; 250 mM NaCl washes are performed in these cases.

V. PITFALLS

1. Wash buffers need to be cold (4°C) to minimize degradation of proteins.

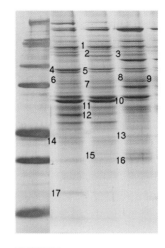

Proteins identified:
1. Glucose regulated protein 94
2. Tec-protein kinase
3. MAPK
4. MAPKK
5. Regulator of G-protein signaling
6. Glycogen synthase kinase
7. MAPKK3
9. Spectrin
10. Protein kinase B
11. FER tyrosine kinase
12. P60-Src
13. MAPK2
14. ZAP70
15. Succinate dehydrogenase
16. HSP90
17. Glucose regulatory protein

FIGURE 2 Staurosporine elution profile of ATP resin charged with human placenta.

2. MC-biotin has to be prepared fresh and is not reusable as it is lost during elution.
3. Its important to know solubility conditions of the drugs used to elute proteins off the resin, e.g., stuarosporine elutions are performed at room temperature, as the inhibitor precipitates at low temperatures.

References

Campos, M., Fadden, P., *et al.* (1996). Identification of protein phosphatase-1-binding proteins by microcystin-biotin affinity chromatography. *J. Biol. Chem.* **271**(45), 28478–28484.

Davies, S. P., Hawley, S. A., *et al.* (1994). Purification of the AMP-activated protein kinase on ATP-gamma-sepharose and analysis of its subunit structure. *Eur. J. Biochem.* **223**(2), 351–357.

Graves, P. R., Kwiek, J. J., *et al.* (2002). Discovery of novel targets of quinoline drugs in the human purine binding proteome. *Mol. Pharmacol.* **62**(6), 1364–1372.

Haystead, C. M., Gregory, P., *et al.* (1993). Gamma-phosphate-linked ATP-sepharose for the affinity purification of protein kinases: Rapid purification to homogeneity of skeletal muscle mitogen-activated protein kinase kinase. *Eur. J. Biochem.* **214**(2), 459–467.

Haystead, T. A., Haystead, C. M., *et al.* (1994). Phosphorylation of PHAS-I by mitogen-activated protein (MAP) kinase: Identification of a site phosphorylated by MAP kinase *in vitro* and in response to insulin in rat adipocytes. *J. Biol. Chem.* **269**(37), 23185–23191.

Shellman, Y. G., Svee, E., *et al.* (1999). Identification and characterization of individual cyclin-dependent kinase complexes from *Saccharomyces cerevisiae*. *Yeast* **15**(4), 295–309.

34

Analysis of Protein–Protein Interactions by Chemical Cross-Linking

Andreas S. Reichert, Dejana Mokranjac, Walter Neupert, and Kai Hell

I. INTRODUCTION

Protein–protein interactions can be studied by a variety of approaches. Frequently employed methods include copurification and coimmunoprecipitation of protein complexes. For studying molecular interactions *in vivo*, yeast two-hybrid assays (Toby and Golemis, 2001) and fluorescence-based approaches using fluorescence resonance energy transfer and bioluminescence resonance energy transfer have been developed (reviewed in Boute *et al.*, 2002; Lippincott-Schwartz *et al.*, 2001). A rather versatile method capable of detecting even weak and transient interactions is chemical cross-linking. The basis of this technique is to covalently link closely apposed proteins or protein domains by chemical cross-linkers. A large number of cross-linkers are available that differ in their selectivity of reactive groups, spacer arm length, bi- or trifunctionality, membrane permeability, solubility, cleavability, use of iodination, and the presence of affinity tags such as biotin (Pierce Biotechnology, Inc. provides a list of a wide choice of available reagents). In order to select a suitable cross-linker, these characterisitics have to be considered. Chemical properties of the protein, such as the presence of functional groups and its chemical environment, have to be taken into account. Most importantly, cross-links between two proteins can only be established when suitable side chains are present in the right distance and are accessible. Therefore, the characterization of protein–protein interactions by chemical cross-linking often requires empirical testing of a variety of different cross-linkers and reaction conditions. Photoreactive derivates can be used independent of the presence of functional side chains in the proteins to be cross-

linked; these can be incorporated co- or posttranslationally (reviewed in Brunner, 1996).

Another important issue is to identify an unknown protein that was cross-linked to the protein of interest. When a known protein is suspected to be cross-linked, several approaches can be employed. One possibility is to check whether the cross-link can be immunoprecipitated by using an antibody against the partner protein in question. Another way is to perform the cross-linking experiment in a mutant background that either lacks the candidate protein or contains a variant that can be distinguished from the endogenous protein (e.g., by the presence of a protein tag or by a variation of its apparent size). For identification of unknown binding partners, it is advisable to enrich the cross-link by affinity chromatography and subsequently identify the protein, e.g., by mass spectrometry (reviewed in Farmer and Caprioli, 1998). In that respect, one may consider the use of proteins that contain an affinity tag (e.g., GST or His_{6-10}) or to use biotinylated and/or cleavable cross-linkers.

In the following example the interaction of a mitochondrially targeted precursor protein with components of the mitochondrial inner membrane translocase complex TIM23 is investigated by chemical cross-linking. To this end, the precursor protein $pb_2\Delta19(167)DHFR_{K5}$ (Schneider *et al.*, 1994) is radiolabelled upon translation *in vitro* and incubated with isolated energized mitochondria from *Neurospora crassa* (for a review about mitochondrial import, refer to Neupert, 1997). The precursor protein is arrested during the *in vitro* import reaction at a certain stage spanning both the outer and the inner membrane. Such a trapping step is often necessary to increase the specificity and yield of the cross-linking reaction. After addition of the homobifunctional,

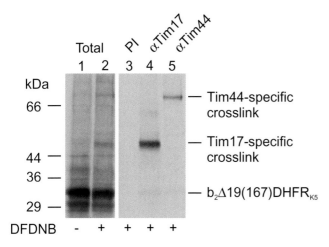

FIGURE. 1 Cross-linking of arrested b2Δ19(167)DHFR$_{K5}$ to Tim17 and Tim44. Radiolabelled b2Δ19(167)DHFR$_{K5}$ was arrested during import into isolated mitochondria from *Neurospora crassa*. After import, a portion (1/12) of mitochondria were mock treated in the absence of the cross-linker DFDNB (lane 1). The rest was subjected to cross-linking with 200 μ*M* DFDNB for 30 min on ice and another portion (1/12) of the sample was withdrawn (lane 2). Mitochondria were reisolated and used for SDS–PAGE directly (lanes 1 and 2). For immunoprecipitation of cross-linked species, the remaining portion (10/12) of the sample was lysed, split in three aliquots, and incubated with preimmune serum (lane 3) and antibodies raised against Tim17 (lane 4) or Tim44 (lane 5). Immunoprecipitates were harvested and subjected to SDS–PAGE and autoradiography. Exposure times for lanes 1 and 2 were approximately 10 times shorter than for lanes 3 to 5.

noncleavable, amine-specific cross-linker 1,5-difluoro-2,4-dinitrobenzere (DFDNB) cross-links can be detected by SDS–PAGE and autoradiography (Fig. 1). In order to identify to which of the known proteins of the TIM23 complex the precursor was cross-linked, the samples were immunoprecipited using antibodies against Tim17 and Tim44, subunits of the TIM23 complex.

II. MATERIALS AND INSTRUMENTATION

NADPH (Cat. No. 1045), NADH (Cat. No. 1051), and HEPES (Cat. No. 1009) are from Gerbu, Germany. Methotrexate (MTX, Cat. No. A6770), sorbitol (Cat. No. S-1876), and glycerol (Cat. No. G-7757) are from Sigma-Aldrich. KCl (Cat. No. 104936), Mg/acetate · 4H$_2$O (Cat. No. 105819), KH$_2$PO$_4$ (Cat. No. 104873), Na$_2$EDTA · 2H$_2$O (Titriplex, Cat. No. 108418), glycine (Cat. No. 104201), MnCl$_2$ · 4H$_2$O (Cat. No. 105927), KOH (Cat. No. 105033), Tris (Cat. No. 108382), NaCl (Cat. No. 106404), NaH$_2$PO$_4$ · H$_2$O (Cat. No. 106349), Triton X-100 (Cat. No. 108603), dimethyl sulfoxide

(DMSO, Cat. No. 102952), ethanol (Cat. No. 100983), and NaOH (Cat. No. 106467) are all from Merck KG, Germany. Sodium dodecyl sulfate (SDS, Cat. No. 20760), bromphenol blue (15375), and phenylmethyl sulforyl fluoride (PMSF, Cat. No. 32395) are from Serva, Germany. ATP · 3H$_2$O (Cat. No. 635316) and protein A–Sepharose CL-4B (Cat. No. 170963) are from Roche, Switzerland, and Amersham Biosciences, Sweden, respectively. DFDNB (Cat. No. 21525) is from Pierce Biotechnology, Inc.

III. PROCEDURES

A. Arrest of a Radiolabelled Precursor Protein as a Translocation Intermediate

Solutions

1. *0.1 mM methotrexate (MTX)*: First make a 10 m*M* stock solution by dissolving 4.54 mg in 1 ml DMSO. Store at −20°C. Dilute the stock solution 100-fold with distilled water to 0.1 m*M* MTX prior to use.

2. *100 mM NADPH*: Dissolve 8.87 mg in 100 μl distilled water. Make fresh each time.

3. *200 mM ATP*: Dissolve 60.5 mg ATP in 500 μl distilled water and adjust with 10 *M* KOH to pH 7. Make aliquots and store at −20°C.

4. *200 mM NADH*: Dissolve 14.2 mg NADH in 100 μl distilled water. Make fresh each time.

5. *2× import buffer*: Dissolve 2.38 g HEPES, 18.2 g sorbitol, 1.19 g KCl, 0.43 g Mg acetate · 4H$_2$O, 54.4 mg KH$_2$PO$_4$, 186.12 mg Na$_2$EDTA · 2H$_2$O, and 99 mg MnCl$_2$ · 4H$_2$O in 90 ml distilled water. Adjust pH to 7.2 with KOH and fill up with distilled water to 100 ml. Store at −20°C and thaw each time before use.

Steps

1. The precursor protein was synthesized in the presence of [^{35}S]methionine using reticulocyte lysate from Promega, USA, according to the manufacturer's instructions. Preincubate 50 μl of the obtained lysate containing the radiolabelled precursor protein with 0.5 μl of methotrexate (0.1 m*M*) and 1.25 μl of 100 m*M* NADPH for 10 min at 25°C.

2. Mix 600 μl 2× import buffer with 24 μl methotrexate (MTX, 0.1 m*M*), 60 μl NADPH (100 m*M*), 24 μl ATP (200 m*M*), 30 μl NADH (200 m*M*), and 390 μl distilled H$_2$O thoroughly. Add 60 μl mitochondria (10 mg/ml) freshly prepared from *N. crassa* (Sebald *et al.*, 1979), 12 μl of pretreated lysate, and mix gently.

3. Incubate for 15 min at 25°C. Stop the import reaction by placing the sample on ice.

B. Cross-Linking

Solutions

1. *20 mM DFDNB*: Dissolve 2.04 mg in 500 µl DMSO. Make fresh immediately before use.

2. *1 M glycine, pH 8.0*: Dissolve 7.51 g in 90 ml distilled water, adjust with KOH to pH 8.0, and fill up to a total volume of 100 ml with distilled water. Store at −20°C.

3. *SHKCl*: Dissolve 10.93 g sorbitol, 476 mg HEPES, and 596 mg KCl in 90 ml distilled water, adjust with KOH to pH 7.2, and fill up to 100 ml with distilled water. Store at 4°C.

4. *Laemmli buffer (without β-mercaptoethanol)*: Dissolve 1 g SDS, 5 ml glycerol, and 0.36 g Tris in 40 ml of distilled water. Adjust to pH 6.8 with HCl, add 4 mg bromphenol blue, and fill up to 50 ml with distilled water. Store at room temperature.

Steps

1. Split the import reaction into a 100-µl aliquot and an 1100-µl aliquot, representing 50 and 550 µg mitochondrial protein, respectively.

2. Add 1 µl DMSO to the 50-µg aliquot for mock treatment (total, no cross-link reagent) and 11 µl DFDNB (20 m*M* stock in DMSO) to the latter aliquot.

3. Incubate the samples for 30 min on ice.

4. Stop the cross-linking reaction by the addition of 1 *M* glycine, pH 8.0, to a final concentration of 0.1 *M* and incubate 10 min on ice.

5. Remove a 100-µl aliquot from the sample containing the DFDNB (total, plus cross-link reagent). The rest of the sample (1000-µl aliquot) can be used for identification of the cross-linked protein. As an example, see Section IIIC. Centrifuge the samples (totals, plus and minus cross-link reagent) for 10 min at 21,000 *g* at 4°C to reisolate mitochondria. Wash samples once carefully with 1 ml SHKCl and centrifuge again as described earlier.

6. Remove the supernatant and resuspend the pellet in 25 µl Laemmli buffer.

7. Resolve mitochondrial proteins and cross-linked products by SDS–PAGE. The radiolabelled cross-linked products can be visualized by autoradiography.

C. Identification of the Cross-Linked Product by Immunoprecipitation

Solutions

1. *1 M NaPi buffer, pH 8.0*: Dissolve 13.8 g $NaH_2PO_4 \cdot H_2O$ in 90 ml distilled water. Adjust pH to 8.0 with NaOH and fill up with distilled water to 100 ml. Store at 4°C.

2. *1 M NaCl*: Dissolve 58.44 g NaCl in 1 liter distilled water. Store at room temperature.

3. *10% (w/v) SDS*: Dissolve 10 g SDS in distilled water and bring to a final volume of 100 ml. Store at room temperature.

4. *20% (w/v) Triton X-100*: Dissolve 10 g Triton X-100 in distilled water and bring to a final volume of 50 ml. Store light protected at room temperature.

5. *TBS buffer*: Dissolve 1.21 g Tris and 9 g NaCl in 950 ml distilled water. Adjust pH to 7.4 with HCl and fill up with distilled water to 1 liter.

6. *SHKCl*: See solutions in Section IIIB.

7. *0.2 M PMSF*: Dissolve 34.8 mg PMSF in 1 ml ethanol. Prepare fresh each time.

8. *SDS lysis buffer*: Mix 20 µl 1 *M* NaPi buffer, pH 8.0, 100 µl 1 *M* NaCl, 100 µl 10% (w/v) SDS, and 775 µl distilled water. Finally add 5 µl 0.2 *M* PMSF.

9. *IP buffer*: Add 0.5 ml 20% (w/v) Triton X-100 to 50 ml TBS buffer.

Steps

1. For three immunoprecipitation reactions, take 75 µl protein A–Sepharose CL-4B (PAS) beads and wash three times with 1.5 ml TBS buffer.

2. Take 25 µl of PAS beads in 500 µl TBS buffer per reaction and add affinity-purified antibodies (approximately 10 µg of IgGs) against Tim17 or Tim44, or preimmune serum, respectively.

3. Incubate for at least 30 min at 4°C under gentle shaking.

4. Before immunoprecipitation, wash the PAS beads with the bound antibodies twice with 1 ml TBS and once with 500 µl IP buffer.

5. Take the 1000-µl aliquot from the cross-link reaction (Section IIIB, step 5) for immunoprecipitation and centrifuge for 10 min at 21,000 *g* at 4°C.

6. Wash isolated mitochondria with 1 ml SHKCl and centrifuge again as in step 5.

7. Lyse mitochondria by resuspension in 50 µl SDS lysis buffer and shake gently for 15 min at room temperature.

8. Dilute with Triton X-100-containing IP buffer to 1 ml.

9. Centrifuge the sample for 30 min at 125,000 *g* in a TLA45 rotor at 2°C to remove nonsolubilized and aggregated material.

10. Take the supernatant and add equal amounts to PAS beads coupled to antibodies against Tim17, Tim44, or preimmune serum, respectively (see earlier). Fill up to 1 ml with IP buffer.

11. Rotate supernatant for 2 h at 4°C.

12. Wash PAS beads twice with 1 ml IP buffer and once with 1 ml TBS.

13. Elute bound material by adding 40 µl Laemmli buffer. Keep sample for 3 min at 95°C and remove eluate from the PAS beads.

14. Analyze the precipitated material by SDS–PAGE and autoradiography.

IV. COMMENTS

One limitation of this technique is that the amounts of cross-linked as related to noncross-linked protein species are normally less than 1% in case of transient interactions but can exceed 10% with stable interactions. Therefore, it is common to halt a biological process at a defined step to increase the cross-linking yield. In the example given earlier, the precursor protein is arrested at a stage, where it spans the outer and the inner mitochondrial membrane via the translocation machineries. Complete import into mitochondria is inhibited by the C-terminal DHFR domain of the precursor protein, which is stably folded due to the presence of the substrate analogue methotrexate and the cosubstrate NADPH. Translocation intermediates can also be generated by depletion of matrix ATP or depletion of membrane potential. Import at low temperature prolongs the time of interaction. Another strategy could be to use mutants, which still have the potential to transport the precursor to a certain intermediate stage but not further.

DFDNB is a homobifunctional aryl halide-containing cross-linker, which contains two reactive fluorine atoms and reacts with amines. It should be noted that DFDNB is not completely specific for amine groups, but can also react with amino acids containing thiol, phenolate, and imidazolyl groups. Commonly used compounds to cross-link amines are the family of *N*-hydroxysuccinimide esters (NHS esters). NHS esters react with deprotonated primary amines present at the N terminus of proteins or in the side chain of lysine residues within proteins. It is important that the cross-linking reaction is performed under alkaline conditions within the pH range of 7.5 to 9 to reduce the protonation of the amine groups.

Cross-link efficiencies between interacting proteins are not predictable. Therefore, cross-link conditions have to be optimised for each protein–protein interaction. In addition to the test of various cross-linkers (e.g., BMH, DSG, DSG, EDC, MBS, and SPDP) with different reactivities and spacer lengths, the concentration of the cross-linker can be titrated and the temperature and reaction time can be modified. We normally test cross-linkers at 12°C, room temperature, or on ice in a concentration range of 0.05 to 1 m*M*.

It is also possible to introduce additional functional groups within the protein to facilitate the cross-link reaction. In the example, additional lysines were added in a position within the precursor protein, where they were likely to be in close contact to the translocation components.

V. PITFALLS

1. As stated earlier, DFDNB and NHS esters react with primary amines. Buffers containing primary amines such as Tris cannot be used as reaction buffers. They would react with the cross-linking reagent and quench the cross-linking reaction. Indeed, Tris at pH 7.5–8 is used as a quenching reagent similar to glycine. In addition, the reaction buffers should not contain high amounts of unspecific proteins, such as bovine serum albumine, whose lysines would compete for the cross-linker. When using thiol-specific reagents, reducing agents such as dithiothreitol or β-mercaptoethanol have to be omitted. For example, some [^{35}S]methionine preparations contain β-mercaptoethanol for stabilisation. These should not be used for labelling of proteins *in vitro* if the obtained lysate is directly used for a cross-linking reaction with a thiol-specific cross-linker.

2. Hydrolysis of most cross-linking reagents occurs quite rapidly. Therefore the cross-linkers have to be kept dry during storage to prevent hydrolysis. It is recommended to store the cross-linker under nitrogen once the vial is opened. As this is not always practicable, we store them in an exsiccator. To protect the cross-linker against condensing water, make sure that the vial is equilibrated to room temperature before opening.

3. In an immunoprecipition experiment the detection of cross-linked products can be complicated by the coelution of IgG chains. Under reducing conditions the light and heavy chains of IgGs are separated and have apparent molecular masses of approximately 25 and 50–60 kDa, respectively. In contrast, under nonreducing conditions, when light and heavy chains are not separated, the IgGs have a molecular mass of larger than 150 kDa. This has to be considered when a cross-linked protein is in the respective size range.

References

Boute, N., Jockers, R., and Issad, T. (2002). The use of resonance energy transfer in high-throughput screening: BRET versus FRET. *Trends Pharmacol. Sci.* **23**, 351–354.

Brunner, J. (1996). Use of photocrosslinkers in cell biology. *Trends Cell Biol.* **6**, 154–157.

Farmer, T. B., and Caprioli, R. M. (1998). Determination of protein-protein interactions by matrix-assisted laser desorption/ionization mass spectrometry. *J. Mass Spectrom.* **33**, 697–704.

Lippincott-Schwartz, J., Snapp, E., and Kenworthy, A. (2001). Studying protein dynamics in living cells. *Nature Rev. Mol. Cell Biol.* **2**, 444–456.

Neupert, W. (1997). Protein import into mitochondria. *Annu. Rev. Biochem.* **66**, 863–917.

Schneider, H. C., Berthold, J., Bauer, M. F., Dietmeier, K., Guiard, B., Brunner, M., and Neupert, W. (1994). Mitochondrial Hsp70/

MIM44 complex facilitates protein import. *Nature* **371**, 768–774.

Sebald, W., Neupert, W., and Weiss, H. (1979). Preparation of *Neurospora crassa* mitochondria. *Methods Enzymol.* **55**, 144–148.

Toby, G. G., and Golemis, E. A. (2001). Using the yeast interaction trap and other two-hybrid-based approaches to study protein-protein interactions. *Methods* **24**, 201–217.

35

Peroxisomal Targeting as a Tool to Assess Protein–Protein Interactions

Trine Nilsen, Camilla Skiple Skjerpen, and Sjur Olsnes

I. INTRODUCTION

Assessing protein–protein interactions is essential in order to elucidate the molecular mechanisms of the cell. Several methods have been developed to identify, monitor, or confirm such interactions, but they all suffer from limitations, such as being purely *in vitro* methods, not adaptable for mammalian systems, or laborious to implement. By exploiting the peroxisomal targeting machinery it is possible to assay for protein–protein interactions in living mammalian cells in a simple and affordable way. Two peroxisomal targeting signals (PTSs) have been described where PTS-1 consists of a C-terminal tripeptide (typically Ser-Lys-Leu) and PTS-2 of an N-terminal nonapeptide (Gould *et al.*, 1989; Swinkels *et al.*, 1991). Protein oligomers can be imported into the lumen of peroxisomes and, as a consequence, proteins lacking a PTS can be imported in a "piggyback fashion" (McNew and Goodman, 1994; Titorenko *et al.*, 2002).

This specific feature is exploited in the protein–protein interaction assay described here. Initially the protein of interest is targeted to the peroxisomes by adding PTS-1 to its extreme C terminus. This targeting is confirmed by colocalising the expressed PTS-tagged protein with the peroxisomal marker catalase (Fig. 1). After confirming peroxisomal targeting, cells are cotransfected with the PTS-tagged protein and a potential interacting partner lacking a PTS. Following coexpression and binding of the two overexpressed proteins in the cytosol, interacting partners can be colocalised in the peroxisomal lumen. Such colocalisation indicates that the two proteins in question bind to each other *in vivo* (Skjerpen *et al.*, 2002).

II. MATERIALS AND INSTRUMENTATION

NH_4Cl (Cat. No. 11145-1), $NaH_2PO_4 \cdot H_2O$ (Cat. No. 17157-1), $Na_2HPO_4 \cdot 12H_2O$ (Cat. No 16579-1), NaCl (Cat. No. 16404-1), and glycerol (Cat. No. 14094-1) are from VWR International.

Tris–HCl (Cat. No. T-1503), digitonin (Cat. No. D-1407), and Triton X-100 (Cat. No. T-9284) are from Sigma. The FuGENE 6 transfection reagent (Cat. No. 1 815 091) is from Roche Molecular Biochemicals, and paraformaldehyde (Cat. No. 762 40) is from Fluka Chemika. Mowiol (Cat. No. 475904) is from Calbiochem. The confocal microscope used for collecting images is a Leica TCS NT from Wezlar, Germany, and the software used for image processing is Adobe Photoshop 5.0 (Mountain View, CA).

III. PROCEDURES

A. Plasmid Preparation

Steps

1. Clone the cDNAs encoding the proteins to be analysed into plasmid vectors suitable for transient transfection.
2. Purify the plasmid DNA. Anion-exchange chromatography, using disposable columns such as Qiagen, is recommended in order to produce high-quality DNA.

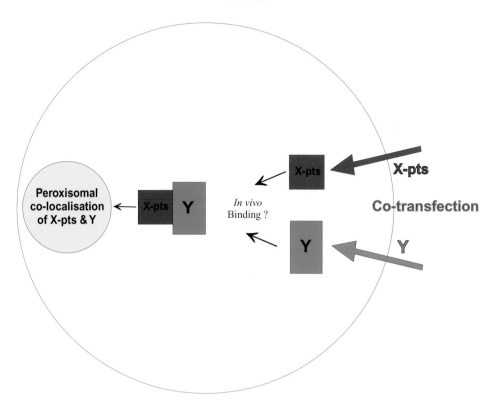

FIGURE 1 Peroxisomal colocalisation of two interacting proteins, X and Y. Protein X (red) is targeted to the peroxisomes by the addition of a peroxisomal-targeting sequence (PTS). X-pts and its potential interacting partner, Y (green), are then cotransfected into cells. If there is an *in vivo* interaction between X-pts and Y in the cytosol, Y can be cotransported into the peroxisomes after the binding takes place in a "piggyback" fashion. In this case the two proteins will colocalise in the peroxisomes as seen when merging the immuno-fluorescent images of X-pts (red) and Y (green), which will yield yellow-coloured peroxisomes.

B. Cotransfection and Confocal Microscopy Analysis

Solutions

1. *Phosphate-buffered saline (PBS)*: To make 1 litre, dissolve 0.16 g NaH$_2$PO$_4 \cdot$ H$_2$O, 1.98 g Na$_2$HPO$_4 \cdot$ 12H$_2$O, and 8.1 g NaCl in water

2. *40 µg/ml digitonin*: To make a stock solution of 400 µg/ml, dissolve 16 mg digitonin in 40 ml PBS. Sterile filter and store in aliquots at −20°C. Dilute 1 : 10 to get a working concentration of 40 µg/ml.

3. *3% paraformaldehyde*: To make 150 ml of 3% paraformaldehyde, heat 90 ml H$_2$O to 60°C and add 4.50 g paraformaldehyde. Stir for 3 h in a sterile hood and make sure not to overheat the solution. Add 2 N NaOH drop wise until the solution is clear. Add 50 ml 3× PBS (to make 250 ml 3×PBS, dissolve 0.12 g Na$_2$HPO$_4 \cdot$ 12H$_2$O, 1.49 g Na$_2$HPO$_4 \cdot$ 12H$_2$O, and 6.08 g NaCl in water, no pH adjustment) and adjust the pH to 7.2 with HCl. Add water until the total volume is 150 ml. Filter the solution and store aliquots at −20°C.

4. *50 mM NH$_4$Cl*: Make a 0.5 M stock solution by dissolving 4.01 g NH$_4$Cl in 150 ml PBS. Store at room temperature and dilute 1 : 10 to obtain a working solution of 50 mM.

5. *0.1% Triton X-100*: To make a stock solution of 10%, add 1 ml Triton X-100 to 9 ml H$_2$O. Sterile filter and store the aliquots at −20°C. Dilute 1 : 100 to get a working solution of 0.1%.

6. *Mowiol*: Add 6.7 g 87% glycerol and 2.4 g Mowiol to 6 ml H$_2$O and 12 ml 0.2 M Tris–HCl, pH 8.5. Mix for 10 min at 50°C. Centrifuge the solution for 15 min at 5000 rpm in a Sorvall RC5C centrifuge. Sterile filter and store the aliquots at −20°C.

Steps

1. Plate cells on glass coverslips in a plastic dish (typically ~2 × 10^5 cells/3.5-cm^2 well, depending on cell type) and incubate overnight at 37°C.

2. Cotransfect the cells with approximately 0.5–1.0 µg DNA of each construct by conventional methods for transient transfection, such as the FuGENE 6 transfection reagent. Allow expression for 20–24 h at 37°C.

3. All the following steps are performed at room temperature. Wash the cells in PBS for 5 min on the bench.

4. Incubate with digitonin dissolved in PBS (20–40 µg/ml, depending on cell density and cells type) for 10 min and wash the cells carefully in PBS for 5 min.

5. Fix the cells in 3% paraformaldehyde in PBS for 20–50 min at room temperature or at 4°C overnight.

6. Quench the autofluorescence with 50 mM NH$_4$Cl in PBS for 10 min.

7. Incubate the cells in 0.1% Triton X-100 in PBS for 5 min.

8. Block unspecific antibody binding sites with 5% FCS in PBS for 20 min.

9. Incubate the cells with primary antibodies. Place the coverslips with cells facing down for 20 min on a piece of Parafilm where 15 µl of the primary antibody solution has been placed. Ensure that catalase detection is included to be able to confirm peroxisomal targeting of the complex.

10. Transfer the coverslips back to the plastic dish and wash three times in PBS for 5 min.

11. Incubate with secondary antibodies as described in step 9.

12. Wash twice for 5 min in PBS.

13. Rinse the coverslips briefly in water and mount on a clean glass slide with Mowiol.

14. Collect separate immunofluorescence images of the two proteins and catalase in cotransfected cells using a confocal microscope.

15. Process the images using suitable software such as Adobe Photoshop 5.0. Merge images representing the PTS-tagged protein and catalase to confirm peroxisomal targeting in the relevant cell. Then merge the image of the protein lacking a PTS with either catalase or its potential interacting partner to analyse for colocalisation.

IV. COMMENTS

The immunofluorescent protocol described here is designed specifically to visualise the peroxisomes

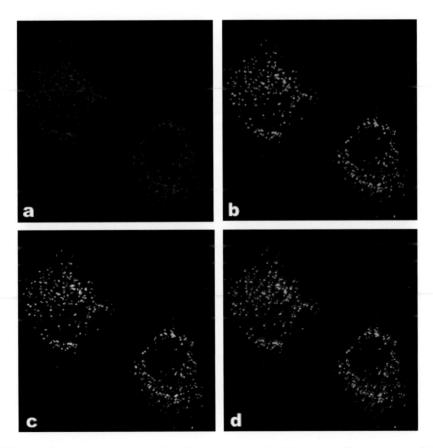

FIGURE 2 Askew imaging ensures detection of protein colocalisation specificity in merged images. (A) Confocal image representing protein X-pts. (B) Confocal image representing protein Y. (C) Merged confocal image of A and B. (D) Confocal image merged askew.

against a low background of cytosolic material. Digitonin is a cholesterol-specific detergent and will therefore leave noncholesterol-containing compartments such as the peroxisomes intact. Because the plasma membrane contains a substantial amount of cholesterol, the cytosolic material can be depleted from the cell by digitonin treatment prior to fixation. The antibodies used in immunoflourescence are given access to the organelle lumen by Triton X-100 treatment.

V. PITFALLS

1. Before initiating cloning, ensure that detection of both expressed proteins by immunofluorescence will be feasible. If no antibodies suitable for immunofluorescence microscopy are available, tags such as Myc or green fluorescence protein should be added to the construct. Adding an autofluorescent marker to at least one of the proteins will facilitate the immunofluorescent procedure as it requires triple staining.

2. It is recommended to use cell lines that are tranfected easily, as this assay is dependent on cotransfection. In principle, any transfectable cell line can be used depending on the purpose of the study.

3. Some cells lines, such as COS-1 cells, are more susceptible to digitonin treatment than others and tend to loosen from the coverslip. In such cases it is recommended to decrease the digitonin concentration.

4. If one or both of the proteins in question are associated with intracellular membranes, it may be difficult, if not impossible, to redirect these proteins to the peroxisomes. A solution to this problem can be to delete the protein domain responsible for the attachment and use the truncated soluble form in this assay. Such an approach has been successful in the detection of protein–protein interaction studies (Nilsen *et al.*, 2004).

5. It may be difficult to determine whether the colocalisation between two proteins is specific, especially if the confocal images have a high background. In order to ensure specificity, merge the two images askew by the approximate diameter of one peroxisome in the relevant image. Using Adobe Photoshop 5.0, select and copy an image approximately one peroxisomal diameter smaller on two sides of the rectangle than the full size image. Then merge it onto its corresponding image of the interacting protein. The two images will now be askew, as merging is centered automatically. When merging a red (X-pts) and a green (Y) image askew, yellow-stained peroxisomes as observed in conventional merged images will appear as red and green (Fig. 2).

Acknowledgment

T.N and C.S.S are fellows of the Norwegian Cancer Society.

References

Gould, S. J., Keller, G. A., Hosken, N., Wilkinson, J., and Subramani, S. (1989). A conserved tripeptide sorts proteins to peroxisomes. *J. Cell Biol.* **108**, 1657–1664.

McNew, J. A., and Goodman, J. M. (1994). An oligomeric protein is imported into peroxisomes in vivo. *J. Cell Biol.* **127**, 1245–1257.

Nilsen, T., Skjerpen, C. S., and Olsnes, S. (2004). Peroxisomal targeting as a tool to assay protein-protein interactions into living cell. *J. Biol. Chem.* **279**, 4794–4801.

Skjerpen, C. S., Nilsen, T., Wesche, J., and Olsnes, S. (2002). Binding of FGF-1 variants to protein kinase CK2 correlates with mitogenicity. *EMBO J.* **21**, 4058–4069.

Swinkels, B. W., Gould, S. J., Bodnar, A. G., Rachubinski, R. A., and Subramani, S. (1991). A novel, cleavable peroxisomal targeting signal at the amino-terminus of the rat 3-ketoacyl-CoA thiolase. *EMBO J.* **10**, 3255–3262.

Titorenko, V. I., Nicaud, J. M., Wang, H., Chan, H., and Rachubinski, R. A. (2002). Acyl-CoA oxidase is imported as a heteropentameric, cofactor-containing complex into peroxisomes of *Yarrowia lipolytica*. *J. Cell Biol.* **156**, 481–494.

Biomolecular Interaction Analysis Mass Spectrometry

Dobrin Nedelkov and Randall W. Nelson

I. INTRODUCTION

Biomolecular interaction analysis mass spectrometry (BIA/MS) is multidimensional methodology for functional and structural protein analysis (Krone *et al.*, 1997; Nelson *et al.*, 1997a,b, 2000a,b). In essence, BIA/MS represents a synergy of two individual technologies: surface plasmon resonance (SPR) sensing and matrix-assisted laser desorption/ionization time-of-flight (MALDI-TOF) mass spectrometry. SPR is employed for quantification, whereas MS is utilized to delineate the structural features of the analyzed proteins. Proteins are affinity captured and quantified from solution via ligands covalently attached on the SPR sensor surface. Because the SPR detection is nondestructive, proteins retrieved on the SPR sensing surface can be further analyzed via mass spectrometry, either directly from the sensor/chip surface (as described in most of the publications from our laboratory) or separately, following elution and microrecovery (Gilligan *et al.*, 2002; Nelson *et al.*, 1999; Sonksen *et al.*, 1998, 2001). The combination of SPR with MS overcomes the limitation of nondiscriminatory SPR detection and allows for elucidation of nonspecific binding (NSB) of other biomolecules to the surface-immobilized biomolecules (or to the underivatized sensor surface itself), binding of protein fragments, protein variants (existing due to posttranslational modifications and point mutations), and complexed (with other molecules) proteins. The BIA/MS approach has been utilized for isolation, detection, and identification of epitope-tagged proteins (Nelson *et al.*, 1999), detection of food pathogens (Nedelkov *et al.*, 2000), analysis of human urine protein biomarkers (Nedelkov and Nelson, 2001a), delineation of *in vivo*

assembled multiprotein complexes (Nedelkov and Nelson, 2001c; Nedelkov *et al.*, 2003), and screening for protein functionalities (Nedelkov and Nelson, 2003). Detection of attomole amounts of proteins from complex biological mixtures is possible via the combined SPR-MS approach (Nedelkov and Nelson, 2000a). The modus operandi of the BIA/MS approach is illustrated via the analysis of apolipoproteins A-I and A-II (apoA-I and apoA-II) from human plasma sample.

II. MATERIALS AND INSTRUMENTATION

Biacore X Biosensor (Biacore AB, Uppsala, Sweden) is utilized for the protein affinity retrieval and SPR analysis. Ligands (e.g., antibodies) are immobilized on the carboxymethyldextran surface of a CM5 research grade sensor chip (Biacore AB, Cat. No. BR. 1000-14) using amine-coupling kit chemicals (Biacore AB, Cat. No. BR-1000-50).

A chip cutter with a circular heated cutter head (made in our laboratory) is used for excising a chip/plastic mount of a defined circular shape that fits into an appropriately configured MALDI mass spectrometer target (Nedelkov and Nelson, 2000b).

The MALDI matrix is applied to the chip via an aerosol-spraying device (also developed in our laboratory). The device consists of an aspirating/sheath gas needle (~30-μm orifice), backed by ~30 psi of compressed air (Nedelkov and Nelson, 2000b). The air/matrix solution ratio can be adjusted to produce a fine mist of matrix solution that is aimed at the entire surface of the cutout chip. The matrix of choice is α-cyano-4-hydroxycinnamic acid (ACCA, Aldrich,

Milwaukee, WI, Cat. No. 47,687-0), which is further processed by powder-flash recrystallization from a low-heat-saturated acetone solution of the original stock.

MALDI-TOF mass spectrometry analysis from the chip surface is performed on a custom-made MALDI-TOF mass spectrometer (Intrinsic Bioprobes, Inc.). The instrument consists of a linear translation stage/ion source capable of precise targeting of each of the flow cells under a focused laser spot. Ions generated during a 4-ns laser pulse (357 nm, nitrogen) are accelerated to a potential of 30 kV over a single-stage ion extraction source distance of ~2 cm before entering a 1.5-m field-free drift region. The ion signals are detected using a two-stage hybrid (channel plate/discrete dynode) electron multiplier. Time-of-flight spectra are produced by signal averaging of individual spectra from 50 to 100 laser pulses (using a 500-MHz; 500-Ms/s digital transient recorder). Custom software is used in acquisition and analysis of mass spectra. All spectra are obtained in the positive ion mode.

III. PROCEDURES

A. SPR Analysis and Protein Affinity Retrieval

Ligands, Solutions, and Buffers

1. *Ligands*: When using antibodies (or other proteins) as ligands, dilute the original stock solutions with 10 mM acetate buffer, pH 5.0, to a final ligand concentration of 0.01–0.1 mg/mL.

2. *Buffers and solutions*: Ultrapure, molecular biology grade, sterile water (American Bioanalytical, Natick, MA, Cat. No. 7732-18-5) is used for solution making. HBS-EP [0.01 M HEPES, pH 7.4, 0.15 M NaCl, 0.005% (v/v) polysorbate 20, 3 mM EDTA] is typically used as a running buffer in the SPR biosensor. From the amine-coupling kit, make 400 mM solution of EDC [N-ethyl-N'-(dimethylaminopropyl)carbodiimide] in the ultrapure water, 100 mM solution of N-hydroxysuccinimide (NHS), and use the 1 M ethanolamine (pH 8.5) solution as supplied. Make 60 mM HCl for ligand surface regeneration.

Steps

1. Remove a new CM5 sensor chip from its packaging, take out the chip from the plastic housing cassette, wash it with five 200-µl aliquots of water, dry, and put the chip back in the cassette.

2. Insert the chip into the Biosensor (via the dock command), prime, set the flow rate at 5 µl/min, and let it equilibrate with the HBS-EP running buffer at 5 µl/min for 10–20 min.

3. Set the flow to a single flow cell and start the amine-coupling surface preparation procedure from the Biosensor software. Mix 35 µl of the EDC solution and 35 µl of the NHS solution in a small vial, and inject 35 µl of the mixture over the flow cell surface to activate the carboxyl groups of the dextran matrix.

4. After the surface activation injection has ended, inject 70 µl of the antibody solution. Monitor the SPR response for a sharp increase that will indicate binding of the antibody to the chip surface.

5. Inject 35 µl of the ethanolamine solution to block free (unreacted) esters.

6. Inject 20 µl of the HCl solution to release any noncovalently attached antibody.

7. Measure the SPR response (in resonance units, RU) at the end of the EDC/NHS injection and subtract it from the final SPR response measured after the HCl injection to yield an accurate estimate on the total amount of antibody immobilized on the surface of the flow cell. Because 1 RU equates to 1 pg of proteinacious material per 1 mm^2 of the flow cell surface (the FC dimensions are 0.5 × 2 mm), a response of ~15,000 RUs at the end of the injection, indicating the immobilization of ~100 fmol antibody (MW$_{antibody}$ ~ 150,000, 1 RU = 1 pg protein), is generally satisfactory.

8. Repeat steps 2–6 for immobilization of the other antibody in the second flow cell. Alternatively, leave the second flow cell underivatized and use as control.

9. Switch the flow to both flow cells, and let the antibody surface equilibrate with the running buffer for 10–20 min.

10. Inject a 50-µl aliquot of sample (e.g., plasma sample diluted appropriately) and record the SPR response at the end of the injection.

11. Stop the flow of the buffer (set the flow rate at 0 µl/min) and quickly remove (undock) the chip from the biosensor.

12. Wash the chip with three 200-µl aliquots of ultrapure water and dry it under a stream of nitrogen.

B. MALDI-TOF Mass Spectrometry Analysis of Chips

Solution

Make a fresh solution of the MALDI matrix [aqueous solution of ACCA in 33% (v/v) acetonitrile and 0.4% (v/v) trifluoroacetic acid]

Steps

1. Place a piece of transparent tape on the glass side of the chip so that it covers both the glass chip itself and the surrounding plastic support area.

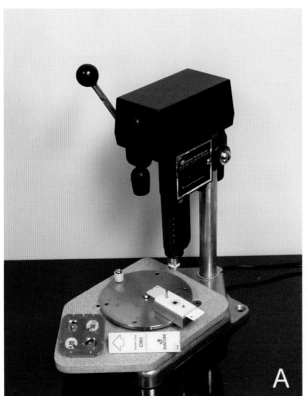
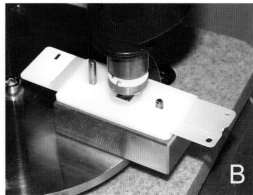

FIGURE 1. (A and B) IBI's chip cutter device. (C) The cutout chip fits into appropriately configured MALDI targets.

2. Warm up the chip cutter and place the chip in the positioning holder, with the active side facing down (**Fig. 1A**). Rotate the holder to position the chip underneath the heater head (**Fig. 1B**) and gently push down the circular heated head onto the chip. After few seconds, release back the cutter head, rotate the holder, and remove the chip.

3. To separate the round chip/support piece from the rest of the plastic support, gently press on one side of the circle cutout.

4. Remove the tape from the back of the chip cutout and position the chip on a flat surface in a chemical hood, with the active side facing up.

5. Fill in the solution holder on the sprayer device with matrix solution and adjust appropriately the air-to-matrix solution ratio on a test surface. Then, position the device ~5 cm from the chip surface and, in one swift motion, spray the matrix evenly over the entire chip surface (**Fig. 2A**). The matrix mist should moisten, but not completely wet the chip surface (i.e., the tiny matrix droplets should stay as individual drops on the surface and not be connected into one large liquid drop). The matrix droplets will desorb the proteins from their respective capturing affinity ligand and,

upon rapid drying, the matrix/protein mixture will be redeposited on the same area from where the proteins were captured originally in the SPR analysis (**Figs. 2B and 2C**). To prevent fast initial evaporation, cover briefly (for 10–20 s) the chip with a small cap and then let it go to dryness.

6. Place a double-stick tape on the back of the chip and place the chip in the MALDI target so that it is firmly positioned and attached in the probe holder (**Fig. 1C**). Carefully insert the probe into the mass spectrometer. Target each flow cell individually with the laser and acquire mass spectra.

7. Following the MS analysis, chips can be stored shortly at room temperature for further analysis and reevaluation. After a week or so, the quality of mass spectra obtained from the stored chips deteriorates significantly.

D. Example

As an example of concerted BIA/MS analysis, we show the investigation of two human apolipoproteins from human plasma. Antibodies to apolipoprotein A-I (Cat. No. 11A-G2B, 1 mg/ml) and apolipoprotein

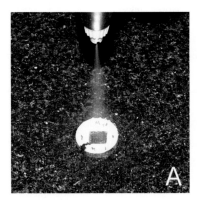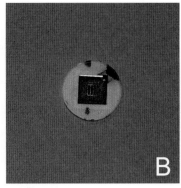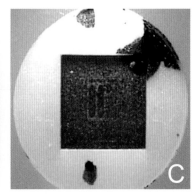

FIGURE 2.   (A) Post-SPR spraying of the chip with MALDI matrix. (B and C) Dried-out protein/matrix mix chip surface showing the preserved spatial resolution between the two flow cells.

A-II (Cat. No. 12A-G1B, 1 mg/ml) were purchased from Academy Biomedical (Houston, TX). Human blood was obtained from a single subject recruited within Intrinsic Bioprobes Inc. (IBI), following a procedure approved by the IBI's Institutional Review Board (IRB), and after signing of an informed consent form. Human blood (45 µl) was drawn under sterile conditions from a lancet-punctured finger with a heparinized microcolumn (Drummond Scientific Co., Broomall, PA), mixed with 200 µl of HEPES-buffered saline (HBS-EP) buffer [0.01 M HEPES, pH 7.4, 0.15 M NaCl, 0.005% (v/v) polysorbate 20, 3 mM EDTA], and centrifuged for 30 s (at 7000 rpm, 2500 g) to pellet red blood cells. The supernatant (plasma) was further diluted 10-fold with HBS-EP buffer to yield plasma sample diluted 100-fold.

Figure 3a shows the immobilization of anti-apoA-I and anti-apoA-II in flow cell 1 (FC1) and flow cell 2 (FC2), respectively. The SPR responses indicate immobilization of ~100 fmol of antibody in each flow cell. Figure 3b shows a sensorgram resulting from the injection of a 50-µl aliquot of the 100-fold diluted plasma over both flow cells in series. Responses of ~500 RUs are indicated in both flow cells. Figures 3c and 3d show mass spectra obtained from the surfaces of the two flow cells following plasma sample injection. The mass spectrum obtained from FC1 contains multiply charged ions from apoA-I, in line with the immobilized antibody specificity. Two minor signals due to apoC-I and apoC-I′ are also seen and can be attributed to nonspecific binding to the carboxymethyldextran surface (Nedelkov and Nelson, 2001b). The spectrum obtained from the surface of FC2 contains multiply charged signals from apoA-II. Three major peaks are observed for each charge state: (1) cysteinylated form of apoA-II (cys-apoA-II, MW 8827) (apoA-II contains a single cysteine residue, which is cysteinylated readily

in vivo), (2) cys-apoA-II missing one terminal glutamine residue [cys-apoA-II (-Gln), MW 8699] (both the C and the N apoA-II termini residues are glutamines), and (3) an apoA-II missing one terminal Gln residue [apoA-II (-Gln), MW 8580]. The native apoA-II (MW 8708) signal is most likely concealed in the mass spectrum by the strong cys-apoA-II (-Gln) peak. The apoA-II homodimer was barely observed in the spectrum (region not shown). Similar results were obtained in another study using a different approach to apolipoprotein extraction and affinity retrieval (Niederkofler *et al.*, 2003).

IV. COMMENTS

Protein modifications (at the native MW level) can be assessed rapidly via BIA/MS, as shown in the apoA-II example. Similarly, protein complexes can be delineated by the observance of signals from the constituent complex components in the MS analysis, and SPR detection can be utilized to monitor specific protein–protein interactions.

V. PITFALLS

1. Contrary to typical kinetic SPR analysis, it is recommended that high ligand densities are utilized in the initial stage of BIA/MS so that ample amounts of analyte are captured for subsequent MALDI-TOF MS analysis.

2. Lower flow rates (promoting mass transfer effects) should also be utilized to increase the binding of the analyte to the immobilized ligands. At high flow

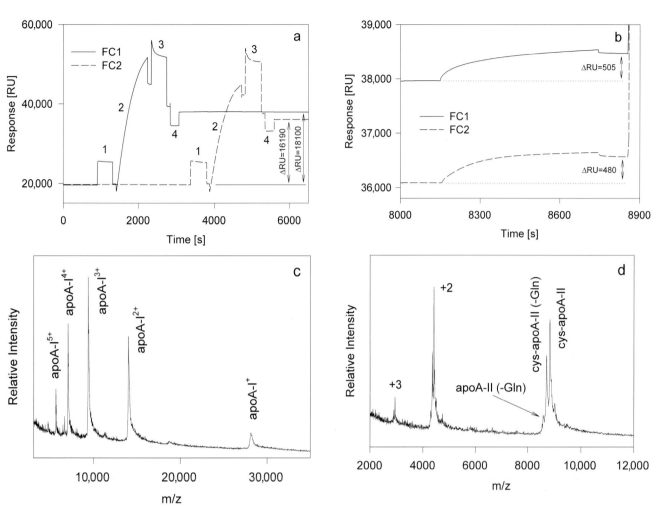

FIGURE 3. **(a)** SPR sensorgrams showing the immobilization of anti-apoA-I and apoA-II antibodies in FC1 and FC2, respectively, via (1) EDC/NHS surface activation, (2) antibody injection, (3) ethanolamine blocking, and (4) HCl for noncovalently attached antibody stripping. **(b)** SPR sensorgram showing the injection of 50 μl of 100-fold diluted human plasma over both flow cells. **(c)** Mass spectrum taken from surface FC1 showing the presence of multiply charged apoA-I ions. **(d)** Mass spectrum taken from the surface of FC2 showing the presence of multiple forms of apoA-II.

rates (60–100 μl/min) low-concentration analytes will not be captured in amounts permissible to downstream MS analysis.

3. Following sample analysis, the chip should be removed promptly from the biosensor to avoid losses of the captured proteins (especially when the interaction system under study exhibits fast dissociation phase). When possible, the undocking should be executed from within the root control software of the biosensor (OS9). For the same reasons, lengthy postcapture washes of the chip should be avoided.

4. Generally, higher quality mass spectra are obtained with ACCA for proteins smaller than ~25 kDa, whereas sinapic acid yields better results for

higher molecular mass (>25 kDa) proteins. For the BIA/MS, however, ACCA is superior in that it is a better energy-absorbing matrix and, consequently, requires less laser power. Lower laser power means that more spectra can be obtained from a single spot and the fast burning through the matrix/sample layer is avoided. Generally, the appearance and the intensity of the multicharged ion analyte signals obtained with ACCA are better indicators of the presence and the mass of the on-chip retained analyte during the BIA/MS analysis.

5. Reapplication of more matrix (following initial application and MS analysis) does not yield better signals and generally results in a decreased signal-to-noise ratio.

References

Gilligan, J. J., Schuck, P., and Yergey, A. L. (2002). Mass spectrometry after capture and small-volume elution of analyte from a surface plasmon resonance biosensor. *Anal. Chem.* **74**, 2041–2047.

Krone, J. R., Nelson, R. W., Dogruel, D., Williams, P., and Granzow, R. (1997). BIA/MS: Interfacing biomolecular interaction analysis with mass spectrometry. *Anal. Biochem.* **244**, 124–132.

Nedelkov, D., and Nelson, R. W. (2000a). Exploring the limit of detection in biomolecular interaction analysis mass spectrometry (BIA/MS): Detection of attomole amounts of native proteins present in complex biological mixtures. *Anal. Chim. Acta* **423**, 1–7.

Nedelkov, D., and Nelson, R. W. (2000b). Practical considerations in BIA/MS: Optimizing the biosensor-mass spectrometry interface. *J. Mol. Recogn.* **13**, 140–145.

Nedelkov, D., and Nelson, R. W. (2001a). Analysis of human urine protein biomarkers via biomolecular interaction analysis mass spectrometry. *Am. J. Kidney Dis.* **38**, 481–487.

Nedelkov, D., and Nelson, R. W. (2001b). Analysis of native proteins from biological fluids by biomolecular interaction analysis mass spectrometry (BIA/MS): Exploring the limit of detection, identification of non-specific binding and detection of multi-protein complexes. *Biosens. Bioelectron.* **16**, 1071–1078.

Nedelkov, D., and Nelson, R. W. (2001c). Delineation of *in vivo* assembled multiprotein complexes via biomolecular interaction analysis mass spectrometry. *Proteomics* **1**, 1441–1446.

Nedelkov, D., and Nelson, R. W. (2003). Delineating protein-protein interactions via biomolecular interaction analysis-mass spectrometry. *J. Mol. Recogn.* **16**, 9–14.

Nedelkov, D., Nelson, R. W., Kiernan, U. A., Niederkofler, E. E., and Tubbs, K. A. (2003). Detection of bound and free IGF-1 and IGF-2 in human plasma via biomolecular interaction analysis mass spectrometry. *FEBS Lett.* **536**, 130–134.

Nedelkov, D., Rasooly, A., and Nelson, R. W. (2000). Multitoxin biosensor-mass spectrometry analysis: A new approach for rapid, real-time, sensitive analysis of Staphylococcal toxins in food. *Int. J. Food Microbiol.* **60**, 1–13.

Nelson, R. W., Jarvik, J. W., Taillon, B. E., and Tubbs, K. A. (1999). BIA/MS of epitope-tagged peptides directly from *E. coli* lysate: Multiplex detection and protein identification at low-femtomole to subfemtomole levels. *Anal. Chem.* **71**, 2858–2865.

Nelson, R. W., Krone, J. R., and Jansson, O. (1997a). Surface plasmon resonance biomolecular interaction analysis mass spectrometry. 1. Chip-based analysis. *Anal. Chem.* **69**, 4363–4368.

Nelson, R. W., Krone, J. R., and Jansson, O. (1997b). Surface plasmon resonance biomolecular interaction analysis mass spectrometry. 2. Fiber optic-based analysis. *Anal. Chem.* **69**, 4369–4374.

Nelson, R. W., Nedelkov, D., and Tubbs, K. A. (2000a). Biomolecular interaction analysis mass spectrometry: BIA/MS can detect and characterize proteins in complex biological fluids at the low- to subfemtomole level. *Anal. Chem.* **72**, 404A–411A.

Nelson, R. W., Nedelkov, D., and Tubbs, K. A. (2000b). Biosensor chip mass spectrometry: A chip-based proteomics approach. *Electrophoresis* **21**, 1155–1163.

Niederkofler, E. E., Tubbs, K. A., Kiernan, U. A., Nedelkov, D., and Nelson, R. W. (2003). Novel mass spectrometric immunoassays for the rapid structural characterization of plasma apolipoproteins. *J. Lipid Res.* **44**, 630–639.

Sonksen, C. P., Nordhoff, E., Jansson, O., Malmqvist, M., and Roepstorff, P. (1998). Combining MALDI mass spectrometry and biomolecular interaction analysis using a biomolecular interaction analysis instrument. *Anal. Chem.* **70**, 2731–2736.

Sonksen, C. P., Roepstorff, P., Markgren, P. O., Danielson, U. H., Hamalainen, M. D., and Jansson, O. (2001). Capture and analysis of low molecular weight ligands by surface plasmon resonance combined with mass spectrometry. *Eur. J. Mass Spectrom.* **7**, 385–391.

37

Blot Overlays with ^{32}P-Labeled GST-Ras Fusion Proteins: Application to Mapping Protein–Protein Interaction Sites

Zhuo-shen Zhao and Edward Manser

I. INTRODUCTION

A well-characterized system includes ligand–receptor interactions, cell adhesion events, antigen recognition, and virus–host recognition: inside the cell the formation of multiple protein complexes during the assembly of cytoskeletal elements and interplay of proteins in signal transduction pathways are some of the best studied. In recent years, a number of popular techniques have evolved to analyze protein–protein interactions. Among these, immunoprecipitation, yeast two-hybrid analysis (Bartel and Fields, 1995; Fields and Song, 1989) and Western overlay assays are the most commonly used.

Protein overlays allow the researcher to visualize protein interactions on a blot and only require that the target be in the correct conformation for binding following SDS electrophoresis and transfer to a solid-phase membrane. The primary challenge is to produce a sensitive protein "probe" by recombinant techniques. This article describes the generation and use of such a probe for overlay binding in which the small GTP-binding protein Ras is used as an acceptor domain that can be labeled rapidly with [γ-^{32}P]GTP. In brief, target proteins of interest are first separated by SDS–polyacrylamide electrophoresis and are immobilized on a PVDF membrane. A radiolabeled Ras fusion protein is incubated with this membrane and, after a suitable time, the blot is processed to detect specific interaction (Fig. 1) as first described (Zhao *et al.*, 2000a). This assay can be used to identify specific binding partners in a complex mixture of proteins or for examining interactions between two proteins to determine a target site. In some cases where protein solubility is a problem, SDS–PAGE is the only route of protein isolation, as immunoprecipitation can only tolerate mild detergents. When protein expression in yeast is toxic, the protein overlay technique provides a useful adjunct to two-hybrid methods. A key advantage of protein overlays is their flexibility. Proteins expressed by a variety of means can be subject to SDS electrophoresis and overlay: total cell extracts, relatively pure recombinant proteins, or even conjugated peptides can be used as targets. Finally, proteins or peptides can be synthesized with specific posttranslational modifications or subsequently modified (e.g., by phosphorylation) and then assessed for changes to the protein–protein interaction in a quantitative manner.

II. MATERIALS AND INSTRUMENTATION

pGEX-2TK vector and glutathione–Sepharose 4B beads are from Amersham Pharmacia Biotech (Piscataway, NJ; Cat. Nos. 27-4587-01 and 17-0756-01,). [γ-^{32}P]GTP (~6000 Ci/mmol) is from NEN Life Science Products (Boston, MA, Cat. No. BLU/NEG/

285

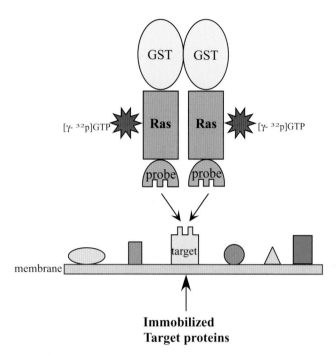

FIGURE 1 A schematic diagram showing the Ras overlay assay. The GST–Ras–probe fusion protein forms a dimer through the GST tag (in light yellow), which further stabilizes the probe–target interaction. The GTPase-deficient Ras mutant (in orange) allows high-affinity radiolabelled GTP (in red) binding.

004Z). The PVDF membrane is from Perkin Elmer NEN (Cat. No. NEF1002). Acrylamide/bisacrylamide is from Genomic Solutions Inc. (Ann Arbor, MI, Cat. No. 80-0084), Ammonium persulfate and N,N,N',N'-tetramethylenediamine (TEMED) are from Bio-Rad (Cat. No. 161-0700 and Cat. No. 161-0800, respectively). Glycine (Cat. No. G-7126) and Trizma base (Cat. No. T-1503) are from Sigma, and SDS (Cat. No. 44215HN) is from BDH. Mini-PROTEAN II module including gel casts, glass plates, spacers (1.5 mm thickness), combs, and running tank are from Bio-Rad (Cat. No. 165-2944).

III. PROCEDURES

A. Construction of Expression Vector for Probe Labeling

The cDNA fragment encoding GTPase-deficient mutant H-Ras^{G12V} residues 1–185 was amplified by polymerase chain reaction (PCR) with forward primer: 5' gaagatctatgacggaatataagctggtgg 3' and reverse primer: 5' gggaattcggggatccccttgcagctcatgcagccggg 3'. The PCR product was digested with BglII (introduced by forward primer) and EcoRI (introduced by reverse

primer) and ligated to a BamHI- and EcoRI-linearized pGEX-4T-1 vector. The resulting vector pGEX-Ras allows us to express a given cDNA downstream of the multiple cloning sites as a GST-Ras^{G12V} fusion protein with the cDNA product C-terminal to the other two domains. The fusion protein is labeled with $[\gamma\text{-}^{32}P]GTP$ via the Ras GTP-binding module.

B. Purification of GST-Ras-Probe Fusion Protein for Labeling

Solutions

1. *Bacteria lysis buffer*: 50 mM Tris (pH 8.0), 0.5% (w/v) Triton X-100, 0.5 mM MgCl$_2$, and 133 mM NaCl. This solution can be kept at 4°C for at least 6 months. Add 1 mg/ml lysozyme, 0.5 mM phenylmethylsulforyl fluoride, 5 mM dithiothreitol (DTT) and 1 tablet of Roche's complete protease inhibitor cocktail (Cat. No. 1697498) freshly before use.
2. *Washing buffer*: Phosphate-buffered saline (PBS) with 0.1% (w/v) Triton-X 100
3. *Elution buffer*: 50 mM Tris–Cl (pH 8.5), 10% (v/v) glycerol in PBS. Add freshly glutathione (reduced form) to 1.5 mg/ml and DTT to 5 mM before use.
4. *Bradford solution*: Dilute 1 : 5 of Bio-Rad protein assay solution (Cat. No. 500-0002)

Steps

1. Subclone the cDNA fragment encoding protein to be used as a probe to the pGEX-Ras vector in frame with the GST-Ras^{G12V} cassette. Transform the resultant plasmid into *Escherichia coli* BL21 strain (from Stratagene, Cat. No. 200133) and plate on a LB agar plate containing 100 μg/ml ampicillin. Incubate the transformation plate at 37°C overnight.
2. Inoculate five colonies into a 1-liter flask containing 400 ml of LB broth with 100 μg/ml ampicillin. Grow the culture in a 37°C incubator with shaking for approximately 4 h until the absorbance of OD$_{600\text{ nm}}$ reaches ~ 1.0.
3. Induce protein expression by adding IPTG to a final concentration of 0.25 mg/ml (4 ml of a 25-mg/ml stock per 400-ml culture flask). Incubate with shaking for 4 h at room temperature.
4. Collect cells by centrifugation (Sorvall RC 5C and GS3 rotor) at 6000 rpm for 10 min and decant the supernatant. The pellet can be processed immediately or be stored at −80°C for future use.
5. Resuspend the cell pellet completely in 40 ml of ice-cold bacteria cell lysis buffer and keep the cell suspension on ice for 5 min.
6. Sonicate the cell suspension (3 × 20 s), allowing 20 s for cooling between each ultrasonic burst of the

XL sonicator (Heat System-Ultrasonics Inc., Farmingdale, NY). During sonication, keep the suspension on ice and avoid foaming.

7. Remove cell debris by centrifugation for 30 min at 40,000 rpm, 4°C using a Ti50.2 rotor (Beckman).

8. Load 400 µl of glutathione–Sepharose 4B into a Poly-Prep chromatography column (Bio-Rad Cat. No. 731-1550). Wash the column twice with 10 ml of washing buffer.

9. Pass the centrifuge cleared supernatant through the glutathione–Sepharose 4B column. Wash the column two times with 10 ml of ice-cold washing buffer.

10. Elute the GST fusion protein in stepwise elution buffer containing reduced glutathione in 200-µl fractions.

11. Determine the protein concentration using the Bradford assay and aliquot and store in aliquots of 1 mg/ml at −80°C.

C. Labeling Fusion Protein Probe with [γ-^{32}P]GTP

Solution

Nucleotide exchange buffer (2×): 100 mM NaCl, 50 mM HEPES (pH 7.3), 10 mM EDTA, 1 mg/ml bovine serum albumin (BSA), and 0.1% (w/v) Triton X-100. Filter the buffer through a 0.2-µm syringe filter (Sartorius, Cat. No. 11107-25-N) and store in aliquots at −20°C.

Step

Add 10 µg of GST-Ras^{G12V}-probe fusion protein into a 1.5-ml Eppendorf tube containing 25 µl of 2× exchange buffer and 5 µCi of [γ-^{32}P]GTP. Make up the volume to 50 µl. Incubate the reaction at room temperature for 5 min and return to ice until use.

D. Preparation and Transfer of Proteins to Membrane

Solution

Semidry transfer buffer: Per 1 liter, add 6 g Trizma base (Sigma, Cat. No. T1503), 3 g glycine, and 10% (v/v) methanol in distilled water

Steps

1. Separate protein samples containing target proteins on a SDS–PAGE gel and blot onto a PVDF membrane using a Bio-Rad semidry blotter.
2. Stain the membrane with 0.05% Coomassie blue R250 for 1 min and destain with 40% methanol and 10% acetic acid until protein bands on the membrane are clear.

3. Record the image of protein bands on the membrane with a digital camera or a scanner for reference purpose.
4. Block the membrane with 5% skimmed milk at least 1 h before probing.

E. Overlay and Autoradiography

Solutions

1. *Binding buffer (2×)*: 100 mM HEPES, pH 7.3, 200 mM NaCl, 10 mM MgCl$_2$, 1 µg/µl BSA, and 0.1% Triton X-100. Filter the buffer through a 0.2-µm syringe filter and store at −20°C.
2. *Washing buffer (1×)*: Phosphate buffer saline containing 10 mM HEPES (pH 7.3), 5 mM MgCl$_2$, and 0.05% Triton X-100 (keep at 4°C)

Steps

1. Transfer the blocked membrane to a 50-ml BD Falcon high-clarity polypropylene conical centrifuge tubes (Cat. No. 352070) and rinse the membrane once with 1× binding buffer. Check that there are no air bubbles between the membrane and the wall of the tube. Dilute the radioactive labeled probe in 4 ml of 1× binding buffer and transfer into the tube containing the membrane: incubate for 2 h on a standard clinical spiral mixer behind a radioactive shield (at 4°C or room temperature).

2. Decant the probe carefully into a radioactive waste container and wash the membrane three times (5 min each) with ice-cold washing buffer. Blot residual liquid with Whatman 3MM filter paper, wrap it with Saran warp, and either expose to Hyperfilm MP (Amersham Biosciences, Cat. No. RPN6K) or a PhosphorImager screen at −20°C for 30 min. For optimal signals, additional exposures may be required. We illustrate the use of this technique to map the PIX-binding site within GIT1 (Fig. 2). In the absence of a specific interaction, we do not observe "background binding." Note that although GST is a dimer in solution, there is no signal when overlaid onto GST immobilized on the target membrane.

IV. COMMENTS

The protein-labeling method described in this article has a number of advantages over others: (a) the process is rapid, it takes ~5 min to label a probe; (b) the only reagent required is [γ-^{32}P]GTP; (c) over 60% labeling efficiency is obtained easily; the free probe does not need to be removed; (d) less waste is generated, a

A

GIT1 deletion constructs

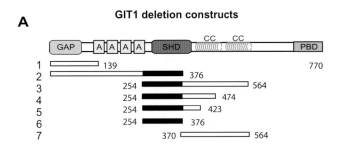

B

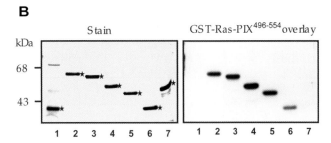

FIGURE 2 Mapping the PIX-binding domain on GIT1 by the overlay assay. (A) Schematic diagram showing the domain structure of GIT1 and deletion constructs used in the overlay assay to determine the PIX-binding region. (B) An overlay assay using the GIT1-binding sequence in PIX as a probe. GST fusion proteins (0.5 µg) of each GIT1 deletion construct shown in A were separated on SDS–PAGE and transferred onto a PVDF membrane. Proteins on the membrane were stained with Coomassie blue (left) and overlaid with [γ-^{32}P]GTP-labeled GST–Ras–PIX$^{496–554}$ (right). The minimal PIX-binding region determined by this assay corresponds to the SHD (Spa2 homologous domain) (lane 6). PAK and GIT1 can simultaneously bind PIX at its N-terminal and C-terminal, respectively, and form a PAK–PIX–GIT1 trimeric complex. This complex, through interaction with its effectors, plays an important role in regulating cell focal adhesion dynamics and cell motility.

single 5-min step with no additional processing compared to other radiolabeling methods such as phosphorylation of chemical modification; (e) as fusion proteins can be stored at −80°C for many years in aliquots, labeling is reproducible and does not rely on

the quality of any other biological; and (f) use of the ^{32}P radioisotope provides very high sensitivity. We have used this method to characterize interactions between PAK, PIX, and GIT1 in detail (Zhao *et al.*, 2000b). In some instances, the sensitivity is such that protein partners can be visualized directly in total cell or tissue lysates. The GST moiety mediates probe dimerization, which in fact provides a stabilization of probe binding (in the same manner as bivalent antibodies), which increases sensitivity of the overlay assay.

V. PITFALLS

1. Use recently purchased radioactive [γ-^{32}P]GTP for best sensitivity.
2. Seal the Falcon tube with Parafilm to prevent any leakage of the radioactive probe.
3. Make sure that the incubation tube is level: because the Falcon tube lid is wider than the tube, one option is to use another Falcon tube cut across the body about 2 cm from the lid and insert the end of probe tube into this.

References

Bartel, P. L., and Fields, S. (1995). Analyzing protein-protein interactions using two-hybrid system. *Methods Enzymol.* **254**, 241–263.

Fields, S., and Song, O. (1989). A novel genetic system to detect protein-protein interactions. *Nature* **340**, 245–246.

Zhao, Z. S., Manser, E., and Lim, L. (2000a). Interaction between PAK and nck: A template for Nck targets and role of PAK autophosphorylation. *Mol. Cell. Biol.* **20**, 3906–3917.

Zhao, Z. S., Manser, E., Loo, T. H., and Lim, L. (2000b). Coupling of PAK-interacting exchange factor PIX to GIT1 promotes focal complex disassembly. *Mol. Cell. Biol.* **20**, 6354–6363.

Ligand Blot Overlay Assay: Detection of Ca^{+2}- and Small GTP-Binding Proteins

Pavel Gromov and Julio E. Celis

I. INTRODUCTION

Protein-targeting interactions play a central role in most biological processes. Their detection and analysis *in vitro* can provide important information on specificity, affinity, and structure/function relationships that are realized *via* these interactions. Protein blot overlay assays, also known as "Western–Western," "Far–Western," "ligand," or "affinity" blotting, are powerful techniques for detecting and analyzing proteins or protein motifs involved in cellular-targeting processes (Clegg *et al.*, 1998 and references therein).

These methods are based on the principle that proteins, or protein fragments resolved by electrophoresis and transferred to an immobilizing matrix such as nitrocellulose or a nylon membrane, can react with putative binding ligands. This article describes protocols for identifying Ca^{+2}- and GTP-binding proteins using whole cellular protein extracts from noncultured human psoriatic keratinocytes and COS-1 cells.

II. MATERIALS AND INSTRUMENTATION

A. ^{45}Ca Overlay Assay

Imidasole (Cat. No. I-0250) is from Sigma, MgCl$_2$ · 6H$_2$O (Cat. No. 105832) is from Merk. ^{45}CaCl$_2$ (Cat. No. CES3) is from Amersham. Nitrocellulose membrane sheets (Hybond C, Cat. No. RPN. 203C) are from Amersham. X-ray films (X-Omat UV, 18 × 24 cm, Cat. No. 524 9792) are from Kodak.

B. α-^{32}GTP Overlay Assay

Tween 20 (Cat. No. 822 184) and MgCl$_2$ · 6H$_2$O(Cat. No. 105832) are from Merk. Dithioltreitol (DTT, Cat. No. D-0632) and ATP (Cat. No. A-2383) are from Sigma. Tris base (Cat. No. 648311) is from Calbiochem. [α-^{32}P]GTP (10 mCi/ml, Cat. No. PB 10201) and nitrocellulose membranes (Hybond C, Cat. No. RPN. 203C) are from Amersham. X-ray films (X-Omat UV, 18 × 24 cm, Cat. No. 524 9792) are from Kodak. All other reagents and materials are as described elsewhere (Celis and Celis, 1997).

III. PROCEDURES

The protocol for blot overlay detection of Ca^{+2}- and GTP-binding proteins is exemplified using whole cellular protein extracts from noncultured human psoriatic keratinocytes and COS-1 cells, but can be applied to a variety of other cultured cells, tissue samples, and biological fluids.

A. ^{45}Ca Overlay Assay

Calcium ion is a universal intracellular signal that acts as a important second messenger for many cellular processes and whose effect is modulated by specific calcium-binding proteins (Berridge *et al.*, 1998). According to well-conserved structural elements, these proteins can be grouped into different families, including annexins, C2 domain proteins, and EF-hand proteins (Celio *et al.*, 1996; Maki *et al.*, 2002;

Heizmann *et al.*, 2002). The calcium overlay assay (Maruyama *et al.*, 1984) as described here is essentially a specific application of a general metal ion-binding assay (Aoki *et al.*, 1986) and is widely used for studying various calcium-binding proteins (Son *et al.*, 1993), including those containing EF hands (Hoffmann *et al.*, 1993). Proteins are separated by means of one (1D)- or two-dimensional (2D) gel electrophoresis (see article by Celis *et al.* and article by Görg and Weiss in this volume), transferred to a nitrocellulose membrane, and overlaid with radioactive ^{45}Ca. Calcium-binding proteins are detected by autoradiography or phosphorimaging.

Solutions

The volumes given in the protocol are for 2D gel nitrocellulose blots (14 × 16 cm). The volumes for 1D gel blots (0.6 × 16 cm) are given in parentheses.

1. *Washing buffer*: 60 mM KCl, 5 mM MgCl$_2$, and 10 mM imidazole–HCl, pH 6.8. Prepare stock solutions in glass-distilled water. To make 1 liter of 3 M KCl, dissolve 223.6 g in 700 ml of water and bring to 1 liter. To make 1 liter of 1 M MgCl$_2$, dissolve 203.1 g in 700 ml of water and bring to 1 liter. Just before use, prepare 40 ml of the 1 M imidazole solution by dissolving 2.72 g of imidazole in 20 ml of distilled water and adjust to pH 6.8 with HCl. Complete to 40 ml with distilled water. To make 500 ml (20 ml) of washing buffer, enough for one 14 × 16-cm (0.6 × 14-cm) membrane, combine the following: 10 ml (0.4 ml) of 3 M KCl, 2.5 ml (0.1 ml) of 1 M MgCl$_2$, 5 ml (0.2 ml) of 1 M imidazole, and complete to 500 ml (20 ml) with distilled water.

2. *Probing buffer*: Add 15 μl (2 μl) of ^{45}CaCl$_2$ to 15 ml (2 ml) of washing buffer (final concentration of 1 μCi/ml)

3. *Aqueous ethanol*: Prepare 150 ml (5 ml) of 67% aqueous ethanol per membrane. Add 50 ml (1.67 ml) of distilled water to 100 ml (3.33 ml) of 96% ethanol.

Steps

1. Work with radioactivity according to the safety procedures enforced in your laboratory.

2. Transfer proteins from the gels to the nitrocellulose membrane as described in the article by Celis *et al.* in Volume 1.

3. Place the nitrocellulose sheet in a rectangular glass container (19 × 24 cm) containing 100 ml of washing buffer. Wear gloves when handling the membrane. The nitrocellulose sheet should be placed with the protein-bearing side facing upward. Rinse the nitrocellulose sheet twice with washing buffer.

4. Remove the membrane from the blotting chambers and wash briefly with the blotting buffer. The membrane can be dried and stored in a plastic bag at room temperature until further use.

5. Soak the membrane in 150 ml (2 ml) of washing buffer and shake gently for 20 min.

6. Remove the washing buffer and add 150 ml (2 ml) of fresh washing buffer. Shake gently for 20 min.

7. Repeat step 6.

8. Place the membrane or strip in a plastic bag and add 15 ml (2 ml) of probing buffer. Seal the bag and incubate for 10 min with gentle agitation. Alternatively, incubate the membrane in an appropriate rectangular glass or plastic container containing 150 ml (2 ml) of probing buffer. The incubation can be extended up to 1 h without adverse effects. Handle radioactive material with care.

9. Transfer the membrane to an appropriate glass or plastic container and add 150 ml (2 ml) of aqueous ethanol. Shake for 5 min with gentle agitation. Dispose of the radioactive solutions according to the safety procedure in your laboratory.

10. Carefully remove the nitrocellulose sheet or strip from the container using plastic tweezers and air dry for at least 4 h at room temperature.

11. Expose the dried membrane to X-ray film for 1 h to 3 days at room temperature. The membrane can be stained with amido black after autoradiography to facilitate protein identification.

Figure 1 shows an isoelectric focusing two-dimensional gel of total protein extracts from psoriatic keratinocytes transferred to nitrocellulose membrane and probed with ^{45}CaCl$_2$.

Comments

No significant differences in calcium binding are observed when using membranes that had been dried before probing or membranes that had been used directly after blotting.

To decrease nonspecific binding, competing metal ions such as Mg^{+2} may be added in steps 5–8 to a final concentration of 10 mM.

The use of a rotating roller system (see Fig. 1 in the article by Celis *et al.* in Volume 1) for incubation of nitrocellulose membranes facilitates the handling of the membrane, reduces considerably the volume of reagents used, and provides even detection of the calcium-binding proteins.

Pitfalls

1. Use a fresh imidazole solution.

2. Do not dry the membrane between pieces of filter paper as it leads to severe background problems.

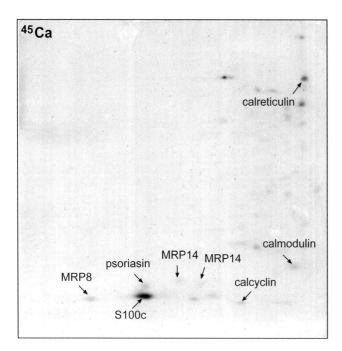

FIGURE 1 Calcium-binding proteins expressed by psoriatic keratinocytes. Autoradiograph of a $^{45}Ca^{2+}$ blot overlay of a 2D gel of whole protein extracts from noncultured, psoriatic keratinocytes.

B. α-^{32}GTP Overlay Assay

Small GTP-binding proteins constitute a rapidly increasing family of monomeric regulatory switches that have been adopted to control various cellular activities. These include proliferation and differentiation (Ras), protein transport and secretion (Rab), cytoskeletal (Rho), and nuclear assembly (Ran) (Manser, 2002; Dasso 2002; Pruitt and Der, 2001; Jaffe and Hall 2002). Unlike oligomeric G proteins, which are composed of α, β, and γ subunits, these proteins are able to bind GTP specifically when separated by SDS–PAGE and blotted onto nitrocellulose or nylon membranes. The blot overlay nucleotide-binding assay allows detection of small GTP-binding proteins in polypeptide mixture separated by one or 2D PAGE. The protocol presented here follows closely those described by McGrath *et al.* (1984), Doucet and Tuana (1991), and Gromov and Celis (1994).

Solutions

The volumes given are for nitrocellulose blot sheets of 14 × 16 cm in size.

1. *10% Tween 20*: To make 50 ml, weigh 5 g of Tween 20 and complete to 50 ml with distilled H$_2$O. Mix carefully. Store at room temperature.

2. *100 mM ATP*: To make 1 ml, add 55.1 mg of ATP and complete to 1 ml with distilled water. Aliquot in 100-μl portions and store at −20°C.

3. *0.1 M MgCl$_2$*: To make 50 ml, weigh 0.102 g of MgCl$_2$ · 6H$_2$O and complete to 50 ml with distilled water. Store at 4°C.

4. *Washing buffer*: 50 mM Tris–HCl, pH 7.6, 10 μM MgCl$_2$, and 0.3% Tween 20. To make 1 liter, add 6.055 g of Tris base to 800 ml of distilled H$_2$O and titrate with HCl. Add 30 ml of a 10% stock solution of Tween 20 and 100 ml of a 0.1 M stock solution of MgCl$_2$. After dissolving, complete to 1 liter with distilled H$_2$O and store at 4°C.

5. *ATP overlay buffer*: 50 mM Tris–HCl, pH 7.6, 10 μM MgCl$_2$, 0.3% Tween 20, 100 mM DTT, and 100 μM ATP. To make 100 ml, dissolve 1.54 g DTT in 90 ml of washing buffer and add 100 μl of a 0.1 M stock solution of ATP. After dissolving, complete to 100 ml with washing buffer and store at 4°C.

6. *Binding buffer*: 50 mM Tris–HCl, pH 7.6, 10 μM MgCl$_2$, 0.3% Tween 20, 100 mM DTT, 100 μM ATP, and 1 nM [α-^{32}P]GTP [1 μCi [α-^{32}P]GTP/ml]. To make 100 ml, add 10 μl of a 10-mCi/ml solution of [α-^{32}P]GTP to the ATP overlay buffer. Prepare directly in the binding container prior to use (see step 4).

Steps

1. Transfer proteins from the gels to the nitrocellulose membrane as described in the article by Celis *et al.* in volume 1.

2. Place the nitrocellulose sheet in a rectangular glass container (19 × 24 cm) containing 100 ml of washing buffer. Wear gloves when handling the membrane. The nitrocellulose sheet should be placed with the protein-bearing side facing upward. Rinse the nitrocellulose sheet twice with washing buffer.

3. Remove the washing buffer from the container and fill it with 50 ml of ATP overlay buffer. Place the container on an orbital shaker and incubate for 10 min with gentle agitation at room temperature.

4. Remove the blotting container from the orbital shaker platform and add 10 μl of a 10-mCi/ml solution of [α-^{32}P]GTP. Incubate with gentle agitation for 60 min at room temperature. Handle radioactive material with care.

5. Remove the binding buffer from the container and fill it with 50 ml of washing buffer. Soak the nitrocellulose sheet for 10 min at room temperature. Dispose of radioactive solutions according to the safety regulations in your laboratory.

6. Repeat the washing procedure twice at room temperature (10 min per wash). Use as much washing buffer as possible.

7. Carefully remove the nitrocellulose sheet from the container using plastic tweezers and air dry for at least 4 h at room temperature.

8. Place the air-dried nitrocellulose sheet into a cassette for autoradiography (12–72 h at –80°C) using an X-ray film and an intensifying screen.

Comments

The use of a rotating roller system for incubating the membranes is recommended.

DTT (100 mM final concentration in steps 3 and 4) enhances the GTP-binding ability of small GTP-binding proteins. This improves considerably the signal-to-noise ratio as well as the sensitivity of the procedure (Gromov and Celis, 1994).

To reduce background, we recommend ATP as an effective competitor for nonspecific binding of GTP to the nitrocellulose. Alternatively, use bovine serum albumin (0.3% final concentration) in the overlay and binding buffer as recommended by McGrath *et al.* (1984) and Doucet and Tuana (1991).

Bound [α-^{32}P]GTP can be removed from the nitrocellulose blot without detectable loss of proteins by incubating the membrane in a solution containing 50 mM Tris–HCl, pH 7.4, and 1% SDS for 30 min at room temperature. The blot can be then reprobed with [α-^{32}P]GTP. In this case do not dry the membrane after probing and perform autoradiography, keeping the membrane in the plastic bag.

To facilitate the identification of [α-^{32}P]GTP-binding proteins on whole protein extracts separated by 2D gel electrophoresis, we recommend adding a small amount of [^{35}S]methionine-labeled proteins from an appropriate source (e.g., keratinocytes) to the protein mixture prior to electrophoresis. Following the [α-^{32}P]GTP-binding overlay assay, the nitrocellulose blot is subjected to autoradiography using two films placed on top each other. The first film, which is placed in direct contact with the blot, visualizes both ^{35}S and ^{32}P isotopes, while the second one reveals only ^{32}P. The positions of detected spots may be compared with those in the master keratinocyte database (http://proteomics.cancer.dk). Using the protocol described here, it is possible to detect many small GTP-binding proteins in various cell types and tissues. Representative 2D autoradiographs of [α-^{32}P]GTP-binding proteins from COS-1 cells that transiently express ADP-ribosylation factor, rab11a, and p21-ras are shown in Fig 2 (see also article by Gromov *et al.* in this volume).

Pitfalls

1. Do not dry the nitrocellulose membrane after protein transfer as it substantially reduces the efficiency of GTP binding.

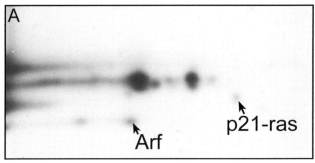
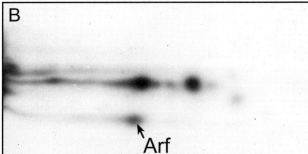
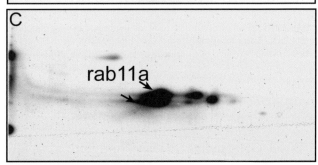
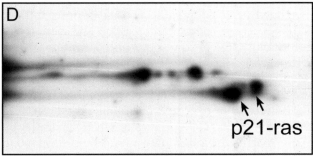

FIGURE 2 Two-dimensional blot autoradiographs of [α-^{32}P]GTP-binding proteins from COS-1 cells that transiently express several small GTP-binding proteins. (A) Control, nontransfected cells, (B) ADP-ribosylation factor (Arf), (C) rab11a, and (D) p21-ras (see also article by Gromov *et al.* in this volume).

2. Make sure that the solution covers the whole nitrocellulose sheet surface during agitation. Avoid scratching or tearing of the nitrocellulose membrane during manipulation.

3. Use high-grade ATP. Small traces of cold GTP, which may contaminate commercial ATP, decrease the efficiency of [α-^{32}P]GTP binding.

References

Aoki, Y., Kunimoto, M., Shibata, Y. Y., and Suzuki, K. T. (1986). Detection of metallothionein on nitrocellulose membrane using Western blotting technique and its application to identification of cadmium-binding proteins. *Anal. Biochem.* **157**, 117–122.

Berridge, M. J., Bootman, M. D., and Lipp, P. (1998). Calcium—a life and death signal. *Nature* **395**, 645–648

Celio, M., Pauls, T., and Schwaller, B. (1996). *"Guidebook to the Calcium-Binding Proteins"*, pp. 1–238. Oxford Univ. Press., Oxford.

Celis, A., and Celis, J. E. (1997). General procedures for tissue culture. *In "Cell Biology: A Laboratory Handbook"* (J. E. Celis, N. Carter, T. Hunter, D. Shotton, K. Simons, and J. V. Small eds.), Vol. 1, pp 5–16. Academic Press.

Clegg, R. A. (ed.) (1998). Protein targeting protocols. *In "Methods in Molecular Biology,"* Vol. 88. Humana Press, Totowa, NJ.

Dasso, M. (2002). The Ran GTPase: Theme and variations. *Curr. Biol.* **12**, R502–R508.

Daucet, J.-P., and Tuana, B. S. (1991). Identification of low molecular weight GTP-binding proteins and their sites of interaction insubcellular fractions from skeletal muscle. *J. Biol. Chem.* **266**, 17613–17620.

Gromov, P. S., and Celis, J. E. (1994). Some small GTP-binding proteins are strongly downregulated in SV40 transformed human keratinocytes. *Electrophoreis* **15**, 474–481.

Hoffmann, H. J., Olsen, E., Etzerodt, M., Madsen, P., Thogersen, H.-G., Kruse, T., and Celis, J. E. (1994). Psoriasin binds calcium and is differentially regulated with respect to other members of the S100 protein family. *J. Invest. Dermatol.* **103**, 370–375.

Jaffe, A. B., and Hall, A. (2002). Rho GTPases in transformation and metastasis. *Adv. Cancer Res.* **84**, 57–80.

Heizmann, C. W., Fritz, G., and Schafer, B. W. (2002). S100 proteins: Structure, functions and pathology. *Front Biosci.* **7**, 1356–1368.

Maki, M,, Kitaura, Y., Satoh, H., Ohkouchi, S., and Shibata, H. (2002). Structures, functions and molecular evolution of the penta-EF-hand Ca2+-binding proteins. *Biochim. Biophys. Acta* **1600**, 51–60.

Manser, E. (2002). Small GTPases take the stage. *Dev. Cell* **3**, 323–328.

Maruyama, K., Mikawa, T., and Ebashi, S. (1984). Detection of calcium-binding proteins by ^{45}Ca autoradiography on nitrocellulose membrane after sodium dodecyl sulfate gel electrophoesis. *J. Biochem.* **95**, 511–519.

McGrath, J. P., Capon, D. J., Goeddel, D. V., and Levinson, A. D. (1984). Comparative biochemical properties of normal and activated human ras p21 protein. *Nature* **310**, 644–649.

Pruitt, K., and Der, C. J. (2001). Ras and Rho regulation of the cell cycle and oncogenesis. *Cancer Lett.* **171**, 1–10.

Son, M., Gunderson, R. E., and Nelson, D. L. (1993). A 2nd member of the novel Ca^{2+}-dependent protein-kinase family from paramecium-tetraurelia; purification and characterization. *J. Biol. Chem.* **268**, 5840–5948.

39

Modular Scale Yeast Two-Hybrid Screening

Christopher M. Armstrong, Siming Li, and Marc Vidal

I. INTRODUCTION

The observation that most transcription factors can be separated into a DNA-binding domain (DB) and a transcriptional activation domain (AD) led to the development of the yeast two-hybrid (Y2H) system as an *in vivo* screen or selection to identify and characterize protein interactions (Fields and Song, 1989). Using the Y2H, one can identify potentially interacting proteins (X–Y heterodimers or X–X homodimers) by generating two different hybrid proteins: one with protein X fused to DB and the other with protein Y fused to AD (see Fig. 1). If protein X and Y interact, the AD can be brought to the promoter by DB-X and thereby activate the gene driven by that promoter (usually a selectable or screenable marker). By fusing a library to the AD (Fields and Song, 1989) and, in some special cases, to DB (Du *et al.*, 1996), it is possible to screen for proteins and identify potential interactors.

Two-hybrid screens have been used quite successfully by scientists interested in identifying potential interacting partners to their protein of interest. While this one gene at a time approach has been fruitful, the ease of use of Y2H has allowed it to be scaled up to perform screens on more of a proteome scale (Ito *et al.*, 2000; Uetz *et al.*, 2000; Walhout *et al.*, 2000). While whole genome two-hybrids may be beyond the scope or interest of most laboratories, medium size screens (on the level of 20 to 50 genes) can be done easily by one or a few scientists (Davy *et al.*, 2001; Drees *et al.*,

2001; Boulton *et al.*, 2002). Screens of this size allow scientists to approach problems on a more modular scale, i.e., studying most of the genes involved in a process rather than a few at a time (Hartwell *et al.*, 1999), thereby addressing more global questions than traditional single gene approaches allow.

While the Y2H is quite powerful, it is important to remember that genes identified in a Y2H screen are only potential interactors and further experiments are necessary to validate the relevance of the interactions. Techniques such as GST-pulldowns or immunoprecipitations can help confirm that the interaction exists, while analysis of expression patterns and phenotypes can help establish the functional relevance of the predicted interactions. The power of the Y2H is its ability to act as a starting point for the identification of interacting proteins.

Many variations of the method exist, but the fundamentals are the same for all of them. Here we use the strains and vectors described in Vidal *et al.* (1996), but the protocols work just as well with other strains so long as you take into account any changes in selectable markers and reporter assays that may be an issue. The strain used is MaV203, which has three screening markers: *GAL1::lacZ*, *SPAL10::URA3*, and *GAL1::HIS3* (Vidal *et al.*, 1996). Initial screening uses the *HIS3* marker to identify potential positives. Secondary screening is then done with other markers to test the strength of the interactions. The techniques here are derived in part from the techniques described in Walhout and Vidal (2001), but have been streamlined to make it possible to work with many baits in parallel.

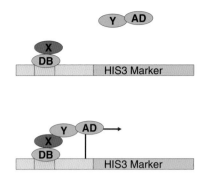

FIGURE 1 Two-hybrid interactions. (Top) Two fusion proteins are created, DB-X and AD-Y. If protein X fails to interact with protein Y, then the activation domain will not be brought to the promoter and the marker gene (in this case HIS3) will fail to activate. (Bottom) Protein X and protein Y interact successfully, bringing the activation domain to the promoter and activating the marker gene successfully.

II. MATERIALS AND INSTRUMENTATION

Bacto agar is obtained from Labscientific (Cat. No. A466). Bactopeptone (Cat. No. DF0118170), yeast extract (Cat. No. DF0127-17-9), ammonium sulfate (Cat. No. A702-500), N,N-dimethylformamide (Cat. No. D119-1), 3-mm glass beads (Cat. No. 11-312A), and 125-mm Whatman filter papers (Cat. No. 09-868C) are from Fisher Scientific. 3-Amino-1,2,4-triazole (3-AT, Cat. No. A-8056), salmon testes DNA (Cat. No. D9156), glucose (Cat. No. G8270), polyethylene glycol MW 3350 (Cat. No. P-4338), lithium acetate (Cat. No. L-4158), 5-bromo-4-chloro-3-indolyl-β-D-galactopyranoside (X-Gal, Cat. No. B-4252), 2-mercaptoethanol (Cat. No. M-3148), yeast synthetic drop-out medium amino acid supplement (without histidine, leucine, tryptophan, and uracil) (Cat. No. Y-2001), tryptophan (Cat. No. T-0254), uracil (Cat. No. U-0750), and histidine (Cat. No. H-6034) are from Sigma. 5-Fluoroorotic acid (5FOA, Cat. No. 1555) is from BioVectra. Zymolyase-20T (Cat. No. 120491) is from Seikagaku. Airpore tape sheets (Cat. No. 19571) are from Qiagen. Nitrocellulose filters (Cat. No. WP4HY13750) are from Osmonics. Eight-and-a-half-inch replica velvets (#2008) and 6-in. replica blocks (#4006) can be obtained from Cora Styles Needles 'N Blocks (www. corastyles.com).

MaV203 yeast strain (MATα, leu2-3,112, trp-901, his3Δ200, ade2-1, gal4Δ, gal80Δ, SPAL10::URA3, GAL1::lacZ, GAL1::HIS3-@LYS2, can1^R, cyh2^R) can be obtained from Invitrogen as part of the ProQuest two-hybrid system with Gateway Technology (Cat. No. 10835–031) or directly from the Vidal laboratory (http://vidal.dfci.harvard.edu/).

III. PROCEDURES

A. Preparation of Yeast Culture Plates and Medium

1. *YPD liquid media*: Add 10 g of yeast extract and 20 g of Bactopeptone to 1 liter of dH$_2$O and autoclave. Add 50 ml of 40% glucose before use.

2. *20 mM uracil*: Dissolve 1.21 g of uracil in 500 ml of ddH$_2$O. Filter to sterilize. Note that it may be necessary to heat the water to get the uracil into solution.

3. *100 mM histidine*: Dissolve 7.76 g of histidine in 500 ml of ddH$_2$O. Filter to sterilize and wrap the bottle in foil.

4. *40 mM tryptophan*: Dissolve 4.08 g of tryptophan in 500 ml of ddH$_2$O. Filter to sterilize and wrap the bottle in foil and keep at 4°.

5. *Synthetic complete (SC) liquid medium*: To prepare a liter of SC medium, add 1.4 g of yeast synthetic drop-out medium amino acid supplement, 1.7 g of yeast nitrogen base, and 5 g of ammonium sulfate into 925 ml of water. Adjust the pH with 10 N NaOH to a final pH of 5.9. Autoclave media to sterilize. Add 50 ml 40% glucose before use. This media is -Leu, -Trp, -Ura, -His. To supplement with a missing amino acid/nucleotide, add 8 ml of the appropriate solution (e.g., to make SC-Leu-Trp, add 8 ml of 20 mM uracil and 8 ml of 100 mM histidine).

6. *Synthetic complete (SC) agar plates*: To prepare a liter of SC plates (ten to twelve 15-cm petri dishes), add 1.4 g of yeast synthetic drop-out medium amino acid supplement, 1.7 g of yeast nitrogen base, and 5 g of ammonium sulfate into 425 ml of water. Adjust pH to 5.9 with 10 N NaOH. At the same time, in a separate flask, add 20 g of bacto agar into 500 ml of water. Autoclave both agar and SC solutions. After autoclaving, mix the flasks. Cool to 55°C in a water bath. Add 50 ml of 40% glucose and 8 ml of the appropriate amino acid solution before pouring.

7. *SC-Leu plates*: Prepare 1 liter of SC/agar and add 8 ml of 20 mM uracil, 8 ml of 100 mM histidine, and 8 ml of 40 mM tryptophan before pouring. Pour into ten to twelve 15-cm petri plates.

8. *SC-Leu-Trp plates*: Prepare 1 liter of SC/agar and add 8 ml of 20 mM uracil and 8 ml of 100 mM histidine before pouring. Pour into ten to twelve 15-cm petri plates.

9. *SC-Leu-Trp-His+3-AT plates (3-AT plates)*: Prepare 1 liter of SC/agar and add 8 ml of 20 mM uracil and 1.18 g of 3-AT powder before pouring (this makes 20 mM 3-AT plates). Pour into ten to twelve 15-cm petri plates.

10. *SC-Leu-Trp+5-FOA plates (5-FOA plates)*: Prepare 1 liter of SC/agar and add 8 ml of 20 mM uracil and 8

ml of 100 m*M* histidine before pouring. Add 2 g of 5-FOA powder and stir to dissolve. It will take a while for 5-FOA to go into solution. Pour into ten to twelve 15-cm petri plates.

11. *YPD plates*: Add 10 g of yeast extract and 20 g of Bactopeptone to 500 ml of dH$_2$O. At the same time, in a separate flask, add 20 g of bacto agar to 500 ml of dH$_2$O. Autoclave both flasks and mix afterward. Cool to 55°C in a water bath. Add 50 ml of 40% glucose and pour into ten to twelve 15-cm petri plates.

B. Preparation of DB-ORF Bait Strains

Any cloning strategy can be used to generate DB-ORF fusion vectors. The Gateway recombination method (Walhout *et al.*, 2000) is useful for the cloning of large numbers of DB-ORF baits. For the strains and plasmids that we describe, the DB plasmid has a *LEU2* marker and the AD plasmid has a *TRP1* marker, but other variations of the two-hybrid system use other markers. It is important to make sure that the selective plates you use match the markers of the vectors you are using. After obtaining the constructs, the next step is to introduce them into the two-hybrid strain (MaV203 in this method). Several transformation protocols that are optimized for the scale of the transformation to be performed are given throughout the article, but any yeast transformation protocol should work. The following protocol is optimized for transforming large numbers of baits in a 96-well plate.

Solutions

1. *1 M lithium acetate stock solution*: Add 51 g of lithium acetate into 500 ml of ddH$_2$O. Autoclave to sterilize.

2. *10× TE stock solution*: 100 m*M* Tris–HCl (pH 7.5), 10 m*M* EDTA, autoclave

3. *50% PEG stock solution*: Dissolve 125 g of polyethylene glycol in warm ddH$_2$O and finalize to 250 ml. Sterilize by filtration (the PEG solution is very viscous and takes a long time to filter).

4. *TE/LiAc*: To make 50 ml, add 5 ml of 10× TE and 5 ml of 1 *M* LiAc into 40 ml of sterile ddH$_2$O

5. *TE/LiAc/PEG*: To make 50 ml, add 5 ml of 10× TE and 5 ml of 1 *M* LiAc into 40 ml of 50% PEG

6. *Boiled ssDNA*: Boil the 10 mg/ml of salmon testes DNA for 5 to 10 min and chill on ice before transformation

Steps

1. Start an overnight culture of MaV203 yeast by scratching a small clump of cells from a patch into at least 0.5 ml of media for each bait you plan to transform (a minimum of 5 ml of media should be used).

2. The next day, take 0.5 ml of the overnight culture for each transformation.

3. Spin down the cells at 2000 rpm for 5 min.

4. Wash the cells by adding 0.25 ml of ddH$_2$O for each transformation.

5. Spin down the cells and wash in 100 μl of TE/LiAc for each transformation.

6. Spin down the cells and resuspend the cell pellet in 20 μl of TE/LiAc for each transformation.

7. Add 2 μl of boiled ssDNA for each tansformation.

8. Aliquot 22 μl of yeast into the wells a 96-well plate.

9. Add 50–100 ng of the appropriate DB-ORF DNA to each well. A transformation without DB-ORF DNA serves as a negative control.

10. Add 100 μl of TE/LiAc/PEG to each well and mix by pippetting.

11. Incubate the plates at 30°C for 30 min.

12. Heat shock at 42°C for 15 min.

13. Spin down and remove the TE/LiAc/PEG solution with a multichannel pipette.

14. Add 120 μl of ddH$_2$0 to each well, but be careful not to resuspend the cells. Remove 105 μl of ddH$_2$O from each well and resuspend the cells in the remaining 15 μl of liquid.

15. Using a multichannel pipette, spot 6–7 μl onto two 15-cm SC-Leu plates. You should be able to spot all 96 wells onto a 15-cm plate (see Fig. 2). The second plate is to have a copy in case spots run together or one plate is contaminated. You can spot onto additional plates if necessary.

16. Incubate for 2 to 3 days at 30°C.

17. Plates can be stored at 4°C for up to 2 months. Fifteen percent glycerol stocks can also be created and stored at −80°C indefinitely.

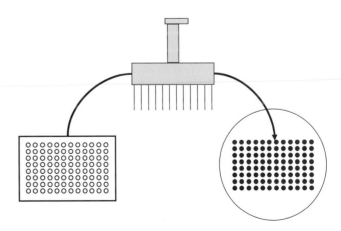

FIGURE 2 Spotting a 96-well plate onto a 15-cm agar plate. If a 12-tip multichannel pipette is used, all 8 rows of a 96-well plate can fit easily onto an agar plate as diagrammed.

C. Introduction of cDNA Libraries

Once the bait strains have been created, an AD-cDNA library needs to be transformed into the yeast. Typically, for most cDNA libraries, at least 10^6 independent clones need to be screened to get good coverage of a library, thereby requiring a high efficiency transformation procedure. If a large number of baits are being screened or if you are using a normalized ORFeome library (Reboul *et al.*, 2003), it may be preferable to screen as few as 2×10^5 colonies. While more potential interactions will be identified if you screen more colonies, screening a smaller number can still identify many interesting interactions and will allow you to screen a larger number of baits as well. We give protocols for both large and small numbers of colonies to be screened. Positives are screened for their ability to activate the *HIS3* marker. In MaV203, *HIS3* has a low level of activity, leading to moderate growth on SC-His media. To overcome this, the medium is supplemented with 3-AT, an inhibitor of *HIS3*, to reduce the background growth (Durfee *et al.*, 1993).

Solutions and Media

See transformation solutions in Section IIIB. SC-Leu liquid media, SC-Leu-Trp plates, and SC-Leu-Trp-His + 3-AT plates (referred to as 3-AT plates from here on) are needed.

Protocol 1. Introduction of AD Library into Y2H Strain (30 Plate Scale)

Steps

1. Grow the DB-ORF baits in 3 ml of SC-Leu yeast liquid media at 30°C for approximately 24 h.

2. Resuspend the cells well by vortexing and inoculate 80–100 μl in 250 ml of YPD and incubate at 30°C for 15 to 18 h until the OD_{600} reaches 0.3 to 0.6. *Note*: It may be necessary to vary the amount of cells added and the time of incubation for the cells to be ready at the right time.

3. Harvest the cells by centrifuging for 5 min at 1800 rpm.

4. Wash the cell pellet in 50 ml of ddH$_2$O.

5. Wash the cells in 10 ml of TE/LiAc.

6. Resuspend the cells in $5 \times OD_{600}$ ml of TE/LiAc (e.g., if the $OD_{600} = 0.5$, then add 2.5 ml of TE/LiAc).

7. Transfer 1.6 ml of resuspended cells into a 15-ml Falcon tube and add 160 μl of boiled ssDNA and 20–30 μg of cDNA library.

8. Add TE/LiAc/PEG to final volume of 9 ml. Mix by inverting several times and aliquot at 1 ml into 1.5-ml Eppendorfs.

9. Incubate at 30°C for 30 min to 1 h.

10. Heat shock at 42°C for 15 min.

11. Transfer the cells into a 15-ml Falcon tube and spin down at 1800 rpm for 5 min.

12. Remove the supernatant and resuspend the cells in 9 ml of sterile ddH$_2$O.

13. Take 10 μl and add to 10 ml of ddH$_2$O to create a 1 : 1000 dilution.

14. To about thirty 15-cm 3-AT plates, add approximately twenty-five 3-mm glass beads. Plate 300 μl of cells from step 12 on each 15-cm 3-AT plate. Spread the cells evenly by shaking the plates with glass beads. Remove beads when done. The beads can be washed, autoclaved, and reused.

15. To measure transformation efficiency, plate 300 μl of the 1 : 1000 diluted cells on a SC–Leu-Trp plates. Count the colonies 2 to 3 days after plating.

16. Incubate the 3-AT plates for 4 to 5 days at 30°C and take to Section III,D.

Protocol 2. Introduction of AD Library in Y2H Strain (Three Plate Scale)

Steps

1. Grow the DB-ORF baits in 3 ml of SC-Leu yeast medium at 30°C for approximately 24 h.

2. Resuspend the cells well by vortexing and inoculate 10–15 μl in 35 ml of YPD and incubate for 15 to 18 h at 30°C until the OD_{600} reach 0.3 to 0.6. This can be done in 50-ml Falcon tubes. *Note*: It may be necessary to vary the amount of cells added and the time of incubation for the cells to be ready at the right time.

3. Harvest the cells by centrifuging for 5 min at 1800 rpm.

4. Wash the cell pellet in 1.5 ml of ddH$_2$O by vortexing and transfer to a 2.0-ml Eppendorf.

5. Spin at highest speed for 5 s in a microcentrifuge.

6. Wash the cells in 1 ml TE/LiAc and spin again.

7. Resuspend the cells in 275 μl of TE/LiAc.

8. Add 30 μl of boiled ssDNA and 3–5 μg of the normalized AD-library.

9. Add 1.5 ml TE/LiAc/PEG and mix by inverting several times.

10. Incubate at 30°C for 30 min to 1 h.

11. Heat shock at 42°C for 15 min.

12. Spin down the cells.

13. Remove the supernatant and resuspend the cells in 900 μl of sterile ddH$_2$O.

14. Take 10 μl and add to 10 ml of ddH$_2$O to create a 1 : 1000 dilution

15. To three 15-cm 3-AT plates, add approximately twenty-five 3-mm glass beads. Plate 300 μl of cells from step 12 on each 15-cm 3-AT plate. Spread the cells evenly by shaking the plates with glass beads. Remove beads when done. The beads can be washed, autoclaved, and reused.

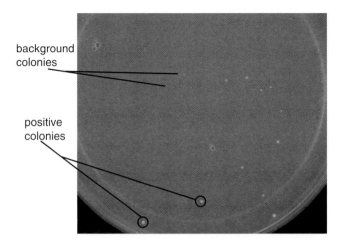

FIGURE 3 An example of a 3-AT screening plate before colonies have been picked. Two of the colonies that are likely two-hybrid positives have been circled.

FIGURE 4 An example of positive colonies that have been picked and patched onto a 3-AT plate. By patching them in the same pattern and dimensions as a 96-well plate, it makes it easier to manipulate the samples at later stages.

16. Calculate the transformation efficiency by plating 300 µl of 1 : 1000 diluted cells on SC-Leu-Trp plates.

D. Isolation of Two-Hybrid Positives

After 4 to 5 days of growth, colonies with interacting proteins should have grown enough to isolate them. At times it is difficult to differentiate a true positive from a large background colony, but later in Section III,E you can use the phenotypic assays to help differentiate them. Colonies are initially grown on 3-AT plates after picking to allow true positives to outgrow any background cells that might have been picked up accidentally.

Solutions

1. SC-Leu-Trp-His + 3-AT plates
2. SC-Leu-Trp liquid media

Steps

1. Get the plates from Section II,C. See Fig. 3 for an example of what a screening plate might look like (with positive colonies growing in a field of background colonies).

2. Use a toothpick to pick colonies that grow above the background. Patch the colony in a small streak onto a 15-cm 3-AT plate. Typically we divide the plate into 96 sectors that correspond to a 96-well plate so that the positives can be stored and manipulated in a 96-well format in the future (see Fig. 4). If you only have a small number of positives you will not fill up a plate, but if you have many baits and a large number of positives then you could fill up many plates with potential positive clones.

3. Incubate the plates at 30°C for 2 to 3 days.

4. When the patches have grown up, use a multichannel pipette with tips to scrape a small clump of cells into 120 µl of SC–Leu-Trp medium in U-bottom 96-well plates (see Fig. 5). Seal the plates with airpore tape.

5. Incubate the plates at 30°C for 2 days before making a 15% glycerol stock.

6. Take the remaining culture to Section III,E for phenotypic assays.

E. Phenotypic Assays

Picking positive colonies is only the first step in identifying potential interacting proteins. It is frequently difficult to determine which of the colonies are true positives and which are merely large growing background colonies. It is important to retest the potential positives and see if they are able to activate expression of a variety of different two-hybrid reporters. We generally look for colonies that are able to activate at least two of the three two-hybrid reporters. It is also important to include at least one positive and one negative control on the plates. A good negative control would be the empty AD and DB vectors, whereas a strong positive control would be the full-length Gal4 transcription factor. For a description of a larger set of two-hybrid controls, see Walhout and Vidal (2001).

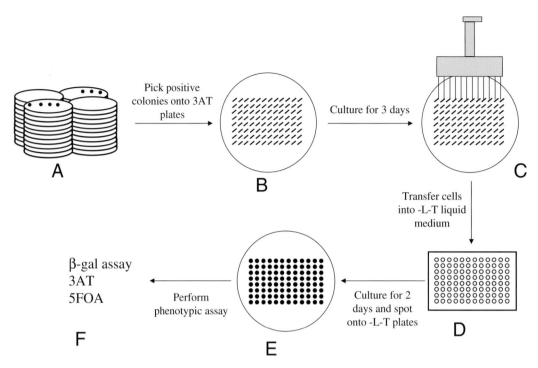

FIGURE 5   Diagram of picking positives. First gather the library transformation screening plates (A) and pick the positives onto a fresh 3-AT plate (B). After 3 days of growth, use a multichannel pipette with tips (C) to transfer the positives to liquid SC-Leu-Trp in a 96-well plate (D). After 2 days of growth, spot the cells onto a SC-Leu-Trp plate (E) and use to replica plate for the phenotypic assays (F).

Solutions and Materials

You need SC-Leu-Trp plates, 3-AT plates, 5-FOA plates, and YPD plates. Note that 5-FOA selects against URA3-positive strains, i.e., positive interactors will not grow on 5-FOA. You also need circular nitrocellulose filters.

Steps

1. Get the 96-well plates from Section III,D. Spot 5 µl of culture onto a SC-Leu-Trp plate. Grow at 30°C for 1 to 2 days.

2. You will use three different plates to assay the three two-hybrid reporters: YPD for lacZ, 3-AT for HIS3, and 5-FOA for URA3. The 5-FOA and 3-AT plates can be used as is, but the YPD plate needs a nitrocellulose filter placed on it prior to replica plating. This is necessary to remove the yeast from the YPD plate and perform the β-Gal filter lift assay (see Section III,F). If there are bubbles between the filter and the agar or if the filter is misaligned, use forceps to move the filter or remove the bubble.

3. Using replica velvets and a replica block, replica plate the yeast from the SC–Leu-Trp growth plate to the YPD/filter, 3-AT, and 5-FOA plates. Use the same velvet for each of the assay plates; there should be enough yeast on it for all three plates.

4. Replica clean the 3-AT and 5-FOA plates as follows. Use a clean velvet to remove excess yeast from the 3-AT plate (press firmly but not harshly). Repeat with a clean velvet until there is no longer any visible sign of yeast on the plate. Then repeat one more time (we typically clean with four velvets). Repeat the procedure with the 5-FOA plate. Replica cleaning is necessary to decrease the background growth and to ensure that you start out with comparable amounts of cells in each spot; it is not necessary to perform on the YPD/filter plates.

5. Culture all three assay plates at 30°C. After 1 day the YPD plate should have large spots of yeast on the filter. Take the YPD plate to Section III,F to perform β-Gal filter lift assays.

6. Examine 3-AT and 5-FOA plate. If the negative control shows growth after 1 day, replica clean again.

7. When the controls on the 3-AT and 5-FOA plates have grown to the appropriate levels (it normally takes 2 days for 5-FOA and 3 days for 3-AT), remove plates

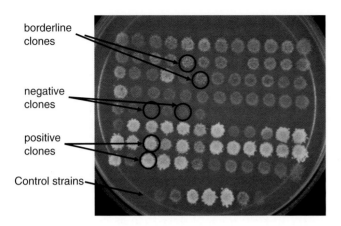

borderline clones

negative clones

positive clones

Control strains

FIGURE 6 An example of a 3-AT assay plate. The bottom seven patches are a range of controls. A range of strengths of interactions can be seen by the strength of growth on the spots. Clones that are indicative of positive and negative interactions have been circled as well as a couple of clones that are on the border between being positive and negative.

and score the results. (See Fig. 6 for an example of what a 3-AT assay plate can look like.)

8. Score the 3-AT and 5-FOA plates along with the β-Gal filters from Section III,F. Any strain that passes at least two of the three tests is considered positive. Consolidate all of the positives into fresh 96-well plates.

9. Grow plates for 2 days and make glycerol stocks of the strains.

F. β-Gal Filter Lift Assay

Solutions

1. *Z buffer*: To make a liter, add 16.1 g $Na_2HPO_4 \cdot 7H_2O$, 5.5 g $NaH_2PO_4 \cdot H_2O$, 0.75 g KCl, and 0.246 g $MgSO_4 \cdot 7H_2O$. Autoclave to sterilize.
2. *4% X-Gal*: Dissolve 40 mg in 1 ml of *N,N*-dimethyl-formamide. Store at −20°C wrapped with foil.
3. *β-Gal solution*: For each plate, prepare 5 ml of Z buffer with 120 μl of 4% X-Gal and 13 μl of 2-mercaptoethanol. Make fresh each time.

Steps

1. Retrieve the YPD/filter plates from Section III,E.
2. Make up the β-Gal solution according to the number of plates that are needed to assay.
3. For each for each plate to be assayed, get one empty 15-cm petri plate. Put two pieces of Whatman filter paper in the plate. Add 5 ml of β-Gal solution to each plate. Let the paper soak up the solution and make sure there are no bubbles under the Whatman paper.

4. Remove the nitrocellulose filter from the YPD (with yeast on the filter) and place in liquid nitrogen for at least 30 s. This lyses the cells.

5. Remove the filter from liquid nitrogen and allow it to thaw in air (this should take ~30 s). Once the filter is flexible again, place in a petri plate with β-Gal solution-soaked Whatman paper. Use forceps to remove any bubbles that may be under the filter.

6. Put the β-Gal assay plates at 37°C overnight. The next day you can read the results with positives being blue; the stronger the positive, the stronger the blue.

7. Remove the Whatman filter paper. The filters can be stored dry for at least 6 months.

G. Yeast PCR to Identify Preys

Once you have identified the clones that pass the phenotypic assays, it is necessary to isolate prey DNA from the yeast clones and identify them by sequencing. Prey DNA can be obtained by polymerase chain reaction (PCR) using the universal primers on the vector (typically we make primers to the activation domain sequence and the termination sequence; the primer sequences that we use are listed later).

Solutions and Media

1. SC-Leu-Trp plates, YPD plates
2. Lysis solution: To make 1 ml, add 2.5 mg zymolyase to 0.1 *M* sodium phosphate buffer (pH7.4). The solution can be stored at −20°C.

Primer sequences: Activation domain –CGCGTTTG GAATCACTACAGGG

Termination sequence –GGA GAC TTG ACC AAA CCT CTG GCG

Steps

1. Spot 5 μl of the positives from the 96-well plate culture or the glycerol stocks made in Section III,E, steps 8 and 9 onto a SC-Leu-Trp plate. Culture at 30°C for 1 day.

2. Replica plate to YPD. Culture at 30°C for 1 day.

3. Add 15 μl of lysis solution to each well in a 96-well PCR plate.

4. Using a multichannel pipette, scrape some of the yeast cells off the YPD plate (one row at a time as in Section III,D, step 4) and resuspend into the 15-μl lysis solution.

5. Put the yeast at 37°C for 5 min then at 95°C for 5 min (this can be done in a PCR machine).

6. Set up the PCR plate using a final volume of 50 μl for each reaction.

7. Make a 1 : 10 dilution of the yeast lysis from step 5. Add 5 μl to the PCR plate.

8. Perform PCR using 5-min extension times to make sure that the large ORF inserts are isolated.

9. The PCR products can be sent out for sequencing after confirming it works. Note that most PCR products need to be purified from the primers to be sequenced properly. We use the Millipore MultiScreen PCR plates (Fisher, Cat. No. MANU-030-10) to purify PCR products.

H. Retest by Gap Repair

Sometimes a positive isolate will occur not because the two proteins interact, but through a spontaneous self-activation mutation that makes the growth independent of the two plasmids. This makes it important to isolate the AD-ORF from the strain and retest it in the bait strain. The easiest way to do this is by gap repair (Orr-Weaver et al., 1983). This involves introducing into yeast both a vector cut at the insert sites and the insert itself. If there is some homology between the insert and the cut vector, the yeast will repair it by ligating the insert into the vector.

The protocol is essentially the same as the method listed in Section III,B with a few exceptions.

Step 1. Instead of starting a culture with untransformed MaV203 at the beginning, start a culture with MaV203 transformed with the DB-ORF of interest.

Step 9. Instead of using DB-ORF DNA, use 25 ng of linearized AD plasmid DNA and 2.5 μl of the PCR product from Section III,G. The AD plasmid should be cut in the linker region between the AD sequence and the termination sequence (in our plasmids, we use SmaI) as that way the PCR product can have homology to the AD sequence and the termination sequence and gap repair can be efficient.

Step 15. Select on -Leu, -Trp plates instead of -Leu.

Step 17. Perform phenotypic assays as in Section III,E, skipping step 1 and only using the 3-AT and β-Gal assays.

IV. COMMENT

Notes on High-Throughput Two Hybrid

The simplicity of two-hybrid screening allows for the screening of large numbers of baits of interest at a time. This allows a scientist to study a greater number of genes, but it requires the researcher to be more organized from the start. Here are several points to remember when screening large numbers.

1. Work in 96-well formats from the beginning. This will allow you to transform all the baits that you want to study in 1 day.
2. When transforming the bait strains with the library, it is still necessary to do each transformation individually, but you can do six or more in a day.
3. When picking the colonies, try to patch them on a 96-well grid so that it will be easier to transfer your positives to 96-well plates in the future.
4. Try to keep track of data such as gene names and strength of interactions in the phenotypic assay in a spreadsheet such as Microsoft Excel. This can allow you to monitor your positives easily and keep track of which ones pass all the phenotypic tests.

The strength of the two-hybrid system is its ease of use. It can identify many potential interactors and maybe even bring interaction networks to light. It is ultimately, however, a first-pass prediction of interactions and should be thought of as a jumping off point to other more detailed studies.

References

Boulton, S. J., Gartner, A., et al. (2002). Combined functional genomic maps of the C. elegans DNA damage response. Science **295**(5552), 127–131.

Davy, A., Bello, P., et al. (2001). A protein-protein interaction map of the Caenorhabditis elegans 26S proteasome. EMBO Rep. **2**(9), 821–828.

Drees, B. L., Sundin, B., et al. (2001). A protein interaction map for cell polarity development. J. Cell Biol. **154**(3), 549–571.

Du, W., Vidal, M., et al. (1996). RBF, a novel RB-related gene that regulates E2F activity and interacts with cyclin E in Drosophila. Genes Dev. **10**(10), 1206–1218.

Durfee, T., Becherer, K., et al. (1993). The retinoblastoma protein associates with the protein phosphatase type 1 catalytic subunit. Genes Dev. **7**(4), 555–569.

Fields, S., and Song, O. (1989). A novel genetic system to detect protein-protein interactions. Nature **340**(6230), 245–246.

Hartwell, L. H., Hopfield, J. J. et al. (1999). From molecular to modular cell biology. Nature **402**(6761 Suppl.), C47–C52.

Ito, T., Tashiro, K., et al. (2000). Toward a protein-protein interaction map of the budding yeast: A comprehensive system to examine two-hybrid interactions in all possible combinations between the yeast proteins. Proc. Natl. Acad. Sci. USA **97**(3), 1143–1147.

Orr-Weaver, T. L., Szostak, J. W., et al. (1983). Genetic applications of yeast transformation with linear and gapped plasmids. Methods Enzymol. **101**, 228–245.

Reboul, J., Vaglio, P., et al. (2003). C. elegans ORFeome version 1.1: Experimental verification of the genome annotation and resource for proteome-scale protein expression. Nature Genet. **34**(1), 35–41.

Uetz, P., Giot, L., et al. (2000). A comprehensive analysis of protein-protein interactions in Saccharomyces cerevisiae. Nature **403**(6770), 601–603.

Vidal, M., Brachmann, R., *et al.* (1996). Reverse two-hybrid and one-hybrid systems to detect dissociation of protein-protein and DNA-protein interactions. *Proc. Natl. Acad. Sci. USA* **93**(19), 10315–10320.

Walhout, A. J., Sordella, R., *et al.* (2000). Protein interaction mapping in *C. elegans* using proteins involved in vulval development. *Science* **287**(5450): 116–122.

Walhout, A. J., Temple, G. F., *et al.* (2000). GATEWAY recombinational cloning: Application to the cloning of large numbers of open reading frames or ORFeomes. *Methods Enzymol* **328**, 575–592.

Walhout, A. J., and Vidal, M. (2001). High-throughput yeast two-hybrid assays for large-scale protein interaction mapping. *Methods* **24**(3), 297–306.

Functional Proteomics

Chromophore-Assisted Laser Inactivation of Proteins by Antibodies Labeled with Malachite Green

Thomas J. Diefenbach and Daniel G. Jay

I. INTRODUCTION

Chromophore-assisted laser inactivation (CALI) permits direct inactivation of proteins within cells or tissues with high spatial and temporal resolution. CALI converts nonfunction-blocking antibodies into function-blocking antibodies through the use of covalently linked chromophores, such as malachite green isothiocyanate (MGITC) and fluorescein isothiocyanate (FITC). Cells or tissues loaded with labelled antibody are exposed to laser light, which results in the generation of free radicals that locally damage the protein to which the antibody is attached. Using free-radical quenchers, the inactivation mechanism for MG-mediated CALI was found to be dependent on hydroxyl radical generation, with the hydroxyl radicals so generated having a half-maximal radius of inactivation of 15 Å (Jay, 1988; Liao *et al.*, 1994; Linden *et al.*, 1992). In the case of FITC-mediated CALI (FALI) (Beck *et al.*, 2002), singlet oxygen is the reactive species, and evidence suggests a half-maximal inactivation radius of ~40 Å (Beck *et al.*, 2002). Thus, the specificity of CALI and the process of photoactivation make possible a protein knockdown strategy with high spatial and temporal control. Protein function can be perturbed directly during precise periods of development without concerns accompanying the use of molecular genetic approaches, such as lethality or compensation. In addition, CALI can be performed on two scales: CALI and micro-CALI. CALI employs a laser to produce a 2-mm-diameter laser spot useful for irradiation of tissue samples, culture dishes, or microtiter wells. Micro-CALI uses a less powerful laser linked to an inverted microscope. By focussing laser light through an objective lens, spot sizes 2–100 μm in diameter are attainable. Micro-CALI can therefore be used to target individual cells or subdomains within cells. Furthermore, the utility of CALI has been expanded to include targeting of EGFP–protein constructs (Rajfur *et al.*, 2002), RNA sequences (Grate and Wilson, 1999), and the use of a membrane-permeable fluorescein derivative that targets recombinant protein sequences (Adams *et al.*, 2002; Marek and Davis, 2002). For recent, comprehensive reviews of the utility of CALI, see Buchstaller and Jay (2000) and Rubenwolf *et al.* (2002). For application-specific use of CALI, consult Beermann and Jay (1994).

II. MATERIALS AND INSTRUMENTATION

A. Micro-CALI

For micro-CALI, a nitrogen laser (Model VSL-337ND-S, Laser Science, Inc., Franklin, MA; 75 kW peak power; Model 337201-00 for a 110-V circuit) is used in conjunction with a DUO-220 visible tunable dye laser module (Laser Science, Inc.; Model 337220-00 for a 110-V circuit). A more compact, less powerful (30 kW peak power) alternative is the VSL-337 nitrogen laser (Laser Science, Inc.; Model 337000-00 for 110 V) with DYE-120 (Laser Science, Inc.; Model 337120) visible tunable dye laser module. For CALI using malachite green, DCM dye [4(dicyanomethylene)-2-methyl-6-(p-diethylaminostyryl)-4H-pyran](Laser Science, Inc., Cat. No. 337999 PBD) is used to generate 630 nm light (the peak excitation wavelength of malachite green is 620 nm). The dye resides in a cuvette in the dye laser module. Dyes are also available for fluorescein and GFP excitation wavelengths. The laser light is routed through the rear port of a Nikon Diaphot 200-inverted microscope or similar type of microscope (i.e., Zeiss Axiovert) (Fig. 1).

The rear microscope port has attached to it a Nikon TE dual mercury lamp house adapter (Micro Video Instruments Inc.), to which the 100-W mercury arc lamp is mounted opposite the laser input. The lamp house adapter permits rapid switching between fluorescence and laser illumination using a rotating mirror. The equivalent attachment for the Zeiss Axiovert is the dual-mirror lamp housing attachment (Cat. No. 447230, Carl Zeiss MicroImaging, Inc.). The laser light is directed through a dichroic mirror that reflects wavelengths greater than 600 nm (e.g., Cat. No. 590DCLP, Chroma Technology Corp.) mounted in a Nikon epifluorescence filter cube. Thus, no excitation or barrier filters are required. The 590DRSP dichroic reflects greater than 90% of 620-nm laser light. The

laser light then passes through an objective lens, which should be the highest numerical aperture (NA) possible at either 20, 40, 60, or 100× magnification. If the NA is too low, insufficient light will reach the specimen plane. Examples of suitable objective lenses include a Nikon Plan Apo 40× 0.95NA DMPh3 (a phase-contrast objective lens) and a Nikon Plan Apo 60× 1.4NA oil objective lens or equivalent. Low fluorescence immersion oil is important for micro-CALI to prevent light scattering. A recommended oil is Cargille type DF (Cat. No. 04108A-AB, Structure Probe, Inc., 120 ml) and is suited for inverted microscope use only (it has higher viscosity than oil for upright microscopes). The laser light then passes through No. 1.5 coverslips (Cat. No. 12-541B, Fisher Scientific Company, 22 × 22 mm), which are optimized for most microscope objectives that are standardized for use with coverslips of 0.17 mm thickness. Laser light intensity can be measured at the specimen plane using a light meter (see later). Because malachite green is strictly a chromophore, there is no fluorescence emission to observe during micro-CALI. However, the laser light itself is in the visible spectrum and may partially obscure the center of the image field. To block this light, a low-pass filter that transmits wavelengths below 620 nm can be

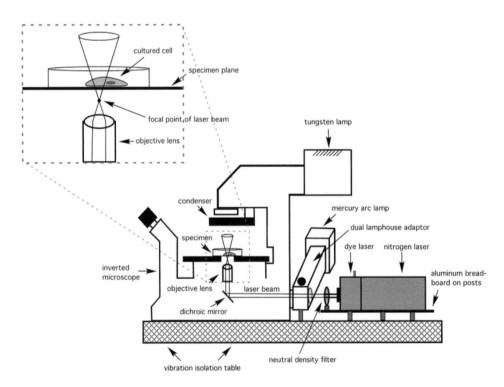

FIGURE 1   A schematic of micro-CALI configuration. Light from a nitrogen dye laser is directed to the rear port of an inverted epifluorescence microscope. An expanded view of the dotted square illustrates that the focal point of the laser beam should be above the specimen plane so that the specimen is exposed to the divergent portion of the beam.

placed in the filter cube or port leading to a CCD digital camera (custom filter from Chroma Technology Corp.). The laser itself is mounted on a solid 300 × 600-mm aluminum breadboard (Cat. No. M-SA-12 and other mounting hardware from Newport Corp.) that stands on six aluminum posts (Cat. No. TPS1, 2, or 3 metric posts, Newport Corp.) held in place with sliding base clamps (Cat. No. SB-TPS, Newport Corp.). To maintain alignment of the laser with the microscope optics, it is important that the laser and microscope be stationed on a breadboard (Cat. No. MIG-33-2-ML or higher quality, Newport Corp.) mounted onto a vibration isolation table (any model from Newport). A neutral density filter set is used to reduce the laser light intensity (and to some extent laser spot size) without altering the wavelength. A set of five filters—0.2, 0.3, 0.5, 0.6, and 1.0—are recommended and can be used singly or in combination (Cat. No. 22000A UV ND, Chroma Technology Corp.). The neutral density filters are mounted on a breadboard between the laser and the microscope using an optical mount (P100-AC "Performa" performance optical mount with 1-in. aperture, Newport). Opticians refer to percentage transmission (the reciprocal of which is percentage absorbance and therefore the percentage reduction of light transmitted). A neutral density filter of ND 0.2 = 63% transmission (37% reduction), ND 0.3 = 50% (50% reduction), ND 0.5 = 32% (68% reduction), ND 0.6 = 25% (75% reduction), and ND 1.0 = 10% (90% reduction). When placed in series, the reduction in light intensity is additive. Without neutral density filters or an objective lens, the light measured at the specimen plane with the VSL-337ND-S laser with dye module is approximately 60 µJ average energy at 30 Hz (3 ns pulse duration), down from the 300-µJ output without the dye module. A recommended laser light meter is the Orion PE with PE10 head (2 µJ–10 mJ sensitivity) from Ophir Optronics Ltd. To align the laser, the center of the laser light birefringence pattern should be at the center of the primary reflecting mirror. The microscope can then be positioned so that the laser light is seen exiting the objective lens. To visualize and confirm laser output and alignment at the specimen plane, irradiate a spot of ink (blue ink for 630 nm light, red ink for 488 nm light) on a glass coverslip at the specimen plane (Fig. 2A) using the desired objective lens.

The ink will absorb the laser light, creating a hole in the ink and delineating the margins of the laser spot. This spot can be imaged digitally on a computer and used as an overlay. Alternatively, a stage micrometer can be used to reference the position of the spot. It is good practice to fire the laser about 10 times to confirm the boundaries of the hole (Fig. 2B) as some burn-in can occur (spot area increases), especially if the laser is

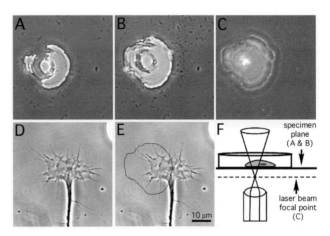

FIGURE 2 (A) Using a 40× objective lens, the shape and position of the laser beam can be determined by irradiating a glass coverslip marked with blue ink. Blue ink absorbs the 630-nm light and burns away the ink, leaving a bright spot. (B) After 10 laser pulses, more ink has been burned away, leaving a slightly larger spot. After about 10 pulses, the spot does not change significantly. (C) By focussing the objective lens below the level of the ink (or from the viewer's perspective, above the specimen plane), the focal point (and therefore the centre) of the laser beam can be located in relation to the spot in the ink. (D) A growing tip of an axon, the growth cone, from a neuron isolated from dorsal root ganglion of embryonic chick. The microspikes, or filopodia, are clearly seen emanating from the growth cone. (E) The perimeter of the laser spot from B is superimposed onto the growth cone (black line) showing how the spot can be used to irradiate a small portion of a cellular structure (scale bar: 10 µm).

not aligned perpendicular to the specimen plane. Depending on the distance between the laser and the objective lens, the focal point of the laser light will likely fall above or below the specimen plane (Fig. 2C). It is important that the focal point of the laser is *not* at the specimen plane as this will cause thermal damage (and even chipping of the glass coverslip) on account of the very high power density of coherent laser light at a focal point (Fig. 2F). It is also important to check manufacturer's specifications of the objective lens used to ensure that the lens coatings can withstand laser light (specifically, peak laser power) without degrading. The duration of continuous irradiation for a single CALI treatment should not exceed 5 min as periods greater than this may cause quenching of the chromophore. To minimize nonspecific effects of long-term laser illumination, periods of irradiation could be interspersed with 5-min rest periods. Because the size of the laser spot is typically 10–20 µm in diameter (although with lower power objectives it can be over 100 µm in diameter), micro-CALI can be used to inactivate proteins in single cells or small regions of cells. For example, we have used micro-CALI to examine the function of distinct myosins (Diefenbach *et al.*, 2002) in the growing tips of nerve cell axons (Figs. 2D and 2E), which are typically 10 μm in diameter.

B. CALI

To avoid stray or direct laser light entering the eyes, approved safety goggles are imperative for use with class IV lasers, in addition to following recommended safety precautions of the laser manufacturer. This includes using light-absorbing material (black cloth or tinted plexiglass) as a shield to prevent reflection of stray laser light. For CALI, a tunable neodymium : yttrium–aluminum–garnet (Nd : YAG) laser (Quanta Ray Model GCR-11, Spectra Physics) is used to pump a custom-fabricated dye laser using the DCM dye as for micro-CALI. Alternatively, a tunable Surelite OPO (optical parametric oscillator) Nd : YAG laser (Continuum Scientific Service) is also suitable (Fig. 3).

The laser output is 630 nm for MGITC (488 nm for FITC), 10 Hz, 6 ns pulse width, and 15 mJ peak energy per pulse. A beam splitter (glass slide) can be used to take a fraction (i.e., 1/6th) of the light and direct it at a 90° angle to a laser light meter (Model JD500 or EM400, Molectron Detector, Inc.). The light is then directed vertically through a right-angle prism and focussed using a 25-mm-diameter plano-convex lens (Cat. No. SPX016AR.14, Newport Corporation) to a 2-mm diameter spot (typically the spot is a flat oval, 2 mm on the short axis and 4 mm on the long axis). Spot size and shape can be recorded, viewed on preexposed photographic paper (600 film, Polaroid Corp.), and altered using an adjustable miniature diaphram (Cat. No. MH-2P, Newport Corporation) placed between the output of the dye laser and the 1/6th beam splitter. To test laser effectiveness at the

specimen plane, control *in vitro* experiments are necessary to establish suitable irradiation times. However, for most applications, the duration of irradiation for a single treatment is typically 2 min and no longer than 5 min. For irradiation of FITC-labelled antibodies, a source of incoherent light can also be used. For example, a 300-W slide projector with a 488-nm narrow band-pass filter has been employed to illuminate microtiter plates (Beck *et al.*, 2002).

Although cost effective, the system has to be configured to ensure uniform illumination over the entire sample, and adequate precautions should be taken to prevent thermal damage, such as cooling of the sample during irradiation or increasing the distance of the sample from the light source.

III. PROCEDURES

A. Antibody Labelling

Antibody labelling involves covalently binding malachite green isothiocyanate (Fig. 4; Cat. No. M-689, Molecular Probes, Inc.) or fluorescein isothiocyanate (Cat. No. F-143, Molecular Probes, Inc.) to nonfunction-blocking, polyclonal or monoclonal IgG antibodies, Fab fragments, or recombinant single chain variable fragments (scFvs).

Solutions

1. *Phosphate-buffered saline (PBS) buffer*: 15 liters of 0.1 M PBS buffer
2. *Hank's balanced salt solution*: Ca^{2+}/Mg^{2+} free, 15 ml (HBSS, Gibco, Cat. No. 24020117, Invitrogen Corporation)
3. *10 mg/ml MGITC*: Weigh 10 mg and resuspend in 1 ml dimethyl sulfoxide (DMSO) in a 1.5-ml centrifuge tube

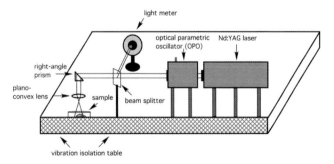

FIGURE 3 A schematic of macro-CALI configuration. An Nd : YAG laser feeds light into an optical parametric oscillator that tunes the wavelength to 630 nm. Light energy is measured from a meter that receives a fraction of the light internally reflected from a beam splitter (glass slide). The light is reflected down to the vibration isolation table surface using a right-angle prism and is focussed to the desired spot size using a plano-convex lens.

FIGURE 4 Chemical structure of malachite green isothiocyanate (MGITC) and fluorescein isothiocyanate (FITC).

4. *Bovine serum albumin (BSA)*: 5 mg/ml (500 µl; weigh 5 mg and resuspend in 1 ml PBS) for blocking the membrane of the concentrating tube and 20 mg/ml (weigh 20 mg and resuspend in 1 ml PBS) to combine with the antibody solution

5. *Antibody solution*: Typically 100 to 200 µg in physiological solution suited to the target cells (i.e., Hank's balanced salt solution without Ca^{2+}/Mg^{2+})

6. *0.5 M NaHCO₃ stock*: To make 100 ml, take 4.2 g of $NaHCO_3$ and dissolve in less than 100 ml of distilled, deionized water. Add 1 M NaOH dropwise to bring the pH to 9.5 and bring the volume up to 100 ml with water

Steps

1. Dialyze antibodies (10,000 molecular weight cutoff) against PBS buffer (pH 7.4) overnight at 4°C.

2. Antibodies can be concentrated by centrifugation (see step 10).

3. Prepare MGITC as a 10-mg/ml solution in DMSO. Because MGITC is very hygroscopic, it should be prepared fresh from powder and kept in a sealed tube just prior to labelling.

4. Add an equal (1 : 1) amount of BSA (Cat. No. B2518, Sigma-Aldrich Corporation) to the antibody solution as a stabilizing reagent.

5. Add a 10-fold molar excess of MGITC to the antibody solution to yield an optimal labelling ratio of 5 MGITC : 1 antibody molecule (or 1 MGITC : 1 Fab or scFv). Typically, between 100 and 200 µg of antibody is labelled at a time. The volume of antibody should be between 20 and 500 µl, as too large a volume of reaction mixture will reduce labelling reaction efficiency.

6. Add 500 µl of 0.5 M NaHCO₃ (pH 9.5) to the antibody solution.

7. Add MGITC to the antibody solution in a stepwise fashion, which favors the labelling reaction over the hydrolysis reaction. Divide the total volume of antibody into three equal aliquots. Add the first aliquot and mix gently for 5 min, followed by the second aliquot for another 5 min, and incubate the third aliquot for 15 min. A rocker (Cat. No. 14-512-28, Lab-Line Maxi Rotator, Fisher Scientific Company) is used to gently mix the reaction mixture during incubation.

8. The free MGITC label in solution is an emerald green color. As the labelling reaction proceeds (hydroxylation at the central carbon of the triarylmethane of amino groups, forming a thioester), the color changes to a deep sea-blue color. The hydrolysis product is purple and forms an insoluble precipitate.

9. After incubation, centrifuge the mixture at high speed (6400 rpm) on a minicentrifuge (Cat. No. EF4241A, Costar WX4241A, Daigger Laboratory Supplies) for 30 s to precipitate any hydrolyzed dye.

10. Separate unbound MGITC and buffer salts from the reaction mixture using a prepacked, 5-cc Sephadex G25 gel filtration column (PD-10, Cat. No. 17-0851-01, Amersham Biosciences Corporation) using a medium suitable for cell loading as an eluant, such as HBSS (Cat. No. 24020117, GIBCO, Invitrogen Corporation) with or without calcium and magnesium.

11. Usually 1–1.5 ml of eluant is collected. Depending on the initial concentration, concentrate the antibody solution to 1.0 mg/ml in Centricon tubes (30,000 MW cutoff; preblocked with 5 mg/ml BSA) using a centrifuge (Sorval RC-5, or equivalent) at 4200 rpm and 20°C. The eluant should be kept above 15°C to avoid precipitation of residual, free MGITC, which is toxic to cells.

12. Once concentrated, determine the molar labelling ratio by measuring the optical density of a diluted sample of the solution at 620 nm based on a molar aborptivity of MGITC of 150,000 $M^{-1}cm^{-1}$ and dividing the concentration of dye by the original concentration of the mixture as follows:

MGITC to protein ratio = [MGITC]/[IgG protein]

[MGITC] = [(optical density @ 620 nm) × dilution factor of sample]/150,000 $M^{-1}cm^{-1}$

[IgG protein] = [(mass of antibodies/ molecular weight of IgG) + (mass of BSA/ molecular weight of BSA)]/total volume of labelled antibody recovered from column

The molecular weight of IgG is approximately 150,000 and for BSA it is 66,000.

13. Aliquot the concentrated antibody solution, quick freeze on dry ice, and store at −80°C, which will keep it viable for up to 6 months, after which time some of the dye dissociates from the antibody.

B. Antibody Loading

Steps

1. It is advisable to centrifuge the aliquots of MGITC-labelled antibody prior to cell loading to precipitate and thus separate any free dye from the labelled antibody solution. Free MGITC is toxic to cells, whereas bound MGITC is not.

2. Using the supernatant collected, perform loading as per the following suggestions. Labelled antibodies raised against extracellular or cell surface proteins can be applied directly to cells in culture or injected into tissue *in vivo*. For intracellular protein targets, MGITC-labelled antibodies can be loaded using trituration (Borasio *et al.*, 1989; see article by Clarke *et al.*), microinjection, lipofection, or electroporation. Generally, the cell type dictates the type of loading permissible, with

neurons requiring microinjection or trituration, and nonneuronal cells being more tolerant of lipofection or electroporation. Cells sensitive to increases in intracellular calcium (i.e., neurons) can be loaded and washed in calcium/magnesium-free HBSS. To confirm loading, cells can be coloaded with either nonimmune FITC-IgG (i.e., Cat. No. F6397, Sigma) or FITC-dextran (Cat. No. D-7168, Molecular Probes). Because free, unbound dye in the MGITC–antibody solution is toxic, loading should not take any longer than 5 min to minimize the exposure of cells to the antibody solution.

3. For trituration loading, treat cells with 0.25% trypsin (Sigma) in HBSS for 15 to 30 min at 37°C, although this step will depend on cell type.

4. Wash cells in HBSS or Ca^{2+}/Mg^{2+}-free HBSS.

5. Remove most of the HBSS and add the MGITC-labelled antibody solution (50 μl or less). Cells are trituration loaded by gently but rapidly passing the cells through either a P200 pipetteman tip about 50 times or through a fire-polished (tip diameter 0.5 mm or less) glass pipette about 5–10 times. It is important that bubbles are not formed during trituration, as exposure of cells to air will reduce their viability.

6. Wash cells immediately after loading via centrifugation at low rpm (30 s at $1400\,g$ on a Costar WX4241A minicentrifuge, Cat. No. EF4241A, Daigger Laboratory Supplies) to remove the antibody solution and gently resuspend in HBSS or Ca^{2+}/Mg^{2+}-free HBSS.

7. Wash cells again in HBSS and resuspend in cell culture medium prior to plating.

8. To confirm loading, use epifluorescence microscopy to locate cells that contain FITC-dextran or FITC-IgG as the loading marker. If cell viability is a concern, viability can be determined using the trypan blue dye exclusion test (de Costa et al., 1999). The half-life of MGITC-labelled antibodies in loaded cells will depend on the cell type. Six hundred and twenty-nanometer laser light penetration can be up to 5 mm; however, this depth will be highly dependent on tissue type, with shorter wavelengths penetrating less. To confirm effectiveness of CALI or penetration of laser light in previously loaded tissues in vivo, laser irradiation followed by immunohistochemistry of the target protein can be used to demonstrate loss of antigenicity through CALI-induced damage of the target protein (Sakuri et al., 2002).

IV. COMMENTS

Control Experiments for CALI

To establish a specific effect of CALI on a protein target, the following control experiments can be performed.

1. Cells that have not been loaded with antibody are exposed to laser light (no antibody, laser control) to exclude an effect of laser light alone.

2. Cells are loaded with preimmune MGITC-labelled IgG and exposed to laser light (IgG control) as an antibody specificity control and MG-labelled antibody control.

3. Cells are loaded with MG-labelled specific antibody without exposure to laser light (MG–antibody, no laser control) as a control for the non-function-blocking nature of the antibody and an MG toxicity control.

4. As a control for nonspecific free radical damage of neighboring proteins during CALI, cells can be loaded with an MG-labelled Fab fragment derived from an anti-Fc antibody (MG–anti-Fc). By directing this MG–anti-Fc against the unlabelled primary antibody, the MG is effectively positioned ~100 Å away from the antibody-binding site (Sakuri et al., 2002). Because the half-maximal radius for inactivation by MG is 15 Å (Liao et al., 1994), the additional distance from the antigenic site should yield no significant effect with laser irradiation.

5. In vitro CALI of an isolated protein preparation using an assay of protein function can be used to demonstrate loss of protein function that complements results of CALI from cell-based assays.

6. An assay of β-galactosidase function can be used as a positive control for CALI (Liao et al., 1994).

V. PITFALLS

1. The utility of CALI is determined in part by the retention of damaged protein in the cell region irradiated. Diffusion of unirradiated protein, degradation of damaged protein, or synthesis of new protein will determine the half-life of the effect of CALI in cells and tissues. Retention of labelled antibody will also determine the utility of CALI and will be cell type specific. For proteins subjected to CALI thus far, the effectiveness of CALI has been observed typically up to 10–15 h after loading (Diamond et al., 1993; Sakurai et al., 2002), while over 15 h after loading there can be considerable reduction in effectiveness (Sakurai et al., 2002).

2. An important consideration in the efficiency of CALI is the specificity of the antibody. The antibody may be highly specific for its intended epitope, sufficiently abundant inside or outside the cell, and optimally labelled with MGITC or FITC, but may still not have an effect on protein function if it binds to a site on the protein that is not essential for the biological function being tested. Polyclonal antibodies that rec-

ognize a greater number of epitopes on a single protein can partially circumvent this limitation, as could the use of more than one monoclonal antibody. In addition, addressing protein function at the level of more than one epitope can be helpful in elucidating the function of distinct protein domains.

3. Abundance of the native protein should be considered when determining the effectiveness of CALI. Because CALI is a knockdown strategy employing a limited number of antibodies per cell, typically less than 100% of the protein targeted is inactivated in an entire cell. Thus, it is technically difficult to inactivate a highly abundant intracellular protein, such as actin. Despite this limitation, most proteins will occupy a small percentage of the total protein in a given cell type, and antibodies targeted against extracellular domains of integral membrane proteins would not be so limited.

4. There are a number of considerations when selecting MGITC or FITC for applications utilizing CALI. First, FITC requires a 488-nm excitation wavelength of light. This wavelength has higher energy and is absorbed more readily by cells and tissues, which necessitates brief illumination periods with laser light, and cell viability tests with longer illumination periods using incandescent light. MGITC requires a longer excitation wavelength (620 nm), which not only has lower energy, but is also a wavelength not significantly absorbed by biological material. Second, FITC has been reported to be 50 times more efficient than MGITC in *in vitro* assays (Surrey *et al.*, 1998). The half-maximal radius of inactivation of FITC (~40 Å; Beck *et al.*, 2002) is 2.7 times greater than that of MGITC (15 Å). However, the inactivation radius of FITC is still half the average distance between proteins in a cell (80 Å; Linden *et al.*, 1992). Therefore, although there is a greater chance of nonspecific effects of FITC-induced radical generation on neighboring proteins, appropriate control experiments can be used to rule out such effects. Third, unlike *in vitro* assays, cell-based assays contend with the scavenging of free radicals by endogenous scavengers. Ascorbate and glutathione scavenge hydroxyl radicals, and ergothioneine and carotenoids scavenge singlet oxygen (Chaudiere and Ferrari-Iliou, 1999). The relative abundance of these scavengers will determine the lifetime, and therefore the radius, of generated free radicals. Furthermore, scavenging ability is dependent on the antioxidant capacity of the cell (Beiswanger *et al.*, 1995), with proliferating or tumorigenic cells in general displaying a greater capacity to deal with oxidative stress (for review, see Das, 2002). Fourth, oxidative damage resulting from irradiation of MGITC (hydroxyl radical) or FITC (singlet oxygen) occurs through distinct mechanisms. For the hydroxyl radical, hydroxylation

of aromatic amino acids likely occurs through an addition reaction to form a hydroxycyclohexadienyl radical on the aromatic ring, which is then oxidized to form a phenol (Halliwell and Gutteridge, 1989). Singlet oxygen likely reacts with tryptophan, histidine, methionine, and cysteine. In general, the reactions are very complicated and can proceed through several mechanisms, yielding distinct decomposition products (Halliwell and Gutteridge, 1989). Finally, hydroxyl radical and singlet oxygen differ in their ability to react with amino acids. The second order rate constant (from Halliwell and Gutteridge, 1989) for the reaction of histidine with singlet oxygen is 30 times less ($10^8 \, M^{-1} \, s^{-1}$) than the reaction with hydroxyl radical ($3 \times 10^9 \, M^{-1} \, s^{-1}$), 300 times less for methionine (OH' = $5.1 \times 10^9 \, M^{-1} \, s^{-1}$, singlet oxygen = $1.7 \times 10^7 \, M^{-1} \, s^{-1}$), and 283 times less for tryptophan (OH' = $8.5 \times 10^9 \, M^{-1} \, s^{-1}$, singlet oxygen = $3 \times 10^7 \, M^{-1} \, s^{-1}$). Thus, singlet oxygen shows much less reactivity than hydroxyl radical to amino acids targeted by either radical, and the presence of endogenous scavengers may amplify these differences. In short, comparison of results of *in vitro*-based CALI assays and cell-based CALI assays, or of experiments using MGITC or FITC, should keep in mind the differences inherent to these applications.

References

Adams, S. R., Campbell, R. E., Gross, L. A., Martin, B. R., Walkup, G. K., Yao, Y., Llopis, J., and Tsien, R. Y. (2002). New biarsenical ligands and tetracysteine motifs for protein labelling *in vitro* and *in vivo*: Synthesis and biological applications. *J. Am. Chem. Soc.* **124**, 6063–6076.

Beck, S., Sakurai, T., Eustace, B. K., Beste, G., Schier, R., Rudert, F., and Jay, D. G. (2002). Fluorophore-assisted light inactivation: A high-throughput tool for direct target validation of proteins. *Proteomics* **2**, 247–255.

Beermann, A. E., and Jay, D. G. (1994). Chromophore-assisted laser inactivation of cellular proteins. *Methods Cell Biol.* **44**, 715–732.

Beiswanger, C. M., Diegmann, M. H., Novak, R. F., Philbert, M. A., Graessle, T. L., Reuhl, K. R., and Lowndes, H. E. (1995). Developmental changes in the cellular distribution of glutathione and glutathione S-transferases in the murine nervous system. *Neurotoxicology* **16**, 425–440.

Buchstaller, A., and Jay, D. G. (2000). Micro-scale chromophore-assisted laser inactivation of nerve growth cone proteins. *Microsc. Res. Tech.* **48**, 97–106.

Chaudiere, J., and Ferrari-Iliou, R. (1999). Intracellular antioxidants: From chemical to biochemical mechanisms. *Food Chem. Toxicol.* **37**, 949–962.

da Costa, A. O., de Assis, M. C., Marques Ede, A., and Plotkowski, M. C. (1999). Comparative analysis of three methods to assess viability of mammalian cells in culture. *Biocell* **23**, 65–72.

Das, U. N. (2002). A radical approach to cancer. *Med. Sci. Monit.* **8**(4), RA79–RA92.

Diamond, P., Mallavarapu, A., Schnipper, J., Booth, J., Park, L., O'Connor, T. P., and Jay, D. G. (1993). Fasciclin I and II have distinct roles in the development of grasshopper pioneer neurons. *Neuron* **11**, 409–421.

Diefenbach, T. J., Latham, V. M., Yimlamai, D., Liu, C. A., Herman, I. M., and Jay, D. G. (2002). Myosin 1c and myosin IIB serve opposing roles in lamellipodial dynamics of the neuronal growth cone. *J. Cell. Biol.* **158**, 1207–1217.

Grate, D., and Wilson, C. (1999). Laser-mediated, site-specific inactivation of RNA transcripts. *Proc. Natl. Acad. Sci. USA* **96**, 6131–6136.

Halliwell, B., and Gutteridge, J. M. C. (1989). "Free Radicals in Biology and Medicine," 2nd Ed. Clarendon Press, Oxford.

Jay, D. G. (1988). Selective destruction of protein function by chromophore-assisted laser inactivation. *Proc. Natl. Acad. Sci. USA* **85**, 5454–5458.

Liao, J. C., Roider, J., and Jay, D. G. (1994). Chromophore-assisted laser inactivation of proteins is mediated by the photogeneration of free radicals. *Proc. Natl. Acad. Sci. USA* **91**, 2659–2663.

Linden, K. G., Liao, J. C., and Jay, D. G. (1992). Spatial specificity of chromophore-assisted laser inactivation of protein function. *Biophys. J.* **61**, 956–962.

Marek, K. W., and Davis, G. W. (2002). Transgenically encoded protein photoinactivation (FlAsH-FALI): Acute inactivation of synaptotagmin I. *Neuron* **36**, 805–813.

Rajfur, Z., Roy, P., Otey, C., Romer, L., and Jacobson, K. (2002). Dissecting the link between stress fibres and focal adhesions by CALI with EGFP fusion proteins. *Nature Cell Biol.* **4**, 286–293.

Rubenwolf, S., Niewohner, J., Meyer, E., Petit-Frere, C., Rudert, F., Hoffmann, P. R., and Ilag, L. L. (2002). Functional proteomics using chromophore-assisted laser inactivation. *Proteomics* **2**, 241–246.

Sakurai, T., Wong, E., Drescher, U., Tanaka, H., and Jay, D. G. (2002). Ephrin-A5 restricts topographically specific arborization in the chick retinotectal projection *in vivo*. *Proc. Natl. Acad. Sci. USA* **99**, 10795–10800.

Surrey, T., Elowitz, M. B., Wolf, P. E., Yang, F., Nedelec, F., Shokat, K., and Leibler, S. (1998). Chromophore-assisted light inactivation and self-organization of microtubules and motors. *Proc. Natl. Acad. Sci. USA* **95**, 4293–4298.

Protein/DNA Interactions

41

Chromatin Immunoprecipitation (ChIP)

Valerio Orlando

I. INTRODUCTION

Current notion of chromatin *in vivo* tends to include any component, DNA, RNA, histone and non-histone proteins that by interacting with defined chromosomal regions contribute to gene specific, as well as global aspects of gene regulation and in general all DNA-dependent processes that take place inside the nucleus of eukaryotic cells.

Chromatin immunoprecipitation (ChIP) is a deductive *in vivo* method that allows the investigation of any chromatin component in its natural context (Orlando, 2000). By this method, living cells are fixed with formaldehyde and fragmented chromatin is subjected to immunoprecipitation. After reversal of crosslinks, specific genomic fractions obtained by immunopurification can be analyzed by semiquantitave methods like conventional PCR, real-time PCR and Southern hybridization. This method can be used to analyze the *in vivo* distribution of any factor and relate this to the activity of a particular gene or locus. Recent applications of the ChIP technology have also allowed the investigation of binding profiles of specific families of factors and core chromatin modifications on a genome wide scale. The completion of several genome projects and the development of microchips containing whole genome sequence arrays, in combination with the ChIP technology (ChIP on CHIP) has been giving a tremendous impulse to our understanding of genome biology.

II. MATERIALS AND INSTRUMENTATION

Tissue culture media: Serum free2 insect cell culture medium, (Hyclone HYQ-SFX, SH30278.02), Penicillin/Streptomycin 100 mg/ml (EuroClone ECB3001D)

Chemicals: Boric acid (BDH, 20185.291); CsCl, Biotechnology grade (EuroClone EMR 016001), 100% Ethanol (Merck); Ethylendiamine tetraacetic acid Sodium salt (Na-EDTA, Serva, 11278), Ethylene Glycol-bis (β-aminoethyl ether)-N,N,N′,N′-tetraacetic acid (EGTA, Calbiochem, 324626), Formaldehyde 37% solution (Sigma F1268), Hepes (Serva, 25245), Glycerol (J. T. Baker, 7044), Glycine (Serva 23390), Glycogen (New England Biolabs); LiCl (Serva, 28053), NaCl (J. T. Baker 1764), Na-acetate 3-hydrate (BDH, 10235 5P); (Phenol/Chloroform saturated with 100 mM Tris-HCl pH 8 (EuroClone, EMR 187100); Sodium Dodecyl Sulphate (SDS) (BDH 444464T), Tris (hydroxymethyl)methylamine (BDH 443866G)

Detergents: Nonidet P40 (NP40, Euroclone 181250), Triton X-100 (TX-100, Serva 37240), Na-deoxycholate Sodium-salt (Serva, 18330), Sarkosyl (N-Lauroyl-Sarkosine, Sigma L-9150)

Enzymes: Proteinase K (stock 20 mg/ml New England Biolabs); RNase DNase free (stock 10 mg/ml New England Biolabs); Taq Polymerase: EuroTaq (EuroClone, EME010001).

Glass beads (150–212 microns, acid washed, Sigma G-1145); Dialysis tubes (1/4 in., GIBCO, 15961)

Tubes (0.45 mm); Capillary tubes (50 μl Corning 7099S-50).

Instrumentation: Lab top centrifuge (Eppendorf 5415 R); Ultracentrifuge (Beckman L7-65); Rotor: Beckman SW55; Sonifier (Branson 250 or Misonix 200) equipped with a microtip (diam.: 0.2 cm) Refractometer (Horizon 2000); Peristaltic pump (Watson Marlow "Dirrel" 4661); Power supplies, (Pharmacia EPS200); Rotor wheel (FALC F200); Gel Doc 2000 (BIORAD); PCR (MJ PT200); Analytical balance (Sartorius).

III. PROCEDURES

The following protocol was readapted for *Drosophila* tissue culture cells from the one described for mammalian cells by Peggy Farnham and collegues (Weinmann *et al.*, 2002).

Solutions:

1. Fixing solution: 11% Formaldehyde (from a 37% stock equilibrated with methanol), 100 mM NaCl, 1 mM Na-EDTA, 0.5 mM EGTA, 50 mM HEPES pH 8.0. To make 100 ml, 29.7 ml of 37% formaldehyde, 2 ml of a 5 M stock solution of NaCl, 1 ml of a 100 mM stock solution of Na-EDTA, 500 μl of a 0.1 M stock of EGTA and 5 ml of a stock solution of 1 M HEPES pH 8.

2. Phosphate Buffer Saline (PBS) pH 7.4

3. 2. Cell lysis buffer: 10 mM PIPES pH 8, 85 mM KCl, 0.5% NP40, proteinase inhibitors. To make 100 ml, 2 ml of a 0.5 M stock solution of HEPES pH8, 2.83 ml of a 3 M stock solution of KCl, 5 ml of a 10% NP40 stock solution and proteinase inhibitors

4. Nuclei lysis buffer: 50 mM Tris-HCl pH 8, 10 mM EDTA, 0.8% SDS, proteinase inhibitors. To make 100 ml, 5 ml of a 1 M stock solution of Tris-HCl pH8, 2 ml of a 0.5 M stock solution of Na-EDTA pH8, 0.4 ml of a 20% stock solution of SDS and protinase inhibitors.

5. 14. Dilution Buffer: 10 mM Tris-HCl pH 8, 0.5 mM EGTA, 1% Triton X-100, 140 mM NaCl, proteinase inhibitors. To make 100 ml, 1 ml of a 1 M stock solution of Tris-HCl pH8, 0.5 ml of a stock solution of EGTA, 10 ml of 10% Triton X-100 and proteinase inhibitors.

Steps

1. Grow 100 ml *Drosophila* Schneider SL-2 tissue culture cells (25°C, in serum free-medium, Hyclone, 100 U/ml penicillin, 100 μg/ml streptomycin) in cell culture bottles to a density of 3–6 × 10^6 per ml. See Section III, step 1.

2. Pour fixing solution (1/10th of volume of cells, 11 ml) directly into the flask and gently mix. Leave at 4°C for decided fixation time. Fixation time ranges between 5 min up to 1 hr and has to be adjusted empirically by checking the immunoprecipitation efficiency.

1. Stop HCHO fixation by adding solid glycine to 125 mM and mix well. Pellet cells into a 50 ml Falcon tubes (2000 rpm, 5 min., Hereaus Minifuge or eq.) and wash once with cold PBS.

2. Take up cell pellet into 15 ml of cell lysis buffer and leave on ice for 10 minutes. Spin down cells and take up pellet into 2 ml of nuclei lysis buffer, transfer 15 ml Falcon tubes and leave on ice for 10 minutes.

3. Add ca. 0.5 ml glass beads (100–212 microns, acid-washed). Store on ice.

4. Sonicate each aliquot, 3–6 × 30 sec. in 1 min intervals (output near microtip limit), using a high power sonicator (e.g. Branson 250 or Misonix XL). Each tube is cooled in a beaker with an ice/water mix. The sonicator-tip should be immersed roughly 1/4. Avoid foaming.

5. Transfer the suspension into 2 × 1.5 ml Eppendorf tubes (leaving most of the glass beads in the old tube) and spin for 10 min. at mix.-speed in a tabletop centrifuge in the cold room. Transfer supernatants to a 15 ml Falcon tube and add dilution buffer up to 8 ml. Rotate at 4°C for 10 min. on a wheel. Take an aliquot of 50 μl for DNA-analysis (Procedure B). Of the rest prepare aliquots of 1 ml and store at −80°C ore use the chromatin directly for immunoprecipitation (Procedure C).

B. Reversal of Crosslinks

In order to estimate the average size of the DNA and trace it along the gradient, remove about 1/10 vol (50 μl) of each fraction to a microfuge tube containing 50 μl and proceed for DNA purification (reversal of crosslinks). The same procedure (except for gel analysis) is performed after chromatin immunoprecipitation, before PCR analysis (Section III C, steps 6–7).

Solutions

1. TE: 10 mM Tris pH 8, 1 mM EDTA. To make 100 ml, 1 ml of a 1 M stock solution of Tris-HCl pH8, 200 μl of a 0.5 M stock solution of Na-EDTA pH8.

2. 20% SDS, Phenol/Chloroform, Chloroform, Na-acetate pH 5.2.

3. TBE 1× (agarose gel running buffer); to make 1 L of 5× TBE, add to 800 ml of autoclaved distilled water

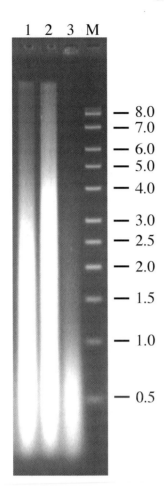

1 2 3 M

— 8.0
— 7.0
— 6.0
— 5.0
— 4.0
— 3.0
— 2.5
— 2.0
— 1.5
— 1.0
— 0.5

FIGURE 1. Determination of chromatin DNA average size after sonication. An aliquot of decrosslinked, purified *Drosophila* SL2 tissue culture cell chromatin DNA was analyzed on a 0.8% agarose gel. A comparison between samples sonicated with or without glass beads is shown. The presence of glass beads ensures a highly efficient and reproducible chromatin DNA shearing. Lane 1: 1% formaldehyde, 10 min crosslinking, no Glass beads. Lane 2: 1% formaldehyde, 20 min crosslink, no Glass beads; Lane 3: 1% formaldehyde, 10 min crosslinking, Glass beads (150–212 microns, acid washed, Sigma G-1145). Similar results were obtained with Drosophila embryos as well as with several mammalian tissue culture cells. Note that a considerable variability can be observed among various preparations depending on crosslinking conditions, sonicator device and operator.

27 g of Tris-HCl, 13.75 g of Boric acid, 10 ml of a 0.5 M stock solution of Na-EDTA. Fill up to 1 L with autoclaved distilled water.

4. 6× DNA loading buffer (Promega, DV 4371).

Steps

1. Add 50 μl of TE and incubate tubes overnight at 65°C.

2. Add Proteinase K (Sigma) to 500 μg/ml and SDS to 0.5%. Incubate at 50°C for ca. 3 h. Extract once with phenol-chloroform and once with chloroform. Precip-

itate chromatin DNA in the presence 0.3 M Na-acetate pH 5.2 with three volumes of ethanol.

3. Spin 30 min wash in 70% ethanol. During centrifugation pour a 0.8% agarose gel.

4. Take up pellets in 10 μl of TE. Add RNase A to 10 μg/ml and incubate 30 min at 37°C. Add 2 μl of 6× loading buffer (Promega). Load and run the gel in 1× TBE 10–15 cm.

5. Stain gel by 0.5 μg/ml ethidium bromide. Take a picture with a GelDoc apparatus (Biorad) or Polaroid film.

C. Immunoprecipitation

Prior to immunoprecipitation of fixed chromatin it is absolutely necessary to test the antibody to check its compatibility with crosslinked material and certain detergents. Normally, "ChIP quality" of the antibody maybe anticipated by its performance in immuno-histochemistry, as the crosslinking agent is often the same. (formaldehyde or para-formaldehyde). Pilot immunoprecipitation experiments using protein extracts or a purified antigen may be performed to check compatibility with certain detergents. The conditions described below work well, especially for rabbit polyclonal antibodies and are considered to be very stringent. Noteworthy, due to the growing interest for histone modifications, and the intrinsic problems of immuno-responsiveness, particular care should be taken in the choice of the commercially available antisera (Perez-Burgos *et al.*, 2003). Particularly for the analysis of histone modifications, the use of non-crosslinked chromatin immuno-precipitation approaches (N-ChIP) and specific immuno-reagents that would recognize non-fixed epitopes, may be considered as a highly valuable alternative to X-ChIP (O'Neill and Turner, 2003).

Solutions

1. RIPA buffer (10 mM Tris-HCl pH 8, 1 mM EDTA pH 8.0, 0.5 mM EGTA, 1% Triton X-100, 0.1% Na-Deoxycholate, 0.1% SDS, 140 mM NaCl, 1 mM PMSF). To make 100 ml, 1 ml of a stock solution of 1 M Tris-HCl, pH 8, 200 μl of a stock solution of 0.5 M Na-EDTA, 0.5 ml of a 0.1 M EGTA stock solution, 10 ml of a 10% Triton X100 stock solution, 1 ml of 10% stock solution of Na-Deoxycholate, 0.5 ml of a 20% SDS stock solution.

6. LiCl-buffer (0.25 M LiCl, 0.5% NP 40, 0.5% Na-Deoxycholate, 1 mM Na-EDTA, 10 mM Tris-HCl, pH 8). To make 100 ml, 5 ml of a 5 M stock solution of LiCl, 5 ml of a 10% stock solution of NP40, 5 ml of a 10% stock solution of Na-Deoxycholate, 1 ml of Tris-HCl pH8 and 200 μl of EDTA.

7. TE (1 mM Na-EDTA, 10 mM Tris-HCl, pH 8), To make 100 ml, 1 ml of a stock solution of 1 M Tris-HCl pH 8, 200 µl of a stock solution of 0.5 M Na-EDTA.

8. Protein A Sepharose beads CL4B (PAS) (Sigma, P 3391) or Protein A/G-PLUS Agarose (Santa Cruz Biotechnology, sc 2003).

Steps

1. Swell PAS by adding 1 vol of RIPA buffer to 1 vol (e.g. 200 µl) of PAS. Rotate on a wheel for 10–20 min at RT. Spin for 5 sec in Eppendorf centrifuge. Take up supernatant and re-suspend PAS in 1 vol of RIPA.

2. Adjust one aliquot (500 µl) of purified chromatin to immunoprecipitation conditions (RIPA buffer). To make a 800-µl suspension add sequentially: 80 µl of a 10% stock solution of Triton X-100, 8 µl of a 10% stock solution of Na-Deoxycholate, 4 µl of 20% stock solution of SDS, 22.4 µl of 5 M stock solution of NaCl, 4 µl a 0.1 M stock solution of EGTA.

3. As a preclearing step, add 20 µl of 50% PAS previously swollen and equilibrated in RIPA (step 1). Incubate for 1–2 hour at 4°C for pre-clearing and spin for 5 minutes in a tabletop centrifuge (14,000 rpm, 4°C).

4. Transfer supernatant to a new tube and add the appropriate amount of antibody (usually dilutions of 1 : 100, 1 : 500). Do not overload with antibody: this will result in higher non-specific immunoprecipitation. The same amount of precleared chromatin is used as negative control, without the addition of antibody (mock-control). Incubate the samples overnight at 4°C on a wheel.

5. Spin samples 10 min. in a tabletop centrifuge (14,000 rpm, 4°C). Transfer supernatants to 1.5 ml new tubes. Add 20 µl of the 50% PAS suspension. After incubation for further 2–4 hours wash the PAS-pellet 5 times with 600 µl of RIPA buffer, 1× with 600 µl LiCl-buffer and 1× with 600 µl TE (pH 8). Always pellet the beads with <u>short</u> spins (15" max speed) with a table-top centrifuge. At the end, take up PAS pellet in 100 µl of TE.

6. Add 1 µg of RNAse (DNAse-free), incubate at 37°C for 30 min. Incubate samples overnight at 65°C.

7. Next day adjust samples to 0.5% SDS and 0.5 mg/ml proteinase K and incubate for 3 more hours at 50°C. Phenol-Chloroform extract the samples. Back-extract the phenol phase by adding an equal volume of TE (pH 8). Combine the aqueous phases and chloroform extract. Precipitate DNA by adding Glycogen to 100 µg/ml as carrier, 1/10 volume of 3 M sodium acetate pH 5.2 and 3 volumes of 100% ethanol. Incubate at –20°C for some hours up to overnight. Spin down DNA and wash pellet in 70% ethanol, air dry briefly and re-suspend the precipitated DNA in 15–50 µl of TE

(depends on initial input material) and store at 4°C (do not freeze) for PCR analysis (Section V).

IV. PURIFICATION OF *IN VIVO* FIXED CHROMATIN FROM DROSOPHILA SL2 (OR MAMMALIAN) TISSUE CULTURE CELLS VIA CSCL GRADIENT

This procedure was described in Solomon and Varshavsky, 1988 and readapted by Orlando and Paro, 1993). Its application is now limited to particular cases in which insoluble macromolecular complexes may interfere with immunoprecipitation efficiency (Schwartz et al., 2005)

Solutions

5. Fixing solution: 11% Formaldehyde (from a 37% stock equilibrated with methanol), 100 mM NaCl 1 mM Na-EDTA, 0.5 mM EGTA, 50 mM HEPES pH 8.0. To make 100 ml, 29.7 ml of 37% formaldehyde, 2 ml of a 5 M stock solution of NaCl, 1 ml of a 100 mM stock solution of Na-EDTA, 500 µl of a 0.1 M stock of EGTA and 5 ml of a stock solution of 1 M HEPES pH 8.

6. Phosphate Buffer Saline (PBS) pH 7.4

7. Wash solution A: 10 mM Tris-HCl pH 8.0, 10 mM Na-EDTA pH 8.0, 0.5 mM EGTA pH 8.0, 0.25% Triton X100. To make 200 ml, 2 ml of a stock solution of 1 M Tris-HCl pH 8, 8 ml of a stock solution of 0.5 M Na-EDTA, 1 µl of a 0.1 M EGTA stock solution, 5 ml of a 10% Triton X100 stock solution.

8. Wash solution B: 10 mM Tris-HCl pH 8.0, 1 mM Na-EDTA pH 8.0, 0.5 mM EGTA pH 8.0, 200 mM NaCl, 0.01% Triton X100. To make 200 ml, 2 ml of a stock solution of 1 M Tris-HCl pH 8, 400 µl of a stock solution of 0.5 M Na-EDTA, 1 ml of a 0.1 M EGTA stock solution, 8 ml of a 5 M NaCl stock solution.

9. Wash solution C: 10 mM Tris-HCl pH 8.0, 1 mM Na-EDTA pH 8.0, 0.5 mM EGTA pH 8.0. To make 200 ml, 2 ml of a stock solution of 1 M Tris-HCl pH 8, 400 µl of a stock solution of 0.5 M Na-EDTA, 1 ml of a 0.1 M EGTA stock solution.

Steps

1. Grow cells at a convenient density (3–6 × 10^6 for *Drosophila* SL2 cells or similar for mammalian cells). The criterion to calculate the amount of cells to be grown is that one needs at least 20–30 microgram DNA per immunoprecipitation. If one assumes that the DNA content of a diploid Drosophila cell nucleus is 0.4 pg, then 1×10^9 cells would contain about 0.4 mg of total DNA. The *Drosophila* genome is ten times less

complex than mammalian genome. Thus, if mammalian cells are to be used, the initial cell input may be changed accordingly, though depending on the antigen and efficiency of immunoprecipitation this may also be unnecessary.

2. Add 1/10 vol/vol of fixing solution to obtain a final concentration of 1% formaldehyde. Fixation time ranges between 5 min up to 1 hour and has to be adjusted empirically.

3. Stop fixation by adding Glycine powder to a final concentration of 0.125 M. Solubilize glycine by gentle shaking.

4. Pellet cells in 50 ml Falcon tubes spun at 700 xg for 5 min.

5. Take up pellets in 15 ml PBS and repeat centrifugation as in step 4.

6. Take up pellets in 15 ml Solution A and gently shake or rotate for 10 min.

7. Centrifuge as in step 4.

8. Take up pellets in 15 ml of Solution B and procede as in steps 6–7.

9. Take up pellet in 5 ml of solution C.

10. Add approximately 1/3 vol/vol of acid-washed glass beads.

11. Sonicate sample (chilled on ice) by 3–5 × 30 sec bursts. Let the tubes rest on ice for approximately 1 min between each burst.

12. Add 250 µl of 10% Sarkosyl (0.5% final concentration).

13. Transfer suspension in 1.5 ml Eppendorf tubes and let rotate on a wheel for 20 min.

14. Spin debris for 5 min in Eppendorf centrifuge at maximum speed. Take up supernatant and transfer to new tubes.

B. Chromatin Purification by Isopycnic Centrifugation

Solutions

3. Dialysis buffer (TEE): 10 mM Tris pH 8, 1 mM EDTA, 0.1 mM EGTA, 5% glycerol. To make 1 L, 10 ml of a 1 M stock solution of Tris-HCl pH 8, 2 ml of a 0.5 M stock solution of Na-EDTA, 5 ml of a 0.1 M stock solution of EGTA.

4. TE: 10 mM Tris pH 8, 1 mM EDTA. To make 100 ml, 1 ml of a 1 M stock solution of Tris-HCl pH8, 200 µl of a 0.5 M stock solution of Na-EDTA pH8.

Steps

1. Add 2.84 g (0.568 g per ml of cell lysate) of CsCl powder to cell lysate (approximately 4 ml) in 15 ml Falcon tube. Mix gently until the salt has dissolved. Adjust volume to 5 ml with buffer TEE supplemented

with 0.5% Sarkosyl. Final density should be 1.42 g/cm^3.

2. Check density with a refractometer or by weighing the suspension on a analytical balance.

3. Transfer the solution into a 5 ml polyallomere Beckman tube (for Beckman SW55 rotor). Spin at 40,000 rpm (Beckman L7-65 ultracentrifuge, SW55 rotor) at 22°C for at least 24 hours.

4. Elute 10 × 50 µl fractions per gradient with a peristaltic pump at about 1 ml/min with tubings of 0.045 inches internal diameter (#1 bottom—#10 top). Alternatively, carefully pipette from the top layer of the gradient 10 × 500 µl aliquots with a Gilson pipette. Check the density profile of fractions with a Refractometer or by weighing each fraction on an analytical balance. The peak-fraction of crosslinked chromatin should have a density of 1.39 g/cm^3. Routinely, chromatin is found at gradient fractions spanning density values between 1.350 and 1.450 g/cm^3.

5. Dialyse fractions in dialysis bags against 300 volumes (approx. 1.5 L) of dialysis buffer. After 2 h, change the buffer and continue dialysis overnight. For chromatin-IPs, fractions that contain the crosslinked chromatin (usually fractions 3–4) are pooled and re-aliquoted in 500 µl aliquots. Chromatin suspension can be either directly processed for immunoprecipitation of stored at −80°C (stable for several months).

V. ANALYSIS OF IMMUNOPRECIPITATED CHROMATIN AND BINDING SITES IDENTIFICATION

A. PCR analysis

When the sequence of the target region of the protein of interest is known, the immunopurified DNA can be directly used as template for a semi-quantitative PCR using primer pairs that span the putative binding sites. In this case amplification is obtained if protein binding occurs, otherwise no amplification will be obtained. Primer pairs (melting temperature 64–68°C, around 25 bp) amplifying 400–500 bp fragments in the target region of interest are designed with the appropriate software, e.g., Proligo®, www.proligo.com). For each primer pair the optimal magnesium concentration (1–2 mM MgCl$_2$) has to be assayed with 200 ng of total genomic DNA from SL-2 culture cells.

Perform PCR in 25–40 µl reactions with the immunoprecipitated material and the genomic control with the optimal magnesium concentration for each primer pair using 1–3 µl of the immunoprecipitated DNA (in 1× reaction buffer, 0,25 mM NTPs, 1 µM primer, 0,5 U Taq polymerase). PCR scheme: 1. 94°C, 3 min, 1×, 2. 94°C, 1 min, 60–65°C, 1 min, 72°C, 1 min, 28×; 3. 94°C, 1 min, 60–65°C, 1 min, 72°C, 7 min; 1×.

For individual primer pairs the annealing temperature and number of cycles may have to be adjusted until no signal is detected for the mock-ip DNA, but the amplification on the genomic template is not altered. Signals obtained with the antibody-ip DNA under these conditions are considered as significant. After the amplification add gel loading buffer and separate half of the reaction on a 1.5% agarose gel and visualize amplified DNA with ethidium bromide.

In order to allow decisions whether the protein of interest or its enzymatic products are significantly presenta a given site in the genome, the measurement of target DNA in the ChIP fraction is analyzed by PCR.

The enrichment for a given ChIP product is function of the relative amount of selected PCR product detected and it should always be higher than the control IP product. This can be measured by quantifying each of the PCR band products by conventional software (e.g., ImageQuant®, or QuantityONE®). The intensities of resolved bands are quantified and plotted in a diagram as percentage of the total input, the total of chromatin DNA used for the immunoprecipitation. In order to have an internal control, it is convenient to have on the same gel serial dilutions of the input. In some cases, a useful control is to include in the analysis promoter sequences of another genes that are likely not to contain the same factors.

When possible, RealTime PCR methodology should be used. The two methods do not differ in terms of "biological significance" of the data produced; RealTime PCR is way less time consuming as it measures the amount of PCR product present in the reaction at a given time point (in particular during the exponential phase) and also gives a direct graphic representation of the measurements as compared to control reactions and input DNA standards.

In all cases, it is imperative to compare ChIP results obtained in functionally distinct contexts, e.g., the same ip performed on the same regulatory region in two different transcription states of the same gene.

Finally, it has to be emphasized that, when comparing ChIP results between different antisera, as each antibody would work different, the ChIP method per se cannot be considered quantitative. This is due to the different precipitation efficiencies of the diverse antisera, thus it is not possible to compare signal intensities from one immunoprecipitation to another when ana-

lyzing the same PCR amplification product obtained with different antisera. Thus, conclusions or comparisons regarding the amounts of the various proteins present in the same region are not feasible.

Current models in gene regulation favor a highly dynamic organization of the eukaryotic nucleus (Osborne et al., 2004). Thus, the intensities obtained from the quantification of the PCR-amplified immuno-purified DNA, may not be considered as an indication for the amount of protein or number of molecules bound to a certain region, but rather as a rate of turnover, a measure for the presence of the molecules of a protein at a certain region at the timepoint of fixation. High intensities would correspond to proteins (or their enzymatic product) present at a genomic region in most of the cells at the timepoint of fixation and viceversa. However, the on/off rate of regulatory factors and multienzymatic complexes from DNA in vivo is in the order of milliseconds. That leads to the conclusion that ChIP enrichments simply testify the mean/event situation represented by million cells at the time of fixation and applied to one ideal cells. Indeed, as ChIP technology works in the range of minutes, the ultimate limit of this technology if talking about single cell reality, may remain the resolution and determination of which factor would be actually present on a particular sequence at the exact time when a given event is taking place.

VI. SINGLE LOCUS UP TO GENOME WIDE ANALYSIS OF CHIP DNA

There are further ways to analyze of immunoprecipitated chromatin. ChIP anaylsis can be performed over a large genomic region or entire chromosomes, by using the immunoprecipitated DNA as a probe in a Southern analysis. This approach has been successfully used in Drosophila (Orlando and Paro, 1993; Orlando et al., 1997; Strutt et al., 1997; Orlando et al., 1998).

An extension of this is the ChIP on CHIP approach, in which the ChIP probe is used to hybridize a genomic microchip. This approach appears to be a highly powerful way to gain insights on direct targets and genome wide distribution of any chromosome associated regulatory protein factor. Details about these applications can be found in the web sites of the Rick Young and Peggy Farnham laboratories (see below and Ren et al., 2000; Lee et al., 2002; Weinmann et al., 2002; Cawley et al., 2004). There procedures will not be described here as working protocols.

Amplification of Immunoprecipitated DNA

This procedure can be used to amplify ChIP DNA for both Southern analysis and ChIP on CHIP experiments in *Drosophila* or mammalian cells. Due to the heterogeneity of the immunoprecipitated chromatin DNA, a linker modified DNA PCR strategy is carried out (Strutt *et al.*, 1997; Orlando *et al.*, 1998). The adapter is prepared as follows: two oligonucleotides, a 24-mer of sequence 5'AGAAGCTTGAATTCGAGCAGTCAG, and a 20-mer of sequence 5'CTGCTCGAATTCAA GCTTCT, are synthesized. Only the 24-mer should be 5' phosphorylated. Equimolar amounts of the two oligonucleotides are mixed and allowed to anneal. Immunopurified-chromatin DNA (approx. 1 ng) is resuspended in 9 μl ligase buffer (12.5 mM MgCl₂, 25 mM DTT, 1.25 mM ATP, 50 mM Tris-HCl, pH 7.6) containing 0.1 μM final concentration of linker. Ligation is carried out by the addition of 4U of T4 DNA ligase (Boehringer) incubation at 4°C for 24 hr. Pilot experiments with digested or sonicated genomic DNA should be performed in advance to test all reagents. To this aim, a fixed amount of ChIP DNA (ranging between 0.1 to 10 ng of DNA) should be tested with various concentrations of linkers (0.1 to 10 micromolar).

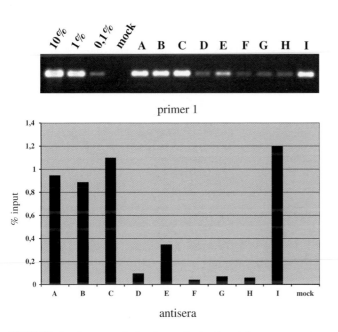

FIGURE 2 An example for the analysis of protein presence at a specific site using X-ChIP. A gel resulting from precipitations of chromatin with various antibodies (A–I) is shown on top. PCR reactions with 10, 1 and 0.1% of the input DNA are loaded to determine the linear range of amplification. Below the gel, a quantitative analysis of ethidium bromide-stained bands is shown. Values on the *y* axis represent the amount of immunoprecipitated DNA as a percentage of the input.

The ligated mixture is directly used as a template in a 100 μl PCR reaction using Taq I polymerase and 1× corresponding buffer. The primer used is the 20-mer oligonucleotide described above added to a final concentration of 1 μM. Amplifications is performed using one cycle of 94°C, 2 min; 35 cycles of 94°C, 1 min, 55°C, 1 min, 72°C, 1 min; 1 cycle of 94°C, 1 min, 55°C, 1 min, 72°C, 10 min.

After PCR, PCR products should be analyzed on a 1.2% agarose gel. Linkers are designed in a way that they can be eliminated by HindIII digestion. ChIP DNA is then purified through a PCR purification kit (Qiagen).

Southern analysis and DNA binding sites mapping

When the immunopurified DNA is used a probe, prior to radiolabeling the amplified DNA is freed from linkers. After PCR samples are extracted once with pheno-chloroform and once with chloroform-isoamylalcohol, and ethanol precipitated. Linkers are removed by digesting the DNA with HindIII and separated by gel filtration (Chroma-spinTE100, Clontech or QiaQuick PCR purification kit (Qiagen)). An aliquot with linkers may be saved to be used as reservoir.

DNA probes are routinely labeled by oligonucleotide random-primed DNA synthesis with α (³²P)-dCTP (Specific activity 3000 Ci/mmol, Amersham).

The hybridization procedure is the one described by Church and Gilbert (1984) with nylon membranes (Gene Screen Plus) preferably baked at 80°C for 2 hr. Briefly, prehybridization is done for 3 hr at 65°C in 7% SDS, 1 mM EDTA, 1% BSA, 0.5 M Na₂HPO₄ (pH 7.2), heat-denatured DNA directly added to prehybridization solution. Filters are washed at 65°C once in 5% SDS, 1 mM Na-EDTA, 0.5% BSA, 40 mM Na₂HPO₄ for 10 min and at least 4 times for 5 min in the same buffer but containing 1% SDS. This protocol seems to guarantee practically no background. This allows, when necessary, the use of higher quantities of probe (up to 50–100 ng/ml) in order to partially compensate for dilution due to its peculiar complexity.

For a more sensitive and quantitative analysis of the hybridization signals a Phosphorimager is used.

PhosphorImager analysis

The identification of a binding site *in vivo* is based on the analysis of the enrichment of specific genomic regions obtained by ChIP. Therefore, determining the baseline as background value above which a given fragment is "elected" as an *in vivo* binding site is crucial. To this aim Southern analysis results are carefully processed by PhosphorImager analysis.

While analyzing Southern hybridization results, specific criteria have to be followed. In particular, the intensity of a hybridization signal is proportional to M_r, and the resulting values should be normalized with respect to M_r.

The amount of signal is given in arbitrary units by the ImageQuant® software or equivalent (e.g., QuantityOne® by Biorad), as a result of the integration of a selected area on the screen corresponding to the band. The absolute value is obtained by dividing the quantitation value by the kb of the fragment. The choice of the Mr set as a reference may reflect the average size of the chromatin fragments (0.5–1 kb). The relative intensity value will be calculated by subtraction of the "background" value obtained from the hybridization and quantification of the corresponding signal on the same fragment obtained by using a mock DNA probe (control/non-enriched). The mock IP probe should hybridize approximately uniformly to all fragments (dependent on M_r).

It has to be mentioned that repetitive elements are always strongly enriched in immunoprecipitations and therefore hybridize strongly to all immunoprecipitated DNA probes. These elements can be identified by their strong hybridisation obtained when genomic DNA is to be used as a probe.

Finally, all relative values of individual genomic restriction fragments may be aligned and plotted against the genomic region of interest, representing the distribution *in vivo* of the protein of interest.

CONCLUSION

In general, ChIP should not be considered as a confirmatory technique. Conversely, ChIP should be used as diagnostic method challenging the *in vivo* distribution of proteins in chromosomes. As more and more novel mechanisms and genetic pathways controlling gene expression are being discovered, several surprises coming from ChIP studies can be anticipated. Moreover, the completion of genome projects and the construction of high resolution genomic microarrays combined with ChIP (ChIP on CHIP) provide unique tools for genome wide target gene identification, epigenetic landscaping and genetic network decoding. However, before embarking on a ChIP-trip, clear cut functional experiments should be planned to help to validate any anticipated result and in most cases try

learn directly from chromosomes, what we do not know yet.

References

Cawley, S., Bekiranov, S., Ng, H., Kapranov, P., Sekinger, E. A., Kampa, D., Piccolboni, A., Sementchenko, V., Cheng, J., Williams, A. J., Wheeler, R., Wong, B., Drenkow, J., Yamanaka, M., Patel, S., Brubaker, S., Tammana, S., Helt, G., Struhl, K., and Gingeras, T. R. (2004). Unbiased Mapping of Transcription Factor Binding Sites along Human Chromosomes 21 and 22 Points to Widespread Regulation of Noncoding RNAs. *Cell* **116**, 499–509.

Church, M. G., and Gilbert, W. (1984). Genomic sequencing. *Proc. Natl. Acad. Scie. USA* **81**, 1991–1995.

Lee, T. I., Rinaldi, N. J., Robert, F., Odom, D. T., Bar-Joseph, Z., Gerber, G. K., Hannett, N. M., Harbison, C. R., Thompson, C. M., Simon I., Zeitlinger J., Jennings, E. G., Murray, H. L., Gordon, D. B., Ren, B., Wyrick, J. J., Tagne, J., Volkert T. L., Fraenkel, E., Gifford, D. K., and Yong, R. A. (2002). Transcriptional Regulatory Networks in *Saccharomyces cerevisiae*. *Science* **298**, 799–804.

O'Neill, L. P., and Turner, B. M. (2003). Immunoprecipitation of native chromatin: N-ChIP. *Methods* **31**, 76–82.

Orlando, V., and Paro, R. (1993). Mapping Polycomb represseddomains in the bithorax complex using *in vivo* formaldehyde cross-linked chromatin. *Cell* **75**, 1187–1198.

Orlando, V., Strutt, H., and Paro, R. (1997). Analysis of chromatin structure by *in vivo* formaldehyde cross-linking. *Methods* **11**, 205–214.

Orlando V., Jane, E., Chinwalla, V., Harte, P. J., and Paro R. (1998). Binding of Trithorax and Polycomb proteins to the bithorax complex: dynamic changes during early embryogenesis. *EMBO J.* **17**, 5141–5150.

Orlando, V. (2000). Mapping chromosomal proteins *in vivo* by formaldehyde crosslinked-chromatin immunoprecipitation. *Trends Biochem Scie.* **25**, 99–104.

Osborne, C. S., Chakalova, L., Brown, K. E., Carter, D., Horton, A., Debrand, E., Goyenechea, B., Mitchell, J. A., Lopes, S., Reik, W., Fraser, P. (2004). Active genes dynamically colocalize to shared sites of ongoing transcription. *Nat. Genet.*, **36**, 1065–1071.

Perez-Burgos, L., Peters, A., Opravil, S., Kauer, M., Mechtler, K., and Jenuwein, T. (2003). Generation and characterization of methyl-lysine histone antibodies. *Methods in Enzymology* **376**, 234–254.

Ren, B., Robert, F., Wyrick, J. J., Aparicio, O., Jennings, E. G., Simon, I., Zeitlinger, J., Schreiber, J., Hannett, N., Kanin, E., Volkert, T. L., Wilson, C. J., Bell, S. P., and Young, R. A. (2000). Genome-wide location and function of DNA binding proteins. *Science* **22**, 2306–2309.

Schwartz, Y. B., Kahn, T. G., and Pirrotta, V. (2005). Characteristic low density and shear sensitivity of cross-linked chromatin containing polycomb complexes. *Mol. Cell. Biol.* **25**, 432–439.

Solomon, M. J., Larsen, P. L., and Varshavsky, A. (1987). Mapping protein-DNA interactions *in vivo* with formaldehyde: evidence that histone H4 is retained at a highly transcribed gene. *Cell* **37**, 937–947.

Weinmann, A. S., Yan, P. S., Oberley, M. J., Huang, T. H., and Farnham, P. J. (2002). Isolating human transcription factor targets by coupling chromatin immunoprecipitation and CpG island microarray analysis. *Genes Dev.* **16**, 235–244.

42

Gel Mobility Shift Assay

Peter L. Molloy

I. INTRODUCTION

Because they are conceptually simple and also relatively straightforward to perform practically, gel mobility shift assays (otherwise known as gel retardation or electrophoretic mobility shift assays) have become one of the most widely used techniques in molecular and cell biology. They provide a key point of entry for identification of protein–DNA interactions important for regulation of gene expression. The discussion in this article focuses on DNA-binding proteins in relation to transcriptional control, but similar principles apply to the use of gel mobility shift assays to study protein/DNA interactions in other processes (replication, recombination, and repair), as well as proteins involved in RNA metabolism. The assay relies on the increased molecular size and decreased charge : mass ratio of a protein–DNA complex compared to free DNA and the observation that many protein–DNA complexes are stable through electrophoresis. Specific protein–DNA complexes can therefore be readily distinguished from free DNA by their slower mobility during electrophoresis. The use of labelled DNA probes in gel mobility shift assays enables visualisation of the specific complexes even in complex protein mixtures. The format of the assays allows both characterisation of DNA sequence requirements for protein binding and characterisation or identification of the proteins involved in complex formation, linking back to the cellular regulatory networks controlling gene expression.

II. MATERIALS AND INSTRUMENTATION

Electrophoresis requires a power supply capable of supplying approximately 30 mA at 200 to 300 V. Conditions described are for a vertical gel apparatus with 20×20-cm plates and 0.75- to 1-mm spacers. Conditions can be scaled down for a minigel apparatus, e.g., Bio-Rad 8×7.3 cm.

General chemicals should all be analytical reagent grade. Solutions for DNA-binding reactions should be prepared using nuclease-free water and reagents. HEPES (Cat. No. H4034), Nonidet P-40 (identical to Igepal CA-630, Cat. No. I8896), and dithiothreitol (DTT, Cat. No. D5545) can be purchased from Sigma Chemical Co. Nonspecific competitor polynucleotides, poly(dI-dC) · poly(dI-dC), poly(dG-dC) · poly(dG-dC), and poly(dA-dT) · poly(dA-dT) are available from Amersham Biosciences, (Cat. Nos. 27-7880-02, 27-7910-02, and 27-7870-02, respectively). Nuclease-free bovine serum albumin (BSA) can be purchased from Promega. Restriction enzymes, Klenow fragment of DNA polymerase 1, T4 polynucleotide kinase, and premixed acry-lamide solutions are available from a number of suppliers.

Radioactive nucleotides $[\alpha\text{-}^{32}\text{P}]\text{dATP}$ or dCTP (>3000 Ci/mmol, 10 mCi/ml) and $[\gamma\text{-}^{32}\text{P}]\text{ATP}$ (>3000 Ci/mmol, 10 mCi/ml) can be purchased from Amersham Biosciences. Reagents for nonradioactive labelling of DNA probes are also available, e.g., digoxygenin-labelling kit from Roche Applied Science and chemiluminescent kit from Pierce. Direct post-electrophoresis staining for DNA and protein components of complexes using sensitive dyes are now also possible (Jing *et al.*, 2002)

Nuclear extracts and related reagents, including specific oligonucleotide sets, are available from Promega, and a kit for nuclear extract preparation can be purchased Pierce. A wide range of antibodies targeted to transcription factors and other DNA-inding proteins is available from Santa Cruz Biotechnology, Inc. and also Chemicon; both catalogues indicate which antibodies are suitable for "supershifting." A range of individual protease inhibitors and prepared cocktails are available from Roche Applied Science and Calbiochem.

III. PROCEDURES

A. Preparation of Probes and Competitor DNAs

Procedures described in this article utilise ^{32}P-labelled probes that have traditionally been used in gel shift assays to allow ready visualisation and potential quantification by autoradiography or phosphorimaging. It is important to use proper shielding (e.g., Perspex screens) to limit exposure to ^{32}P radiation. There are also a number of nonradioactive methods available for visualisation of complexes, including fluorescent probes and biotinylated or digoxygenin-labelled probes. Nonradioactively labelled probes have advantages in safety and length of storage but also have disadvantages, such as the greater number of handling steps needed for visualisation, sensitivity levels, and linearity of signals. Probes are normally prepared by either restriction enzyme digestion and labelling or by labelling of oligonucleotides.

1. Restriction Digestion

Labelling of restriction digests of 1 to 2 µg of plasmid DNA should provide sufficient DNA probe to perform 50 to 100 binding reactions (for a 3-kb plasmid, 2 µg yields about 1 pmol of fragments). To minimise the effect of nonspecific binding to the probe and the number of potential binding sites, restriction fragments should be relatively short (ideally less than 100 bp). If possible, it is best to have binding sites located centrally within the fragment; because affinity is often enhanced by nonspecific interaction of proteins with DNA surrounding their specific binding site, a fragment length of at least 30 to 40 bp is advisable. Restriction enzymes that leave 5' overhangs are most convenient as they can readily be filled in and radiolabelled using the Klenow fragment of DNA

polymerase I and an appropriate [α-^{32}P]dNTP in the presence of other unlabelled deoxynucleotides.

For end labelling, we digest 2 µg of plasmid DNA in a 20-µl reaction in the recommended enzyme buffer. To this is added 20 µCi of suitable ^{32}P-labeled deoxynucleotide for end labelling along with the other three unlabelled dNTPs to a concentration of 100 µM each and 1 unit of Klenow fragment of DNA polymerase I. After a 15-min incubation at room temperature, the fourth unlabelled dNTP is added to 100 µM and incubation is continued for 5 min. Chasing the reaction with the unlabelled nucleotide is important, as single-stranded ends can provide avid binding sites for some proteins.

For optimal gel shift results it is important to gel purify restriction fragment probes. Depending on their size, probes can be separated on 5 to 10% acrylamide gels. For digests of 1–2 ug of DNA, load digest in a 2.5-cm-wide well on a 1-mm-thick gel in TBE buffer and electrophorese at 10 V/cm until the bromphenol blue dye is near the bottom of the gel. For probes 50 bp or less, we routinely run gels and elute small DNA fragments in the cold room. This avoids DNA melting, as single-strand DNAs can produce artefactual results. After the gel apparatus is dismantled, leaving the gel on one of the glass plates, it is covered with plastic wrap and exposed to X-ray film for 2 to 5 min, marking the film for alignment with the gel. The position of the labelled band is identified from the autoradiograph, and the gel slice is excised with a scalpel blade. Fragments can be recovered by elution overnight on a rocking platform or shaker. Depending on its concentration, DNA probes may be used directly or ethanol precipitated and resuspended in TE buffer. Storage of radiolabelled fragments in 0.5 mM DTT or 1 mM β-mercaptoethanol is recommended to limit radiolytic breakdown.

Unlabelled competitor DNA fragments are prepared similarly except that the quantity of DNA is increased to 5 µg, restriction digestion is done in 50 µl, and all four deoxynucleoside triphosphates are added for end filling. A thicker, 2-mm gel should be used and the gel stained lightly with ethidium bromide (soaking in 0.5 µg/ml solution for 5 to 10 min) for fragment visualisation and isolation. The unlabelled competitor needs to be concentrated by ethanol precipitation.

2. Oligonucleotide Probes

Once a target region for protein binding within a gene regulatory region is identified, oligonucleotide probes provide a powerful way to characterise DNA sequences responsible for DNA–protein complex formation. Sets of mutations within the putative binding site can be prepared and assayed readily and

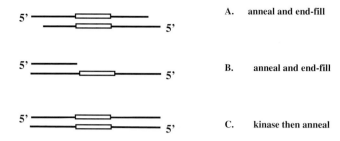

	A.	anneal and end-fill
	B.	anneal and end-fill
	C.	kinase then anneal

FIGURE 1 Preparation of oligonucleotide probes. Three methods as described in the text for preparation of oligonucleotide probes are shown. The boxed region indicates sequences required for protein–DNA binding. (A and B) Oligonucleotides are extended from 3′ ends to flush-ended double-stranded oligonucleotides. (C) Fully two complementary oligonucleotides are annealed after kinase labelling.

binding and migration of complexes can be compared with complexes formed on binding sites for well-characterised proteins. Probes may be prepared by end filling using deoxynucleoside triphosphates (A and B in Fig. 1) or by kinasing one or both strands. For methods A and C, two complementary 25–30 base oligonucleotides must be prepared for each sequence to be studied. For method B, a single 12-base primer adjacent to the binding sequence to be studied can be used to prime on separate oligonucleotides containing variants of the target sequence.

a. End-Filling Reactions. Mix 2 pmol of each oligonucleotide (for method A) or 2 pmol of the long oligonucleotide and 10 pmol of primer for method B in 10 µl of medium salt restriction enzyme buffer, warm to 60°C, and allow to cool slowly to room temperature. Adjust the reaction volume to 20 µl with additional 10× restriction enzyme buffer, 20 µCi of ^{32}P-labeled deoxynucleotide, the other three unlabelled dNTPs to a concentration of 100 µM, and 1 unit of Klenow fragment of DNA polymerase I. Incubation and chase with unlabelled dNTP are as for labelling restriction fragments.

b. Kinase Reactions. Incubate oligonucleotides for 30 min at 37°C in 20 µl final volume reactions containing 2 pmol of oligonucleotide, 2 µL of 10× kinase buffer, 3 µl (30 µCi) of [γ-^{32}P]ATP, and 1 µl of polynucleotide kinase. After kinasing one or both complementary oligonucleotides, mix 20-µl reactions containing 2 pmol of each and add 5 µl of 100 mM MgCl$_2$. Heat the mix to 60°C and allow oligonucleotides to anneal by cooling slowly to room temperature.

After labelling by either method, load reactions on a 10% acrylamide/0.5% bisacrylamide gel run at 4°C in TBE buffer and purify fragments as described

earlier. Gel purification is essential to remove any single-stranded oligonucleotide.

B. Protein–DNA-Binding Reactions

1. Protein Source

Nucleic acid-binding proteins used in gel shift assays may be either purified proteins (endogenous or expressed recombinant proteins) or contained in relatively crude cellular or nuclear extracts. *In vitro* expression in coupled transcription/translation extracts is a convenient way to produce proteins if the gene has been cloned. For many studies, nuclear extracts from cells or tissues in which a gene is expressed provide the starting point for the identification of transcription factors relevant to the expression of a gene. The nuclear extract preparation is based on the principle that elevated levels of salt release specific DNA-binding proteins from chromatin; it is critical that the salt level remains below that which will begin to dissociate histones that bind DNA strongly and non-specifically and interfere with gel shift assays. Preparation of nuclear extracts is usually based on the method of Dignam *et al.* (1984), which is applicable to quantities of 10^8 to 10^9 cells in culture. After isolation of a crude nuclear fraction, proteins are extracted in 0.42 M NaCl and the extract is dialysed against buffer containing 0.1 M KCl. A number of variations of the method have been published that allow for extract preparation from tissue sources (Gorski *et al.*, 1986; Fei *et al.*, 1995) or rapid miniextract preparation from small numbers of cultured cells (e.g., Schreiber *et al.*, 1989). The rapid ammonium sulfate nuclear extract protocol of Slomiany *et al.* (2000) provides a convenient approach for the isolation of high activity extracts from a range of starting sources. When preparing extracts from tissue sources, it is especially important to minimise proteolytic degradation. Early methods included the serine protease inhibitor phenylmethylsulfonyl fluoride, but use of an inhibitor cocktail is recommended to obtain extracts of maximal activity and to avoid potential confusion caused by probes binding to different proportions of intact proteins and proteolytic fragments in extracts from different tissues or cell types.

2. Reaction Setup and Parameters

For each protein–DNA interaction studied it is necessary to optimise the binding conditions. For nuclear extracts and a range of binding proteins, the standard conditions shown in Table I provide a good starting point. Because the amount of a specific protein in an extract and the amount of nonspecific DNA-binding

TABLE I   Protein–DNA-Binding Reaction Conditions with Nuclear Extracts

Component[a]	Standard conditions	Range/alternates
HEPES (pH to 7.9 with 2 M KOH)	12 mM	Tris–HCl pH range 6.5–8.5
KCl	60 mM	0–200 mM sodium or ammonium salts
EDTA	0.6 mM	0.1 mM EGTA for selective inhibition of Ca-dependent proteases
Glycerol	12%	0–12%
Dithiothreitol (DTT)	1.2 mM	0 to 5 mM. *Note*: Binding of some proteins is redox sensitive, whereas for others binding may be enhanced by higher levels of DTT
NP40	0.1%	Up to 2% Tween or Triton X-100 as alternates
poly(dI-dC) : poly(dI-dC).	1 µg/20 µl Rn.	200 ng to 4 µg; other synthetic DNA polymers or mixed sequence DNA such as calf thymus or *E. coli*
Nuclease-free BSA	10 µg/20 µl Rn	Up to 20 µg
MgCl$_2$		Up to 10 mM
Labelled DNA probe	10 fmol	
Temperature	30°C	0–37°C
Time	30 min	5–30 min
Protease inhibitors		Use as required. Cocktail of inhibitors or individual inhibitors: AEBSF (Pefabloc), leupeptin, aprotinin, calpain inhibitors I and II, soybean trypsin inhibitor, chymostatin, pepstatin

[a] HEPES, KCl, EDTA, glycerol, and DTT are all contained within binding buffer A, which comprises 20 mM HEPES, pH 7.9, 100 mM KCl, 1 mM EDTA, 2 mM DTT, and 20% glycerol. Prepare by dissolving the HEPES base, KCl, and EDTA and adjusting the pH to 7.9 using 2 M KOH. Add glycerol and adjust the volume prior to autoclaving. Buffer without DTT can be stored at room temperature. Buffer with DTT should be stored in aliquots at –20°C.

proteins will vary widely, it is important to initially survey a range of protein and nonspecific DNA competitor concentrations to identify levels that allow clear distinction of the protein–DNA complexes.

A setup for a typical exploratory experiment is shown in Table II. Reactions should be set up on ice in 1.5- or 0.5-ml microfuge tubes, first adding all components except the nuclear extract and DNA probe. For nuclear protein extracts, a wide range of protein concentrations, e.g., from 2 to 20 µg in a 20-µl reaction, should be assayed. A typical nuclear extract contains 2 to 4 µg of protein per microliter. Levels of purified proteins need to be titrated to determine optimal levels. It can be preferable to add the nuclear extract prior to the probe if the extract contains significant levels of avid, nonspecific DNA-binding proteins, but in practice the order of addition does not normally make a significant difference. The important point is that the protein and DNA probe should not come into contact until the final mixing step. Binding reactions are typically incubated at 30°C for 30 min. Binding buffer (4 µl) containing bromphenol blue dye is added to aid in gel loading. Provided the binding buffer contains >5% glycerol, there is no need for addition of a density reagent. At this early stage it is also a good idea to test the separation of complexes on different gel systems (see later), as the buffer type and ionic strength can significantly affect separation and band quality.

The exploratory conditions of Table II were applied to the binding site for an ets-related protein (GGAA core sequence) found in the N-ras gene promoter (Fig. 2). The amount of specific complex formed increases with increasing levels of nuclear extract, but at higher levels of extract, a significant fraction of the labelled probe is trapped in the well. A clear signal with minimal background is achieved with 0.5 to 1 µl of nuclear extract and 1 µg of poly(dI-dC) competitor. Replacement of poly(dI-dC) with either poly(dA-dT) or poly(dG-dC) reveals the presence of additional complexes, indicating the value of testing alternate nonspecific competitor DNAs.

Table I indicates a range of possible variations of binding conditions, and the effect of reaction conditions on complex formation should be explored systematically.

TABLE II Exploration of Binding Conditions

Reaction #	1	2	3	4	5	6	7	8	9	10	11	12
Nuclear extract[a]	0	1	2	4	8	12	4	4	4	4	4	4
Binding buffer A[b]	12	11	10	8	4	0	8	8	8	8	8	8
BSA, 10 mg/ml	1	1	1	1	1	1	1	1	1	1	1	1
poly(dI-dC) : poly(dI-dC), 2 mg/ml[c]	1	1	1	1	1	1	0.5	1.5	2	1	1 dAT[e]	1 dGC[e]
NP-40, 2%	1	1	1	1	1	1	1	1	1	1	1	1
MgCl₂, 100 mM[d]											1	
H₂O	3	3	3	3	3	3	3.5	2.5	2	2		
DNA probe (10 fmol)	2	2	2	2	2	2	2	2	2	2	2	2

[a] Conditions based on nuclear extract being dialysed against binding buffer A.

[b] Composition of buffer is in footnote to Table I.

[c] Polynucleotides are dissolved in 10 mM Tris–HCl, 0.1 mM EDTA, pH 8, and stored in aliquots at –20°C.

[d] MgCl₂ should be prepared from fresh solid using sterile, nuclease-free water and filter sterilised.

[e] dAT = poly(dA-dT) · poly(dA-dT)poly and dGC = (dG-dC) · poly(dG-dC), both prepared as in footnote c.

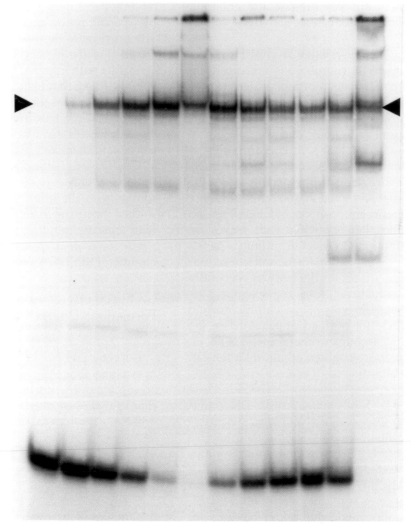

	1	2	3	4	5	6	7	8	9	10	11	12
Nuc. extract (μl)	0	0.2	0.5	1.0	2.0	4.0	1.0	1.0	1.0	1.0	1.0	1.0
poly dI-dC (μg)	1.0	1.0	1.0	1.0	1.0	1.0	0.5	1.5	2.0	1.0		
poly dX-dY, 1μg											dA-dT	dG.dC
MgCl₂, 5 mM	-	-	-	-	-	-	-	-	-	+	-	-

FIGURE 2 **Analysis of protein–DNA-binding conditions.** Binding reactions were set up using a 50 bp fragment containing a binding site for an ets-related protein. Reaction components were as in Tables I and II; the variation in specific components is indicated in the table beneath the gel lanes. Free and bound complexes were separated on a 5% acrylamide gel(30:1 acrylamide: bisacrylamide) run in TNAE buffer. The major specific complex is indicated by the arrow.

a. Time and Temperature. While reactions are often incubated for 30 min, all that is necessary is sufficient time for complex formation to have reached equilibrium, which can often be as short as 5–10 min. Binding reactions are done most commonly at 30°C or room temperature. Lower temperatures sometimes improve binding, e.g., Sp1 binds better at 20°C than at 30°C. Lower temperatures and/or shorter times of incubation can also limit the effect of phosphatases that may be present in some extracts (critical if the active form of a protein is phosphorylated or if probe 5' is end labelled) (Laniel and Guerin, 1998). Similar considerations apply to the inhibition of nucleases if divalent ions are present in the binding reaction. A characteristic downward smearing of the unbound DNA band is indicative of nuclease activity.

b. Ionic Strength. The relative level of formation of specific and nonspecific complexes is influenced significantly by the concentration of monovalent cation; increased salt concentrations can favour the formation of specific complexes. Most nonspecific protein–DNA interactions are principally ionic and the strength of interaction declines as the salt concentration is increased. For sequence-specific complexes, key interactions involve hydrogen bonding and hydrophobic interactions between bases and amino acids, although the interaction between positively charged amino acids and the phosphate backbone still makes an important contribution to overall affinity. When working with crude protein extracts, salt concentrations in the range of 50 to 100 mM are generally optimal, but individual complexes can sometimes be differentiated by their stability at concentrations up to 200 mM. Particularly for purified proteins, when the issue of nonspecific binding by other proteins is not an issue, use of a lower or no added monovalent ion can improve binding affinity.

c. Divalent Ions. While many protein–DNA complexes form in the absence of divalent ions, it is advisable to always evaluate binding in both the presence and the absence of Mg^{2+}, as sometimes binding properties can be affected significantly. An illustrative example is binding of the helix–loop–helix protein USF. In the presence of Mg^{2+}, the rates of both association and dissociation and equilibrium binding are affected significantly and even the sequence specificity of binding is altered (Chodosh *et al.*, 1986; Bendall and Molloy, 1994). In some cases, a specific metal ion can be necessary for DNA binding and as well as inclusion of the metal ion in the binding reaction it may be necessary to omit EDTA from the electrophoresis buffer (Anderson *et al.*, 1990). Examples are the metal response element-binding factor that requires Cd^{2+} and Zn finger proteins, which require low levels of Zn^{2+} for proper protein folding.

d. Nonionic Detergents and BSA. The addition of a nonionic detergent and a carrier protein such as NP-40 and BSA helps minimise protein aggregation and generally results in less smearing of bands and less trapping of the DNA probe in the wells. This is beneficial when working with crude nuclear extracts or to minimise protein denaturation and sticking to tube walls when working with highly purified proteins

e. Nonspecific Competitor DNA. All DNA-binding proteins will bind to a certain extent to both the specific probe and nonspecific competitor DNA. The purpose of the nonspecific competitor is to bind as much of the nonspecific DNA-binding proteins in an extract with minimal binding of the specific protein being studied. Synthetic copolymers have become the preferred choice for competing nonspecific DNA binding because they are less likely to be able to bind efficiently with sequence-specific DNA-binding proteins. Natural, mixed sequence competitor DNAs such as *Escherichia coli* or salmon sperm DNA can also prove to be very effective. However, competitors that have similarity to the binding site of the protein(s) being studied have the potential to interfere significantly with specific binding. For example, poly(dA-dT) would be a poor choice of competitor for studying binding of the TATA-binding protein.

C. Specificity of Protein–DNA Complex Formation: Characteristics of Target DNA Site

Gel retardation assays can be used effectively both to establish that formation of a complex is dependent on specific DNA-sequences and to define the sequence characteristics of the DNA binding site. This can be achieved through a combination of DNA-binding assays using probes of different sequence and competition experiments using unlabelled competitor DNAs.

1. Competitor DNA

Competition assays are based on the principle that the amount of a sequence-specific complex formed on a labelled probe will be reduced in proportion to the amount of competitor DNA if the competitor DNA has specificity for the same sequence. In contrast, if the competitor lacks the specific binding site the additional DNA will compete generally for binding of proteins in the reaction and will add to the overall level of nonspecific competitor DNA. Competitor DNAs may

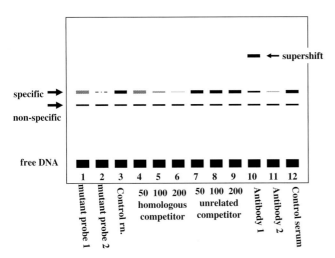

FIGURE 3   Gel mobility shift schematic. The control binding reaction that produces a specific as well as a nonspecific complex is shown in track 3. Reactions are identical to that of track 3 except for a single factor as indicated below each track. Tracks 3 to 12 all contain the same labelled DNA probe. Tracks 1 and 2 contain labelled mutant probes of different binding affinity. In tracks 4 to 6, competitor unlabelled DNA identical in sequence to the probe has been added at 10-, 50-, and 200-fold excess. In tracks 7–9 an unrelated competitor DNA of the same length as the labelled probe has been added at 10-, 50-, and 200-fold excess. In track 10, an antibody that specifically binds the protein in the specific complex without disrupting the DNA-binding interaction has been added. In track 11, antiserum that binds the protein in the specific complex and disrupts complex formation has been preincubated in the reaction prior to probe addition. In track 12, nonimmune control serum has been added.

be either restriction fragments or oligonucleotides (see earlier discussion for preparation), but for comparison between competitors it is important to use DNAs of equivalent size as affinity can increase significantly with increasing fragment size, particularly with oligonucleotides of about 20 bp. Competitor DNAs should minimally include a DNA fragment identical to the probe and a control of equivalent length but unrelated sequence (Fig. 3, tracks 4–9). Competitors with specific mutations can be used to identify or confirm the location of bases important for binding; this approach can be extended to a series of mutants to provide a detailed analysis of the relative importance of different bases to binding affinity. If binding of a known protein is suspected, competition with a DNA fragment that contains a known high-affinity binding site surrounded by unrelated DNA can provide good supporting evidence.

For competition assays the level of DNA-binding protein should not be saturating on the labelled probe; binding of 20 to 30% of the labelled probe in the absence of competitor is a good starting point. It is

reasonable to use up to 1 pmol of competitor in a 20-ul reaction, allowing, for example, 10-, 50-, and 200-fold ratios if 5 fmol of probe is included in the reaction. For proper comparison of relative binding affinities the probe and competitor should be added together. It is important to remember, especially with impure protein preparations, that binding kinetics are not simple and that the final level of binding seen depends not just on the specific binding parameters of the protein with its specific DNA target, but also the binding of the protein to both specific and nonspecific competitor DNAs and to other proteins in the reaction, as well of the binding of the DNA probe to other proteins.

2. Direct Analysis of Binding to Probes of Different Sequences

Complementary to the aforementioned competitor approach, DNA fragments of different sequence can be used as labelled probes. Probes covering sequence variants in the same way as for competitor analysis can be run in parallel. Formation of complexes with equivalent migration in the gel provides evidence that both sequences can bind the same protein (Fig. 3, tracks 1–3). This approach is well suited to analysis of series of mutations within a fragment and identification of bases critical for binding. An approach that allows relative binding affinities to be estimated is to include a reference-labelled DNA probe of different size in binding reactions; if probe lengths are sufficiently different, it can be feasible to separate complexes formed on the separate probes and quantify binding in relation to the reference probe (Bendall and Molloy, 1994). The principle of reacting proteins with sequence variants can be extended to incubating proteins with populations of oligonucleotides containing segments of random sequence DNA. Complexes can be isolated from gels and successive rounds of PCR and gel retardation used to isolate enriched populations of DNA-binding sites (Oliphant et al., 1989). For studies of a limited number of base positions, bound populations can be studied by direct sequencing (Luo et al., 1997)

D. Characterisation of DNA-Binding Proteins

Gel retardation assays can provide an entry point for identification and characterisation of proteins regulating gene expression. Antibodies to a wide range of characterised transcription factors are available and can be used to determine if a specific protein is contained in a complex (see Fig. 3). If an antibody binds to or near the DNA-binding domain it may disrupt complex formation, reducing the amount of complex formed (track 11). Conversely, if the target epitope is distant from the DNA-binding domain, antibody binding will not be disrupted and formation of a ternary DNA–protein–antibody complex will produce a supershifted complex of lower mobility (track 10).

In many instances the nature of the binding sequence provides a guide to possible binding proteins. Comparison can then be made of the migration of protein complexes formed on the sequence under study and known high affinity sites for the candidate protein. Protease clipping can be further used to confirm the equivalence of complexes formed on the two target DNA sequences (e.g., Watt and Molloy, 1993). For many proteins, cDNA clones can be used to express candidate proteins and binding and complex migration compared between nuclear extracts and expressed proteins and between complexes formed by the expressed proteins on known binding sites and the sequence being analysed.

For complexes where the protein involved is unknown, gel retardation can be used to determine the molecular weight of a protein. This can be done indirectly through plotting mobility vs gel concentration (Orchard and May, 1993) or by cross-linking protein DNA complexes and subsequent electrophoresis on SDS–polyacrylamide gels (Miyamoto et al., 1995). With the advent of increasingly sensitive proteomic technology, the direct identification of proteins using mass spectrometry has become feasible (see Woo et al., 2002).

E. Kinetic Analysis

For purified proteins, gel retardation assays provide a convenient approach to the determination of association and dissociation rates and equilibrium-binding constants of protein–DNA complexes (Meisterernst et al., 1988; Chodosh et al., 1986). Accurate data can be obtained if the distribution between complex and free DNA and protein is not altered during gel loading and running. Bain and Ackers (1998) have described a cryogenic gel system where the rapid quenching of reactions by transfer to –40°C and electrophoresis at this temperature results in almost complete stabilisation of species distribution. As well as its application in kinetic analysis, this gel system may allow visualisation of complexes that are present in solution but are not stable under conditions normally used in gel retardation experiments. With crude nuclear or cell extracts, the number of competing reactions for binding of proteins and DNA means that it is difficult to obtain good kinetic data (for discussion, see Cann, 1989), although measurement of dissociation rates is possible.

F. Gel Preparation and Running

1. Gel Matrix

Acrylamide : bisacrylamide gels with acrylamide concentration 5% and ratios to cross-linker of 30 : 1 to 80 : 1 are used most commonly. Concentrations can be adjusted readily to provide the best resolution for individual complexes. Agarose gels can also be used in gel retardation assays; they can be useful for separation of large complexes such as nucleosomes or when using large DNA fragments as probes. With gel additives, agarose can also provide similar resolution to acrylamide gels (Chandrasekhar et al., 1998)

2. Electrophoresis Conditions

Gel retardation assays rely on the continued association of protein–DNA complexes during electrophoresis. To minimise dissociation of complexes, it is best to conduct electrophoresis at 4°C (in a cold room or refrigerated cabinet) and in low ionic strength buffers. Commonly used buffers and electrophoretic conditions are provided in Table III. Because buffer conditions can differentially affect the migration and/or stability of complexes, it is valuable to test the different systems and choose the one that proves best for the complexes being studied.

Complex dissociation is minimised in the low salt TNAE buffer, but its low buffering capacity means that the running buffer needs to be recirculated. The higher

TABLE III   Electrophoresis Buffers and Conditions

Buffer	Components	Running conditions
TNAE	6.7 mM Tris–HCl, pH 7.9 3.3 mM sodium acetate 1 mM Na₂EDTA Adjust the pH of a 50× stock buffer to 7.9	Gel 5% acrylamide, 40 : 1 acrylamide : bisacrylamide Prerun at 10–15 V/cm for at least 30 min. Gel run at 10–15 V/cm for 2–3 h
TG	50 mM Tris base 380 mM glycine 1.67 mM Na₂EDTA Prepare a ×5 stock solution	Gel 5% acrylamide, 30 : 1 or 80 : 1 acrylamide : bisacrylamide. Prerun at 10 V/cm for at least 30 min. Gel run at 15 V/cm for 2–3 h
TBE	89 mM Tris base 89 mM boric acid 2 mM EDTA, pH 8.3 Prepare a 5× stock solution	Gel 5% acrylamide, 40 : 1 or 30 : 1 acrylamide : bisacrylamide, prepared and run in either ½× or ¼× TBE Prerun at 10–15 V/cm for at least 30 min. Gel run at 10–15 V/cm for 2–3 h

salt buffers are more convenient in that they do not require recirculation and background nonspecific binding to the probe can be reduced. However, not all complexes are stable in the higher salt buffers. For convenience, electrophoresis at room temperature should also be tested, taking care to lower the voltage to about two-thirds that used in the cold in order to prevent the gel from overheating.

References

Anderson, R. D., Taplitz, S. J., Oberbauer, A. M., Calame, K. L., and Herschman, H. R. (1990). Metal-dependent binding of a nuclear factor to the rat metallothionein-I promoter. *Nucleic Acids Res.* **18**, 6049–6055.

Bain, D. L., and Ackers, G. K. (1998). A quantitative cryogenic gel-shift technique for analysis of protein-DNA binding. *Anal. Biochem.* **258**, 240–245.

Bendall, A. J., and Molloy, P. L., (1994). Base preferences for binding by the bHLH-Zip protein USF: Effects of MgCl2 on specificity and comparison with binding of Myc family members. *Nucleic Acids Res.* **22**, 2801–2810.

Cann, J. R. (1989). Phenomenological theory of gel electrophoresis of protein-nucleic acid complexes. *J. Biol. Chem.* **264**, 17032–17040.

Chandrasekhar, S., Soubar, W. W., and Abcouwer, S. F. (1998). Use of modified agarose gel electrophoresis to resolve protein-DNA complexes for electrophoretic mobility shift assay. *Biotechniques* **24**, 217–218.

Chodosh, L. A., Carthew, R. W., and Sharp, P. A. (1986). A single polypeptide possesses the binding and transcrption activities of the adenovirus major late transcription factor. *Mol. Cell. Biol.* **6**, 4723–4733.

Dignam, J. D., Lebowitz, R. M., and Roeder, R. G. (1983). Accurate transcription initiation by RNA polymerase II in a soluble extract from isolated mammalian nuclei. *Nucleic Acids Res.* **11**, 1475–1489.

Fei, Y., Matragoon, S., and Liou, G. I. (1995). Simple and efficient method for the preparation of nuclear extracts. *Biotechniques* **18**, 984–987.

Gorski, K., Carneiro, M., and Schibler, U. (1986). Tissue-specific *in vitro* transcription from the mouse albumin promoter. *Cell* **47**, 767–776.

Jing, D., Agnew, J., Patton, W. F., Hendrickson, J., and Beechem, J. M. (2003) A sensitive two-color electrophoretic mobility shift assay for detecting both nucleic acids and proteins in gels. *Proteomics* **3**, 1172–1180.

Laniel, M.-A., and Guerin, S. L. (1998). Improving sensitivity of the electrophoretic mobility shift assay by restricting tissue phosphatase activities. *Biotechniques* **24**, 964–969.

Luo, B., Perry, D. J., Zhang, L., Kharat, I., Basic, M., and Fagan, J. B. (1997). Mapping sequence specific DNA-protein interactions: A versatile quantitative method and its application to transcription factor XF1. *J Mol. Biol.* **266**, 479–492.

Meisterernst, M., Gander, I., Rogge, L., and Winnacker, E.-L. (1988). A quantitative analysis of nuclear factor I/DNA interactions. *Nucleic Acids Res.* **16**, 4419–4435.

Miyamoto, S., Cauley, K., and Verma, I. M. (1995). Ultraviolet cross-linking of DNA binding proteins. *Methods Enzymol.* **254**, 632–641.

Oliphant, A. R., Brandl, C. J., and Struhl, K. (1989). Defining the sequence-specificity of DNA-binding proteins by selecting binding sites from random-sequence oligonucleotides: Analysis of yeast GCN4 protein. *Mol. Cell. Biol.* **9**, 2944–2949.

Orchard, K., and May, G. E. (1993). An EMSA-based method for determining the molecular weight of a protein-DNA complex. *Nucleic Acids Res.* **21**, 3335–3336

Schreiber, E., Matthias, P., Muller, M. M., and Schaffner, W. (1989). Rapid detection of octamer binding proteins with 'mini-extracts' prepared from a small number of cells. *Nucleic Acids Res.* **17**, 6419

Slomiany, B. A., Kelly, M. M., and Kurtz, D. T. (2000). Extraction of nuclear proteins with increased DNA binding activity. *Biotechniques* **28**, 938–942

Watt, F., and Molloy, P. L. (1993). Specific cleavage of transcription factors by the thiol protease, m-calpain. *Nucleic Acids Res.* **21**, 5092–5100.

Woo, A. J., Dods, J. D., Susanto, E., Ulgiati, D., and Abraham, L. J. (2002). A proteomics approach for the identification of DNA binding activities observed in the electrophoretic mobility shift assay. *Mol. Cell. Proteomics* **1**, 472–478.

43

DNA Affinity Chromatography of Transcription Factors: The Oligonucleotide Trapping Approach

Suchareeta Mitra, Robert A. Moxley, and Harry W. Jarrett

I. INTRODUCTION

The purification of transcription factors is a complex topic. A brief search of genetic databases reveals over 10,000 transcription factor entries. Of these, several hundred have now been purified. Typically these purifications have involved some form of DNA affinity chromatography. Most recent purifications use specific, double-stranded oligonucleotide sequences attached covalently to a suitable chromatographic support. This topic has been reviewed elsewhere (Gadgil *et al.*, 2000).

We described a variant of this technique called oligonucleotide trapping (Gadgil and Jarrett, 2002), depicted in Fig. 1. A column ('the trap") is prepared with the single-stranded oligonucleotide ACACACA CAC attached to CNBr-activated Sepharose through an aminoalkyl linker. Another DNA ("the probe") is prepared, which has a double-stranded region containing the element bound by a transcription factor and additionally has a GTGTGTGTGT single-stranded tail. The probe is mixed with a cell extract containing a transcription factor to be purified along with salt and various competitors and surfactants, which improve selectivity. The mixture is then applied to the trapping column and, after washing thoroughly, the column can be eluted either using high salt concentrations or using low salt and elevated temperatures.

The method has several advantages. The same "trap" column can be used with a variety of probes to purify different transcription factors and other DNA-binding proteins. The probe can be 5' end labeled to test new trap columns and, during protein purification, to follow the efficiency of column trapping of the protein–DNA complex. The labeled probe can be used directly in an electrophoretic mobility shift assay (EMSA) to follow transcription factor binding, to measure the amount of transcription factor present in various cell fractions, and to assess the effect of various competitors, detergents, and so on on this binding. The trapping method also has much higher capacity for transcription factors than traditional, covalent DNA affinity chromatography. This is because virtually all of the probe DNA is active and productively binds the transcription factor while much of covalently coupled DNA is usually inactive or inaccessible for binding (Gadgil and Jarrett, 2002; Massom and Jarrett, 1992). This article describes the current oligonucleotide trapping method practiced in our laboratory.

II. MATERIALS AND INSTRUMENTATION

CNBr-activated Sepharose (C-9142), heparin (H-3393), *N,N,N',N'*-tetramethylethylenediamine (T-9281), polydeoxyinosinic-deoxycytidylic acid [poly(dI), dC, P-4929], and igepal CA-630 (I-3021) are from Sigma. Acrylamide (161-0101) and *N,N'*-methylene-bisacrylamide (161-0201) are from Bio-Rad. Oligonucleotides are purchased from Integrated DNA Technologies (IDT), but other suppliers have also been used with good results. For oligonucleotides that will be coupled covalently to Sepharose (e.g.,

335

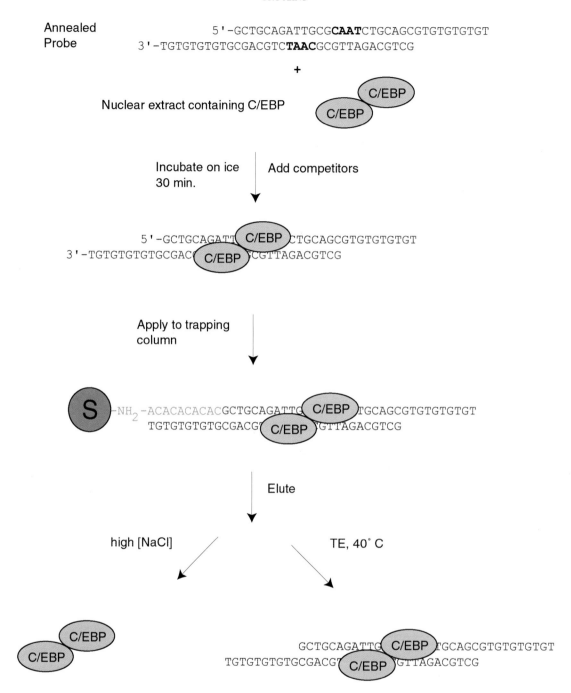

FIGURE 1 Oligonucleotide-trapping purification of the CAAT enhancer-binding protein (C/EBP) is shown schematically.

ACACACACAC), the "5′ Amino Modifier C6" is added on the last (5′) cycle. We have also used the Applied Biosystems Aminolink II reagent for this purpose and it also works well. We purchase unpurified oligonucleotides at the 1-μmol scale which have "trityl off" and are deblocked.

Chromatography is either by simply using gravity flow with a fraction collector or using a Bio-Rad Bio-logic LP Chromatograph (Cat. No. 731-8300). Electrophoresis is with the Bio-Rad Mini-Protean II apparatus (Cat. No. 165-2944) and the PowerPac 200 power supply (Cat. No. 165-5050).

III. PROCEDURES

A. Preparation of Oligonucleotides

Solutions

1. *0.5 M EDTA*: 15 g Na$_4$EDTA, 3.2 g Na$_2$EDTA, 81 ml distilled water; autoclave for 45 min
2. *1 M Tris, pH 7.5*: 60.6 g Tris base, 31.25 ml fresh 37% HCl, 441 ml distilled water, 0.5 ml 0.5 M EDTA. Check the pH of a 1 : 10 dilution. Autoclave for 1 h.
3. *TE*: 10 mM Tris, 1 mM EDTA, pH 7.5. To make 100 ml, add 1 ml 1 M Tris, pH 7.5, 0.2 ml 0.5 M EDTA, and sterile distilled water to complete 100 ml.
4. *3 M sodium acetate*: For 100 ml, add 40.8 g sodium acetate trihydrate, 5 ml glacial acetic acid, 0.2 ml 0.5 M EDTA, and 66 ml H$_2$O. Autoclave for 45 min.
5. *70% ethanol*: For 100 ml, mix 70 ml absolute ethanol with 30 ml distilled water. Keep in a −20°C freezer and use while cold.

Steps

1. Dissolve a dried, unpurified oligonucleotide (1 μmol) in 300 μl of TE, 30 μl 3 M sodium acetate, and add 1 ml of absolute ethanol for oligonucleotides longer than 12-mers and 1.32 ml ethanol for those between 6- and 11-mers.
2. Place the mixture in a −85°C freezer for 1 h and then centrifuge at 4°C at 16,000 g for 12 min. Aspirate the supernatant carefully away.
3. Wash the pellet with 500 μl 70% ethanol (80% is used for shorter oligonucleotides), centrifuge again, and discard the supernatant.
4. Dry the oligonucleotide for 5 min in a Speed-Vac SC110 (Savant) at room temperature. Oligonucleotides purified by this simple ethanol precipitation are used for all our DNA-affinity chromatography experiments and we have found them to be of adequate purity. For long-term storage, the oligonucleotides are either left dry or dissolved in 500 μl TE, adjusted to 1 mM using their absorption at 260 nm (we use an Excel spreadsheet we configured to calculate molar absorptivities based upon a sequence; a similar calculator is available on the internet at http://www.basic.nwu.edu/biotools/oligocalc.html), and stored at −20°C. If stored in TE, oligonucleotides must be again precipitated and washed with ethanol prior to coupling, as Tris interferes with coupling.

B. Preparing the ACACACACAC-Sepharose Column

Solutions

1. *Coupling buffer*: 0.1 M NaHCO$_3$, pH 8.3, 0.5 M NaCl. To prepare 1 liter, dissolve 8.40 g NaHCO$_3$ in 800 ml distilled water, adjust the pH to 8.3 with HCl, add 29.22 g NaCl, and complete to 1 liter. Keep in a 4°C refrigerator and put on ice an hour before use.
2. *1 mM HCl*: 43 μl of concentrated (11.6 M) HCl in a total volume of 500 ml distilled water. Keep in a 4°C refrigerator and put on ice an hour before use.
3. *Blocking buffer*: 0.1 M Tris, pH 8.0, 0.5 M NaCl. For 1 liter, dissolve 12.1 g Tris base in 800 ml distilled water, adjust to pH 8 with HCl, add 29.22 g NaCl, and complete to 1 liter. Store at 4°C.
4. *1 M NaN$_3$*: Dissolve 6.50 g sodium azide to a total volume of 100 ml with distilled water.
5. *TE/azide*: 10 mM Tris, 1 mM EDTA, 10 mM NaN$_3$, pH 7.5. To make 100 ml, add 1 ml 1 M Tris, pH 7.5, 0.2 ml 0.5 M EDTA, 1 ml 1 M NaN$_3$, and sterile distilled water to complete 100 ml.

Steps

1. Dissolve the dried, ethanol-precipitated 5′-[amino modifier C6]-ACACACACAC oligonucleotide, 200 nmol, in 2 ml coupling buffer, transfer to a 15-ml polypropylene screw-cap tube (Sarstedt # 62.554.002), and keep on ice.
2. Suspend 1.2 g CNBr-preactivated Sepharose in 100 ml ice-cold 1 mM HCl, cover with Parafilm, and invert occasionally while keeping in an ice-water slurry for 15 min.
3. Filter on a coarse sintered glass funnel (Pyrex brand, available from Fisher #36060-150C) and wash with an additional 100 ml ice-cold 1 mM HCl. Filter under vacuum until the filtered cake of resin just begins to form cracks and immediately scrape the Sepharose into the DNA-coupling buffer solution and mix with a vortex mixer.
4. The slurry should be watery enough to mix easily; if not, add an additional milliliters of coupling buffer. Mix the slurry in a cold room (4°C) overnight on a tube rotator (Cole Parmer Roto-Torque A-07637-00) at low speed just sufficient to keep the resin from settling.
5. The next day, filter the resin again, and wash five times with 5-ml portions of blocking buffer. Determine the absorption at 260 nm of the combined washes and measure their volume (typically 25–30 ml) either

gravimetrically or volumetrically. The amount of coupling is determined by difference using

$$\% \text{ coupling} = 100\% - \frac{A_{260} \times \text{volume (ml)}}{200 \times 10^{-9} \times 111,250 \times 10}$$

$$= 100 - \frac{A_{260} \times \text{volume (ml)}}{0.2225}$$

where $111,250 \, M^{-1}cm^{-1}$ is the calculated molar absorptivity of CACACACACA. For example, if the total washes were 30 ml, the absorbance would be 0.742 if none couples. If the washes actually have $A_{260} = 0.186$, the equation calculates 75% coupling. To convert this to nanomoles DNA per milliliter Sepharose, the 1.2-g preactivated Sepharose yields 3.5 ml/g or about 4.2 ml DNA–Sepharose. Thus 75% coupling would give $0.75 \times 200 \, \text{nmol}/4.2 \, \text{ml} = 36 \, \text{nmol DNA/ml}$ Sepharose.

6. Leave the resin as a watery slurry in blocking buffer with mixing overnight in the cold room (4°C) and then put in TE/azide (10 mM Tris, pH 7.5, 1 mM EDTA, 10 mM NaN₃) and store at 4°C. We store resins in graduated 15-ml polypropylene tubes as 50% settled resin/50% liquid mixtures for a year or more with no loss of activity.

7. Then pack the (AC)₅-Sepharose into a 1-ml syringe column (Bio-Rad Poly-Prep empty columns, Cat. No. 731-1550 or Alltech Extract-Clean, Cat. No. 211101). This is done simply by adding 1 ml of the well-mixed 50% slurry to the column and allowing it to drain and overlaying with fresh buffer. We usually place an extra column frit on top of the 0.5-ml packed support bed to prevent disturbing it and keep the column in TE/azide when not in use. Assuming 75% coupling as described earlier, this column should bind 18 nmol of probe.

Pitfalls

There are essentially three kinds of failure we have observed as we help others do trapping.

1. Tris cannot be present during DNA coupling to CNBr-activated Sepharose. It can be removed effectively from DNA stored in TE by careful ethanol precipitation.

2. After hydration in 1 mM HCl and filtration, care must be taken not to introduce air into the Sepharose. Filter only until cracks begin to appear in the filtrate and stop, leaving a moist cake. Once air is introduced, the Sepharose particles will float, making column packing difficult or impossible.

3. Columns should usually only be used when EDTA is present. DNases all require divalent cations and, in the absence of EDTA, can destroy columns. In

some cases, using divalent cations is unavoidable though and this will lessen column lifetime.

C. Preparing the Probe

The oligonucleotides, usually one for each strand, are obtained commercially and separately ethanol precipitated as described in Section III,A. In the case of C/EBP, an internally symmetric sequence [i.e., EP24(TG)₅] was used (Gadgil and Jarrett, 2002), requiring a single oligonucleotide. Probes are usually stored at −20°C as 1 mM solutions in TE. A portion of either strand can be labeled using the 5′ polynucleotide kinase reaction and [γ-³²P]ATP (Sambrook et al., 1989). In our laboratory, 5′ end labeling is performed with oligonucleotide in excess over ATP. Just prior to use, probe strands are annealed by diluting to 10–100 μM in TE0.1 (10 mM Tris, pH 7.5, 1 mM EDTA, 0.1 M NaCl) and heating to 95°C for 5 min followed by a linear decrease to 4°C at 1 h in a thermocycler. They can also be annealed by boiling 50 ml water in a 100-ml beaker, removing it from the heat, and adding the sealed tube containing the DNA. After allowing the beaker to cool for 30 min, the tube is removed to an ice bath. In the case of EP24(TG)₅, after annealing we have

5′-GCTGCAGATTGCG**CAAT**CTGCAGCGTGT GTGTGT-3′
3′-TGTGTGTGTGCGACGTC**TAAC**GCGTTAG ACGTCG-5′

When first using a new probe or trap, we generally test the apparatus by 5′ end labeling the probe (about 10,000 counts per minute of ³²P in 1–2 nmol DNA is sufficient) and applying it to the trap column in TE0.1 at 4°C. If all of the ³²P is attached to the DNA (i.e., proper care was taken to remove any excess [γ-³²P]ATP used for labeling) and if the trap column has sufficient capacity (e.g., 18 nmol in the example given for 75% coupling), greater than 95% of the counts should be retained by the trapping column. The column is then placed in a 37°C water bath and eluted with TE. All of the bound counts should elute. This test can also be included during purification itself, i.e., the 5′ end-labeled probe can be used during purification to monitor column performance.

D. Preparing the Cell Extract

Most transcription factors are in the nucleus and are prepared from nuclear extracts. However, for a new transcription factor, this is not at all certain and should be ascertained. C/EBP is found in the nucleus of rat liver. In this case, a nuclear extract prepared by the method of Gorski et al. (1986) was used as the starting point. Other methods for preparing nuclear extracts can also be used. Other fractions should also be checked for activity, however, to be certain of localization and to follow the success of the preparation. In working with other transcription factors, we have found that extracting the nucleus with 0.36 M (NH$_4$)SO$_4$ as in the usual method (Gorski *et al.*, 1986) did not effectively release the transcription factor and higher salt, i.e., 0.6 M (NH$_4$)SO$_4$, was required. Therefore, monitoring each fraction as cells are fractionated and perhaps altering the fractionation procedure to accommodate the protein of interest are usually worthwhile.

Monitoring is accomplished by preparing serial dilutions of each cell fraction and adding these to an electrophoretic gel shift assay (EMSA). This is usually a qualitative analysis. For example, we typically use three-fold serial dilutions and note for each fraction (cytosol, nuclear extract) the dilution that gave an approximate 50% of the maximal shift obtained. This 50% shift is thus a rough unit of activity. Where the bulk of activity resides is determined from the dilution required for 50% shift and accounting for the volume of each fraction.

Alternatively, several kits are available for preparing nuclear extracts (i.e., the ones from Panomics, Inc., Cat. No. AY2002 or Active Motif, Cat. No. 40010) and nuclear extracts already prepared from a variety of cell lines are available commercially (see Active Motif's catalogue). While these may not be suitable in all cases, they can certainly save time when applicable.

For C/EBP, the rat liver nuclear extracts prepared (Gadgil and Jarrett, 2002) are dialyzed into 25 mM HEPES, pH 7.6, 40 mM KCl, 0.1 mM EDTA, I mM dithiothreitol, and 10% glycerol, adjusted to 5 mg/ml protein, and stored in 0.2-ml aliquots at −85°C.

E. Electrophoretic Mobility Shift Assay

Here we describe the complete assay in use in our laboratory, including directions for preparing the gel. Alternatively, we have also used 4–12% Bio-Rad Tris/glycine ready gels (with 0.25× TBE as the running buffer) with good results.

Solutions

1. *5× TBE*: To make 1 liter, dissolve 30.03 g Tris base, 15.25 g boric acid, 10 ml *0.5 M* EDTA, and add distilled water to 1000 ml and store at room temperature

2. *5 M NaCl*: Combine 29.2 g NaCl, 89 ml distilled water, and 0.1 ml *0.5 M* EDTA. Autoclave for 45 min and store at room temperature.

3. *5× incubation buffer (for 8 ml)*: Combine 400 µl I M Tris–Cl (pH 7.5), 80 µl 0.5 M EDTA (pH 8.0), 320 µl 5 M NaCl, 1.6 ml glycerol, and 3.2 µl β-mercaptoethanol. Add distilled water to 8 ml and store at 4°C for up to 1 week.

4. *Acrylamide*: Combine 29 g acrylamide, 1 g *N,N'*-methylene-bisacrylamide, and distilled water to a total of 100 ml. Store at 4°C for up to 1 month.

5. *10% ammonium persulfate*: Weigh 0.1 g ammonium persulfate and add 1 ml distilled water. Discard after 1 day.

6. *N,N,N',N'-Tetramethylethylenediamine (TEMED)*

7. *1 µg/µl dI.dC*: Dissolve 1 mg polydeoxyinosinic–deoxycytidylic acid in 1 ml TE

8. *10× TE0.1*: For 1 liter, combine 12.1 g Tris base, 20 ml 0.5 M EDTA, and 800 ml distilled water. Titrate to pH 7.5 with HCl, dissolve 58.44 g NaCl, and complete to 1000 ml with distilled water.

9. *TE0.4*: For 100 ml, mix 10 ml 10× TE.1, 6 ml 5 M NaCl, and complete to 100 ml

10. *TE1.2*: For 100 ml, mix 10 ml 10× TE.1, 22 ml 5 M NaCl, and complete to 100 ml. Store solutions 8–10 at room temperature.

1. Gel Preparation

Steps

1. Prepare the gel (5% acrylamide) in an 16 × 100-mm test tube by combining 7.7 ml distilled water, 0.5 ml 5× TBE, 1.7 ml acrylamide, 0.1 ml 10% ammonium persulfate, and 20 µl TEMED.

2. Mix well and quickly pour into the assembled plates and place the comb of a Bio-Rad protein II minigel apparatus.

3. Let the gel polymerize.

4. Fill the gel tank with 0.25× TBE (15 ml 5× TBE completed to 300 ml with distilled water). Remove the comb, wash, and fill the wells with 0.25× TBE.

5. For some transcription factors, preelectrophorese the gel for 30 min at 100 V.

2. Mobility Assay

Steps

1. In a microtube, combine 5 µl 5× incubation buffer, I µl 1 µg/µl poly(dI-dC), 1–5 µg transcription factor or nuclear extract, 20,000 cpm ^{32}P-labeled annealed oligonucleotides, and distilled water for 25 µl total.

2. Mix gently and incubate at room temperature for 20 min.

3. Add 2 µl of bromphenol blue (0.1% in distilled water) to each sample and mix gently.

4. Load 20–25 µl onto gel.

5. Run gel at 150 V for 1 h until the dye front is ~1 mm from bottom.

6. Dry gel overnight or place in a water-tight Zip-Lock bag.
7. Expose film overnight at −85°C using an intensifying screen or use a Phosphorimager.

F. Oligonucleotide Trapping

By adding different dilutions of nuclear extract to a gel shift, performed using a known amount of DNA, usually a total shift of the DNA can be obtained. By observing what dilution was just sufficient to give a total shift (or, better, use densitometry to determine what gives a 50% shift), the approximate concentration of the DNA-binding activity can be obtained. This method should be applied to a new transcription factor to estimate its concentration. We have experience with only a small number of transcription factors, but the highest concentration we have found of a transcription factor in a nuclear extract is about 40 fmol/μl (40 nM). This number may not be meaningful to all transcription factor/nuclear extract combinations but does provide a starting point for estimation. If a nuclear extract is diluted 10-fold into TE0.1 and an annealed probe is added to give a concentration of 40 nM or greater, the DNA should be in 10-fold or greater excess over the transcription factor. At this very low concentration of DNA, very little nonspecific binding should occur and yet all of the transcription factor should be bound. To test if the conditions are correct, EMSA can be performed under the conditions of chromatography to test whether binding is complete. Thus, if the nuclear extract is diluted 10-fold into TE0.1 and increasing amounts of 5′-end-labeled DNA probe are added (from 0.1 to 1000 nM) and the mixture is applied to EMSA, all of the DNA should be shifted at the lower concentrations but eventually a point will be reached where all of the DNA is not shifted because it is now in excess. An excess of about 10-fold is appropriate.

We also use a similar approach to determine which competitors and detergents we can use during trapping. We have found that heparin [an inhibitor of double-stranded DNA binding (Gadgil and Jarrett, 1999)], in concentrations as high as 4 mg/ml, and (dT)$_{18}$ (a competitor of single-stranded DNA binding), as high as 20 μM, have little effect on the DNA binding of many transcription factors but can diminish nonspecific binding greatly. However, we have found some transcription factors, notably some members of the E2F family, do not bind DNA even when very small amounts of heparin are used (unpublished data). Poly(dI,dC) at 0.05 mg/ml can greatly diminish nonspecific binding greatly. Similarly, we find that 0.1% igepal, Triton, or Tween detergents do not adversely

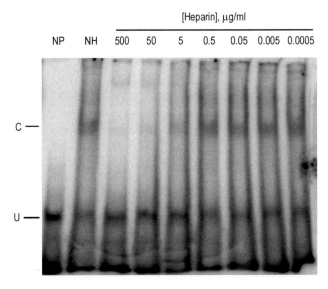

FIGURE 2 Electrophoretic mobility shift assay (EMSA) of C/EBP in the presence of heparin. EMSA was performed as in Section III,E in the presence of the heparin concentrations shown. NP, no protein added showing the position of the unshifted (U) oligonucleotide. NH, no heparin added to the complete assay mixture showing the position of the shifted complex (C).

affect most gel shifts but we have found exceptions. We usually test various concentrations of each of these reagents in an EMSA experiment before using them in chromatography.

Such an experiment to determine the concentration of heparin to use in the purification of rat liver C/EBP is shown in Fig. 2. Concentrations of heparin of 0.5 μg/ml or less have no effect on the gel shift observed and 0.5 μg/ml would then be chosen tentatively for use in trapping. The same experiment is then performed with each component to be tested (detergents, dIdC, etc.) to find the highest concentration that can be used safely. As a last test, a mixture containing all the components is then tested to ascertain that there is no synergism between the reagents that could affect DNA binding adversely.

Trapping is then performed using this tested combination. What follows is a set of conditions that work well for rat liver C/EBP.

Steps

1. Combine 1 ml 10X TE0.1, 1 μl 5 mg/ml heparin (0.5 μg/ml final), 90 μl 1 mM (dT)$_{18}$ (9 μM final), 0.5 ml 1 mg/ml dI,dC (50 μg/ml final), and distilled water to complete 8 ml. Cool on ice 30 min.

2. Add 1 ml rat liver nuclear extract and mix gently. Add 20 μl 100 μM annealed EP24GT trapping oligonucleotide (final is 200 nM strand, 100 nM duplex) and mix gently. Incubate on ice for 30 min.

3. Chromatography. All operations take place in a 4°C cold room. All solutions are kept on ice. Apply the trapping mixture to the 0.5 ml (AC)$_5$-Sepharose column at about 0.3 ml/min.

4. Wash the column with 20 ml TE0.4. Elute the column by either of two procedures (see Fig. 1)

5a. High salt: Elute the column with 10 ml TE1.2 (10 mM Tris, pH 7.5, 1 mM EDTA, 1.2 M NaCl).

5b. Temperature: Place the column in a 37°C water bath and elute with 10 ml 37°C TE (no salt).

6. Column fractions are then assayed by EMSA. In either case, the conditions for EMSA are not ideal. For high salt elution, the salt can interfere with electrophoresis and DNA binding; therefore, only 1–2 µl can be added safely to the 25-µl gel-shift mixture. For temperature, the protein–DNA complex elutes and the DNA acts as a "cold competitor" in EMSA and diminishes the amount of shift observed. However, in either case, a gel shift is observed sufficiently to identify the active fractions. Our experience has been that temperature elution gives somewhat higher purity than salt elution (Gadgil and Jarrett, 2002).

G. Other Applications

We have extended the trapping method in two directions. We have developed a CNBr-activated HPLC silica (Jurado et al., 2002) and produced (AC)$_5$-silica. This allows trapping to be performed using the much higher mass transfer characteristics of 7-µm, 300-Å pore silica. Second, we have extended the method to restriction enzymes and used catalytic means for elution. An example of this technique is shown in Fig. 3. In this case, the trapping oligonucleotide is

Ecogt 5'-GCATGC**GAATTC**GCATGTGTGTGTGT

3'-CGTACG**CTTAAG**CGTA

which of course binds to the *Eco*RI restriction endonuclease. This oligonucleotide was used to trap *Eco*RI from a crude bacterial (*E. coli* strain RY13) extract (400 µl) in the presence of EDTA. *Eco*RI binds DNA in the absence of Mg^{2+} (presence of EDTA) but is catalytically inactive (Jurado et al., 2000). Once bound to the column and washed thoroughly, the enzyme was eluted catalytically by adding Mg^{2+} (50 mM) to the column buffer. The enzyme binds Mg^{2+}, digests the column DNA, and elutes. It is detected in Fig. 3 by its characteristic digestion of λ phage DNA (in fractions 11–23).

Once columns are eluted catalytically, the DNA is converted to product (in this case, digested DNA) and are no longer useful for affinity chromatography. However, by using the trapping approach, the digested DNA is simply removed (by washing with 37°C TE or 70°C water) to return to (AC)$_5$-silica and then the EcoGT DNA is replaced (i.e., trapped) for the next purification. This strategy allows DNA, rather than columns, to become a consumable reagent in the catalytic chromatography of enzymes such as endonucleases.

IV. CONCLUSIONS

Trapping has proven to be a useful technique for the purification of transcription factors. It allows the protein–DNA complex to form in solution at low concentrations and under conditions that can be tested beforehand using the EMSA assay with the same DNA.

It has now been extended to DNases and can probably also be applied to DNA or RNA polymerases and to DNA repair enzymes, although we have not yet done so.

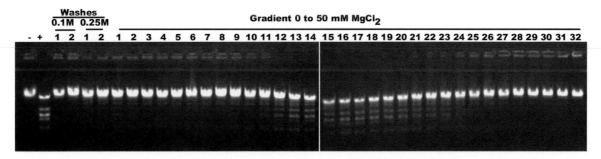

FIGURE 3 Catalytic chromatography of *Eco*RI using the trapping approach on a 4.6 × 50-mm (AC)$_5$-silica column. Fractions were assayed as described previously (Jurado et al., 2000) by digestion of λ phage. A minus sign shows control reaction without *Eco*RI, and a plus sign shows digestion with pure *Eco*RI. After loading 0.4 ml of the RY13 extract, the column was washed with 10 ml of a buffer containing 1 mM EDTA and 0.1 M KCl and then with 0.25 M. Following washing, the column was eluted with a linear gradient to 50 mM MgCl$_2$. The chromatography conditions are described in more detail elsewhere (Jurado et al., 2000).

Acknowledgment

This work was supported by the National Institutes of Health (GM43609).

References

Gadgil, H., and Jarrett, H. W. (1999). Heparin elution of transcription factors from DNA-Sepharose columns. *J. Chromatogr. A* **848**, 131–138.

Gadgil, H., and Jarrett, H. W. (2002). An oligonucleotide trapping method for purification of transcription factors. *J. Chromatogr. A* **966**, 99–110.

Gadgil, H., Jurado, L. A., and Jarrett, H. W. (2000). DNA affinity chromatography of transcription factors. *Anal. Biochem.* **290**, 147–178.

Gorski, K., Carneiro, M., and Schibler, U. (1986). Tissue-specific *in vitro* transcription from the mouse albumin promoter. *Cell* **47**, 767–776.

Jurado, L. A., Drummond, J. T., and Jarrett, H. W. (2000). Catalytic chromatography. *Anal. Biochem.* **282**, 39–45.

Jurado, L. A., Mosley, J., and Jarrett, H. W. (2002). Cyanogen bromide activation and coupling of ligands to diol-containing silica for high performance affinity chromatography: Optimization of conditions. *J. Chromatogr. A* **971**, 95–104.

Massom, L. R., and Jarrett, H. W. (1992). High-performance affinity chromatography of DNA. II. Porosity effects. *J. Chromatogr. A* **600**, 221–228.

Sambrook, J., Fritsch, E. F., and Maniatis, T. (1989). "Molecular Cloning, a Laboratory Manual," 2nd Ed. Cold Spring Harbor Laboratory Press, Cold Spring Harbor, NY.

Protein Degradation

44

Protein Degradation Methods: Chaperone-Mediated Autophagy

Patrick F. Finn, Nicholas T. Mesires, and James Fred Dice

I. INTRODUCTION

Proteins can be degraded within lysosomes by several different pathways (Dice, 2000). Chaperone-mediated autophagy (CMA) is a lysosomal pathway of proteolysis that degrades 30% of cytosolic proteins during nutritional deprivation. Substrates for this pathway of proteolysis have targeting sequences biochemically related to the pentapeptide KFERQ (Dice, 1990). Molecular chaperones, such as the constitutively expressed heat shock 70-kDa protein (hsc70), along with cochaperones (Agarraberes and Dice, 2001), bind to substrate proteins and may unfold them, which is a necessary prerequisite to import (Salvador *et al.*, 2000). A molecular chaperone in the lysosomal lumen (lyhsc70) is required for import of the substrate protein (Agarraberes *et al.*, 1997) and for optimal membrane insertion of the receptor for this proteolytic pathway, lysosome-associated membrane protein 2a (lamp 2a) (Cuervo and Dice, 2000).

Another lysosomal pathway of proteolysis, macroautophagy, is induced in rat liver for the first day of starvation, but macroautophagy declines to basal levels by the second day (Dice, 2000). CMA is only activated slightly during the first day of starvation but is activated greatly during longer starvation times (Cuervo *et al.*, 1995). This pattern suggests that the relatively nonselective macroautophagy provides amino acids during short starvation times but that the substrate selective CMA is used for prolonged starvation. Presumably, the substrates for CMA are dispensable under these conditions.

Primary human fibroblasts in culture stimulate macroautophagy when they become confluent. Withdrawal of serum growth factors from confluent fibrob-lasts activates CMA (Chiang and Dice, 1988). However, not all cells in culture behave in this orderly manner. For example, hepatocytes and kidney tubule cells may simultaneously activate macroautophagy and CMA in response to withdrawal of growth factors.

We have reproduced CMA using isolated lysosomes and defined components, such as molecular chaperones, ATP, and an ATP-regenerating system (Chiang *et al.*, 1989; Cuervo *et al.*, 2000; Terlecky and Dice, 1993). We have used lysosomes isolated from rat liver or from a variety of cultured cells in such assays. Others have used yeast vacuole preparations to study CMA (Horst *et al.*, 1999). It is with these *in vitro* preparations that many molecular details of CMA have been established.

II. MATERIALS AND INSTRUMENTATION

There are no special materials or instrumentation required to reproduce CMA *in vitro*. Standard laboratory equipment, including a low-speed centrifuge, an ultracentrifuge, a scintillation counter, a fluorimeter, and materials for sodium dodecylsulfate polyacrylamide electrophoresis (SDS–PAGE), are all that are needed. We give specific sources for materials that we use, but except when stated in Section IV, this does not imply that products from other venders are inferior.

III. PROCEDURES

A. Cell Culture

In order to obtain enough lysosomes for experimentation, you should grow cells on several 500-cm^2

diameter plates (Corning Science Products, Acton, MA 01720, #431110). The cell culture conditions for IMR-90 fibroblasts (Coriell Institute for Medical Research, Camden, NJ 08103) are as follows: high glucose Dulbecco's modified Eagle's medium (DMEM) (Sigma Chemical Company, St. Louis, MO 63178, #D5648), 10% newborn calf serum (NCS; Atlanta Biologicals, Norcross, GA 30093, #S11250), and 1% penicillin/ streptomycin with fungicide (GIBCO/Invitrogen, Rockville, MD 20849, #15240-062) grown at 37°C in 5% CO_2. Once the cells reach confluence, cells can be trypsinized (GIBCO/Invitrogen, Rockville, MD 20849, #25300-054) and replated following standard cell culture procedures.

B. Harvesting Cells

The procedures for isolating lysosomes are shown schematically in Fig. 1. Once the cells are ready to harvest, usually at confluence, media can be removed from the plate using a vacuum flask and Pasteur pipette. The cells should then be washed twice with cold (4°C) phosphate-buffered saline (PBS; 720 mM Na_2HPO_4/28 mM NaH_2PO_4/26 mM NaCl, pH 7.2). Next, add 1 ml of PBS to the plate and gently rock the plate to ensure that cells are covered with PBS. Collect the cells by scraping the monolayer with a rubber policeman. Repeat with other 500-cm^2 plates of cells and add the collected cells to an ice-cold 15-ml corex

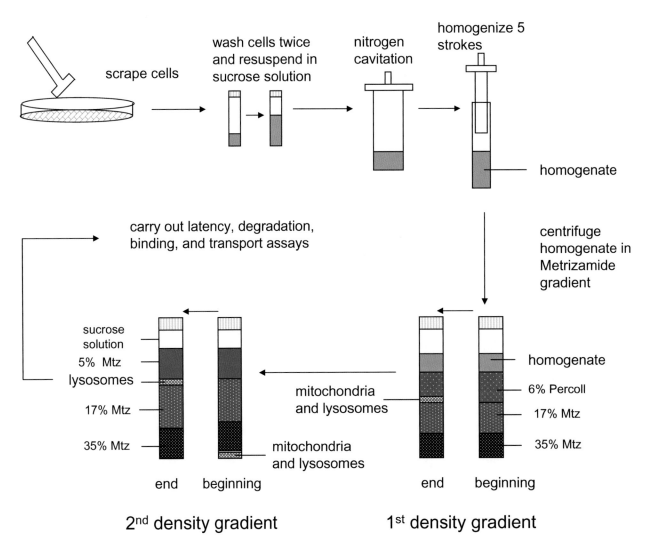

FIGURE 1 Preparation of lysosomes from cultured cells. Procedures are as described in the text.

tube (VWR, Marlboro, MA 01752, #8441-15; see Comment 1). Centrifuge the cells at 500 g for 5 min at 4°C and discard the supernatant. Gently resuspend the pellet in 2 ml 0.25 M sucrose (Sigma, #S-9378), pH 7.2 (sucrose solution), centrifuge again, and discard the supernatant.

C. Preparation of Lysosomes

Resuspend the pelleted cells in 1 ml of sucrose solution and place them into a nitrogen cavitation chamber (Kontes Glass Company, Vineland, NJ 08360, #08363-1502). Attach the cavitation chamber to a nitrogen tank and equilibrate the cavitation chamber to 30 pounds per square inch over a period of 1 min and cavitate for an additional 5 min. We have shown that this cavitation pressure and time are optimal for IMR-90 fibroblasts, human embryo kidney (HEK 293), and Chinese hamster ovary (CHO) cells. For other cell types you should optimize the pressure and time to achieve maximal cell disruption with minimal lysosomal disruption. To increase lysosomal yield, homogenize with five strokes of a loose-fitting Teflon pestle (Thomas Scientific, Swedesboro, NJ 08085, #AA22). Transfer the homogenate back to the corex tubes and centrifuge at 2500 g for 15 min at 4°C to separate nuclei from the cytoplasm.

While you perform this centrifugation you should pour the first density gradient. In a Beckman centrifuge tube (Beckman Instruments, Fullerton, CA 92634, #344059) pipet 2 ml of 35% metrizamide (Mtz; Accurate Chemical and Scientific Corporation, Westbury, NY 11590, #AN-6300KG) in sucrose solution (see Comment 2). Next layer 2 ml of 17% Mtz in sucrose solution, followed by 3 ml of 6% Percoll (Amersham Biosciences Corporation, Piscataway, NJ 08855, #17-0891-01) in sucrose solution. Layer the postnuclear supernatant on top of the gradient. This gradient should be centrifuged in a swinging bucket rotor such as the Beckman SW40.1 using a Beckman L8-M or equivalent ultracentrifuge at 69,000 g for 35 min at 4°C with the rate of acceleration and deceleration set to 7. The band of interest containing lysosomes and mitochondria will migrate at the 17% Mtz/6% Percoll interface. Aspirate and discard the upper bands and collect 1.2 ml of the lysosomal/mitochondria band. Add the 1.2 ml of lysosomes/mitochondria to 0.875 ml of 80% Mtz resulting in a 35% Mtz solution into another Beckman centrifuge tube. This will be the bottom portion of the second gradient. Keep this tube on ice while pouring the gradient. Place the tube into a larger tube or beaker containing ice at the bottom. Next, layer 2 ml of 17% Mtz in sucrose solution followed by

2 ml of 5% Mtz in sucrose solution. Fill the centrifuge tube to the top with sucrose solution, place the tube in the swinging bucket, and centrifuge using the same settings as the previous gradient. The pure lysosomal fraction will float to the 17% Mtz/5% Mtz interface (Storrie and Madden, 1990). Collect the lysosomes in 0.8–1.2 ml. These lysosomes can be used directly for transport, latency, or binding and uptake assays.

For SDS–PAGE or to separate lysosomal membrane from the matrix, further centrifugation is necessary to concentrate the lysosomes or to separate the Mtz from lysosomes. Take the lysosomal fraction from the second gradient and divide it into two polycarbonate ultracentrifuge tubes (Beckman Instruments, Spinco Division, Palo Alto, CA 94304, #343778). Dilute in 4 volumes of sucrose solution and centrifuge at 100,000 g for 45 min. You will notice a small pellet of lysosomes at the bottom of the tube. This pellet can be resuspended in SDS–PAGE buffer, boiled for 5 min, and subjected to SDS–PAGE. In order to separate membrane from matrix the pellet should be resuspended in 0.1 ml of water. Freeze and thaw the lysosomal resuspension five times and then centrifuge at 150,000 g for 30 min. The pellet from this centrifugation is the lysosomal membrane and the supernatant is the matrix fraction. The supernatant can be removed with a pipette and the pellet resuspended in the same hypotonic buffer. The membrane and matrix can then be subject to SDS–PAGE or other analytical methods.

D. Lysosomal Enzyme Latency Assay

The latency assay is necessary to determine the percentage of broken lysosomes after isolation. The latency assay is conducted using a 96-well 0.22-μm filter plate (available from Millipore, Bedford, MA 01730, #MAGVN2210) along with a 96-well plate. The first step is to wet the desired wells of the filter plate with distilled water for 5 min at room temperature. Shake the water out of the plate and then block the wells with 20 mg/ml of bovine serum albumin (BSA) in water. Next, filter the blocking solution with a vacuum manifold (Millipore, Bedford, MA 01730, #MAVM0960R). Add 42.5 μl of 10 mM MOPS solution [3-(N-morpholino)propanesulfonic acid]/0.25 M sucrose, pH 7.2 (USB Corporation, Cleveland, OH 44128, #19256), per well and 7.5 μl of lysosomes directly from the gradient. Apply the vacuum for 30 s or until all the solution is filtered. Take the flow through and assay for β-hexosaminidase as described below. For total β-hexosaminidase activity, add 7.5 μl of lysosomes and 42.5 µl of MOPS solution. These samples can also be stored at −20°C until assaying.

For the β-hexosaminidase assay, prepare glass test tubes with the flow through from the filter plate in MOPS solution, leaving one tube as a blank. Next, add 0.2 ml of reaction mixture [10 ml acetate buffer, 10 ml β-hexosaminidase substrate solution, 0.5 ml Triton X-100, 19.5 ml of water (see Comment 3)]. Incubate at 37°C for 45 min. Stop the assay by adding 0.5 ml of 0.5 M glycine/0.5 M Na_2CO_3. Measure the fluorescence using a fluorimeter with an excitation of 364 nm and an emission of 448 nm with the slit set at 3 nm. Calculate the percentage of broken lysosomes using the formula: fluorescence of filtrate/fluorescence of total lysosomes ×100. Generally, lysosomes preparations are considered successful if less then 5% of lysosomes are broken. Lysosome preparations with greater than 10% breakage should not be used.

E. Reductive Methylation of CMA Substrate Proteins

Glyceraldhyde-3-phosphate dehydrogenase (GAPDH; Sigma, #G-2267) and ribonuclease A (RNase A; Worthington Chemical Company, Lakewood, NJ 08701, #LS003433) are the most commonly used substrate proteins to study the transport of proteins into isolated lysosomes during CMA due to their commercial availability, as well as the ability to label them by reductive methylation without denaturing the proteins (see Comment 4). Labeling reactions are typically done using 250 µCi of [14C]formaldehyde (Perkin-Elmer Life Sciences, Boston, MA 02118, #NEC-039H) and 1 mg of either GAPDH or RNase A.

Before the labeling reaction is performed, prepare disposable G-50 Sephadex (Sigma, #G-50-80) columns in 1-ml syringes that have had the plungers removed and the tips plugged with glass fiber. Place empty columns into glass tubes, fill the columns with resin to the 1-ml mark on the syringe barrel, and allow to drain. Draining is facilitated by centrifugation at 300 g for 2 min (see Comment 5). Next, equilibrate columns with 5 ml of 10 mM MOPS (pH, 7.2) and centrifuge at 300 g for 1 min. Prepare six columns for the labeling reaction (see Comment 6).

Equilbrate GAPDH or RNase A (1 mg) into MOPS buffer by adding 100 µl of a 10-mg/ml solution to one of the prepared G-50 Sephadex columns. The flow through containing the protein equilibrated in MOPS buffer will be used in the subsequent labeling reaction. Add 50 µl of a 5-µCi/µl solution (250 µCi total) of [14C]formaldehyde to the equilibrated protein solution followed by 50 µl of a 9-mg/ml solution of NaCNBr (Sigma, #AS-8638) and incubate the reaction at 25°C

for 1 h. During this incubation, block the remaining G-50 Sephadex columns to prevent nonspecific binding by adding 2 ml of 10 mg/ml BSA followed by centrifugation as described earlier. Equilibrate the columns with another 5 ml of MOPS buffer. After incubation, add 75 µl of the labeling reaction to 3 of the blocked columns, place in clean glass tubes, and centrifuge for 2 min. Flow through from the centrifugation step contains the labeled protein, which is then collected and pooled from the three columns. Unreacted [14C]formaldehyde remains in the column. To determine labeling efficiency, place 2 µl of the protein flow through into 98 µl of MOPS buffer and determine acid-soluble and insoluble radioactivity as described later. The labeled protein solution should be diluted 1 : 50 in MOPS buffer before use for transport assays.

F. Protein Transport and Degradation Assay

CMA substrate proteins are transported across the lysosomal membrane and are degraded by hydrolases present in the lysosomal lumen. Transport of [14C]GAPDH into isolated lysosomes can be quantified by an increase in acid-soluble radioactivity as it is degraded into amino acids and small peptides. Data are expressed as the percentage of total radioactivity that is converted to acid-soluble radioactivity during the incubation period.

Use 10 to 40 µl of the lysosomal fraction from the second Mtz gradient in a total reaction volume of 100 µl of 0.25 M sucrose, 10 mM MOPS, pH 7.2, and 5 mM dithiothreitol. Add 10 µl of a 1 : 50 dilution of [14C]GAPDH stock solution to lysosomes and allow transport and degradation to proceed for 90 min at 25°C. To stop the transport assay, filter the reaction mixture using Millipore multiscreen assay filter plates (96-well format, 0.22-µm pore size) and a vacuum manifold. Filters should be wetted and blocked as described in Section III,D. The acid-insoluble protein is then precipitated from the filtrate using trichloroacetic acid (TCA) as outlined below.

To precipitate acid-insoluble radioactivity, add 100 µl of 20% TCA to the filtrate followed by 10 µl of 20 mg/ml BSA as carrier protein. Incubate for 1 h to overnight at 4°C. Separate precipitated proteins by centrifugation at 13,000 g for 10 min. Remove the resulting supernatant, which contains acid-soluble radioactivity, from the pellet, which contains acid-insoluble radioactivity. Wash the pellets with an additional 100 µl of 20% TCA and, after centrifugation, pool the supernatants. Solubilize the pellets with 100 µl of 1 N NaOH followed by neutralization with 8.3 µl of 12 N HCl. Count radioactivity in the pooled acid-soluble fraction as well as the solubilized and neutralized acid-insoluble fraction by scintillation counting. The percentage protein degraded is acid-soluble radioactivity/acid-soluble + acid-insoluble radioactivity × 100. CMA activity is stimulated by the inclusion of hsc70, ATP, and an ATP-regenerating system (Chiang *et al.*, 1989, Terlecky and Dice, 1993).

G. Protein Binding and Transport Assay

Substrates of CMA specifically bind to lamp2a in the lysosomal membrane (Cuervo and Dice, 1996). Chymostatin-treated lysosomes do not significantly degrade proteins that are transported into the lysosomal lumen. To measure the amount of binding and transport of CMA substrate proteins by isolated lysosomes, treat 198 µl of lysosomes from the second Mtz gradients with 2 µl of a 10 mM chymostatin (Sigma, #C-7268) stock dissolved in dimethyl sulfoxide for 10 min on ice. After chymostatin treatment, dilute the lysosomes three-fold in 10 mM MOPS/0.25 sucrose, pH 7.2, buffer for a final chymostatin concentration of 33 µM. Use 100 µl of diluted lysosomes for the binding/transport assay. Add 20 µg of nonradioactive GAPDH or RNase A to chymostatin-treated lysosomes and incubate the reaction for 20 min at 37°C. Treat one replicate of each with 2.5 µl of a 2-mg/ml stock of proteinase K (Sigma, #P6556) and incubate on ice for 15 min. Inhibit proteinase K by the addition of 10 µl of a 24 mM 4-(2-aminoethyl)-benzene sulfonyl fluoride (Sigma, #A8456) stock solution dissolved in dimethyl sulfoxide. Collect lysosomes by centrifugation at 100,000 g for 30 min. Resuspend the pelleted lysosomes in SDS–PAGE sample buffer and separate the proteins by SDS–PAGE. Alternatively, you may collect lysosomes by filtration as described earlier followed by punching and boiling the filters in electrophoresis sample buffer. The binding and uptake of CMA substrates are then assessed via immunoblotting using anti-GAPDH (Biodesign International, Saco, ME 04072, #H86504M) or anti-RNase A antibodies (Rockland Immunochemicals, Gilbertsville, PA 19525, #200-4188). Substrate that is susceptible to proteinase K cleavage is bound to the surface of lysosomes. Sunstrate that is not susceptible to proteinase K is in the lysosomal matrix. You may substitute [^{14}C]GAPDH or [^{14}C]RNase A in the aforementioned assay to provide a quantitative measure of substrate binding and uptake into isolated lysosomes.

IV. COMMENTS

1. Throughout the entire lysosomal preparation the samples should be kept on ice. Failure to keep the samples on ice will significantly decrease lysosomal recovery and increase the percentage of lysosomes that are broken.

2. An 80% Mtz stock solution (w/v) should be made several days prior to the lysosomal preparation. Dissolve the Mtz very slowly in sucrose solution using a stir bar and stir plate mixer. Generally, making 120 ml will take 8 h. Once the Mtz stock solution is made it should be divided into 40-ml aliquots and frozen in disposable 50 ml conical tubes wrapped with aluminum foil at −20°C.

3. Make the β-hexosaminidase reaction mixture using the following reagents: 0.4 M sodium acetate (Sigma, #S-8628), pH 4.4, 10% Triton X-100 (Sigma, #X-100), and β-hexosaminidase substrate solution (4 mM 4-methylumbelliferyl-N-acetyl-β-D-glucosaminide (Sigma, #M-2133) in water. Sonicate to dissolve and, for immediate use, keep at 37°C. For storage, keep at −20°C wrapped with aluminum foil.

4. Some preparations of RNase A from Sigma Chemical Company contain trace amounts of RNase B. RNase A preparations from Worthington Chemical Company should be used.

5. The Sepadex-50 compacts by approximately one-third with this centrifugation.

6. Extra G-50 Sephadex columns in addition to the four required are prepared in case some columns have bubbles in the resin layer. Bubbles will impede the flow of fluid through the column and may affect recovery of labeled protein and should not be used.

V. PITFALLS

It is crucial that the lysosomal preparation contain few broken lysosomes. In our experience, preparations with <5% broken lysosomes give reproducible results. Preparations with >10% broken lysosomes should be discarded. Results using preparations of lysosomes with breakage between 5 and 10% should be regarded with suspicion.

References

Agarraberes, F. A., and Dice, J. F. (2001). A molecular chaperone complex at the lysosomal membrane is required for protein translocation. *J. Cell Sci.* **114**, 2491–2499.

Agarraberes, F. A., Terlecky, S. R., and Dice, J. F. (1997). An intralysosomal hsp70 is required for a selective pathway of lysosomal protein degradation. *J. Cell Biol.* **137**, 825–834.

Chiang, H.-L., and Dice, J. F. (1988). Peptide sequences that target proteins for enhanced degradation during serum withdrawal. *J. Biol. Chem.* **263**, 6797–6805.

Chiang, H.-L., Terlecky, S. R., Plant, C. P., and Dice, J. F. (1989). A role for a 70-kilodalton heat shock protein in lysosomal degradation of intracellular proteins. *Science* **246**, 382–385.

Cuervo, A. M., and Dice, J. F. (1996). A receptor for the selective uptake and degradation of proteins by lysosomes. *Science* **273**, 501–503.

Cuervo, A. M., and Dice, J. F. (2000). Regulation of lamp2a levels in the lysosomal membrane. *Traffic* **1**, 570–583.

Cuervo, A. M., Gomes, A. V., Barnes, J. A., and Dice, J. F. (2000). Selective degradation of annexins by chaperone-mediated autophagy. *J. Biol. Chem.* **275**, 33329–33335.

Cuervo, A. M., Knecht, E., Terlecky, S. R., and Dice, J. F. (1995). Activation of a selective pathway of lysosomal proteolysis in rat liver by prolonged starvation. *Am. J. Physiol.* **269**, C1200–C1208.

Dice, J. F. (1990). Peptide sequences that target cytosolic proteins for lysosomal proteolysis. *Trends Biochem. Sci.* **15**, 305–309.

Dice, J. F. (2000). "*Lysosomal Pathways of Protein Degradation.*" Landes Bioscience, Austin, TX.

Horst, M., Knecht, E., and Schu, P. (1999). Import into and degradation of cytosolic proteins by isolated yeast vacuoles. *Mol. Biol. Cell* **10**, 2879–2889.

Storrie, B., and E. A. Madden (1990). Isolation of subcellular organelles. *Methods Enzymol.* **182**, 203–225.

Terlecky, S. R., and Dice, J. F. (1993). Polypeptide import and degradation by isolated lysosomes. *J. Biol. Chem.* **268**, 23490–23495.

45

Methods in Protein Ubiquitination

Aaron Ciechanover

I. INTRODUCTION

Degradation of a protein via the ubiquitin-proteasome (UPS) pathway involves two discrete and successive steps: **(a)** tagging of the substrate by covalent attachment of multiple ubiquitin molecules and **(b)** degradation of the tagged protein by the 26S proteasome complex with release of free and reusable ubiquitin. This last process is mediated by ubiquitin recycling enzymes (isopeptidases; deubiquitinating enzymes; DUBs)(for a scheme of the UPS, see Fig. 1).

Conjugation of ubiquitin, a highly conserved 76 residue polypeptide, to the protein substrate proceeds via a three-step cascade mechanism. Initially, the ubiquitin-activating enzyme, E1, activates ubiquitin in an ATP-requiring reaction to generate a high-energy thiol ester intermediate, E1-S~ubiquitin. One of several E2 enzymes (ubiquitin-carrier proteins or **ub**iquitin-**c**onjugating enzymes, UBCs) transfers the activated ubiquitin from E1 via an additional high-energy thiol ester intermediate, E2-S~ubiquitin, to the substrate that is specifically bound to a member of the ubiquitin-protein ligase family, E3. There are several families of E3 enzymes. Members of the RING finger-containing E3s, the largest family of E3s, catalyze direct transfer of the activated ubiquitin from E2 to the E3-bound substrate. For HECT (**H**omologous to the **E6-AP C T**erminus) domain E3s, the ubiquitin is transferred from the E2 to an active site Cys residue on the E3 to generate a third high-energy thiol ester intermediate, ubiquitin-S~E3, prior to its transfer to the ligase-bound substrate.

E3s catalyze the last step in the conjugation process: covalent attachment of ubiquitin to the substrate. Ubiquitin is generally transferred to an ε-NH$_2$ group of an internal lysine residue in the substrate to generate a covalent isopeptide bond. In some cases however, ubiquitin is conjugated to the N-terminal amino group of the substrate. By successively adding activated ubiquitin moieties to internal lysine residues on the previously conjugated ubiquitin molecule, a polyubiquitin chain is synthesized. The chain is recognized by the downstream 26S proteasome complex. Thus, E3s play a key role in the ubiquitin-mediated proteolytic cascade, as they serve as the specific substrate-recognition elements of the system. Approximately 1,000 different E3s have been identified in the human genome based on specific, commonly shared structural motifs. A single modification by ubiquitin or by ubiquitin-like proteins (UBLs), such as the Small Ubiquitin MOdifier (SUMO) or NEDD8, serves other, nonproteolytic purposes, such as routing cellular proteins to subcellular destinations or to the lysosome/vacule for degradation. UBLs are also conjugated via their C-terminal residue to an internal lysine residue in the acceptor protein. The specific enzymes that catalyze modification by ubiquitin-like proteins are somewhat different from those involved in conjugation of ubiquitin, although they utilize a similar mechanism. SUMO, for example, is conjugated by a heterodimeric E1, Aosl•Uba2, and the E2-conjugating enzyme Ubc9. Although Ubc9 can recognize the SUMOylation motif and transfer SUMO to certain substrates, for other proteins, specific E3 enzymes have been described (for reviews on poly- and oligoubiquitination and on UBLs, see Pickart, 2001; Weissman, 2001; Glickman and Cicchanover, 2002; Schwartz and Hochstrasser, 2003; Hicke and Dunn, 2003; Huang *et al.*, 2004).

Degradation of the polyubiquitinated substrates is carried out by the 26S proteasome that does not

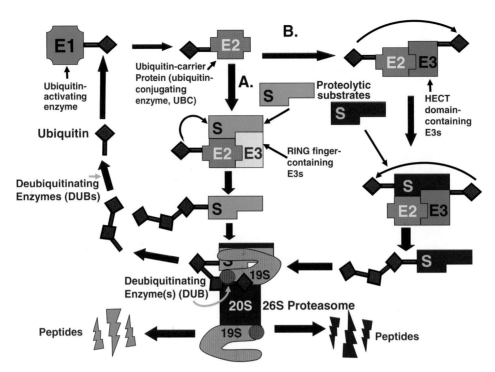

FIGURE 1 Ubiquitin is first activated to a high-energy intermediate by E1. It is then transferred to a member of the E2 family of enzymes. From E2 it can be transferred directly to the substrate (S; light blue) that is bound specifically to a member of the ubiquitin ligase family of proteins, E3. This occurs when the E3 belongs to the RING finger family of ligases (A). In the case of an HECT domain-containing ligase (B), the activated ubiquitin is transferred first to the E3 before it is conjugated to the E3-bound substrate (S; dark purple). Additional ubiquitin moieties are added successively to the previously conjugated moiety to generate a polyubiquitin chain. The polyubiquitinated substrate binds to the 26S proteasome complex (constituted of 19S and 20S subcomplexes): the substrate is degraded to short peptides, and free and reusable ubiquitin is released via the activity of deubiquitinating enzymes. These enzymes are both proteasomal and soluble.

recognize, in the vast majority of cases, nonmodified substrates. The proteasome is a multicatalytic protease that degrades polyubiquitinated proteins to short peptides. It is composed of two subcomplexes: a 20S core particle (CP) that carries the catalytic activity and a regulatory 19S regulatory particle (RP). The 20S CP is a barrel-shaped structure composed of four stacked rings, two identical outer α rings, and two identical inner β rings. The eukaryotic α and β rings are composed each of seven distinct subunits, giving the 20S complex the general structure of $\alpha_{1-7}\beta_{1-7}\beta_{1-7}\alpha_{1-7}$. The catalytic sites are localized to some of the β subunits. Each extremity of the 20S barrel can be capped by a 19S RP. One important function of the 19S RP is to recognize ubiquitinated proteins and other potential substrates of the proteasome. A ubiquitin-binding subunit of the 19S RP has indeed been identified; however, its biological function and mode of action have remained enigmatic. A second function of the 19S RP is to open an orifice in the α ring that will allow entry of the substrate into the proteolytic chamber.

Also, because a folded protein would not be able to fit through the narrow proteasomal channel, it is assumed that the 19S particle unfolds substrates and inserts them into the 20S CP. Both the channel opening function and the unfolding of the substrate require metabolic energy, and indeed, the 19S RP contains six different ATPase subunits. Following degradation of the substrate, short peptides derived from the substrate are released, as well as reusable ubiquitin. These peptides are further degraded into free amino acids by cytosolic amino- and carboxypeptidases. A small fraction of the peptides is transported across the ER membrane, binds to the MHC class I complex, and is carried to the cell surface to be presented to cytotoxic T cells. In case the peptides are derived from a "nonself" antigen, the T cell lyses the presenting cell. Proteasomal degradation is not always complete. In some cases, the proteasome processes the ubiquitinated substrate in a limited manner, releasing a truncated product. In the case of the NF-κB transcriptional regulator, an active subunit (p50 or p52) is thus released from a longer inactive precursor (p105 or p100). For reviews on the proteasome, see Zwickl *et al.* (2002), Gröll and Huber (2003), Adams (2003), and Glickman and Adir (2004).

A major, yet unresolved problem is how the UPS achieves its high specificity and selectivity toward its innumerable substrates. Why are certain proteins extremely stable in the cell, whereas others are extremely short-lived? Why are certain proteins degraded only at a particular time point during the cell cycle or only following specific extracellular stimuli, yet they are stable under most other conditions? It appears that specificity of the ubiquitin system is determined by two distinct and unrelated groups of proteins: (i) E3s and (ii) modifying enzymes and ancillary proteins. Within the ubiquitin system, substrates must be recognized and bind to a specific E3 as a prerequisite to their ubiquitination. In most cases, however, the substrates are not recognized in a constitutive manner and they must undergo a posttranslational modification such as specific phosphorylation or oxidation that renders them susceptible for recognition. In some other cases the target proteins are not recognized directly by the E3, and their recognition depends on association with ancillary proteins such as molecular chaperones that act as recognition elements in *trans* and serve as a link to the appropriate ligase. Other proteins, such as certain transcription factors, have to dissociate from the specific DNA sequence to which they bind in order to be recognized by the system. Stability of yet other proteins depends on oligomerization. Thus, in addition to E3s, modifying enzymes, such as kinases, ancillary proteins, or DNA sequences to which substrates bind, also play an important role in the recognition process. In some instances, it is the E3 that must be switched on by undergoing posttranslational modification in order to yield an active form that recognizes the target substrate.

II. MATERIALS AND INSTRUMENTATION

ATP (adenosine-5'-triphosphate disodium salt)(Sigma A-7699)

ATP-γ-S (adenosine-5'-[γ-thio]triphosphate tetralithium salt)(Sigma A-1388)

Chloramine-T (Sigma C-9887)

Creatine phosphokinase (CPK; from rabbit muscle) (Sigma C-3755)

Creatine phosphate (CP; phosphocreatine, disodium salt) (Sigma P-7936)

Cultured cells (HeLa, 293, Cos; E1 WT and its mutant cells—CHO-E36 and CHO-ts20) for preparation of cell extract and for pulse–chase labeling and immunoprecipitation experiments

Cycloheximide (Sigma C-1988)

2-Deoxyglucose (2-DOG; Sigma D-6134)

Diethylaminoethyl cellulose (DEAE; DE52; Whatman 4057-050)

2,4-Dinitrophenol (DNP; Aldrich D198501)

1,4-Dithiothreitol (DTT)(Sigma D-8255)

Dulbecco's modified Eagle medium (DMEM) with high glucose, L-glutamine, and sodium pyruvate (GIBCO 11995-065)

DMEM with high glucose, without L-glutamine, sodium pyruvate, and L-methionine (Biological Industries, Beit Ha'emek, Israel 01-054-1)

Epoxomicin (BIOMOL International L.P. PI-127)

N-Ethylmaleimide (NEM; Sigma E-1271)

Foetal bovine serum (certified, heat inactivated; US)(GIBCO 10082-147)

Foetal calf serum, dialyzed (Biological Industries, Beit Ha'emek, Israel 04-011-1)

Anti-HA antibody (Roche; clone 3F10; 1867423)

HEPES buffer (Sigma H-4034)

Hexokinase (~1,500 units/ml as ammonium sulfate slurry, Roche 1426362)

6 × Anti-His antibody (Qiagen; RGS; 34610)

[^{125}I]Na (350–600 mCi/ml; Amersham Biosciences IMS 300)

clasto-Lactacystin β-lactone (BIOMOL International L.P. PI-108)

Z-Leu-Leu-Leu-H (MG132; BIOMOL International L.P. PI-102)

L-[^{35}S]methionine (>1000 Ci/mmol; *in vitro* translation grade; Amersham biosciences SJ1515)

L-[^{35}S]methionine (×1000 Ci/mmol; cell labeling grade; Amersham biosciences SJ1015)

Methylated ubiquitin (MeUb; BIOMOL International L.P. UW8555)

Anti-Myc antibody (Roche; clone 9E10; 1667149)

New Zealand white rabbits (preferably females) of ~2 kg body weight (2-3 months old; for preparation of reticulocyte lysate)

Phenylhydrazine-HCl (Sigma P-6926)

Ribonuclease inhibitor (recombinant RNasin; Promega N2511)

Salts (KCl, NaCl, MgCl$_2$, NaF, NaN$_3$, ammonium sulfate, NaPi, KPi) (various suppliers; of highest analytical grade)

Sodium bisulfite (sodium metabisulfite; Sigma S-9000)

Translation-transcription kit (rabbit reticulocyte lysate based; L4600, L4610, or L4950, dependent on the RNA polymerase promoter in the vector, SP6, T3 or T7, respectively; TNT; Promega)

Translation-transcription kit (wheat germ extract based; L4120, L4130, or L4140, dependent on the RNA polymerase promoter in the vector, T3, SP6, or T7, respectively; TNT; Promega)

Trichloroacetic acid (TCA; 100% solution; Sigma T-0699)

Tris buffer (Trizma base; Tris base; THAM; Trometamol; 2-Amino-2-(hydroxymethyl)-1,3-propanediol; Sigma T-1378)

Ubiquitin (Sigma U-6253)

Ubiquitin (His-tagged; BIOMOL International L.P. UW8610)

Ubiquitin aldehyde (UbAl; BIOMOL International L.P. UW8450)

Ubiquitin, antibody (BIOMOL International L.P. PW8810)

Water (nuclease free; Promega P1193)

Dialysis tubing (SnakeSkin, Pierce, 68100)

Desalting column (PD-10, Amersham Biosciences 17-0851-01)

Nitrogen cavitation bomb (Parr Instrument Company)

III. PROCEDURES

A. Preparation of Cell Extracts for Monitoring Conjugation and Degradation

To conjugate or degrade a protein substrate *in vitro*, one has to utilize the appropriate cell extract. Rabbit reticulocyte lysate contains the enzymes required for degradation of most proteins and can be therefore used in most cases. Reticulocytes have several advantages. They contain a relatively small number of proteins and do not have lysosomes from which proteases can leak during preparation of the extract. Unlike cultured cells lysate, one can obtain reticulocyte lysate in a relatively large amount. Also, the lack of requirement for tissue culture media and sera makes this lysate significantly less expansive then its nucleated cultured cells counterpart. All these attributes make this lysate an ideal extract in which one can test conjugation and proteolysis of the studied protein. For monitoring conjugation and degradation of labeled proteins in the crude extract, it is not necessary to deplete ATP from the cells prior to the preparation of the extract. This will be necessary, however, in order to reconstitute the cell-free proteolytic system and to monitor dependence of the proteolytic process on the addition of exogenous ubiquitin. It will also be necessary in order to monitor conjugation of labeled or tagged ubiquitin to different substrates. Depletion of ATP from cells leads to deubiquitination of most proteins. Once such an ATP-depleted lysate is fractionated over the anion-exchange resin diethylaminoehtyl (DEAE)-cellulose, ubiquitin is eluted in the unadsorbed, flow-through material (fraction I) that also contains several E2 enzymes. Fraction II, the high salt eluate, contains E1, the remaining E2s, all the E3s, and the 26S proteasome.

1. Preparation of Reticulocyte Lysate

Steps

1. Inject rabbits subcutaneously with 10 mg/kg of phenylhydrazine (freshly dissolved in phosphate-buffered saline, PBS) on days 1, 2,4, and 6.

2. Bleed the rabbits from the ear artery or vein or from the heart (following anaesthesia) on day 8. Induction of reticulocytosis is dramatic, and more then 90% of the circulating red blood cells are reticulocytes as determined by methylene blue or brilliant cresyl blue staining.

3. Wash the cells three times with ice-cold PBS and, using a Pasteur pipette, aspirate carefully the thin layer of white blood cells ("buffy coat") that overlays the pelleted red blood cells.

4. Lyse the cells in 1.6 volumes (of pelleted cells volume) of ice-cold $H_2O \times 2$ (double distilled water) containing 1 mM DTT (diluted from 1 M solution).

5. Centrifuge at 80,000 g for 1 h at 4°C to remove particulate material.

6. Collect the supernatant and freeze in aliquots at −70°C.

7. To deplete ATP, wash cells twice in PBS and resuspend in 1 volume of Krebs Ringer phosphate buffer (120 mM NaCl, 4.8 mM KCl, 1 mM CaCl$_2$, 1.2 mM MgSO$_4$, and 16.5 mM NaPi, pH 7.4) containing 20 mM 2-deoxyglucose and 0.2 mM 2,4-dinitrophenol. Following incubation accompanied by gentle shaking for 90 min at 37°C, wash cells twice in PBS, lyse, and centrifuge as described previously.

2. Preparation of Extract from Cultured Cells

All procedures are carried out at 4°C.

Steps

1. Wash cells three times in HEPES (20 mM pH 7.5)-saline buffer and resuspend to a concentration of 10^7–10^8/ml in 20 mM HEPES, pH 7.5, that contains also 1 mM DTT.

2. Cavitate cells in a high-pressure nitrogen chamber. For HeLa cells, the best conditions are 1000 psi for 30 min. However, these conditions may vary among different cell species. For example, it may be necessary to repeat the pressure cycle twice. Make sure that most of the cells are disrupted by visualizing the suspension in a light microscope before and after cavitation. Following disruption, one should observe intact nuclei (that are not broken) and cell debris.

3. Centrifuge the homogenate successively at 3,000 g and 10,000 g for 15 min each time and then at 80,000 g for 60 min. Collect and freeze the supernatant at −70°C.

4. To deplete ATP, wash cells twice in HEPES-saline buffer and resuspend in Krebs Ringer phosphate buffer (to a density of 10^7 cells/ml) in the presence of 2-deoxyglucose, 2,4-dinitrophenol (as described earlier), 20 mM NaF, and 10 mM of NaN$_3$. Following incubation for 60 min at 37°C, wash cells twice in HEPES-saline, resuspend in HEPES-DTT, lyse, and centrifuge as described earlier.

B. Fractionation of Cell Extract to Fraction I and Fraction II

As described earlier, fractionation of the lysate into fraction I and fraction II separates ubiquitin from many of the other components of the system, thus enabling one to monitor the dependence of conjugation and degradation upon the addition of exogenous ubiquitin and certain E2 enzymes. To fractionate the lysate, ATP-depleted lysate is resolved on a DEAE-cellulose column. In the ATP-depleted lysate, all the ubiquitin is free: it was released from conjugates by isopeptidases during incubation in the presence of the glycolysis and respiration inhibitors. In the absence of ATP, reconjugation cannot occur. Under these conditions, ubiquitin is resolved in fraction I, and fraction II is dependent for its conjugating and proteolytic activities upon the addition of exogenous ubiquitin. In cell extracts from which ATP was not depleted, the ubiquitin that is still conjugated to endogenous protein substrates will adsorb to the anion-exchange resin DEAE (via the protein substrate moiety) and will elute in fraction II. During incubation, the bound ubiquitin will be released by the activity of isopeptidases and will be available for conjugation to other proteins, including the test substrate examined. Therefore, it will be difficult to demonstrate ubiquitin-dependent conjugation and degradation in fraction II that is prepared from an extract from which ATP was not depleted. In addition, the bound ubiquitin fraction, when released, will dilute any added labeled or tagged ubiquitin, thus decreasing the detectable signal in the biosynthesized ubiquitin adducts.

All procedures are carried out at 4°C.

Steps

1. Swell the DEAE resin in 0.3 M KPi, pH 7.0, for several hours. Use enough resin to adsorb all the proteins in the extract that can be bound. As a rule, use 0.6 resin volume per volume of reticulocyte lysate, or 1 ml resin/~5 mg protein of nucleated cell extract [in principle, one can use also a chromatographic system such as the AKTÄ FPLC (Amersham Biosciences) with a Mono Q column, although for resolution of large quantities, the DEAE resin procedure is advantageous].

2. Load the resin onto a column and wash with 10 column volumes of a buffer containing 5 mM KPi, pH 7.0, and 1 mM DTT (buffer A).

3. Load the extract. Once all the material is loaded, elute fraction I with buffer A. When resolving reticulocyte lysate, collect only the dark red fraction. When resolving nucleated cell extract, collect only the fractions with the highest absorption at 280 nm. Freeze fraction in aliquots at −70°C.

4. Wash the column extensively with a buffer containing 20 mM KCl in buffer A. When resolving reticulocyte lysate, make sure all the hemoglobin is eluted. When resolving nucleated cell extract, wash until the absorbency at 280 returns to baseline.

5. Elute fraction II with 2.5 column volumes of a buffer containing 20 mM Tris–HCl, pH 7.2, 1 mM DTT, and 1 M KCl.

6. Add to the eluted fraction II ammonium sulfate to saturation (~70 g/li solution) and swirl on ice for 30 min.

7. Centrifuge at 15,000 rpm for 15 min.

8. Resuspend pellet in 0.2–0.3 of the volume of the original extract in a buffer containing 20 mM Tris–HCl, pH 7.2, and 1 mM DTT. At times, it will be difficult to dissolve all the proteins. This is not essential. They will be dissolved during dialysis.

9. Dialyze against two changes of a large volume of a buffer containing 20 mM Tris–HCl, pH 7.2, and 1 mM DTT. Dialysis should be carried out on ice. Remove particulate material by centrifugation at 15,000 rpm for 15 min. Freeze in aliquots at −70°C.

C. Labeling of Proteolytic Substrates

In most cases, monitoring the conjugation and/or degradation of a specific protein requires its labeling. The fate of the protein can also be followed via Western blot analysis using specific antibodies directed against the test protein [Western blot analysis will not be described here; however, the conjugation and degradation assays for labeled proteins (described later) can be applied in an almost identical manner for unlabeled proteins, using immune detection]. Two methods of labeling have proven to be useful: iodination and biosynthetic incorporation of [^{35}S]methionine. Iodination is utilized mostly when a purified recombinant/commercial protein is available. The main advantage of the method is the high specific radioactivity that can be attained. The disadvantage of the method is that one needs a pure protein. Also, during iodination, unless it is carried out using the Bolton–

Hunter reagent, the protein can be damaged from the chloramine T used to oxidize the iodide. In addition, during storage, the labeled substrate may be subjected to radiochemical damage. A different method of labeling utilizes incorporation of ^{35}S-labeled methionine to a protein that is synthesized in a cell-free system from its corresponding cDNA/mRNA. The generated protein is native; however, the specific activity obtained is relatively low. Also, the labeled protein is contained in the crude extract in which it is synthesized and it is not pure. This extract contains, among other proteins, enzymes of the ubiquitin system that may interfere in the reconstitution of a cell-free system from purified components.

Despite these shortcomings, metabolic labeling of proteins is the most frequently used procedure to label them and follow their fate *in vitro*. To label proteins biosynthetically, one can first synthesize the mRNA on the cDNA template, using the appropriate RNA polymerase. Following digestion of the cDNA, the extracted mRNA can be translated *in vitro* in reticulocyte or wheat germ extracts. Alternatively, one can use a coupled transcription–translation cell-free extract that synthesizes the mRNA and translates it simultaneously. Such systems are available commercially (TNT; Promega). An even more advanced system (Quick TNT; Promega) contains all the necessary components for biosynthesis, except for the template cDNA and the labeling amino acid. Biosynthesis is carried out basically according to the manufacturer's instructions. In principle, it is preferable to use wheat germ extract. This extract lacks many, although not all, of the mammalian E3 enzymes. Therefore, in most cases, a protein synthesized in this extract can be used in experiments in which a cell-free system is reconstituted from purified enzymes and, in particular, when the role of a specific E3 is tested. A protein synthesized in reticulocyte lysate may be "contaminated" in many cases with its cognate E2 and/or E3 enzyme(s). This enzyme(s), which is being carried to the reconstituted system, may interfere with the examination of the role of an exogenously added E2 and/or E3 in the conjugation/degradation of the translated protein. However, at times, one must use the reticulocyte lysate, as the translation efficiency in the wheat germ extract is extremely low. In that case, the "contaminating" E2 or E3 in the reticulocyte lysate can be inactivated, if necessary, by NEM (10-min incubation at room temperature in a final concentration of 10 mM of freshly prepared solution). Because E1, all known E2s, and some of the E3s (HECT domain-containing) have an essential –SH group, the alkylating agent inactivates them. The NEM is then neutralized by the addition of DTT (final concentration of 7.5 mM). It should be noted that this procedure can

also denature/inactivate the substrate. Alternatively, the labeled substrate can be immunoprecipitated immobilized from the translation mixture, and the system can be reconstituted using the isolated protein. In most cases, however, the substrate can still be utilized and reproduces faithfully the behavior of the native substrate. The cDNA template coding for the test protein should be driven by one of the following RNA polymerase promoters: SP6, T7, or T3.

Radioiodination of Proteins

Steps

1. Add the following reagents in the following order to 1.5-ml microcentrifuge (Eppendorf) tube. The volume of the reaction mixture can vary from 20 to 100 µl.
   a. NaPi, pH 7.5, final concentration of 100 mM
   b. Protein substrate, 10–500 µg
   c. Unlabeled NaI, 50 nmol (use a stock solution of 10 mM in H$_2$O)
   d. Radiolabeled Na^{125}I, 0.1–2.0 mCi
   e. Chloramine-T solution (10 mg/ml in 10 mM NaPi, pH 7.4, freshly dissolved), 10–50 µg
2. Mix once (vortex) and incubate for 1–2 min at room temperature.
3. Add 20–100 µg Na-metabisulfite solution (20 mg/ml in 10 mM NaPi, pH 7.4, freshly dissolved) and mix. Add two-fold the amount of the added chlovamine T.
4. To remove unreacted radioactive iodine, resolve the mixture over a desalting column (PD10) equilibrated with 10 mM Tris–HCl, pH 7.6, and 150 mM NaCl. Collect fractions (in a fraction collector or manually) of ~10% of column volume each. The radioactive protein is typically eluted in fraction 4 (void volume of the column, which is ~35% of the total volume of the column). To keep a relatively small elution volume, it is recommended not to follow the one step elution procedure suggested by the manufacturer.
5. Store in aliquots at –18°C.

When monitoring conjugation, the *in vitro*-translated substrate can be used without further processing. This is also true in many cases when the degradation of the labeled substrate is followed by monitoring its disappearance in PhosphorImager-analyzed SDS–PA electrophoresed gels. However, as the degradation of certain proteins is not always efficient, it may be difficult to follow with accuracy the disappearance of 10–30% of a labeled protein band in a gel. In this case it will be necessary to monitor the release of radioactive material into a TCA-soluble fraction. For such preparations, in order to decrease background, the excess of unincorporated labeled methionine (or iodine) in the preparation of the translated protein must be first removed. This can be achieved via chromatography over DEAE exactly as described earlier for fractionation of lysate into resin-unadsorbed (fraction I) and adsorbed fractions (fraction II). The vast majority of the labeled proteins will resolve in fraction II, whereas the labeled amino acid (or iodine) will be eluted in fraction I. If the labeled protein is eluted in fraction I, changing the pH may lead to its adsorption. Alternatively, extensive dialysis of the labeled protein against a solution of 20 mM Tris–HCl, pH 7.6, 150 mM NaCl that also contains 1 mM of unlabeled methionine (or No. I) will also remove efficiently the labeling amino acid.

D. Conjugation of Proteolytic Substrates *in vitro*

To demonstrate that the degradation of a certain protein proceeds in a ubiquitin-dependent manner, it is essential to demonstrate the intermediates in the process, ubiquitin–protein adducts. Typically, incubation of the labeled protein in a complete cell extract supplemented with ubiquitin and in the presence of ATP will lead to the formation of high molecular mass adducts that can be detected following resolution of the mixture in SDS–PAGE. To increase the amount of the adducts generated, one can use two approaches, independently or simultaneously. The nonhydrolyzable ATP analog, adenosine-5′-O-(3-thiotriphosphate), ATP-γ-S, can be used instead of ATP (Johnston and Cohen 1991). The ubiquitin-activating enzyme, E1, can catalyze activation of ubiquitin in the presence of the analog, as it utilizes the α–β high energy bond of the nucleotide that is cleavable also in the analog. In contrast, assembly and activity of the 26S proteasome complex require the β–γ bond, which cannot be cleaved in the analog. Caution should be exercised, however, when utilizing the ATP analog. Often, phosphorylation of the target protein is required in order for the ubiquitin ligase to recognize it (for example, see Yaron *et al.*,1997). In these cases, the analog cannot substitute for the hydrolyzable native nucleotide, ATP. An additional approach to increase the amount of generated conjugates in a cell-free system is to use ubiquitin aldehyde (UbAl), a specific inhibitor of certain ubiquitin C-terminal hydrolases, isopeptidases (Hershko and Rose, 1987).

Steps

1. Add the following reagents to a 0.5-ml microcentrifuge (Eppendorf) tube. The volume of the reaction mixture can vary from 10.0 to 50.0 µl. Addition of all the reagents should be carried out on ice:

   a. 50 mM Tris–HCl, pH 7.6 (1 M stock solution)

   b. 5 mM MgCl$_2$ (1 M stock solution)

   c. 2 mM DTT (1 M stock solution)

   d. 5.0–30 µl of reticulocyte lysate or 50–200 µg of complete mucleated cell extract protein

   e. 2.5–10 µg ubiquitin (10 mg/ml stock solution in H$_2$O)

   f. 0.5–2.0 µg UbAl (1 mg/ml stock solution in H$_2$O)

   g. ATP and ATP-regenerating system [0.5 mM ATP (0.1 M stock solution), 10 mM CP (0.5 M stock solution), and 2.5–10 µg CPK (10 mg/ml stock solution in 10 mM Tris, pH 7.2) or 2 mM ATP-γ-S (0.1 M stock solution)]

   h. For depletion of endogenous ATP, the system should contain instead of ATP and the regenerating system 10 mM 2-DOG (1 M stock solution) and 0.2–1.0 unit hexokinase (ammonium sulfate slurry; centrifuge an aliquot of the slurry and resuspend to the same volume in 20 mM Tris–HCl buffer, pH 7.6. Dilute in the same buffer. Stock solution in the buffer can be stored at 4°C for at least 4 weeks)

   i. For the substrate, use either a labeled protein (25,000–100,000 cpm) or an unlabeled substrate in an amount that is sufficient for detection by Western blot analysis (100–2000 ng)

Typically we prepare three reaction mixtures: (i) one that contains Tris, MgCl$_2$, DTT, ubiquitin, and UbAl, (ii) one that contains ATP, CP, and CPK, and (iii) one that contains 2-DOG and hexokinase.

2. Incubate the mixture for 30 min at 37°C and resolve via SDS–PAGE (7.5–10% acrylamide).

3. Detect high molecular mass conjugates by PhosphorImager analysis (labeled proteins) or via enhanced chemiluminescence (ECL) following Western blot (for unlabeled substrates) using a specific primary antibody against the test protein and a secondary tagged antibody.

There are several ways to demonstrate that the high molecular mass adducts generated are indeed ubiquitin conjugates of the test protein.

a. It is expected that the adducts will not be generated in an ATP-depleted system.

b. Generation of conjugates of the specific substrate should be inhibited reversibly by the addition of increasing amount of MeUb (Hershko and Heller, 1985). This reductively methylated derivative of ubiquitin lacks free amino groups and therefore cannot generate polyubiquitin chains. It serves therefore as a chain terminator in the polyubiquitination reaction, and consequently as an inhibitor in this reaction.

c. Adducts can be precipitated from the reaction mixture with an antibody directed against the test protein and, following SDS–PAGE, can be further detected with an antiubiquitin antibody. Alternatively, the reaction can be carried out in the presence of His-, HA-, or Myc-tagged ubiquitin (His-tagged ubiquitin is available commercially; HA- and Myc-tagged ubiquitins or the bacterial expression cDNA clones that code for them can be obtained from different researchers), and the immunoprecipitate can be detected with an antibody against the appropriate tag.

d. A cell-free system can be reconstituted from purified or isolated components of the ubiquitin system and the formation of conjugates can be followed, dependent upon the addition of these components. Instead of adding a complete cell extract, it is possible to add fraction II (50–200 μg; derived from ATP-depleted cells) and free or tagged ubiquitin (2.5–10 μg: same amount as added to supplement the complete extract; see earlier discussion). Because fraction II is devoid of ubiquitin, the formation of conjugates that is dependent upon the addition of exogenous ubiquitin will strongly suggest that the high molecular mass derivatives generated slurring the reaction are indeed ubiquitin adducts of the test substrate. Because not all E2 enzymes are present in fraction II, it may be necessary, at times, to add to the reconstituted system purified UbcH5a, b, or c, UbcH7, or UbcH8 (~0.5–2.0 μg; available from BIOMOL International L.P.). In most cases, one of the UbcH5 (typically b or c) enzymes will be able to reconstitute activity and support conjugation.

E. Degradation of Proteolytic Substrates in Vitro

With several exceptions, cell-free systems for monitoring the degradation of proteolytic substrates are similar to those used for monitoring their conjugation. In the proteolytic assays however, unlike in the conjugation assays, ATP (and not ATP-γ-S) must be used, as activity of the 26S proteasome complex is dependent upon cleavage of the high-energy β–γ bond (see earlier discussion). ATP is added along with ATP-regenerating system as described earlier. Also, UbAl is not added. Following incubation for 2–3 h at 37°C, the reaction mixture is resolved via SDS–PAGE and disappearance of the substrate can be monitored either via PhosphorImager analysis (in case the protein substrate is radioactively labeled) or via Western blot analysis (in case of unlabeled substrate). Control reactions are complete mixtures that have been incubated on ice and/or mixtures that were incubated at 37°C in the absence of ATP. At times, degradation efficiency is low and it is difficult to follow the reduction in the amount of a protein band in gel analysis. In these cases, it is necessary to monitor the appearance of radioactivity in the TCA-soluble fraction. Here, only a radioactive substrate can be used. In moA cases, radioiodinated proteins can be used directly. At times, excess unincorporated iodine

must be removed. *In vitro*-translated proteins must first undergo DEAE fractionation or extensive dialysis in order to remove excess unincorporated labeled methionine (see earlier discussion). At the end of the incubation, a carrier protein (10–25 μl of 100 mg/ml solution of BSA) is added, followed by the addition of 0.5 ml of ice-cold TCA (20%). Following mixing, the reaction is incubated on ice for 10 min and centrifuged (5 min at 15,000 g). The supernatant is collected and the radioactivity is determined in either a β-scintillation counter (for methionine) or a γ-counter (for iodine-labeled substrates). Again, control reactions are complete mixtures that have been incubated on ice and/or mixtures that were incubated at 37°C in the absence of ATP.

F. Involvement of Ubiquitin System in Degradation of Proteins in vivo: Effect of Specific Proteasomal Inhibitors and Inactivation of E1 on Stability of Proteins in Intact Cells

All the known proteolytic substrates of the ubiquitin system are degraded, following generation of the polyubiquitin chain, by the 26S proteasome. The opposite notion, that all substrates of the 26S proteasome must be ubiquitinated prior to their recognition by the enzyme, is true in all but one established case, that of ornithine decarboxylase (ODC) (Murakami et al., 1992). This enzyme is degraded by the 26S complex without prior ubiquitination. A noncovalent association with a specific binding protein, antizyme, renders ODC susceptible to degradation. The core catalytic subunit of the 26S enzyme is the 20S proteasome complex and inhibition of this complex inhibits all proteolytic activities of the 26S proteasome. To test whether a certain protein is degraded by the proteasome, it is possible to inhibit the enzyme, both *in vitro* and *in vivo*. Inhibition of the proteasome in a cell-free system requires higher concentrations of the inhibitor (two- to five-fold) compared to the concentrations required to inhibit it in cultured cells. Also, as noted earlier, for accumulation of ubiquitin adducts in cell-free systems, it is possible to inhibit the proteasome by using ATP-γ-S, the nonhydrolyzable ATP analog (see earlier discussion). Stabilization of a protein following inhibition of the proteasome is a strong indication that the protein is indeed degraded by this enzyme. Furthermore, inhibition of the 20S proteasome may lead to accumulation of ubiquitin adducts of the test protein that cannot be detected under conditions of rapid degradation when the proteasome is active. Detection of such intermediates serves as strong support for the notion that the protein is degraded by the 26S proteasome complex following its tagging by ubiquitin. The adducts can be detected by probing the specific immunoprecipitate that was resolved via SDS–PAGE and blotted onto the membrane with an anti-ubiquitin antibody. Alternatively, to increase the sensitivity of the signal, the cells can be transfected with a cDNA coding for HA-, His-, or Myc-tagged ubiquitin, and the immunoprecipitate can be detected with the appropriate anti-tag antibody.

Determination of Stability (Half-Life; $t_{1/2}$) of a Protein in Cells; Effect of Proteasome Inhibitors

Steps

1. Wash cells (in case the fate of a naturally occurring protein is monitored; or transfected cells in case a cDNA coding for a specific protein was transfected to them) twice in a methionine-free medium at 4°C.

2. Add methionine-free medium that contains dialyzed serum (serum is added in the concentration used for growing the cells).

3. Incubate for 30 min (to remove endogenous methionine), remove the medium (by aspiration for adherent cells and following centrifugation at $800\,g$ for 10 min for cells in suspension), and add fresh methionine-free medium with dialyzed serum. To save on labeled methionine, for adherent cells add medium to barely cover the cells (1–1.5 ml to a 60-mm dish). For cells in suspension, resuspend cells to 2×10^6/ml.

4. Add labeled methionine (50–250 µCi/ml) and continue the incubation for 1 h (pulse).

5. Add the inhibitor to the experimental dishes. Lactacystin or its lactone inhibitor (which penetrates cells better) or epoxomicin should be added to a final concentration of 5–20 µM, whereas MG132 should be added to a final concentration of 20–50 µM. The inhibitor should be added for 0.5 h (the last 30 min of the labeling period).

6. Remove the labeling medium.

7. Add ice-cold complete medium that contains, in addition to the inhibitor, also 2 mM of unlabeled methionine, and wash the cells twice. The complete medium should contain also 10% untreated complete FLS.

8. Add prewarmed complete medium (that contains the inhbitior and 2 mM of unlabeled methionine) and continue the incubation for the desired time periods (chase).

9. Withdraw samples at various time points and monitor degradation/stabilization of the target protein by immunoprecipitation, SDS–PAGE, and PhosphorImaging analysis. High molecular mass conjugates of the labeled protein should be precipitated by a specific antibody directed against the target protein under study. To avoid proteolysis of the conjugates by

ubiquitin C-terminal hydrolases, it is recommended to dissolve the cells in a detergent-containing lysis buffer at 100°C. Also, the buffer should contain 10 mM NEM to inhibit ubiquitin hydrolases. The NEM can be later neutralized by DTT (7.5 mµ) or β-mercaptoethenol (15 mµ).

Instead of using pulse–chase labeling and immunoprecipitation, one can use cycloheximide (20–100 µg/ml diluted from 20 to 100 mg/ml freshly water-dissolved solution) to stop protein synthesis and follow its degradation via Western blot analysis. The advantage of this approach is that it does not necessitate the use of radioactive material and immunoprecipitation, and one can load a whole cell extract onto the gel. Utilization of the proteasome inhibitors in this system is similar to that described previously for the pulse–chase labeling approach. The disadvantage of the method is the potential interference of the drug in the proteolytic process. Thus, if cycloheximide inhibits the synthesis of a short-lived ubiquitin ligase, E3, involved in the process, the test protein can be stabilized or further destabilized, dependent on the role of the ligase in its degradation.

A complementary approach to the utilization of proteasome inhibitors, stabilization of the protein and accumulation of ubiquitin adducts, is the use of cells that harbor a temperature-sensitive mutation in the ubiquitin-activating enzyme E1, the first enzyme in the ubiquitin proteolytic cascade. At the nonpermissive temperature, the cells fail to conjugate the target proteins, which are consequently stabilized. Such cells can be, for example, the CHO-E36 (WT) and CHO-ts20 (E1 ts mutant) (used, for example, in Aviel et al., 2000). The experimental approach used with these cells can be either pulse–chase labeling and immunoprecipitation or a cycloheximide chase.

Acknowledgments

Research in the laboratory of A.C. is supported by grants from Prostate Cancer Foundation (PCF) Israel—Centers of Excellence Program, the Israel Science Foundation founded by the Israeli Academy of Sciences and Humanities—Centers of Excellence Program, and the Foundation for Promotion of Research in the Technion. A.C. is an Israel Cancer Research Fund (ICRF) Professor. Infrastructural equipment was purchased with support of the Wolfson Charitable Fund Center of Excellence for studies on *Turnover of Cellular Proteins and its Implications to Human Diseases*.

References

Adams, J. (2003). The proteasome: Structure, function, and role in the cell. *Cancer Treat. Rev.* **29** (Suppl. 1), 3–9.

Aviel, S., Winberg, G., Massucci, M., and Ciechanover, A. (2000). Degradation of the Epstein–Barr virus latent membrane protein 1 (LMP1) by the ubiquitin-proteasome pathway: Targeting via ubiquitination of the N-terminal residue. *J. Biol Chem.* **275**, 23491–23499.

Glickman, M. H., and Adir, N. (2004). The proteasome and the delicate balance between destruction and rescue. *PLoS Biol.* **2**, E13.

Glickman, M. H., and Ciechanover, A. (2002). The ubiquitin proteasome pathway: Destruction for the sake of construction. *Physiol. Rev.* **82**, 373–428.

Gröll, M., and Huber, R. (2003). Substrate access and processing by the 20S proteasome core particle. *Int. J. Biochem. Cell Biol.* **35**, 606–616.

Hershko, A., and Heller, H. (1985). Occurrence of a polyubiquitin structure in ubiquitin-protein conjugates. *Biochem. Biophys. Res. Commun.* **128**, 1079–1086.

Hershko, A., and Rose, I. A. (1987). Ubiquitin-aldehyde: A general inhibitor of ubiquitin-recycling processes. *Proc. Natl. Acad. Sci. USA* **84**, 1829–1833.

Hicke, L., and Dunn, R. (2003). Regulation of membrane protein transport by ubiquitin and ubiquitin-binding proteins. *Annu. Rev. Cell Dev. Bio.* **19**, 141–172.

Huang, D. T., Walden, H., Duda, D., and Schulman, B. A. (2004). Ubiquitin-like protein activation. *Oncogene* **23**, 1958–1971.

Johnston, N. L., and Cohen, R. E. (1991). Uncoupling ubiquitin-protein conjugation from ubiquitin-dependent proteolysis by use of β, γ-nonhydrolyzable ATP analogues. *Biochemistry* **30**, 7514–7522.

Murakami, Y., Matsufuji, S., Kameji, T., Hayashi, S.-I., Igarashi, K., Tamura, T., Tanaka, K., and Ichihara, A. (1992). Ornithine decarboxylase is degraded by the 26S proteasome without ubiquitination. *Nature* **380**, 597–599.

Pickart, C. M. (2001). Mechanisms of ubiquitination. *Annu. Rev. Biochem.* **70**, 503–533.

Schwartz, D. C., and Hochstrasser, M. (2003). A superfamily of protein tags: Ubiquitin, SUMO and related modifiers. *Trends Biochem. Sci.* **28**, 321–328.

Weissman, A. M. (2001). Themes and variations on ubiquitylation. *Nature Rev. Cell Mol. Biol.* **2**, 169–179.

Yaron, A., Gonen, H., Alkalay, I., Hatzubai, A., Jung, S., Beyth, S., Mercurio, F., Manning, A. M., Ciechanover, A., and Ben-Neriah, Y. (1997). Inhibition of NF-κB cellular function via specific targeting of the IκBα-ubiquitin ligase. *EMBO J.* **16**, 6486–6494.

Zwickl, P., Seemüller, E., Kapelari, B., and Baumeister W. (2001). The proteasome: A supramolecular assembly designed for controlled proteolysis. *Adv. Protein Chem.* 59, 187–222.

Mass Spectrometry: Protein Identification and Interactions

46

Protein Identification and Sequencing by Mass Spectrometry

Leonard J. Foster and Matthias Mann

I. INTRODUCTION

In the postgenomic world, research priorities are shifting toward understanding the function of gene products, and proteomics is the term given to large-scale determination of gene product function, starting with where and when the products are expressed. The underlying technology required for mass spectrometry-based proteomic experiments is very young and still undergoing rapid development in both hardware and software areas (Aebersold and Goodlett, 2001; Aebersold and Mann, 2003; Mann *et al.*, 2001; Washburn *et al.*, 2002). The most powerful and most popular method for elucidating the protein composition of highly complex samples is proteolytic digestion of the proteins to peptides followed by single- or multidimensional high-pressure liquid chromatography (HPLC or LC) with on-line coupling to electrospray ionization tandem mass spectrometry (MS/MS) to generate peptide fragmentation spectra . These fragments are then compared to theoretical fragments predicted from amino acid sequence databases to arrive and protein identifications. We prefer quadrupole time-of-flight hybrid mass spectrometers, sacrificing the ease of use of ion trap-type spectrometers for higher resolution data. This article describes the general procedures used in our laboratory for sequencing and identifying proteins from highly complex samples. Because this is not a literature review, the reference list is not comprehensive and does not necessarily refer to the original description of a given technique.

II. MATERIALS AND INSTRUMENTATION

Urea (Cat. No. U5128), thiourea (Cat. No. T8656), dithiothreitol (DTT, Cat. No. D9163), absolute ethanol (EtOH, Cat. No. E7023), iodoacetamide (Cat. No. I1149), and heptafluorobutyric acid (HFBA, Cat. No. H7133) are from Sigma. LysC is from Wako (Osaka, Japan, Cat. No. 12502543). Sequencing-grade porcine trypsin (Cat. No. V511C) is from Promega, acetonitrile (Cat. No. 34881) is from Riedel-da Haën, methanol (Cat. No. M/4056/17) is from Fisher, and acetic acid (Cat. No. 6052) is from J. T. Baker. All water used here is "MilliQ"-quality distilled, deionized water. The following consumables are all obtained from the indicated sources: C18 Empore extraction disks (3M, Cat. No. 2215), P200 pipette tips (Gilson but any laboratory plastics supplier will suffice), 22-gauge flat-tip syringes (Hamilton, Cat. No. 90134), LC columns (New Objective, FS 360-100-8-N-20-C15), and 50- and 20-μm-inner-diameter fused silica capillary tubing (Polymicro Technologies LLC, Cat. Nos. 020375 and 050375). Vydac prototype 3-μm C18 beads (Cat. No. 218MSB3) are a kind gift from Grace Vydac (Hesperia, CA). The helium pressure cells used in our laboratory are custom made by a local metal workshop, but similar instruments can be purchased from Brechbühler AG (http://home.flash.net/~massevo). The HPLC system used in these protocols is the Agilent 1100 Series with 0.2- to 20-μl/min flow rate. The hybrid quadrupole TOF mass spectrometer is from MDS Sciex and Applied Biosystems. All peptide fragmentation data are searched against the appropriate databases using a dual processor LinuxOS Mascot search engine (Matrix Science).

III. PROCEDURES

A. Sample Preparation

1. In-Solution Digestion

In-solution digestion can be used where the protein sample contains no detergent and is relatively simple (i.e., <300 proteins).

Solutions

1. *8 M urea*: 8 M urea in 10 mM HEPES, pH 8.0. To make 10 ml, dissolve 4.80 g urea and 23.8 mg HEPES in 10 ml water. Adjust pH with NaOH. Store at room temperature.

2. *6 M urea/2 M thiourea*: 6 M urea, 2 M thiourea in 10 mM HEPES, pH 8.0. To make 10 ml, dissolve 3.60 g urea, 23.8 mg HEPES, and 1.52 g thiourea in water. Adjust pH as necessary with NaOH. Solutions of thiourea often contain insoluble particles so it is a good idea to centrifuge this sample at 5000 g for 10 min to clarify it. Store at 4°C.

3. *8 M guanidine*: 8 M guanidine HCl, pH 1.5. To make 10 ml, dissolve 7.65 g guanidine HCl in water. Adjust pH with HCl.

4. *Digestion buffer*: 50 mM NH_4HCO_3. To make 10 ml, dissolve 40 mg NH_4HCO_3 in 10 ml water. Store at room temperature.

5. *Iodoacetamide stock solution*: 0.5 μg/μl iodoacetamide in digestion buffer. To make 10 ml, dissolve 5 mg iodoacetamide in 10 ml digestion buffer. Separate into 100-μl aliquots and store at −20°C.

6. *DTT stock solution*: 0.5 μg/μl DTT in water. To make 10 ml, dissolve 5 mg DTT in 10 ml water and store at −20°C.

7. *LysC stock solution*: 0.5 μg/μl. To make 1 ml, dilute 0.5 μg in 1 ml digestion buffer. Separate into small aliquots and store at −20°C.

Steps

1. Solubilize the protein pellet (either precipitated from solution or a centrifuge pellet from a subcellular fractionation) in 8 M urea, 6 M urea/2 M thiourea or 8 M guanidine. Selection of the particular denaturant will be sample specific, but urea/thiourea works best for membrane proteins. Keep the volume as small as possible.

2. Add 1 μg DTT/50 μg sample protein and incubate for 30 min at room temperature. If material is lim-

iting, then the protein mass needs only to be roughly estimated.

3. Add 5 μg iodoacetamide/50 μg sample protein and incubate for 20 min at room temperature.

4. Add 1 μg LysC/50 μg sample protein and incubate for 3 h or overnight, at room temperature.

5. Dilute sample with 4 volumes digestion buffer, add 1 μg trypsin/50 μg sample protein, and incubate overnight at room temperature. Digested peptides may be stored at −20°C indefinitely.

2. In Gel Digestion

This procedure is as described by Shevchenko *et al.*, (1996). Single-dimension SDS–PAGE can be used to reduce sample complexity but sample recovery is likely much less than in-solution digestions and so should only be used for complex (>300 proteins) samples where sample amount is not limited.

Solutions

1. *Digestion buffer*: 50 mM NH_4HCO_3. To make 10 ml, dissolve 40 mg NH_4HCO_3 in 10 ml water. Store at room temperature.

2. *NH_4HCO_3/EtOH*: 50% digestion buffer/50% EtOH. Combine equal volumes of digestion buffer and neat ethanol. Store at room temperature.

3. Trypsin solution: Dilute trypsin stock solution to 12.5 ng/μl with digestion buffer. Prepare *immediately* prior to use to minimize autocatalysis.

4. *Iodoacetamide*: 55 mM iodoacetamide in digestion buffer. To make 1 ml, dissolve 10.2 mg iodoacetamide in 1 ml digestion buffer. Prepare fresh.

5. *DTT*: 10 mM DTT in water. To make 1 ml, dilute 10 μl of a 1 M DTT solution in 990 μl water and store at −20°C.

6. *Extraction solution*: 3% trifluoroacetic acid (TFA), 30% acetonitrile. To make 1 ml, dilute 300 μl acetonitrile and 30 μl TFA in 670 μl water.

Steps

1. Excise individual stained gel bands and/or molecular weight ranges and place gel pieces in clean 1.5-ml microfuge tubes.

2. Chop each gel piece into smaller pieces approximately 1 mm per side using a scalpel.

3. Wash the gel pieces twice with NH_4HCO_3/EtOH for 20 min each at room temperature. Discard the supernatant each time. For each of the steps described here, enough solution should be used to cover the gel pieces.

4. Dehydrate the gel pieces by incubating for 10 min in absolute EtOH. Discard solution afterward.

5. Reduce the proteins by incubating for 45 min in DTT at 56°C. Discard solution afterward.

6. Block free sulphydryl groups by incubating for 30 min in iodoacetamide at room temperature. Discard solution afterward.

7. Wash gel pieces once with digestion buffer for 20 min at room temperature. Discard supernatant afterward.

8. Dehydrate pieces as in step 4.

9. Wash gel pieces once as in step 7.

10. Dehydrate gel pieces twice as in step 4.

11. Remove remaining ethanol from gel pieces by vacuum centrifugation.

12. Digest proteins with trypsin overnight at 37°C. Add enough trypsin solution to cover the dehydrated gel pieces. When the gel has swelled as much as possible (~20 min), remove excess trypsin solution and cover gels in digestion buffer.

13. Add 2 µl TFA to the digestion, quickly finger vortex the solution, and separate the liquid from the gel pieces, storing the liquid in a clean microfuge tube.

14. Extract the gel pieces by adding extraction solution to cover the gel. Shake the mixture vigorously for 5 min at room temperature. Remove the liquid and combine with that from step 13.

15. Dehydrate gel pieces in acetonitrile for 10 min at room temperature. Combine supernatant from this step with that from steps 13 and 14.

3. Desalting, Filtering, and Concentration

This procedure is identical to that described by Rappsilber *et al.* (2003).

Solutions

1. *Sample buffer*: 1% TFA, 5% acetonitrile. To prepare 1 ml, dilute 10 µl TFA and 50 µl acetonitrile in 940 µl water.
2. *Buffer B*: 0.02% HFBA, 0.5% acetic acid, 80% acetonitrile in water. To make 500 ml, combine 400 ml acetonitrile, 2.5 ml acetic acid, and 50 µl HFBA and top to 500 ml with water.

Steps

1. Prepare as many desalting columns as necessary by punching out small disks of C18 Empore filter using a 22-gauge flat-tipped syringe and ejecting the disks into P200 pipette tips. Ensure that the disk is securely wedged in the bottom of the tip.

2. Condition a column by forcing 5 µl of methanol through the Empore disk with a syringe fitted to the end of the pipette tip.

3. Remove any remaining organic solvent in the column by forcing 5 µl of sample buffer through.

4. Prepare the peptide sample for binding to reverse-phase material. For in-solution digestions, dilute the sample 3× with buffer A and pH the resulting solution to <2.5 with acetic acid. For in-gel digestions, dry down the extracted peptides in a vacuum centrifuge, resuspend in 100 µl sample buffer, and pH the resulting solution to <2.5 with acetic acid.

5. Force the acidified peptide sample through the C18 column.

6. Wash the column with 10 µl of sample buffer.

7. Elute the peptides from the C18 material using 5 µl buffer B. Elute directly into a microfuge tube.

B. LC/MS/MS

1. Column Preparation

This procedure is identical to that described by Rappsilber *et al.* (2003).

Solutions

1. *Matrix slurry*: Place a few cubic millimeters of dry Vydac matrix, a small magnetic stirbar, and 300 µl methanol in a 1.5-ml flat-bottomed sample vial.
2. *Buffer B as described earlier.*
3. *Buffer A*: 0.5% acetic acid, 0.02% HFBA. To make 500 ml, dilute 2.5 ml acetic acid and 50 µl HFBA in 500 ml water.

Steps

1. Insert the vial of matrix slurry in the helium pressure cell and position of the cell on a magnetic stir plate. Ensure that the plate is capable of rotating the bar inside the vial.

2. Insert an empty fused silica column into the pressure cell and tighten the seal. The back end of the column should rest in the matrix slurry but high enough that it is not bumped by the rotating stirbar.

3. Apply 30 to 50 bar helium to the system. The beads should be visible collecting in the column tip.

4. Allow the packing to proceed until the desired column length is achieved. Usually we use columns of between 7 and 10 cm in length.

5. Slowly release the pressure from the system. The column should remain tightly packed. If the packing separates, then it usually indicates that the tip of the column is blocked in some manner. Columns can be emptied by reversing the column orientation in the pressure cell and forcing methanol through the column backward.

6. Before loading any peptides onto the column, wash any methanol from the matrix by forcing buffer A through the column for 5 min using the pressure cell with 50 bar helium. At this pressure the flow rate should be 0.3 to 0.5 µl/min.

7. Column resolution can be checked using the mass spectrometer as a readout by analyzing a known amount (we use 100 fmol) of trypsin-digested bovine serum albumin. In a 20-min linear gradient from 5 to 80% buffer B, peptides should elute in approximately 20 s (width at half-maximum).

2. Liquid Chromatography

This procedure is similar to that described by Rappsilber et al. (2002) and Foster et al. (2003) with some modifications. For a comprehensive review of more advanced LC plumping, see Rappsilber et al. (2003).

Solutions

Buffer B and sample buffer as described earlier.

Steps

1. Remove all liquid from the desalted peptide sample by vacuum centrifugation.

2. Resuspend the peptide sample in 3 µl sample buffer. Alternatively, if only a small fraction of the desalted sample is to be analyzed, then instead of evaporating and resuspending, it can be directly diluted sufficiently with sample buffer to reduce the acetonitrile content to <5%. The high organic solvent content here will prevent peptides from binding to the reverse-phase matrix.

3. Measure the output rate of the HPLC pump required to achieve a flow rate through the column of 200 nl/min. Volumes this low can be measured by collecting the flow through in a graduated glass capillary.

4. Load the peptide sample onto the analytical column prepared earlier using the pressure cell again. Place the microfuge tube containing the sample in the cell, insert the analytical column, tighten the seal, and apply 50 bar pressure. Semiaccurate volumes can be loaded in this way by simply measuring the volume of liquid collecting at the column tip.

5. During sample loading, prepare the HPLC plumping (Fig. 1) and gradient program (Fig. 2). The HPLC system should be situated as close to the mass spectrometer orifice as possible (50 cm separating the two allows enough working space).

6. Flush the HPLC plumbing for 5 min with 7% buffer B in buffer A to remove air bubbles from the system.

7. Reset the flow to 0.00 µl/min and insert the analytical column containing the sample peptides into the plumbing system (Fig. 1).

3. Acquiring Mass Spectra

This procedure is similar to that described previously (Blagoev et al., 2003; Foster et al., 2003; Rappsilber et al., 2002).

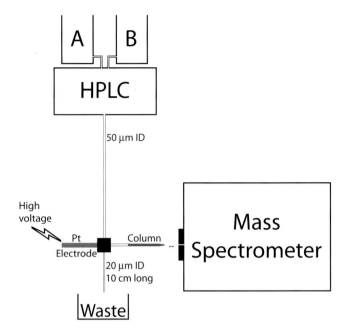

FIGURE 1 **Simplest HPLC plumbing arrangement.** The HPLC pump forces a mixture of buffers A and B out of the pump at a high flow rate. This high flow rate is then split, with most of the volume going through 20-µm capillary tubing (to provide back pressure) to a waste collection vessel and the remaining flow (200 nl/min) being forced through the analytical column. Electrospray voltage (1800 V) is applied to a platinum (Pt) electrode in the fourth port of the splitter, spraying the sample into the spectrometer for analysis.

Steps

1. Calibrate the mass spectrometer using a calibration standard containing at least two ions of known masses spanning the mass region to be measured.

2. Set up an acquisition method for the LC/MS experiment. Typically the method should be approximately 30 min longer than the HPLC elution program to allow time for the gradient delay and setup (see Fig. 2a for a typical elution gradient profile). Current mass spectrometers work by performing a few short experiments that are repeated over and over for the entire length of the LC run. Each cycle typically involves one survey scan (MS, see Fig. 2b for an example) from which the software chooses between two and four of the most intense peptides for product ion scans (called an MS–MS scan or fragmentation scan, see Fig. 2c) before starting the cycle again. The cycle length and scan times are defined by the user and should be adjusted to get sufficient sampling across an LC peak. For instance, if peptides elute from the LC column in 10 s and a cycle time of 15 s was in use, then it would be possible for a peptide to elute from the column and never be recorded by a survey scan (and thus never be

a

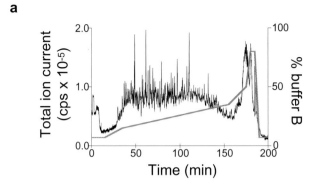

b

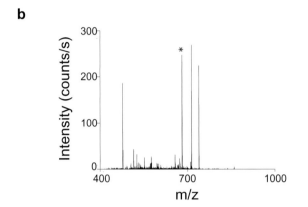

c

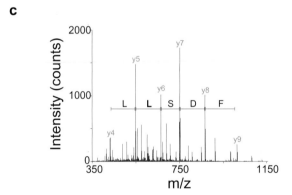

FIGURE 2 **Sample LC, MS, and MS/MS traces.** (a) Typical elution profile of a reversed-phase HPLC program and the associated total ion chromatogram registered by the mass spectrometer detector. Most of the peptides will elute from the analytical column between 15 and 35% B and so the majority of analysis time is spent in this range. (b) A typical survey or MS scan. Multiply charged ions such as the one indicated by an asterisk (*) detected in this spectra would then be selected for subsequent fragmentation. (c) A typical fragmentation or MS-MS scan. The Y-ions used to identify this peptide are labeled.

chosen for sequencing). The other major factor in deciding cycle times is sensitivity. The scan times should be long enough such that sufficient ion statistics are obtained for confident identifications. Cycle times are largely dependent on the duty cycle and the sensitivity of the mass spectrometer. For the QSTAR

Pulsars in use in our laboratory, the following settings are typical: one 1.0-s MS time-of-flight (TOF) experiment followed by three or four 1.5-s information-dependent MS/MS TOF experiments, only multiply charged ions selected for sequencing, 3 min and 0.9 Da exclusion windows, enhancement around 700 Da, MS and MS/MS scan windows of 400 to 1200 Da, 35 counts/s IDA threshold, 1800 to 2000 V spray voltage.

3. Position the analytical column in front of the orifice plate and use the tuning mode of the mass spectrometer software to adjust the column position for maximum sensitivity.

4. Start the HPLC pump to the predetermined flow rate (see Section III,B,2) using a mix of 5% B in A.

5. Add a sample to the acquisition queue that calls the method described in step 2.

6. Start this sample and open the file to check that the detector is receiving a stable signal.

7. Start the elution program on the HPLC described in Section III,B,2.

C. Protein Identification

1. Mascot Searching

There are a number of commercial products for searching measured fragmentation spectra against theoretical spectra predicted from sequence databases. We use the Mascot search engine in our laboratory.

Steps

1. Mascot cannot interpret raw mass spectra itself and must by given a peak list to work with. Different mass spectrometer manufacturers use different software to do this but the output is all generally similar. For spectra acquired with ABI Analyst, fragmentation spectra are processed from the LC/MS file using the IDA Processor supplied with Analyst. If the mass spectrometer is tuned properly, then the default settings in IDA Processor are a good starting point for optimal extraction.

2. Use Mascot to search the peak list file against the protein sequence database of interest with. Standard settings for Mascot v1.8 are shown in Fig. 3. We typically search any human samples against the IPI database (to download, go to http://srs.ebi.ac.uk/srs6bin/cgi-bin/wgetz?-page+LibInfo+-id+1hEHl1KZ36s+-lib+IPI) rather than NCBI due to the high redundancy of the latter. The UniProt database from the creators of IPI and with financial backing from NIH will likely replace all general databases as the library of choice for such searches.

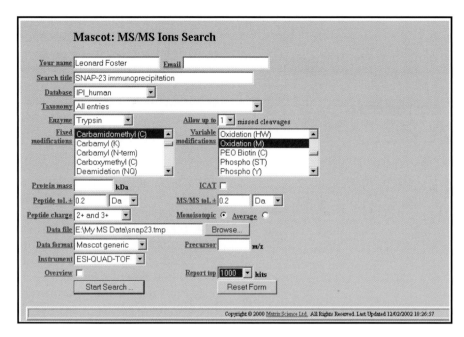

FIGURE 3 Typical parameters for a Mascot search.

2. Understanding Mascot Output, Confirming Identifications, and Internal Calibration

Steps

1. Peptides are coded in Mascot according to the rules listed in Table I. Only checked queries should be further considered for positive identifications. Thus, any protein with no checked queries can and should be immediately discarded from the "hit list", as their identifications were based on spectra who fit different peptides more closely.

2. With the remaining queries some guidelines can be established to minimize time-consuming manual verification. Depending on the database searched and the conditions used for the search, the significance level in Mascot scores will shift. With the conditions specified in Fig. 3 we typically accept any protein with a single IonsScore <45 or two IonsScores < 30 with no further verification.

3. Any remaining proteins in the list should be verified manually by direct examination of the fragmentation spectra of the peptides used for the identification. For a quadrupole-TOF hybrid instrument, we look for a Y-ion series of at least three fragments and a Mascot score that is distinct from all other matches to that spectra. Many additional sequence-specific factors can be considered but are beyond the scope of this article (Rappsilber *et al.*, 2003).

4. While certain types of mass spectrometers have inherently high mass accuracy, spectra obtained from these instruments can often be improved even further.

Table I Coding Scheme for Queries (Peptides) in Mascot

Coding method	Meaning
Bold type	First use of spectra in the output list
Light type	Subsequent use of spectra
Red type	Highest-scoring match of spectra
Black type	Less than the highest-scoring match of spectra
Checked query	Highest-scoring use of spectra

By simply calculating the mass errors for each identified peptide (the difference between the measured mass and the calculated mass for the amino acid sequence), a regression correction can be calculated that, when applied back against the whole data set, can greatly improve the mass accuracy. Using the highest-scoring peptides from a Mascot search, we have obtained mass accuracies <10 ppm, representing up to 10-fold improvements over the raw data. While there will always be outliers, peptides with a mass deviation larger than twice the average mass deviation should be inspected carefully.

IV. COMMENTS

A. Keratin

There are, of course, many biological studies where the identification of keratins by mass spectrometry would be interesting (i.e., studies of intermediate filaments), but the vast majority of keratin identifications are unwanted. Fortunately, keratin contamination does not have to be accepted as a *fait d'accompli*, as careful sample handling can severely reduce and sometimes eliminate keratins altogether.

1. Check all solutions before use—with any sign of precipitates, or "floaties," new solutions should be prepared.
2. Use Milli-Q-quality water for everything.
3. Work in a laminar flow hood where possible. Remember that flow hoods are *not* fume hoods, however, so do not use large volumes of volatile or hazardous chemicals in them.
4. *Never* bring the sample into contact with anything that may have been touched by fingers or had dust fall on it. Implicit in this is to keep unused sample tubes and pipette tips in covered containers.
5. When working with solutions, take care not to pass anything over the top of the open container, as particles may fall off. Beyond this, always enter a container with a pipettor or forceps with the container tilted so that any dust that might fall does not end up in the container.

B. Gloves

We have not yet resolved the gloves/no gloves issue in our laboratory. While they certainly help prevent flakes of skin from falling off our hands, they may increase static charges on sample tubes, thereby attracting air-borne particles. Regardless, by following the general rules described earlier, any additional benefits of gloves seem to be minimal, as those who use them in our laboratory generally see no more or less keratin contamination than those who do not.

V. PITFALLS

While much of the sample handling and data acquisition can be automated fairly easily, there is currently no commercially available software designed to handle the enormous volumes of data that this procedure is capable of generating. For single experiments the manual verification of spectra assignments and data compilation is a relatively simple task but one that grows exponentially more complex with each additional set of results added to the data set. Interpretation and analysis are additionally hampered by the lack of standards in the field for what constitutes an "identified protein." It is hoped that open source or public domain packages will become available soon to make these processes more efficient, transparent, and comparable between groups.

An additional pitfall is the observation that two analyses of the same sample will often not give identical lists of protein components. This is not due to the database search software but rather to the data the software has available. For unknown reasons the mass spectrometer does not choose the same set of peptides for sequencing in duplicate runs, even under identical conditions using the very simplest samples (i.e., purified bovine serum albumin). Proteins that are abundant enough to have three or four (or more) peptides selected for sequencing will likely be found in parallel analyses, but in complexity-limited samples where a large fraction of identifications are based on single peptides, an investigator should not expect to identify the same proteins in a second analysis. To get around this problem, we typically analyze a single sample multiple times and/or attempt to fractionate the sample prior to reversed-phase HPLC (either at protein or peptide levels).

References

Aebersold, R., and Goodlett, D. R. (2001). Mass spectrometry in proteomics. *Chem. Rev.* **101**, 269–295.

Aebersold, R., and Mann, M. (2003). Mass spectrometry-based proteomics. *Nature* **422**(6928), 198–207.

Blagoev, B., Kratchmarova, I., Ong, S. E., Nielsen, M., Foster, L. J., and Mann, M. (2003). A proteomics strategy to elucidate functional protein-protein interactions applied to EGF signaling. *Nature Biotechnol.* **21**.

Foster, L. J., de Hoog, C. L., and Mann, M. (2003). Unbiased quantitative proteomic analysis of lipid rafts reveals high specificity for signalling factors. Submitted for publication.

Mann, M., Hendrickson, R. C., and Pandey, A. (2001). Analysis of proteins and proteomes by mass spectrometry. *Annu. Rev. Biochem.* **70**, 437–473.

Rappsilber, J., Andersen, J. S., Ishihama, Y., Ong, S. E., Foster, L. J., and Mann, M. (2003). A recipe collection for the identification of peptides in complex mixtures. *Sci. STKE*.

Rappsilber, J., Ishihama, Y., and Mann, M. (2003). Stage (STop And Go Extraction) tips for MALDO, nanoelectrospray, and LC/MS sample pre-treatment in proteomics. *Anal. Chem.* **175**, 663–670.

Rappsilber, J., Ryder, U., Lamond, A. I., and Mann, M. (2002). Large-scale proteomic analysis of the human spliceosome. *Genome Res.* **12**, 1231–1245.

Shevchenko, A., Wilm, M., Vorm, O., and Mann, M. (1996). Mass spectrometric sequencing of proteins silver-stained polyacrylamide gels. *Anal. Chem.* **68**, 850–858.

Washburn, M. P., Ulaszek, R., Deciu, C., Schieltz, D. M., and Yates, J. R., 3rd (2002). Analysis of quantitative proteomic data generated via multidimensional protein identification technology. *Anal. Chem.* **74**, 1650–1657.

47

Proteome Specific Sample Preparation Methods for Matrix-Assisted Laser Desorption/Ionization Mass Spectrometry

Martin R. Larsen, Sabrina Laugesen, and Peter Roepstorff

I. INTRODUCTION

Early in the beginning of the 1990s a number of groups demonstrated that it was possible to identify proteins in databases based on peptide mass maps obtained by mass spectrometric (MS) analysis of peptides derived by specific enzymatic proteolysis of a given protein (Henzel *et al.*, 1993; Mann *et al.*, 1993; Pappin *et al.*, 1993; James *et al.*, 1993). Shortly after that, the field of proteomics started to expand concurrent with the development of more sensitive mass spectrometers so that even very low amounts of gel-separated proteins could be identified. Presently, most mass spectrometers routinely allow analysis of peptides or proteins in the low femtomole level.

With the development of sensitive mass spectrometers, the real limitation for the analysis of complex and contaminated peptide and protein mixtures is sample preparation prior to mass spectrometric analysis. Therefore, major efforts have been invested into developing new methods for sample preparation prior to MS.

In the traditional sample preparation method for matrix-assisted laser desorption/ionization (MALDI) MS, the analyte and matrix solutions are mixed directly on the MALDI target, referred to as the dried droplet method (Karas and Hillenkamp, 1988; Hillenkamp *et al.* 1991). Alternatively, the matrix and analyte solutions can be mixed in a test tube prior

to application onto the target. The tolerance of this method toward nonvolatile contaminants is limited to the efficiency of the crystallization process and to sample washing postmatrix crystallization. Additionally, it has been reported that the matrix solution conditions (especially the solvent) have a large influence on the quality of the MALDI-MS analysis (Cohen and Chait, 1996). With the introduction of the fast evaporation or thin-layer method (Vorm *et al.*, 1994), an increase in sensitivity and resolution was obtained. Here, the matrix is dissolved in a highly volatile organic solvent, which is applied onto the target to create a thin homogeneous layer of matrix crystals. The analyte solution is placed on top of the film and dried at ambient temperature followed by rinsing, resulting in decreased alkali metal adduct formation and a concomitant increase in signal intensity and sensitivity. The fast evaporation method was further developed with the inclusion of nitrocellulose (NC) in the matrix (Jensen *et al.*, 1996; Kussmann *et al.*, 1997). The presence of NC reduced the intensity of the alkali metal ion adducts, presumably because it binds the alkali metal ions very strongly, thereby preventing them from entering the gas phase. A noteworthy improvement in sample preparation for both MALDI-MS and electrospray ionization (ESI) MS came with the introduction of custom-made disposable microcolumns as a fast cleanup step prior to MS, first introduced by Wilm and Mann (1996). They used a pulled capillary needle in which they packed a small volume of reversed-phase chromatographic material. The analyte molecules were eluted directly into a nano-ESI capillary needle by centrifugation prior to mass analysis. Later, this sample preparation was simplified by using Eppendorf GELoader tips packed with chromatographic material or prepacked microcolumns (e.g., ZipTips, Millipore) (Gobom et al., 1997, Kussmann et al., 1997; Erdjument-Bromage et al., 1998). Elution of the analyte from the microcolumns with matrix solution was later demonstrated to be a very efficient sample preparation method for MALDI (Gobom et al., 1999). The use of microcolumns allows upconcentration and desalting of highly diluted samples prior to MS analysis, resulting in an increase in the sensitivity and of the overall quality of the mass spectra. In addition, significantly higher sequence coverage from peptide mass maps is observed, which is important not only for unambiguous protein identification, but especially for complete characterization of posttranslational modifications in proteomics.

This article reports a number of protocols currently used in our laboratory for MALDI-MS sample preparation optimised for proteomic research, i.e., high sensitivity, high tolerance towards low molecular weight contaminants, and high sequence coverage. We focus mainly on sample preparation methods involving sample cleanup using microcolumn technology, as traditional methods such as the dried droplet and the fast evaporation method have been described extensively in Roepstorff et al. (1998).

II. MATERIALS AND INSTRUMENTATION

The GELoader pipette tips are from Eppendorf (Hamburg, Germany). The chromatographic column materials, Poros R2 and Poros Oligo R3, are from PerSeptive Biosystems (Framingham, MA). α-Cyano-4-hydroxycinnamic acid (4HCCA) and activated charcoal (C-5510) are from Sigma (St. Louis, MO). 2,5-Dihydroxybenzoic acid (DHB) and 2-hydroxy-5-methoxybenzoic acid (HMB) are from Aldrich. Acetonitrile, formic acid (FA), and trifluoroacetic acid (TFA) are all analytical grade and obtained from different manufacturers. Disposable syringe (1 ml) are from BD Plastipak. The water is from a Milli-Q system (Millipore, Bedford, MA).

MALDI mass spectra are recorded on a Bruker Reflex III mass spectrometer (384 sample plate inlet), a Perseptive Voyager STR mass spectrometer, or a TOF-Spec 2E (Micromass, Manchester, UK). All are equipped with delayed extraction.

III. PROCEDURES

A. Preparation of Matrix Solutions

1. Matrix Solutions Containing 4HCCA

a. *4HCCA (I)*: Dissolve 10–20 mg 4HCCA in 1 ml 70% acetonitrile/0.1% TFA

b. *4HCCA (II)*: Dissolve 10–20 mg 4HCCA in 1 ml acetone/water (99/liter, v/v)

2. Matrix Solutions Containing 2,4,6-THAP

THAP: Dissolve 10 mg 2,4,6-THAP in 0.5 ml acetonitrile/water (70/30, v/v)

3. Matrix Solutions Containing DHB

DHB: Dissolve 20 mg DHB in 0.5 ml acetonitrile/water (50/50, v/v)

4. Mixed Matrix Solutions

a. *4HCCA/DHB*: Dissolve 20 mg 4HCCA in 1 ml acetonitrile/5% FA (70 : 30, v/v). Dissolve 20 mg DHB in 1 ml acetonitrile/0.1% TFA (70 : 30, v/v). The two solutions are then combined in a 1 : 1 volume ratio

b. *DHB/HMB (super DHB)*: Dissolve 9 mg DHB and 1 mg HMB in 0.5 ml acetonitrile/water (20/80, v/v)

B. Sample Preparation for MALDI-MS Analysis

1. Traditional Dried Droplet Method

The dried droplet method originally introduced by Karas and Hillenkamp (1988) is the oldest sample preparation method and is, in many applications, still the preferred one for MALDI-MS. It is surprisingly simple, it provides good results for different types of samples, and it can tolerate moderate concentrations of low molecular weight contaminants. This matrix preparation method is used preferentially in proteomics for high throughput automated peptide mass mapping of proteins present in relative high abundance. The following matrices are traditionally used with this method: 4HCCA (I), THAP, and DHB.

Steps

1. Mix equal volumes of analyte and matrix solutions and, if appropriate, 0.1% TFA or 2% FA on the sample support.
2. Let the sample dry at ambient temperature. Alternatively, evaporation of the solvent can be assisted by a stream of inert gas.
3. If the matrix solution is 4HCCA or THAP, then the matrix crystals can be washed by depositing 10 μl

of 0.1% TFA on top of the dried preparation and then removing it carefully with a piece of paper tissue.

Note: A common problem with the dried droplet method is the accumulation of analyte/matrix crystals in the periphery of the sample surface, especially in the presence of low molecular weight contaminants. For DHB, the analytes tend to associate with the big crystals that form at the periphery of the sample surface, whereas salts are found predominantly in the smaller crystals formed in the centre of the sample surface. This is the reason why it is often necessary to search for "sweet" spots on the sample surface. Inclusion of acid in the first step increases tolerance towards low molecular weight contaminants and helps in the crystallization process.

2. Mixed Matrix Dried Droplet Method

The use of matrix mixtures was reported in the 1990s by Karas and coworkers (Tsarbopoulos *et al.*, 1994). They described a matrix mixture consisting of DHB and HMB (super DHB), which was used pre-dominantly for the analysis of glycosylated proteins. Mixed matrices presumably combine the different properties of the individual matrices to a property superior to the individual matrices. Our group developed a mixed matrix solution consisting of 4HCCA and DHB, which has proven to be very useful in analysis of low amounts of peptides and proteins (Laugesen and Roepstorff, 2003). Here the hydrophilic matrix (DHB) forms crystals in the outer edge of the spot, whereas the hydrophobic 4HCCA matrix forms crystals in the centre of the spot. An increased tolerance towards salts is observed, presumably because the peptides cocrystalize with matrix in the centre, whereas the salt molecules are upconcentrated in the hydrophilic DHB edge. An example of the performance of the 4HCCA/DHB matrix compared to traditional dried droplet using only one of the matrices is shown in Fig. 1.

Steps

1. Mix the two matrix solutions in a small Eppendorf tube in a proper ratio (typically 1 : 1) and vortex intensively.

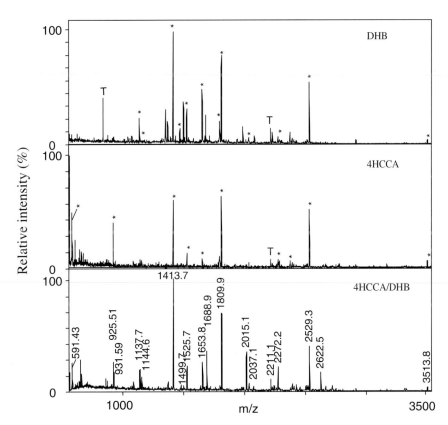

FIGURE 1   MALDI-MS peptide mass maps of peptides derived by tryptic digestion of a silver-stained spot obtained from 2D-PAGE of proteins extracted from barley grain. Spectra were obtained with DHB, CHCA, and the matrix mixture preparations. Asterisks indicate signals matching protein z-type serpin from barley. Commonly observed trypsin autodigest signals are indicated by T.

2. Mix an equal volume of analyte, matrix mix solution, and, if appropriate, 0.1% TFA or 2% formic acid on the sample support.
3. Let the solution dry at ambient temperature.

Note: These matrix preparations are not washable, as DHB is soluble in water/0.1% TFA solution. Desorption with the laser is performed in the centre of the preparation.

3. Fast Evaporation Thin-Layer/Sandwich Method

The fast evaporation method is used less frequently in proteome analysis nowadays. However, several laboratories prepare a layer of matrix prior to spotting sample and matrix on top. The method is especially useful to ensure even crystallization when anchor chip targets are used.

Steps

1. Place a small droplet (0.5 μl) of the matrix solution [4HCCA (II)] onto the MALDI target so that it spreads out. Evaporation of the solvent results in a homogeneous layer of matrix crystals.
2. Apply 0.5 μl 2% FA or 0.1% TFA, 0.5 μl analyte solution, and 0.5 μl 4HCCA (I) on top of the thin matrix layer. The addition of FA or TFA solution is not necessary if the analyte solution is already acidified.
3. Let the solution dry at ambient temperature.
4. The matrix crystals can be washed as described earlier.

Note: The fast evaporation methods are used exclusively with 4HCCA and THAP as matrices.

4. Micropurification Method

The dominating sample preparation method used for proteomics in our laboratory and many other laboratories involves the use of microcolumns, either homemade GELoader tip (Eppendorf) columns or prepacked microcolumns (e.g., ZipTips, Millipore). This sample preparation method is compatible with 4HCCA, DHB, HMB, and the DHB/4HCCA mixture.

This section describes the use of GELoader tip microcolumns and refers to the manufacturer's protocol for the ZipTips.

Steps

1. Make a partially constricted GELoader pipette tip by squeezing the narrow end. The two most common ways to generate a partially constricted GELoader pipette tip are illustrated in Fig. 2. Put the narrow end of a GELoader tip flat on a hard surface. Then roll a 1.5-ml microfuge tube over the final 1 mm of the tip. Alternatively, fix the narrow end of the

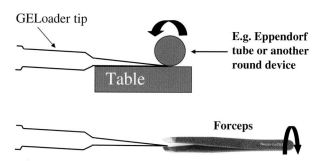

FIGURE 2 Preparation of a constricted GELoader tip can be performed in a number of different ways. One way is to roll a 1.5-ml Eppendorf tube over the narrow end of the tip with gentle pressure to squeeze the approximately last 0.5 mm of the tip flat (top). The other strategy is to use a forceps to twist the last 1 mm of the tip (bottom).

GELoader tip using a flat-surface forceps and twist the tip to close the end.

2. Prepare a slurry of 100–200 μl chromatographic material, e.g., Poros R2, Oligo R3, or Graphite powder, in 70% acetonitrile (approximately 1.5 mg/100 μl).

3. Load 20 μl 70% acetonitrile in the top of the constricted GELoader tip followed by approximately 0.5 μl of the chromatographic material slurry. Use a 1-ml syringe fitted to the diameter of the GELoader tip via a disposable pipette tip to press the liquid down gently, thereby generating a small column at the end of the constricted tip. Push tall liquid out of the column before performing the next step.

4. Apply 20 μl of 2% FA or 0.1% TFA to the top of the column. Press approximately 10 μl through the column for equilibration. Leave the remaining 10 μl on top of the column bed.

5. Mix the analyte with the remaining acid solution and use gentle air pressure to press the liquid through the column. Leave approximately 3 μl of solvent on top of the column. Do not dry the column in this step!

6. Apple 10–20 μl of 2% FA or 0.1% TFA on top of the remaining solution and use air pressure to wash the column. In this step, all liquid should be pushed out of the column.

7. Elute the peptides with 0.5 μl matrix solution directly onto the MALDI MS target. The eluent should be spotted in several small droplets on the target for a further concentration. Most peptides are only present in the two first droplets, but some may come later depending on the size and hydrophobicity. As an alternative the peptides can be eluted with increasing concentrations of organic solvent and applied by the dried droplet method.

8. Depending on the size of the column and the abundance/concentration of the analyte molecules that have been loaded, the column can be reused from

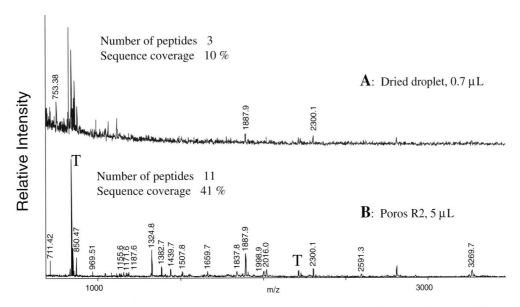

FIGURE 3 MALDI-MS peptide mass maps of peptides derived by tryptic digestion of a protein from a 2D gel of a membrane preparation from *Pseudomonas aeruginosa* (hypothetical protein, NC_002516). (A) Spectrum obtained using the dried droplet method with 0.7 μl peptide solution. Three peptides were detected covering 10% of the protein. The protein could not be identified using this peptide mass map. (B) Spectrum obtained after desalting and concentration of 5 μl peptide solution using a Poros R2 GELoader tip microcolumn. Eleven peptides were detected, and the protein could be identified easily with 41% sequence coverage. T indicates peptides originating from tryptic autodigestion.

2 to 10 times after extensive washing with 70–100% acetonitrile.

Notes: Using microcolumns, the low molecular weight contaminants can be removed easily from the analyte molecules and diluted samples are concentrated readily with a resulting increase in sensitivity and sequence coverage. An example is shown in Fig. 3. Several types of reversed-phase material can be used for column packing. Poros R2 and Poros Oligo R3 have proven excellent in our hands. Graphite powder has been used as an alternative to reversed-phase material, giving superior recovery of hydrophilic peptides and higher sequence coverage (Larsen *et al.*, 2002). Examples of the use of Poros R2 and graphite column materials in proteome analysis are given in Fig. 4.

The same procedure as described for GELoader tip microcolumns can be used with commercially available ZipTip microcolumns, resulting in increased sensitivity compared to the protocol described by the manufacturer.

The performance of the homemade microcolumn is superior to that of the commercially available ZipTips, especially for analysis of very low abundant proteins from 2D gels. An example of the performance of Poros R2 versus ZipTip C_{18} is shown in Fig. 5.

C. Column Material for GELoader Tip Microcolumns Used in Proteomics

The most common chromatographic material used for microcolumns in proteomics is reversed-phase material. However, other kind of materials can be applied with success.

i. Poros R2 and Oligo R3 are the main reversed-phase material used in proteomics to desalt and concentrate peptides prior to mass spectrometry. Poros R2 is designed for the general separation of proteins, peptides, and nucleic acids. The binding strength is similar to low carbon-loading C8 or C18 supports. The Oligo R3 medium is designed for hydrophilic peptides and nucleic acids and is similar to high carbon-loading C18 supports.

ii. Graphite powder can be used as an alternative to RP material as a peptide cleanup medium, especially for small hydrophilic peptides or phosphopeptides (Chin and Papac, 1999; Larsen *et al.*, 2002). However, it also works very well for all other peptides if they are eluted with the 4HCCA matrix (Larsen *et al.*, 2002). Alternatively, it can be used to desalt and concentrate carbohydrates prior to mass spectrometric analysis.

iii. Anion (e.g., Poros HQ) and cation (e.g., Poros S) exchange columns are used less frequently for protein

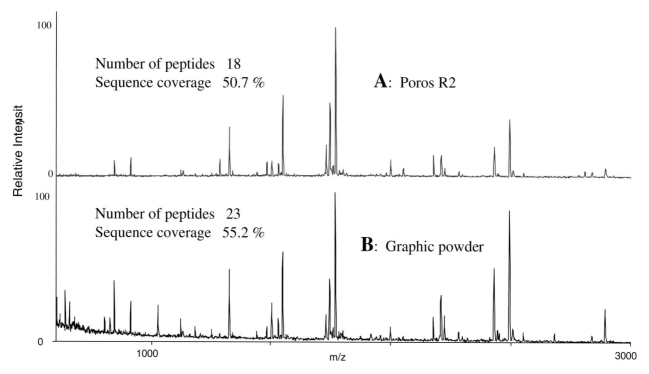

FIGURE 4 MALDI-MS peptide mass maps of peptides derived by tryptic digestion of a protein from a 2D gel of a membrane preparation from *Pseudomonas aeruginosa* (probable porin, NC_002516). An aliquot of the peptide solution was desalted and concentrated using a Poros R2 and a graphite powder GELoader tip microcolumn, respectively. (A) Spectrum obtained from the Poros R2 column. A total of 18 peptides could be assigned to the sequence of the probable porin covering 50.7% of the sequence. (B) Spectrum obtained from the graphite powder column. Twenty-three peptides covering 55.2% of the sequence could be assigned.

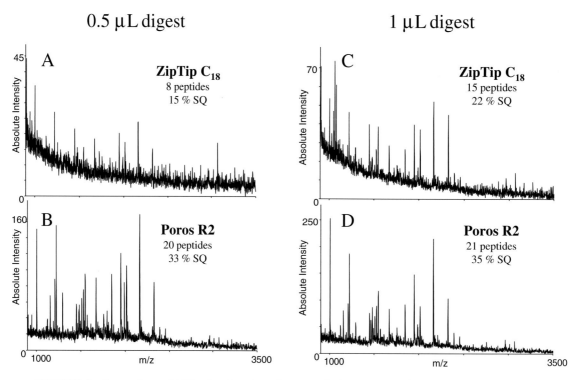

FIGURE 5 Comparison of GELoader tip microcolumns with the commercially available ZipTip C_{18}. A weak Coomassie-stained spot on a 2D gel of a membrane preparation from *Pseudomonas aeruginosa* [Fe(III)-pyochelin receptor precursor, NP_252911] was submitted to in-gel digestion using trypsin in a total volume of 20 µl buffer. Aliquots (0.5 and 1 µl) were desalted and concentrated on the two different microcolumns. Spectra obtained from the ZipTip C_{18} microcolumn purification of 0.5- and 1-µl aliquots are shown in A and C, respectively. Spectra obtained from the GELoader tip microcolumn purification of 0.5- and 1-µl aliquots are shown in B and D, respectively. As judged from the intensity of the peptide signals (absolute intensities), GELoader tip microcolumns are between 5 and 10 times more sensitive than the commercially available ZipTip columns.

identification purposes in proteomics but can be advantageous for specific purposes, such as the isolation of specific peptides when the goal is full characterization of proteins. Protocols for the use of these types of column materials are available from the resin manufacturer's home page (www.applied biosystems.com).

Note: Many other types of column material can be packed into the small GELoader tip microcolumns, including immobilized metal affinity material, immobilized enzymes, and immobilized DNA/RNA.

D. Selection of Sample Preparation Method in Proteomics

The choice of matrix and sample preparation method for MALDI-MS analysis of peptides in proteome analysis is very dependent on the amount and complexity of the sample to be analysed. In general, peptides derived by proteolytic digestion of proteins visible by Coomassie blue require less sensitive sample preparation methods than proteins visible by more sensitive staining procedures, such as silver staining or fluorescent staining (e.g., Sypro Ruby). For the first category of peptides, a standard dried droplet or sandwich method is fast and straightforward and results in detection of sufficient peptides for ambiguous protein identification. In addition, these methods are automated easily and therefore suitable for high-throughput proteomics. For a lower amount of starting material, the dried droplet or sandwich methods will not provide enough peptide signals due to suppression effects by low molecular weight contaminants or simply because the sample is too diluted. In such cases, desalting and concentration using microcolumns will be needed. As judged from the intensity of the peptides in the example shown in Fig. 5, the homemade GELoader tip microcolumns are 5–10 times more sensitive than commercially available tips. However, commercially available microcolumns are easy to handle and are compatible with several different liquid-handling robots, making them suitable for high-throughput proteomics. The combined use of different matrices and several types of microcolumns may be needed if the goal is to obtain the highest possible sequence coverage. This is relevant for observation of posttranslationally modified peptides and for determination of the processing sites when the proteins are truncated.

E. Final Comments

The sample preparation procedures are under constant development and may also vary according to the individual experimentalist. Therefore, this article should merely serve as a source of inspiration. It must, however, be kept in mind that the quality of all solvents and chemicals is of paramount importance for success. Therefore, all new batches of solvents and chemicals must be checked with a standard protein digest before being used. In addition, we always aliquot the solvents immediately after positive testing, as any solvent will be contaminated after opening the bottle a few times in the laboratory.

References

Chin, E. T., and Papac, D. I. (1999). The use of a porous graphitic carbon column for desalting hydrophilic peptides prior to matrix-assisted laser desorption/ionization time-of-flight mass spectrometry. *Anal. Biochem.* **273**(2), 179–185.

Cohen, S. L., and Chait, B. T. (1996). Influence of matrix solution conditions on the MALDI-MS analysis of peptides and proteins. *Anal. Chem.* **68**(1), 31–37.

Erdjument-Bromage, H., Lui, M., Lacomis, L., Grewal, A., Annan, R. S., McNulty, D. E., Carr, S. A., and Tempst, P. (1998). Examination of micro-tip reversed-phase liquid chromatographic extraction of peptide pools for mass spectrometric analysis. *J. Chromatogr. A* **826**(2), 167–181.

Gobom, J., Nordhoff, E., Ekman, R., and Roepstorff, P. (1997). Rapid micro-scale proteolysis of proteins for MALDI-MS peptide mapping using immobilized trypsin. *Int. J. Mass Spectrom. Ion Proc.* **169/170**, 153–163.

Gobom, J., Nordhoff, E., Mirgorodskaya, E., Ekman, R., and Roepstorff, P. (1999). Sample purification and preparation technique based on nano-scale reversed-phase columns for the sensitive analysis of complex peptide mixtures by matrix-assisted laser desorption/ionization mass spectrometry. *J. Mass Spectrom.* **34**, 105–116.

Henzel, W. J., Billeci, T. M., Stults, J. T., and Wong, S. C. (1993). Identifying proteins from two-dimensional gels by molecular mass searching of peptide fragments in protein sequence databases. *Proc. Natl. Acad. Sci. USA* **90**, 5011–5015

Hillenkamp, F., Karas, M., Beavis, R. C., and Chait, B. T. (1991). Matrix-assisted laser desorption/ionization mass spectrometry of biopolymers. *Anal. Chem.* **63**(24), 1193A–1203A.

James, P., Quadroni, M., Carafoli, E., and Gonnet, G. (1993). Protein identification by mass profile fingerprinting. *Biochem. Biophys. Res. Commun.* **195**, 58–64

Jensen, O. N., Podtelejnikov, A., and Mann, M. (1996). Delayed extraction improves specificity in database searches by matrix-assisted laser desorption/ionization peptide maps. *Rapid Commun. Mass Spectrom.* **10**(11), 1371–1378.

Karas, M., and Hillenkamp, F. (1988). Laser desorption ionization of proteins with molecular masses exceeding 10,000 daltons. *Anal. Chem.* **60**, 2299–2301.

Kussmann, M., Nordhoff, E., Nielsen, H. R., Haebel, S., Larsen, M. R., Jacobsen, L., Jensen, C., Gobom, J., Mirgorodskaya, E., Kristensen, A. K., Palm, L., and Roepstorff, P. (1997). MALDI-MS sample preparation techniques designed for various peptide and protein analytes. *J. Mass Spectrom.* **32**, 593–601.

Larsen, M. R., Cordwell, S. J., and Roepstorff, P. (2002). Graphite powder as an alternative or supplement to reversed-phase material for desalting and concentration of peptide mixtures prior to matrix-assisted laser desorption/ionization mass spectrometry. *Proteomics* **2**, 1277–1287.

Laugesen, S., and Roepstorff, P. (2003). Combination of Two Matrices Results in Improved Performance of MALDI MS for Peptide Mass Mapping and Protein Analysis. *J. Am. Soc. Mass. Spectrom.* **14**: 992–1002.

Mann, M., Højrup, P., and Roepstorff, P. (1993). Use of mass spectrometric molecular weight information to identify proteins in sequence databases. *Biol. Mass Spectrom.* **22**, 338–345.

Pappin, D. J. C., Højrup, P., and Bleasby, A. J. (1993). Rapid identification of proteins by peptide-mass finger printing. *Curr. Biol.* **3**, 327–332.

Roepstorff, P., Larsen, M. R., Rahbek-Nielsen, H., and Nordhoff, E. (1998). Sample preparation methods for matrix assisted laser desorption/ionization mass spectrometry of peptides, proteins and nucleic acids. *In* "Cell Biology: A Laboratory Handbook," 2rd Ed. Vol. 4, pp. 556–565.

Tsarbopoulos, A., Karas, M., Strupat, K., Pramanik, B. N., Nagabhushan, T. L., and Hillenkamp, F. (1994). Comparative mapping of recombinant proteins and glycoproteins by plasma desorption and matrix-assisted laser desorption/ionization mass spectrometry. *Anal. Chem.* **66**(13), 2062–2070.

Vorm, O., Roepstorff, P., and Mann, M. (1994). Matrix, surfaces made by fast evaporation yield improved resolution and very high sensitivity in MALDI TOF. *Anal. Chem.* **66**, 3281–3287.

Wilm, M., and Mann, M. (1996). Analytical properties of the nanoelectrospray ion source. *Anal. Chem.* **68**(1), 1–8.

48

In-Gel Digestion of Protein Spots for Mass Spectrometry

Kris Gevaert and Joël Vandekerckhove

I. INTRODUCTION

Modern techniques for studying complex protein mixtures utilize two main analytical techniques: (a) one- or two-dimensional polyacrylamide gel electrophoresis (1D or 2D PAGE) for dividing the protein mixture into its individual components and (b) accurate mass spectrometry (MS) for identifying these proteins. Different gel-staining procedures are available for visualizing gel-separated proteins, of which most are compatible with further MS analysis. Different types of Coomassie brilliant blue staining (Bennett and Scott, 1971; Wilson, 1983) and silver-staining procedures (Merril *et al.*, 1979) are by far the most used visualisation protocols prior to mass spectrometric analysis. Proteins spots of interest, e.g., protein spots of which the staining intensity (thus concentration) and/or the position (modification status) differs between two compared samples, are excised from the gel and cleaved into peptides after protease addition. The generated peptide mixture elutes out of the gel passively and is analyzed by a variety of chromatographic and mass spectrometric methods (e.g., reviewed by Aebersold and Mann, 2003).

Frequently, trypsin is used to generate peptides, cleaving at the carboxy-terminal side of lysine and arginine, generating peptides of an average size of about 10 amino acids. For most proteins, this leads to a sufficient number of analyzable peptides and thus to unambiguous identification. This article describes a standard protocol for *in-gel* tryptic protein digestion leading to MS-based protein identification and characterization.

II. MATERIALS AND INSTRUMENTATION

A Sentry ionizing air blower (product number 4003143) is from Simco, Deerlijk, Belgium. Singly wrapped sterile stainless steel scalpels (product number 10.295.10MN) are from BCB Ltd., Cardiff, UK. Singly wrapped Eppendorf Biopur Safe-Lock micro test tubes (product number 0030 121.589) are from VWR International Belgium, Leuven, Belgium. Water (product number 4218) and acetonitrile (product number 9017) used to prepare the different buffers are of the best quality available and are both from Malinckrodt Baker B.V., Deventer, The Netherlands. Sequencing grade modified trypsin (product number V5111) is from Promega Benelux BV, Leiden, The Netherlands. Ammonium hydrogen carbonate (product number A-6141) is from Sigma-Aldrich Corp., St. Louis, MO. Recombinant, proteomics grade trypsin from *Pichia pastoris* (product number 3 357 228) can be obtained from Roche Diagnostics GmbH, Penzberg, Germany. Peptide synthesizer graded tri-fluoroacetic acid (product number PTS6045) is from Rathburn Chemicals Ltd., Walkerburn, Scotland, UK.

III. PROCEDURE

In-Gel Digestion of Proteins Separated in Polyacrylamide Gels

The following procedure is derived from a general procedure that has been described previously (Rosenfeld *et al.*, 1992), but to which important modifications are added. Generally, this approach can be used to digest proteins *in gel* independent from the procedure that was used to visualize them, although care must be taken, as some silver-staining protocols use amino acid cross-linkers that render most of the proteins inaccessible for the employed protease. In the following procedure, trypsin is used to digest the protein into peptides. In principle, every available protease can be used as long as the final buffer conditions (pH, additives, etc.) are met.

Solutions

1. *Wash solution*: Prior to *in-gel* digestion, excised gel pieces containing the proteins are washed extensively with an aqueous acetonitrile solution (1/1, v/v) of the highest purity available. This solution is preferentially freshly prepared in small volumes (e.g., 10 ml).

2. *Digestion buffer*: This buffer must always be freshly prepared and consists of 50 m*M* ammonium bicarbonate dissolved in water of the highest purity available. The pH of this buffer does not need any adjustment and will be around 7.8.

Steps

1. Following Coomassie or silver staining, wash the polyacrylamide gels in water to remove excess substances, derived from the electrophoresis and staining procedure, from the gel matrix. Repeat this washing step at least three times (for 30 min each), each time using fresh volumes of water.

2. Then transfer the gel to a clean container and keep moist during the excision of protein spots. Immediately prior to gel spot excision, open the container and blow a stream of ionized air over it. This air stream almost completely eliminates charging of plastic material by static electricity by neutralising the surface of the working area. When a neutralizing air stream is not used, dust particles will be attracted to the gel container, the polyacrylamide gel, gloves, scalpel, and microtest tubes.

3. Excise protein spots using a sterile scalpel or a hollow stainless steel needle, and excise only the heart of the protein spot (i.e., the most intensely coloured gel area). Using the scalpel blade or a clean forceps, transfer each individual protein spot to a Biopur microtest tube. Prior to excising the next protein spot, wash the scalpel (and the forceps) extensively with methanol to remove small gel particles that may stick to it and thus cross-contaminate other protein spots.

4. Cover each protein spot with wash solution and incubate for 15 min at room temperature. During this wash step, gel pieces will shrink and buffer components (and Coomassie) are efficiently extracted from the gel.

5. Centrifuge the tubes briefly, remove the wash solution, and submerge the gel pieces for a second time in this solution for 15 min at room temperature.

6. Following a second centrifugation step, remove the wash solution and transfer the tubes to a centrifugal vacuum concentrator in which the gel pieces are dried to complete dryness (this takes about 10 min).

7. Rehydrate the dried gel pieces with 10 μl (corresponding to 0.1 μg) of a freshly prepared 0.001% (w/v) trypsin solution in 50 m*M* ammonium bicarbonate (pH 7.8) for 10 min at room temperature, after which excess trypsin solution is removed.

8. Subsequently submerge the rehydrated gel pieces in 50 m*M* ammonium bicarbonate buffer (pH 7.8); depending up the size of the excised gel pieces, between 50 and 100 μl of buffer must be added.

9. Close the tubes and place into a thermostatically controlled incubator, and tryptic digestion proceeds overnight at 37°C.

10. Terminate protein digestion by adding 10 μl of 10% (v/v) trifluoroacetic acid (TFA).

11. Following a brief centrifugation step, remove and transfer the supernatant containing the peptide mixture to a new tube, which is frozen at −20°C until further analysis (e.g., by mass spectrometric techniques).

IV. COMMENTS

1. It has been shown that the removal of metallic silver from silver-stained proteins by oxidation with hydrogen peroxide prior to *in-gel* digestion not only increases the overall sensitivity of MS-based protein identification, but also augments the coverage of the sequence of the analyzed protein (Sumner *et al.*, 2002). One of the side effects is that methionines and carbamidomethylated cysteines will be (at least partly) oxidized to their sulfoxide derivatives (Gevaert *et al.*, 2002). This will increase the complexity of the generated peptide mixtures in two ways: (a) all peptides containing methionine and/or cysteine will be present as couples (oxidized and nonoxidized forms) and (b) neutral losses of the side chains of the oxidized amino acids give rise to satellite peaks (Steen and Mann, 2001), which may cause problems during MALDI-MS analysis. In contrast, this added complexity might help in identifying proteins, as peptides containing rare amino acids (methionine and cysteine) can be recognized in the peptide mass maps and verify the database findings of search algorithms.

2. When analyzing Coomassie-stained protein spots by MALDI-MS-based peptide mass fingerprint-

ing (reviewed by Cottrell, 1994), it is important to remove as much of the noncovalently bound Coomassie molecules as possible during the washing steps (steps 4, 5, and 6 of the procedure). Coomassie molecules ionize rather efficiently in MALDI mode and thereby suppress ionization of peptide molecules. Even more severe, in some cases, Coomassie ions will saturate the detector of the mass spectrometer, making it less sensitive for peptide ions striking it. Furthermore, these Coomassie ions disturb the lower mass section of the peptide mass maps, possibly masking some important small peptide ions.

3. The accuracy of protein identification largely depends on the total number of peptides that are available for analysis. Proteins that have many hydrophobic patches are sometimes difficult to digest, as either the protease has limited access to the chain of the protein or the liberated peptides are too hydrophobic and thus do not elute well out of the gel pieces. For such proteins, the efficiency of protein digestion is improved by adding detergents or chaotropes that partly unfold the protein in the gel, thereby exposing more cleavage sites. Unfortunately, the choice of such additional components is rather limited, as many of them block the action of the protease or interfere with further reversed-phase high-performance liquid chromatographic (RP-HPLC) analysis or with mass spectrometric analyses. One detergent that has proven to be particularly useful for the digestion of membrane proteins is octyl-β-glucopyranoside (van Montfort et al., 2002). This detergent increases the recovery of hydrophobic peptides following digestion and does not interfere with subsequent MS analysis. From our experience, we have found that, especially when working with trypsin, fairly high concentrations of urea (up to 4 M) in the protein digestion buffer can be tolerated without any severe effect on the activity of the protease. Urea notably increases the total coverage of the analyzed protein and thus simplifies protein identification by mass spectrometry. The main drawback when using urea is carbamylation of free amino groups by ammonium cyanate present in urea solution, which, in addition to making less sites available for tryptic cleavage, makes the peptide mixtures more complex.

4. Particularly when employing MALDI-MS for protein identification, it is important to remove any substances that might interfere with the matrix crystallization and the ionization processes. Such substances include salts, buffering substances, chaotropes, and detergents. Several procedures have been described (Gevaert et al., 1997) and are compatible with the in-gel digestion procedure described here.

5. Recombinant trypsin made in the yeast Pichia pastoris has been made available commercially. The main advantage of this protease over other commercially available proteases is the complete absence of nontryptic proteolytic activity. Other types of proteases, although treated with specific inhibitors of commonly copurifying proteases, still lead to nontryptic peptides, which complicate peptide mass maps and make unambiguous protein identification more complex. For example, the most prominent contamination found in trypsin isolated from mammalian organs is chymotrypsin. Because chymotrypsin cleaves at the C-terminal side of five different amino acids, upon prolonged incubation of protein substrates with "chymotrypsin-contaminated" trypsin, this leads to very complex peptide mass maps.

6. Transient or permanent modifications of amino acid side chain are molecular switches to control the activity of a great number of proteins/enzymes. Mass spectrometric analysis is the sole technology available for the in-depth analysis of the vastness of modifications that occur in vivo. However, when digesting a gel-separated protein with only one enzyme, too few ionisable peptides will be generated so as to cover the complete sequence of the analyzed protein and thus many possible modifications may escape analysis. Therefore, it is advantageous to use multiple enzymes either in sequence or together in order to generate many protein fragments and thus cover as much of the protein sequence as possible. Combinations of site-specific and unspecific proteases have been used for that purpose (e.g., MacCoss et al., 2002).

V. PITFALLS

1. One of the major pitfalls encountered frequently is contamination of protein spots with human epidermal keratins. During in-gel protein digestion, keratins that are present digest as well and will give a characteristic and easily recognizable pattern of peptide masses in the mass maps obtained (Parker et al., 1998). To our experience, it is very important to wear clean gloves during all steps of the procedure, i.e., when mounting and handling polyacrylamide gels, when excising stained proteins out of the gel, and when handling the test tubes containing these spots prior to digestion. Next to this, the staining procedures are preferentially kept as short as possible and in the smallest volume possible, as it has been shown that a long exposure of gels to air increases the risk for keratin contamination (Sinha et al., 2001). In our laboratory, we minimize this problem by working in a

laminar air flow hood during the excision of protein spots and during preparation of the protein digests. Furthermore, the gels and the test tubes are constantly in a flow of neutralizing air by which electrostatic charging of plastic surface, which attracts dust particles (and thus keratins), is reduced.

2. Because it is very difficult to estimate the amount of protein present in a gel, in some cases too much protease will be added. Although many of the proteases that are available commercially are so-called "proteomics grade", during typical incubation times, they give rise to autodigestion products. These peptides can, in some cases, be used as internal standards, by which the peptide mass will be determined more accurately (e.g., Li *et al.*, 1997) or, when overrepresented, suppress ionization of peptides from the analyzed proteins, thus hampering protein identification.

Acknowledgment

K.G. is a Postdoctoral Fellow of the Fund for Scientific Research—Flanders (Belgium) (F.W.O.-Vlaanderen).

References

Aebersold, R., and Mann, M. (2003). Mass spectrometry-based proteomics. *Nature* **422**, 198–207.

Bennett, J., and Scott, K. J. (1971). Quantitative staining of fraction I protein in polyacrylamide gels using Coomassie brillant blue. *Anal. Biochem.* **43**, 173–182.

Cottrell, J. S. (1994). Protein identification by peptide mass fingerprinting. *Pept. Res.* **7**, 115–124.

Gevaert, K., Demol, H., Puype, M., Broekaert, D., De Boeck, S., Houthaeve, T., and Vandekerckhove, J. (1997). Peptides adsorbed on reverse-phase chromatographic beads as targets for femtomole sequencing by post-source decay matrix assisted laser desorption ionization-reflectron time of flight mass spectrometry (MALDI-RETOF-MS). *Electrophoresis* **18**, 2950–2960.

Gevaert, K., Van Damme, J., Goethals, M., Thomas, G. R., Hoorelbeke, B., Demol, H., Martens, L., Puype, M., Staes, A., and Vandekerckhove, J. (2002). Chromatographic isolation of methionine-containing peptides for gel-free proteome analysis: Identification of more than 800 *Escherichia coli* proteins. *Mol. Cell. Proteomics* **1**, 896–903.

Li, G., Waltham, M., Anderson, N. L., Unsworth, E., Treston, A., and Weinstein, J. N. (1997). Rapid mass spectrometric identification of proteins from two-dimensional polyacrylamide gels after in gel proteolytic digestion. *Electrophoresis* **18**, 391–402.

MacCoss, M. J., McDonald, W. H., Saraf, A., Sadygov, R., Clark, J. M., Tasto, J. J., Gould, K. L., Wolters, D., Washburn, M., Weiss, A., Clark, J. I., and Yates, J. R., 3rd (2002). Shotgun identification of protein modifications from protein complexes and lens tissue. *Proc. Natl. Acad. Sci. USA* **99**, 7900–7905.

Merril, C. R., Switzer, R. C., and Van Keuren, M. L. (1979). Trace polypeptides in cellular extracts and human body fluids detected by two-dimensional electrophoresis and a highly sensitive silver stain. *Proc. Natl. Acad. Sci. USA* **76**, 4335–4339.

Parker, K. C., Garrels, J. I., Hines, W., Butler, E. M., McKee, A. H., Patterson, D., and Martin, S. (1998). Identification of yeast proteins from two-dimensional gels: Working out spot cross-contamination. *Electrophoresis* **19**, 1920–1932.

Rosenfeld, J., Capdevielle, J., Guillemot, J. C., and Ferrara, P. (1992). In-gel digestion of proteins for internal sequence analysis after one- or two-dimensional gel electrophoresis. *Anal. Biochem.* **203**, 173–179.

Sinha, P., Poland, J., Schnölzer, M., and Rabilloud, T. (2001). A new silver staining apparatus and procedure for matrix-assisted laser desorption/ionization-time of flight analysis of proteins after two-dimensional electrophoresis. *Proteomics* **1**, 835–840.

Steen, H., and Mann, M. (2001). Similarity between condensed phase and gas phase chemistry: Fragmentation of peptides containing oxidized cysteine residues and its implications for proteomics. *J. Am. Soc. Mass Spectrom.* **12**, 228–232.

Sumner, L. W., Wolf-Sumner, B., White, S. P., and Asirvatham, V. S. (2002). Silver stain removal using H_2O_2 for enhanced peptide mass mapping by matrix-assisted laser desorption/ionization time-of-flight mass spectrometry. *Rapid Commun. Mass Spectrom.* **16**, 160–168.

van Montfort, B. A., Canas, B., Duurkens, R., Godovac-Zimmermann, J., and Robillard, G. T. (2002). Improved in-gel approaches to generate peptide maps of integral membrane proteins with matrix-assisted laser desorption/ionization time-of-flight mass spectrometry. *J. Mass Spectrom.* **37**, 322–330.

Wilson, C. M. (1983). Staining of proteins on gels: Comparisons of dyes and procedures. *Methods Enzymol.* **91**, 236–247.

49

Peptide Sequencing by Tandem Mass Spectrometry

John R. Yates, III, David Schieltz, Antonius Koller, and John Venable

I. INTRODUCTION

Microcolumn reversed-phase HPLC electrospray ionization tandem mass spectrometry (ESI-MS/MS) is a rapid and sensitive technique for the analysis of complex mixtures of peptides (Hunt *et al.*, 1986; Griffin *et al.*, 1991). This technique can be used to determine the amino acid sequence of unknown peptides, to verify the structure of proteins, and to determine post-translational modifications. In particular, the strength of this approach is the analysis of peptides in complicated mixtures. Two approaches for sequence analysis of peptides are described: low flow rate infusion (microelectrospray ionization) and reversed-phase microcapillary liquid chromatography tandem mass spectrometry (Gale and Smith, 1993; Emmett and Caprioli, 1994; Andren *et al.*, 1994; Wilm and Mann, 1994; Davis *et al.*, 1995). Algorithms have been developed to help interpret tandem mass spectra to derive a sequence *de novo* and an example of their use is illustrated.

II. MATERIALS AND INSTRUMENTATION

Solvents are from Fisher Scientific (Springfield, NJ): J. T. Baker, 2-propanol 9095-02, acetonitrile OPTIMA A996-4, J. T. Baker, acetic acid 6903-05. Fused silica capillaries are from Polymicro Technologies (Tucson, AZ): 375 µm o.d. × 20 µm i.d., TSP375020, 365 µm o.d. × 100 µm i.d., TSP200100. Reversed-phase chromatographic supports are from Zorbax Eclipse XDB. The column packing device is homemade and is shown in

Fig. 1. The microelectrospray platform was built in the The Scripps Research Institutes instrument shop (Fig. 2). The XYZ manipulator is from Newport Corp. (Irvine, CA). MT-XYZ and the high voltage connector suitable for the ThermoFinnigan electrospray voltage are from Lemo USA (www.lemo.com) (San Jose, CA; P/N 4-89626). A P-2000 laser puller is from Sutter Instruments (Novato, CA).

The electrospray ionization tandem mass spectrometers, LCQ and Q-TOF mass spectrometers, are from ThermoFinnigan (San Jose, CA) and Waters (Milford, MA), respectively. Pumps for HPLC and HP1100 are from Agilent (Palo Alto, CA). DeNovoX is provided with the BioWorks software from ThermoFinnigan.

III. PROCEDURES

A. Preparation of Microcolumns for Reversed-Phase HPLC

Solutions

1. *HPLC solvent A*: To make 1 liter, add 5 ml of formic acid to 995 ml of distilled and deionized water
2. *HPLC solvent B*: To make 1 liter, add 5 ml of formic acid to a solution of 200 ml of distilled and deionized water and 795 ml of acetonitrile

Steps

1. Construct microcapillary columns according to Gatlin *et al.* Rinse 50-cm piece of fused silica capillary (365 µm o.d. × 100 µm i.d.) with 2-propanol and dry. Remove the polyimide coating with a low temperature flame in the middle of the capillary. Position the

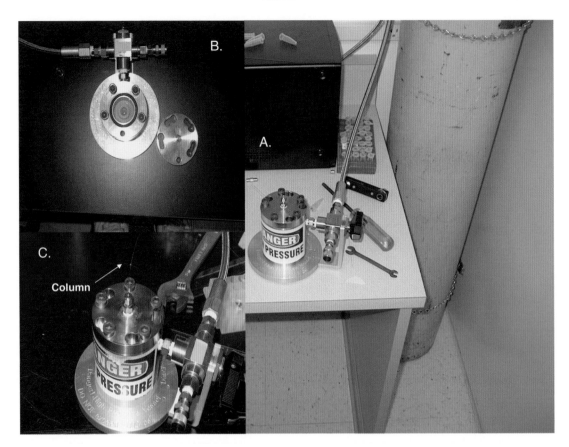

FIGURE 1   Pneumatic device for packing capillary columns and loading samples. A: Device is shown connected to gas line. B: Device is shown with the top removed showing the inside. A slot can be seen where an eppendorf tube can be inserted for either column packing or sample loading. C: Pneumatic device with a fused silica capillary column.

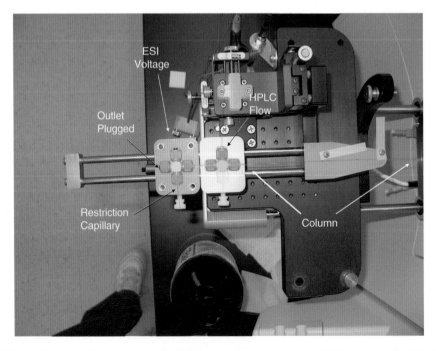

FIGURE 2   Microelectrospray stage for capillary liquid chromatography. HPLC solvent is initially split using a peak microTee to another microTee. This microTee contains a gold electrode for application of the ESI voltage and a restriction capillary to create back pressure and direct flow to waste.

capillary on a laser puller and pull to create two columns each with tips containing 5-μm openings. The settings on the laser puller for 365 × 100-μm i.d. capillary are heat 270, filament 0, velocity 30, delay 128, and pull 0. Repeat this program three times before the capillary is pulled. This program produces a 5-μm tip, although the exact program may depend on mirror alignment and capillary positioning. *Caution*: Conditions to pull a capillary may vary from puller to puller.

2. The packing device is depicted in Fig. 1A. Insert the blunt end of the column into the swagelock fitting in the top of the packing device. To pack the column, fill a polypropylene Eppendorf centrifuge tube (1.5 ml) with ~100 μg of packing material and 1 ml of methanol and sonicate briefly to suspend the material and minimize aggregation. Place the solution in a high-pressure packing device (Fig. 1B), insert the column, and place the end of the column in the solution (Fig. 1C). Use helium gas at a pressure of ~500 psi to drive the packing material into the column. Continue packing until the material fills a length of the capillary corresponding to 10–20 cm. Allow the pressure to slowly drop to zero. Condition the column by rinsing with 100% solvent B and slowly reducing the percentage of solvent B until it reaches initial HPLC conditions (100% solvent A). Use a linear gradient of 100% solvent A to 20% solvent A over 30 min to finish conditioning the column.

3. Configuration for microcolumn HPLC is shown in Fig. 2 and 3.

Gradient: linear, 30–60 min, 0 to 100% (80 : 20 acetonitrile: 0.5% acetic acid)
Flow: 100 μl/min, flow split to final flow rate 200–300 nl/min
Columns: 100 μm i.d. × 20–25 cm, Phenomonex AQUA, 5-μm particles

B. Low Flow Rate Infusion Tandem Mass Spectrometry of Peptides

Media and Solutions

1. *1 : 1 0.5% acetic acid/ methanol*: To make 100 ml, add 2.5 ml of glacial acetic acid to 47.5 ml of distilled and deionized water and add 50 ml of HPLC grade methanol. Store in closed container at room temperature.

Steps

1. To accomplish low flow rate infusion, create a liquid metal junction. Valco fitting (P/N MU1XCTI), titanium is preferred because it is a biologically inert

material. The fused silica capillary is 375 μm o.d. ×20 μm i.d.

2. To infuse a sample, connect an entrance line of fused silica capillary to a syringe pump or to a device to pneumatically drive the liquid at a flow rate of 10–200 nl/min. Strip a fused silica microelectrospray needle (4 cm of 375 μm o.d. × 20 μm i.d.) of its polyimide coating near the exit and then connect to the Valco fitting. Place a voltage (900–2000 V) on the fitting to transfer the potential to the liquid and form the electrospray. Place the exit end of the electrospray needle close (0.1–0.5 mm) to the entrance of the heated capillary.

3. The advantage of low flow rate infusion is that a small volume of sample can be infused over a long time period (5–30 min). This allows time to record a mass spectrum and then to begin acquiring tandem mass spectra of ions observed in the mass spectrum. Low flow rate infusion works quite well for reasonably uncomplicated peptide mixtures. Complicated peptide mixtures, such as those derived from proteolytic digestion of large proteins or mixtures of proteins, can suffer from ion suppression. The tandem mass spectrum shown in Fig. 4 was acquired through infusion and subjected to *de novo* sequence analysis as discussed later.

C. Analysis of Peptides by Microcolumn Liquid Chromatography Microelectrospray Ion Trap Mass Spectrometry (LCQ)

Steps

1. Insert the packed 10- to 20-cm column into an Upchurch PEEK tee. Insert the blunt end of the column into the tee and direct the solvent from the HPLC into one of the other arms of the tee. Insert a length of fused silica into the remaining arm of the tee and connect to a second Upchurch PEEK tee. Insert into one of the arms a gold wire so that a length of the wire protrudes approximately 1–2 cm (Fig. 2). Attach the electrospray voltage to the gold wire. In the remaining arm of the tee, insert a length of fused silica 365 × 5 μm. The fused silica tubing is used to create a restriction to the HPLC flow to force 100–300 nl/min through the microcolumn and this solvent is directed to waste. The microelectrospray source connected to an ion trap mass spectrometer is depicted in Fig. 2.

Electrospray voltage: −1800–2000 V

2. An aliquot of sample is injected to record the molecular weight of the peptides.

Scan mass range: 400–1500 amu
Electron multiplier voltage: 1000 V

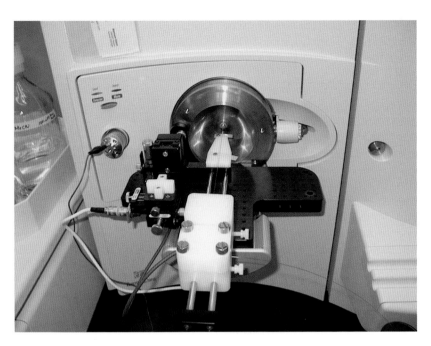

FIGURE 3   Microelectrospray source is shown connected to the mass spectrometer. The column is aligned with the opening in the heated capillary using the XYZ manipulator.

3. Sequence analysis of peptides is performed during a second HPLC analysis by selecting the precursor ion with a 3 u Full Width at Half Height (FWHH) wide notch and exciting the ions within the ion trap to induce fragmentation. Collision energies are on the order of 35 eV. The fragment ions are then scanned out of the ion trap to an electron multiplier set at 1000 V.

4. Alternatively, the LCQ mass spectrometer can automatically acquire tandem mass spectra over the course of an LC analysis. The analysis incorporates a unit mass resolution scan to find the most abundant ion above a preset ion abundance threshold. This ion is isolated and a high mass resolution scan is acquired. By resolving the isotopic peaks of the isolated ion, the charge state can be determined. For example, if the m/z distance between the isotope peaks is 1, then the charge state is +1. If the m/z distance is 0.5, then the charge state is +2. The LCQ can resolve a +4 charge state up to an m/z value of 1500. The tandem mass spectrum for Glu-fibrinopeptide is shown in Fig. 4.

IV.   Comments

Under low-energy, multiple collision conditions, peptides fragment primarily at the amide bonds, producing sequence-specific fragmentation patterns. When fragment ions are produced the charge can be retained on the N terminus of the fragment of the ion

to form ions of type b. If the charge is retained on the C-terminal fragment, the ion is of type y. There is no easy method to distinguish a b-ion from a y-ion without derivatization of the peptide. The mass spectrum shown in Fig. 6 was obtained by a ThermoFinnigan LCQ Deca ion trap mass spectrometer. A hallmark of this instrument is the ability to automatically acquire tandem mass spectra. After a mass spectrum is obtained, the data system identifies the most abundant ion and then acquires a tandem mass spectra in the next scan.

To demonstrate the sequencing process, the MS/MS spectrum for the peptide shown in Fig. 4 is interpreted. The process is demonstrated on a tandem mass spectrum acquired on an LCQ ion trap mass spectrometer. The process is identical to the one followed for triple quadrupole mass spectrometers as described in Yates *et al*. (1994) and McCormack *et al*. (1994). The mass spectrum is produced by collisional activation of a doubly charged ion, $(M + 2H)^{2+}$, at m/z 786.10. The protonated molecular weight of this peptide is calculated by multiplying the m/z value by 2 and subtracting the weight of one proton to give a value of 1571.2 (average mass). Subtraction of the highest mass fragment ion, m/z 1396.2 (1571.2 − 1396.2 = 175.0), from the $(M + H)+$ value yields a mass of 175. The mass fits C-terminal Arg. If this peptide was created through trypsin digestion, then this is a good assignment. Subtraction of the ion at 1396.2 from the next highest ion, 1285.4, yields a mass difference of 110.8, which does not correspond

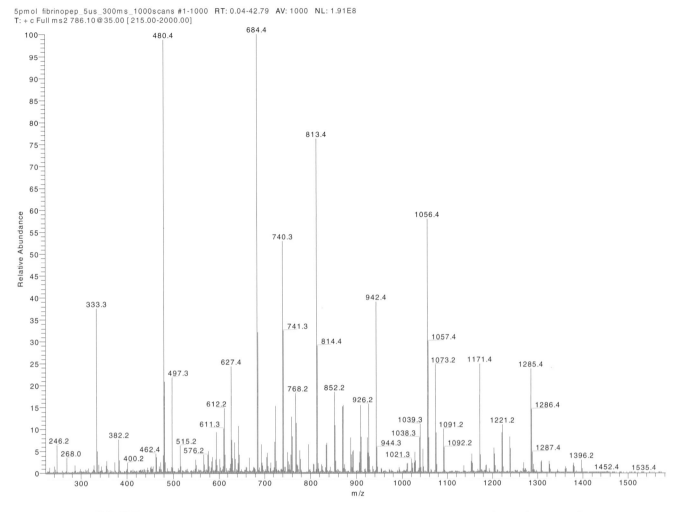

5pmol fibrinopep_5us_300ms_1000scans #1-1000 RT: 0.04-42.79 AV: 1000 NL: 1.91E8
T: + c Full ms2 786.10@35.00 [215.00-2000.00]

FIGURE 4 A tandem mass spectrum of the doubly charged ion, (M+2H)+2 786.10, of Glu-Fibrinopeptide.

to a mass for one of the common amino acids. This situation strongly suggests the ion at 1285.4 is of a different type. Assuming the ion at 1285.4 is of type y and that y-ions tend to predominate in tandem mass spectra of doubly charged peptide ions created through trypsin digestion, then it would be worthwhile to start from this point in the spectrum. A window can be set for where the next ion should reside in the spectrum using a range between a mass of 57 (Gly) and 186 (Trp). This window assumes no modifications to amino acids that would be larger then Trp. The next ion that fits the aforementioned criteria is 1171.4, which corresponds to a mass of 114 or Asn. The difference between the next ion at 1056.4 and 1171.4 is 115 for Asp and then the difference between this ion and 942.4 is 114. Thus far a sequence of Asn–Asp–Asn has been determined. The next abundant ion in the spectrum is 813.4, which nets a difference of 129 or Glu. An abundant ion is observed at 740.3, which

yields a difference of 73. This value does not correspond to the weight of a common amino acid and thus can be temporarily discounted. Proceeding to the next ion to determine if a mass difference exists with the ion at 813.4 that provides a better fit, the ion at 684.4 is tried. This ion produces a mass difference of 129 and we now have a sequence of Asn–Asp–Asn–Glu–Glu. The next set of ions are less abundant and the largest of this group 627.4 gives a mass difference of 57 corresponding to Gly. If the other ions are tried, none produce as close a fit with an amino acid residue mass. The next two ions are rather obvious at 480.4 and 333.3. These ions yield a sequence of Phe–Phe. Two ions remain in the low end of the mass spectrum but only 246.2 produces a good fit and the amino acid Ser is obtained. The sequence now corresponds to Asn–Asp–Asn–Glu–Glu–Gly–Phe–Phe–Ser. By using the presumption the ion series used to interpret the sequence was of type y, this sequence corresponds to the N-terminal sequence of the peptide. We know the C-terminal residue of the peptide is Arg so a calculation of the difference between 175 and 246.2 produces a mass of 71 and thus Ala. Completion of the sequence will require closer scrutiny of ions at the ends of the spectrum. At the high mass end of the spectrum between ions at 1285.4 and 1396.2 there are several ions. The ion at 1384.6 is 99 Da from the ion at 1285.4 and could be a Val residue. The corresponding ion at 286 is present at the low *m/z* end of the spectrum, confirming this residue as the likely sequence. The difference between 1384.6 and 1571.2 is 186.6. This corresponds to a Trp residue or a combination of two or more amino acids. Several combinations of amino acids will fit this mass: Glu–Gly, Asp–Ala, or Ser–Val. There is insufficient information in the spectrum to differentiate these possible sequences. The Ser–Val sequence can be eliminated by converting the peptide

to its methyl ester as neither Ser nor Val should convert to a methyl ester with a *resulting* mass increase of 14 Da. The final sequence for the peptide derived from this spectrum is [EG,DA,SV]NDNEEGFFSAR.

De Novo Sequencing Algorithm

Results from two different *de novo* sequencing algorithms are shown in Fig. 5 and 6. The DeNovoX program automatically determines a complete or partial amino acid sequence of an unknown peptide by interpreting its MS/MS spectra. It is capable of identifying peptides with posttranslational modifications, mutations, or single amino acid deletion in the peptide. The probability-based program, which is optimized for data collected with the ThermoFinnigan LCQ ion trap instrument, takes data files that contain the *m/z* values and intensities of ions in MS/MS

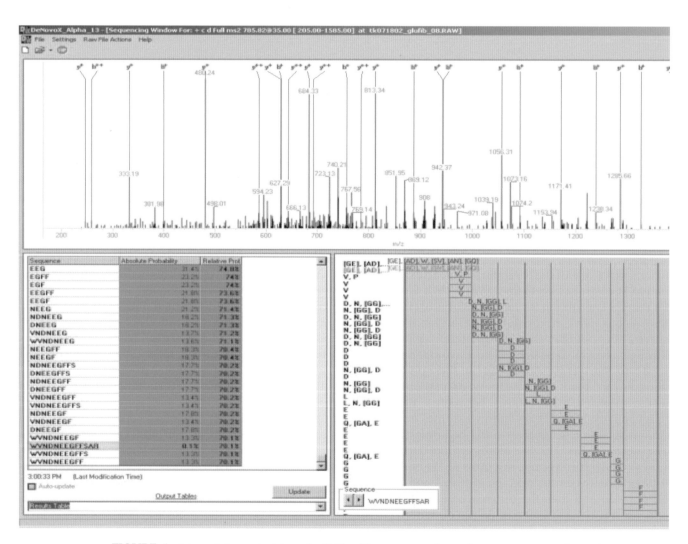

FIGURE 5   Interpretation output from the DeNovoX program on the tandem mass spectrum shown in Figure 4.

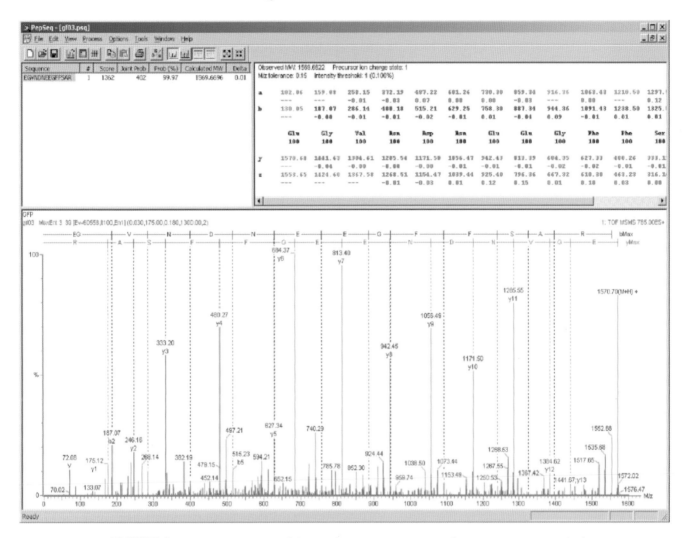

FIGURE 6 Interpretation output of the MassSeq program using a tandem mass spectrum acquired on a Q-TOF mass spectrometer (data not shown).

spectra together with the information on the respective parent ion mass and charge state as input. The data files are processed by the DeNovoX program in a fully automated mode. The program takes into account over 25 different factors relevant to *de novo* peptide sequencing. Amino acid tables that specify the amino acids considered for the search can be selected by the user and can include all major post-translational modifications. All other parameters are set by default in the program, including the mass tolerance of 0.8 Da for fragment ions. A list of suggested sequence tags and completed sequences with respective probability values are included in the output of the program, in addition to assigned spectra. The sequence interpretation with the highest probability is WVNDNEEGFFSAR and is shown in Fig. 5. The correct sequence is EGNDNEEGFFSAR. Clearly the

first two amino acid residues are incorrect for reasons similar to those encountered in the manual interpretation. Ion trap tandem mass spectra are often lacking the necessary sequences to definitively identify the N-terminal amino acid residues. These sequences can then be used for further database searches (e.g., BLAST search).

Peptide tandem mass spectrometry data obtained by Q-TOF mass spectrometers can be sequenced using the MassSeq program from MicroMass. *De novo* sequencing is performed by using the MaxEnt 3/ MassSeq programs in the Waters-Micromass Masslynx 3.5 software package. The raw tandem mass spectrum is first processed with the Massive Inference algorithm (MaxEnt 3) (Jaynes, 1957; Skilling, 1989), which produces a spectrum where any higher charge state ion is collapsed into a single charge state and the isotopic distribution of each peak is folded into the lowest mass member (^{12}C isotope). Interpretation of the processed spectrum is performed by the MassSeq program, which generated candidate sequences based on the molecular mass of the peptide. Because the total number of possible sequences is very large, a terminated Markov Chain Monte Carlo algorithm is employed to generate a smaller, yet representative set of sequences. Predicted spectra are generated from each of the candidate sequences using probabilistic fragmentation and then each are compared to the MaxEnt 3 processed spectrum. The final result is a list of sequences, which are scored based on their probability of a correct match to the MaxEnt 3 spectrum. Within each resulting peptide sequence, each amino acid is assigned a probability as to how likely the residue is correct assigned to that position in the sequence. Therefore, this method provides the ability to sequencing peptides *de novo* and have information on the degree of likelihood that the sequence is correct. Q-TOF spectra often contain more information about the sequence and the complete sequence can often be inferred. Interpretation of a tandem mass spectrum of Glu-fibrinopeptide is shown in Fig. 6. Both algorithms produce a nearly complete sequence.

V. PITFALLS

1. Results from sequencing should be checked by derivatizing the peptide to ensure that the appropriate mass shift is observed. One derivatization method consists of forming the peptide methyl ester by adding 2 *N* methanolic HCl to the dry peptide. A second tandem mass spectrum can be acquired for the derivatized peptide to determine if the fragment ions also shift by the correct mass increment.

2. Distilled and deionized water must be used in the HPLC solvents to avoid the production of sodium or potassium adducts.

References

Andren, P. E., Emmett, M. R., and Caprioli, R. M. (1994). Micro-electrospray—zeptomole-attomole per microliter sensitivity for peptides. *J. Am. Soc. Mass Spectrom.* **5**, 867–869.

Davis, M. T., Stahl, D. C., Hefta, S. A., and Lee T. D. (1995). A microscale electrospray interface for online, capillary liquid-chromatography tandem mass-spectrometry of complex peptide mixtures. *Anal. Chem.* **67**, 4549–4556.

Emmett, M. R., and Caprioli, R. M. (1994). Micro-electrospray mass spectrometry: Ultra-high-sensivity analysis of peptides and proteins. *J. Am. Soc. Mass Spectrom.* **5**, 605–613.

Gale, D. C., and Smith, R. D. (1993). Small volume and low flow rate electrospray ionization mass spectrometry of aqueous samples. *Rapid Commun. Mass Spectrom.* **7**, 1017–1021.

Gatlin, C. L., Kleemann, G. R., Hays, L. G., Link, A. J. & Yates, J. R. (1998). Protein identification at the low femtomole level from silver-stained gels using a new fritless electrospray interface for liquid chromatography-microspray and nanospray masss spectrometry. *Anal Biochem* **263**, 93–101.

Griffin, P. R., Coffman, J. A., Hood, L. E., and Yates, J. R. III (1991). Structural analysis of proteins by HPLC-MS and HPLC-MS/MS using electrospray ionization on a triple quadrupole mass spectrometer. *Int. J. Mass Spectrom. Ion Proc.* **111**, 131–149.

Hunt, D. F., Yates, J. R. III, Shabanowitz, J., Winston, S., and Hauer, C. R. (1986). Protein sequencing by tandem mass spectrometry. *Proc. Natl. Acad. Sci. USA* **84**, 620–623.

Jaynes, E. T. (1957). Information theory and statistical mechanics, I. *Phys. Rev.* **106**, 620–630.

Kennedy, R. T., and Jorgenson, J. W. (1989). Preparation and evaluation of packed capillary liquid chromatography columns with inner diameters from 20 to 50 μm. *Anal. Chem.* **56**, 1128–1135.

McCormack, A. L., Eng, J. K., and Yates, J. R. III (1994). Peptide sequence analysis on quadrupole mass spectrometers. *Methods* **6**, 274–283.

Skilling, J. (1989). *In* "Maximum Entropy and Bayesian Methods" (J. Skilling, ed.), p. 45. Kluwer Academic, Dordrecht.

Wilm, M. S., and Mann, M. (1994). Electrospray and taylor-cone theory, doles beam of macromolecules at last. *In. J. Mass Spectrom. Ion Proc.* **136**, 167–180.

Yates, J. R. III, McCormack, A. L., Hayden, J., and Davey, M. (1994). Sequencing peptides derived from the class II major histocompatibility complex by tandem mass spectrometry. *In* "Cell Biology: A Laboratory Handbook" (J. E. Celis, ed.), pp. 380–388. Academic Press, San Diego.

50

Direct Database Searching Using Tandem Mass Spectra of Peptides

John R. Yates III and William Hayes McDonald

I. INTRODUCTION

Whole genome sequencing has provided a sequence infrastructure to the enormous benefit of mass spectrometry and protein biochemistry. The ability to match tandem mass spectra to sequences in protein or nucleotide databases allows accurate and high throughput identification of peptides and proteins (Eng *et al.*, 1994; Yates *et al.*, 1995a,b, 1996). Furthermore, this data analysis capability, when combined with automated acquisition of tandem mass spectra, allows direct analysis and identification of proteins in mixtures. Applications of this procedure to rapidly survey the identities of proteins in protein complexes, subcellular compartments, cells, or tissues are made possible because of the combination of tandem mass spectrometry and database searching (McCormack *et al.*, 1995; Link *et al.*, 1997, 1999; Washburn *et al.*, 2001; Florens *et al.*, 2002). The method allows both automated and accurate searches of protein and nucleotide databases, as well as the use of spectra of posttranslationally modified peptides. Modifications such as phosphorylation, methylation, and acetylation can be identified through the analysis of digested protein complexes (Cheeseman *et al.*, 2002; MacCoss *et al.*, 2002). A detailed description of the manner in which tandem mass spectra are processed and then used to search databases is described.

II. MATERIALS AND INSTRUMENTATION

Tandem mass spectra of peptides were obtained by electrospray ionization tandem mass spectrometry (Finnigan MAT, San Jose, CA) as described previously (Gatlin *et al.*, 1998; Verma *et al.*, 2000; Washburn *et al.*, 2001).

Several computer algorithms now exist to search tandem mass spectra through sequence databases. Information about these programs can be accessed through the following URLs: http://www.matrix science.com, http://prospector.ucsf.edu, http://pepsea.protana.com, http://fields.scripps.edu/sequest, and http://www.proteometrics.com. This article discusses the use of SEQUEST, as we have the most experience with this program. SEQUEST searches were performed on a PC Linux computing cluster described elsewhere using a parallelized version of the SEQUEST algorithm (Sadygov, 2002). Methods for data reduction, preliminary scoring, and cross-correlation analysis used in the computer algorithm have been described in detail elsewhere (Eng *et al.*, 1994; Yates *et al.*, 1995a,b). Search results are analyzed with the program DTAselect (Tabb *et al.*, 2002). Information on how to obtain the DTAselect program can be found at http://www.sequest.org. This program can accommodate output from the

SEQUEST and Mascot search programs. Databases can be obtained over the Internet using anonymous ftp to the National Center for Biotechnology Information (NCBI) (ftp://ncbi.nlm.nih.gov). Databases such as the GenBank database of nucleotide sequences, the NRP (nonredundant protein) database, and dbEST—the collection of expressed sequence tag sequences—can be obtained from the NCBI site. The complete *Saccharomyces cerevisiae* sequence can be obtained from Stanford Genomic Resources (http://genome-www.stanford.edu).

III. PROCEDURES

A. Method for Database Search with Tandem Mass Spectra

Steps

1. Convert raw, binary data from an MS/MS file to ASCII in the following format:

```
1734.9      2
 110.3      49422.8
 112.3      32433.3
 112.9      65452.1
   :            :
```

The first line in the file contains (M + H)+ and charge state information. Conversion of binary data is accomplished automatically by SEQUEST with data files from ThermoFinnigan mass spectrometers. Most mass spectrometers can convert proprietary data file formats to a form that can be used by the database search programs or the search programs can read the formats directly. The next task calculates the molecular weight of the peptide based on its precursor *m/z* value. This calculation is accomplished with the program *2to3*, which identifies the charge state of the precursor ion by identifying related fragment ion pairs (Sadygov *et al.*, 2002). Fragment ion pairs sum to the molecular weight of the peptide and by initially assuming that the precursor ion is doubly charged and checking that related fragment ion pairs exist in the spectrum will verify this assumption. A molecular weight based on a +3 charge state is then calculated and the existence of fragment ion pairs. The correct molecular weight will produce a much larger number of related ion pairs. This program is designed to use nominal resolution spectra where the charge state based on isotope spread cannot be calculated. If the spectrum does not contain a minimum amount of fragmentation information to identify the charge state, then that spectrum is not used for database searching.

2. For SEQUEST searches, search parameters can be set in the param.h header file. This includes the database to be searched, whether it is a nucleotide or protein database, and whether to perform calculations using average or monoisotopic masses. The set of sequence ions (types b and y) to be considered can also be selected as well as the relative abundance values to be used during the theoretical reconstruction of tandem mass spectra. The masses of amino acids to consider for modified peptides can be input. The mass tolerance for the peptide (M + H)+ can be set as well. Enzyme or chemical cleavage specificity can be entered from the following list: no enzyme, trypsin, chymotrypsin, clostripain, cyanogen bromide, iodoso benzoate, proline endopeptidase, *Staphylococcus aureus* V8 protease, lysine endoproteinase, arginine endoproteinase, AspN-endoproteinase, and elastase.

3. The tandem mass spectrometry data file is then processed in two ways. First the precursor ion is removed and the remaining ions are normalized. This process is not required when analyzing ion trap mass spectra. All but the top 200 most abundant ions are then removed from the search file. This processed spectrum is used to search the database. To perform the cross-correlation analysis described later, a second file from the tandem mass spectrum is created. The spectrum is divided into 10 equal regions and within each region the ions are normalized to the most abundant ion. The molecular weight of the peptide in the tandem mass spectrum is calculated directly from the precursor ion.

4. A search of the database involves scanning each entry to find linear combinations of amino acids, proceeding from the N to the C terminus, that are within some tolerance of the mass of the peptide represented by the tandem mass spectrum. Sequence selection can also be guided by the cleavage specificity of the protease used to create the peptide, including consideration of incompletely digested peptides from either side of the primary sites, or it can be performed with no assumptions about how the peptide was created. If a nucleotide database is searched, the nucleotide sequences are translated "on the fly" to protein sequences in 6-reading frames (Yates *et al.*, 1995b). Chemical modifications can be considered by changing the amino acid mass used to calculate the masses of the peptides. The modified amino acid is then considered at every occurrence in the sequence (Yates *et al.*, 1995a,b).

5. Once an amino acid sequence is within the defined mass tolerance, a preliminary evaluation is performed (Eng *et al.*, 1994). First, the number (n_i) of predicted fragment ions that match ions observed in the spectrum within the fragment ion mass tolerance and their abundances (i_m) are summed. If an ion series is continuous, i.e., if consecutive sequence ions are

present, then a component of the score, β, is incremented. A sequence that matches a continuous set of ions is weighted more heavily then one that matches a few sequence ions randomly. If an immonium ion is present in the spectrum (not usually the case with Ion Trap MS/MS data), then the associated amino acid must be present in the sequence under consideration or an additional component of the score, ρ, is increased or decreased correspondingly. The total number of predicted sequence ions is also noted (n_t). A score is calculated for each amino acid sequence by using the following relationship in Eq. (1),

$$Sp = \left(\sum im\right) * ni * (1+\beta) * (1+\rho) / n_x. \quad (1)$$

6. Each of the top 500 scoring sequences are subjected to a cross-correlation analysis. This is performed by reconstructing a model tandem mass spectrum for each of the amino acid sequences in the list of 500 and comparing each one to the processed experimental tandem mass spectrum (step 2) by using a cross-correlation function. The cross-correlation function is a very sensitive signal processing method used to compare the coherence of two signals (Owens, 1992). This is performed, in effect, by translating one signal across another. If two signals are the same or very similar, the correlation function should maximize when there is no offset between the signals. A cross correlation score is computed for each of the 500 amino acid sequences.

7. Cross-correlation values are normalized by dividing the XCorr value by the autocorrelation of the experimental tandem mass spectrum. The formula used to calculate a normalized cross-correlation value is shown by Eq. (2). By normalizing correlation scores for tandem mass spectra of +1, +2, and +3 ions, the scores are roughly equivalent and a statistical confidence for each sequence assignment can be assessed (MacCoss *et al.*, 2002).

$$XCorr = \frac{Corr(Exp, Theo)}{\sqrt{Corr(Exp, Exp) * Corr(Theo, Theo)}} \quad (2)$$

B. Data Assembly and Filtering

Steps

1. For very large analyses, such as multidimensional liquid chromatography experiments (MudPIT), extremely large data sets can be produced. A program, DTASelect, is used to assemble, filter, and display the results from search (Tabb *et al.*, 2002). The program is especially suited for the analysis of data produced in multidimensional liquid chromatography separations together. DTASelect can be used with output from SEQUEST or Mascot searches. Figure 1 shows output from DTASelect from a multidimensional analysis of the *S. cerevisiae* proteasome. In the first column, the probability the match is correct is shown. The closer the value is to 1 the higher the probability the match is correct. This entry in the table is linked to the tandem mass spectrum and a representation of the quality of the sequences is fit to the spectrum. How well other sequences fit to the spectrum can be viewed from this page as well. Each of the predicted b- and y-ions are aligned with matching fragment ions in the spectrum. Mass differences between predicted fragment ions and fragments ions observed in the spectrum are shown in the upper right hand side of the page. At the bottom of the figures are check boxes to indicate the validation status of the fit of the spectrum to the sequence (Fig. 2). Access to individual search data allows the inspection of search results to look for sequence similarity between sequences.

2. In the second and third columns marked XCorr and DeltaCn, respectively, are scores for the tandem mass spectrum's match to the sequence. The values under XCorr are obtained from the cross-correlation analysis and are normalized as described earlier. The larger the value, the closer the fit between the experimental tandem mass spectrum and the model tandem mass spectrum constructed from the sequence. The DeltaCn value is the difference between the normalized XCorr values for the first and second search results. The larger DeltaCn is the more dissimilar the first and second answers will be.

3. The fourth and fifth columns show the (M + H)+ value calculated from the precursor ion chosen for MS/MS and the (M + H)+ value calculated from the sequence, respectively.

4. Columns SpR, SpScore, and Ion% show the rank of the peptide match based on the preliminary score, the preliminary score and the percentage of ions matched between the experimental spectrum, and sequence ions predicted from the sequence. Generally, short peptides produce an SpR that is less than one. The correct result usually matches a large percentage of ions, Ion%, between those predicted and those observed. Larger peptides can result in a low percentage of ion matches because of the limited mass range of the ion trap mass spectrometer.

5. In the last column the matched sequence is shown. The actual matching sequence is preceded and terminated by a period. Amino acid residues before the preceding period and after the terminating period are shown to illustrate the enzymatic cleavage sites.

6. Peptide matches are grouped by gene locus, and information pertaining to the protein is shown in the header. From left to right the information shown in the header describes the validation status of peptides matched to that protein, the locus name for the gene,

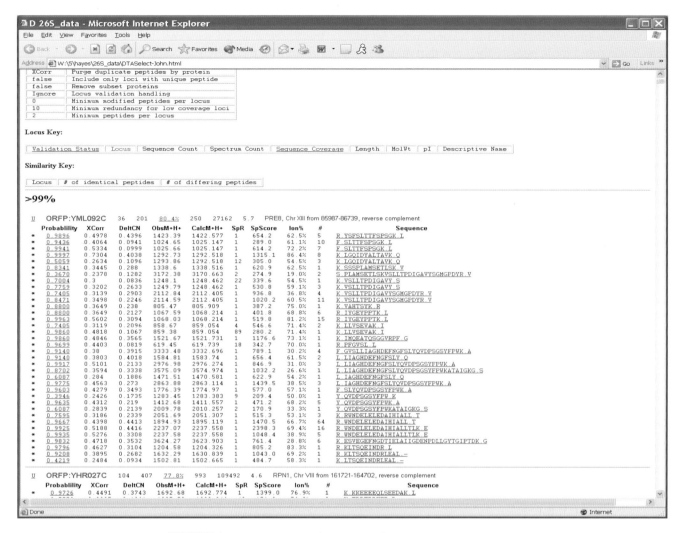

FIGURE 1   DTASelect output for analysis of the *S. cerevisiae* proteasome.

the number of peptides identified, the spectrum count indicating the total number of spectra matched (redundant spectra are included), percentage sequence coverage, which is linked to a visual display of coverage and overlap (Fig. 3), protein length, molecular weight, pI, and the descriptive name for the protein.

IV. COMMENTS

SEQUEST software can provide highly accurate protein identifications. The highly specific information represented in a tandem mass spectrum allows proteins present in mixtures to be identified, as each tandem mass spectrum is a specific address to a protein in the same manner that an amino acid sequence can be highly specific for a protein. Gener-

ally, several or more spectra are obtained for each protein, but a protein can be identified on the basis of one tandem mass spectrum if the following criteria are met: the amino acid sequence represented by the tandem mass spectrum is at least seven amino acids in length, the tandem mass spectrum contains a sufficient number of sequence ions to allow validation of the identified sequence, and the amino acid sequence is unique to a single protein within the organism from which the protein was derived.

When evaluating a match between a spectrum and a sequence, several scores should be considered. The greater the probability value, the more statistically significant the match will be. These probability scores are based on empirical measures. The DTASelect program does retain search results with lower probabilities as their significance may increase if many higher scoring spectra match to the same protein. The normalized

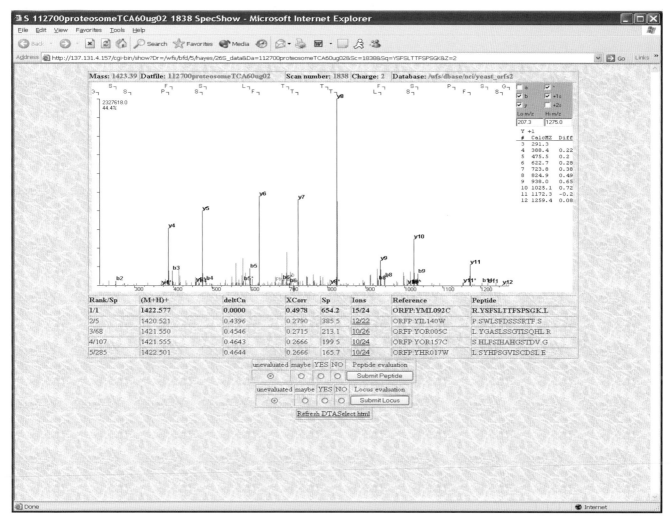

FIGURE 2 Following the link from the probability value displayed in the first column (Fig. 1), information for individual searches of tandem mass spectra can be found. This is useful to determine if a close score is the result of identifying two closely related sequences. The tandem mass spectrum is shown as well as the match to the sequence.

cross-correlation value provides a mathematical evaluation of how close the experimental MS/MS fits to a model derived from the sequence. The limitations of this method relate to the ability to model a sequence as a spectrum and the predictability of peptide fragmentation. The DeltaCn value reflects how well the spectrum matches the model spectrum in comparison to background matches from the database. The XCorr value is an inherent property of the spectrum and changes as a function of spectral quality, whereas DeltaCn changes as a function of spectral quality and database size (e.g., background). Thus these two values provide a good quantitative measure of the match quality.

SEQUEST is ideal for creating a high-throughput, automated system for the identification of proteins from two-dimensional gels. By combining an autosampler with an HPLC protein, digests can be automatically injected and separated and tandem mass spectra automatically acquired. Tandem mass spectra can be acquired through an instrument control program on the ThermoFinnigan LCQ series mass spectrometer allowing unattended acquisition. Proteins present in mixtures can also be identified through the acquisition of tandem mass spectra and matching spectra to their respective proteins. Database sequence errors can be tolerated by the adjustment of parameters such as the ion series and mass tolerance to be used in a search. Database searching using tandem mass spectra enables "shotgun proteomics" or the analysis of digested protein mixtures. Unlike a 2 Dimensional Gel Electrophoresis analysis, it is only after a database search

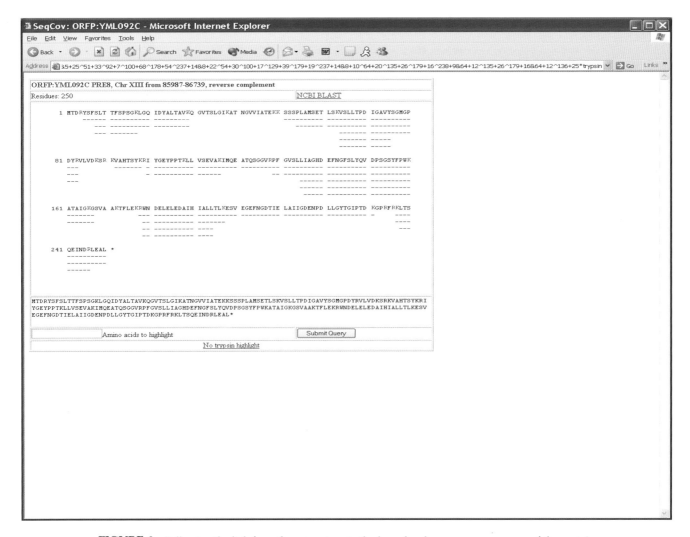

FIGURE 3 Following the link from the percentage in the locus header, sequence coverage of the protein can be observed. This figure shows regions of the protein where peptides were identified. This information is useful when a search of modification is desired. Regions lacking sequence coverage could represent sites of modification, areas that were difficult to digest, or areas where digestion yields too many short peptides. By using the input box at the bottom of the page, specific amino acid residues can be highlighted. Sites of K and R in the sequence are shown.

can relationships among spectra be determined. For relatively simple protein mixtures, combining the "shotgun proteomics" method with multidimensional liquid chromatography can yield reasonably good protein sequence coverage.

V. PITFALLS

1. All tandem mass spectra should be validated against the sequence matched in the database.
2. SEQUEST will match a tandem mass spectrum to a similar, mass conserved sequence if the correct sequence is not present in the database.

3. SEQUEST can match tandem mass spectra with a relatively poor signal-to-noise ratio to the correct sequence, but the spectrum should be of sufficient quality for validation or the match should be considered tentative.

References

Cheeseman, I. M., Anderson, S., *et al.* (2002). Phospho-regulation of kinetochore-microtubule attachments by the *Aurora kinase* Ipl1p. *Cell* **111**(2), 163–172.

Eng, J. K., McCormack, A. L., *et al.* (1994). An approach to correlate tandem mass spectral data of peptides with amino acid sequences in a protein database. *J. Am. Soc. Mass Spectrom.* **5**, 976–989.

Florens, L., Washburn, M. P., *et al.* (2002). A proteomic view of the *Plasmodium falciparum* life cycle. *Nature* **419**(6906), 520–526.

Gatlin, C. L., Kleemann, G. R., *et al.* (1998). Protein identification at the low femtomole level from silver-stained gels using a new fritless electrospray interface for liquid chromatography-microspray and nanospray mass spectrometry. *Anal Biochem* **263**(1), 93–101.

Link, A. J., Carmack, E., *et al.* (1997). A strategy for the identification of proteins localized to subcellular spaces: Application to *E. coli* periplasmic proteins. *Int. J. Mass Spectrom. Ion Proc.* **160**, 303–316.

Link, A. J., Eng, J., *et al.* (1999). Direct analysis of protein complexes using mass spectrometry. *Nature Biotechnol.* **17**(7), 676–682.

MacCoss, M. J., McDonald, W. H., *et al.* (2002). Shotgun identification of protein modifications from protein complexes and lens tissue. *Proc. Natl. Acad. Sci. USA* **99**(12), 7900–7905.

MacCoss, M. J., Wu, C. C., *et al.* (2002). Probability-based validation of protein identifications using a modified SEQUEST algorithm. *Anal. Chem.* **74**(21), 5593–5599.

McCormack, A. L., Eng, J. K., *et al.* (1995). Microcolumn liquid chromatography-electropray ionization tandem mass spectrometry. *Biochem. Biotech. Appli. Electrospray Ionization Mass Spectrom.* **619**, 207–225.

Owens, K. G. (1992). Application of correlation analysis techniques to mass spectral data. *Appl. Spectrosc. Rev.* **27**(1), 1–49.

Sadygov, R. G., Eng, J., Durr, E., Saraf, A., McDonald, H., MacCoss, M. J., and Yates, J. R. III (2002). Code developments to improve the efficiency of automated MS/MS spectra interpretation. *J. Proteome Res.* **2**, 211–215.

Tabb, D. L., McDonald, H. W., Yates, J. R., III (2002). DTASelect and Contrast: Tools for aseembling and comparing protein identifications from shotgun proteomics. *J. Proteome Res.* **1**(1), 21–36.

Verma, R., Chen, S., *et al.* (2000). Proteasomal proteomics: Identification of nucleotide-sensitive proteasome-interacting proteins by mass spectrometric analysis of affinity-purified proteasomes. *Mol. Biol. Cell* **11**(10), 3425–3439.

Washburn, M. P., Wolters, D., *et al.* (2001). Large-scale analysis of the yeast proteome by multidimensional protein identification technology. *Nature Biotechnol.* **19**(3), 242–247.

Yates, J. R., 3rd, Eng, J. K., *et al.* (1995a). Mining genomes: Correlating tandem mass spectra of modified and unmodified peptides to sequences in nucleotide databases. *Anal. Chem.* **67**(18), 3202–3210.

Yates, J. R., 3rd, Eng, J. K., *et al.* (1995b). Method to correlate tandem mass spectra of modified peptides to amino acid sequences in the protein database. *Anal. Chem.* **67**(8), 1426–1436.

Yates, J. R., 3rd, McCormack, A. L., *et al.* (1996). Mining genomes with MS. *Anal. Chem.* **68**(17), 534A–540A.

Identification of Proteins from Organisms with Unsequenced Genomes by Tandem Mass Spectrometry and Sequence-Similarity Database Searching Tools

Adam J. Liska and Andrej Shevchenko

I. INTRODUCTION

The analysis of proteomes by mass spectrometric methods that correlate peptide fragments from proteins with database entries *in silico* has been dependent on the sequencing of genomes. Mass spectrometry and database sequences have enabled the analysis of the human, mouse, and *Arabidopsis* proteomes, among others. Due to the high homology between living organisms at the molecular level, it is possible to use the available protein sequences accumulated in databases from a range of organisms as a reference for the identification of proteins from organisms with unsequenced genomes by sequence-similarity database searching. As research continues in organisms such as *Xenopus*, maize, cow, and others with limited database sequence resources, sequence-similarity searching is a powerful method for protein identification. This article focuses on MS BLAST (Shevchenko *et al.*, 2001) and MultiTag (Sunyaev *et al.*, 2003) as bioinformatic methods for the identification of proteins by the interpretation of tandem mass spectra of peptides and sequence-similarity searching.

II. MATERIALS AND INSTRUMENTATION

In analyses where MS BLAST is utilized for protein identification, tandem mass spectra of peptides can be acquired with any ionization source and mass spectrometer that enables *de novo* sequence prediction: nanoelectrospray, LC/MS/MS, o-MALDI (MALDI quadrupole TOF), MALDI TOF-TOF, QqTOF, triple quad, PSD MALDI-TOF, and ion trap. Alternatively, in analyses where MultiTag is applied, tandem mass spectra of peptides can be acquired with any ionization source and tandem mass spectrometer that enables the creation of peptide sequence tags (Mann and Wilm, 1994): nanoelectrospray, LC/MS/MS, QqTOF, triple quad, or other novel system.

Software for *de novo* sequence prediction from tandem mass spectra is often included in the software packages associated with mass spectrometers: BioMultiview, BioAnalyst (both are from MDS Sciex, Canada), BioMassLynx (Micromass Ltd, UK), BioTools (Bruker Daltonics, Germany), and DeNovoX (ThermoFinnigan). The Lutefisk program (Johnson and Taylor, 2000) can be acquired from http://www.

hairyfatguy.com/Sherpa/. BioAnalyst with the Pro-Blast processing script can generate a complete MS BLAST query automatically from multiple-spectra files acquired by nanoelectrospray or LC/MS/MS (Nimkar and Loo, 2002). A web browser such as Internet Explorer or Netscape is also required to gain access to the MS BLAST web interface located at http://dove.embl-heidelberg.de/Blast2/msblast.html. An independent BLAST computer (Paracel BlastMachine system) may also be purchased and installed for rapid and private MS BLAST operation.

Sequence tags that contain a few confidently designated amino acid residues and mass values that lock the short sequence stretch into the length of the peptide can often be generated from tandem mass spectra with the software packages associated with mass spectrometers. Microsoft Excel or an alternative spreadsheet program is required for compiling of search results before submission to MultiTag. BioAnalyst software has an associated processing script that produces a list of database search results from a generated list of sequence tags to accelerate spectra processing with MultiTag.

III. PROCEDURES

A. Identification of Proteins by MS BLAST Database Searching

1. MS BLAST is a specialized BLAST-based tool for the identification of proteins by sequence-similarity searching that utilizes peptide sequences produced by the interpretation of tandem mass spectra (Shevchenko *et al.*, 2001). The algorithm and principles of BLAST sequence-similarity searching are reported in detail elsewhere (Altschul *et al.*, 1997). A useful list of BLAST servers accessible on the web is provided in Gaeta (2000).

2. Peptide sequences are generated from the interpretation of tandem mass spectra from the analysis of a single in-gel or in-solution digest of an unknown protein, edited and assembled into a query list for the MS BLAST search. If tandem mass spectra were interpreted by *de novo* sequencing software, disregard relative scores and use the entire list of candidate sequences (or some 50–100 top scoring sequence proposals per fragmented peptide precursor) (Fig. 1). Automated interpretation of tandem mass spectra often requires adjustment of parameters that affect the quality of predicted sequences. It is therefore advisable to test the settings in advance using digests of standard proteins and to adjust them if necessary. Note that the settings may depend on a charge state of the fragmented precursor ion. Use only the standard single-letter symbols for amino acid residues. If the software introduces special symbols for modified amino acid residues, replace them with standard symbols.

3. When interpreting MS/MS spectra manually, try making the longest possible sequence stretches, although their accuracy may be compromised. For example, it is usually difficult to interpret unambiguously fragment ion series at the low *m/z* range because of abundant peaks of chemical noise and numerous fragment ions from other series. In this case, it is better to include many complete (albeit low confidence) sequence proposals into the query rather than using a single (although accurate) three or four amino acid sequence stretch deduced from a noise-free high *m/z* segment of the spectrum.

4. Gaps and ambiguities in peptide sequences can occur due to the fragmentary nature of tandem mass spectra of peptides. Some *de novo* sequencing programs may suggest a gap in the peptide sequence that can be filled with various isobaric combinations of amino acid residues. For example,

DTPS[...]HYNAR, [...] = [S, V] or [D, A]

If one or two combinations were suggested, include all variants into a searching string:

-DTPSSVHYNAR-DTPSVSHYNAR-
DTPSDAHYNAR-DTPSADHYNAR-

If more combinations were possible, the symbol X can be used instead to fill the gap. Zero score is assigned to X symbol in PAM30MS scoring matrix and therefore it matches weakly any amino acid residue:

-DTPSXXHYNAR-

Note that MS BLAST is sensitive to the number of amino acid residues that are filling the gap. If the gap could be filled by a combination of two and three amino acid residues, consider both options in the query

-DTPSXXHYNAR-DTPSXXXHYNAR-

5. Isobaric amino acids need to be altered in the MS BLAST query. L stands for Leu (L) and Ile (I). Z stands for Gln (Q) and Lys (K), if undistinguishable in the spectrum. Use Q or K if the amino acid residue can be determined. The query string needs to be further altered for cleavage site specificity. If the proposed sequence is complete, a putative trypsin cleavage site symbol B is added prior to the peptide sequence:

...-BDTPSVDHYNAR-

It is often difficult to determine two amino acid residues located at the N terminus of the peptide. In this case, present them as

...-BXXPSVDHYNAR-...

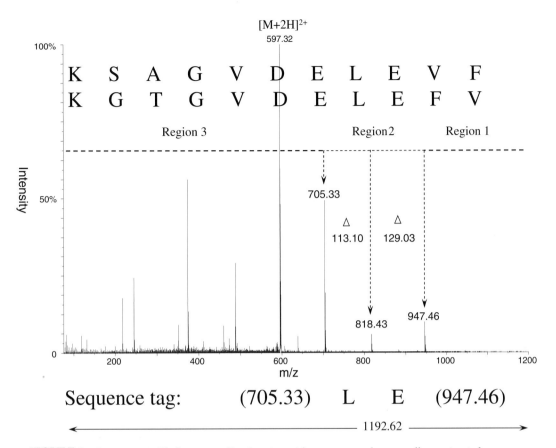

FIGURE 1 A spectrum with *de novo*-predicted amino acid sequences and a manually constructed sequence tag. Multiple candidate peptide sequences can be generated from a single spectrum for MS BLAST analysis, whereas MultiTag requires one sequence tag per spectrum.

MS BLAST will then consider BXX residues in possible sequence alignments.

6. The regular BLAST search must be altered in options and settings for an MS BLAST query:

NOGAP is absolutely essential, it turns off gapped alignment method so that only high-scoring pairs (HSPs) with no internal gaps are reported.

SPAN1 is absolutely essential, it identifies and fetches the best matching peptide sequence among similar peptide sequences in the query. Therefore the query may contain multiple partially redundant variants of the same peptide sequence without affecting the total score of the protein hit.

HSPMAX 100 limits the total number of reported HSPs to 100. Set it to a higher number (e.g., 200) if a large query is submitted and a complete list of protein hits (including low confidence hits) is required in the output.

SORT_BY_TOTALSCORE places the hits with multiple high scoring pairs to the top of the list.

Note that the total score is not displayed, but can be calculated, if necessary, by adding up scores of individual HSPs.

EXPECT: It is usually sufficient to set EXPECT at 100. Searching with higher EXPECT (as, 1000) will report many short low-scoring HSPs, thus increasing the sequence coverage by matching more fragmented peptides to the protein sequence. Note that low scoring HSPs do not increase statistical confidence of protein identification. The EXPECT setting also does not affect the scores of retrieved HSPs.

MATRIX: PAM30MS is a specifically modified scoring matrix. It is not used for conventional BLAST searching.

PROGRAM: blast2p.

DATABASE: nrdb95 are default settings of the MS BLAST interface.

FILTER: Filtering is set to "none" default. However, if the sequence query contains many repeating stretches (as . . . EQEQEQ . . .), filtering should be set to "default."

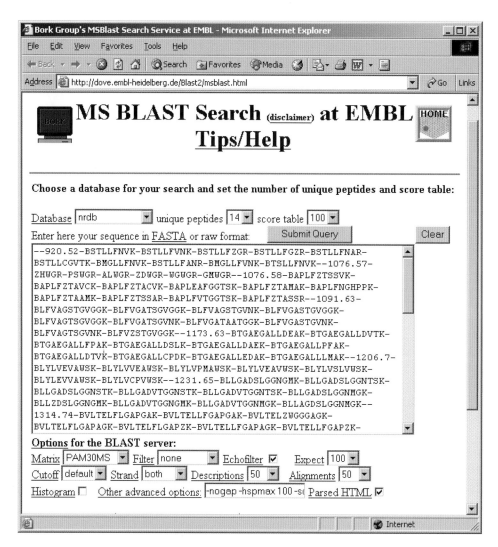

FIGURE 2   The MS BLAST web interface. A generated query is pasted in the input window, the number of tandem mass spectra from which sequences were derived is input in "unique peptides," and all other settings are set automatically according to MS BLAST parameters.

At the EMBL web interface, all parameters are preset and only the number of fragmented peptides and query sequences need to be input.

7. Space all candidate sequence proposals obtained from MS/MS spectra with a "−" (minus) symbol and merge them into a single text string that can be pasted directly into the query window at the MS BLAST web interface (Fig. 2). The query may contain space symbols, hard returns, numbers, and so on, as the server ignores them. For example, it is convenient to keep masses of precursor ions in the query, as it makes retrospective analysis of data much easier. Statistical evaluation is a very important element of MS BLAST protocol, as the query typically comprises many incorrect and partially redundant peptide sequences. Note that the statistics of conventional BLAST searching are

not applicable and therefore ignore reported E-values and P-values. Thresholds of statistical significance of MS BLAST hits were estimated in a computational experiment and scoring thresholds were set conditionally on the number of reported HSPs and the size of the database searched. Experimental MS BLAST hits are evaluated based on the number of fragmented precursors (this value is entered in the search parameters), and confident hits appear in red, borderline hits in green, and random matches in black at the web interface cited earlier (Fig. 3).

8. MS BLAST can, in principle, be used to search protein, EST, and genomic databases. The EMBL site only supports protein BLAST searching due to available computational capacity. A script can be written to retrieve specific genomic sequences that lie within the

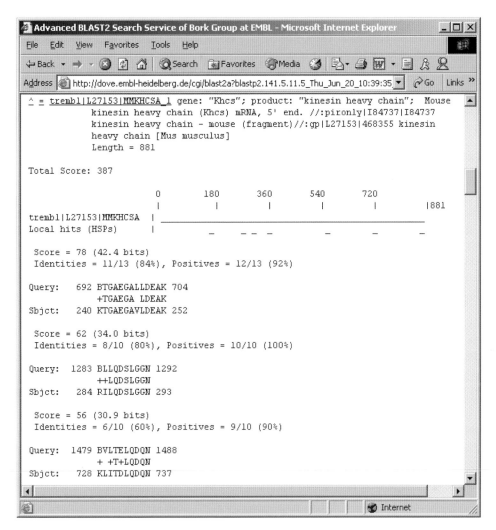

FIGURE 3 A section of the MS BLAST output where three of seven matched peptides to one database entry are shown. A list of matching database entries with the highest significance match at the top of the list is generated. Significant hits are color coded for easy data interpretation.

aligned peptides from a *tblastn* search against specific genomic databases. This search enables the use of unannotated genomic sequences and makes it possible to identify novel genes in large genomes. However, in both EST and genomic searches, different scoring schemes would need to be developed and installed locally. To set up a local BLAST searching engine, WU-BLAST 2.0 can be acquired from http://blast.wustl.edu/.

B. Identification of Proteins by MultiTag Database Searching

1. MultiTag is a software program that sorts compiled results from database searches with partial and complete sequence tags and calculates the significance of matches that align multiple sequence tags (for a

complete description, see Sunyaev *et al.*, 2003). MultiTag is based on error-tolerant searching with multiple partial sequence tags. This technique enables the correlation of search results from multiple searches with sequence tags representative of numerous spectra and gauges the significance of those matches.

2. Sequence tags should be generated manually from tandem mass spectra acquired in the analysis of a single in-gel or in-solution digest of an unknown protein or proteins. Some mass spectrometer software enables the automatic prediction of sequence tags; however, for best results it is advisable to make sequence tags by manual interpretation or gauge the accuracy of the automatic prediction software with a standard protein prior to the analysis of unknown samples. One sequence tag per spectrum should be made using prominent Y-ions, usually larger

than the multiply charged precursor. Sequence tags made with two to four amino acids each from multiple MS/MS spectra should be compiled in a text file list that includes the tag followed by the parent mass.

(360.20)FLL(733.44)918.64

(561.27)LA(745.40)935.48

(866.41)DEA(1181.52)1422.59

3. Each sequence tag is used to search a protein database, and the results from four searches are compiled in a spreadsheet. For the most specific results, mass tolerances should be narrow, taking into consideration the best accuracy of the mass spectrometer employed. The database is first searched using the complete sequence tag:

(866.41)DEA(1181.52)

This search will only find proteins that contain peptides with exactly these amino acids, spaced with exact amino acids residues that give mass combinations to make up the gaps to the peptide's termini. The second search allows for one error within the amino acid representation itself:

(866.41)D?A(1181.52)

The third search allows for errors between the analyzed peptide and the database at the C terminus of the peptide (searching with regions 1 and 2 only):

DEA(1181.52)

The fourth search allows for errors between the analyzed peptide and the database at the N terminus of the peptide (searching with regions 2 and 3 only):

(866.41)DEA

In the first column of the results table, the parent mass should be followed by an "NC", "E", "N", or "C" for sequence tag search results from complete, one-error, regions 1 and 2 (N-terminal match), and regions 2 and 3 (C-terminal match), respectively. The second, third, fourth, fifth, and six columns should include the amino acid sequence of the peptide matched, the molecular weight of the protein the tag matched, the accession number of the entry, the name of the protein, and the species name, respectively (Fig. 4). Compile the search results in a spreadsheet from all searches with the sequence tags generated, and save results as a text file.

4. Submit the search results table to MultiTag. Designate the mass tolerance used for searching (in daltons), input the approximate number of entries in the database searched, input the list of tags used to generate search results, and compute significance. Results will be sorted, with the database entry containing the most unlikely correlation event at the top of the list, and probabilities are calculated (Fig. 5). Results can be evaluated based on the number of sequence tags matched, E-values and P-values [count (predicted)]. E-values lower than 1×10^{-3} and P-values lower than 1×10^{-4} can be considered significant matches. Final E-values are highly dependent on the number of tags submitted for database searching; more tags will tend to diminish the significance of the alignment of multiple tags to one database entry. Reported P-values are less affected by the number sequence tags in the query. P-values reflect an approximation of the probability that the tags that are aligned will match randomly to an entry in a database of a specific size and at a specific mass accuracy, while neglecting the query size (number of sequence tags). Low P-values (but with higher E-values) are good indicators of alignments having borderline significance that need further manual evaluation to conclude a confident identification. Three partial sequence tags are normally specific enough to identify one entry in a database of 1,000,000 entries, at a mass accuracy of 0.1 Da.

IV. COMMENTS

It is not known in advance if the sequence of the analyzed protein is already present in a database. Therefore, conventional database-searching routines based on stringent matching of peptide sequences should be applied first (Mann and Wilm, 1994; Perkin et al., 1999). Only if the protein is unknown and no convincing cross-species matches can be obtained is it recommended to proceed with *de novo* interpretation of tandem mass spectra and sequence-similarity searching.

The success of MS BLAST and MultiTag identification depends on the size of a query and the corresponding database, the number of peptides aligned, the quality of peptide sequences or sequence tags, and the sequence similarity between the protein of interest and its homologues available in a database. On average, candidate sequences determined for five tryptic peptides should be submitted to MS BLAST or MultiTag searching to identify the protein by matching to a homologous sequence. With 10 sequences submitted and aligned, MS BLAST can identify 50% of homologues containing 50% sequence similarity, and MultiTag can identify 50% of homologues at 70% sequence similarity. However, because MultiTag can utilize less

Nr	Tag Mass	Sequence	Mass(kDa)	DB Accession	Protein name	Species
0	1173.628E	SAAKKVKNAEK	47.465796	gi|14210646	(AY033620) putative RNA-binding pro	Unknown
1	1173.628E	TGAEHLWLTR	27.823135	gi|7019377	(NM_013393) cell division protein Ft	Human
2	1173.628E	TGAEHLWLTR	27.487108	gi|13386002	(NM_026510) RIKEN cDNA 2310037B18 [	Mouse
3	1173.628E	SAANKALNDKK	15.032396	gi|12744797	AF323725_1 (AF323725) PsaN precurso	Unknown
4	1173.628E	ASAEILSVDRV	47.31585	gi|15611313	(NC_000921) EXODEOXYRIBONUCLEASE LA	Helicobacter Pylori
5	1173.628E	ASAEILSVDRV	47.455069	gi|15644887	(NC_000915) exonuclease VII, large	Helicobacter Pylori
6	1173.628E	TGAETLVEEAK	12.263558	gi|401181	THGF_TOBAC FLOWER-SPECIFIC GAMMA-THIO	Unknown
7	1173.628E	SAAERKRQEK	79.528375	gi|18375979	(AL356173) conserved hypothetical p	Unknown
8	1173.628E	SAAERKRQEK	82.102377	gi|11359450	T49456 hypothetical protein B14D6.8	Unknown
9	1173.628E	SANEKKSINVK	143.453267	gi|17224297	AF218388_1 (AF218388) apoptotic pro	Rat
10	1173.628E	SANEKKSINVK	143.435205	gi|13027436	(NM_023979) apoptotic protease acti	Rat
11	1173.628E	SAAEAQATRGR	21.850893	gi|13162112	(AL512667) putative tetR-family tra	some Streptomyces
12	1173.628N	GTAEQPRLFVG	32.418744	gi|7294725	(AE003544) CG7547 gene product [Dros	Fruit Fly
13	1173.628N	SAAEQWKQDL	74.867942	gi|17227422	(NC_003267) ORF_ID:all8048~unknown	Unknown
14	1173.628N	ASAEQRATQTI	36.747253	gi|17481280	(AB062896) vomeronasal receptor 1 A	Mouse
15	1173.628N	ASAEQRATQTI	35.05824	gi|3892596	(Y12724) pheromone receptor 2 [Mus m	Mouse
16	1173.628N	ASAEQRATQTI	35.84985	gi|17481276	(AB062895) vomeronasal receptor 1 A	Mouse
17	1173.628N	ASAEQRATQTI	36.402747	gi|18558569	(AY065464) vomeronasal receptor V1R	Mouse
18	1173.628N	ASAEQRATQTI	34.605305	gi|16716523	(NM_053218) vomeronasal 1 receptor,	Mouse
19	1173.628N	ASAEKGIASVRS	13.48812	gi|15802561	(NC_002655) orf, hypothetical prote	Escherichia Coli
20	1173.628N	SAAEQSGLDKNG	35.440047	gi|12620486	AF322012_67 (AF322013) ID142 [Brady	Unknown
21	1173.628N	ASAEKKRQATS	56.223343	gi|8570440	AC020622_1 (AC020622) Contains simil	Human
22	1173.628N	ASAEKKRQATS	64.846084	gi|15223502	(NM_100069) hypothetical protein [A	Mouse-Ear Cress
23	1173.628N	SAAEKLSEETL	272.279457	gi|4874311	AC006053_15 (AC006053) unknown prote	Mouse-Ear Cress
24	1173.628N	SAAEKLSEETL	60.547133	gi|15081785	(AY048285) At2g25730/F3N11.18 [Arab	Mouse-Ear Cress
25	1173.628N	SAAEKLSEETL	277.4555	gi|18400918	(NM_128132) unknown protein [Arabid	Mouse-Ear Cress
26	1173.628N	ASAEKKAEKSE	105.810124	gi|15900468	(NC_003028) translation initiation	Streptococcus Pyogenes
27	1173.628N	ASAEKYPHEF	31.188912	gi|17988633	(NC_003318) DIPEPTIDE TRANSPORT ATP	Unknown
28	1173.628N	GTAEKMPTTSR	13.777599	gi|3204328	(AJ008500) gag protein [Human immuno	Human
29	1173.628N	GTAEKMPSTTR	13.71449	gi|3204368	(AJ008521) gag protein [Human immuno	Human
30	1173.628N	TGAEKRSFVAD	80.784195	gi|7302767	(AE003803) CG4878 gene product [alt	Fruit Fly
31	1173.628N	SAAEKIVVYSGG	50.711733	gi|17231331	(NC_003272) unknown protein [Nostoc	Unknown
32	1173.628N	SAAEKAVSAPPR	55.045677	gi|13471797	(NC_002678) ATP-binding protein of	Unknown
33	1173.628N	SAAEKFDVSMT	24.872427	gi|10956719	(NC_002490) conjugal transfer prote	Unknown
34	1173.628N	ASAEKEQIAQI	144.720391	gi|16555336	(AY056833) chitin synthase [Anophel	Unknown
35	1173.628N	GTAEQHGRNVK	46.230479	gi|16759094	(NC_003198) putative IS element tra	Unknown
36	1173.628N	GTAEQHIKEGK	51.501952	gi|695769	(X84038) transposase [Xanthobacter au	Unknown
37	1173.628N	GTAEKGGLAIGDT	86.792704	gi|8894820	(AL360055) putative ABC transport sy	some Streptomyces
38	1173.628N	SAAEKDKGKQE	10.64305	gi|18550306	(XM_103535) hypothetical protein XP	Human
39	1173.628N	TGAEKAPKSPSK	13.977534	gi|6009909	(AB018242) histone H2A-like protein	Unknown
40	1173.628N	ASAEQCGRQAGG	33.741525	gi|7798662	AF135145_1 (AF135145) class I chitin	Unknown
41	1173.628N	GTAEKMPNTSR	13.580314	gi|3204322	(AJ008497) gag protein [Human immuno	Human
42	1173.628N	GTAEKMPNTSR	13.697865	gi|3355417	(AJ011213) gag protein [Human immuno	Human
43	1173.628N	GTAEKMPNTSR	13.807781	gi|3204271	(AJ008470) gag protein [Human immuno	Human
44	1173.628N	GTAEKMPNTSR	13.637407	gi|3204303	(AJ008487) gag protein [Human immuno	Human
45	1173.628N	GTAEKMPNTSR	13.552304	gi|3204338	(AJ008505) gag protein [Human immuno	Human

FIGURE 4 Compiled and formatted search results from sequence tag searching are input in the MultiTag software via opening a text-formatted results file.

intense and noisy spectra, it can outperform MS BLAST in many cases (for a more thorough discussion on homologue identification specificity, see Sunyaev *et al.*, 2003).

Both MS BLAST and MultiTag can identify proteins present in mixtures. Usually two or three components per sample can be identified easily. The sensitivity of both methods is determined primarily on the quality of the *de novo* sequences or sequence tags.

V. PITFALLS

1. Poor sample preparation can frequently deteriorate the quality of tandem mass spectra of peptides. The digestion of proteins with trypsin or other proteases should be carried out with chemicals of the highest degree of purity available. Plasticware (pipette tips, gloves, dishes, etc.) may acquire a static charge and attract dust, thus leading to contamination of samples with human and sheep (wool) keratin during in-gel or liquid digestion. Any polymeric detergents (Tween, Triton) should not be used for cleaning the laboratory materials.

2. When generating *de novo* sequences or sequence tags, if the software automatically extrapolates the parent mass from the precursor isotope cluster in the MS/MS spectra or the proceeding survey scan in a LC/MS/MS run, it is advisable to manually calculate this value, as software may determine the parent mass incorrectly by designating an incorrect charge state or ^{12}C monoisotopic peak of the parent ion isotope cluster, thus disabling correct *de novo* sequence prediction and sequence tag prediction.

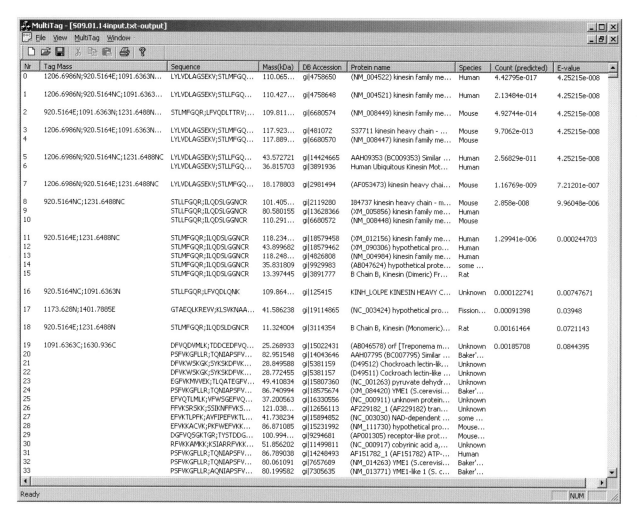

FIGURE 5 A section of MultiTag output. Margins may be adjusted to see the full list of tags matched, as well as full peptide sequences aligned, names, and so on. Results may be saved as text files to be viewed in an appropriate spreadsheet application.

3. MultiTag is laborious. Without scripted sequence tag database searching and processing of search results, manual data processing can demand extended effort; however, in cases where conventional methods fail to identify analyzed proteins, positive identifications are of a high value to cell biological studies.

4. Poor queries tend to obscure protein identification by both MS BLAST and MultiTag. It is best to submit fewer higher quality sequences than numerous lower quality sequences to MS BLAST. MS BLAST is particularly susceptible to low-complexity glycine- and proline-rich sequences generated incorrectly by *de novo* software. These low-complexity sequences tend to mask correct alignments. MultiTag functions best with sequence tags containing multiple (2–4) amino acids that have a low prevalence, such as tryptophan

(W) or methionine (M), whereas common amino acids such as leucine (L) in the tag tend to be of less significance and are likely to produce more false positives. Sequence tags generated from larger peptides also have more significance in a database search than those generated from smaller peptides.

References

Altschul, S. F., Madden, T. L., Schaffer, A. A., Zhang, J., Zhang, Z., Miller, W., and Lipman, D. J. (1997). Gapped BLAST and PSI-BLAST: A new generation of protein database search programs. *Nucleic Acids Res.* **25**, 3389–3402.

Gaeta, B. A. (2000). BLAST on the Web. *Biotechniques* **28**, 436–440.

Johnson, R. S., and Taylor, J. A. (2000). Searching sequence databases via de novo peptide sequencing by tandem mass spectrometry. *Methods Mol. Biol.* **146**, 41–61.

Mann, M., and Wilm, M. (1994). Error-tolerant identification of peptides in sequence databases by peptide sequence tags. *Anal. Chem.* **66**, 4390–4399.

Nimkar, S., and Loo, J. A. (2002). Orlando FL. Application of a new algorithm for automated database searching of MS sequence data to identify proteins. Abstract 334.

Perkins, D. N., Pappin, D. J., Creasy, D. M., and Cottrell, J. S. (1999). Probability-based protein identification by searching sequence databases using mass spectrometry data. *Electrophoresis* **20**, 3551–3567.

Shevchenko, A., Sunyaev, S., Loboda, A., Bork, P., Ens, W., and Standing, K. G. (2001). Charting the proteomes of organisms with unsequenced genomes by MALDI-quadrupole time-of-flight mass spectrometry and BLAST homology searching. *Anal. Chem.* **73**, 1917–1926.

Sunyaev, S., Liska, A., Golod, A., Shevchenko, A., and Shevchenko, A. (2003). MultiTag: Multiple error-tolerant sequence tag search for the sequence-similarity identification of proteins by mass spectrometry. Submitted for publication.

52

Identification of Protein Phosphorylation Sites by Mass Spectrometry

Rhys C. Roberts and Ole N. Jensen

I. INTRODUCTION

Tight regulation of cellular proteins is a prerequisite for viable cell function. Cells require various mechanisms whereby intracellular pathways can be activated and inactivated in a reversible manner depending on the prevailing environment at a particular moment in time. Reversible phosphorylation of proteins is one such mechanism. The importance of phosphorylation is exemplified by the finding that at least 2% of the human genome encodes proteins with predicted kinase domains (Lander *et al.*, 2001). Furthermore, other workers predict that approximately a third of all proteins expressed in vertebrates can be phosphorylated at some point in their lifetime (Hunter, 1998).

Many phosphoproteins have been identified through incorporation of radioactively labelled ATP (usually [γ-^{32}P]ATP) either in cells or following incubation with specific kinases *in vitro*. Once identified, the significance of phosphorylation can then be addressed. In eukaryotes, phosphorylation occurs predominantly on serine, threonine, and tyrosine residues. Ideally, the specific phosphorylated residue should be pinpointed in order to gain further important insights into the consequences of the addition and removal of a phosphate group.

Identifying specific phosphorylated residues can be fraught with difficulties. Most strategies, as stated previously, have used radioactively labelled ATP to phosphorylate the protein of interest. The purified phosphorylated protein can then be cleaved using specific chemical or enzymatic processes (e.g., CNBr or

trypsin) before separation of the peptides by an appropriate method (e.g., SDS–PAGE, HPLC, or thin-layer chromatography). The radioactively labelled peptide, which should contain the phosphorylated residue, is detected, isolated, and the sequence determined, usually by Edman degradation (Moyers *et al.*, 1995). In addition, phosphoamino acid analysis can be used to determine the nature of the phosphorylated residue (Sefton, 1995). This approach relies on a high degree of protein phosphorylation with radiolabelled phosphate. Handling large amounts of radiolabelled ^{32}P is clearly a potential hazard and, in practice, the efficiency of protein phosphorylation can vary significantly from case to case. Furthermore, the amount of protein required for identification using this approach means that only those proteins available in significant quantities, with high incorporation of radiolabelled phosphate, are feasible for phosphorylation site identification. Because phosphorylated residues are not detected by conventional Edman sequencing, many sites are determined indirectly. In many cases, subsequent mutational analysis is used to support the identification of the putative phosphorylation site.

Following many developments and improvements in biological mass spectrometry, unambiguous identification of specific phosphorylation sites is possible without the use of radioactive isotopes (Larsen *et al.*, 2001; Stensballe *et al.*, 2001). The main principle behind phosphorylation site identification by mass spectrometry is the fact that the mass of a specific residue increases by 80 Da upon phosphorylation. Therefore, by accurately measuring the masses of a mixture of peptides from a phosphorylated protein (whose sequence is known) following specific enzymatic or chemical cleavage, potentially phosphorylated species with a mass shift of 80 Da can be observed. Phosphorylation can be confirmed by dephosphorylating the peptide mix with alkaline phosphatase and observing the disappearance of the 80-Da peptide signal. The specific site of phosphorylation can then be determined by tandem mass spectrometry using, for example, a quadrupole time-of-flight (TOF) mass spectrometer.

To maximise the yield of phosphorylated peptides, techniques such as Fe^{3+}-IMAC (immobilised metal affinity chromatography) micropurification can be used prior to analysis by mass spectrometry. This technique is based on the relative affinity of Fe^{3+} for phosphorylated residues. By using this method, it is possible to identify specific phosphorylation sites in proteins phosphorylated to a low degree.

This article describes a stepwise approach for the identification of phosphorylation sites by mass spectrometry. First, phosphopeptide enrichment by Fe^{3+}-IMAC micropurification is described. This is followed by a method to confirm phosphorylation by using alkaline phosphatase to dephosphorylate the peptide mixture already analysed on a MALDI-TOF mass spectrometer. Finally, unambiguous determination of a phosphorylation site using nanoelectrospray ionisation and tandem mass spectrometry is described.

II. MATERIALS AND INSTRUMENTATION

Milli-Q H_2O (Millipore) is used throughout this method. Formic acid, acetic acid, trifluoroacetic acid, methanol and trifluoroacetic acid (TFA), and acetonitrile (all HPLC grade) are from Sigma. Ethylenediaminetetraacetic acid (EDTA), NaCl, and $FeCl_3$ are from Sigma. Ni^{2+}-NTA (nitrilotriacetic acid)-silica (16–24 μm particle size) is supplied by Qiagen, and OligoR3 reverse-phase resin is from PE biosystems. 2,5-Dihydroxybenzoic acid (DHB) is supplied by Sigma. Calf intestinal alkaline phosphatase is from Roche (0108138). GELoader tips (1–10 μl) from Eppendorf (0030 001.222) are used for making micropurification columns.

MALDI time-of-flight analysis of peptides is performed using a REFLEX II time-of-flight mass spectrometer with delayed extraction (Bruker-Daltonics). The samples are ionised with a nitrogen laser (λ = 337 nm) and data acquired in the positive ion mode. "moverz" (Proteometrics Ltd.) is used to analyse the spectra obtained.

Electrospray mass spectrometry is performed on a Q-TOF hybrid mass spectrometer fitted with a nano-ESI Z-spray interface (Micromass). Nanoelectrospray needles (gold/pallidum precoated borosilicate glass) are from MDS Protana, Odense Denmark. The instrument is used in positive ion mode with the following typical settings: needle 700–1000 V; cone 55 V; collision gas (Argon) pressure 5.5–6.0 × 10^{-5} atm; collision energy 4–34 eV. Commonly, the nanoelectrospray flow rate would be in the order of 15–50 nl/min. Selected ions are subject to collision-induced dissociation (CID) with argon gas, and fragment ions are detected by the orthogonal time-of-flight analyser. Resulting data are analysed using the MassLynx software supplied by Micromass.

III. PROCEDURES

A. Phosphopeptide Enrichment by Fe^{3+} IMAC Micropurification

Following separation and purification of the protein of interest (e.g., by SDS–PAGE, reverse-phase HPLC), the phosphorylated protein is cleaved into a peptide fragment either by enzymatic digest or by chemical cleavage. For separation by SDS-PAGE followed by in-gel digestion of a protein with trypsin, followed by the extraction of peptides, see article by Gevaert and Vandekerckhove.

1. Preparation of Fe^{3+}-NTA

Steps

1. Resuspend 15 mg Ni^{2+}-NTA silica in 200 μl H_2O in a 1.5-ml microcentrifuge tube.

2. Pellet the resin in a benchtop centrifuge and remove the supernatant. Wash the resin with 200 μl H_2O.

3. Incubate the resin with 50 mM EDTA in 1 M NaCl for 2 min at room temperature. Centrifuge and discard the supernatant. Repeat once.

4. Wash once with 200 μl H_2O and twice with 200 μl 100 mM acetic acid.

5. Incubate resin with 400 μl 50 mM $FeCl_3$: 50 mM acetic acid for 5 min at room temperature, mixing gently. Pellet the resin and discard the supernatant. Repeat once.

6. Wash with 200 μl 100 mM acetic acid and once with 200 μl 3 : 1 100 mM acetic acid : 100% acetonitrile. Wash again with 200 μl 100 mM acetic acid.

7. Resuspend the resin in 100 μl 100 mM acetic acid. The resin is now ready for use and can be stored at 4°C for 3–4 weeks.

2. Preparation of Microcolumn

Steps

1. Take a 1- to 10-μl GELoader tip (Eppendorf) and gently twist the tip end with fingers and thumb. Care must be taken to make the tip outflow sufficiently narrow in order to pack the resin, whilst avoiding complete occlusion.

2. Add 10 μl of the prepared Fe^{3+}-NTA resin to the microcolumn and pack into the tip using a 10 ml syringe for pressure. The column should measure 20–25 mm in height.

3. The Fe^{3+}-NTA IMAC microcolumn is now ready for use.

3. Purification of Phosphopeptides by Fe^{3+}-IMAC

Steps

1. Dilute the peptide mixture with 30x excess 100 mM acetic acid (e.g., from a Coomassie-stained band resuspended in a final volume of 20 μl, take 1–2 μl and dilute in 60 μl of 100 mM acetic acid).

2. Add the resuspended petide mixture to the Fe^{3+}-NTA IMAC microcolumn and load very slowly using the 10-ml syringe for pressure. For optimum binding of phosphopeptides, this step should take 15–20 min.

3. Wash the column sequentially with 20 μl 100 mM acetic acid, 20 μl 3 : 1 100 mM acetic acid : 100% acetonitrile, and 20 μl 100 mM acetic acid.

4. Elute the bound phosphopeptides with 2× 5 μl H_2O, pH 10.5 (NH_3). To concentrate the peptides prior to analysis, the eluted peptides can be applied directly into 60 μl of 5% formic acid in a preequilibrated OligoR3 microcolumn. Prepare the microcolumn in a similar manner to the Fe^{3+}-IMAC microcolumn but using OligoR3 reversed-phase resin (PE Biosystems) in 5% formic acid and forming a column height of 2–3 mm.

5. Wash the bound peptides with 20 μl 5% formic acid.

6. Elute the peptides directly onto the MALDI target with 1 μl of DHB [DHB (Sigma) 20 μg/μl in 70% acetonitrile/0.1% TFA]. Allow the eluate to crystalise before analysis by MALDI mass spectrometry (Fig. 1).

4. Analysis of Purified Phosphopeptides by MALDI

Steps

1. Measure the masses of the purified peptides by MALDI mass spectrometry as described previously (see article by Roepstorff et al.).

2. Identify peptide peaks corresponding to predicted peptide masses +80 Da. Candidate phosphopeptides are identified in this way and by also looking for phosphorylated partially digested peptides. In

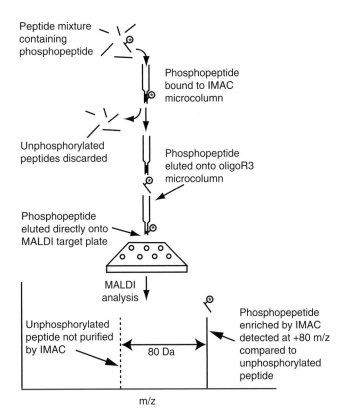

FIGURE 1 Schematic diagram illustrating the method of selective purification of phosphopeptides by Fe^{3+}-IMAC followed by analysis by MALDI mass spectrometry. A mixture of peptides is applied to the IMAC microcolumn and the phosphorylated peptide (labelled with asterisks) is selectively purified prior to desalting and concentrating using an OligoR3 microcolumn. The peptides are eluted directly onto the MALDI target for analysis by MALDI mass spectrometry. The phosphorylated peptide corresponds to a signal seen at 80 Da with respect to its unphosphorylated counterpart.

addition, the presence of a metastable ion, corresponding to a phosphopeptide which has undergone β elimination (detected ~98 Da lower than the predicted phosphorylated peptide), is highly suggestive of a phosphorylated species.

B. Confirmation of Phosphorylation Using Alkaline Phosphatase

1. Remove the MALDI target from the mass spectrometer.

2. Resuspend the crystallised peptides in 0.5 μl 70% acetonitrile in 100 mM NH_4HCO_3.

3. Transfer to a new OligoR3 microcolumn containing 20 μl dephosphorylation mixture (19 μl 100 mM NH_4Cl with 1 unit alkaline phosphatase).

4. Seal the column with Parafilm and incubate at 37°C for 1 h.

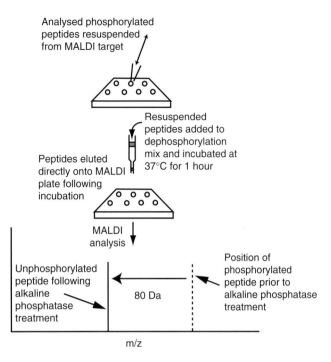

FIGURE 2 Schematic diagram illustrating the use of alkaline phosphatase to confirms that a candidate phosphopeptide is phosphorylated. Following initial MALDI analysis, the crystallised peptides are resuspended and added to a dephosphorylation mixture containing alkaline phosphatase. The mixture is present in a preequilibrated OligoR3 microcolumn. Following incubation for 1 h at 37°C, the peptides are purified and desalted on the microcolumn and eluted directly onto the MALDI target plate for further analysis. The disappearance of the signal corresponding to the phosphopeptide and the appearance of a signal 80 Da smaller confirm that the peptide of interest is indeed phosphorylated.

5. Acidify the enzyme mixture by adding 20 µl 5% formic acid and purify the peptides using the OligoR3 microcolumn as described earlier.
6. Elute the peptides using DHB onto the MALDI target and analyse as before.
7. Peaks corresponding to dephosphorylated peptides should now be seen, confirming that these peptides are indeed phosphorylated (Fig. 2).

C. Identification of Specific Phosphorylation Sites by Tandem Mass Spectrometry Using Nanoelectrospray Ionisation

Once candidate phosphopeptides are detected by IMAC and MALDI mass spectrometry, the precise residue should be identified. This can be achieved by selecting and sequencing the phosphopeptide using a nanoelectrospray quadrupole time-of-flight mass spectrometer. Sample preparation is vital to ensure successful analysis.

For analysis by nanoelectrospray, 5x the amount of peptide mixture is typically required, which will need to be determined on an individual basis. To minimise nonspecificc binding of phosphopeptides to the Fe^{3+}-IMAC microcolumn, three to four microcolumns are used simultaneously, and the bound peptides are concentrated on a single OligoR3 column prior to analysis.

Steps

1. Apply the phosphopeptide mixture to three to four Fe^{3+}-IMAC microcolumns simultaneously and wash as described in Section III,A.
2. Elute the peptides onto a single OligoR3 microcolumn and wash with 5% formic acid.
3. Elute the peptides directly into a nanospray needle with 1 µl 50% methanol : 1% formic acid.
4. Analyse the peptide mixture on the quadrupole time-of-flight mass spectrometer and select the relevant candidate phosphopeptides for further analysis.
5. Sequence the phosphopeptide by collision-induced dissociation.
6. The phosphorylated residue can be identified by observing a mass difference equalling the mass of an amino acid +80 Da between two sequential ions (Fig. 3) or by observing a series of dephosphorylated ions resulting from β elimination (with masses corresponding to 98 Da lower than the predicted phosphorylated ion or 18 Da lower than the nonphosphorylated counterpart). By careful analysis of data obtained, the phosphorylated residue can be deduced.

IV. COMMENTS AND PITFALLS

A. Sample Preparation

The importance of sample preparation cannot be overemphasised. Often, the proportion of the protein sample that is phosphorylated is very low. Therefore, without meticulous sample handling from protein purification to analysing peptides, the chances of success diminish rapidly. Many potential problems have been described previously when analysing peptides by mass spectrometry, such as contamination with exogenous proteins (e.g., keratin) and inadequate desalting prior to ionisation (a particular problem with electrospray ionisation). To maximise the success rate for identifying phosphorylation sites, these factors must be optimised.

Purification of phosphopeptides by Fe^{3+}-IMAC is a method to select and concentrate the phosphorylated species prior to analysis by mass spectrometry. It is an

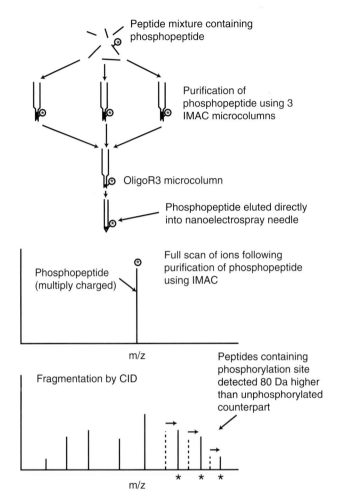

FIGURE 3 Schematic diagram illustrating the identification of specific sites of phosphorylation by IMAC micropurification and nanospray tandem mass spectrometry. The phosphopeptide is purified by using Fe^{3+}-IMAC microcolumns in parallel, prior to concentration and desalting on a single OligoR3 column. Once the multiply charged ion corresponding to the phosphopeptide is identified, the ion is fragmented by collision-induced dissociation (CID) with the production of specific fragment ions. The specific site of phosphorylation can be determined by careful analysis of this spectrum, looking particularly for phosphorylated and nonphosphorylated ions. In this simple schematic, a complete y-ion series is shown from a hypothetical nine residue phosphopeptide. The four largest y-ions are detected 80 Da higher than the predicted masses for the unphosphorylated peptide (indicated with dashed lines). This confirms that the fourth residue from the C-terminal is the site of phosphorylation. Further confirmation can be obtained by analysing the b-ion series and identifying the corresponding confirmatory ions.

extremely effective technique to assist in the identification of phosphorylation sites. However, this technique can yield a very high background of nonspecific binding if a number of key points are not followed closely. First, it is important to ensure that the peptide sample is well diluted. Second, the sample should be applied to the resin for at least 10 min, if not longer,

with the minimum amount of back pressure applied from the syringe. Both these steps reduce the amount of nonspecific binding to the Fe^{3+}-IMAC microcolumn and facilitate successful outcomes.

Because the phosphorylated peptide might exist in a small quantity, it is sometimes worth purifying three to four aliquots of peptide sample on separate Fe^{3+}-IMAC microcolumns. The phosphopeptides can then be eluted onto a single OligoR3 microcolumn. This last step concentrates the peptides prior to analysis and also desalts the sample effectively, ensuring optimal analysis by mass spectrometry.

B. The Phosphoprotein

In addition to the amount of phosphorylation of an individual protein, specific factors relating to the phosphoprotein itself may determine the ease to which a phosphorylation site is identified. For example, the molecular mass of the peptide containing the phosphorylation site should ideally be between 500 and 4000 Da. Peptides with masses outside this range are very difficult to analyse using the methods described in this chapter. Trypsin is usually the enzyme of choice for producing measurable peptides. However, other enzymes, such as chymotrypsin, Lys-C, or Arg-N, may be required to isolate a phosphopeptide in the detectable range. Chemical cleavage with CNBr can also be used. A rare difficulty arises when a nonphosphorylated peptide has the same molecular weight as the phosphopeptide of interest.

In many cases, phosphorylation sites are flanked by polar residues. Generally, peptides containing polar residues are ionised more efficiently than their nonpolar counterparts. However, occasionally, a hydrophobic sequence surrounding the phosphorylation site may decrease the efficiency of ionisation, resulting in low detection rates. In addition, some peptides prove difficult to sequence by CID and this problem is exacerbated when the amount of phosphorylated peptide is small.

With respect to Fe^{3+}-IMAC micropurification, a particular difficulty arises when a protein contains a number of acidic repeats. The acidic residues, glutamate and aspartate, have signficant affinities to Fe^{3+} and hence may result in a high background when analysed.

A further obstacle when attempting to pinpoint sites of phosphorylation occurs when the specific phosphate group is lost upon ionisation. Although this property can lead to difficulty in identifying the phosphorylated species, the same property can be used to identify a specific site. If the phosphate group is lost through β elimination, a signal can be seen corresponding to 98 Da less than the predicted phosphopeptide. Conversely, in some cases, the intact phosphorylated peptide remains undetected while its degraded product becomes the principal ion seen on analysis.

C. Ionisation and Analysis

As discussed earlier, a balance is required when analysing a phosphopeptide mixture to ensure efficient ionisation on the one hand whilst also protecting the phosphate group. At lower energies, the phosphopeptide will not be seen, whilst if the energy of ionisation is too high, the labile phosphate group will be lost. The optimal settings and conditions may need to be determined for different phosphoproteins. We have found that DHB is the matrix of choice for the analysis of most phosphopeptides.

An interesting phenomenon relating to phosphopeptides is the "suppression effect." This is a term describing the low detection of phosphorylated peptides when ionised in the presence of their nonphosphorylated counterparts. Fe^{3+}-IMAC is used to discard the nonphosphorylated species and can result in a phosphoprotein signal that was not detectable before micropurification.

This article describes sequencing using a quadruplole time-of flight mass spectrometer. It is possible to obtain very high resolution data using these instruments. However, these mass spectrometers are less sensitive than their MALDI-TOF counterparts and, in practice, approximately 5x as much sample is required for sequence analysis by electrospray ionisation. In addition, the Q-TOF is less tolerant than a MALDI-TOF mass spectrometer to salt and samples should always be desalted thoroughly before analysis. We routinely micropurify all our peptide samples on self-made reverse-phase microcolumns prior to both MALDI and electrospray ionisation to ensure optimal results at all times.

References

Hunter, T. (1998). The Croonian Lecture 1997. The phosphorylation of proteins on tyrosine: Its role in cell growth and disease. *Philos. Trans. R. Soc. Lond. B. Biol. Sci.* **353**, 583–605.

Lander, E. S., Linton, L. M., Birren, B., Nusbaum, C., Zody, M. C., Baldwin, J., Devon, K., Dewar, K., Doyle, M., FitzHugh, W., Funke, R., Gage, D., Harris, K., Heaford, A., Howland, J., Kann, L., Lehoczky, J., LeVine, R., McEwan, P., McKernan, K., Meldrim, J., Mesirov, J. P., Miranda, C., Morris, W., Naylor, J., Raymond, C., Rosetti, M., Santos, R., Sheridan, A., Sougnez, C., Stange-Thomann, N., Stojanovic, N., Subramanian, A., Wyman, D., Rogers, J., Sulston, J., Ainscough, R., Beck, S., Bentley, D., Burton, J., Clee, C., Carter, N., Coulson, A., Deadman, R., Deloukas, P., Dunham, A., Dunham, I., Durbin, R., French, L., Grafham, D., Gregory, S., Hubbard, T., Humphray, S., Hunt, A., Jones, M., Lloyd, C., McMurray, A., Matthews, L., Mercer, S., Milne, S., Mullikin, J. C., Mungall, A., Plumb, R., Ross, M., Shownkeen, R., Sims, S., Waterston, R. H., Wilson, R. K., Hillier, L. W., McPherson, J. D., Marra, M. A., Mardis, E. R., Fulton, L. A., Chinwalla, A. T., Pepin, K. H., Gish, W. R., Chissoe, S. L., Wendl, M. C., Delehaunty, K. D., Miner, T. L., Delehaunty, A., Kramer, J. B., Cook, L. L., Fulton, R. S., Johnson, D. L., Minx, P. J., Clifton, S. W., Hawkins, T., Branscomb, E., Predki, P., Richardson, P., Wenning, S., Slezak, T., Doggett, N., Cheng, J. F., Olsen, A., Lucas, S., Elkin, C., Uberbacher, E., Frazier, M., *et al.*, (2001). Initial sequencing and analysis of the human genome. *Nature* **409**, 860–921.

Larsen, M. R., Sorensen, G. L., Fey, S. J., Larsen, P. M., and Roepstorff, P. (2001). Phospho-proteomics: Evaluation of the use of enzymatic de-phosphorylation and differential mass spectrometric peptide mass mapping for site specific phosphorylation assignment in proteins separated by gel electrophoresis. *Proteomics* **1**, 223–238.

Moyers, J. S., Linder, M. E., Shannon, J. D., and Parsons, S. J. (1995). Identification of the in vitro phosphorylation sites on Gs alpha mediated by pp60c-src. *Biochem. J.* **305**, 411–417.

Sefton, B. M. (1995). Phosphoamino acid analysis, In *"Current Protocols in Protein Science"* (J. E. Coligan, B. M. Dunn, D. W. Speicher, and P. T. Wingfield, Eds.), pp. 13.3.1–13.3.8. Wiley, New York.

Stensballe, A., Andersen, S., and Jensen, O. N. (2001). Characterization of phosphoproteins from electrophoretic gels by nanoscale Fe(III) affinity chromatography with off-line mass spectrometry analysis. *Proteomics* **1**, 207-222.

Analysis of Carbohydrates/ Glycoproteins by Mass Spectrometry

Mark Sutton-Smith and Anne Dell

I. INTRODUCTION

Electron impact mass spectrometry (EI-MS) has been employed in carbohydrate analysis since the early 1960s and is still used for defining sugar compositions and for linkage analysis (Albersheim *et al.*, 1967). The introduction of fast atom bombardment-mass spectrometry (FAB-MS) at the beginning of the 1980s (Morris, 1980; Barber *et al.*, 1981; Dell *et al.*, 1983) revolutionised the structure determination of a very wide range of carbohydrate-containing biopolymers (Fukuda *et al.*, 1985; Laferte *et al.*, 1987; Dell *et al.*, 1990; McConville *et al.*, 1990). A decade later, electrospray ionisation (ES-MS) (Fenn *et al.*, 1990) and matrix-assisted laser desorption ionisation (MALDI-MS) technologies (Karas and Hillenkamp, 1988; Karas *et al.*, 1989) expanded the range of glycobiological structural problems amenable to mass spectrometry because of their higher sensitivity and applicability to much larger molecules (Lopez *et al.*, 1997). The lessons learnt from FAB-MS investigations in the 1980s (Fukuda *et al.*, 1984, 1985; Dell, 1987) have turned out to be equally applicable to ES-MS and MALDI-MS. Notably, it has been found that although native samples are amenable to MS analysis, it is often desirable to prepare derivatives prior to analysis. As a general rule, derivatisation vastly improves sensitivity and derivatised glycans yield fragment ions much more reliably than their native counterparts. Permethylation is the most important type of derivatisation employed in carbohydrate MS.

Broadly speaking, MS can be exploited in two general ways in the analysis of carbohydrates and glycoproteins.

i. Detailed characterisation of purified individual glycopolymers or mixtures of glycopolymers. This usually requires acquisition of MS data from both intact material and chemical, and enzymatic digests. Case studies that exemplify strategies applicable to a range of glycopolymers are described elsewhere (Sasaki, *et al.*, 1987; Dell *et al.*, 1995).

ii. Glycomics analyses that involve screening of cell and tissue extracts for their overall glycan content. Examples of such strategies are given elsewhere (Sutton-Smith, *et al.*, 2000; Manzi, *et al.*, 2000) and at the NIH Functional Glycomics Consortium Web site http://web.mit.edu/glycomics/consortium.

This article documents protocols for isolating, derivatising, and digesting glycans and glycopeptides in preparation for MS analysis. The emphasis is on glycoprotein analysis, but many of the methodologies are also applicable to other glycopolymers, such as glycolipids and polysaccharides. Also documented are procedures for high-sensitivity MS and MS/MS analyses of glycans and glycopeptides. The generic strategy is outlined below and in Fig. 1.

1. Preparation of biological matrix for analysis
2. Purification of glycoprotein(s)
3. Reduction/carboxymethylation
4. Tryptic digestion or, in some instances, cynanogen bromide degradation
5. *N*-Glycosidase F digestion
6. Separation of *N*-glycans from peptides/ *O*-glycopeptides
7. Reductive elimination of *O*-glycans from *O*-glycopeptides
8. Dowex purification of *O*-glycans
9. Permethylation of *N*- and *O*-glycans
10. MALDI-TOF MS profiling of *N*- and *O*-glycans

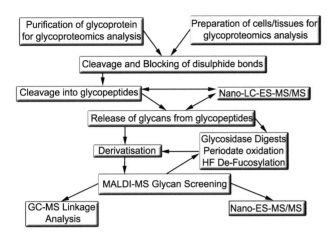

FIGURE 1 Glycomics strategy for screening mammalian cells, tissues, and glycoproteins.

11. ES-MS/MS on key selected peaks observed in the MALDI profile

The initial steps for glycoproteome analysis of mammalian biological matrices, such as cells, biological fluids, tissues, and organs, involve homogenisation and sample cleanup by dialysis. For screening glycans in mammalian tissues, 100–400 mg of tissue is sufficient for a series of MS analyses, including glycan profiling, sequential exoglycosidase digestions, specific chemical degradations, linkage analyses, and MS/MS studies. As a rough guide, high-quality mapping data can be obtained from 10% of a typical extract from a single mouse kidney. For specific glycoproteins, initial steps usually involve some form of immunoprecipatation or affinity chromatography with, or without liquid chromatography purification or electrophoresis.

Once initial steps have been performed, the disulphide bridges are split by reduction and blocked by carboxymethylation. This allows the glycoproteins to be efficiently cleaved by trypsin, or another protease (or by cyanogen bromide if enzymatic digestion is problematical due to poor solubility in the digestion buffer). Once the glycopeptides are generated, these can be purified by reversed-phase chromatography (e.g., Sep-Pak or MicroTrap purification), or be analysed directly by nano-ES-LC/MS/MS before, or after *N*-glycosidase F digestion. Alternatively, the glycopeptides are digested with *N*-glycosidase F and separated into *N*-glycan and *O*-glycopeptide fractions, with the latter being subject to reductive elimination. The released *N*- and *O*-glycans are permethylated and analysed by MALDI and nano-ES-MS/MS. To complement the screening experiments, GC-MS linkage analysis and sugar analysis are performed to provide details of the linkages and the sugar compositions in

the sample. In detailed studies the released glycans are often treated with chemical and/or enzymatic reagents to further characterize ambiguous glycan assignments, but this is outside the scope of this article. Initial assignments of relevant signals in MALDI-MS spectra are based on compositions that take into account biosynthetic considerations.

II. MATERIALS AND INSTRUMENTATION

Ammonium acetate (Cat. No. 100134T), ammonium hydrogen carbonate (Ambic, Cat. No. 103025E), hydrochloric acid (HCl, Cat. No. 101254H), and ion-exchanger Dowex 50 W-X8 (H+ form, Cat No. 105221) are from VWR. High-quality solvents: acetonitrile (UPS, ultra purity solvent, Cat. No. H050), dimethyl sulfoxide (DMSO) (Hi Dry, anhydrous solvent, D4281), methanol (UPS, ultra purity solvent, Cat No. H411), and propan-1-ol (SPS, super purity solvent, H624) are from Romil. Acetic acid (SpS, super purity reagent, Cat. No. H014), ammonia solution (SpS, super purity reagent, Cat No. H058), and sodium hydroxide are also from Romil. Acetic anhydride (Cat. No.4295) and methyl iodide (Cat. No. 0347) are from Lancaster. Acetyl chloride (Cat. No. 23,957-7), ammonium formate (Cat. No. F-2004), adrenocorticotropic hormone fragment (ACTH)1–17 (Cat. No. A-2407), ACTH fragment 18–39 (Cat. No. A-0673), ACTH fragment 7–38 (Cat. No. A-1527), ACTH fragment 1–39 (Cat. No. A-0423), angiotensin I (Cat. No. A9650), bradykinin (B-4764), calcium hydride lumps, +4 mesh (Cat. No. 213322), cyanogen bromide (Cat. No. 16774), 2,5-dihydroxybenzoic acid (Cat. No. G-5254), dithiothreitol (DTT, Cleland's reagent, Cat. No. D-5545), ethelyene glycol (Cat. No. 10,246-8), EDTA (Cat. No. 43,178-8), formic acid (Cat No. 94318), hexanes (Cat. No. 15,617-5), hydrogen chloride gas (Cat. No. 29,542-6), hydrofluoric acid 48% (HF, Cat. No. 33,926-1), iodoacetic acid (IAA, Cat. No. I-4386), insulin (bovine pancreas, Cat. No. I-5500), leucine enkephalin (Cat. No. L-9133), exo-β-mannosidase (*Helix pomatia*, Cat. No. M9400), neurotensin (Cat. No. N-6383), sodium acetate (Cat. No. 24,124-5), sodium m-periodate (Cat. No. S-1878), sodium borodeuteride (NaBH₄ Cat. No. 20,559-1), potassium borohydride (KBH₄ Cat. No. 438472), sodium chloride (Cat. No. 20,443-9), sodium dodecyl sulfate (SDS, Cat. No. 436143), Tris(hydroxymethyl)aminomethane (Tris, Cat. No. 154563) and trypsin (bovine pancreas, TPCK-treated, T-1426) are from Sigma-Aldrich. 3-((3-Cholamidopropyl)dimethyl-ammonio)-1-propane sulfonate (CHAPS, Cat. No. 810126) and N-glycosidase F (*Escherichea coli*, Cat. No. 1365177) are from Roche Diagnostics. Exo-β-galactosidase (bovine testes, Cat. No. EG02), exo-β-N-acetylhexosaminidase (*Streptomyces pneumoniae*, Cat. No. E-GL01), exo-α-mannosidase (jack bean, Cat. No. E-AM01), exo-neuraminidase (*Clostridium perfringes*, Cat. No.), and exo-neuraminidase (*Streptococcus pneumoniae*, Cat, No. E-S007) are from Qa-Bio. Endo-β-galactosidase (*Escherichia freundii*, 100455) is from Seikagaku corporation. Tri-sil "Z" derivatizing agent (Cat. No. 49230) and Snakeskin Pleated dialysis tubing (Cat. No. 68700) are from Perbio Science UK Ltd. Ultra pure water is generated from an Analytical Purite Neptune ultrapure water purification system from Purite Ltd.

Homogenisation of cells and tissues is achieved with a compact electric CAT homogeniser (×120)-fitted *T6.1* dispersion shaft from Ingenieurburo CAT. Sonications of cells are achieved by a VC 130 PB (130 W) Vibra-Cell ultrasonic processor within a sound-abating enclosure from Sonics & Materials Inc. Screw-cap style Pyrex disposable culture tubes (Cat. No. 99449-13, 7.5 ml) capped with disposable phenolic lids (Cat. No. 99999-13) and plasticware are from Corning. Caps are lined with Teflon inserts (Cat. No. 0402) from Owens Polyscience Ltd. Sep-Pak Classic C_{18} cartridges (Cat. No. WAT051910) are from Waters Ltd. Medium NanoES capillaries (Cat. No. ES387) for the Micromass Q-Tof are from Proxeon Biosytems. The Hamilton syringe (Cat. No. 002520) is from SGE. MicroTrap peptide cartridges (Cat. No. 004/25108/02) and the Manual Trap Holder kit (Cat. No. 004/25111/01) are from Michrom BioResources, Inc.

III. PROCEDURES

A. Preparation of Homogenates/Cell Lystates

Wear suitable protective clothing, including safety glasses, and work in a fume hood when preparing all solutions and performing various steps of the protocols.

Solutions

1. *Solution A:* 80% (v/v) methanol in H_2O. To make 25 ml, add 5 ml of ultrapure water to 20 ml methanol to make a 80% (v/v) methanol solution.
2. *Solution B:* 33.33% ultrapure water, 33.33% formic, and 33.33% methanol. To make 30 ml, add 10 ml of formic acid and 10 ml of methanol to 10 ml of ultrapure water.
3. *Dilute acetic acid:* To make a 5% acetic acid solution, add 25 ml of acetic acid and complete to 500 ml with ultrapure water. Store at room temperature.
4. *Dialysis buffer:* 50 mM Ambic buffer. To make 4.5 liters, add 17.6 g of Ambic and complete with 4.5 liters of water. Adjust to pH 7.5 with dilute acetic acid. Prepare fresh on day of use.
5. *10% SDS:* To make 100 ml, add 10 g of SDS and complete with 100 ml of ultrapure water. Store at room temperature.
6. *Homogenisation buffer:* 0.5% SDS (w/v) in 50 mM Tris. To make 50 ml, add 0.3029 g of Tris and complete to 50 ml with ultrapure water. Adjust the pH to 7.4 with dilute acetic acid. Withdraw 2.5 ml of the solution and add 2.5 ml of 10% SDS.
7. *Cell lysis buffer:* 25 mM Tris, 150 mM NaCl, 5 mM EDTA, and 1% CHAPS at pH 7.4. To make 50 ml, add 0.1514 g of Tris, 0.4383 g NaCl, 0.5 g CHAPS, and 0.0731 g EDTA. After dissolving in a small volume of ultrapure water, complete to 50 ml and adjust to pH 7.4 with dilute acetic acid.

1. Homogenisation

Steps

1. Immerse the tip of the dispersion shaft into solution A; activate the drive motor at a low to intermediate setting for 60 s.
2. Examine the tip and carefully remove any residual debris with a 3-mm hypodermic needle. Avoid scratching the dispersion shaft.
3. Put the dispersion shaft in solution B and sonicate in a sonicator bath for 10 min.
4. Repeat step 1 with fresh homogenisation buffer to clean the dispersion shaft.
5. Add 2–3 ml of ice-cold homogenisation buffer to the sample.
6. Homogenise on ice for 10 s. Repeat two or three times, pausing for 15 s between each homogenisation step.
7. Transfer the homogenate to high-quality dialysis tubing and seal.
8. Place the sample into cool dialysis buffer and dialyse at 4°C. Change the buffer regularly with fresh buffer over a period of 48 h.
9. Lyophilise the sample in a clean screw-capped glass culture tube covered with perforated Parafilm.

2. Sonication of Cells

Steps

1. Immerse the probe of the sonicator into solution A; activate the ultrasonic processor in continuous mode at 20 A for 10 s. Sonicate the probe again with ultrapure water and then the cell lysis buffer.

2. Add enough ice-cold cell lysis buffer (1–2 ml) to completely suspend the cell pellet.

3. Sonicate on ice in continuous mode at 40 A for 10 s. Repeat two or three times, pausing for 15 s between each sonication.

4. Transfer the homogenate to high-quality dialysis tubing and seal.

5. Place the sample into cool dialysis buffer and dialyse at 4°C. Change the buffer regularly with fresh buffer over a period of 48 h.

6. Lyophilise the sample in a clean screw-capped glass culture tube covered with perforated Parafilm.

B. Cleavage and Blocking of Disulphide Bridges

Wear suitable protective clothing, including safety glasses, and work in a fume hood when preparing all solutions and performing various steps of the protocols.

Solutions

1. *Tris buffer:* 0.6 M Tris. To make 50 ml, weigh 3.63 g of Tris and complete to 50 ml with ultrapure water. Adjust to pH 8.5 with acetic acid. Degas by passing a slow stream of nitrogen (oxygen free) through the solution via a Pasteur pipette.

2. *Dilute acetic acid:* To make a 5% acetic acid solution, add 25 ml of acetic acid and complete to 500 ml with ultrapure water. Store at room temperature.

3. *Dialysis buffer:* 50 mM Ambic buffer. To make 4.5 liters, add 17.6 g of Ambic and complete with 4.5 liters of water. Adjust to pH 8.5 with dilute acetic acid. Prepare fresh on day of use.

1. Reduction and Carboxymethylation of Homogenates/Cell Lysates

Steps

1. Weigh 10 mg of DTT and dissolve in 5 ml of degassed Tris buffer to make a 2-mg/ml DTT solution.

2. Add 0.5 ml of the DTT solution to the sample.

3. Incubate for 60 min at 37°C and then centrifuge.

4. Weigh 60 mg of IAA and dissolve in 1 ml of degassed Tris buffer.

5. Add 0.5 ml of fresh IAA solution to the sample.

6. Incubate in the dark at room temperature for 90 min.

7. Transfer the homogenate to high-quality dialysis tubing and seal.

8. Place the sample into cool dialysis buffer and dialyse at 4°C. Change the buffer regularly with fresh buffer over a period of 48 h.

9. Lyophilise the sample in a clean screw-capped glass culture tube covered with perforated Parafilm.

2. Reduction and Carboxymethylation of Glycoproteins

Steps

1. Weigh out 1 mg of DTT and add 1 ml of Tris buffer to make a 1 μg/μl DTT solution.

2. Predict the number disulphide bridges in the glycoprotein if the disulphide bridges have not been mapped previously, i.e., (No. of cysteines ÷ 2).

3. Add enough Tris buffer (~200 μl) to completely dissolve sample and add the appropriate volume of DTT to the sample by using the following equation: Volume of DTT (L) = (No. of disulphide bridges) × (No. of moles of glycoprotein) × (4 × 154.3).

4. Incubate for 60 min at 37°C and then centrifuge.

5. Weigh out 1 mg of IAA and add 1 ml of Tris buffer to make a 1 μg/μl IAA solution.

6. Add the appropriate volume of IAA to the sample using the following equation: Volume of IAA (L) = (No. of moles of DTT in sample) × (5 × 186.0).

7. Incubate in the dark at room temperature for 90 min.

8. Transfer the homogenate to high-quality dialysis tubing and seal.

9. Place the sample into cool dialysis buffer and dialyse at 4°C. Change the buffer regularly with fresh buffer over a period of 48 h.

10. Lyophilise the sample in a clean screw-capped glass culture tube covered with perforated Parafilm.

C. Cleavage into Glycopeptides

Wear suitable protective clothing, including safety glasses, and work in a fume hood when preparing all solutions and performing various steps of the protocols.

Solutions

1. *Dilute acetic acid:* To make a 5% acetic acid solution, add 25 ml of acetic acid and complete to 500 ml with ultrapure water. Store at room temperature.

2. *Ambic buffer:* To make 50 ml, weigh 0.1977 g of ammonium hydrogen carbonate and complete with 50 ml ultrapure water. Adjust to pH 8.4 with ammonia solution. Prepare fresh on day of use.

3. *20 and 40% Propan-1-ol:* To make 50 ml of each solution, add 10 ml and 20 ml of propan-1-ol to separate clean glass bottles, and complete to 50 ml with ultrapure water. Prepare fresh on day of use.

1. Tryptic Digestion of Homogenates/Cell Lysates

Steps

1. Weigh out 1 mg of trypsin and add 1 ml of Ambic buffer to make a 1 µg/µl trypsin solution.
2. Add sufficient solution (300–600 µl) to completely cover the sample.
3. Incubate at 37°C for 14 h and then centrifuge.
4. Terminate the reaction by dispensing 2 drops of acetic acid with a Pasteur pipette.
5. Attach a 10-ml glass syringe to a Sep-Pak C_{18} cartridge and condition by eluting successively with methanol (5 ml), 5% acetic acid (5 ml), propan-1-ol (5 ml), and 5% acetic acid (3×5 ml).
6. Load the sample onto a cartridge.
7. Elute stepwise with 5% acetic acid (20 ml), 20% propan-1-ol solution (4 ml), 40% propan-1-ol solution (4 ml), and 100% propan-1-ol (4 ml). Collect all the fractions apart from the 5% acetic acid fraction (hydrophilic contaminants).
8. Reduce the volume with a Speed-Vac and lyophilise the sample in a clean screw-capped glass culture tube covered with perforated Parafilm.

2. Tryptic Digestion of Glycoproteins

Steps

1. Weigh out 1 mg of trypsin and add 1 ml of Ambic buffer to make a 1 µg/µl trypsin solution.
2. Estimate the approximate amount of glycoprotein.
3. Add sufficient solution (100–200 µl) to completely cover the sample, incubate at 37°C for 5 h and then centrifuge.
4. Dispense 3 drops of acetic acid from a Pasteur pipette to terminate the reaction.

3. Cyanogen Bromide Cleavage of Glycoproteins

Wear suitable protective clothing, including safety glasses, and work in a fume hood due to the toxic cyanogen bromide.

Steps

1. Add 700 µl of formic acid to 300 µl of water in an Eppendorf to make a 70% formic acid solution.
2. Add 8–10 good quality white cyanogen bromide crystals to the Eppendorf.
3. Vortex the mixture until the crystals dissolve completely.
4. Ensure that the glycoprotein sample is dry. Dissolve the sample in minimum amount of the cyanogen bromide solution, typically 100–200 µl.
5. Incubate at room temperature for 14-20 h.
6. Terminate the reaction by adding 4 volumes of ultrapure water.

7. Reduce the volume with a Savant Speed-Vac and lyophilise the sample in a clean screw-capped glass culture tube covered with perforated Parafilm.

D. Cleavage of Glycans From Glycopeptides

Wear suitable protective clothing, including safety glasses, and work in a fume hood when preparing all solutions and performing various steps of the protocols.

Solutions

1. *Dilute acetic acid:* To make a 5% acetic acid solution, add 25 ml of acetic acid and complete to 500 ml with ultrapure water. Store at room temperature.
2. *Ambic buffer:* To make 50 ml, weigh 0.1977 g of ammonium hydrogen carbonate and complete with 50 ml ultrapure water. Adjust to pH 8.4 with ammonia solution. Prepare fresh on day of use.
3. *20 and 40% propan-1-ol:* To make 50 ml of each solution, add 10 ml and 20 ml of propan-1-ol to separate clean glass bottles and complete to 50 ml with ultrapure water. Prepare fresh on day of use.
4. *0.1 M KOH:* To make 100 ml, weigh 0.56 g of KOH and complete to 100 ml with ultrapure water. Store at room temperature.
5. *Dowex solution:* To prepare 100 g of Dowex beads, add 100 g of Dowex beads and complete with 100 ml of 4 M HCl and decant. Repeat this twice more and then wash beads by adding, agitating, and decanting with ultrapure water until the pH does not change (usually 10–15 times). Wash the beads three times with 150 ml of dilute acetic acid and leave the beads immersed in dilute acetic acid. The treated beads can be kept equilibrated in this state for many months at room temperature.
6. *10% methanolic acetic acid:* To make 25 ml, add 2.5 ml of acetic acid and complete to 25 ml with methanol. Prepare fresh on day of use.

1. N-Glycosidase F Digest

Steps

1. Dissolve the 20% and 40% Sep-Pak fractions of the tryptic digest each in 150 µl of Ambic buffer and combine.
2. Add 5 U of *N*-glycosidase F, incubate at 37°C for 20–24 h, and then centrifuge.
3. Lyophilise the sample.
4. Separate *N*-glycans from the mixture using the Sep-Pak C_{18} propanol-1-ol / 5% acetic acid system.

2. Separation of N-Glycans from Peptides/O-Glycopeptides by Sep-Pak C₁₈ Purification

Steps

1. Attach a 10-ml glass syringe to a Sep-Pak C_{18} cartridge and condition by eluting successively with methanol (5 ml), 5% acetic acid (5 ml), propan-1-ol (5 ml), and 5% acetic acid (3×5 ml).
2. Load the sample on to a cartridge.
3. Elute stepwise with 5% acetic acid (5 ml), 20% propan-1-ol solution (4 ml), 40% propan-1-ol solution (4 ml), and 100% propan-1-ol (4 ml). Collect all fractions.
4. Reduce the volume with a Speed-Vac and lyophilise the sample in a clean screw-capped glass culture tube covered with perforated Parafilm.
5. Permethylate the 5% acetic acid fraction (the N-glycan fraction).
6. Perform reductive elimination on the 20% propan-1-ol fraction.

3. Reductive Elimination

Steps

1. Add 54–55 mg of KBH_4 to 1 ml of 0.1 M KOH.
2. Add 400 μl of the $NaBH_4$ solution to the sample in a Teflon-lined screw-capped culture tube.
3. Incubate at 45°C overnight for 20–24 h and then centrifuge.
4. Terminate the reaction by adding 5 drops of acetic acid from a Pasteur pipette until fizzing stops.
5. Assemble the desalting column by packing a Pasteur pipette fitted with a piece of silicone tubing at its tapered end (to control the flow) with Dowex beads.
6. Elute the desalting column with 15 ml of acetic acid.
7. Load the sample onto the top of the desalting column.
8. Elute with 5% acetic acid and collect 5 ml in a glass culture tube.
9. Reduce the volume on a Savant Speed-Vac and lyophilise.
10. Remove excess borate by coevaporating with 10% methanolic acetic acid (4 × 0.5 ml) under a stream of nitrogen at room temperature. Repeat twice more.

E. Derivatisation of Glycans and Sep-Pak Cleanup

Wear suitable protective clothing, including safety glasses, and work in a fume hood when preparing all solutions and performing various steps of the protocols.

Solutions

1. *Anhydrous DMSO solution:* To make 400 ml, using a steel spatula carefully place good sized calcium hydride lumps into a clean 500-ml round-bottomed Quick-fit flask. Avoid putting powder into the flask. Add 400 ml of DMSO and stand overnight or longer until all powder has settled to the bottom. Keep the stopper tight and replace immediately each time after the DMSO has been dispensed. This DMSO stock solution may be kept as a stock solution at room temperature.
2. *15, 35, 50, 75% Acetonitrile solutions:* To make 100 ml of each solution, add 15, 35, 50, and 75 ml of acetonitrile to separate clean glass bottles and complete them to 100 ml with ultrapure water. Store at room temperature.

1. NaOH Permethylation

Wear suitable protective clothing, including safety glasses, and work in a fume hood due to the caustic NaOH and toxic methyl iodide.

Steps

1. Place 5 pellets of NaOH in a dry mortar and add 3 ml of anhydrous DMSO to the pellets.
2. Grind the NaOH pellets with a pestle to form a slurry. This should be done fairly swiftly to avoid excessive absorption of moisture from the atmosphere. Ensure the sample to be permethylated is completely dry.
3. Add 0.5–1 ml of the slurry to the sample and then add 0.2–0.5 ml of methyl iodide (or deuteromethyl iodide).
4. Vortex the sample and agitate the reaction mixture on an automatic shaker for 10 min at room temperature.
5. Quench the reaction by slow dropwise additions of ultrapure water (~1 ml) with constant shaking to lessen the effects of the highly exothermic reaction.
6. Add 1–2 ml of chloroform and make up to 5 ml with ultrapure water.
7. Mix thoroughly and allow the mixture to settle into two layers. Remove and discard the upper aqueous layer.
8. Wash the lower chloroform layer several times with ultrapure water.
9. Dry down the chloroform layer under a gentle stream of nitrogen.
10. Purify the mixture by Sep-Pak C_{18} purification.

2. Purification of Permethylated Samples by Sep-Pak C₁₈ Purification

Steps

1. Condition the Sep-Pak cartridge by eluting successively with methanol (5 ml), ultrapure water (5 ml), acetonitrile (5 ml) and ultrapure water (3 × 5 ml).
2. Dissolve the sample in 1 : 1 methanol : ultrapure water (200 µl).
3. Load it onto the Sep-Pak cartridge.
4. Elute stepwise with 5 ml of ultrapure water and 3 ml each of 15, 35, 50, and 75% aqueous acetonitrile. Collect the fractions in culture tubes.
5. Reduce the volume by Savant Speed-Vac and lyophilise.

F. Useful Glycan Degradation Procedures

Wear suitable protective clothing, including safety glasses, and work in a fume hood when preparing all solutions and performing various steps of the protocols.

Solutions

1. *50 mM ammonium acetate:* To make 50 ml, add 0.1927 g of ammonium acetate and complete with 50 ml of ultrapure water. Adjust the pH to optimum range of the glycosidase.
2. *50 mM ammonium formate:* To make 50 ml, add 0.1577 g of ammonium formate and complete with 50 ml of ultrapure water. Adjust the pH to the optimum range of the glycosidase.

1. Hydrofluoric Acid 2-, 3-, and 4-Linked Fucose Removal

Wear suitable protective clothing, including safety glasses, and work in a fume hood due to toxic HF.

Steps

1. Add 50 µl of 48% HF to sample using a plastic micropipette tip. Ensure the sample is in a clean Eppendorf and is completely dry before adding the HF.
2. Incubate the sample on ice at 4°C for 20 h.
3. Terminate the reaction by drying under a gentle stream of nitrogen.
4. Dissolve the sample in 5% acetic acid and transfer to a screw-capped glass culture tube for subsequent glycosidase digestion or permethylation.

2. Mild Periodate Oxidation for O-Glycan Core Definition

Steps

1. Add 100 µl of 2–20 mM sodium m-periodate in ammonium acetate buffer (100 mM, pH 6.5) and incubate in the dark at 0°C for 20 h.
2. Quench the reaction by adding 2–3 µl of ethylene glycol.
3. Stand at room temperature for 1 h and then lyophilise.
4. Incubate the sample with 400 µl of NaBH₄ in 2 M NH₄OH (10 mg/ml) for 2 h.
5. Terminate the reaction by adding a few drops of acetic acid dispensed from a Pasteur pipette until fizzing stops.
6. Assemble the desalting column by packing a Pasteur pipette fitted with a piece of silicone tubing at its tapered end (to control the flow) with Dowex beads.
7. Elute the desalting column with 15 ml of acetic acid.
8. Load the sample onto the top of the desalting column.
9. Elute with 5% acetic acid and collect 5 ml in a glass culture tube.
10. Reduce the volume on a Savant Speed-Vac and lyophilise.
11. Remove excess borates by coevaporating with 10% methanolic acetic acid (4 × 0.5 ml) under a stream of nitrogen at room temperature. Repeat twice more.

3. Exo-glycosidase Digestion

Perform glycosidase digestions on released or partially digested glycans.

Steps

1. Add up to 1 nmol of the glycan in an Eppendorf or a screw-capped glass culture tube.
2. Adjust the pH of ammonium acetate to the pH optimum of the glycosidase and then add 50 µl to the sample. Suitable pH values are as follows: α-mannosidase (jack bean, pH 5), β-mannosidase (*Helix pomatia*, pH 4.6), β-galactosidase (bovine testes, pH 4), neuraminidase (*Clostridium perfringes*, pH 6), neuraminidase (*Streptococcus pneumoniae*, pH 6), and endo-β-galactosidase (*Bacteroides fragilis*, pH 5.8).
3. Add enough glycosidase to digest the sample. As a rough guide use the following amounts: α-mannosidase (jack bean, 20 mU), β-mannosidase (*H. pomatia*, 10 mU), β-galactosidase (bovine testes, 20 mU), neuraminidase (*C. perfringes*, 60 mU),

neuraminidase (*S. pneumoniae*, 10 mU), and endo-β-galactosidase (*B. fragilis*, 50 mU).

4. Incubate at 37°C for 20 h and then centrifuge.

5. Add a second aliquot and incubate at 37°C for a further 12 h.

6. Lyophilise the sample in a clean screw-capped glass culture tube covered with perforated Parafilm.

7. Remove an appropriate aliquot and permethylate for subsequent analysis by MALDI-MS. *NB*: for β-*N*-acetylhexosaminidase (*S. pneumoniae* 80 mU, pH 5), use ammonium formate

G. Defining Linkages and Sugar Compositions

Wear suitable protective clothing, including safety glasses, and work in a fume hood when preparing all solutions and performing various steps of the protocols.

Solutions

1. *2.0 M TFA:* Add 200 µl TFA to 1.1 ml of water. Prepare fresh on day of use.

2. *2.0 M NH₄OH:* Add 200 µl NH₃ to 1.62 ml of water. Prepare fresh on day of use.

3. *10% methanolic acetic acid:* To make 25 ml, add 2.5 ml of acetic acid and complete to 25 ml with methanol. Prepare fresh on day of use.

4. *Methanolic HCl:* To make a 1.0 M solution, add dropwise 100 µl of acetyl chloride to 1.3 ml of ice-cold methanol with constant shaking between each addition. Prepare fresh on day of use. Alternatively, for a rapid way of preparing a solution of approximately 1.0 M concentration, connect a Pasteur pipette to tubing attached to a cylinder of HCl gas and bubble the gas into 1 ml of methanol at room temperature until the bottom of the glass tube is hot to the touch.

1. Linkage Analysis of Permethylated Glycans

Steps

1. Add 200 µl of 2 M TFA to the permethylated sample and incubate at 121°C for 2 h.

2. Allow to cool, centrifuge, and dry under nitrogen.

3. Weigh out 10–20 mg NaBD₄ and add the appropriate volume of 2.0 M NH₄OH to make a 10-mg/ml solution.

4. Add 200 µl of the reducing reagent to the hydrolysates and stand at room temperature for 2 h.

5. Add 5 drops of acetic acid dropwise with a Pasteur pipette until fizzing stops and dry under nitrogen. It is not necessary to wait for complete dryness before proceeding to the next step.

6. Add 1 ml of 10% acetic acid in methanol and evaporate the solution under nitrogen until dry. Repeat twice more.

7. Add 200 µl acetic anhydride and incubate at 100°C for 1 h. Dry down under a stream of nitrogen.

8. Add 1 ml of chloroform and wash with ultrapure water.

9. Vortex the mixture and allow the two layers to separate. Discard the upper water layer.

10. Repeat two more times and dry under nitrogen. The resulting partially methylated alditol acetates (PMAA) can now be dissolved in a small volume of hexanes and analysed by GC-MS.

2. TMS Sugar Analysis

Steps

1. Add 200 µl of the methanolic–HCl reagent to the underivatised sample in a Teflon-lined screw-capped glass culture tube.

2. Incubate overnight (14–16 h) at 80°C and centrifuge.

3. Remove the reagent by blowing down under a stream of nitrogen.

4. Add the following reagents sequentially with thorough mixing in between: 500 µl of methanol, 10 µl (one drop from Pasteur pipette) of pyridine, and 50 µl of acetic anhydride.

5. Stand at room temperature for 15 min and remove the reagent by blowing down under a stream of nitrogen.

6. Break a 1-ml Tri-sil Z ampoule and add 100–200 µl of the TMS derivatisation reagent to the sample.

7. Stand at room temperature for about 15 min and dry down under a gentle stream of nitrogen.

8. Add a small squirt of hexane (~500 µl) and dry down again.

9. Add 1 ml of hexanes, mix thoroughly, and centrifuge at 3000 rpm for about 5–10 min.

10. Transfer the clear hexane supernatant to a new tube.

H. Mass Spectrometry

Solutions

1. *1% TFA solution:* To make 100 ml, add 1 ml of TFA to 99 ml of ultrapure water. Store at room temperature.

2. *0.1% TFA solution:* To make 100 ml, add 100 µl of TFA to 99.9 ml of ultrapure water. Store at room temperature.

3. *30% and 60% acetonitrile TFA solutions:* To make 50 ml of each solution, add 15 ml and 30 ml of acetonitrile to clean glass bottles and complete to 50 ml with 0.1% TFA to make 30 and 60% acetonitrile solutions.

I. Gas Chromotography–Mass Spectrometry

Analyse derivatised monosaccharide mixtures on a PerkinElmer's Clarus 500 gas chromograph / mass spectrometer (GC-MS) fitted with a RTX-5 column (30 × 0.32 mm internal diameter, Restek Corp.). Dry down the hexanes and redissolve the derivatives in ~50 μl of hexanes. Samples may be injected directly onto the column or by the autosampler. For PMAA analysis, the oven is held at 90°C for 1 min and subsequently ramped to 290°C at a rate of 8°C/min, held at 290°C for 5 min and finally to 300°C at a rate of 10°C/min. For standards, various synthetic glycans can be permethylated and taken through the linkage protocol to generate retention information for various linkages. As a rough guide, the elution times are ordered as follows: terminal fucose, terminal hexoses, linked hexoses, terminal *N*-acetylhexosamines, and finally linked *N*-acetylhexosamines. Note that sialic acids are not observed in this analysis. For TMS derivatives the oven is held at 90°C for 1 min and subsequently ramped to 140°C at a rate of 8°C/min, then to 200°C at a rate of 5°C/min, and finally to 300°C at a rate of 10°C/min. For TMS derivatives, arabitol is used as an internal standard to track the reproducibility of retention times and for quantitative information. Monosaccharide standards usually include 1 nmol/μl solutions of arabitol, fucose, glucose, galactose, mannose, GlcNAc, GalNAc, and NeuAc.

J. Matrix-Assisted Laser Desorption Ionisation Mass Spectrometry

MALDI MS is generally performed in positive reflectron mode using a Perspective Biosystems Voyager-DE STR MALDI workstation equipped with delayed extraction technology. Data acquisition is performed using Voyager 5 Instrument Control Software, and data processing by Data Explorer MS processing software. Calibration is performed by external calibration of a mixture of leucine, enkephalin, bradykinin, bradykinin (fragment 1–8), angiotensin I, neurotensin, adrenocorticotropic hormone fragment (ACTH) 1–17, ACTH fragment 18–39, ACTH fragment 7–38, ACTH fragment 1–39, and insulin. Typical mass ranges are as follows: permethylated *N*-glycans in the range of *m/z* 1000-7000, and permethylated *O*-glycans in the range of *m/z* 500–4000. Permethylated glycans are usually in the 35%, 50%, and 75% fractions. The relevant fractions for *N*- and *O*-glycan studies are the 5% acetic acid fraction (*N*-glycans) and the 20% propan-1-ol fraction (*O*-glycopeptides) obtained after N-glycosidase.

1. MALDI Sample Preparation of Derivatised Glycans

Steps

1. Weigh 10 mg of DHB in an Eppendorf.
2. Add 200 μl of water and 800 μl methanol to the DHB
3. Vortex the matrix solution until the DHB completely dissolves.
4. Dissolve the derivatised glycans in a small volume of methanol ~5–10 μl.
5. Mix a 1-μl aliquot of the derivatised glycans with 1 μl or the fresh DHB solution.
6. Spot 1 μl of the sample matrix mixture on a clean stainless steel target and allow it to dry under vacuum. Perform MALDI-TOF experiments.

2. MALDI Sample Preparation of Peptides and Glycopeptides

If the sample has not been purified by Sep-Pak, perform steps 1–4, otherwise go to step 5.

Steps

1. Assemble a manual Microtrap Holder with a clean MicroTrap peptide cartridge.
2. Condition the cartridge by eluting successively with acetonitrile (5 × 20 μl) and then 0.1% TFA (8 × 20 μl).
3. Load the sample onto the column and then elute with 0.1% TFA (20 μl × 5) without collection.
4. Elute stepwise with 20 μl each of 30% and 60% acetonitrile in 0.1% TFA. Collect the fractions in Eppendorfs.
5. Weigh 10 mg of α-cyanocinnamic acid in an Eppendorf.
6. Add 1 ml of 30% acetonitrile in 0.3% TFA to the α-cyanocinnamic a acid and mix.
7. Dissolve the sample in an appropriate volume of 0.1% TFA to make a picomole per microliter solution if the sample is dry.
8. Mix a 1-μl aliquot of the peptides with 1 μl fresh α-cyanocinnamic acid solution.
9. Spot 1 μl of the sample matrix mixture on a clean stainless steel target and allow it to dry under vacuum. Perform MALDI-TOF experiments.

K. Nanoflow Electrospray Ionisation

Nano-ES MS and nano-ES MS/MS data are acquired using positive ion mode of a quadrupole orthogonal acceleration time-of-flight, Q-Tof 1, mass spectrometer (Waters Ltd, UK) fitted with a Z-spray atmospheric ion source. In MS mode the quadrupole is used in RF-only mode and transmits about two decades in mass to the TOF. In MS/MS the quadrupole is in resolving mode, allowing selection of precursors for collision in the hexapole gas cell. A voltage of 1.5 kV is applied to the NanoES capillary tip, generating a nanoflow in the range of 10–30 nl/min. Argon and nitrogen are used as the collision and bathing gases, respectively. Collision gas pressure is maintained at 10^{-4} mbar, and collision energies up to 90eV are used for large glycopeptides, but do not usually exceed 50 eV for most peptides. As a rough guide, collision energies for doubly charged peptides range from <18 eV for m/z values of <500 to collision energies >35 eV for species with m/z values >1200. Collision energies are usually varied between 40 and 70 eV for permethylated glycans. Data are acquired and processed using Masslynx 4 software (Waters Ltd., UK). The instrument is calibrated with a 0.1 to 1 pmol/µl solution [Glu[1]]-fibrinopeptide B in methanol / 5% (v/v) aqueous acetic acid [1 : 3, (v/v)]. Experiments may also be performed in a similar fashion on other quadrupole orthogonal acceleration time-of-flight instruments, such as the QSTAR Pulsar I mass spectrometer (AB/MDS Sciex, Toronto, Canada) fitted with a nanoelectrospray ion source (MDS Proteomics, Odense, Denmark) controlled by Analyst QS software.

1. ES Sample Preparation of Derivatised Glycans

Steps

1. Dissolve derivatised glycans in a suitable volume of methanol (~10 µl).
2. Withdraw a few microliters of the sample into a clean Hamilton syringe and inject into a nano-ES capillary.
3. Open the tip of the needle and load into the nano-ES interface of the mass spectrometer. Perform MS and MS/MS experiments.

2. ES Sample Preparation of Glycopeptides

If the sample has not been purified by Sep-Pak, perform steps 1–4, otherwise go to step 5.

Steps

1. Assemble a manual Microtrap Holder with a clean MicroTrap peptide cartridge.
2. Condition the cartridge by eluting successively with acetonitrile (5 × 20 µl) and then 0.1% TFA (8 × 20 µl).
3. Load the sample onto the column and then elute with 0.1% TFA (20 µl × 5) without collection.

4. Elute stepwise with 20 µl each of 30% and 60% acetonitrile in 0.1% TFA. Collect the fractions in Eppendorfs.
5. Withdraw a few microliters of the sample into a clean syringe and inject into a nano-ES capillary.
6. Open the tip of the needle and load into the nano-ES interface of the mass spectrometer. Perform MS and MS/MS experiments.

L. Nano-LCMS/MS Q-Star Hybrid MS/MS

Complex tryptic digests are analysed by nano-LC-MS/MS using a reversed-phase nano-HPLC system (Dionex, Sunnyvale) connected to a quadrupole TOF mass spectrometer (QSTAR Pulsar I, MDS Sciex, Canada). The digests are separated by a binary nano-HPLC gradient generated by an Ultimate pump fitted with a Famos autosampler and a Switchos microcolumn-switching module (LC Packings, Amsterdam, Netherlands). An analytical C18 nanocapillary (75 µm i.d. × 15 cm, PepMap) and a micro precolumn C18 cartridge (300 µm i.d. × 1 mm) are employed for on-line peptide separation. Digests are injected onto the precolumn by a Famos autosampler with volumes from volumes typically ranging from 0.5 to 5 µl. The digests are first loaded onto the precolumn and eluted with 0.1% formic acid in water (HPLC grade, Purite) for 2–4 min. The sample is then transferred onto an analytical C18 nanocapillary HPLC column and eluted at a flow rate of 150–200 nl/min using the following gradient: 0–5 min 99% A, 5–10 min 99–90% A, 10–70 min 90–60% A, 70–71 min 60–50% A, 71–75 min 50–5% A, 75–85 min 5% A, 85–86 min 5–95% A, 86-90 min and 95% A. Solvent A = 0.05% (v/v) formic acid in 95 : 5 (v/v) water:acetonitrile; solvent B = 0.04% formic acid in 95 : 5 (v/v) acetonitrile:water. Data acquisition was performed using Analyst QS software with an automatic information-dependent acquisition (IDA) function. Similar experiments may also be performed on other quadrupole orthogonal acceleration time-of-flight instruments such as the MicroMass Q-ToF MS (Waters Ltd., UK) fitted with a CapLC system (Waters Ltd., UK) or with the same nano-HPLC system (Dionex, Sunnyvale) described earlier.

IV. COMMENTS

The protocols outlined in this article are suitable for profiling the major N- and O-glycan populations on a wide range of mammalian cells, fluids, tissues, and organs. In addition, these methodologies are used routinely to complement detailed studies of specific biologically active glycopolymers.

V. PITFALLS

1. Use ion-free ultrapure water at all times for washing plasticware and glassware. Always use ion-free ultrapure water to prepare solutions.

2. Clean all glassware thoroughly before use with water before drying in a 90°C oven. Do not use detergents to clean glassware.

3. Use screw-capped disposable Pyrex culture tubes with caps that have Teflon inserts as much as possible, i.e., incubations, reactions, and drying steps.

4. Use prepierced Parafilm (American National Can) to cover glass culture tubes during vacuum centrifugation and lyophilisation to minimise potential cross contamination.

5. Where possible, use glassware rather than plasticware in all steps apart from steps involving HF, as HF reacts with glass, e.g. use sterile nonplugged Pasteur pipettes to dispense solutions >0.2 ml and use sterile glass micropipettes to dispense solutions < 0.2 ml except for HF transfers.

6. Prior to reduction/carboxymethylation, take appropriate measures to ensure that extracts or samples are as free as possible of detergents and involatile salts.

7. TMS derivatives are volatile so do not leave for extended periods under nitrogen.

8. For detergent solutions, add the detergents after adjusting the pH in order to prevent precipitation.

9. When using gloves, wear the powder-free type.

10. For bulky tissues, it is advisable to split the sample, as large amounts of material can be difficult to process. As a rough guide, a mouse kidney should be split into two, whereas a mouse small intestine should be split into at least four samples.

References

Albersheim, P., Nevins, D. J., English, P. D., and Karr, A. (1967). A method for the analysis of sugars in plant cell wall polysaccharides by gas-liquid chromatography. *Carbohydr. Res.* **5**, 340–345.

Barber, M., Bordoli, R. S., Sedgwick, R. D., and Tyler, A. N. (1981). Fast atom bombardment of solids (F.A.B): A new ion source for mass spectrometry. *J. Chem. Soc. Chem. Commun.* **7**, 325–327.

Dell, A. (1987). F.A.B.: Mass spectrometry of carbohydrates. *Adv. Carbohydr. Chem. Biochem.* **45**, 19–72.

Dell, A., Azadi, P., Tiller, P., Thomas-Oates, J., Jennings, H. J., Beurret, M., and Michon, F. (1990). Analysis of oligosaccharides epitopes of meningococcal lipopolysaccharides by fast-atom-bombardment mass spectrometry. *Carbohydr. Res.* **200**, 59–76.

Dell, A., Morris, H. R., Easton, R. L., Panico, M., Patankar, M., Oehniger, S., Koistinen, H., Seppala, M., and Clark, C. F. (1995). Structure analysis of the oligosaccharides derived from glycodelin, a human glycoprotein with potent immunosuppressive and contraceptive activities. *J. Biol. Chem.* **270**, 24116–24124.

Dell, A., Morris, H. R., Egge, H., von Nicolai, and Strecker, G. (1983). Fast-atom-bombardment mass spectrometry for carbohydrate structure determination. *Carbohyr. Res.* **115**, 41–52.

Fenn, J. B., Mann, M., Meng, C. K., Wong, S. K., and Whitehouse, C. M (1990). Electrospray ionization-principles and practice. *Mass Spectrom. Rev.* **9**, 37–70.

Fukuda, M., Bothner, B., Ramsamooj, P., Dell, A., Tiller, P. R., Varki, A., and Klock, J. C. (1985). Structures of sialylated fucosyl polylactoaminioglycans isolated from chronic myelogenous leukemia cells. *J. Biol. Chem.* **260**, 12957–12967.

Fukuda, M., Dell, A., and Fukuda, M. N. (1984). Structure of fetal lactosaminoglycan. The carbohydrate moiety of band 3 isolated from human umbilical cord erythrocytes. *J. Biol. Chem.* **259**, 4782–4791.

Karas, M., and Hillenkamp, F. (1988). Laser desorption ionization of proteins with molecular masses exceeding 10,000 daltons. *Anal. Biochem.* **191**, 332–336.

Karas, M., Igendoh, A., Bahr, U., and Hillenkamp, F. (1989). Ultraviolet-laser desorption ionization mass-spectrometry of femtomolar amounts of large proteins. *Biomed. Environ. Mass Spectrom.* **18**, 841–843.

Laferte, S., Fukuda, M. N., Fukuda, M., Dell, A. and Dennis J. W. (1987). Glycosphingolipids of lectin-resistant mutants of the highly metastatic mouse tumour cell line, MDAY-D2. *Cancer Res.* **47**, 150–159.

Lopez, M., Coddeville, B., Langridge, Plancke, Y., Sautierre, P., Chaabihi, H., Chirat, F., Harduin-Lepers, A., Cerutti, M., Verbert, A., and Delannoy, P. (1997). Microheterogeneity of the oligosaccharides carried by the recombinant bovine lactoferrin expressed in *Mamestra brassicae* cells. *Glycobiology* **7**, 635–651.

Manzi, A. E, Norgard-Sumnicht, K., Argade, S., Marth, J. D., van Halbeek, H., and Varki, A. (2000). Exploring the glycans repertoire of genetically modified mice by isolation and profiling of the major glycan classes and nano-NMR analysos of glycans mixtures. *Glycobiology* **10**, 669–689.

McConville, M. J., Homans, S. W., Thomas-Oates, J. E., Dell, A., and Bacic, A. (1990). Structures of the glycoinositolphospholipids from *Leishmania major*: A family of novel galactofuranose-containing glycolipids. *J. Biol. Chem.* **265**, 7285–7294.

Morris, H. R. (1980). Biomolecular structure determination by mass spectrometry. *Nature* **286**, 447–452.

Saski, H., Bothner, B., Dell, A., and Fukuda (1987). Carbohydrate structure of erythropoietin exopressed in Chinese hamster ovary cells by a human erythropoietin cDNA. *J. Biol. Chem.* **262**, 12059–12076.

Sutton-Smith, M., Morris, H. R., and Dell, A. (2000). A rapid mass spectrometric strategy suitable for the investigation of glycan alterations in knockout mice. *Tetrahedron-Asymmetry* **11**, 363–369.

54

Stable Isotope Labeling by Amino Acids in Cell Culture for Quantitative Proteomics

Shao-En Ong, Blagoy Blagoev, Irina Kratchmarova, Leonard J. Foster, Jens S. Andersen, and Matthias Mann

I. INTRODUCTION

Analysis by combined liquid chromatographic separation and mass spectrometry (LC-MS) is rapidly becoming the most popular and effective approach for large-scale protein identification (Aebersold and Mann, 2003; Yates, 2000) (see also article by Foster and Mann). The increasing sensitivity and the development of better protein/peptide separation methods will surely extend the catalogues of proteins emerging from proteomics projects. However, the significance of such data is further amplified if such identifications are coupled with quantitative abundance information. Indeed, it has been argued that almost all proteomics experiments should soon be converted to a quantitative format as qualitative lists of proteins may not be sufficient in the future (Aebersold and Mann, 2003).

Quantitative methods in mass spectrometry (MS) rely largely on the principle of stable isotope labelling (SIL). Differential incorporation of stable isotopic nuclei (^{2}H, ^{13}C, ^{15}N, ^{18}O) allows the "light" and a "heavy" form of the same peptide to be resolved in the mass spectrometer due to their mass difference. The ratio of signal intensities of the two peptides measures peptide abundance and, correspondingly, the relative abundance of the two proteins.

The main strategies for stable isotope incorporation can be broadly classified into two groups: "postharvest" incorporation and metabolic incorporation. The prototypical example of the former approach is the "isotope-coded affinity tag" (ICAT) reagent described in 1999 (Gygi *et al.*, 1999). In this approach, biotinylated tags are differentially mass encoded with stable isotopes (either 0 or 8 deuteriums) and bear a chemical moiety targeted at cysteine sulphydryl groups. The protein samples to be compared are derivatised separately with the two forms of ICAT, digested, and purified over an avidin column to enrich ICAT-labelled peptides for subsequent mass spectrometric analyses. Several other variations on this chemical derivatisation theme have been applied since then (Goodlett *et al.*, 2001; Regnier *et al.*, 2002; Olsen *et al.*, 2004).

Metabolic incorporation of stable isotopes for MS quantitation of proteins was first introduced in 1998 for bacteria (Langen *et al.*, 2000) and in 1999 in yeast (Oda *et al.*, 1999). These first reports relied on growing simpler organisms on ^{15}N-enriched media. The growth of mammalian cells in ^{15}N-enriched media has also been described (Conrads *et al.*, 2001) but is largely untenable for most cell culture protocols due to the lack of a source of ^{15}N-enriched serum. In our laboratory, we have developed a method (SILAC) for stable isotope labelling by amino acids specifically suited for mammalian cell culture (Ong *et al.*, 2002). Independently, other groups have also described similar experiments for use in yeast (Jiang and English, 2002) and in mammalian systems (Zhu *et al.*, 2002).

The name SILAC (for stable isotope labelling by amino acids in cell culture) is apt as it accurately describes the mode in which the stable isotope labels are delivered, and the acronym is an extension of SIL already popularised from the small molecule field and in NMR. Rather than introducing the mass differential in all nitrogen atoms as in ^{15}N labelling, the SILAC method relies on stable isotope containing amino acids. This makes the incorporation of stable isotopes sequence dependent and inherently predictable. Furthermore, by introducing the stable isotope label in the form of amino acids in growth media, the label is effectively "encoded" into each newly synthesized protein. This differs significantly from chemical derivatisation methods that target specific amino acid residues. For instance, a limitation of the ICAT approach is that not all proteins contain cysteine. Often, SILAC labelling is more practical, as beyond preparation of labelling media, there are no chemical modifications required to "encode" proteins with quantitative information. Indeed, the chemical derivatisations required in postharvest labelling methods are not usually compatible with high-sensitivity MS measurements.

Importantly, SILAC provides investigators with two cell populations, grown in parallel and thus identical save for the fact that the proteins from each individual population are distinguishable by MS. The "encoded" cells can be mixed at harvest and can be subjected to any biophysical separation or protein purification directly. This contrasts with postharvest labelling approaches that require separate processing of protein samples until the label has been incorporated. Derivatisation methods that incorporate the label at the peptide level (i.e., after proteins have been processed enzymatically) demand that these purification steps and protein digestion steps be done individually, thus significantly compromising the accuracy of quantitation.

This article describes the study of C2C12 muscle differentiation with SILAC as an example. One set of cells are kept at the myoblast stage whilst the other set is induced to differentiate into myotubes (Ong *et al.*, 2002). Cells from both samples are mixed 1 to 1 and lysed. Proteins extracted from the mixture are then analysed by standard MS techniques (Fig. 1). Many muscle-specific proteins, such as myosin, were found to be highly upregulated whilst other proteins, such as histones, were found at similar levels in both samples. This illustrates the ability of the SILAC approach to accurately measure changes in protein expression during a cell differentiation experiment.

Another intrinsic benefit of "coded" cell populations is apparent when SILAC is coupled to standard

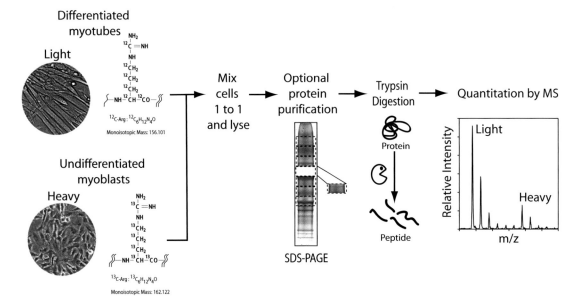

FIGURE 1   **An overview of a quantitative study of C2C12 muscle differentiation with SILAC.** Myoblasts are grown to confluence separately in "light" and "heavy" media. Myoblasts in "light" $^{12}C_6$-Arg media are induced to differentiate to form myotubes by reducing serum concentration to 2% dialysed FBS. Myoblasts in $^{13}C_6$-Arg are left undifferentiated. Cells are harvested and mixed 1 to 1 after protein concentration determination. Nuclei and cytoplasmic fractions from mixed cells are obtained. Proteins were separated on a SDS–PAGE gel, and gel slices were excised and digested with trypsin. Peptides were extracted from gel slices and analysed by LC-MS. The ratio of relative intensities of the "light" over the "heavy" peak cluster gives the fold increase in relative abundance of the protein during muscle differentiation.

biochemical purifications of proteins with antibodies or bait proteins (Fig. 2). Pull-down experiments and subsequent MS identification have been used for large-scale analysis of protein complexes and protein–protein interactions (Gavin *et al.*, 2002; Ho *et al.*, 2002). The exquisite sensitivity of mass spectrometers generates large lists of identified proteins that are purportedly true binders to the bait. Although informative, it is not always easy to judge if a particular protein identified in this way is a real interaction partner or a nonspecific "background" interaction with the antibody/bait. This necessitates time-consuming and often difficult follow-up experiments to validate one's findings. Such uncertainty can be appropriately addressed with the following approach that combines SILAC with affinity purifications.

SILAC-labelled cells are used to generate separate pools of proteins or protein complexes that are distinct to an affinity bait. Cells may be differentially treated with a drug or growth factor; alternatively, cells may express the wild-type or mutant form of a component of a protein complex. Proteins from both states can be mixed in equal proportions and then purified together over the affinity bait. In this manner, peptides from proteins specific to the differential treatment will give a large differential ratio whilst background, nonspecific interactions with the bait will be close to the 1-to-1 ratio of mixing.

In one example, $^{13}C_6$-Arg-labelled cells were stimulated briefly with EGF alongside untreated, unlabelled cells. Cells were lysed and equal amounts of protein from each sample were combined for subsequent affinity purification with the SH2 domain from Grb2 as a bait (Blagoev *et al.*, 2003). As the SH2 domain of Grb2 binds to phosphorylated tyrosines on the EGF receptor upon stimulation, we observed specific enrichment of the phosphorylated EGF receptor along with its associated proteins. Direct interaction partners exhibited a large (> fivefold) enrichment of one peptide form versus the other and were easily distinguished from background-binding proteins, which showed a 1-to-1 ratio (see Fig. 1). This discrimination between specific interaction partners and background binding is extremely powerful and addresses many of the inadequacies of previous approaches.

Using three different forms of arginine with SILAC makes it possible to functionally encode three cellular states in a single experiment. We have used $^{13}C_6$-Arg, $^{13}C_6{}^{15}N_4$-Arg in addition to normal arginine to label all arginine-containing peptides in the proteome (Blagoev *et al.*, 2004). This is especially useful in cases where various time intervals of drug treatment are examined or where multiple cellular conditions are presented to an affinity bait. Using immunoprecipitation as an example, applying the "two-state" approach with only two distinct quantitative labels would require two separate pull downs followed by separate MS identifications and quantitative analyses. With triple encoding, this is reduced to a single experiment with a shared affinity purification step.

Like other methods of labelling that rely on metabolic incorporation of stable isotopes, SILAC can only be used with live cells. In dynamic studies of protein abundance in live cells (Pratt *et al.*, 2002), SILAC has obvious advantages over chemical methods. Clearly the two approaches (metabolic incorporation and postharvest derivatisation) are complementary, each with their accompanying strengths and limitations.

Stable isotopes for quantitative measurements in proteomics are becoming increasingly important, as MS has grown to become a cornerstone technology in this field. The sensitivity and high mass accuracy of mass spectrometric readouts will likely mean that these tools will become indispensable in the life science field for years to come.

II. MATERIALS AND INSTRUMENTATION

Dialysed foetal bovine serum (FBS) (Cat. No. 12480-026), L-glutamine (Cat. No. 25030-024), penicillin/streptomycin (Cat. No. 15070-063), custom-synthesised media—Dulbecco's modified Eagle medium (like Cat. No. 21969 but without amino acids) and RPMI 1640 (like Cat. No. 61870 but without amino acids) are from Invitrogen. MEM Eagle's deficient with Earle's salts and L-glutamine, without L-leucine, L-lysine, L-methionine (Cat No. M7270), L-arginine (Cat. No. A-6969), L-leucine (Cat No. L-8912), L-lysine (Cat. No. L-9037), and all remaining unlabelled L-amino acids were from Sigma. L-$^{13}C_6$-Arginine (Cat. No. CLM-2265) and L-$^{13}C_6$-lysine (CLM-2247) are from Cambridge Isotope Labs. Leucine-5,5,5-d3 (Cat. No. 48,682-5) is from Sigma-Isotec. All water is "Milli-Q" quality distilled, deionised water.

Sequencing-grade porcine trypsin (Cat. No. V511C) is from Promega. Dithiothreitol (DTT, Cat. No. D-9163) and iodoacetamide (Cat No. I-1149) are from Sigma. The HPLC system used in LC-MS analyses is the Agilent 1100. The hybrid quadrupole time-of-flight mass spectrometer is supplied by MDS-SCIEX-Applied Biosystems. The search engine for protein and peptide identification from LC-MS data is the Mascot search program from Matrix Science.

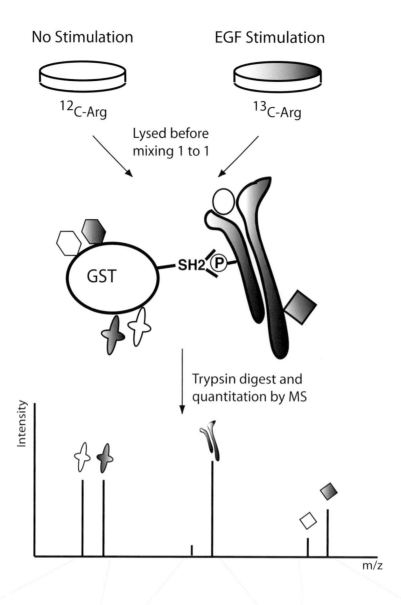

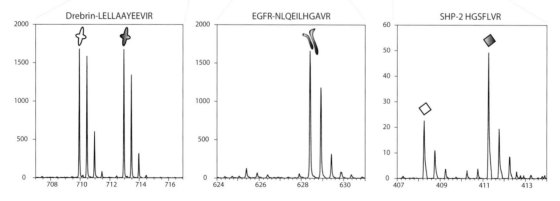

FIGURE 2 SILAC discriminates between true interaction partners in affinity purifications—posttranslational modification-dependent enrichment of epidermal growth factor receptor (EGFR) and associated proteins by binding to GST-SH2(Grb2) upon EGF stimulation. Two cell populations are differentially labelled with ^{13}C-Arg-SILAC. After treatment of "heavy" cells with EGF, lysates from both cell populations are mixed 1 to 1 and incubated with GST-SH2(Grb2). Proteins that interact specifically with the bait in conjunction with EGF stimulation will exhibit a large differential ratio in their peptide ratios. In contrast, nonspecific interactions with GST beads will bind equally from both cell populations with ratios similar to the original mixing ratio. Drebrin, an actin-binding protein, shows up with a peptide ratio of 1 (left spectrum), indicating that it was not pulled down by specific interaction with the bait. Quantitation of EGFR (middle spectrum) and Shp-2 (right spectrum) peptides gives enrichment ratios of >10 and 2.5, respectively, strongly indicative that they were affinity purified by the bait in response to EGF treatment.

III. PROCEDURES

A. Labelling Media Preparation

Cell culture media can be prepared exactly according to the experimenter's specifications to suit any particular cell type (refer to Volume 1 for cell culture techniques). The only requirement is that the labelling AA of choice is substituted with an isotopically "heavy" form to distinguish it from unlabeled, "light" media. We have applied SILAC extensively in a variety of cell types in human and rodent systems and with an assortment of stable isotope nuclei-containing amino acids and found that cell growth, nutrient use, and general morphology are similar to cells grown in standard cell culture media.

Media preparations depleted in certain amino acids are available in powdered form. We have previously used MEM with Earle's salts deficient in lysine, methionine, and leucine (powdered media). Presently, we obtain standard media formulations (DMEM and RPMI) deficient in specific amino acids as custom-synthesised products from Invitrogen. A primary concern may be the availability of certain types of dialysed serum (horse, calf), even though most cell culture media companies routinely supply dialysed FBS. A limitation of the use of dialysed serum is the potential loss of low molecular mass (below 10 kDa) components, which may be important for cell growth, through the dialysis process but the importance of using dialysed serum cannot be overstated. It is absolutely necessary to avoid the contribution of unlabelled AAs in labelling media and is critical for accurate quantitation (Ong *et al.*, 2002).

The choice of a labelling amino acids for use in SILAC is likewise important. Common amino acids such as leucine (70% of all tryptic peptides contains at least one leucine residue—human IPI database, EBI) are preferable, generating several quantitatable peptides per protein. Alternatively, an enzymatic digest with trypsin for mass spectrometric analysis produces peptides with lysine or arginine at the C-terminal of the peptide. Labelling with arginine or lysine therefore results in incorporation of a single labelled residue in half of the tryptic peptides. If both AAs are used in SILAC, essentially all tryptic peptides would be labelled and quantitatable—a critical advantage in the quantitation of posttranslational modifications.

Essential amino acids are an obvious first choice in SILAC experiments. However, not all cell lines are capable of *de novo* synthesis of nonessential amino acids such as arginine and the empirical evaluation of a particular amino acid may well be worthwhile.

The use of $^{13}C_6$-arginine (where all ^{12}C atoms are substituted with ^{13}C) in SILAC is particularly desirable because of the suitable mass differential encoded, as well as the coelution of peptide pairs (Zhang and Regnier, 2002)(see later). ^{13}C-containing amino acids cost more in comparison to deuterated amino acids. By testing cell lines with growth media containing less arginine or lysine, we find that significantly reducing the amount of these amino acids does not adversely affect the growth of cells. Because both control and experimental cell populations are grown on identical media compositions (other than the form of the labelling AA), the reduction of a particular amino acid concentration does not compromise the validity of the experiment. However, it is important to test each separate cell line for the nutritional requirements for a particular amino acid. For example, a cell line may begin to synthesize arginine *de novo* when the arginine concentration is reduced by a third whilst others (e.g., HeLa) only begin to do so at one-tenth the normal arginine concentration (DMEM media). Reducing the levels of labelled amino acids may be cost effective and, in some cases (see later), even necessary.

Separately, we find that arginine is metabolically converted to proline in certain cell lines when supplied at concentrations described in standard media formulations. This observation was made as peptides containing [^{13}C]-proline were detected in MS analyses in the samples from human adenocarcinoma (HeLa) cells but not in a mouse fibroblast cell line (NIH 3T3). This proline conversion was undetectable when one-fifth of the original arginine concentration was used with HeLa cells.

As an example, this article describes the preparation of DMEM (a common media preparation for many commonly used cell types) for use with SILAC and arginine labelling (Ong *et al.*, 2003), but we reiterate that the general approach is directly applicable to one's custom media preparations (Ong *et al.*, 2004). An additional resource for SILAC information is available on our laboratory's Web page at http://www.cebi.sdu.dk/silac.html.

Solutions

1. *Base media preparation*: The formulation of labelling media should be prepared according to the needs of the cell lines used. Labelling media should only be distinct in the form of labelling amino acid used. A base media should be prepared in the same manner right up to the final addition of normal or labelling amino acid. The custom media formulations we purchase only require the addition of the labelling amino acid, dialysed serum (10% FBS), penicillin (50 units/ml), streptomycin (50 μg/ml), and glutamine (2 m*M* final) before use.

2. *Preparation of amino acid stock solutions*: Concentrated stock solutions of amino acids are dissolved in phosphate-buffered saline and filter sterilized with a 0.22-μm filter. These stocks should be prepared at as high a concentration as possible to minimise the dilution of other media components, e.g., arginine and lysine are prepared as 1000× (84 and 146 mg/ml, respectively) stocks. With labelled amino acids, it may be necessary to prepare unfiltered stock solutions and to filter media only after addition of the labelled amino acid in order to avoid losses due to filtration.

3. *Dialysed serum*: We use dialysed serum obtained from a commercial source. It is also possible to dialyse existing serum stocks in order to employ lower molecular weight cutoff dialysis filters and/or to reduce costs but it can be difficult to maintain consistency and to avoid contamination.

Steps

1. Work in a sterile environment. To two separate lots of base media, add the appropriate amount of arginine (either normal L-arginine to give "normal" media or L-^{13}C$_6$-arginine to give "labelling" media) to make up a full complement of amino acids, according to the manufacturer's specifications. Filter media through a 0.22-μm filter if unfiltered amino acid stocks are used.

2. Add antibiotics and glutamine as required along with 10% (v/v) dialysed serum to media containing the full complement of amino acids. Media are now ready for use and can be stored at 4°C like standard cell culture media.

B. Incorporation of Labelled AA in Growing Cells

Cells growing in normal cell culture media (in our example, murine C2C12 myoblasts) are passaged into dishes containing either normal or labelling media. C2C12 cells are allowed to undergo at least five cell doublings in SILAC media to ensure full incorporation of the labelled amino acid (Fig. 3). After five cell doublings, only a minimal amount of the original unlabelled AA should exist in the entire protein population—a theoretical maximum of $(1/2)^5$ or 3.125% could remain. In actuality, the cells would incorporate the label much sooner through normal protein turnover and in addition to novel synthesis. Furthermore, during cultivation to obtain sufficient cell numbers for the intended experiment, continual passages will result in faster incorporation of the

label (Fig. 3). When working with a new cell line or if certain experimental parameters have been changed (i.e., a different lot of dialysed serum is used), it is best to assay the state of incorporation by obtaining a protein sample for MS analysis before beginning the experiment. Ideally, small aliquots of unmixed cells from each condition should be saved from each experiment in order to check for incorporation state. MS analysis of proteins from the labelled lysates will reveal if the unlabelled amino acid is present in the protein sample. A correction factor for the protein ratios may be applied if necessary or the experiment can be repeated with cells after a longer period of adaptation.

Steps

1. From a dish of cells grown in standard cell culture media, passage cells into two separate lots, one containing the unlabelled "light" SILAC media and the other with labelled "heavy" SILAC media.
2. Grow cells in respective labelling media for a minimum of five cell doublings. If working with immortalised cells, passage the cells to deplete the cell populations of normal amino acid and to increase rate of incorporation of label.
3. Perform differential treatment of cell populations, i.e., differentiation protocol, and drug treatment protocol, and growth factor treatment.

C. Cell Harvesting and Protein Purification

Steps (Assuming SILAC-Labelled Cells Have Been Treated Differentially)

1. Harvest cells from the tissue culture dish as in normal protocols. Save a small aliquot of unmixed cell populations to check levels of incorporation of the labelled AA if necessary.
2. Mix the two cell populations in a specific ratio (e.g., 1 to 1). This should be based on cell number (measured with a haemocytometer or Coulter counter) or protein concentration (determined by the Bradford method or the equivalent). In some instances and with sufficient sample, a combination of several experiments with different mixing ratios may be advantageous to expand the dynamic range of quantitation.
3. Optional protein purification steps may be included at this point. Examples include subcellular fractionation, gel filtration, immunoprecipitations, and one- or two-dimensional gel electrophoresis.
4. Digest proteins with a protease with high cleavage specificity, e.g., trypsin. With ^{13}C-Arg SILAC,

SILAC labelling of cells

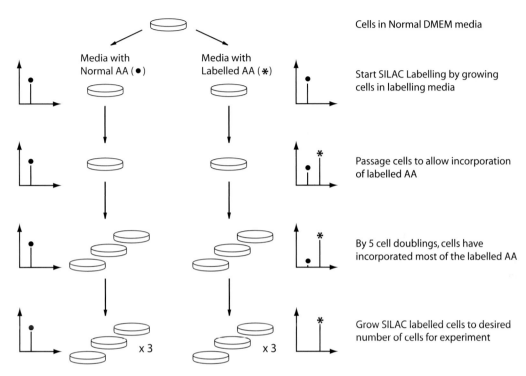

FIGURE 3   SILAC cell labelling and incorporation of labelled amino acid with continual passage of cells. Cells in standard DMEM supplemented with normal FBS, antibiotics, and glutamine are passaged into two separate dishes containing either isotopically "light" or "heavy" media. X–Y plots illustrate the presence of "light" and "heavy" peak clusters for a quantitatable peptide pair. The filled dot (•) indicates the peptide containing a normal amino acid, whilst asterisks (*) mark the peptide bearing the labelled, isotopically heavy amino acid. Cells grown in "heavy" media incorporate the labelled amino acid rapidly, as seen by the presence of the characteristic peptide pair even after the first passage. The intensity of the normal amino acid-containing peptide diminishes over time and, after sufficient cell doublings and passages, only the fully labelled peptide is detectable.

approximately half of the tryptic peptides contain arginine as the C-terminal amino acid and are quantitatable. Analyse samples with MS.

D. Analysis and Quantitation of SILAC Samples by Mass Spectrometry

A main benefit of SILAC is the option to mix cells prior to any further treatment of the sample. It is therefore possible to perform a subcellular fractionation to obtain organelles such as the nuclei without having to worry about introducing quantitative errors by way of differential treatment of samples. The downstream processing of protein samples for mass spectrometric analyses are similarly straightforward. Reduction,

alkylation, and enzymatic digestion of protein samples (as described in Shevchenko *et al.*, 1996) (see article by Foster and Mann) are performed on the same sample. Biases that may arise through labelling efficiency and enzymatic cleavage are thus avoided.

Mass spectrometric analyses can be performed with any of the standard MS instruments, but best results are obtained with a high mass accuracy, high-resolution instrument to resolve the natural isotope clusters of peptides. In our laboratory, we use the combination of nano-flow capillary liquid chromatography for the separation of complex peptide mixtures with subsequent detection and identification by a hybrid quadrupole time-of-flight mass spectrometer (QSTAR PULSAR—ABI).

In a typical LC-MS experiment, peptides eluting from the reversed-phase column are ionised and electrosprayed directly into the mass spectrometer. The mass spectrometer first obtains a survey MS scan where the entire mass range is analysed. From the survey scan, suitable peptides (of charge state $z = 2, 3$, or 4) can be selected for fragmentation in the collision cell in the MS/MS mode. The fragmentation of peptides in MS/MS spectra produces characteristic fragments, which give sequence-specific information for subsequent protein identification (Aebersold and Goodlett, 2001; Mann et al., 2001). This acquisition cycle, comprising of a survey scan followed by several MS/MS scans, is repeated throughout the LC-MS run. Ion intensities of monoisotopic peaks from each survey scan MS are used to quantitate peptide pairs. The ratios obtained from peptides used to identify a particular protein can then be averaged to give the relative ratio of protein abundance. Different laboratories approach MS analyses in subtly different ways depending on available instrumentation; we present here some guiding principles for quantitative analyses that should be applicable to anyone working in this field.

Guidelines

1. In quantitative LC-MS, a balance has to be struck between the goal of peptide identification (MS/MS sequencing events) and accurate quantitation (MS survey scans across eluting peaks). If the instrument is configured to perform a MS survey scan (taking 1 s) followed by four MS/MS spectra (each 1.5 s), the machine spends only one-seventh of the total run time acquiring the MS spectrum from which peptide ratios are acquired. Therefore, it is important that sufficient data points are collected (at least nine points for a Gaussian curve) to accurately plot the ion intensities across the eluting peptide peak.

2. In all cases where quantitation is performed, one should check that no unrelated peak clusters overlap the peptides in question. This is especially important where adequate protein/peptide separation steps do not precede MS analyses. Quantitation should be based on distinct peptide signals, and best results are obtained from peptides with a minimum signal-to-noise (S/N) ratio of 10.

3. The selection of a suitable amino acid and the mass shift that will be incorporated by the stable isotopes is important. A common amino acid such as leucine is present in about 70% of unique tryptic peptides in the human proteome and is thus a good choice as a labelling amino acid. We also favour arginine and lysine, as enzymatic cleavage with trypsin generates peptides with these amino acids at C termini. The mass shift generated from the labelling should ideally be large enough (4 Da or greater) to avoid overlapping of isotope clusters (Fig. 4A). If peptide isotope clusters overlap, calculation of the isotopic envelope of each peptide based on sequence composition can still provide accurate quantitation data.

4. Care should be taken when quantitating different proteins that have regions of sequence identity (a protein family for, e.g., the histones), as some peptides can be shared across several proteins. Quantitation based on a shared peptide may not accurately reflect the quantitation of any single protein, but might instead be an average across multiple proteins. Quantitation is best based on peptides unique to the protein of interest.

5. Many labelling reagents make use of deuterium (^{2}H) instead of ^{15}N and ^{13}C. There is a tendency for peptides labelled with deuterated reagents to elute earlier than the corresponding unlabelled peptide in reversed-phase chromatography due to isotope effects. As peptide quantitation is obtained from the MS survey scan at various time points across the elution of a peptide (Fig. 4B, left), accurate determination of the peptide ratio is possible at each sampling point of the MS. More data points are required with partially resolved peaks in order to accurately plot each eluting peptide peak (Fig. 4B, right) and quantitation should be based on peak area of the extracted ion chromatogram or a sampling of MS spectra over the retention times of the peptide pair. Although generally more expensive, using ^{13}C-substituted amino acids in SILAC helps reduce errors in quantitation caused by separation of the peptide pair. It also simplifies the process of quantitation, as the peptide ratio can be obtained from each MS scan directly. Having said that, it is very straightforward to obtain good quantitation data from deuterated reagents by simply being aware of the potential pitfalls.

6. Adequate ion statistics for quantitation peaks. Regardless of the type of mass spectrometer used to acquire quantitative data, it is imperative that mass spectra acquired should comprise enough data collection events in order to accurately describe the peptide peaks. For peptide identification, this requirement may not be as critical, but it should be apparent that accurate quantitation is only achievable where sufficient data points can be averaged. In cases where highly differential peptide ratios are observed, one of the peaks may not be detectable above noise. Here, it may be sufficient to assign the smaller peak at the level of noise and describe a lower limit of the peptide ratio rather than to give some arbitrary (and most likely incorrect) value to the smaller peak.

A

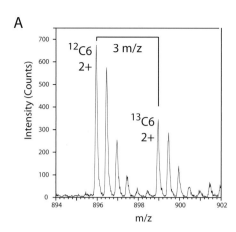

B

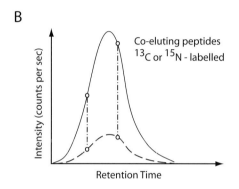
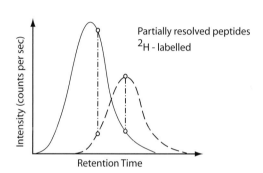

FIGURE 4   **Peptide quantitation in MS.** (A) A doubly charged SILAC-$^{13}C_6$-arginine-labelled peptide pair in the MS survey scan (acquisition time of 1 s) with 3 *m/z* separation between the "light" and the "heavy" peak. There is clear separation between peak clusters and unrelated peaks are absent. If peptides elute from the reversed-phase column over a period of 40 s, about 7 to 15 (depends on MS acquisition parameters) of these MS spectra may be acquired by the mass spectrometer. Each of these spectra give peak ratios, which are averaged to give a peptide ratio and standard deviation. (B) An example of coeluting peptides in an LC-MS run (left) and noncoeluting peptides (right). The peaks are extracted ion chromatograms (XICs) that monitor a particular mass range over time (see text for discussion on the adequate MS acquisition events to accurately describe a peak). Vertical dotted lines are examples of two individual MS acquisition events across the eluting peak. From the coeluting pair (left), it is apparent that similar ratios will be obtained regardless of the point when the peak is sampled. However, the partially resolved peaks (right) show inverse peptide ratios at the two time points indicated.

References

Aebersold, R., and Goodlett, D. R. (2001). Mass spectrometry in proteomics. *Chem Rev.* **101**(2), 269–295.

Aebersold, R., and Mann, M. (2003). Mass spectrometry-based proteomics. *Nature* **422**(6928). 198–207.

Blagoev, B., Kratchmarova, I., Ong, S. E., Nielsen, M., Foster, L. J., and Mann, M. (2003). A proteomics strategy to elucidate functional protein-protein interactions applied to EGF signaling. *Nature Biotechnol.* **21**(3), 315–318.

Blagoev, B., Ong, S.E., Kratchmarova, I., and Mann, M. (2004). Temporal analysis of phosphotyrosine-dependent signaling networks by quantitative proteomics. *Nat Biotechnol.* **22**(9), 1139–45.

Conrads, T. P., Alving, K., Veenstra, T. D., Belov, M. E., Anderson, G. A., Anderson, D. J., Lipton, M. S., Pasa-Tolic, L., Udseth, H. R., Chrisler, W. B., Thrall, B. D., and Smith, R. D. (2001). Quantitative analysis of bacterial and mammalian proteomes using a combination of cysteine affinity tags and 15N-metabolic labeling. *Anal. Chem.* **73**(9), 2132–2139.

Gavin, A. C., Bosche, M., Krause, R., Grandi, P., Marzioch, M., Bauer, A., Schultz, J., Rick, J. M., Michon, A. M., Cruciat, C. M., Remor, M., Hofert, C., Schelder, M., Brajenovic, M., Ruffner, H., Merino, A., Klein, K., Hudak, M., Dickson, D., Rudi, T., Gnau, V., Bauch, A., Bastuck, S., Huhse, B., Leutwein, C., Heurtier, M. A., Copley, R. R., Edelmann, A., Querfurth, E., Rybin, V., Drewes, G., Raida, M., Bouwmeester, T., Bork, P., Seraphin, B., Kuster, B., Neubauer, G., and Superti-Furga, G. (2002). Functional organization of the yeast proteome by systematic analysis of protein complexes. *Nature* **415**(6868), 141–147.

Goodlett, D. R., Keller, A., Watts, J. D., Newitt, R., Yi, E. C., Purvine, S., Eng, J. K., von Haller, P., Aebersold, R., and Kolker, E. (2001). Differential stable isotope labeling of peptides for quantitation and de novo sequence derivation. *Rapid Commun. Mass Spectrom.* **15**(14), 1214–1221.

Gygi, S. P., Rist, B., Gerber, S. A., Turecek, F., Gelb, M. H., and Aebersold, R. (1999). Quantitative analysis of complex protein mixtures using isotope-coded affinity tags. *Nature Biotechnol.* **17**(10), 994–999.

Ho, Y., Gruhler, A., Heilbut, A., Bader, G. D., Moore, L., Adams, S. L., Millar, A., Taylor, P., Bennett, K., Boutilier, K., Yang, L., Wolting, C., Donaldson, I., Schandorff, S., Shewnarane, J., Vo, M., Taggart, J., Goudreault, M., Muskat, B., Alfarano, C., Dewar, D., Lin, Z., Michalickova, K., Willems, A. R., Sassi, H., Nielsen, P. A., Rasmussen, K. J., Andersen, J. R., Johansen, L. E., Hansen, L. H., Jespersen, H., Podtelejnikov, A., Nielsen, E., Crawford, J., Poulsen, V., Sorensen, B. D., Matthiesen, J., Hendrickson, R. C., Gleeson, F., Pawson, T., Moran, M. F., Durocher, D., Mann, M., Hogue, C. W., Figeys, D., and Tyers, M. (2002). Systematic identification of protein complexes in *Saccharomyces cerevisiae* by mass spectrometry. *Nature* **415**(6868), 180–183.

Jiang, H., and English, A. M. (2002). Quantitative analysis of the yeast proteome by incorporation of isotopically labeled leucine. *J. Proteome Res.* **1**(4), 345–350.

Langen, H., Takacs, B., Evers, S., Berndt, P., Lahm, H. W., Wipf, B., Gray, C., and Fountoulakis, M. (2000). Two-dimensional map of the proteome of *Haemophilus influenzae. Electrophoresis* **21**(2), 411–429.

Mann, M., Hendrickson, R. C., and Pandey, A. (2001). Analysis of proteins and proteomes by mass spectrometry. *Annu. Rev. Biochem.* **70**, 437–473.

Oda, Y., Huang, K., Cross, F. R., Cowburn, D., and Chait, B. T. (1999). Accurate quantitation of protein expression and site-specific phosphorylation. *Proc. Natl. Acad. Sci. USA* **96**(12), 6591–6596.

Olsen, J. V., Andersen, J. R., Nielsen, P. A., Nielsen, M. L., Figeys, D., Mann, M., and Wisniewski, J. R. (2004). HysTag–a novel proteomic quantification tool applied to differential display analysis of membrane proteins from distinct areas of mouse brain. *Mol Cell Proteomics.* **3**(1), 82–92.

Ong, S. E., Mittler, G., and Mann, M. (2004). Identifying and quantifying in vivo methylation sites by heavy methyl SILAC. *Nat. Methods.* **1**, 119–126.

Ong, S. E., Blagoev, B., Kratchmarova, I., Kristensen, D. B., Steen, H., Pandey, A., and Mann, M. (2002). Stable isotope labeling by amino acids in cell culture, SILAC, as a simple and accurate approach to expression proteomics. *Mol. Cell Proteomics* **1**(5), 376–386.

Ong, S. E., Kratchmarova, I., and Mann, M. (2003). Properties of 13C-substituted arginine in stable isotope labeling by amino acids in cell culture (SILAC). *J. Proteome Res.* **2**(2), 173–181.

Pratt, J. M., Petty, J., Riba-Garcia, I., Robertson, D. H., Gaskell, S. J., Oliver, S. G., and Beynon, R. J. (2002). Dynamics of protein turnover, a missing dimension in proteomics. *Mol. Cell Proteomics* **1**(8), 579–591.

Regnier, F. E., Riggs, L., Zhang, R., Xiong, L., Liu, P., Chakraborty, A., Seeley, E., Sioma, C., and Thompson, R. A. (2002). Comparative proteomics based on stable isotope labeling and affinity selection. *J. Mass Spectrom.* **37**(2), 133–145.

Shevchenko, A., Wilm, M., Vorm, O., and Mann, M. (1996). Mass spectrometric sequencing of proteins silver-stained polyacrylamide gels. *Anal. Chem.* **68**(5), 850–858.

Yates, J. R., 3rd (2000). Mass spectrometry: From genomics to proteomics. *Trends Genet.* **16**(1), 5–8.

Zhang, R., and Regnier, F. E. (2002). Minimizing resolution of isotopically coded peptides in comparative proteomics. *J. Proteome Res.* **1**(2), 139–147.

Zhu, H., Pan, S., Gu, S., Bradbury, E. M., and Chen, X. (2002). Amino acid residue specific stable isotope labeling for quantitative proteomics. *Rapid Commun Mass Spectrom.* **16**(22), 2115–2123.

Site-Specific, Stable Isotope Labeling of Cysteinyl Peptides in Complex Peptide Mixtures

Huilin Zhou, Rosemary Boyle, and Ruedi Aebersold

I. INTRODUCTION

Relative quantification of proteins from different samples by mass spectrometry (MS) is based on the stable isotope dilution approach (Gygi *et al.*, 1999; Han *et al.*, 2001; Smolka *et al.*, 2002; Zhou *et al.*, 2002). Proteins or peptides are labeled with chemically identical tags that differ in mass due to a stable isotope content. In a typical experiment, proteins (or peptides derived from proteolytic digestion of proteins) from one sample are labeled with an isotopically heavy mass tag, whereas the light isotope tag is used to label the sample to be compared. The isotopically labeled peptides are combined, purified or separated into fractions, and analyzed by mass spectrometry, which measures the mass and ion abundance of peptides. Because isotopically heavy and light forms of a peptide of the same amino acid sequence are chemically identical, they generate responses with identical sensitivity from the mass spectrometer and are readily distinguished based on their mass differences. Therefore the measured ion abundance ratio between heavy- and light-labeled peptides by the mass spectrometer is the actual abundance ratio of this peptide from two different samples. In this way, the relative abundance of peptides, and thus proteins, in two different samples can be determined accurately. Quantitative and comparative analysis of protein abundance from different samples has many applications, including large-scale protein expression profiling from different cell states to identify proteins that are unique in one cell condition, not the other, or comparative analysis

of protein complexes derived from different cell states to reveal their compositional differences. In this case, conventional protein purification techniques would be used to enrich the protein complexes of interest. Many cell biological processes are carried out by large, multisubunits protein complexes. It would be an essential step to identify their protein components and dynamic changes in protein compositions under a different cell context. Comparative protein analysis by mass spectrometry would be a powerful tool toward this goal.

This article describes a method for site-specific stable isotope labeling of cysteinyl peptides in complex peptide mixtures via a solid-phase capture and release process, and the concomitant isolation of the labeled peptides (Zhou *et al.*, 2002). The recovered, tagged peptides were analyzed by microcapillary liquid chromatography and tandem mass spectrometry (μLC-MS/MS) to determine their sequences and relative abundance.

II. MATERIALS AND INSTRUMENTATION

1. Amino propyl glass beads, 200–400 mesh, pore size 170 Å (Sigma, St. Louis, MO., G4518)
2. Organic solvents: anhydrous dimethylformamide (DMF, Aldrich, 22705-6) and dichloromethane (DCM, Aldrich, 220997-1L)
3. 1-Hydroxybenzotriazole (HOBt) (Nova Biochem, Laufelfingen, Switzerland, 01-62-0008)

4. Fmoc-protected amino acids: Fmoc-aminoethyl photolinker (Nova Biochem, 01-60-0042) and Fmoc-γ-amino butyric acid (Fmoc-GABA) (Nova Biochem, 04-12-1088)

5. Diisopropyl carbodiimide (DIC) (Aldrich, D12540-7)

6. Acetic anhydride (Aldrich, 24284-5)

7. Pyridine (Aldrich, 36057-0)

8. Piperidine (Aldrich, 104094)

9. D6-γ-aminobutyric acid (d6-GABA) (Isotec, Inc., 82-222-02-7)

10. Fmoc-N-hydroxysuccinimide (Fmoc-Osu) (Nova Biochem, 01-63-0001)

11. Diisopropyl ethyl amine (DIPEA) (Aldrich, D12580-6)

12. Iodoacetic anhydride (Aldrich, 28426-2)

13. Micro Bio-Spin columns (Bio-Rad Labs, Hercules, CA, 732-6204)

14. Blak-Ray long-wave UV lamp (100 W, VWR Scientific, Inc., 36595-020)

15. Tri(caboxyl ethyl)phosphine (TCEP) (Pierce, 20490)

16. Trypsin, sequencing grade (Promega, V5111)

Stable isotope-labeled peptides can be analyzed by all types of mass spectrometry instrumentation, including electrospray ionization (ESI) and matrix-assisted laser desorption ionization (MALDI)-based techniques. We typically use liquid chromatography (LC) and ESI-tandem mass spectrometry to analyze complex peptide mixtures. A tandem mass spectrometer allows identification of peptide sequences as well as their relative ion abundance of the isotopically related peptides in the same experiment. The Finnigan LCQ ion-trap instrument is used in combination with the Hewlett-Packard HP1100 series HPLC system.

III. PROCEDURES

A. Synthesis of Solid-Phase Isotope-Labeling Reagents

A schematic diagram of the chemical structure of the solid-phase reagent is shown in Fig. 1. Synthesis of solid-phase reagents is based on a method that has been published previously (Holmes and Jones, 1995). For synthesis of beads with a heavy isotope, Fmoc-d6-GABA was prepared from d6-GABA and Fmoc-OSu, as described in Section III,B because the Fmoc-protected, deuterated amino acid is not available commercially. The methods described involve standard peptide chemistry, making it possible to use other amino acids with isotopically heavy or light forms.

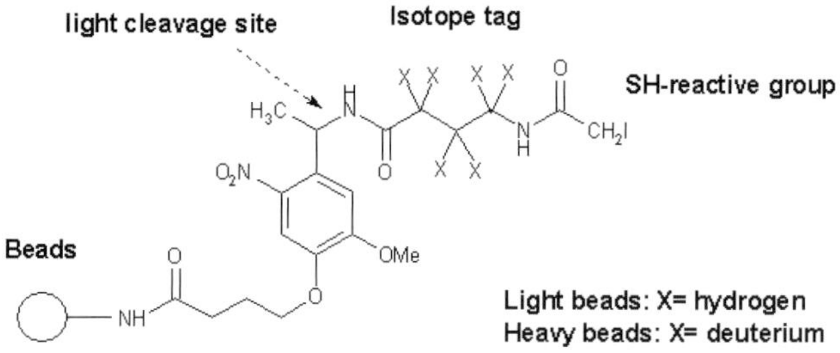

FIGURE 1 A schematic diagram of the solid-phase isotope-tagging reagent showing the chemical structure. The amino-propyl glass bead is first coated by a photocleavable linker molecule. Peripheral to the photocleavable linker, an amino acid, γ-aminobutyric acid, is used as an isotope-encoding mass tag that can be either nondeuterated (d0) or deuterated (d6). Following the isotope mass tag, an iodoacetyl group is used as a SH-reactive group to capture cysteinyl peptides. Following capture, photocleaving will lead to the recovery of cysteinyl peptides with the isotope tags attached to their cysteine residues.

Steps

1. Load 100 mg of aminopropyl glass beads in an empty Bio-Rad column or other column of suitable size. Wash beads once with 1 column volume of anhydrous DMF.

2. Form amino acid ester: Dissolve 120 μmol each of HoBt and Fmoc-aminoethyl photolinker in 0.8 ml of dry DMF completely. Add to this solution 120 μmol of DIC for 30 min. (Keep light-sensitive reagents from direct room light.)

3. Add the amino acid ester to the beads, mixing the beads by pipetting a few strokes. Incubate for 90 min.

4. Wash beads with 3 column volumes of DMF and 2 column volumes of dry dichloromethane. Always remove excess solvent between washes by applying a little pressure (squeeze a Pasteur pipettor bulb or apply house vacuum).

5. Block: Prepare a 1-ml mixture of 20% acetic anhydride, 30% pyridine, and 50% dichloromethane. Add this mixture to the beads for 30 min to block residual free amines on the beads.

6. Wash beads with 2 column volumes of dichloromethane and 3 column volumes of DMF. Remove excess DMF.

7. Deprotect: Prepare 3 ml of 20% (v/v) piperidine/DMF solution. Add 1 ml to the beads and incubate for 30 min. Collect all of the 1-ml flow through containing Fmoc released from the photolinker. Calculate the capacity of the beads by measuring the absorbance of the released Fmoc. Use 20% piperidine/DMF solution as a blank solution and measure absorbance at 290 nm (A_{290}) of the 1/100 dilution (by 20% piperdine/DMF) of the flow through. The A_{290} should be between

0.6 and 0.8. Calculate capacity according to the formula: $[A_{290} \times$ dilution factor $\times$ flow through volume (ml)]/[1.65 $\times$ weight of beads (mg)] = capacity (mmol/g).

8. Wash beads with 5 column volumes of dry DMF.

9. Repeat steps 2 to 8 with Fmoc-d0-GABA or its heavy form. The calculated capacity for GABA should be close to that for the photolinker.

10. Attach iodoacetyl group to the beads (Zhou *et al.*, 2001): Dissolve 120 μmol of iodoacetic anhydride in 0.8 ml dry DMF and add to the beads. Immediately add 132 μmol of diisopropyl ethyl amine (DIPEA) to the beads and mix well by pipetting a few strokes. Let it incubate for 90 min.

11. Wash the beads with 5 column volumes of DMF and excess methanol; dry the beads in a Speed-Vac (covered in foil). The beads can be stored in the dark at room temperature or in the refrigerator indefinitely.

B. Synthesis of Fmoc-Protected Amino Acid

The following procedure permits custom synthesis of Fmoc-protected amino acid for attachment to solid phase as described previously.

Steps

1. Dissolve 600 μmol d6-GABA in 3 ml 9% sodium carbonate in water in a vial under stirring.

2. Dissolve 900 μmol Fmoc-OSu in 3 ml DMF. Add to d6-GABA solution in one proportion. Continue to stir for 30 min at room temperature.

3. Divide the sample into six 1.5-ml Eppendorf tubes and dry out DMF under reduced pressure in a Speed-Vac.

4. Add 1 ml H_2O to dissolve the white powder as much as possible. Spin down and collect the supernatant. Repeat the water wash. Combine all of the supernatant, adding water until the final volume is 15 ml. Discard insoluble material.

5. Add 600 μl concentrated HCl to the supernatant very slowly. The solution should become cloudy immediately with foam due to carbon dioxide and precipitation of reagent. Check by pH paper that the final pH is approximately 2.

6. Extract Fmoc-d6-GABA: Add 4 ml or more ethyl acetate to the acidified aqueous solution, wait for phase separation, and collect the ethyl acetate phase that contains Fmoc-d6-GABA. Repeat this extraction procedure three times and combine the extracts.

7. Wash the ethyl acetate extract once with 3 ml 0.1% HCl in H_2O and once more with 3 ml of water. Dry the extract completely in a Speed-Vac. The dried

sample can be used directly with the assumption of >90% yield.

C. Preparation of Protein Digest

For labeling with SH-specific solid-phase reagents, it is advantageous to label peptides instead of proteins because proteins may possess tertiary structures that render some cysteine residues inaccessible to the solid-phase reagent. Protein digestion can be performed with any commercially available and suitable protease. Trypsin is used most frequently. For example, proteins in 100 μl 0.2 M Tris, pH, 8.0, can be digested by 1/50 (w/w) trypsin at 37°C overnight.

D. Capture and Release of Isotope-Labeled Peptides

Keep light-sensitive beads out of direct light as much as possible. Brief exposure to room light should not significantly affect the performance of the reagents.

1. Reduce protein digest with 5 mM TCEP for 30 min at room temperature. Because TCEP is quite acidic, it is essential that there is sufficient buffering capacity; 200 mM Tris, pH, 8.0, in the buffer should be sufficient to maintain the pH. TCEP is usually prepared as a 250 mM solution in water and kept at −20°C prior to use. We found that the TCEP stock solution is stable for many months at −20°C.

2. Weigh 5 mg each of isotopically light and heavy beads into tubes that are covered with foil to protect against light.

3. Add reduced protein digests to the beads (or the beads to the protein digest) and shake immediately for 15 min on a vortex mixer at a speed such that the beads should be suspended in the solution, not settled in the bottom of the tube. Efficient mixing of the beads with peptides is important for binding to occur.

4. After 15 min of binding, quench the labeling reaction with 2 μl β-mercaptoethanol for 1–2 min. The solid-phase isotope-labeling reagent should have excess capacity compared to the cysteinyl peptides. The addition of β-mercaptoethanol will block any remaining iodoacetyl group on the beads and prevent any possible side reactions to occur.

5. Combine the beads by loading onto a foil-wrapped Bio-Spin column, rinsing with water and methanol to transfer all beads. (Retain the flow through of each labeling reaction separately for analysis of noncysteine-containing peptides if so desired.) Wash with

    a. 2 × 1 ml 2 M NaCl

    b. 2 × 1 ml 0.1% TFA

   c. 2 × 1 ml 80% ACN/0.1% TFA
   d. 2 × 1 ml MeOH
   e. 2 × 1 ml 28% NH₄OH : MeOH (1 : 9 v/v)
   f. 2 × 1 ml MeOH
   g. 2 × 1 ml water

6. Seal bottom of Bio-Spin column with a cap. Suspend beads in 200 µl 20 mM Tris/1 mM EDTA, pH 8, and 4 µl β-mercaptoethanol with a magnetic stirrer on a stir plate.

7. Expose beads to UV light for 2 h and then collect the supernatant through the Bio-Spin column.

8. Wash the remaining beads with 5 × 100 µl 80% ACN/0.4% acetic acid, combining with the previous supernatant.

9. Reduce the sample volume in Speed-Vac to approximately 200 µl. Check that the pH is acidic.

E. Sample Cleanup and Mass Spectrometric Analysis

Although the labeled peptides appear to be highly pure, free from side reactions of the peptides themselves, we have observed side products other than peptides following photocleaving of the beads. These residual products are likely to be impurities generated during synthesis of the solid-phase reagent. Because these products interfere with MS analysis of peptide samples, it is necessary to remove them prior to MS analysis. Additionally, we have observed that these side products are not positively charged under acidic pH, whereas peptides are positively charged due to protonation of basic residues such as the N terminus, histidine, lysine, and arginine. We therefore devised a strategy using cation-exchange chromatography to remove adducts from peptides. A disposable mixed cation-exchange (MCX) cartridge can be used.

Steps

1. Load sample onto MCX column (30-mg beads).
2. Wash with 3 column volumes of 0.1% TFA in water.
3. Wash with 3 column volumes of 80% ACN/0.1% TFA in water.
4. Wash by 1 column volume of water to prevent salt formation.
5. Elute in 500 µl of elution solvent consisting of 1 volume of ammonia solution (28% NH₄OH stock) and 9 volume of methanol.
6. Dry out ammonia and methanol in a Speed-Vac and resuspend the sample in 10 µl water for MS analysis.

There are several advantages to this solid-phase approach for isotopic labeling of peptides. First, isolation of cysteine-containing peptides and stable isotope

incorporation are achieved in a single step. Therefore, the solid-phase method is rather simple. Second, the covalent attachment of peptides to a solid phase allows for the use of stringent wash conditions to remove noncovalently associated molecules. Third, this procedure is unaffected by the presence of proteolytic enzymes, such as trypsin, or strong denaturants or detergents, such as urea or SDS. There is no need for additional steps for their removal prior to peptide capture by the solid-phase beads and it is easy to remove them by washing. Fourth, the standard solid-phase peptide chemistry involved in the coupling process enables the use of a range of natural or unnatural amino acids in place of the d0/d6-GABA to function as the isotopic mass tag. This allows for synthesis of beads with a range of mass tags for analysis of multiple samples (i.e., more than two) in a single experiment if desired.

F. An Example of Protein Quantitation

We show an example of protein quantitation by this approach (see Table I). Three proteins—glyceraldehyde 3-phosphate dehydrogenase from rabbit, bovine lactoalbumin, and ovalbumin from chicken—were prepared in different amounts and labeled by the solid-phase isotope-tagging reagents. Following light cleavage, the recovered peptides were analyzed by mass spectrometry, and the isotopically labeled peptides were identified and quantified as described (Gygi *et al.*, 1999; Han *et al.*, 2001; Eng *et al.*, 1994). The agreement with expected values was generally within 20% and, for any given protein, consistent ratios were observed. Additional application of this method can be found elsewhere (Zhou *et al.*, 2002).

IV. NOTES

1. The amine capacity of the aminopropyl glass beads should be measured, despite the value quoted by the manufacturer. Other derivatized beads may be used in place of the glass beads if desired, provided that they have good swelling properties under aqueous condition. Clearly, the use of the newly synthesized reagents should be tested with standard cysteinyl peptides.

2. During synthesis of the Fmoc-protected amino acid, it is important to acidify the sample very slowly and to shake well. This should alleviate foaming due to the release of carbon dioxide following the change of pH. Also, one could use more ethyl acetate than pre-

TABLE I Quantitation of Protein Mixture by Solid-Phase Isotope Tagging and MS

Gene name	Cys-containing peptides found	Observed ratio (light/heavy)	Expected ratio (light/heavy)
G3P_rabit	VPTPNVSVVDLTC*R	4.6	4.0
	IVSNASC*TTNC*LAPLAK	4.3	
LCA_bovin	DDQNPHSSNIC*NISC*DK	1.8	2.0
	FLDDDLTDDIMC*VK	1.9	
	LDQWLC*EK	2.1	
	ALC*SEK	2.0	
	C*EVFR	1.9	
Oval_chick	YPILPEYLQC*VK	1.0	1.0
	LPGFGDSIEAQC*GTSVNVHSSLR	0.9	
	ADHPFLFC*IK	1.1	

^a Isotopically labeled cysteine residues are marked by asterisks.

scribed in order to achieve better phase separation during the extraction step.

3. For protein digestion by trypsin, proteins can be denatured by boiling for a few minutes if the protein concentration is not so high as to cause precipitation. In the current protocol, proteins are not reduced prior to digestion; however, it is possible that one could reduce proteins prior to digestion.

4. TCEP is quite acidic and the optimal pH for solid-phase capture of cysteinyl peptides is 8.0. Therefore, it is essential that there is sufficient buffering capacity, such as 200 mM Tris at pH 8.0. In this case, the pH of the solution would be not strongly affected by the addition of 5 mM TCEP.

5. It is necessary to quench the capturing reaction after 15 min by mercaptoethanol or other excess SH-containing reagent because histidine side chains or other nucleophilic functional groups could suffer potential side reactions with the iodoacetyl group on the solid-phase beads. The protocol for washing beads following the quenching reaction can be altered by individual investigators, as we found that solid-phase-captured peptides are stable to a variety of washing conditions.

6. The UV light can be filtered through a copper sulfate solution that passes light of 300 to 400 nm. Although we used a long-wave UV lamp, the cleaving reaction can be accelerated by using a more powerful mercury arc lamp according to Holmes and Jones (1995).

7. When small amounts of peptides are expected, it is particularly important to remove labeling contaminants.

8. The use of β-mercaptoethanol in the photocleaving buffer prevents methionine oxidation.

9. Although very stringent washing steps were used to remove nonspecifically associated molecules from the solid phase after capturing, they may not be entirely necessary for all applications. The readers are encouraged to test different washing conditions for their own applications.

10. Some of the materials used here were published previously (Zhou et al., 2002).

References

Eng, J., McCormack, A.L., and Yates, J.R., 3rd (1994). An approach to correlate tandem mass spectral data of peptides with amino acid sequences in a protein database. *J. Am. Soc. Mass Spectrom.* **5**, 976–989.

Gygi, S.P., *et al.* (1999). Quantitative analysis of complex protein mixtures using isotope-coded affinity tags. *Nature Biotechnol.* **17**, 994–999.

Han, D., Eng, J., Zhou, H., and Aebersold, R. (2001). Quantitative profiling of differentiation induced membrane associated proteins using isotope coded affinity tags and mass spectrometry. *Nature Biotechnol.* **19**, 946–951.

Holmes, C.P., and Jones, D.G. (1995). Reagents for combinatorial organic synthesis: Development of a new o-nitrobenzyl photolabile linker for solid phase synthesis. *J. Org. Chem.* **60**, 2318–2319.

Smolka, M., Zhou, H., and Aebersold, R. (2002). Quantitative protein profiling using two-dimensional gel electrophoresis, isotope-coded affinity tag labeling, and mass spectrometry. *Mol. Cell Proteomics* **1**, 19–29.

Zhou, H., Boyle, R., and Aebersold, R. (2002). Quantitative protein analysis by solid phase isotope tagging and mass spectrometry. In "Protein-Protein Interactions" (H. Fu, ed.), Humana Press, New Jersey.

Zhou, H., Ranish, J.A., Watts, J.D., and Aebersold, R. (2002). Quantitative proteome analysis by solid-phase isotope tagging and mass spectrometry. *Nature Biotechnol.* **5**, 512–515.

Zhou, H., Watts, J.D., and Aebersold, R. (2001). A systematic approach to the analysis of protein phosphorylation. *Nature Biotechnol.* **19**, 375–378.

56

Protein Hydrogen Exchange Measured by Electrospray Ionization Mass Spectrometry

Thomas Lee, Andrew N. Hoofnagle, Katheryn A. Resing, and Natalie G. Ahn

I. INTRODUCTION

Hydrogen exchange-mass spectrometry (HX-MS) is a technique that measures the rate of exchange between protons on macromolecules and isotopically labeled water. It is commonly applied to the exchange of protein backbone amide hydrogens with deuterium oxide, where each exchange event leads to a mass increase of 1 Da, which can be monitored by mass spectrometry. The measurement can be used to obtain information about aspects of solution structure, protein folding, and conformational mobility (Chowdhury *et al.*, 1990; Zhang and Smith, 1993; Johnson and Walsh, 1994; Resing and Ahn, 1998; Hoofnagle *et al.*, 2001) and can also be used to analyze protein solvent accessibility and ligand-binding sites (Neubert *et al.*, 1997; Mandell *et al.*, 1998; Andersen *et al.*, 2001). Both electrospray ionization (ESI) and matrix-assisted laser desorption ionization (MALDI) methods can be used for HX-MS (Chowdhury *et al.*, 1990; Mandell *et al.*, 1998), although each ionization method has its own advantages (discussed by Hoofnagle *et al.*, 2003). This article outlines a practical protocol for hydrogen exchange measurements on proteins using ESI-MS, which provides high protein sequence coverage and low back-exchange, thus improving the sensitivity of HX-MS for monitoring changes in solvent accessibility and conformational mobility.

II. MATERIALS AND INSTRUMENTATION

Pepsin (Cat. No. P6887), succinic acid (Cat. No. S5047), sodium citrate (Cat. No. S4641), and deuterium oxide (Cat. No. 15188-2) are from Sigma-Aldrich, trifluoroacetic acid (Cat. No. 28904) is from Pierce, and HPLC grade water (Cat. No. 26830-0025) and HPLC grade acetonitrile (Cat. No. 32573-0025) are from Fisher. Capillary HPLC columns (10–15 cm × 500 µm i.d.) are constructed from fused silica tubing (320 µm i.d. for outlet, 500 µm i.d. for inlet, Cat. No. TSP320450, TSP530700, respectively, Polymicro Technologies), assembled with epoxy glue (Cat. No. 302 part A, 302 part B, Epoxy Technology), and hand packed with reversed-phase POROS 20 R1 resin (Cat. No. 1-1028-02, Applied Biosystems Inc.) as described by Resing and Ahn (1997). Modifications to the previously described HPLC system added polyetheretherketone (PEEK) loading loops for sample injection (1 ml) and solvent precooling (2 ml) from Upchurch (Cat. No. 1820, 1821) and a PEEK HPLC injector apparatus from Rheodyne (Cat. No. 9010). Hamilton syringes are from SGE (50 µl, 250 µl, 1 ml, Cat. No. 004312, 006312, 008105, respectively), sample tubes and caps are from Bio-Rad (Cat. No. 223-9391, 223-9393), and screw cap tubes are from CLP (Cat. No. 3463). A stainless steel pan and an ice bucket serving, respectively, as ice bath and dry ice/ethanol/water bath are from Fisher Scientific (Cat. No. 13-361A, 11-676).

For data collection we obtain excellent results with a quadrupole time-of-flight (TOF) mass spectrometer (QStar Pulsar, Applied Biosystems Inc.) with standard electrospray source and AnalystQS software, interfaced with any HPLC capable of delivering a steady flow rate of 10–40 µl/min (e.g., Agilent Model 1100 capillary HPLC system or Eldex MicroPro HPLC-2g). Incubations are carried out in a circulating water bath (Cat. No. 13271-036, VWR). Useful for data reduction are software for nonlinear least squares (e.g., Datafit 7.1, Oakdale Engineering Inc.) and spreadsheet analyses (e.g., Microsoft Excel).

III. PROCEDURES

A. HX-MS Data Collection

The basic protocol involves timed incubations of protein in 90% (v/v) D_2O at neutral pH, which allows in-exchange of deuterons for protons within timescales ranging from seconds to hours. At the end of each incubation, the in-exchange reaction is quenched by rapidly lowering pH and temperature. Pepsin is added in amounts that enable rapid protein digestion (1–5 min), and the resulting peptides are separated by reversed-phase HPLC coupled to LC/MS. In order to minimize back-exchange of deuterium for water, all steps following the quench are carried out at 0°C. Applications of this protocol have been reported by Resing *et al.* (1998, 1999) and Hoofnagle *et al.* (2001).

Solutions

1. D_2O (99.9% atom D), stored at room temperature.

2. *Pepsin solution*: Dissolve lyophilized pepsin in HPLC grade water to 2 mg/ml and store in 50-µl aliquots at −80°C. Each day, thaw a new aliquot, dilute to 0.5 mg/ml in HPLC grade water, clarify by centrifugation for 20 min × 12,000 rpm, and store on ice.

3. *Citrate/succinate solution*: 25 mM sodium citrate + 25 mM sodium succinate is titrated with HCl to pH 2.40, filtered through a 0.22-µm membrane, and stored on ice.

4. *HPLC buffers*: Make buffer A [0.05% trifluoroacetic acid (TFA) in HPLC grade water] and buffer B [0.05% TFA in 100% (v/v) HPLC grade acetonitrile] fresh daily.

5. *Step gradient solutions*: Mix buffers A and B in appropriate ratios to yield solutions of 5, 7.5, 10, 12.5, 15, 17.5, 20, 22.5, 25, 30, 35, 40, and 50% (v/v) acetonitrile, 0.05% TFA. Prepare 2-ml aliquots of each in screw-cap vials and store on ice.

6. *HPLC grade water*: ~100 ml for washing the sample loop, stored on ice.

7. *Protein sample*: Ideally > 90% pure at ~10 µM stored long term at −80°C in 50-µl aliquots. Thaw new aliquots each day, clarify by centrifugation for 20 min × 12,000 rpm, and store on ice.

8. *Dry ice bath*: Add dry ice to ethanol : water (23 : 77) in an ice bucket to form a slurry at −10°C. This is used to quickly lower sample temperature after the in-exchange reaction.

Steps

1. A recommended configuration is shown in Fig. 1. Place a 2-ml PEEK loop between pump and injector to facilitate solvent cooling. Immerse this, the PEEK injector, a 1-ml sample loop, and the reversed-phase column at 0°C in an ice/water slurry. Run the HPLC pump isocratically in buffer A at 40 µl/min and equilibrate the reversed-phase column in this solution. Maintain protein sample, citrate/succinate, step gradient solutions, and HPLC water on ice and maintain D_2O at room temperature.

2. Equilibrate D_2O in a water bath at 10°C. *Note*: Performing the in-exchange reaction at 10°C reduces exchange rates to levels measurable in the dead time of the experiment (~5 s).

3. Aliquot 10 µl protein (~100 pmol) to a sample tube and equilibrate in the 10°C water bath for 30 s.

4. Initiate the in-exchange reaction by transferring 90 µl D_2O to protein. Incubate protein + D_2O at 10°C for varying times (e.g., 5–18,000 s). *Note*: In order to minimize sample heating in these and subsequent steps, use P200 pipettors attached to tips that have been prechilled by storing the pipettor + tip in a 15-ml tube on ice.

5. After incubation, remove the sample tube from the water bath to the −10°C dry ice bath and begin timing the postincubation period at t = 00:00 (min:sec). Incubate the sample in the −10°C bath briefly enough to cool rapidly to 0°C but not long enough to freeze the solution.

6. At t = 00:05, add 90 µl citrate/succinate solution to the sample tube and gently mix by tapping the tube against the walls of the dry ice bath.

7. At t = 00:20, remove the sample tube to ice. Immediately add 10 µl × 0.5 mg/ml pepsin to the sample and mix gently.

8. Between t = 00:20 and t = 00:40, load the 200-µl volume into the sample loop using a 250-µl Hamilton syringe, prechilled on ice and insulated by wrapping the syringe barrel with Parafilm to 0.5 cm thickness in order to minimize heat transfer from handling.

9. At t = 01:20, inject the digest onto the column, running isocratically in buffer A at 40 µl/min.

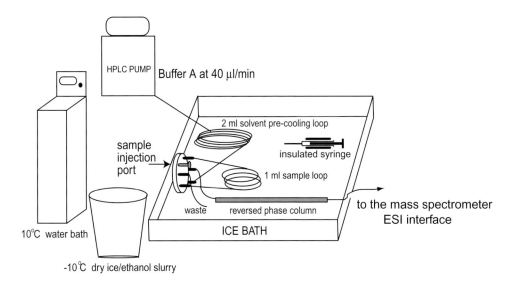

FIGURE 1 Experimental setup for hydrogen exchange-mass spectrometry. The reversed-phase column, injector, 2-ml solvent precooling loop, 1-ml injection loop, and Hamilton syringes wrapped with Parafilm are immersed in ice to minimize back-exchange during chromatography. The HPLC pump washes the column with buffer A (0.05% TFA in water) at a flow rate of 40 µl/min. Before buffer A reaches the injector apparatus, it is cooled to 0°C as it passes through the 2-ml solvent precooling loop. The protein sample, citrate/succinate solution, pepsin, and step gradients are stored separately on ice. The water bath is set at the desired temperature (e.g., 10°C) for in-exchange reactions, and a −10 C° bath (dry ice/23% ethanol/77% water) for quenching the in-exchange reaction is prepared.

10. At $t = 06:20$, switch the injector back to the load position and allow the column to desalt at 40 µl/min. While the column is washing, rinse the sample loop with ≥2 ml cold water and then load a step gradient into the sample loop by injecting 40 µl of buffer B followed successively with 17.5-µl aliquots of 50, 40, 35, 30, 25, 22.5, 20, 17.5, 15, 12.5, 10, 7.5, and 5% (v/v) acetonitrile, 0.05% TFA. Forming the HPLC gradient in the sample loop minimizes the dead time of the gradient.

11. Configure the mass spectrometer computer for data collection. At $t = 12:00$, attach the column to the electrospray source.

12. At $t = 12:20$, start data collection on the mass spectrometer. Immediately set the HPLC flow rate to 20 µl.

13. At $t = 13:20$, inject the gradient onto the column. Peptide elution is usually complete within 10 min.

B. Data Analysis

Peptic peptides are identified by LC/MS/MS sequencing, run under conditions outlined in steps 1–13, except without D₂O. An example of peptide identification by MS/MS and data reduction is presented by Resing *et al.* (1999). The following discussion of HX data analysis specifies the AnalystQS software available with the ABI QStar Pulsar for analysis of quadru-

pole TOF data (WIFF files). Other programs with equivalent features can be substituted.

1. Calculation of Weighted Average Mass

Steps

a. Open the data file with AnalystQS and open the "extract ion" dialog box.

b. Enter the mass/charge (m/z) for an ion and view the extract ion chromatogram (XIC). Select the scan range corresponding to the extract ion and view the mass spectrum. Smooth the spectrum three times and then adjust the threshold to view the m/z values of all isotopic peaks.

c. List the m/z and signal intensities of each isotopic peak for each ion. Save data in a text file, generating separate text files for each peptide at each time point.

d. Open text files with a spreadsheet program (e.g., Microsoft Excel). List the m/z and signal intensity in the first and second columns, respectively. The weighted average mass for each ion of each peptide ($M_{t,wa}$) is then calculated by Eq. (1):

$$M_{t,wa} = \left(\left[\sum (m/z \times \text{intensity}) / \sum (\text{intensity}) \right] \times z \right) - z \tag{1}$$

where $\sum (m/z \times \text{intensity})$ is the mass/charge of each isotopic form (column 1) multiplied by its corre-

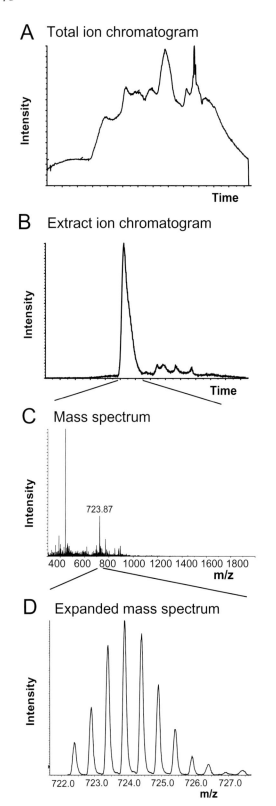

A Total ion chromatogram

Intensity

Time

B Extract ion chromatogram

Intensity

Time

C Mass spectrum

Intensity

723.87

400 600 800 1000 1200 1400 1600 1800

m/z

D Expanded mass spectrum

Intensity

722.0   723.0   724.0   725.0   726.0   727.0

m/z

FIGURE 2 Processing mass spectra for a peptide (EETARFQPGYRS, undeuterated MH_2^{+2} =721.2). (A) Total ion chromatogram showing the data set from a single time point. (B) Extract ion chromatogram identifying the subset of ions with m/z between 721.0 and 723.0 showing maximal peak intensity between 5 and

6 min. (C) Mass spectrum displaying all ions eluting between 5 and 6 min. (D) Expanded spectrum of ions between m/z 721.0 and 726.0, where isotopic peaks are separated by 0.5 Da for MH_2^{+2} ions. Each peak is labeled, and the m/z values and intensities are imported in text file format into a spreadsheet for calculating weighted average mass.

sponding signal intensity (column 2) and summed over all observed isotopic forms for the ion, and $\Sigma(\text{intensity})$ is the sum of intensities for all isotopic forms (Fig. 2).

2. Correction for Artifactual In-Exchange

Artifactual in-exchange occurs after quenching the in-exchange and is facilitated by the partial denaturation of protein in acidic solution. This correction requires measuring the weighted average mass at $t = 0$ (M_0), which is measured by reversing the order of adding citrate/succinate and D_2O in place of steps 4–6 in Section III,A.

The peptide mass corrected for artifactual in-exchange can be calculated by Eqs. (2) and (3):

$$M_{t,\text{corr(IE)}} = (M_{t,\text{wa}} - LM_{\infty,90})/(1-L) \qquad (2)$$

$$L = (M_0 - M_{\text{calc}})/(M_{\infty,90} - M_{\text{calc}}) \qquad (3)$$

where $M_{t,\text{corr(IE)}}$ is the corrected peptide mass at time t, $M_{t,\text{wa}}$ is the observed weighted average mass at time t, L is the fraction of artifactual in-exchange at $t = 0$, M_0 is the observed peptide mass at $t = 0$, M_{calc} is the theoretical average mass of the peptide, and $M_{\infty,90}$ is the theoretical mass of the peptide with complete deuterium exchange at backbone amide hydrogens (for incubation in 90% D_2O). ($M_{\infty,90} - M_{\text{calc}}$) is equal to the total number of nonproline amide residues in the peptide (total number of nonproline residues, minus one), multiplied by 0.9.

3. Back-Exchange Correction

Deuterons at backbone amides will slowly back-exchange to hydrogen during HPLC separation in water. We have used three different methods for estimating the fractional back-exchange for each peptide in the sample and have observed similar results with each.

Use the following steps for direct measurement of back-exchange.

Steps

1. Dilute the protein sample (500 pmol in 50 µl) with 50 µl H_2O. Add 90 µl citrate/succinate solution and digest with 10 µl pepsin as in steps 6 and 7 in Section III,A. Load peptides onto the column and desalt. Inject 30 µl of 40% acetonitrile, 0.05% TFA

onto the column, and collect all peptides into one tube.

2. Lyophilize the peptides, dissolve them in 30 μl of the buffer used to prepare the protein sample, add 270 μl D$_2$O, and heat to 90°C for 90 min in order to completely deuterate the peptides.

3. Cool the sample on ice and load 300 μl volume into the sample loop. Inject the peptide onto the column after 1 min of incubation in the sample loop as in steps 8 and 9 in Section III,A and proceed with steps 10–13. Measure the weighted average mass of each peptide (Section III,B).

4. Fractional back-exchange can be calculated for each ion using Eq. (4):

$$BE = (M_0 - M_{BE})/(M_{\infty,90} - M_{calc}) \qquad (4)$$

where BE is the fractional back-exchange of the peptide and M_{BE} is the observed mass of the peptide in the back-exchange experiment.

The following equation calculating the fractional back-exchange was derived empirically by Resing et al. (1999):

$$\begin{aligned} BE = L &\times (\%H_2O/\%D_2O) + [(\text{peptide elution} \\ &\text{time from HPLC in min} + 6\,\text{min}) \\ &\times 0.01/\text{min}] \end{aligned} \qquad (5)$$

where L is the fraction of artifactual in-exchange at $t = 0$ from Eq.(3) and $(\%H_2O/\%D_2O)$ is the ratio of H$_2$O to D$_2$O during proteolysis (Section III,A, step 7; e.g., 0.55/0.45 for initial incubation with 90% D$_2$O). The calculation is based on an observed back-exchange of approximately 1% for each minute the peptide is on the column prior to elution.

The exchange rates for amide backbone hydrogens have been measured empirically, accounting for inductive and steric blocking effects within different primary sequences (Bai et al., 1993). This study presents tables and equation that can be used to calculate predicted back-exchange rates at 0°C. The program HXPep (Zhang et al., 1997), written by Dr. Zhongqi Zhang (Amgen Inc., Thousand Oaks, CA), calculates exchange rates using the derivation of Bai et al. (1993) and can be used to calculate back-exchange rates by entering the peptide sequence, choosing "NH/D$_2$O" exchange, "oligo" peptide size, "low salt," "pH/pD read of 2.400," "0°C," and "C-terminal considered." For each backbone amide hydrogen,

$$BE_{amide} = k_{HXPep} \times (\text{elution time} + \text{wash time}) \qquad (6)$$

where BE_{amide} is the average back-exchange for each backbone amide hydrogen and k_{HXPep} is the rate of exchange calculated for the amide hydrogen by

HXPep. For the entire peptide, fractional back-exchange may be calculated as

$$BE = \sum (BE_{amide})/(M_{\infty,90} - M_{calc}) \qquad (7)$$

where $\Sigma(BE_{amide})$ equals the sum of BE_{amide} calculated from Eqn.(6) for every amide hydrogen in the peptide.

After estimating fractional back-exchange using any of the aforementional methods, the weighted average mass of each ion is corrected by Eq.(8):

$$M_{t,corr(BE)} = M_{calc} + (M_{t,corr(IE)} - M_{calc})/(1 - BE) \qquad (8)$$

where $M_{t,corr(BE)}$ is the peptide mass at time t after exchange in 90% D$_2$O, corrected for artifactual in-exchange and back-exchange.

C. Curve Fitting

Following correction for artifactual in-exchange and back-exchange, data may be fit using nonlinear least squares to a sum of exponentials (Resing et al., 1999; Hoofnagle et al., 2003). Further discussion and details on curve fitting and modeling are described by Resing and Ahn (1998) and Resing et al., (1999).

The time courses are modeled by a sum of exponentials in which each amide hydrogen exchanges with deuterium at a given rate. While in theory each amide backbone hydrogen is represented by a separate rate constant, in practice, exchange rates are averaged into fast (>1 min^{-1}), intermediate (0.1–1.0 min^{-1}) and slow rates (0.002–0.1 min^{-1}), and time courses can be fit with one, two, or three exponential terms:

$$Y = N - Ae^{-k_1 t} - Be^{-k_2 t} - Ce^{-k_3 t} \qquad (9)$$

where Y is the observed weighted average mass corrected for back-exchange and artifactual in-exchange [i.e., $M_{t,corr(BE)}$], A, B, and C correspond to the number of amides (multiplied by 0.9), respectively, exchanging with average rate constants k_1, k_2, and k_3, and N is the peptide mass after maximal in-exchange of deuterium ($=M_{calc} + A + B + C$).

Some backbone amides are nonexchanging over the experimental time period and cannot be fit to this equation. For instance, in the protein kinase ERK2, 44% of backbone amides show no exchange after the longest time point of 5 h (rate constant < 0.002 min^{-1}) (Hoofnagle et al., 2001). The number of amides in this nonexchanging group (NE) can be estimated by subtracting the number of exchanging amides ($A + B + C$) from the total number of backbone amides in the peptide, excluding proline residues.

Steps

1. Start the DataFit program with a new project using one independent variable.

2. Enter time (minutes) and $M_{t,corr(BE)}$ in columns 1 and 2, respectively.

3. Under the Solve menu, select "Define User Model" and then select "New."

4. Enter the equation describing the sum of up to three exponentials under Model Definition:

"Y = n − a∗Exp(−d∗x) − b∗Exp(−e∗x) − c∗Exp(−f∗x)"

5. Similarly, enter equations describing the sum of two and one exponentials.

6. For each equation, provide initial guesses of parameter values for nonlinear least squares. Test several different initial guesses, which should converge to the same fit.

7. Select Regression under the Solve menu to fit data to the three user-defined models.

8. Record parameter values and standard errors (Fig. 3), choosing the equation fit with lowest variance.

IV. PITFALLS

1. The buffer conditions for the exchange reaction should be optimized to minimize protein denaturation. Salt and buffer concentrations should be low as possible to minimize buffer/salt-catalyzed hydrogen exchange. We have found that 5 mM sodium phosphate, pH 7.0, 50 mM sodium chloride in the exchange reaction produces minimal back-exchange. The sample pH should be identical between experimental conditions in order to minimize effects on hydrogens that exchange via EX2 mechanism, where observed exchange rates vary with pH of the solution (Clarke and Itzhaki, 1998). Higher buffer concentrations may be used when varying solutes that influence pH, e.g., see Andersen et al. (1998).

2. The duration of pepsin digestion and the amount of protease should be optimized to generate the greatest number of peptides in the size range of 8–15 amino acids, which yields optimal resolution for exchange measurements. Peptides should be short enough to enable mass resolution, but long enough to bind the HPLC column and provide some sequence overlap.

3. The acetonitrile concentrations in the step gradient solutions should be varied to optimize peptide resolution from reversed-phase chromatography. Peptide elution should be spread throughout the gradient, while eluting all peptides within 10 min of gradient injection.

4. In order to avoid instrument bias and variations in data collection, randomize the time points and

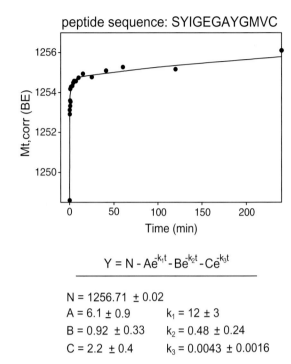

peptide sequence: SYIGEGAYGMVC

$$Y = N - Ae^{-k_1 t} - Be^{-k_2 t} - Ce^{-k_3 t}$$

N = 1256.71 ± 0.02

A = 6.1 ± 0.9 k_1 = 12 ± 3

B = 0.92 ± 0.33 k_2 = 0.48 ± 0.24

C = 2.2 ± 0.4 k_3 = 0.0043 ± 0.0016

FIGURE 3 Curve fitting and modeling. Weighted average masses corrected for artifactual in-exchange and back-exchange [$M_{t,corr(BE)}$] are plotted vs incubation time with D_2O. Time courses are fit to a sum of exponentials by nonlinear least squares. Nonlinear least-square parameters and their standard errors obtained include k_1, k_2, and k_3 (rate constants, min^{-1}), A, B, C (0.9 × number of amides exchanging with apparent rate constants k_1, k_2, and k_3, respectively), and $N = M_{calc} + 0.9(A + B + C)$.

samples during data collection. Perform instrument calibration and measure the $t = 0$ experiment each day.

5. The quality of curve fitting varies with the data quality and the number of time points. Typically, at least 20 data points between 5 s and 3 h are required for curve fitting with modest standard errors. In addition, because the distribution of time points affects curve fitting, best results are achieved by including more time points during the transition between fast and intermediate exchange rates (0–30 min for most peptides) than during the times close to maximal in-exchange.

6. In optimizing the acquisition method for the ABI QStar qTOF mass spectrometer, the orifice voltage, the pulse frequency, and the declustering potential (DP) were optimized to reduce fragmentation at the orifice and to obtain the best signal at minimal peptide sample. Parameters in our acquisition method are: Mass range is 300–2000 amu; accumulation time is 1 s; duration is 50 min; cycles are 3000; delay time is 0 s; cycle time is 1 s; pulser frequency is 7 kHz; pulse 1 duration is 13 μs; pause between mass range is 5 ms;

ionspray voltage is 5200 V; declustering potential is 40 V; and declustering potential 2 is 20 V.

References

Andersen, M. D., Shaffer, J., Jennings, P. A., and Adams, J. A. (2001). Structural characterization of protein kinase A as a function of nucleotide binding: Hydrogen-deuterium exchange studies using matrix-assisted laser desorption ionization-time of flight mass spectrometry detection. *J. Biol. Chem.* **276**, 14204–14211.

Bai, Y., Milne, J. S., Mayne, L., and Englander, S. W. (1993). Primary structure effects on peptide group hydrogen exchange. *Proteins*, **17**, 75–86.

Barksdale, A. D., and Rosenberg, A. (1982). Acquisition and interpretation of hydrogen exchange data from peptides, polymers, and proteins. *Methods Biochem. Anal.* **28**, 1–113.

Chowdhury, S. K., Katta, V., and Chait, B. T. (1990). Probing conformational changes in proteins by mass spectrometry. *J. Am. Chem. Soc.* **112**. 9012–9013.

Clarke, J., and Itzhaki, L. S. (1998). Hydrogen exchange and protein folding. *Curr. Opin. Struct. Biol.* **8**, 112–118.

Hoofnagle, A. N., Resing, K. A., and Ahn, N. G. (2003). Protein analysis by hydrogen exchange and mass spectrometry. *Annu. Rev. Biophys. Biomol. Struct.* **32**, 1–25.

Hoofnagle, A. N., Resing, K. A., Goldsmith, E. J., and Ahn, N. G. (2001). Changes in conformational mobility upon activation of extracellular regulated protein kinase-2 as detected by hydrogen exchange. *Proc. Natl. Acad. Sci. USA* **98**, 956–961.

Johnson, R. S., and Walsh, K. A. (1994). Mass spectrometric measurement of protein amide hydrogen-exchange rates of apomyoglobin and holo-myoglobin. *Protein Sci.* **3**, 2411–2418.

Mandell, J. G., Falick, A. M., and Komives, E. A. (1998). Identification of protein-protein interfaces by decreased amide proton solvent accessibility. *Proc. Natl. Acad. Sci. USA* **95**, 14705–14710.

Neubert, T. A., Walsh, K. A., Hurley, J. B., and Johnson, R. S. (1997). Monitoring calcium-induced conformational changes in recoverin by electrospray mass spectrometry. *Protein Sci.* **6**, 843–850.

Resing, K. A., and Ahn, N. G (1998). Deuterium exchange mass spectrometry as a probe of protein kinase activation: Analysis of wild-type and constitutively active mutants of MAP kinase kinase-1. *Biochemistry* **37**, 463–475.

Resing, K. A., Hoofnagle, A. N., and Ahn, N. G. (1999). Modeling deuterium exchange behavior of ERK2 using pepsin mapping to probe secondary structure. *J. Am. Soc. Mass. Spectrom.* **10**, 685–702.

Zhang, Z. Q., and Smith, D. L. (1993). Determination of amide hydrogen-exchange by mass spectrometry: A new tool for protein structure elucidation. *Protein Sci.* **2**, 522–531.

Nongel-Based Proteomics: Selective Reversed-Phase Chromatographic Isolation of Methionine-Containing Peptides from Complex Peptide Mixtures

Kris Gevaert and Joël Vandekerckhove

I. INTRODUCTION

Contemporary gel-free or nongel proteome analytical techniques are used increasingly as alternatives to highly resolving two-dimensional polyacrylamide gel electrophoresis (2D PAGE) for the analysis of complex protein mixtures. Some of the limitations of 2D PAGE (e.g., its bias to mainly visualize well-soluble, abundant proteins) can be overcome by these novel techniques, explaining their increasing success. A variety of procedures have been described and definitely the most interesting ones are those in which peptides, constituting the original proteins, are analyzed (e.g., Gygi *et al.*, 1999; Geng *et al.*, 2000; Washburn *et al.*, 2001; Zhou *et al.*, 2001). In these techniques, we can distinguish three general steps. First, the protein mixture is digested in solution using a highly specific protease. Second, the generated peptide mixture or an (affinity/covalently) isolated subset of it is separated by a liquid chromatographic step(s) (LC) and analyzed by automated tandem mass spectrometry (MS/MS). Third, the obtained peptide fragmentation spectra are fed into database-searching algorithms and linked to peptide and protein amino acid sequences stored in databases. We have described a technique for reversed-phase (RP) HPLC-based isolation of representative peptides of a proteome (peptides containing rare amino acids that are well distributed over a given proteome) (Gevaert *et al.*, 2002). Because this technique has a number of similarities to the previously described diagonal electrophoresis (Brown and Hartley, 1966) and diagonal chromatography (Cruickshank *et al.*, 1974) methods, we call this technique COmbined FRActional DIagonal Chromatography or COFRADIC. The core technology of COFRADIC is a chromatographic shift, evoked by a chemical or enzymatic reaction, of representative peptides between two identical separations steps. This article describes the different steps involved in the COFRADIC-based isolation of methionine-containing peptides as their sulfoxide derivatives using currently available RP-HPLC instrumentation.

II. MATERIALS AND INSTRUMENTATION

HPLC-graded water (product number 4218) and acetonitrile (product number 9017) used to prepare the HPLC solvents are of the best quality available and are from Malinckrodt Baker B.V., Deventer, The Netherlands. Peptide synthesizer graded trifluoroacetic acid (product number PTS6045) is from Rathburn Chemicals Ltd. (Walkerburn, Scotland, UK). The RP column used for separations is a 2.1 i.d. × 150-mm ZORBAX 300SB-C18 column (product number 883750-902) and is from Agilent Technologies (Waldbronn, Germany). The hydrogen peroxide stock solution [30% (w/w), product number 21,676-3] is from Aldrich Chemical Co., Inc. (Milwaukee, WI).

The HPLC system used to sort the methionine peptides is an Agilent 1100 series capillary LC system, equipped with an Agilent 1100 series capillary pump, thermostatted microwell-plate sampler, thermostatted column compartment, variable wavelength detector, and a thermostatted fraction collector. The system runs under the control of a 3D Agilent ChemStation.

III. PROCEDURES

Chromatographic Isolation of Methionine-Containing Peptides

This procedure describes the isolation of methionine-containing peptides using a porous silica-based C18 RP-HPLC column with an internal diameter of 2.1 mm and a length of 15 cm. When using other brands or types of reversed-phase columns, clearly the amount of material used for the isolation of methionine-peptides, the solvent flow rate, the collection scheme for the primary fractions, and the evoked retention time shifts should be checked and, if necessary, adapted.

Solutions

1. *HPLC solvents:* For the chromatographic sorting of methionine-peptides, HPLC solvent A consists of 0.1% (v/v) trifluoroacetic acid (TFA) in water and solvent B contains 70% (v/v) acetonitrile and 0.1% (v/v) TFA in water. At least 1 litre of each solvent is made so that during the isolation procedure, the same HPLC solvents can be used, thereby increasing the overall reproducibility of the COFRADIC procedure for the isolation of methionine-peptides. These HPLC solvents should be thoroughly degassed prior to use.

2. *Oxidation solution:* The oxidation solution must be freshly prepared prior to the chromatographic isolation of methionine-peptides (secondary runs, see later) and consists of 3% (v/v) of H_2O_2 and 1% (v/v) TFA in water.

Steps

1. Typically, load a total of 1 to 10 nmol of a digested protein mixture, such as a cell lysate, onto a 2.1 i.d. × 150-mm C18 RP column. Following injection, equilibrate the column for 10 min with solvent A. Then, create a binary solvent gradient to 100% of solvent B over a time span of 100 min at a constant flow rate of 80 µl/min. This HPLC separation is referred to as the primary run.

2. During this primary run, peptide fraction collection starts from 40 min on (corresponding to an acetonitrile concentration of 21%) and, in total, 48 primary fractions of 1 min (or 80 µl) are collected in the wells of a microtitreplate (see Fig. 1A).

3. Using this setup, pool primary fractions, which are separated by 12 min, (Table I) and dry to complete dryness in a centrifugal vacuum concentrator. Prior to the methionine-oxidation reaction and the secondary runs, redissolve these pooled peptide fractions in 70 µl of 1% (v/v) TFA in water.

4. The methionine oxidation reaction is done by transferring 14 µl of the freshly made oxidation solution to the vial containing the pooled peptides. The reaction proceeds for 30 min at a 30°C, after which the sample must be immediately injected onto the RP-HPLC column.

5. Separated the oxidized peptides on the same column as the one used for the primary separation and under identical chromatographic conditions. In the experimental setup described here, methionine-sulfoxide containing peptides typically elute in a time frame 3 to 8 min before the elution of the unmodified, methionine-free peptides, which elute in the same interval as during the primary run (see Fig. 1B). Thus, during these secondary runs, the shifted methionine-sulfoxide peptides can be time-based collected in a distinct number of subfractions in a microtitreplate (see Table I).

6. The sorted peptides are analysed most conveniently by a mass spectrometer. For LC-MS/MS experiments using an electrospray ionisation (ESI)-based mass spectrometer, multiple secondary fractions obtained during one secondary run may be pooled, dried to complete dryness, and redissolved in an appropriate solvent for further analysis. Alternatively, matrix-assisted laser desorption ionisation (MALDI)-based analysis combined with collision-induced dissociation can be used for peptide identification. In this case, peptide fractions are not pooled in order to keep the complexity of the samples as low as possible, but are dried and redissolved in a small volume of MALDI matrix solution and finally loaded on a MALDI target for further analysis.

7. Finally, convert the information in the obtained peptide fragmentation spectra to peak lists in which the masses of the observed peptide fragment ions and their relative or absolute intensities are saved. These peak lists are then used to identify the corresponding peptides using commercially available database search engines such as MASCOT (Perkins *et al.*, 1999).

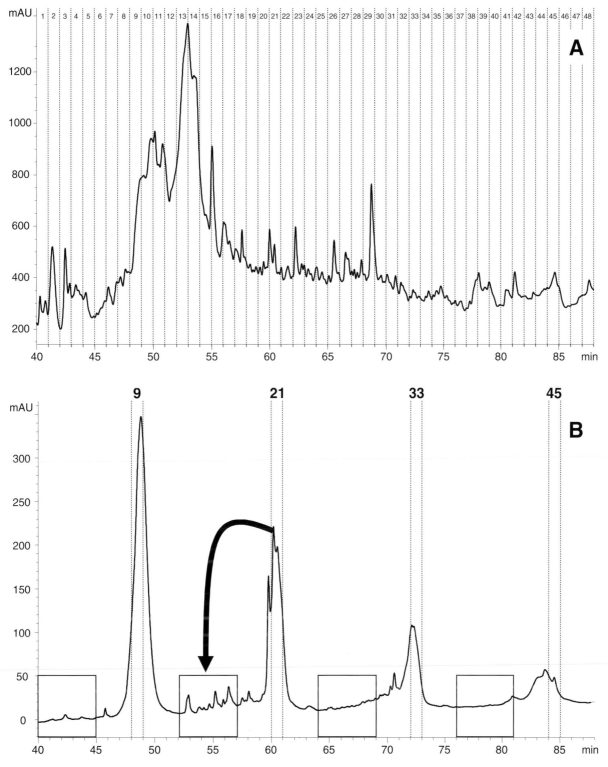

FIGURE 1 (A) UV absorption chromatogram (214 nm) of a reversed-phase HPLC separation of a tryptic digest of a human sputum sample. Peptides are fractionated in 48 distinct fractions. (B) Prior to the isolation of methionine-peptides, four fractions (in this case, fractions 9, 21, 33, and 45) are combined and oxidized to their methionine-sulfoxide counterparts. When rerun on the same column and under identical chromatographic conditions, the oxidized peptides shift out of this primary collection interval (an 8- to 3-min hydrophilic shift) and can be specifically collected (shaded boxes) for further LC-MS/MS analysis.

TABLE I   Primary and Secondary RP-HPLC Fraction Collection Scheme for the Isolation of Methionine-Peptides[a]

Secondary run	Primary fractions				Elution of primary fractions				Elution of methionine-sulfoxide peptides			
A	12	24	36	48	51–52	63–64	75–76	87–88	43–48	55–60	67–72	79–84
B	11	23	35	47	50–51	62–63	74–75	86–87	42–47	54–59	66–71	78–83
C	10	22	34	46	49–50	61–62	73–74	85–86	41–46	53–58	65–70	77–82
D	9	21	33	45	48–49	60–61	72–73	84–85	40–45	52–57	64–69	76–81
E	8	20	32	44	47–48	59–60	71–72	83–84	39–44	51–56	63–68	75–80
F	7	19	31	43	46–47	58–59	70–71	82–83	38–43	50–55	62–67	74–79
G	6	18	30	42	45–46	57–58	69–70	81–82	37–42	49–54	61–66	73–78
H	5	17	29	41	44–45	56–57	68–69	80–81	36–41	48–53	60–65	72–77
I	4	16	28	40	43–44	55–56	67–68	79–80	35–40	47–52	59–64	71–76
J	3	15	27	39	42–43	54–55	66–67	78–79	34–39	46–51	58–63	70–75
K	2	14	26	38	41–42	53–54	65–66	77–78	33–38	45–50	57–62	69–74
L	1	13	25	37	40–41	52–53	64–65	76–77	32–37	44–49	56–61	68–73

[a] The pooling scheme of primary RP-HPLC fractions and the time interval (given in minutes) during which they elute from the column during the primary run are indicated. The collection of methionine-sulfoxide peptides during 12 consecutive secondary runs (A–L) is indicated in the last column (given in minutes). These peptides generally elute in a time interval of 5 min starting 8 min prior to the elution of their respective primary fractions. Primary and secondary fraction collection time intervals corresponding to a particular primary fraction are indicated by an identical colour code.

IV. COMMENTS

Using mass spectrometric analysis, the presence of a methionine-sulfoxide side chain in sorted peptides can be recognized easily by the unstable behaviour of the methionine-sulfoxide side chain. In both MALDI- and ESI-based experiments, the facile loss of methane sulfenic acid (64 amu) has been observed (Lagerwerf et al., 1996). Furthermore, we have noticed that, especially in the postsource decay (PSD) mode (Spengler et al., 1992), peptides baring methionine-sulfoxide residues tend to fragment preferentially at this residue, thereby only a small amount of vibration energy is left for fragmentation of the peptide backbone. Generally, we noticed that MALDI-PSD spectra obtained from peptides containing a methionine-sulfoxide residue are very difficult to interpret, as only a small number of sequence-specific fragment ions (due to fragmentation of peptide bonds) are observed. However, we have not observed a similar behaviour for these type of peptides when analyzed by ESI-MS/MS, meaning that LC-MS/MS can be used routinely to analyze sorted methionine-sulfoxide peptides by fragmentation.

One of the interesting features of COFRADIC is the fact that during secondary runs, methionine-sulfoxide peptides elute in a five times larger time interval compared to the primary run (see Table I, last column). This means that the sorted peptides are delivered in a much less condensed manner for analysis by LC-MS/MS, thus implying that more peptides will be finally analyzed, resulting in a higher coverage of the analyzed proteome.

The aforementioned method for the isolation of methionine-peptides as their sulfoxide derivatives can be highly automated using existing HPLC equipment such as automated injectors and fraction collectors. It should be clear that, in theory, every type of peptide that contains an amino acid that can be specifically altered using chemicals and/or enzymes can be sorted using the COFRADIC technology. For example, similar strategies can be used for selective isolation of cysteine-containing peptides, phosphorylated peptides, and peptides spanning the amino-terminal part of the protein.

V. PITFALLS

1. Do not incubate methionine-peptides for more than 30 min in 0.5% H_2O_2, as this leads to over-oxidation, resulting in the formation of methionine-sulfones. The hydrophilic shift evoked by this sulfone is smaller as compared to the one evoked by the sulfoxide and may thus interfere with the isolation procedure.

2. Likewise, following the oxidation reaction, the altered peptides must be injected immediately onto the reversed-phase column. We have noticed that storing the peptides in the oxidation solvent, even just for a couple of hours, in the freezer leads to almost complete oxidation of methionine to the sulfone derivative, oxi-

dation of cysteine to cystic acid, and almost complete destruction of tryptophane residues.

3. It is important to keep the chromatographic conditions for the primary run and the secondary run as similar as possible. This means that the same HPLC solvents should be used and that the column compartment should be controlled thermostatically.

Acknowledgments

K.G. is a Postdoctoral Fellow of the Fund for Scientific Research - Flanders (Belgium) (F.W.O. - Vlaanderen). The project was further supported by the GBOU-research initiative of the Flanders Institute of Science and Technology (IWT).

References

Brown, J. R., and Hartley, B. S. (1966). Location of disulphide bridges by diagonal paper electrophoresis: The disulphide bridges of bovine chymotrypsinogen. *Biochem. J.* **101**, 214–228.

Cruickshank, W. H., Malchy, B. L., and Kaplan, H. (1974). Diagonal chromatography for the selective purification of tyrosyl peptides. *Can. J. Biochem.* **52**, 1013–1017.

Geng, M., Ji, J., and Regnier, F. E. (2000). Signature-peptide approach to detecting proteins in complex mixtures. *J. Chromatogr. A* **870**, 295–313.

Gevaert, K., Van Damme, J., Goethals, M., Thomas, G. R., Hoorelbeke, B., Demol, H., Martens, L., Puype, M., Staes A., and Vandekerckhove, J. (2002). Chromatographic isolation of methionine-containing peptides for gel-free proteome analysis: Identification of more than 800 *Escherichia coli* proteins. *Mol. Cell. Proteomics* **1**, 896–903.

Gygi, S. P., Rist, B., Gerber, S. A., Turecek, F., Gelb, M. H., and Aebersold, R. (1999). Quantitative analysis of complex protein mixtures using isotope-coded affinity tags. *Nature Biotechnol.* **17**, 994–999.

Lagerwerf, F. M., van de Weert, M., Heerma, W., and Haverkamp, J. (1996). Identification of oxidized methionine in peptides. *Rapid Commun. Mass Spectrom.* **10**, 1905–1910.

Perkins, D. N., Pappin, D. J., Creasy, D. M., and Cottrell, J. S. (1999). Probability-based protein identification by searching sequence databases using mass spectrometry data. *Electrophoresis* **20**, 3551–3567.

Spengler, B., Kirsch, D., Kaufmann, R., and Jaeger, E. (1992). Peptide sequencing by matrix-assisted laser-desorption mass spectrometry. *Rapid Commun. Mass Spectrom.* **6**, 105–108.

Washburn, M. P., Wolters, D., and Yates, J. R., III (2001). Large-scale analysis of the yeast proteome by multidimensional protein identification technology. *Nature Biotechnol.* **19**, 242–248.

Zhou, H., Watts, J. D., and Aebersold, R. (2001). A systematic approach to the analysis of protein phosphorylation. *Nature Biotechnol.* **19**, 375–378.

58

Mass Spectrometry in Noncovalent Protein Interactions and Protein Assemblies

Lynda J. Donald, Harry W. Duckworth, and Kenneth G. Standing

I. INTRODUCTION

Mass spectrometry (MS) has emerged as an important tool in the study of proteins and their interactions over the past several years (Daniel *et al.*, 2002; Robinson, 2002). Electrospray (ESI) provides a gentle ionization method that does not disrupt the weak bonds found in noncovalent complexes, and the adoption of nanospray technology (Wilm and Mann, 1996) has allowed the concentration of the buffer to be varied over a large range. Moreover, the use of time-of-flight (TOF) spectrometers for this purpose (first by Tang *et al.*, 1994) has removed previous limitations on the m/z values that can be examined, particularly after such instruments have been modified in order to study larger and larger assemblies (Chernushevich *et al.*, 1999; Rostom and Robinson, 1999; Rostom *et al.*, 2000; Van Berkel *et al.*, 2000).

Because the mass spectrum shows all the components present in a sample, information may be acquired on quantity, stoichiometry, and equilibria. Problems in the method arise from the limited selection of suitable buffers, interference from inorganic ions (especially Na^+ and K^+), and difficulties inherent in maintaining an uncooperative protein in solution under conditions required for the mass spectrometry.

II. MATERIALS

Ammonium acetate (99.999% Cat. No. 37,233-1), dithiothreitol (99%, Cat. No. 45,777-9), and 1,1,1-trichloroethane (Cat. No. 40,287-7) are from Aldrich. Ammonium bicarbonate is Fisher certified grade (Cat. No. A643). Dimethylchlorosilane is from Pierce (Cat. No. 83410). Methyl alcohol is from Mallinckrodt (Cat. No. 3041). Acetic acid (99+%) (Cat. No. A6283), formic acid (ACS reagent) (Cat. No. F4636), and substance P (Cat. No. S2136) are from Sigma. Dialysis membranes for waterbugs are from Spectra/Por (Cat. No. 132 128 for 50 K MWCO; 132 113 for 8 K MWCO; 132 703 for 12–14 K MWCO). For salt removal from ligands, we use Spectra/Por Dispodialyser MWCO 500, 1 ml capacity (Cat. No. 135 504). Centricon ultrafiltration units are from Amicon Bioseparations (Millipore) (YM10 is Cat. No. 4205; YM30 is Cat. No. 4208; YM50 is Cat. No. 4224). All plasticware is from Fisher Scientific: LDPE drop dispensing bottles (Fisher Cat. No. 03-006), 100-ml beakers (Nalgene Cat. No. 1201-0100), and syringe filters SFCA, 0.2 μm, 25 mm (Nalgene Cat. No. 190-2520). Microcentrifuge tubes, 0.6-ml flat top (Fisher Cat. No. 05-407-16) and 1.5 ml (Fisher Cat. No. 05-406-16), are used for waterbug construction. The 10-ml sterile syringes are latex free with Luer-lok from Becton-Dickinson (Cat. No. 309604). Nanospray capillaries are from Protana (types S and N), and New Objective PicoTip (type Econo12). GELoader tips are from Eppendorf (Cat. No. 22 35 165-6).

Some samples are prepared in a Sorvall RC-5B refrigerated centrifuge with a Sorvall SS34 rotor. Water from a reverse osmosis supply is run through a Barnstead NANOpure II system set at 17.0 MΩ/cm. All buffers are prepared with this grade of water.

457

III. INSTRUMENTATION

Mass spectrometry measures the ratio of the mass (m) to the charge (z) of an ion. Because many commercial instruments have a limited m/z range (usually <4000 for quadrupole mass filters), ions with large m/z values, such as those from buffered proteins, cannot be measured. Consequently, most intact protein and protein–ligand complexes require the "unlimited" m/z range found in TOF mass analyzers, and orthogonal ion injection (Verentchikov *et al.*, 1994) enabled the use of such instruments with ESI sources. Tang *et al.* (1994) pioneered the measurement of noncovalent complexes by ESI-TOF MS. More recently, we have added a heated metal capillary to provide another stage of pumping and desolvation and a new section containing a small RF quadrupole to provide collisional cooling of the ions (Krutchinsky *et al.*, 1998). A schematic diagram of the main elements of the instrument is shown in Fig. 1.

Some of the results reported here were obtained with a conventional electrospray ion source. In this device, solution is delivered to the sharpened tip of a stainless steel needle by a syringe pump. Alternatively, we use a glass nanospray source, which requires less sample, at lower concentration, and is tolerant of high concentrations of buffer. A potential difference of 3–3.5 kV (1 kV for nanospray) between the tip of the needle and the inlet of a heated metal capillary starts the ionization process. A counterflow of hot nitrogen gas helps with desolvation. We usually use nitrogen as this "curtain" gas, but sometimes use SF_6, especially for very large complexes, because it is considerably more efficient than nitrogen both in removing adducts and in dissociating complexes.

Expansion of the ion mixture into the next region produces a supersonic jet. Normally we apply a declustering voltage to the focusing electrode. Up to 300 V potential difference for nitrogen (or 400 V for SF_6, due to its superior insulating properties) can be maintained between the focusing electrode and the

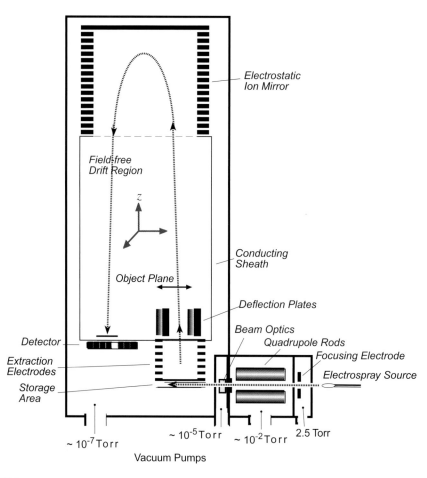

FIGURE 1   Schematic diagram of an electrospray ionization time-of-flight mass spectrometer. The ion path is indicated by the dotted line. A more detailed description can be found in Chernushevich *et al.* (1999).

flat aperture plate further downstream; the voltage is determined empirically for each protein or complex. A small opening in the plate connects this region to the second pumping stage where the ions oscillate in the two-dimensional potential well produced by the RF quadrupole and are cooled to near-thermal energies by collisions with the ambient gas (Krutchinsky et al., 1998). After passing through the quadrupole, the ion beam is focused and directed into the storage region of the modulator, entering perpendicular to the TOF axis at an energy <10 eV. After a group of ions has filled the storage region, it is accelerated along the TOF axis by applying a pulse to the extraction electrodes, which also starts the TOF measurement. The time of flight is recorded for each ion using TOFMA, an in-house software program that is also used for data analysis.

IV. PROCEDURES

A. Finding the Correct Conditions for the Protein

Our successes have come from the two commercially available volatile buffers, ammonium acetate and ammonium bicarbonate, but it is possible to have a small amount of nonvolatile material present in the sample (often unintentionally). Na^+ and K^+ are the worst contaminants, but buffer components such as glucose, glycerol, polyethylene glycol, and detergents are best avoided.

Using the preparative buffer of the protein as a guide, choose ammonium acetate for pH < 7 or ammonium bicarbonate for pH > 7. The pH can be adjusted with ammonium hydroxide or acetic acid. However, these solutions do change pH over time, and it is best to prepare the samples and then acquire spectra as soon as practicable. Choose a concentration that is close to the ionic strength of the preparative buffer and prepare the protein at a high concentration so that both it and the buffer can be diluted during the experiment. Dithiothreitol can be added to limit oxidation. If possible, the exact protein concentration should be determined from the known molar extinction coefficient and measurement of UV absorbance of an aliquot.

In our instrument, the declustering voltage, Vc, is an important component of the successful spectrum. At the "right" voltage, there is a perfect balance between maximum desolvation needed for high resolution and the collisionally induced dissociation of the intact complex. In practice, it is always wise to take spectra at several voltages because each will provide different kinds of information.

B. Desalting with "Waterbugs"

This procedure is modified from Orr et al. (1995). It is an efficient method of salt exchange and can also be used to assess the suitability of different buffer conditions.

1. Preparation of Apparatus

Steps

1. Prepare 2% sialysation solution by adding 10 ml of dimethyldichlorosilane to 500 ml of 1,1,1-trichloroethane. Mix carefully, working in a fume hood.
2. Cut the rim and the attached lid from microcentrifuge tubes using a razor or scalpel. It is very important that the cut edge is smooth or it will puncture the dialysis tubing.
3. Wash the lids with methanol and allow them to air dry.
4. Wearing gloves, transfer the lids to a clean beaker containing sialysation solution. Agitate carefully and decant off the solution.
5. Wash the lids in running tap water and then rinse several times with nanopure water.
6. Tip onto a fresh tissue, cover with a second tissue, and allow to air dry.
7. Minimize handling the waterbugs by using clean forceps for transfer. They should be stored in a covered container, such as a plastic petri dish, to avoid dust contamination.

2. Sample Introduction

Steps

1. Wash hands thoroughly. Do *not* use hand cream. It is much easier to do this without gloves; only step 6 *requires* that no gloves be worn. Hand cream components and the powder used in gloves can all cause adducts in the mass spectrum.
2. Prepare a clean, salt-free work surface.
3. For volumes of 10–50 μl, use the lids from 0.6-ml microcentrifuge tubes. For volumes up to 200 μl, use lids from 1.5-ml microcentrifuge tubes.
4. Prepare and chill the appropriate buffer, using plastic beakers kept specifically for this purpose.
5. Choose dialysis tubing with the appropriate MWCO and cut into squares of about 1.5-cm sides. Immerse them in nanopure water for about 10 min.
6. Rub the dialysis tubing squares between ungloved thumb and finger to separate the two layers. (This should be done in advance of setup if gloves are required for sample handling.)
7. Place a lid assembly on the clean work surface. Using an accurate pipettor, transfer a measured

volume of pure protein into the lid of the former micro-centrifuge tube.

8. Place a single layer of dialysis tubing over the sample. Hold it in place with the rim of the micro-centrifuge tube and carefully press down around the edge. Check the profile carefully in good light to make sure the dialysis membrane is intact. If not, remove the rim, discard the dialysis membrane tubing, and try again.

9. Place the waterbug onto the surface of the buffer, membrane side down. Gentle stirring is optional. Do this at 4°C, especially if the protein is unstable.

10. After at least 4 h, remove the waterbug using clean tongs. Working quickly, invert several times and then transfer to a fresh container of buffer. After a further 4–6 h (or overnight), repeat into a third plastic container of fresh buffer. If using 1-liter beakers, and less than 1 ml total sample, then three changes are sufficient. If using smaller containers in order to use less buffer, then increase the number of buffer changes to five or six.

3. Recovery of Sample

Steps

Set a pipettor at the expected recovery volume. Puncture the dialysis membrane with the tip of the pipettor and withdraw the sample. Put into a clean microcentrifuge tube. If the volume recovered is less than expected, there may be some residue on the underside of the dialysis membrane. Measure the concentration using an aliquot of the recovered sample. The method used will depend on the protein and may require most of the sample.

If the sample has precipitated, add acetic acid by puncturing the dialysis membrane with the end of the pipette tip and then inject the acid, adjust the volume, and remove the solubilized protein. Normally 5 µl of acetic acid is adequate for 100 µl of sample prepared in 5 mM buffer. Formic acid may also be used. This sample can be analysed for mass determination.

C. Desalting by Ultrafiltration

The following is based on the Centricon user guide (Millipore Corp., 2001) that comes with the centrifugal filter devices. Any refrigerated centrifuge may be used, provided it has a fixed-angle rotor, adaptors for 17 × 100-mm tubes, and is capable of running at 1000–7600 g. This requires much less buffer than the water-bug method, but it is more laborious and some proteins may come out of solution when they are concentrated too much.

Steps

1. Prepare and chill the appropriate buffer. The 99.999% ammonium acetate can be used directly. For ammonium bicarbonate buffer, make 100 ml at the desired concentration and then force through a 0.2-µm filter into a plastic container, such as a drop dispensing bottle, which has never been used for anything else.

2. Select a filtration device with a MWCO that will hold back the protein. Assemble the device and rinse the sample reservoir with buffer.

3. Add protein solution to the sample reservoir and fill it with buffer. Centrifuge for an appropriate length of time. Normally, it takes 15–30 min in a Centricon 30 or 50 to reduce the volume to about 100 µl, depending on the concentration of the sample. Buffer components such as glycerol and glucose will slow down the process, and it is best to try and omit these at some earlier stage in the purification. If the retentate vial is attached to the Centricon unit, it may not fit in a closed rotor. We normally use the vial and do not close the rotor.

4. When the volume has been decreased to about 100 µl, add more buffer, put the retentate vial back onto the unit, invert gently several times, and then recentrifuge.

5. Repeat the fill and spin at least six times. The residual salt concentration should be *at least* three orders of magnitude less than that of the protein.

D. Final Preparation of the Protein(s) or Complex

Calibrate the instrument by electrospray using a 10^{-5} M solution of substance P in 50% methanol, 2% acetic acid. Other calibrants can be substituted.

For electrospray, dilute the protein to 5–20 µM in 5 mM buffer. Load into the electrospray needle by back pressure on the syringe pump. Adjust the needle in the holder and turn on the spray voltage to about 3 kV. Adjust the declustering voltage until a clear spectrum appears. Adjust the position of the end of the needle and the voltages until a cone is formed at the end. Collect spectra at various voltages.

For nanospray, cut the capillary to ~3 cm. Using a GELloader tip on a 1- to 10-µl pipettor set at 2.5 µl, rinse the inside of the capillary with an aliquot of the buffer used for the protein preparation. Discard this and reload the tip with sample. Assemble the capillary in its holder and attach to the mass spectrometer. If using a Protana capillary, brush it against the front cone of the instrument to break off the end. (New Objective capillaries are open already.) Establish some back pressure to the assembly to force the sample to the orifice. Gradually turn up the nozzle voltage to 1 kV and adjust the declustering voltage until a clear spectrum emerges. Often the first burst of ions is badly contaminated by salts so it is best to be patient.

V. SPECIFIC EXAMPLES

A. A Small DNA-Binding Protein

The *Escherichia coli trp* apo-repressor protein TrpR is a dimer in the high salt buffer conditions used for nuclear magnetic resonance and X-ray crystallographic measurements (Zhang *et al.*, 1987; Zhao *et al.*, 1993). The pure enzyme is stable when lyophilized in elution buffer (10 mM NaPO$_4$ pH 7.6, 0.1 mM EDTA, 0.45 M NaCl). For our first experiments (Potier *et al.*, 1998), an aliquot of protein was dissolved in water and dialysed in "waterbugs" against 10-ml aliquots of 10 mM ammonium acetate. Just before mass analysis by electrospray ionization, samples were diluted to 10 µM protein in 5 mM ammonium acetate. As shown in Fig. 2A, the spectrum was complex, with two charge envelopes for the monomer, centred at the 7$^+$ and 10$^+$ ions, and good evidence for dimer, trimer, and perhaps higher order aggregates, all of which indicate a par-

tially unfolded or denatured protein. If specific DNA was added to the protein before dialysis, a new ion envelope appeared that corresponded to a complex of one dsDNA and two protein monomers (Potier *et al.*, 1998), in keeping with results from other analytical methods.

More recently, the development of nanospray ionization has expanded the limits of buffer concentration. If the same TrpR protein is prepared in the same way, but with 500 mM ammonium acetate, the spectrum is completely different (Fig. 2B), with one compact ion envelope of four ions, a spectrum typical of that of a properly folded protein. Deconvolution of these data gives a mass of 24452 Da, as expected for a dimer. Other DNA-binding proteins have shown the same sorts of buffer dependence (Donald *et al.*, 2001; Kapur *et al.*, 2002), where a very high concentration of buffer is required to maintain the protein in a native state, yet the DNA–protein complexes are relatively more stable at lower buffer concentrations.

B. A Large Allosterically Regulated Protein

Citrate synthase is a key enzyme for the entry of two-carbon units into the citric acid cycle and is essential in the biosynthesis of amino acids related to glutamate. Although many organisms have a homodimeric citrate synthase, that of *E. coli* crystallizes as a hexamer (Nguyen *et al.*, 2001). This protein cannot be frozen and is most stable when stored at high concentration in buffer (20 mM Tris–Cl, pH 7.8, 1 mM EDTA, 50 mM KCl) and is best desalted with a 50 K MWCO Centricon unit and kept at high concentration until just before mass analysis. When the protein was desalted using 20 mM ammonium bicarbonate and diluted so that the final buffer concentration was 5 mM, the electrospray spectrum was similar to that shown in Fig. 3A. There are three distinct charge envelopes, representing ions from a dimer (96 kDa), a tetramer (192 kDa), and a hexamer (288 kDa). Increasing the concentration of protein changed the spectrum to one with no tetramer and a preponderance of hexamer ions (Ayed *et al.*, 1998; Krutchinsky *et al.*, 2000). Addition of the allosteric inhibitor NADH (also desalted into ammonium bicarbonate) also changed the spectrum, providing data that were used to calculate equilibrium constants. These agreed well enough with results from sedimentation equilibrium measurements, considering that the latter depend on averages, whereas a mass spectrum shows all the components in the mixture. If the protein is desalted into 100 mM ammonium bicarbonate and analysed by nanospray, then the response of the enzyme to salts can be ascertained (Fig. 3) because NH$_4$$^+$, like K$^+$, is an activator of

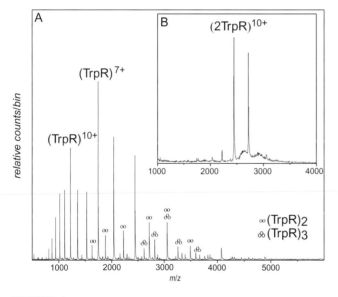

FIGURE 2 Spectra of *E. coli trp* apo-repressor protein. (A) A 10 µM protein prepared in 10 mM NH$_4$OAc by "waterbug" dialysis (8000 MWCO membrane) diluted to 5 mM just before electrospray. Nitrogen curtain gas with an 80-V declustering voltage. (B) A 2 µM protein transferred into 500 mM NH$_4$OAc using a Centricon 10, diluted to 250 mM, and analysed by nanospray in a Protana type S capillary. Nitrogen curtain gas with a 100-V declustering voltage.

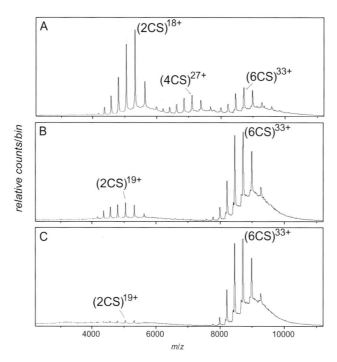

FIGURE 3 Spectra of *E. coli* citrate synthase. Samples were transferred into 100 mM NH$_4$HCO$_3$ using a Centricon 50, diluted to 10 μM protein, and analysed by nanospray ionization from a New Objective PicoTip. For all spectra, the curtain gas was SF$_6$ and the declustering voltage was 200 V. (A) 5 mM NH$_4$HCO$_3$ showing three charge envelopes for dimer (2CS), tetramer (4CS), and hexamer (6CS); (B) 20 mM NH$_4$HCO$_3$; and (C) 100 mM NH$_4$HCO$_3$.

citrate synthase (Faloona and Srere, 1969; Stokell *et al.*, 2003). In this case, changing only the concentration of buffer causes a change in the spectrum from one where ions from dimers are the most abundant, at 5 mM buffer, to one where ions from hexamers are the most abundant, at 100 mM buffer.

C. A Large Very Stable Enzyme with a Haem Residue at Its Active Site

Haem-containing bacterial catalase–peroxidases are large bifunctional enzymes that degrade hydrogen peroxide as part of their oxidative defense system. Haem is an essential component of the active site; mass spectrometry measurements have revealed heterogeneity of haem composition in the tetrameric *E. coli* enzyme HPI (Hiller *et al.*, 2000). However, the comparable enzyme from *Burkholderia pseudomallei*, KatG, is a dimer of 160 kDa (Carpena *et al.*, 2003; Donald *et al.*, 2003) and it seemed reasonable to expect two haems in the folded protein. This enzyme is so stable it can be kept frozen in 5 mM ammonium acetate and it remains a dimer even with very high declustering voltage in the mass spectrometer. However, the ion envelope changes shape as the declustering voltage is increased

(Fig. 4), with a concomitant increase of a singly charged ion at *m/z* 616. The deconvolutions show clearly a loss in mass from the dimer, in two steps of ≈ 600 Da. The high declustering voltage required to dissociate this complex gives an indication of its stability, and the single protein ion seen at lower voltage suggests that the stable form of the enzyme has one haem per subunit. Using a modified commercial instrument, Robinson's group has done similar experiments to fragment the 804-kDa GroEL chaperonin assembly (Rostom and Robinson, 1999) and to dissect intact ribosomes (Rostom *et al.*, 2000).

Acknowledgments

We thank James McNabb and Victor Spicer for their invaluable technical assistance. The pure proteins were supplied by Cheryl Arrowsmith, Peter Loewen, Gillian Sadler, David Stokell, and Jack Switala. This research was funded by grants from the Natural Sciences and Engineering Research Council of Canada (to HWD and KGS), the Canadian Institute of Health Research (to HWD, P. Hultin, and G. Brayer), and the U.S. National Institutes of Health (GM 59240 to KGS).

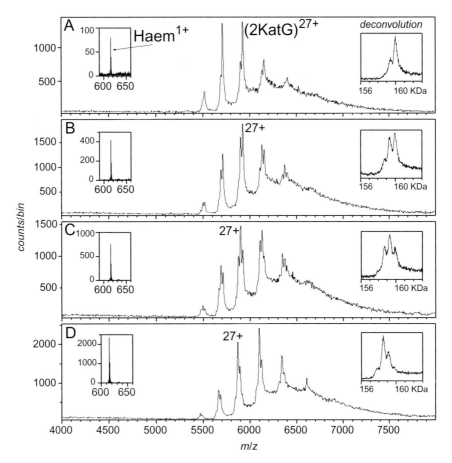

FIGURE 4 Spectra of the *B. pseudomallei* KatG protein prepared in 5 m*M* NH₄OAc using "traditional" dialysis and analysed by electrospray at 10 μ*M* protein. The curtain gas was SF₆, and the declustering voltage was varied in order to dissociate the haem from the complex. (A) 150 V, (B) 200 V, (C) 250 V, and (D) 300 V.

References

Ayed, A., Krutchinsky, A. N., Ens, W., Standing, K. G., and Duckworth, H. W. (1998). Quantitative evaluation of protein-protein and ligand-protein equilibria of a large allosteric enzyme by electrospray ionization time-of-flight mass spectrometry. *Rapid Commun. Mass Spectrom.* **12**, 339–344.

Carpena, X., Loprasert, S., Mongkolsuk, S., Switala, J., Loewen, P. C., and Fita, I. (2003). Catalase-peroxidase KatG of *Burkholderia pseudomallei* at 1.7 Å resolution. *J. Mol. Biol.* **327**, 475–489.

Chapman, J. R. (ed.) (2000). Protein and peptide analysis: New mass spectrometric applications. *In "Methods in Molecular Biology,"* Vol. 146. Human Press, Totawa, NJ.

Chernushevich, I. V., Ens, W., and Standing, K. G. (1999). Orthogonal-injection TOFMS for analysing biomolecules. *Anal. Chem.* **71**, 452A–461A.

Daniel, J. M., Friess, S., Rajagopalan, S., Wendt, S., and Zenobi, R. (2002). Quantitative determination of noncovalent binding interactions using soft ionization mass spectrometry. *Int. J. Mass Spectrom.* **216**, 1–27.

Donald, L. J., Hosfield, D. J., Cuvelier, S. L., Ens, W., Standing, K. G., and Duckworth, H. W. (2001). Mass spectrometric study of the

Escherichia coli repressor proteins, IclR and GclR, and their complexes with DNA. *Protein Sci.* **10**, 1370–1380.

Donald, L. J., Krokhin, O. V., Duckworth, H. W., Wiseman, B., Deemagarn, T., Singh, R., Switala, J., Carpena X., Fita, I., and Loewen, P. C. (2004). Characterization of the catalase-peroxidase KatG from *Burkholderia pseudomallei* by mass spectrometry. *J. Biol. Chem.*

Faloona, G. R., and Srere, P. A. (1969). *Escherichia coli* citrate synthase: Purification and the effect of potassium on some properties. *Biochemistry* **8**, 4497–4503.

Golemis, E. (ed.) (2002). "Protein-Protein Interactions: A Molecular Cloning Manual." Cold Spring Harbor Press, Cold Spring Harbor, NY.

Hiller, A., Peters, B., Pauls, R., Loboda, A., Zhang, H., Mauk, A. G., and Loewen, P. C. (2000). Modulation of the activities of catalase-peroxidase HPI of *Escherichia coli* by site-directed mutagenesis. *Biochemistry* **39**, 5868–5875.

Kapur, A., Beck, J. L., Brown, S. E., Dixon, N. E., and Sheil, M. M. (2002). Use of electrospray ionization mass spectrometry to study binding interactions between a replication terminator protein and DNA. *Protein Sci.* **11**, 147–157.

Krutchinsky, A. N., Ayed, A., Donald, L. J., Ens, W., Duckworth, H. W., and Standing, K. G. (2000). Studies of noncovalent complexes in an electrospray ionization/time-of-flight mass spectrometer. *Methods Mol. Biol.* **146**, 239–249.

Krutchinsky, A. N., Chernushevich, I. V., Spicer, V. L., Ens, W., and Standing, K. G. (1998). A collisional damping interface for an electrospray ionization time-of-flight mass spectrometer. *J. Am. Soc. Mass Spectrom.* **9**, 569–579.

Millipore Corporation (2001). Centricon centrifugal filter devices. User Guide.

Nguyen, N. T., Maurus, R., Stokell, D. J., Ayed, A., Duckworth, H. W., and Brayer, G. D. (2001). Comparative analysis of folding and substrate binding sites between regulated hexameric type II citrate synthases and unregulated dimeric type I enzymes. *Biochemistry* **40**, 13177–13187.

Orr, A., Ivanova, V. S., and Bonner, W. M. (1995). "Waterbug" dialysis. *Biotechniques* **19**, 204–206.

Potier, N., Donald, L. J., Chernushevich, I., Ayed, A., Ens, W., Arrowsmith, C. H., Standing, K. G., and Duckworth, H. W. (1998). Study of a noncovalent *trp* repressor: DNA operator complex by electrospray ionization time-of-flight mass spectrometry. *Protein Sci.* **7**, 1388–1395.

Robinson, C. V. (2002). Characterization of multiprotein complexes by mass spectrometry. *In* "Protein-Protein Interactions: A Molecular Cloning Manual" (E. Golemis, ed.), pp. 227–240. Cold Spring Harbor Press, Cold Spring Harbor, NY.

Rostom, A. A., Fucini, P., Benjamin, D. R., Juenemann, R., Nierhaus, K. H., Hartl, U., Dobson, C. M., and Robinson, C. V. (2000). Detection and selective dissociation of intact ribosomes in a mass spectrometer. *Proc. Natl. Acad. Sci. USA* **97**, 5185–5190.

Rostom, A. A., and Robinson, C. V. (1999). Detection of the intact GroEL chaperonin assembly by mass spectrometry. *J. Am. Chem. Soc.* **121**, 4718–4719.

Stokell, D. J., Donald, L. J., Maurus, R., Nguyen, N. T., Sadler, G., Choudhary, K., Hultin, P. G., Brayer, G. D., and Duckworth, H. W. (2004). Probing the roles of key residues in the unique allosteric NADH binding site of Type II citrate synthase of *E. coli*. *J. Biol. Chem.*

Tang, X. J., Brewer, C. F., Saha, S., Chernushevich, I., Ens, W., and Standing, K. G. (1994). Investigation of protein-protein noncovalent interactions in soybean agglutinin by electrospray ionization time-of-flight mass spectrometry. *Rapid Commun. Mass Spectrom.* **8**, 750–754.

VanBerkel, W. J. H., Vandenheuvel, R. H. H., Versluis, C., and Heck, A. J. R. (2000). Detection of intact megaDalton protein assemblies of vanillyl-alcohol oxidase by mass spectrometry. *Protein Sci.* **9**, 435–439.

Verentchikov, A. N., Ens, W., and Standing, K. G. (1994). Reflecting time-of-flight mass spectrometer with an electrospray ion source and orthogonal extraction. *Anal. Chem.* **66**, 126–133.

Wilm, M., and Mann, M. (1996). Analytical properties of the nanoelectrospray ion source. *Anal. Chem.* **68**, 1–8.

Zhang, R. G., Joachimiak, A., Lawson, C. L., Schevitz, R. W., Otwinowski, Z., and Sigler, P. B. (1987). The crystal structure of *trp* aporepressor at 1.8 Å shows how binding of tryptophan enhances DNA affinity. *Nature* **327**, 591–596.

Zhao, D., Arrowsmith, C. H., and Jardetzky, O. (1993). Refined solution structures of the *Escherichia coli trp* holo- and aporepressor. *J. Mol. Biol.* **229**, 735–746.

APPENDIX

Appendix

59

Bioinformatic Resources for *in Silico* Proteome Analysis

Manuela Pruess and Rolf Apweiler

I. INTRODUCTION

In the last years, the continual advancement in sequencing technology and proteome research has led to the massive rise of numbers of protein sequences from a wide range of species, representing a new challenge in the field of bioinformatics. Structure determination is also proceeding at an increasingly rapid rate, and all these data need to be stored in comprehensive specialised databases. These databases deal with nucleic acid sequences, protein sequences, and protein tertiary structures, respectively. Moreover, the sequences need to be assembled and analysed to represent a solid basis for further comparisons and investigations. Especially databases about model organisms such as *Mus musculus*, *Drosophila melanogaster*, *Caenorhabditis elegans*, *Arabidopsis thaliana*, or *Brachydanio rerio* (zebrafish) are of great interest.

To get as much information out of data as possible and to make use of the combination of different resources are two of the current challenges. An important instrument is the *in silico* analysis of proteomes. The term "proteome" is used to describe the protein equivalent of the genome. Most of the predicted protein sequences lack a documented functional characterisation. The task is to provide statistical and comparative analysis and structural and other information for these sequences as an essential step toward the integrated analysis of organisms at the gene, transcript, protein, and functional levels. Whole proteomes, as they are becoming more and more available, represent an important source for meaningful comparisons between species. To fully exploit the potential of this vast quantity of data, tools for *in silico* proteome analysis are necessary. There are genome

and proteome analysis sites available, which benefit from the information stored in different databases and make use of different protein analysis tools to provide computational analysis of whole proteomes. Other proteomic tools focus on similarity searches, structure analysis and prediction, detection of specific regions, alignments, data mining, two-dimensional PAGE analysis, or protein modelling, respectively.

This article is dedicated to the description of some important sources for proteome analysis such as sequence databases and analysis tools. It aims at enabling the researcher to take advantage of all the different data sources available for optimal evaluation of sequences and structures.

II. RESOURCES AND THEIR USE

Biomolecular databases are, in combination with database search tools and tools for the computational analysis of the data, invaluable resources for biological and medical research. Their importance is ever increasing, especially as much data are no longer published in conventional publications, but are only available from databases. (All databases mentioned in the following, as well as their URLs, are listed in Table I.)

A. Sequence Databases

Sequence databases are comprehensive sources of information on nucleotide and protein sequences.

1. Nucleotide Sequence Databases

Nucleotide sequence databases store data on nucleic acid sequences submitted by genome sequenc-

TABLE I   All Databases and Resources Important for the Analysis of Proteins and
Proteomes Mentioned in the Text

Resource	Database (short name)	URL
Nucleotide sequence databases	DDBJ	http://www.ddbj.nig.ac.jp
	EMBL Bank	http://www.ebi.ac.uk/embl/
	GenBank	http://www.ncbi.nlm.nih.gov
Protein sequence databases	CluSTr	http://www.ebi.ac.uk/clustr
	PIR	http://pir.georgetown.edu
	Swiss-Prot	http://www.ebi.ac.uk/swissprot/, http://www.expasy.org/
	TrEMBL	http://www.ebi.ac.uk/trembl/
Protein tertiary structure databases	CATH	http://www.biochem.ucl.ac.uk/bsm/cath
	CCDC	http://www.ccdc.cam.ac.uk/
	DSSP	http://www.sander.ebi.ac.uk/dssp/
	FSSP	http://www.ebi.ac.uk/dali/fssp/
	HSSP	http://www.sander.ebi.ac.uk/hssp/
	PDB	http://www.rcsb.org/pdb/, http://www.ebi.ac.uk/msd/
	SCOP	http://scop.mrc-lmb.cam.ac.uk/scop/
Genome analysis databases	COG	http://www.ncbi.nlm.nih.gov/COG/
	Ensembl	http://www.ensembl.org/
	KEGG	http://www.genome.ad.jp/kegg/
	RefSeq	http://www.ncbi.nlm.nih.gov/LocusLink/refseq.html
	WIT	http://wit.mcs.anl.gov/WIT2/
Proteome analysis databases and tools	ExPASy	http://www.expasy.org
	Proteome Analysis DB	http://www.ebi.ac.uk/proteome
	InterPro	http://www.ebi.ac.uk/interpro
	IPI	http://www.ebi.ac.uk/IPI
	SRS	http://srs.ebi.ac.uk/
	SWISS-2DPAGE	http://www.expasy.ch/ch2d/

ing projects and individual researchers. Nucleotide sequence data are collected, organised, and distributed by the International Nucleotide Sequence Database Collaboration (Stoesser *et al.*, 2001), which is a joint effort of the nucleotide sequence databases EMBL-Bank (maintained at the European Bioinformatics Institute, Hinxton, UK), the DNA Data Bank of Japan (maintained at the National Institute of Genetics, Mishima, Japan), and GenBank (maintained at the National Center for Biotechnology Information, Bethesda, MD). Nucleotide sequence databases are data repositories, accepting nucleic acid sequence data from the community and making it freely available. The databases strive for completeness, with the aim of recording and making available every publicly known nucleic acid sequence. EMBL-Bank, GenBank, and DDBJ automatically update each other every 24 h with new or updated sequences. Since their conception in the 1980s, nucleic acid sequence databases have experienced constant exponential growth. There is a tremendous increase of sequence data due to techno-

logical advances. At the time of writing, the DDBJ/EMBL/GenBank Nucleotide Sequence Database contains more than 79 billion nucleotides in more than 46 million individual entries and is doubling their size every year. Today, electronic bulk submissions from the major sequencing centers overshadow all other input and it is not uncommon to add in 1 day more than 30,000 new entries to the archives.

2. Protein Sequence Databases

Protein sequence databases store information on proteins. Here it has to be distinguished between universal databases covering proteins from all species and specialised data collections storing information about specific families or groups of proteins or about the proteins of a specific organism. Two categories of universal protein sequence databases can be discerned: simple archives of sequence data and annotated databases where additional information has been added to the sequence record. Especially the latter are of interest for the needs of proteome analysis.

PIR, the Protein Information Resource (Wu *et al.*, 2002), has been the first protein sequence database, which was established in 1984 by the National Biomedical Research Foundation (NBRF) as a successor of the original NBRF Protein Sequence Database. Since 1988 it has been maintained by PIR-International, a collaboration among Georgetown University/NBRF, the Munich Information Center for Protein Sequences (MIPS), and the Japan International Protein Information Database (JIPID). The PIR release 80.00 (Dec-2004) contains 283,416 entries. It presents sequences from a wide range of species.

The PIR stopped this database's activities in 2004. The existing staff that worked at Georgetown University/NBRF on the maintenance of PIR started during 2002 to collaborate with the groups in charge of Swiss-Prot and TrEMBL (see later) and is now contributing to the maintenance of these databases.

Swiss-Prot (Bairoch and Apweiler, 2000) is an annotated protein sequence database established in 1986 and maintained since 1988 collaboratively by the Swiss Institute of Bioinformatics (SIB) and the EMBL Outstation—The European Bioinformatics Institute (EBI). It strives to provide a high level of annotation, such as the description of the function of a protein, its domain structure, posttranslational modifications, and variants, with a minimal level of redundancy. More than 60 biomolecular databases, such as the EMBL/GenBank/DDBJ international nucleotide sequence database, the PDB tertiary structure database (Bhat *et al.*, 2001), or PubMed, are cross-referenced to provide a high level of integration. The links sum up to more than 4,000,000 individual links in total. Swiss-Prot contains data that originate from a wide variety of biological organisms. Release 45.5 (Jan-2005) contains a total of 167,089 annotated sequence entries from 8,811 different species; 11,777 of them are human sequences. The annotation of the human sequences is part of the HPI project, the human proteomics initiative (O'Donovan *et al.*, 2001), which aims at the annotation of all known human proteins, their mammalian orthologues, polymorphisms at the protein sequence level, and posttranslational modifications and at providing tight links to structural information and clustering and classification of all known vertebrate proteins.

TrEMBL (Translation of EMBL nucleotide sequence database) (Bairoch and Apweiler, 2000) is a computer-annotated supplement to Swiss-Prot, created in 1996 with the aim to make new sequences available as quickly as possible. It consists of entries in Swiss-Prot format derived from the translation of all coding sequences (CDSs) in the EMBL Nucleotide Sequence Database, except the CDSs already included in Swiss-Prot. TrEMBL release 28.5 (Jan-2005) contains 1,560,235 entries, which should be eventually incorporated into Swiss-Prot. Before the manual annotation step, automated annotation (Kretschmann *et al.*, 2001) is applied to TrEMBL entries to accelerate the annotation process.

SP_TR_NRDB (or abbreviated SPTR or SWALL) is a database created to overcome the problem of the lack of comprehensiveness of single sequence databases: it comprises both the weekly updated Swiss-Prot work release and the weekly updated TrEMBL work release. The CluSTr (Clusters of Swiss-Prot and TrEMBL proteins) database (Kriventseva *et al.*, 2001) is a specialised database, which offers an automatic classification of Swiss-Prot and TrEMBL proteins into groups of related proteins. The clustering is based on a hierarchical single-linkage clustering of all pairwise comparisons between protein sequences, yielding a hierarchical organisation of clusters.

3. Protein Structure Databases

The number of known protein structures is increasing very rapidly and these are available through PDB, the Protein Data Bank. PDB represents the single worldwide repository for the processing and distribution of three-dimensional biological macromolecular structure data, gained by techniques of X-ray crystal structure determination, nuclear magnetic resonancy, cryoelectron microscopy, and theoretical modeling.

In addition, there are also a number of derived databases, which enable comparative studies of three-dimensional structures as well as to gain insight on the relationships among sequence, secondary structure elements, and three-dimensional structure. DSSP (Dictionary of Secondary Structure in Proteins; Kabsch and Sander, 1983) contains the derived information on the secondary structure and solvent accessibility for the protein structures stored in PDB. HSSP (Homology-derived Secondary Structure of Proteins; Dodge *et al.*, 1998) is a database of alignments of the sequences of proteins with known structure with all their close homologues. FSSP (Families of Structurally Similar Proteins; Holm and Sander, 1996) is a database of structural alignments of proteins. It is based on an all-against-all comparison of the structures stored in PDB. Each database entry contains structural alignments of significantly similar proteins but excludes proteins with high sequence similarity, as these are usually structurally very similar.

The SCOP (Structural Classification of Proteins) database (Lo Conte *et al.*, 2002) has been created by manual inspection and abetted by a battery of automated methods. This resource aims to provide a detailed and comprehensive description of the structural and evolutionary relationships between all proteins whose structure is known. As such, it provides a broad survey of all known protein folds and detailed information about the close relatives of any particular protein.

Another database that attempts to classify protein structures in the PDB is the CATH database (Pearl *et al.*, 2001), a hierarchical domain classification of protein structures in the PDB. There are four major levels in this hierarchy: class, architecture, topology (fold family), and homologous superfamily.

B. Genome Analysis Databases

There are a number of databases that address some aspects of genome comparisons. The Kyoto Encyclopedia of Genes and Genomes (KEGG; Kanehisa *et al.*, 2002) is a knowledge base for systematic analysis of gene functions, linking genomic information with higher order functional information. The WIT project (Overbeek *et al.*, 2000) attempts to produce metabolic reconstructions for sequenced (or partially sequenced) genomes. A metabolic reconstruction is described as a model of the metabolism of the organism derived from sequence, biochemical, and phenotypic data. KEGG and WIT mainly address regulation and metabolic pathways, although the KEGG scheme is being extended to include a number of nonmetabolism-related functions. The Clusters of Orthologous Groups of proteins (COGs; Tatusov *et al.*, 2001) provide a phylogenetic classification of proteins encoded in complete genomes. COGs group together related proteins with similar but sometimes nonidentical functions.

Ensembl (Hubbard *et al.*, 2002) is a project to develop a software system that produces and maintains automatic annotation on metazoan genomes. By presenting up-to-date assembled sequence data and automatic analysis of the genomes, it represents an important resource for the interpretation of metazoan genome data.

Functional annotation of genomes is also provided in the framework of the NCBI Reference Sequence project (RefSeq; Pruitt and Maglott, 2001), which creates reference sequence sets for the naturally occurring molecules of the central dogma, from chromosomes to mRNAs to proteins. Toward this goal, intermediate larger genomic regions are also produced.

C. Proteome Analysis Databases and Tools

Tools and databases for proteome analysis provide algorithmic analysis and information about protein sequences and structures derived from comprehensive protein databases. It is sometimes difficult to distinguish between "database" and "tool," as databases providing precomputed data and search algorithms can both offer a high functionality towards protein analysis.

1. Proteome Analysis Databases

The classic proteomics databases are those of two dimensional gel electrophoresis gels and their analysis is like the SWISS-2DPAGE database (Hoogland *et al.*, 2000) and the human keratinocyte two-dimensional gel protein database from the universities of Aarhus and Ghent. Such databases allow interesting insights into the proteome of a given cell under given experimental conditions. Analysis of the protein spots on a gel by mass spectrometry has enabled the identification of many proteins on such gels, although the majority of proteins are usually still unidentified.

An additional way to gain insights into the variety of proteins produced by an organism is the bioinformatics exploitation of genome sequences. Genome sequencing is proceeding at an increasingly rapid rate, which leads to an equally rapid increase in predicted protein sequences entering the protein sequence databases. Most of these predicted protein sequences are without a documented functional role. The challenge is to bridge the gap until functional data have been gathered through experimental research by providing computational analysis as an essential step towards the integrated analysis of organisms at the gene, transcript, protein, and functional levels. Proteome analysis databases have been set up to provide comprehensive statistical and comparative analyses of the predicted proteomes of fully sequenced organisms.

The Proteome Analysis Database (Apweiler *et al.*, 2001b) has the general aim of integrating information from a variety of sources that will together facilitate the classification of proteins in complete proteome sets. The proteome sets are built from the Swiss-Prot and TrEMBL protein sequence databases that provide reliable, well-annotated data as the basis for the analysis. Proteome analysis data are available for all the completely sequenced organisms present in Swiss-Prot and TrEMBL, spanning archaea, bacteria, and eukaryotes. In the proteome analysis effort, the Inter-Pro (Apweiler *et al.*, 2001a) and CluSTr resources have been used. Links to structural information databases such as PDB, HSSP, and SCOP are provided for individual proteins from each of the proteomes. A functional classification using gene ontology (GO; Ashburner *et al.*, 2000) is also available. The Proteome Analysis Database provides a broad view of the proteome data classified according to signatures describing particular sequence motifs or sequence similarities and, at the same time, affords the option of examining various specific details, such as structure or functional classification. It currently (as of January 2005) contains statistical and analytical data for more than 210 complete proteomes.

The International Protein Index (IPI) provides a top-level guide to the main databases that describe the human and mouse proteome, namely Swiss-Prot, TrEMBL, RefSeq, and EnsEMBL. IPI maintains a database of cross-references between the primary data sources with the aim of providing a minimally redundant yet maximally complete set of human proteins (one sequence per transcript).

2. Proteome Analysis Tools

Traditional proteomics tools, such as those accessible from the EBI's SRS server or from the ExPASy server represent a variety of possibilities to analyse proteins. They help identify and characterise proteins, convert DNA sequences into amino acid sequences, and perform similarity searches, pattern and profile searches, posttranslational modification prediction, primary structure analysis, secondary and tertiary structure prediction, detection of transmembrane regions, alignments, and biological text analysis. Moreover, software is available for two-dimensional PAGE analysis, automated knowledge-based protein modelling, and structure display and analysis.

The analysis of whole proteomes represents an even bigger challenge. Large and comprehensive databases and knowledge bases are developed and used that provide large sets of precomputed data. To gather this comprehensive data, a vast amount of underlying information is necessary. The Proteome Analysis Database, mentioned earlier, uses annotated information about proteins from the Swiss-Prot/TrEMBL database and automated protein classifications from InterPro, CluSTr, HSSP, TMHMM (Sonnhammer *et al.*, 1998), and SignalP (Nielsen *et al.*, 1999). An example of data available for a single organism is shown in Fig. 1. Precomputation permits comparisons of whole proteomes of completely sequenced organisms with those of others. Users of the database can perform their own interactive proteome comparisons between any combinations of organisms in the database. Moreover, structural features of individual proteomes, such as protein length distribution, amino acid composition, and the affiliation of the different proteins to protein families, can be requested. Users are also able to run FASTA similarity searches (Fasta3) of their own sequence against a complete proteome or a set of complete proteomes. It is possible to download a proteome set or a list of InterPro matches for a given organism, to see the current status of all complete proteomes in Swiss-Prot and TrEMBL, and to download GO annotation for the human proteome.

Organism : *Drosophila melanogaster*			
Drosophila melanogaster, the fruit fly, is one of the most valuable organisms in biological research and has been studied for many years. The whole genome of *D. melanogaster* was sequenced in collaboration between Celera Genomics, the Berkeley Drosophila Genome Project (BDGP) and the European Drosophila Genome Project (EDGP).			
Statistical analysis of the *D. melanogaster* proteome is presented in the table below:			
InterPro [help]	CluSTr [help]	Structure [help]	InterPro comparative analysis [help]
General statistics (proteins with InterPro hits)	General statistics	Protein length distribution	*Proteome comparisons vs. A. thaliana, C. elegans, H. sapiens and S. cerevisiae*
Top 30 hits	List of singletons	*Primary* Amino acid composition	*Top 30 hits vs. A. thaliana, C. elegans, H. sapiens and S. cerevisiae*
Top 200 hits	30 biggest clusters		
15 most common families	Clusters without InterPro links	*Secondary* Proteins with HSSP links	*Top 200 hits vs. A. thaliana, C. elegans, H. sapiens and S. cerevisiae*
15 most common domains	Clusters without HSSP links		
15 most common repeats		*Tertiary* Proteins with PDB links	*15 most common families vs. A. thaliana, C. elegans, H. sapiens and S. cerevisiae*
Top 30 proteins with the highest occurrence of different InterPro hits			
			15 most common repeats vs. A. thaliana, C. elegans, H. sapiens and S. cerevisiae

FIGURE 1 Top-level statistical analysis page for the *Drosophila melanogaster* proteome, taken from the Proteome Analysis database.

III. COMMENTS

The last years have seen a tremendous increase in genomic data, particularly in the amount of data contributing to the understanding of the molecular basis of genetic diseases. Every week new discoveries are made that link genetic diseases to defects in specific genes. The Swiss-Prot and TrEMBL databases are in response to these developments gradually enhanced by the addition of a number of features that are specifically intended for researchers working on human genetic diseases and polymorphisms. The latter are of importance as they represent the basis for differences between individuals, which are of particular interest for research on topics such as disease susceptibility and differences in response to drug treatment. The comprehensive capturing of such data in sequence databases is fundamental for proteome analysis tools, such as the proteome analysis database, which combines information about different proteins of a given organism to comprehensive information about a complete proteome. A proteome can be regarded as a whole new unit, analysable according to different points of view (such as distribution of domains and protein families, and secondary and tertiary structure of proteins), and makes it comparable to other proteomes. To use proteomics data in areas such as healthcare and drug development, the characteristics of proteomes of entire species have to be understood before differentiation between individuals can be surveyed.

IV. PITFALLS

The number of proteome analysis tools and databases is increasing and most of them are providing high-quality resources of computational analysis and annotation. However, the user should not forget that *in silico* tools and databases have, like wet lab technologies, their very own pitfalls. The material in databases and the output of tools are trustworthy, but only to a certain point, just as the results of wet lab experiments. It is important to emphasise again that most sequence data today come from large-scale sequencing efforts and lack experimental functional characterisation. Many sequencing centres still annotate the predicted coding sequences just based on automated high-level sequence similarity searches against protein sequence databases. However, this methodology has several drawbacks.

i. Because many proteins are multifunctional, the assignment of a single function, which is still

common in genome projects, results in loss of information and outright errors.

ii. Because the best hit in pairwise sequence similarity searches is frequently a hypothetical protein or poorly annotated or has simply a different function, the propagation of wrong annotation is widespread.

iii. There is no coverage of position-specific annotation such as active sites.

iv. The annotation is not updated constantly and is thus outdated quickly.

It is also worth mentioning that a single sentence describing some predicted properties of an unknown protein should not be regarded as full annotation, but rather more as an attempt to characterise a protein. Full annotation means the combination of extracting experimentally verified information from the literature with sequence analysis to add as much reliable and up-to-date information as possible about properties such as the function(s) of the protein, domains and sites, catalytic activity, cofactors, regulation, induction, subcellular location, quaternary structure, diseases associated with deficiencies in the protein, tissue specificity of a protein, developmental stages in which the protein is expressed, pathways and processes in which the protein may be involved, and similarities to other proteins.

The Swiss-Prot knowledge base strives to provide extensive annotation as defined previously. However, annotation to such a high quality requires careful and detailed analysis of every sequence and of the scientific literature. This is the rate-limiting step in the production of Swiss-Prot and only a relatively small portion of the whole sequence data submitted to the sequence databases can be annotated to such a high standard and gets incorporated into Swiss-Prot.

Users of bioinformatics tools and databases should in no way feel discouraged in using these important resources, but they should bear in mind the potential pitfalls, check all data carefully, and not blindly rely on data. We also stress that all databases appreciate suggestions for improvements and the reporting of errors, which allow data custodians to improve their resources.

References

Apweiler, R., Attwood, T. K., Bairoch, A., Bateman, A., Birney, E., Biswas, M., Bucher, P., Cerutti L., Corpet F., Croning M. D., Durbin R., Falquet L., Fleischmann W., Gouzy J., Hermjakob, H., Hulo, N., Jonassen, I., Kahn, D., Kanapin, A., Karavidopoulou, Y., Lopez, R., Marx, B., Mulder, N. J., Oinn, T. M., Pagni, M., Servant, F., Sigrist, C. J., and Zdobnov, E. M. (2001a). The

InterPro database, an integrated documentation resource for protein families, domains and functional sites. *Nucleic Acids Res.* **29**, 37–40.

Apweiler, R., Biswas, M., Fleischmann, W., Kanapin, A., Karavidopoulou, Y., Kersey, P., Kriventseva, E. V., Mittard, V., Mulder, N., Phan, I., and Zdobnov, E. (2001b). Proteome Analysis Database: Online application of InterPro and CluSTr for the functional classification of proteins in whole genomes. *Nucleic Acids Res.* **29**, 44–48.

Ashburner, M., Ball, C. A., Blake, J. A., Botstein, D., Butler, H., Cherry, J. M., Davis, A. P., Dolinski, K., Dwight, S. S., Eppig, J. T., Harris, M. A., Hill, D. P., Issel-Tarver, L., Kasarskis, A., Lewis, S., Matese, J. C., Richardson, J. E., Ringwald, M., Rubin, G. M., and Sherlock, G. (2000). Gene ontology: Tool for the unification of biology. The Gene Ontology Consortium. *Nature Genet.* **25**, 25–29.

Bairoch, A., and Apweiler, R. (2000). The Swiss-Prot protein sequence database and its supplement TrEMBL in 2000. *Nucleic Acids Res.* **28**, 45–48.

Bhat, T. N., Bourne, P., Feng, Z., Gilliland, G., Jain, S., Ravichandran, V., Schneider, B., Schneider, K., Thanki, N., Weissig, H., Westbrook, J., and Berman, H. M. (2001). The PDB data uniformity project. *Nucleic Acids. Res.* **29**, 214–218.

Dodge, C., Schneider, R., and Sander, C. (1998). The HSSP database of protein structure-sequence alignments and family profiles. *Nucleic Acids Res.* **26**, 313–315.

Holm, L., and Sander, C. (1996). The FSSP database: Fold classification based on structure-structure alignment of proteins. *Nucleic Acids Res.* **24**, 206–209.

Hoogland, C., Sanchez, J.-C., Tonella, L., Binz, P.-A., Bairoch, A., Hochstrasser, D. F., and Appel, R. D. (2000). The 1999 SWISS-2DPAGE database update. *Nucleic Acids Res.* **28**, 286–288.

Hubbard, T., Barker, D., Birney, E., Cameron, G., Chen, Y., Clark, L., Cox, T., Cuff, J., Curwen, V., Down, T., Durbin, R., Eyras, E., Gilbert, J., Hammond, M., Huminiecki, L., Kasprzyk, A., Lehvaslaiho, H., Lijnzaad, P., Melsopp, C., Mongin, E., Pettett, R., Pocock, M., Potter, S., Rust, A., Schmidt, E., Searle, S., Slater, G., Smith, J., Spooner, W., Stabenau, A., Stalker, J., Stupka, E., Ureta-Vidal, A., Vastrik, I., and Clamp, M. (2002). The Ensembl genome database project. *Nucleic Acids Res.* **30**, 38–41.

Kabsch, W., and Sander, C. (1983). Dictionary of protein secondary structure: Pattern recognition of hydrogen-bonded and geometrical features. *Biopolymers* **22**, 2577–2637.

Kanehisa, M., Goto, S., Kawashima, S., and Nakaya, A. (2002). The KEGG databases at GenomeNet. *Nucleic Acids Res.* **30**, 42–46.

Kretschmann, E., Fleischmann, W., and Apweiler, R. (2001). Automatic rule generation for protein annotation with the C4.5 data mining algorith applied on Swiss-Prot. *Bioinformatics* **17**, 920–926.

Kriventseva, E. V., Fleischmann, W., Zdobnov, E. M., and Apweiler, R. (2001). CluSTr: A database of clusters of Swiss-Prot+TrEMBL proteins. *Nucleic Acids Res.* **29**, 33–36.

Lo Conte, L., Brenner, S. E., Hubbard, T. J. P., Chothia, C., and Murzin, A. G. (2002). SCOP database in 2002: Refinements accommodate structural genomics. *Nucleic Acids Res.* **30**, 264–267.

Nielsen, H., Brunak, S., and von Heijne, G. (1999). Machine learning approaches for the prediction of signal peptides and other protein sorting signals. *Protein Eng.* **12**, 3–9.

O'Donovan, C., Apweiler, R., and Bairoch, A. (2001). The human proteomics initiative (HPI). *Trends Biotechnol.* **19**, 178–181.

Overbeek, R., Larsen, N., Pusch, G. D., D'Souza, M., Selkov, E.Jr., Kyrpides, N., Fonstein, M., Maltsev, N., and Selkov, E. (2000). WIT: Integrated system for high-throughput genome sequence analysis and metabolic reconstruction. *Nucleic Acids Res.* **28**, 123–125.

Pearl, F. M. G., Martin, N., Bray, J. E., Buchan, D. W. A., Harrison, A. P., Lee, D., Reeves, G. A., Shepherd, A. J., Sillitoe, I., Todd, A. E., Thornton, J. M., and Orengo, C. A. (2001). A rapid classification protocol for the CATH Domain Database to support structural genomics. *Nucleic Acids Res.* **29**, 223–227.

Pruitt, K. D., and Maglott, D. R. (2001). RefSeq and LocusLink: NCBI gene-centered resources. *Nucleic Acids Res.* **29**, 137–140.

Sonnhammer, E. L. L., von Heijne, G., and Krogh, A. (1998). A hidden Markov model for predicting transmembrane helices in protein sequences. *In* "Proc. of Sixth Int. Conf. on Intelligent Systems for Molecular Biology" (J. Glasgow, T. Littlejohn, F. Major, R. Lathrop, D. Sankoff, and C. Sensen, eds.), pp. 175–182. AAAI Press, Menlo Park, CA.

Stoesser, G., Baker, W., van den Broek, S., Camon, E., Garcia-Pastor, M., Kanz, C., Kulikova, T., Lombard, V., Lopez, R., Parkinson, H., Redaschi, N., Sterk, P., Stoehr, P., and Tuli, M. A. (2001). The EMBL nucleotide sequence database. *Nucleic Acids Res.* **29**, 17–21.

Tatusov, R. L., Natale, D. A., Garkavtsev, I. V., Tatusova, T. A., Shankavaram, U. T., Rao, B. S., Kiryutin, B., Galperin, M. Y., Fedorova, N. D., and Koonin, E. V. (2001). The COG database: New developments in phylogenetic classification of proteins from complete genomes. *Nucleic Acids Res.* **29**, 22–28.

Wu, C. H., Huang, H., Arminski, L., Castro-Alvear, J., Chen, Y., Hu, Z.-Z., Ledley, R. S., Lewis, K. C., Mewes, H.-W., Orcutt, B. C., Suzek, B. E., Tsugita, A., Vinayaka, C. R., Yeh, L.-S., Zhang J., and Barker, W. C. (2002). The Protein Information Resource: An integrated public resource of functional annotation of proteins. *Nucleic Acids Res.* **30**, 35–37.

List of Suppliers

3M AG
Rüschlikon, Switzerland
http://www.3m.com/

AAPER Alcohol
PO Box 339
Shelbyville, KY 40066-0339
USA
TEL 800-456-1017
FAX 502-633-0685

AB
SE-75184 Uppsala
Sweden
TEL 46-018 6121900
FAX 46-018 6121920
http://www.amershambiosciences.com

Abbott Laboratories
100 Abbott Park Road
Abbott Park, IL 60064
http://abbott.com

Abgene
ABgene House
Blenheim Road, Epsom KT19 9AP
UK

Abgene, Inc.
565 Blossom Road
Rochester, NY 14610
USA
TEL 585-654-4800
FAX 585-654-4810

Academy Bioanalytical
1417 Kress Street, Houston
TX 77020
USA

Accurate Chemical and Scientific Corporation
300 Shames Drive
Westbury, NY 11590
http://www.accuratechemical.com

Active Motif
1914 Palomar Oaks Way
Suite 150
Carlsbad California 92008
USA
TEL 877 222 9543
TEL 760 431 1263
FAX 760 431 1351
http://www.activemotif.com

Acufirm
Landsteinerstr 2
Dreieich, D-63303
Germany
TEL +49 6103 9833
FAX +49 6103 983470
E-MAIL ernstkratz@acufirm.de

Affinity Bioreagents
14818 West 6th Avenue
Suite 10A
Golden, Colorado 80401
USA

Affymetrix, Inc.
3380 Central Expressway
Santa Clara, CA 95051
USA
TEL +1-888-362-2447
FAX 408-731-5441
http://www.affymetrix.com

African Reptile Park
30 Easson Rd.
Steenberg 7945
Capetown, South Africa

Agar Scientific Limited
66A Cambridge Road
Stansted Essex, CM24 8DA
England
TEL (+44) 1279 813519
FAX +44 279 815106
http://www.agarscientific.com

Agfa Medical
Agfa-Gevaert Headquarters
Septestraat, 27
2640 Morstel
(Belgium)
http://www.agfa.com/healthcare

Agilent Technologies
395 Page Mill Rd.
P.O. Box #10395
Palo Alto, CA 94303
USA
TEL 800-227-9770
TEL 302-993-5304
FAX 302-633-8901
http://www.agilent.com

AID, Autoimmun Diagnostika GmbH
Ebinger Strasse 4
D-72479 Strassberg
Germany
TEL +49-7434-9364-0
FAX +49-7434-9364-40
http://www.elispot.com

Air Products Europe
Hersham Place
Molesley Road
Walton-on-Thames, Surrey KT12 4RZ
UK
TEL (+44) 1932 249200
FAX (+44) 1932 24 9565

Alcon Laboratories Inc.
6201 S. Freeway
Fort Worth, Texas
USA
http://www.alconlabs.com/

Aldrich Chemical Co., Inc.
1001 West St. Paul Avenue
Milwaukee, WI 53233
USA
TEL 414-273-3850
FAX 414-273-4979

Alexis Biochemicals
6181 Cornerstone Court East
Suite 103
San Diego, CA 92121
TEL 858-658-0065
TEL 800-900-0065

Alfa Aesar Johnson Matthey
Postbox 11 07 65
D-76057 Karlsruhe
Germany
TEL (+49) 721 84007 280
http://www.alfa-chemcat.com

Alltech Associates, Inc.
2051 Waukegan Rd.
Deerfield, IL 60015

Alpha Laboratories, Division of Eurofins
    Scientific, Inc.
1365 Redwood Way
Petaluma—CA 94954
TEL 1-800-92-ALPHA
TEL 707-792-7300
FAX 707-792-7309
http://www.alphalabs.com/index.html

Alpha Laboratories Ltd.
40 Parham Drive
Eastleigh Hampshire SO50 4NU
UK
TEL +44 (0) 23 8048 3000
FAX +44 (0) 23 8064 3701
E-MAIL info@alphalabs.co.uk

Ambion Inc.
2130 Woodward St.
Austin, Texas 78744
USA
http://www.ambion.com

American Bioanalytical
15 Eire Dr.
Natick, MA 01760-1329
USA
TEL (800)443-0600
E-MAIL info@americanbio.com
http://www.americanbio.com/contact.asp

American Type Culture Collection (ATCC)
12301 Parklawn Drive
Rockville, Maryland 20852
USA
TEL (301)-881-2600
FAX (301)-770-1848

American Type Culture Collection (ATCC)
P.O. Box 1549
Manassas, VA 20108
TEL 1 703 365 2700
FAX 1 703 365 2750
http://www.atcc.org/

Amersham Biosciences
800 Centential Avenue
PO Box 1327
Picataway, NJ 08855-1327
USA
TEL 732-457-8000
FAX 732-457-0557
http://www.amershambiosciences.com

Amersham Biosciences
Munzinger Str. 9
D-79111 Freiburg
Germany
TEL +49 761 451 90
FAX +49 761 4519 159
http://www.amershambiosciences.com

Amersham Biosciences Corp.
800 Centennial Ave
Piscataway, NJ 08855
USA
TEL 1-732-457-8000
FAX 1-732-457-0557
TEL 1-800-526-3593
FAX 1-877-295-8102
http://www4.amershambiosciences.com

Amersham Biosciences Europe GmbH,
Branch office Benelux
Bergrand 230
4707 AT Rosendaal,
The Netherlands
http://www.amershambiosciences.com

Amersham Biosciences Europe GmbH
Zweigniederlassung Österreich
Wurzbachgasse 18, A-1152 Wien
AUSTRIA
TEL +43 1 57 606 16 13
FAX +43 1 57 606 16 14
E-MAIL cust.servde@amersham.com
http://www1.amershambiosciences.com

Amersham Biosciences UK Ltd.
International Trading Divison,
Pollards Wood
Nightingales Lane
Chalfont St Giles
Buckinghamshire, HP8 4SP
England
TEL +44 1494 49 8163
FAX +44 1494 49 8235
http://www.amershambiosciences.com

Amersham Pharmacia Biotech
AB
SE-75184 Uppsala
Sweden
TEL 46-018 6121900
FAX 46-018 6121920
http://www.amershambiosciences.com

Amersham Pharmacia Biotech
500 Morgan Blvd.
Baie d'Urfe, PQ H9X 3V1
Canada
TEL 1-800-463-5800

Amersham Pharmacia Biotech
P.O. Box 1327
Piscataway, NJ 08855-1327
TEL 1-800-526-3593
http://www.amershambiosciences.com

Amicon
Millipore
290 Concord Rd.
Billerica, MA 01821
USA
http://www.millipore.com

Amicon Bioseparations (Division of Millipore
    Corporation)
Amicon Ltd.
Upper Mill
Stonehouse, Glos. GL10 2BJ
UK
TEL +44 (0)1453825181
FAX +44 (0)1453826686

Amimed BioConcept
Innovationszentrum Nordwestschweiz
    Gewerbestrasse 14
Postfach 427 CH—4123 Allschwil 1
Switzerland
TEL +41 61 486 80 80
FAX +41 61 486 80 00
E-MAIL info@bioconcept.ch
http://www.bioconcept.com

Amresco/Electra Box Diagnostica AB
Box 2035
Tyreso 2, 13502
Sweden
TEL 46-8-712-3000
FAX 46-8-712-6509

AMRESCO Inc.
30175 Solon Industrial Parkway
Solon, Ohio 44139
TEL (Local): 440-349-1313
TEL 800-366-1313
FAX 440-349-1182
E-MAIL info@amresco-inc.com

Anachemia Chemicals Inc.
3 Lincoln Blvd.
Rouses Point, NY 12979
TEL 800-323-1414
http://www.anachemiachemicals.com

Analis
rue Dewez 14
5000 Namur
TEL ++32 (0)81 25 50 50
FAX ++32 (0)81 23 07 79

Angus Buffers and Biochemicals
2236 Liberty Drive
Niagara falls, New York
USA
TEL 1-800-648-6689

Antec International Ltd.
Chilton Industrial Estate
Sudbury, Suffolk C010 2KD
UK

Apex Microtechnology Corporation
5980 N. Shannon Road
Tucson, AZ 85741-5230
USA
TEL (+01) 520 690-8600
FAX (+01) 520 888-3329
E-MAIL sales@apexmicrotech.com
http://eportal.apexmicrotech.com

Apodan A/S
Lergravsvej 63
Copenhagen, Denmark

Applied Biosystems
850 Lincoln Centre Drive
Foster City, CA 94404
TEL 800-327-3002
TEL +1-650-638-5800
FAX +1-650-638-5884
http://www.appliedbiosystems.com/

Applied Precision, LLC
1040 12th Avenue Northwest
Issaquah, Washington 98027
USA
TEL 425.557.1000
FAX 425.557.1055

AR42J-B13 cells: Professor I Kojima
Department of Cell Biology
Gunma University
Maebashi, Gumma 3718512
Japan

Arcturus Engineering, Inc.
400 Logue Ave.
Mountain View, CA 94043
TEL 888-446-7911
http://www.arctur.com

Asahi Techno Glass
3-7-2 Nihonbashi-honcho
Chuo-ku, Tokyo 103-0023
Japan
TEL 81-3-5625-2751
http://www.atgc.co.jp

Assistent Glasware
Karl Hecht KG
Stettener Str. 22-24
97647 Sondheim, Germany
http://www.hecht-assistent.de

Atlanta Biologicals
1425 Oakbrook Drive
Norcross, GA 30093

Atto Bioscience
15010 Broschart Road
Rockville, MD 20850

Austral Biologicals
125 Ryan Industrial Ct.
Suite 207
San Ramon, CA 94583

Avanti Polar Lipids, Inc.
700 Inductrial Park Drive
Alabaster, AL 35007
TEL 1-800-227-0651
TEL 205-663-2494
FAX 1-800-229-1004
FAX 205-663-0756
http://www.avantilipids.com

Aventis Behring
PO Box 1230.
35002 Marburg
Germany

Avestin Europe GmbH
Weinheimer Str. 64b
D68309 Mannheim
Germany
Axis-Shield PoC AS
Marstrandgata 6
PO Box 6863
Rodeløkka, 0504 Oslo
Norway
TEL +47 22 04 20 00
FAX +47 22 04 20 10
http://www.axis-shield-poc.com

Axon Instruments, Inc.
3280 Whipple Road
Union City, CA 94587
USA
TEL +1 510-675-6200
FAX +1 510-675-6300
http://www.axon.com/

Bachem AG
Hauptstrasse 144
4416 Bubendorf, 19406
Switzerland
TEL +41 61 935 2323
FAX +41 61 935 2325
http://www.bachem.com/

Bachem Bioscience Inc.
3700 Horizon
Drive King of Prussia, PA
TEL +1 800 634 3183
FAX +1 610 239 0800

Bachem, California
3132 Kashiwa Street
Torrance, CA 90505
USA

Bacto Laboratories Pty Lt
Falcon Labware
PO Box 295
Liverpool, NSW 2170
Australia
TEL +61 (0)2 9602 5499
FAX +61 (0)2 9601 8293
E-MAIL info@bacto.com.au
http://www.bacto.com.au

Baker Company
PO Box Drawer E
Sanford, ME 04073
USA
TEL 207-324-8773
FAX 207-324-3869
E-MAIL bakerco@bakerco.com
http://www.bakerco.com

BAL-TEC
Föhrenweg 16
FL-9496 Balzers
Principality of Liechtenstein
TEL +423-388-1212
FAX +423-388-1260
http://www.bal-tec.com/

Bal-TEC AG
Postfach 75
FL-9496 Balzers
Fürstentum Liechtenstein
TEL +41 75 388 5611
FAX +41 75 388 5660

Barnstead International
2555 Kerper Blvd.
PO Box 797
Dubuque, IA 52001-1478
TEL 800-446-6060
FAX 563-589-0516

Baxter Scientific Products
1750 Stone Ridge Dr.
Stone Mountain, GA 30083
USA
TEL 404-270-9645
FAX 800-964-5227

B. Braun Biotech International
Schwarzenberger Weg 73-79
D-34212 Melsungen
Germany
http://www.bbraunbiotech.com

B. Braun Medical AG
Seesatz CH-6203
Sempach-Station
Switzerland
TEL +41 848 83 00 44
FAX +41 800 83 00 43
E-MAIL info.bbmch@bbraun.com
http://www.bbraun.com

BCB Ltd.
Moorland Rd
Cardiff, CF24 2YL
UK
TEL (029) 2046 4464
FAX (029) 2048 1100
http://www.bcb.ltd.uk

BD Biosciences
1-2900 Argentia Road
Mississauga, ON L5N 7X9
Canada
TEL 1-888-259-0187

BD Biosciences
133 Venture Ct.
Lexington, KY 40511-2624

BD Biosciences
2350 Qume Drive
San Jose, CA 95131-1807

BD Biosciences
Two Oak Park Drive
Bedford, MA 01730
USA (Becton, Dickinson and Company)
Tullastrasse 8-12
69126 Heidelberg
Germany
http://www.bdbiosciences.com

BD Biosciences (and Pharmingen)
European Office
Customer Service
Erembodegem—Dorp 86, 9320 Erembodegem
Belgium
TEL +32 53720211
FAX +32 53720450
E-MAIL bdb@europe.bd.com
http://www.bdbiosciences.com

BD Biosciences/Clontech
1020 East Meadow Circle
Palo Alto, CA 94303-4230
TEL (+44) 01865 781688
FAX (+44) 01867 781 627

BD Biosciences Discovery Labware
Two Oak Park Drive
Bedford, MA 01730
USA

BD Biosciences GmbH
Postfach 10 16 29
D-69008 Heidelberg
Germany

BD Biosciences—Immunocytometry Systems
2350 Qume Drive
San Jose, California, 95131-1807
TEL 1 877.232.8995
FAX 1 408.954.2347
E-MAIL facservice@bdis.com
http://www.bd.com

BD Biosciences Pharmingen
10975 Torreyana Road
San Diego, CA 92121
TEL 858-812-8800
FAX 858-812-8888
Toll Free: 877-232-8995
E-MAIL info@pharmingen.com
http://www.bdbiosciences.com.

BD Biosciences—Transduction Laboratories
133 Venture Ct.
Lexington, KY 40511-2624
TEL 859-259-1550
FAX 859-259-1413
TEL 0-227-4063
E-MAIL tlbinfo@translab.com
http://www.bdbiosciences.com

BD Clontech
1020 East Meadow Curcle
Palo Alto, CA 94303

BDH
Merck House
Poole, Dorset, BH15 1TD
England
http://www.bdh.com

BDH, Inc.
350 Evans Avenue
Toronto, Ontario, M8Z 1K5
Canada
TEL 416-255-8521
http://www.bdhinc.com

BDH Chemicals
Broomroad
Poole, Dorset BH12 4NN
England
TEL +44 202 666856
FAX +44 202 660444

BDH Laboratories
Roche Diagnostics GmbH
Roche Applied Science
Sandhofer Straße 116
D-68305 Mannheim
http://www.roche-applied-science.com

BD Pharmingen
10975 Torreyana Road
San Diego, CA, 92121

BD Plastikpak
1 Becton Drive
Franklin Lakes, NJ 07417
USA
http://www.bd.com

Beckman Coulter AB
Archimedesvägen 7
Box 11156
S-16111 Bromma
TEL 46 8 564 859 00
FAX 46 8 564 859 01

Beckman-Coulter
6755 Mississauga Road
Suite 600
Mississauga
Ontario, Canada L5N 7Y2
TEL 800-387-6799
http://www.beckmancoulter.com

Beckman Coulter (U.K.) Limited
Oakley Court
Kingsmead Business Park
London Road
High Wycombe, Buckinghamshire HP11 1JU
England
TEL 44 (0) 1494 441181
FAX 44 (0) 1494 463843
http://www.beckmancoulter.com/

Beckman Coulter, Inc.
4300 N. Harbor Boulevard
P.O. Box 3100
Fullerton, CA 92834-3100
USA
TEL +1 (800) 742-2345
FAX +1 (714) 773-8283
http://www.beckman.com

Beckman Coulter GmbH
Frankfurter Ring 115
80807, München
Germany
TEL 49 89 35870226
FAX 49 89 35870490

Beckman Coulter Inc.
11800 SW 147th Avenue
Miami, FL 33196
http://www.beckmancoulter.com

Beckman Instruments, Inc.
2500 Harbor Boulevard
Box 3100, Fullerton
California, 92634-3100
USA
TEL 800-742-2345
FAX 800-643-4366
http://www.beckman.com

Beckman Instruments Inc.
Spinco Division
P.O. Box 10200
Palo Alto, California 94304
USA
TEL 800-551-1150
TEL 415-859-1694
FAX 800-643-4366

Becton, Dickinson and Company
1 Becton Drive
Franklin Lakes, NJ 07417
TEL 201.847.6800
http://www.bd.com/

Becton Dickinson Biosciences—Transduction
   Laboratories
133 Venture Ct.
Lexington, KY 40511-2624
TEL 859-259-1550
FAX 859-259-1413
TEL 800-227-4063
E-MAIL tlbinfo@translab.com
http://www.bdbiosciences.com

Becton Dickinson Biosciences
2350 Qume Drive
San Jose, CA 95131-1807
USA
TEL +1 800 223-8226
TEL 877.232.8995
FAX +1 408 954-2347
http://www.bdbiosciences.com

Becton Dickinson GmbH
Tullastr. 8-12
69126 Heidelberg
Germany

Becton Dickinson Labware
Two Oak Park
Bedford, MA 01370
TEL 1-800-343-2035
(Outside the USA, 617-275-0004)
FAX 617-275-0043

Becton Dickinson UK Ltd.
Between Towns Road
Cowley, Oxford, Oxfordshire, OX4 3LY
TEL 01865 748844
FAX 01865 781635
E-MAIL bduk customerservice@europe.bd.com

Bellco Glass, Inc.
340 Edrudo Road
P.O. Box B
Vineland, New Jersey, 08360
USA
TEL 800-257-7043
FAX 609-691-3247
E-MAIL cservice@bellcoglass.com
http://www.bellcoglass.com

Berthold Technologies
99 Midway Lane
Oak Ridge, TN 37831
USA
http://www.berthnoldtech.com

BFI Optilas, Germany
Assar-Gabrielsson-Strasse 1
D-63128 Dietzenbach
Germany,—Calbiochem
http://www.calbiochem.com

Biacore AB
Rapsgaten 7
SE-754 50 Uppsala
Sweden

Bibby Sterilin
distributed via Appleton Woods Ltd Lindon
    House
Heeley Road
Selly Oak, Birmingham
West Midlands, B29 6EN
UK

Billups-Rothenberg, Inc.
PO Box 977
Del Mar, CA 92014-0977
TEL 1 877 755-3309 (Toll free in U.S.)
TEL 1 858 535-0545 (International)
FAX 1 858 535-0546
E-MAIL bri@brincubator.com
http://www.brincubator.com

Biochrom AG
Leonorenstraße 2-6 D-12247
Berlin Germany
TEL +49 30 7799 06-0
FAX +49 30 771 0012
http://www.biochrom.com

Biochrom Ltd
Cambridge Science Park
Milton Road, Cambridge, CB4 0FJ
UK
TEL +44 (0)1223 423723
FAX +44 (0)1223 420164
http://www.biochrom.co.uk/contact.htm

BioComp Instruments Inc.
650 Churchill Row
Fredericton, NB E3B 1P6
Canada
TEL 506-453-4812
FAX 506-453-3583
TEL 800-561-4221
E-MAIL dhc@unb.ca
http://www.biocompinsruments.com

BioConcept
Innovationszentrum Nordwestschweiz
    Gewerbestrasse
14 Postfach 427 CH—4123 Allschwil 1
Switzerland
TEL +41 61 486 80 80
FAX +41 61 486 80 00
E-MAIL info@bioconcept.ch
http://www.bioconcept.com

Biocrom KG
Leonorenstr. 2-6
D-12247 Berlin
Germany
TEL +49-30-77 99 06 0
FAX +49-30-77 10 01 2
E-MAIL info@biochrom.de
http://www.biochrom.de

Biodesign Inc. of New York
P.O. BOX 1050
CarmelL, NY 10512
TEL 845-454-6610
FAX 845-454-6077
E-MAIL service@biodesignofny.com
http://www.biodesignofny.com/ny.com

Biodesign International
60 Industrial Park Road
Saco, Maine 04072
USA

Bioline USA Inc.
PMB 311, 28 South Main Street
Randolph, MA 02368-4800
USA
TEL 781 830 0360
FAX 781 830 0205

Bioline GmbH
Im Biotechnologiepark
TGZ 2, D-14943 Luckenwalde
Germany
TEL +49 (0) 3371 68 12 29
FAX +49 (0) 3371 68 12 44
E-MAIL info@bioline.com
http://www.bioline.com/n_distribut2.htm
http://bioline.com

Bioline Ltd.
16 The Edge Business Centre
Humber Road, London NW2 6EW
UK
TEL +44 (0) 20 8830 5300
FAX +44 (0) 20 8452 2822

Biological Indstries Ltd.
Kibbutz Beit Haemek
25115 Israel
TEL 972-(0)4-996-0595
FAX 972-(0)4-996-8896
E-MAIL info@bioind.com
http://www.bioind.com/

Bio-Logic—Science Instruments SA
1, rue de l'Europe
F-38640—CLAIX
France
TEL +33 476 98 68 31
FAX +33 476 98 69 09

BIOLOG Life Science Institute
Forschungslabor und Biochemica-Vertrieb GmbH
Flughafendamm 9a
P.O. Box 107125
D-28071 Bremen
Federal Republic of Germany
http://www.biolog.de

Biomedical Resources International, Inc.
100 Fountain Street Framingham
MA 01702

Biomol Feinchemikalien, Waidmannstr
35, D 22769 Hamburg

Bioresearch Information
4300 N. Harbor Boulevard
P.O. Box 3100
Fullerton, CA 92834-3100
USA
TEL 1-800-742-2345
FAX 1-800-643-4366
http://www.beckmancoulter.com/

BIOMOL Research Laboratories, Inc.
5120 Butler Pike
Plymouth Meeting, PA 19462-1202
N-ethylmaleimide (NEM; Sigma E-1271)
TEL 1-800-942-0430
TEL 610-941-0430
FAX 610-941-9252
http://www.roche.de/

Bioptechs Inc.
3560 Beck Rd.
Butler, PA 16002
TEL 877-LIVE-CELL (5483-2355)
Direct 724-282-7145
FAX 724-282-0745
http://www.bioptechs.com

Bio-Rad Cell Science Division
Bio-Rad House
Maylands Avenue
Hemel Hempstead, Hertfordshire HP2 7TD
England
TEL 1-(800) 4 BIORAD
TEL 1-(800) 424 6723

Bio-Rad Laboratories
2000 Alfred Nobel Drive
Hercules, CA 94547
USA
TEL (800) 424-6723
TEL (510) 741-1000
FAX (510) 741-5800
FAX (800) 879-2289
E-MAIL lsg.orders.us@bio-rad.com
http://www.bio-rad.com

Bio-Rad Laboratories GmbH
Heidemannstrasse 164
D-80939 München
Postfach 45 01 33
D-80901 München
Germany
FAX 49 (89) 31884-100
http://www.bio-rad.de

Bio-Rad Laboratories S.A.
López de Hoyos
245-247
28043 Madrid, Spain
http://www.bio-rad.com

Biosciences AB
SE-751 84 Uppsala
Sweden

BioSource International
542 Flynn Road
Camarillo, CA 93102
USA

BioSource International
Distributed by Medicorp Inc.
5800 Royalmount
Montreal, PQ H4P 1K5
Canada
TEL 1-877-733-1900

Biostatus Limited
56 Charnwood Road
Shepshed, Leicestershire
LE12 9NP
UK
TEL 07957 575402
FAX 01509 651061
E-MAIL enquiry@biostatus.co.uk
http://www.biostatus.co.uk

Bio-Tek Instruments, Inc.
Highland Park
P.O. Box 998
Winooski, Vermont 05404-0998, USA
TEL (888) 451-5171
TEL (802) 655-4740
FAX (802) 655-7941
http://www.biotek.com/

BioVectra
160 Christian Street
Oxford, Connecticut 06478, USA
http://www.biovectra.com/

BioWhittaker (Clonetics)
8830 Biggsford Road
Wallcersville, MD 21793

BioWhittaker/Cambrex
One Meadowlands Plaza
East Rutherford, NJ 07073
USA

Biozym
31840 Hessisch Oldendorf
Germany

Bitplane AG
Badenerstrasse 682
CH-8048 Zürich
Switzerland
http://www.bitplane.com

BMG Labtechnologies Inc.
2415 Presidential Drive
Bldg 204, Suite 118
Durham, North Carolina 27703
http://www.bioresearchonline.com

BOC-Edwards
Manor Royal
Crawley, Sussex, RH10 2LW
UK
TEL +44 (0)1293528844
FAX +44 (0)1293533453

BOCHEM
Industriestraße 3
35781 Weilburg
Germany
http://www.bochem.de/

Boehringer Mannheim Gmbh—now Roche
    Diagnostics
Sandhofer Strasse 116
D-68305 Mannheim
Germany
http://www.roche.de/

Boehringer Manheim—now Roche Diagnostics
201 Boulevard Armand Frappier
Laval PQ H7V 4A2
Canada
TEL 1-800-263-5887
Boehringer Mannheim
9115 Hague Rd
P.O. Box 50414
Indianapolis, IN 46250

Boule (Corning Costar)
Nordic Denmark
Egensvej 25
DK-2770 Kastrup
Denmark

Boule Nordic AB
Box 1080
141 22 Huddinge
Sweden
http://www.boule.se/nordic/

Brain Research Laboratories
Waban PO Box 88
Newton, MA 02468
USA
http://www.brainresearchlab.com

Brand, Germany
11 Bokum Rd.
Essex, CT 06426
http://www.brandtech.com/

Brand GmbH+Co KG
Otto Schott Str. 25
D-97877 Wertheim
Germany
FAX 49 (9342) 808-236
http://www.brand.de

Branson Korea Co., LTD.
8th floor, Dongil Techno Town, #823
Kwanyang-2dong
Dongan-gu, Anyang-si, Kyonggi-do
Korea
TEL (82 31)422-0631
FAX (82 31)422-9572
E-MAIL buc@branson.co.kr

Braun Biotech Int.
34212 Melsungen
Germany
http://www.bbraunbiotech.com

Brechbühler AG
Steinwiesenstrasse 3
8952 Schlieren
Switzerland
TEL +41 1 732 31 31
FAX +41 1 730 61 41
E-MAIL info@sciex.com

Brinkmann Instruments, Inc.
One Cantiague Road
PO Box 1019
Westbury, NY 11590
USA
TEL 800-645-3050
http://www.brinkmann.com/

Bristol-Myers Squibb Canada, Inc.
2365 Cote-de-Liesse
Saint-Laurent, Quebec, H4N 2M7
Canada
TEL 800-267-0005
http://www.bms.com

Brookhaven Instruments Corporation
750 Blue Point Road
Holtsville, NY 11742-1896
USA
http://www.bic.com

Bruker
40 Manning Road
Manning Park
Billerica, MA 01821
http://www.bdal.com/

Brunschwig AG
PO-Kasten CH-4009 Basel
Switzerland
TEL +41 61 308 91 11
FAX +41 61 308 91 19
E-MAIL info@brunschwig-ch.com
http://www.brunschwig-ch.com

BTC engineering
12 Shirley Close
Milton, Cambridge CB4 4BG
UK

BTX
11199 Sorrento Valley Road
San Diego, CA 92121-1334
USA

BTX Instrument Division
Harvard Apparatus, Inc.
84 October Hill Road
Holliston, MA 01746-1388
http://www.btxonline.com

Caenorhabditis Genetics Center
250 Biological Sciences Center
University of Minnesota
1445 Gourtner Ave.
St. Paul, MN 55108
Cairn Research Ltd.
Graveney Road
Faversham, Kent, ME13 8UP
UK
TEL +44 (0)1795 590140
FAX +44 (0)1795 594510
http://www.cairn-research.co.uk

CalBiochem
10394 Pacific Center Court
San Diego, California 92121
USA
Mailing Address:
P.O. Box 12087
La Jolla, California 92039-2087, USA
TEL 1 800 854-3417
TEL 1 858 450-9600
FAX 1 800 776-0999
FAX 1 858 453-3552
E-MAIL orders@calbiochem.com
http://www.calbiochem.com

Caltag
1849 Bayshore Blvd. #200
Burlingame, CA 94010
TEL 650.652.0468
TEL 800.874.4007
FAX 650.652.9030
E-MAIL caltag@caltag.com
http://www.caltag.com

Cambrex (former BioWhittaker Inc.)
8830 Biggs Ford Rd.
Walkersville, MD 21793
USA
TEL +1 800/638-8174
FAX +1 301/845-8338
E-MAIL cs@biowhittaker.com
http://www.cambrex.com/default.asp

Cambridge Bioscience
24-25, Newmarket Road
Cambridge CB5 8LA
UK
TEL (+44) 1223 316855
FAX (+44) 1223 360732

Cambridge Isotope Laboratories
50 Frontage Road
Andover, MA 01810-5413
USA
http://www.isotope.com

Carlo Erba Reagenti
Chaussée du Vexin
27106 Val de Reuil
France
http://www.carloerbareagenti.com

Carl Roth GmbH & Co.
Schoemperlenstrasse 1-5
D-76185 Karlsruhe
Germany
E-MAIL info@carlroth.de
http://www.Carl-Roth.de

Carlson Scientific Inc.
514 S. Third Street
Peotone, IL 60468

Carl Zeiss
Carl-Zeiss-Str. 22
73447 Oberkochen
Germany
http://www.zeiss.com

Carl Zeiss France SAS
60, route de Satrouville
78230 Le Pecq
France
TEL +33 1 34802000
FAX +33 1 34802001
http://www.zeiss.com

Carl Zeiss Ltd.
PO Box 78
Woodfield Road
Welwyn Garden City
Herts, AL7 1LU
UK
TEL 01707 871300
FAX 01707 871289
http://www.zeiss.co.uk

Carl Zeiss MicroImaging, Inc.
One Zeiss Drive
Thornwood, NY 10594
USA
TEL 1-800-233-2343
FAX 1-914-681-7446
E-MAIL micro@zeiss.com
http://www.zeiss.com/micro/

Cartesian Technologies Europe, Ltd.
8 Blackstone Road
Huntingdon, Cambridgeshire
PE296EF, UK
TEL +44 (0) 1480 426700
FAX +44 (0) 1480 426767
http://www.cartesiantech.com/

CBS Scientific
420 South Cedros
Solana Beach, CA 92075

Cellgro, Mediatech, Inc.
13884 Park Center Rd.
Herndon, VA 20171
TEL 800-235-5476
http://www.cellgro.com

Cell Signaling Technology
166B Cummings Center
Beverly, MA 01915
Cell Systems Biotech GmbH
53562 St. Katharinen
Germany

Charles River Laboratories, Inc.
251 Ballardvale Street
Wilmington, MA 01887-1000
http://www.criver.com

Chemicon International
28820 Single Oak Drive
Temecula, CA 92590
USA

Chemicon International
Fischbacher Weg 3A
D-65719 Hofheim

CHIMERx,
6143 North 60th Street
Milwaukee WI 53218

Chiron B.V.
4560 Horton Street
Emeryville, CA 94608
USA
TEL +1 (510) 655-8730
FAX +1 (510) 655-9910
http://www.chiron.com/chironglobal.html

Chroma Technology
Chroma Technology Inc Business Office
74 Cotton Mill Hill
Brattleboro, VT 05302
USA
TEL 1-800-824-7662
TEL 1-802-428-2500
FAX 1-802-428-2525
http://www.chroma.com/

Chroma Technology Corp.
10 Imtec Lane
PO Box 489
Rockingham, VT 05101
USA
2TEL 1-802-257-1800
FAX 1-802-257-9400
E-MAIL info@chroma.com
http://www.chroma.com/index.cfm

Chromatographic Specialities
300 Laurier Blvd.
Brockville, Ontario, K6V 5W1
Canada
TEL 613-342-4678
FAX 613-342-1144
http://www.chromspec.com

Clarkson Chromatorgaphy products Inc
213 Main Street
Sout Williamsport PA 17701
USA

Clay Adams
Div. of Becton Dickinson
Diagnostic Instr. Sys.
383 Hillen Rd.
Towson, MD 21204
USA
TEL 800-638-8656

Clontech
21 Between Towns Road
Cowley, OX4 3LY Oxford
UK

Clontech
1020 East Meadow Circle
Palo Alto, CA 94303-4230
USA
http://www.corning.com/lifesciences

Cocalico Biologicals, Inc.
494 Stevens Road
P.O. Box 265
Reamstown, PA 17567
TEL 717-336-1990

Coherent (U.K.) Ltd.
28 St Thomas
The Cambridgeshire Business Park
Ely, CB7 4EX
UK
TEL 011 44 1353 658833
FAX 011 44 1353 659107
E-MAIL coherent-ltd@cohr.com

Cohesion Technologies, Inc.
2500 Faber Place
Palo Alto, CA 94303
TEL 650-320-5500
FAX 650-320-5511
E-MAIL admin@cohesiontech.com

Cole-Parmer
625 East Bunker Court
Vernon Hills
Illinois 60061-1844
USA
TEL 800-323-4340
FAX 847-247-2929

Color your Enzyme, Inc.
Dr. R. Bowers
Queen's University
Kingston, Ontario, K7L 3N6
Canada
TEL (613)533-6000 ext. 75005

CompuCyte Corporation
12 Emily Street
Cambridge MA 021139
TEL 800-840-1303
TEL 617-492-1300
E-MAIL salesinfo@compucyte.com

Computech
1701 Iron Street
Kansas City, MO 64116
USA

Continental Lab Products
5648 Copley Drive
San Diego, CA92111 USA
Continuum Scientific Service
3150 Central Expressway
Santa Clara, CA 95051-0816
USA
TEL 1-800-532-1064
TEL 1-408-727-3240
FAX 1-408-727-3550
E-MAIL Continuum@ceoi.com
http://www.continuumlasers.com/mainswf.html

Cora Styles Needles 'N Blocks
105 Cypress Point
Hendersonville, NC 28739
USA
http://www.corastyles.com/

Coriell Cell Repositories
403 Haddon Avenue
Camden, NJ 08103
USA

Corning, NY
14831, Sutter Instrument Co
40 Leveroni Court
Novato, CA 94949
Distributed by VWR/Canlab.
8567 Chemin Dalton
Ville Mont-Royal
PQ H4T 1V5
Canada
TEL (514)344-3525
http://www.corning.com

Corning (Life Science) Inc.
P.O. Box 5000
Corning, New York 14830
USA
TEL 800-492-1110
TEL 978-635-2200
FAX 978-635-2476
E-MAIL CLSCustServ@corning.com
http://catalog.corning.com/Lifesciences/

Corning B.V.
Life Sciences
Koolhovenlaan 12
1119 NE Schiphol-Rijk
The Netherlands
TEL 31 (0) 20-659-6051
FAX 31 (0) 20-659-7673
http://www.corning.com

Corning Costar Corporation
One Alewife Center
Cambridge, MA 02140
USA
TEL 617-868-6200
FAX 617-868-2076

Corning Incorporated Life Sciences
45 Nagog Park
Acton, MA 01720
USA
TEL 978-635-2200
TEL 800-492-1110
FAX 978-635-2476
E-MAIL clswebmail@corning.com
http://www.corning.com

Corning Science Products Corning Costar
Am Kuemmerling 21-25
55294 Bodenheim
Germany
TEL ++49 6135 9215-0
FAX ++49 6135 5148
http://www.corning.com/lifesciences
Coulter
http://www.beckmancoulter.com/

Covance Research Products
P.O. Box 7200
Denver, PA 17517

CPG Inc.
3 Borinski Road
Lincoln Park, NJ 07035
USA
TEL (201) 305-8181
FAX (201) 305-0884

Crescent Chemical Company, Inc.
1324 Motor Parkway
Islandia, NY 11749
USA
FAX 1 (631) 348-0913
http://www.creschem.com

CVI Laser Corp.
361 Lindbergh Ave.
Livermore, CA 94550
USA
TEL 510-449-1064
FAX 510-294-7747

Cytoskeleton, Inc.
1650 Fillmore Street
Suite #240
Denver, CO 80206
USA

DACO Corporation
Carpinteria, CA
TEL 800-235-5763
http://www.dakousa.com

Daigger Laboratory Supplies
620 Lakeview Parkway
Vernon Hills, IL 60061
TEL 1-800-621-7193
FAX 1-800-320-7200
E-MAIL daigger@daigger.com
http://www.daigger.com/

DAKO
DakoCytomation Denmark A/S
Produktionsvej 42
DK-2600 Glostrup
Denmark

Dako Cytomation
6392 Via Real
Carpinteria, CA 93013
TEL 805-566-6655
TEL 800-235-5743/800-424-0021
FAX 805-566-6688

DakoCytomation GmbH
Hamburger Strasse 181
D-22083 Hamburg
Germany
E-MAIL info@dakocytomation.de
http://www.dakocytomation.com

DAKO Danmark A/S
Productionsvej 42
DK-2600 Glostrup
Denmark
TEL 45 44 85 95 00
FAX 45 44 85 84 29
E-MAIL contact@dakocytomation.com
http://www.dako.com

Dako UK Ltd.
Denmark House
Angel Drove
Ely CB7 4ET
England

DELTA Acoustics & Vibrations
Building 356
Akademivej
DK-2800 Lyngby
Danemark
2TEL ++45-45-931211
FAX ++45-45-931990
E-MAIL dh@delta.dk

Denley Instruments
Thermo Life Sciences International (UK) Ltd.
Unit 5 The Ringway Centre
Edison Road
Basingstoke, RG21 6YH , Hampshire
UK
TEL +44 (0)1256 817282
FAX +44 (0)1256 817292
http://www.thermo-lifesciences.co.uk

Deville Scientific Inc.
P.O. Box 4588
Metuchen, NJ, 08840

Dianova
20148 Hamburg, Germany

Diatec.com AS
Gaustadalleen 21
0349 Oslo
Norway
TEL +47 22 95 86 25
FAX +47 22 95 86 49
E-MAIL diatec@diatec.com
http://www.diatec.com

DIFCO BD
1 Becton Drive
Franklin Lakes, NJ 07417
USA
TEL 201.847.6800

Difco (Voigt Global Distribution LLC)
P.O. Box 412762
Kansas City, MO 64141-2762
USA
http://www.vgdusa.com/DIFCO.htm

Digital Instruments
Veeco Metrology Group
Santa Barbara, CA
USA
http://www.veeco.com/

Digital Scientific Limited
Sheraton House
Castle Park, Cambridge CB3 0AX
UK
TEL +44 (0) 1223 329993
FAX +44 (0) 1223 460178
E-MAIL digitalscientific.co.uk

Digitial Equipment Corp.
PO Box 9501
Merrimack, NH
USA
TEL 800-344-4825
TEL 800-234-2298
http://www.dec.com

Dionex Corporation
1228 Titan Way
P.O. Box 3603
Sunnyvale, CA 94085-3603

DOJINDO Laboratories
2025-5 Tabaru
Mashiki-machi, Kamimashiki-gun
Kumamoto, 861-2202
Japan
TEL (+81) 96-286-1515
FAX (+81) 96-286-1525
E-MAIL info@dojindo.co.jp
http://www.dojindo.com

Dounce
Wheaton Science Products An Alcan Packaging
   Company
1501 N.10th Street
Millville, NJ 08332-2093
USA
TEL 800-225-1437
FAX 856-825-1368
TEL 1 856-825-1100
FAX 1 856-825-4568

Dow Corning Corporation
PO Box 994
Midland, MI 48686-0994, USA
http://www.dowcorning.com

Dow Corning STI, Inc.
Walnut Site
20832 Currier Road
Walnut CA 91789
USA
TEL +1 909 595 6331
FAX +1 909 595 1946

Drummond Scientific Co.
500 Parkway
Box 700
Broomall, PA, 19008
USA

Dumont, via Fine Sciene Tools
http://www.finescience.com/fst/misc/dumont.html

DuPont NEN
549 Albany Street
Boston, MA 02118

Dupont NEN Research Products—now PerkinElmer
    Canada Inc.
501 Rowntree Dairy Road, Unit #6
Woodbridge, ON L4L 8H1
Canada
TEL 1-800-561-4646

Duxford Cambridge
CB2 4PZ
England

Dynal Biotech
Deutsche Dynal
Bramfelder Chaussee 41
Postbox 710190
D 22177 Hamburg
Germany

Dynal Biotech ASA
PO Box 114
Smestad, 0309 Oslo
Norway
TEL +47 22 06 10 00
FAX +47 22 50 70 15
E-MAIL dynal@dynalbiotech.com
http://www.dynalbiotech.com

DYNEX Technologies, A Capital Genomix Company
14340 Sullyfield Circle Chantilly
Virginia 20151-1683
USA
TEL 703-631-7800
TEL 800-288-2354
FAX 703-631-7816

Eastman Chemical Co.
Laboratory & Research Products
1001 Lee Road
Rochester, NY 14652-3512
USA
TEL 800-225-5352
FAX 800-879-4979

Eastman Kodak Co.
Scientific Imaging Systems
343 State St.
Rochester, NY 14652
TEL 203-786-5657
TEL 877-SIS-HELP
FAX 203-786-5656
E-MAIL sis-info@kodak.com
http://www.kodak.com/go/scientific

EBioscience
6042 Cornerstone Court
Suite B-D
San Diego, CA 92121
USA

ECACC, CAMR
Salisbury
Wiltshire, SP4 OJG
UK

EDWARDS
Manor Royal
Crawley, West Sussex, RH10 9LW
UK
http://www.edwards.boc.com/

Eldex Laboratories, Inc.
30 Executive Court
Napa, CA 94558-6278
USA

Electron Microscopy Sciences
1560 Industry Road
P.O. Box 550
Hatfield, PA 19440
USA
TEL 1-215-412-8400
FAX 1-215-412-8450
http://www.emsdiasum.com/ems/

Electron Microscopy Sciences
P.O. Box 251
321 Morris Road
Fort Washington, PA 19034
TEL 215/646-1566
FAX 215/646-8931
E-MAIL SGKCCK@aol.com
http://www.emsdiasum.comm/home.html

EMBL Workshop
EMBL Heidelberg
Meyerhofstrasse 1
D-69117 Heidelberg
Germany
TEL +49 6221 3870
FAX +49 6221 387306
http://www.embl-heidelberg.de/

EMD Biosciences, Inc.
CALBIOCHEM®
10394 Pacific Center Court
San Diego, California 92121
USA

EMD Biosciences, Inc.
P.O. Box 12087
La Jolla, CA 92039-2087
USA
TEL 858 450 9600
FAX 858 453 3552
http://www.emdbiosciences.com/html/EMD/
    intl_sales_office.html
http://www.emdbiosciences.com/html/CBC/
    home.html

EMD Biosciences Inc.
10394 Pacific Centre Court
San Diego California 92121
USA
TEL 1-800-854-3417

E. Merck
Postfach 4119
D-6100 Darmstadt 1
BRD
TEL +49 6151 700
FAX +49 6151 72 2000

EM Science
480 S. Democrat Road
Gibbstown, NJ 08027
TEL 800-222-0342

Endecotts Limited
9 Lombard Road
London, SW19 3TZ
England
TEL +44 (0) 20 8542 8121
FAX +44 (0) 20 8543 6629
E-MAIL sales@endecotts.com Bellingham

Epoxy Technology
14 Fortune Drive
Billerica, MA01821
USA

Eppendorf
Barkhausenweg 1
22339 Hamburg
Germany
TEL ++49 40 53 8010
FAX ++49 40 53 801-556
http://www.eppendorf.com/en/splash.php

Eppendorf France
Parc des Grillons
60, Route de Sartrouville
78230 Le Pecq, France
TEL +33 1 30 15 67 40
FAX +33 1 30 15 67 45
http://www.eppendorf.com/france

Eppendorf Vertrieb Deutschland GmbH
Peter-Henlein-Strasse 2
D-50389 Wesseling-Berzdorf
Germany

EquiBio
Action Court, Ashford Road
Ashford, Middlesex, TW 151XB-UK
TEL 44-1784-425000
FAX 44-1784-248085

Erie Scientific Company
20 Post Road
Portsmouth, NH 03801
USA
E-MAIL eriesci.com

ESCO
1 South Point Dr.
Lake Forest California 92630
TEL 949-330-3602

Essex Pharma
PO Box 83 03 47
81737 München
Germany

Ethicon Incorporated
Somerville
New Jersey 08876-0151
USA
TEL 1-800-4-ETHICON

Euroclone
Via Figino 20/22
20016 Pero (Milano)
ITALY

Euromedex
24 rue des Tuileries
BP 74684 SOUFFELWEYERSHEIM
67458 MUNDOLSHEIM CEDEX
FRANCE
TEL +33 3 88 18 07 27
FAX +33 3 88 18 07 28

EXFO Burleigh Products Group Inc.
7647 Main St. Fishers
Victor, NY 14564-8909
USA
TEL +1 585 924-9355
FAX +1 585 924-9072
E-MAIL info@burleigh.com

Falcon
Becton Dickinson Labware
2 Bridgewater Lane
Lincoln Park, NJ 07035

Falcon
2350 Qume Drive
San Jose, CA 95131-1807
USA
TEL (800) 223-8226
TEL (877) 232-8995
FAX (408) 954-2347
http://www.bdbiosciences.com/discovery_labware/

Falls Church Store
260 W. Broad Street
Falls Church, Virginia, 22046
TEL 1-888-STOCKS6 (1.888.786.2576)
TEL 1.703.579.4209
FAX 1.703.995.4422
E-MAIL Scripophily.com
http://www.scripophily.net/millabinc1.html

Fermentas
BWI Commerce Park
7520 Connelley Drive
Suite A
Hanover, MD 21076
TEL (800) 340-9026
FAX (800) 472-8322
http://www.fermentas.com/

FERMENTAS UAB
V.Graiciuno 8
Vilnius 2028, Lithuania

F. Hoffmann-La Roche Ltd.
Diagnostics Division
Grenzacherstrasse 124
CH-4070 Basel
Switzerland
TEL 41-61-6881111
FAX 41-61-6919391

F. Hoffmann-La Roche Ltd
Group Headquarters
Grenzacherstrasse 124
CH-4070 Basel, Switzerland
TEL +41-61-688 1111
FAX +41-61-691 9391
E-MAIL Pharma (Rx) Webmaster (non-US residents)
http://www.roche.com

Roche Consumer Health
Wurmisweg, CH-4303 Kaiseraugst
Switzerland
TEL +41-61-688 1111
FAX +41-61-691 9391
E-MAIL Consumer Health Webmaster
http://www.roche.com/home.html

Fine Science Tools GmbH
Fahrtgasse 7-13
D-69117 Heidelberg
Germany
TEL +49 62 21 90 50 50
FAX +49 62 21 60 00 01
E-MAIL europe@finescience.com
http://www.finescience.com

Fischer Chemicals
Bishop Meadows Road
Loughborough, Leicestershire LE110RG
UK
TEL (+44) 01509 231166
FAX (+44) 01509 555111

Fisher Scientific AG
Wilstrasse 57
CH-5610 Wohlen
Switzerland
TEL 41-56-618-41-11
FAX 41-56-618-41-41
E-MAIL info@ch.fishersci.com

Fisher
Becton, Dickson
1 Becton Drive
Franklin Lakes, NJ 07417
USA
http://www.bd.com/

Fisher Bioblock Scientific
Bd Sébastien Brant
BP 50111, 67403 Illkirch Cedex
France
TEL +33 (0)3 88 67 14 14
Fax +33 (0)3 88 67 11 68
http://www.bioblock.com/

Fisher Laboratory Equipment Division
600 Business Center Drive
Pittsburgh, PA 15205
USA
TEL 1-800-926-0505
FAX 1-412-490-7286
http://www.fishersci.com

Fisher Scientific
2000 Park Lane Dr.
Pittsburgh, PA 15275-9943
USA
TEL 800-766-7000
TEL 201-467-6400
FAX 800-926-1166
FAX 201-379-7415
http://www.fishersci.com

Fisher Scientific
2761 Walnut Avene
Tustin, CA 92780
TEL 714-669-4600
TEL (800) 766-7000

Fisher Scientific
3970 John's Creek St.
Ste 500
Suwanee, GA 30024

Fisher Scientific
9999 Veteran's Memorial Dr.
Houston TX 77038
TEL 1-800-766-7000
https://www1.fishersci.com/index.jsp

Fisher Scientific
Customer Service Centre
112 Colonnade Rd.
Nepean, ON, K2E 7L6
Canada
TEL 800-234-7437
http://www.fishersci.ca

FISHER Scientific
Pittsburg, PA
TEL 1-800-766-7000 (USA)

Fisher Scientific
P.O. Box 14989
St. Louis, MO 63178 USA

Fisher Scientific International, Inc.
1200 Denison St.
Unionville, Ontario L3R 8G6
Canada
TEL 416-479-8700
http://www.fisherscientific.com

Fisher Scientific International, Inc.
Liberty Lane
Hampton, NH 03842
TEL (603) 926-5911
FAX (603) 929-2379
E-MAIL webmaster@nh.fishersci.com
http://www.fishersci.com

FJW Optical System, Inc.
629 S.Vermont St.
Palatine, IL 60067-6949
USA
TEL 708-358-2500
FAX 708-358-2533

Fluka
Chemie GmbH
CH-9471, Buchs
Germany

Fluka
Industriestrasse 25
9471 Buchs, Switzerland
TEL ++41(0)81 755 25 11
FAX ++41(0)81 755 28 15
E-MAIL Fluka@sial.com
http://www.sigmaaldrich.com/

FMC BioProducts
191 Thomaston Street
Rockland, ME 04841
USA
TEL 800 341 1574
TEL 207 594 3400
FAX 800 362 5552, 207 594 3491

FMC Corporation
1735 Market Street
Philadelphia, PA 19103
USA
TEL 215-299-6000
FAX 215-299-5998

Fresenius
Else-Kröner-Str. 1
61352 Bad Homburg
Germany

GATAN
5933 Coronado Lane
Pleasanton, CA 94588
USA
http://www.gatan.com/

GE Bayer Silicones
Bergen op Zoom
The Netherlands

Gemini Bio-Products
1301 East Beamer Street
Woodland, CA 95776
TEL 1.800.543.6464
TEL 1.530.668.3636
FAX 1.530.668.3630
http://www.gembio.com

GeneMachines
935 Washington Street
San Carlos, California 94070
USA
E-MAIL genemachines.com

General Mills Inc.
800 Derr Street
Vallejo, CA 94590, USA
http://www.generalmills.com

Geneva Bioinformatics (GeneBio) S.A.
25 Avenue de Champel
CH- 1206 Geneva
Switzerland
TEL +41 22 702 99 00
FAX +41 22 702 99 99
E-MAIL info@genebio.com
http://www.genebio.com/

GERBU Biochemicals GmbH
Am Kirchwald 6
69251 Gaiberg
Germany
http://www.gerbu.de/

Geron Corp.
230 Constitution Drive
Menlo Park, CA 94025
USA
TEL 1-650-473-7700
http://www.geron.com

GIBCO
9800 Medical Cnt Dr.
P.O. Box 6482
Rockville, Maryland

GIBCO
Distributed by Invitrogen Canada Inc.
2270 Industrial St.
Burlington, ON L7P 1A1
Canada
TEL 1-800-263-6236

GIBCO
Invitrogen Ltd
3 Fountain Drive
Inchinnan Business Park , Paisley
UK
TEL 0141 814 6100
FAX 0141 814 6260
http://www.invitrogen.com

GIBCO
P.O. Box 880
Langley, OK 74350-0880
TEL 918-782-4000
FAX 918-782-4002
E-MAIL gibco@worldnet.att.net

GIBCO-BRL
Invitrogen Corporation
1600 Faraday Avenue
PO Box 6482
Carlsbad, California 92008
TEL (760) 603-7200
TEL 800-955-6288
FAX (760) 602-6500
E-MAIL tech_service@invitrogen.com
http://www.gibcobrl.co

GIBCO BRL
Life Technologies
9800 Medical Center Drive
Rockville, MD, 20849

GIBCO BRL Div. of Invitrogen
64271 P.O. Box 9418
Gaithersburg MD 20898
TEL +1-301-840-8000
FAX +1-301-670-8539
http://www.gibcobrl.com

GIBCO BRL / Life Technologies GmbH
Technologiepark Karlsruhe
Emmy-Noether-Strasse 10
D-76131 Karlsruhe
E-MAIL eurocustom@lifetech.com
http://www.lifetech.com

GIBCO Invitrogen Sarl
BP 96
95613 Cergy Pontoise Cedex
France
TEL +33 1 34 32 31 00
FAX +33 1 30 37 50 07
http://www.invitrogen.com

GIBCO Laboratories
3175 Staley Road
Grand Island, NY 14072
TEL (800) 828-6686

GIBCO products from Invitrogen
PO box 3326
4800 DH Breda
The Netherlands
http://www.invitrogen.com/

Gilson, Inc.
3000 W. Beltline Hwy.
P.O. Box 620027
Middleton, WI 53562-0027
USA
TEL 608-836-1551
TEL 800-445-7661
FAX 608-831-4451
http://www.gilson.com/

Glaswarenfabrik Karl Hecht KG
97647 Sondheim
Germany
http://www.hecht-assistent.de/

Glen Research
22825 Davis Drive
Sterling, VA 20164
USA
TEL 1-703-437-6191
FAX 1-703-435-9774
http://www.glenresearch.com/

GLW GmbH
Hüberstrasse 19
D-97084 Würzburg
Germany

Goldschmidt UK Ltd., TEGO House
Chippenham Drive
Kingston, Milton Keynes
Bucks MK10 0AF
UK

Goodfellow Cambridge Ltd.
Ermine Business Park
Huntingdon, Cambridgeshire, PE29 6WR
lUK

Goudoh-Shusei
Tokyo, Japan
http://www.godo.jp

Grace Vydac (Hesperia, CA, USA)
Western US Grace Vydac Representative
Peter R. Krinsky, Western Region Sales Manager
Anaheim, CA
TEL 1-714-518-3353
FAX 1-714-518-3356
E-MAIL pete.krinsky@grace.com

GRATICULES Ltd.
17-19 Morley Road
Tonbridge, Kent TN9 1RN
UK

Greiner-Bio One
Maybachstrasse 2
72636 Frickenhausen
Germany
FAX 49 (7022) 948-514
http://www.greiner-bio-one.com/

Greiner Bio-One Inc.
1205 Sarah Street
Longwood, FL 32750
USA
TEL 407-333-2800
TEL 800-884-4703
FAX 407-333-3001
E-MAIL info@greinerbiooneinc.com
http://www.greinerbioone.com

Hamamatsu
325-6, Sunayama-cho
Hamamatsu City, Shizuoka Pref. 430-8587
Japan
TEL (+81) 53-452-2141
FAX (+81) 53-456-7889
E-MAIL webmaster@hq.hpk.co.jp
http://www.hamamatsu.com

Hamamatsu Photonics
360 Foothill Rd.
Bridgewater, NJ 08807
TEL 908-231-0960
TEL 1-800-524-0504
FAX 908-231-1218
http://www.hamamatsu.com

Hamamatsu Photonics Deutschland GmbH
Arzbergerstr 10
D-82211 Herrsching am Ammersee
Germany

Hamilton Deutschland GmbH
Fraunhoferstr 17
D-82152 Martinsried
Germany
TEL +49-(0)89-5526-49-0
FAX +49-(0)89-5526-49-10

Hansatech Instruments Ltd.
Narborough Road
Pentney
King's Lynn, Norfolk PE32 1JL
England
http://www.hansatech-instruments.com

Harco (Harlow Chemical Company Ltd)
Central Road
Templefields, Harolow, CM20 2BH
Essex
http://www.harlowchem.com

Harlan Bioproducts for Science
P.O. Box 29176
Indianapolis, IN
TEL 1-800-972-4362
FAX 1-317-357-9000
http://www.hbps.com

Harlan Sera Labs Ltd.
Dodgeford Lane
Loughborough
Leicestershire, LE12 9TE
England
TEL 01530 222123
FAX 01530 224970
E-MAIL hslcsd@harlanuk.co.uk
http://www.harlanseralab.co.uk//home.html

Harvard Apparatus, Inc.
84 October Hill Road
Holliston, MA 01746
TEL 508-893-8999
TEL 800-272-2775
FAX 508-429-5732
http://www.harvardapparatus.com

Harvard Apparatus ltd.
Fircroft Way
Edenbridge , Kent TN8 6HE
U.K.
TEL +44 (0)1732 864001
FAX +44 (0)1732 863356

Headway Research, Inc.
3713 Forest Lane
Garland, TX 75042-6928
USA
TEL (972) 272-5431
FAX (972) 272-7817

Henogen SA
Site de Seneffe
14, rue de la Marlette
7180 Seneffe
Belgium

Heraeus
Heraeus Holding GmbH
Heraeusstraße 12-14
D-63450 Hanau

Heraeus Centrifuges, by Kendro Laboratory Products
275 Aiken Road
Asheville, NC 28804

Heraeus Instruments/, Kendro Laboratory Products
   GmbH
Robert-Bosch-Strasse 1
D-63505 Langenselbold
Germany
E-MAIL info@kendro.com
http://www.heraeus-instruments.de

Heraeus S.A.
Manuel Tovar
24, 28034 Madrid
Spain
http://www.sorvall.com

Hettich Zentrifugen
Andreas Hettich GmbH & Co.KG
Gartenstr 100
D-78532 Tuttlingen
TEL +49 7461 705 0
FAX +40 7461 705 125
E-MAIL info@hettichLab.com
http://www.HettichLab.com

Hewlett-Packard
http://www.hewlettpackard.com/
H. Hölzel Laborgeräte GmbH
Bahnhofstraße 23
D-85457 Wörth / Hörlkofen
Germany
E-MAIL info@hoelzel-gmbh.de
http://www.hoelzel-gmbh.de

Hitachi
Tokyo, Japan
http://www.hitachi.com

Hoefer Scientific Instruments
654 Minnesota Street
San Francisco CA 94107
USA

Hoffmann-La Roche AG
Emil-Barell-Str. 1
79639 Grenzach-Wyhlen
Germany

Hoffmann-La Roche Inc.
340 Kingsland Street
Nutley, NJ 07110
USA
TEL +1-973-235 5000
FAX +1-973-235 7605
http://www.rocheusa.com

Hoffmann-La Roche Ltd
Diagnostics Division
Grenzacherstrasse 124
CH-4070 Basel
Switzerland
TEL +41-61-688 1111
FAX +41-61-691 9391
http://www.roche.com

Holm & Halby
Vallensbækvej 35
DK-2605 Brønby
Denmark
TEL +45 43 26 94 00
E-MAIL info@holm-halby.dk
http://www.holm-halby.dk/

Houm AS
PO box 83 Grefsen
TEL +47 22 09 40 00
FAX +47 22 09 40 40
E-MAIL firmapost@houm.no
http://www.houm.no

Huber Kältemaschinenbau GmbH
Werner-von-Siemens-Strasse 1
77656 Offenburg
Germany

HV Skan Ltd.
425-433 Stratford Road
Shirley, Solihull, B90 4AE (Road Map)
West Midlands
TEL 0121 733 3003
FAX 0121 733 1030
http://www.skan.co.ukFisher Scientific

HyClone
925 West 1800 South
Logan, UT 84321 USA
TEL 1-800-492-5663
http://www.HyClone.com

Hydro Systems
PO Box 12137
Research Triangle Park, NC 27709
USA

IBA GmbH
Rudolf-Wissell-Str. 28
D-37079 Goettingen
Germany
E-MAIL info@iba-go.com
http://www.iba-go.com

ICN (now MP Biomedicals)
Biomedica GmbH, Medizinprodukte GmbH & Co. KG
Divischgasse 4, 1210 Wien
AUSTRIA
TEL +43 1 291 0754
FAX +43 1 291 0771
E-MAIL sales@biomedica@bmgrp.at
http://www.biomedica.co.at, www2.icnbiomed.com

ICN Biomedicals, Inc.
1263 South Chillicothe Road
Aurora OH 44202-8064
USA
http://www.icnbiomed.com/

ICN Biomedicals, Inc.
15 Morgan Street
Irvine, California 92618-2005
USA
TEL 800-854-0530
TEL 714-545-0113
FAX 800-334-6999
FAX 714-557-4872
http://www.icnbiomed.com/

ICN Biomedicals Inc.
3300 Hyland Ave.
Costa Mesa, CA 92626
TEL 714-545-0100
TEL 800-854-0530
FAX 800-334-6999
E-MAIL sales@icnbiomed.com
hhttp://www.icnbiomed.com

IDT
1710 Commercial Park
Coralville, IA 52241
TEL 800-328-2661
http://www.idtdna.com/

IKA Works, Inc.
2635 Northchase
PKWY SE Wilmington, NC 28405
USA
TEL 910-452-7059
TEL 800-733-3037
E-MAIL usa@ika.com

Ikemoto Rika
3-25-11, Hongo
Bunkyo-ku, Tokyo 113-8680
JAPAN
http://www.ikemoto.co.jp

ImageJ
Rasband, W., National Institute of
    Health (NIH)
Bethesda, Maryland
USA
http://rsb.info.nih.gov/ij

Industrial and Scientific
The Grip
Hadstock Road
Linton, Cambridge CB1 6NR
UK
TEL (+44) 01223 891953
FAX (+44) 01223 894223

Ingenieurburo CAT
M. Zipperer GmbH
79219, Staufen
Germany

Innomed-Konsult AB
Box 6141
S-102 33 Stockholm
Sweden
http://www.innomed.se/

Instrumedics, Inc.
61 South State Street
Hackensack, NJ 07601
USA
TEL 800-237-2772
TEL 201-343-1313
FAX 201-487-4884
E-MAIL info@instrumedics.com
http://www.instrumedics.com

Integrated DNA Technologies
1710 Commercial Park
Coralville, IA 52241
USA
TEL +1-800-328-2661
FAX 319-626-8444
http://www.idtdna.com

Intelligent Imaging Innovations, Inc.
820 16th Street
Suite 850
Denver, CO 80202

Interferon Sciences Inc.
783 Jersey Avenue
New Brunswick NJ 08901-3660
USA
http://www.interferonsciences.com/

INTERGEN Company
Two Manhattanville Road
Purchase, New York 10577
USA
TEL 1-800-431-4505

Intervet Germany
Feldstr. 1a
D 85716 Unterschleissheim

Intracel
Unit 4
Station Road
Shepreth. Royston, Herts. SG8 6PZ
http://www.intracel.co.uk/

Intrinsic Bioprobes, Inc.
625 S. Smith Rd.
Suite 22
Tempe, AZ 85284
USA

Invitrogen
1600 Faraday Avenue
Carlsbad, CA 92008
TEL (760) 603-7200
FAX (760) 602-6500
http://www.invitrogen.com

Invitrogen
9800 Medical Center Drive
P.O. Box 6482
Rockville, MD 20849-6482
USA
TEL +1-301-610-8709
FAX +1-301-610-8724
http://www.invitrogen.com

Invitrogen AG
Elisabethenstrasse 3
Postfach 533
CH-4019 Basel
Switzerland

Invitrogen Corporation
Thistedgade 6
stuen, 2630 Taastrup
Denmark
http://www.invitrogen.com

Invitrogen Canada, Inc.
2270 Industrial St.
Burlington, Ontario, L7P 1A1
Canada
TEL 800-263-6236
FAX 800-387-1007
http://www.invitrogen.com

Invitrogen GmbH
Technologiepark Karlsruhe
Emmy-Noether Strasse 10
76131 Karlsruhe
Gebührenfreie Bestellungen
TEL 0800 083 09 02
FAX 0800 083 34 35
Technical Information: Euro Tech-LineSM—0800 181
    54 50
E-MAIL eurotech@invitrogen.com

Invitrogen S.A.
Edificio Océano
Parque Mas Blau
C/ Garrotxa 10-12
08820 Prat de Llobregat (Barcelona)
Spain
http://www.invitrogen.com

Irvine Scientific
2511 Daimle Street
Santa Ana, CA 92705
TEL (800) 437-5706

ISCO
PO Box 5347
Lincoln NE 68505, USA
TEL (402) 464-0231

ITC Biotechnologies (Clontech)
Tullastr. 4
D 69126 Heidelberg

IWAKI
Asahi Technoglass Corporation
7-2 Nihonbashi-Honcho3-chome
Chuo-ku, Tokyo 103-0023
Japan

IWAKI, Scitech Division
Asahi Techno Glass
1-50-1 Gyouda
Funabasi, Chiba
Japan

Jackson
P.O. Box 9
872 West Baltimore Pike
West Grove, PA 19390, USA
TEL 1-800-FOR-JAXN (367-5296)
TEL 610-869-4024
FAX 610-869-0171
http://www.jacksonimmuno.com/

The Jackson Laboratory
600 Main Street
Bar Harbor
Maine 04609
USA
TEL 207-288-6000
TEL 207-288-6051
http://www.jax.org/

Jackson Immuno Research Labs Incorporated,
    Stratech Scientific Ltd
61-63 Dudley Street
Luton, Bedfordshire LU2 0NPK4
UK
TEL (+44) 01582 529000
FAX (+44) 01582 481895
Jasco International Co. Ltd.
4-21 Sennin-cho 2-chome
Hachioji, Tokyo 193-0835
Japan
TEL 81-426-66-1322
FAX 81-426-65-6512

Jencons (Scientific) Ltd.
Cherrycourt Way
Stanbridge Road
Leighton Buzzard, Bedfordshire LU7 4UA
UK
TEL (+44) 1525 372010
FAX (+44) 1525 379547

Jencons Scientific Inc.
800 Bursca Drive
Suite 801
Bridgeville, PA 15017
TEL 800-846-9959
TEL 412-257-8861
FAX 412-257-8809
E-MAIL info@jencons.com

JRH Biosciences, Inc.
13804 W. 107th Street
Lenexa, KS
TEL 1-800-255-6032
FAX 1-913-469-5584
http://www.jrhbio.com/

JRH Biosciences Ltd.
Smeaton Road
West Portway
Andover, Hampshire SP10 3LF
UK
http://www.jrhbio.com/

J.T. Baker
Mallinckrodt Baker B.V.
Teugseweg 20
P.O. Box 1
7400 AA, Deventer
The Netherlands
TEL 31-570-687500
FAX 31-570-687574
E-MAIL service.nl@mkg.com
http://www.jtbaker.com

JT Baker
Mallinckrodt Baker, and Inc.
222 Red School Lane
Phillipsburg NJ 08865
USA

JY Horiba
3880 Park Avenue
Edison, NJ 08820-3012
USA
http://www.jobinyvon.com

KEBO
Fagerstagatan 18A
SE-163 94 Stockholm
Sweden
http://www.kebolab.se/

Kendro Laboratory Products
31 Pecks Lane
Newtown, CT 06470-2337 USA
Kimble / Kontes
Vineland, New Jersey
USA
TEL (888) 546-2531 Extension 1
TEL (856) 692-3600 Extension 1
FAX (856) 794-9762
E-MAIL cs@kimkon.com

KIBBUTZ BEIT HAEMEK
25115 ISRAEL
TEL 972-(0)4-996-0595
FAX 972-(0)4-996-8896
E-MAIL info@bioind.com
http://www.bioind.com/

Kinetic Imaging Limited
2 Brunel Road
Croft Business Park
Bromborough, Wirral, Merseyside CH62 3NY
TEL 0151 343 0060
FAX 0151 343 1524
http://www.kineticimaging.com

Kodak
Distributed by VWR/Canlab.
8567 Chemin Dalton
Ville Mont-Royal, PQ H4T 1V5
Canada
TEL (514)344-3525

Kodak Laboratory Chemicals
Building 70
Eastman Kodak Company
343 State St., Rochester, NY
TEL 1-800-225-5352
FAX 1-800-225-5352

Kodak Scientific Imaging Systems
Eastman Kodak Company, 4
Science Park, New Haven, CT 06511
USA
www.kodak.com/US/en/health/scientific

Kojair Tech Oy
Tellollisuusitie 3
35700 Vilpulla, Finland
http://www.kojair.com

Kontes Glass Company
1022 Spruce Street
Vineland, NJ 08360

KVT König
Dietikon, Switzerland
http://www.kvt.ch/

Labassco
Aminogatan 30
431 53 Mölndal
Sweden
TEL +46-31-730 70 00
FAX +46-31-706 30 30

Lab Chem Inc.
200 William Pitt Way
Pittsburgh, PA 15238
Labconco Corporation
8811 Prospect Avenue
Kansas City, Missouri 64132-2696
TEL (800) 821-5525
TEL (816) 333-8811
FAX (816) 363-0130
http://www.labconco.com

Labcor Products Inc.
7309 Governors Way
Frederick, MD, 21704
USA

Lab-Line Instruments, Inc.
15th and Bloomingdale Avenues
Melrose Park, IL 60160

Laboratoires EUROBIO
avenue de Scandinavie 7
91953 Les Ulis Cedex B
France
TEL +33(0)1 69 07 94 77
FAX +33(0)1 69 07 95 34
http://www.eurobio.fr/

Laboratory Products Sales
1655 Buffalo Rd.
Rochester, NY 14624

Laborel
Caspar Stormsvei 2
Postboks 109 Alnabru
0614 Oslo
TEL +47 23 05 19 30
FAX +47 23 05 19 31
E-MAIL laborel@laborel.no
http://www.laborel.no

Labscientific
114 West Mt. Pleasant Avenue
Livingston, New Jersey 07039
USA
http://www.labscientific.com/

Lancaster Synthesis Ltd.
Newgate, White Lund
Morecambe, Lancashire LA3 3DY

Laser Science, Inc.
8E Forge Parkway
Franklin, MA 02038
USA
TEL 1-508-553-2353
FAX 1-508-553-2355
E-MAIL Via the website
http://www.laserscience.com/index.html

LC Laboratories
165 New Boston Street
Woburn, MA 01801

Leica
111 Deer Lake Road
Deerfield, IL 60015 USA
Leica Microsystems
www.light-microscopy.com

Leica Microsystems AG
Ernst-Leitz-Strasse 17-37
Wetzlar, 35578
Germany
http://www.leica-microsystems.com

Leica Microsystems Inc.
2345 Waukegan Road
60015 Bannockburn
TEL +1 800 248 0123
Fax +1 847 405 0164
http://www.leica-microsystems.com

Leica Mikrosysteme Vertrieb GmbH
Lilienthalstrasse 39-45
Bensheim, D-64625
Germany
TEL +49 6251 136 0
FAX +49 6251 136 155
http://www.leica-microsystems.com

Leysop Ltd.
18 Repton Court
Repton Close
Basildon, Essex, SS13 1LN
England
FAX (+44) 1268 522111
E-MAIL sales@leysop.com
http://www.leysop.co.uk

Life Imaging Services, LIS
CH-4153 Reinach BL
Switzerland
http://www.lis.ch

Life Science Headquarters
549 Albany Street
Boston, Massachusetts 02118
USA

Life Science Research
2000 Alfred Nobel Drive
Hercules, California 94547
USA
TEL 1-800-424-6723
FAX 510-741-5800
E-MAIL lsg.orders.us@bio-rad.com
http://www.bio-rad.com

Life Science Products Inc.
P.O. Box 1150
Frederick, CO 80530
USA
TEL 1-800-245-5774
http://www.e-LSPI.com

Life Technologies
76344 Eggenstein-Leopoldshafen
Contact Invitrogen Corp.
1600 Faraday Ave.
Carlsbad, CA 92008, USA
TEL 1-800-955-6288
http://www.invitrogen.com

Life Technologies, Inc.
P.O. Box 68
Grand Island, NY 14072-0068
USA
TEL 800-828-6686
FAX 800-331-2286

Life Technologies, Inc.
P.O. Box 6009
Gaithersburg, MD 20884-9980

LKB-Wallac
PerkinElmer Analytical Instruments
Chalfont Rd.
Seer Green
Beaconsfield, Bucks HP9 2FX
UK
TEL +44 (0)1494 874515
FAX +44 (0)1494 679335
http://uk.instruments.perkinelmer.com

Lochhamer Schlag 19
D—82166 Gräfelfing
Germany
TEL 49 89 895 662 0
FAX 49 89 895 662 101
http://www.till-photonics.de/home_e.htm

Ludl Electronic Products Ltd.
171 Brady Avenue
Hawthorne, NY 10532
USA

Luminex Corporation
12212 Technology Blvd.
Austin, TX 78727
USA

Luxo Corporation
200 Clearbrook Road
Elmsford, NY 10523
TEL 914-345-0067
FAX 914-345-0068

Luxo Schweiz GmbH
Oberebenestrasse 67
CH—5620 Bremgarten AG
TEL +41 56 633 88 28
FAX +41 56 633 99 04
http://www.luxo.com/

Maag Technic AG
Birsfelden
Switzerland
http://www.maagtechnic.ch/

MABTECH
Gamla Värmdöv. 2
SE-131 37 Nacka
Sweden

Mallinckrodt
675 McDonnell Blvd.
Hazelwood, MO 63042
USA

Mallinckrodt Baker
Im Leuschnerpark 4
D-64347 Griesheim
Germany

Mallinckrodt Baker, Inc.
222 Red School Lane
Phillipsburg NJ 08865
USA
TEL 908-859-2151
TEL 1-800-582-2537 (within U.S.)
FAX 908-859-6905
http://www.jtbaker.com/

Mallinckrodt Baker B.V.
Teugseweg 20
P.O. Box 1
7400 AA, Deventer
The Netherlands
TEL 31-570-687500
FAX 31-570-687574
E-MAIL service.nl@mkg.com

Mallinckrodt Laboratory Chemicals
A Division of Mallinckrodt Baker, Inc.
222 Red School Lane
Phillipsburg, NJ 08865
TEL (800) 582-2537
TEL (908) 859-2151 (outside U.S.)
FAX (908) 859-6905
E-MAIL infombi@mkg.com
http://www.mallchem.com/prodlit/prod_lit.asp

Marcherey-Nagel GmbH
Postfach 10 13 52
D-52313 Duren
Germany
TEL +49 (0) 2421-9690
FAX +44 (0) 2421-969 199
http://www.macherey-nagel.com/

Marienfeld Laboratory Glassware
Paul Marienfeld GmbH & Co. KG
Am Wöllerspfad 4
97922 Lauda-Königshofen
Germany
TEL +49 (0) 9343 6272-0
FAX +49 (0) 9343 6272-25
E-MAIL info@marienfeld-superior.com
http://www.marienfeld-superior.com

Marysol
Tokyo, Japan

Matrix
Lower Meadow Road
Brooke Park
Handforth, Wilmslow, Cheshire, SK9 3LP
UK

Matrix Science Ltd.
8 Wyndham Place
London W1H 1PP
UK
http://www.matrixscience.com

Matrix Technologies Corp
22 Friars Drive
Hudson, NH 03051
USA
TEL (866) 229 9770
http://www.matrixtechcorp.com

Matsunami Glass Ind., Ltd.
2-1-10 Yasaka-cho
Kishiwada City, Osaka 596-0049
Japan
TEL 81-724-33-1163
FAX 81-0724-36-2265

MatTek Corp.
200 Homer Ave.
Ashland, MA 01721
USA
TEL (508) 881-6771
FAX (508) 879-1532
E-MAIL information@mattek.com
http://www.mattek.com

Max F. Keller GmbH
Elsteinstrasse 14a
D-68169 Mannheim
Postfach 121036
TEL 0621 / 3227932
TEL 0621 / 3227927

MCI Optonix
Division of USR Optonix, Inc.
P.O. Box 509
Cedar Knolls, NJ 07927
USA
http://www.mcio.com

McMaster
http://www.mcmaster.com/

MDS, Inc.
100 International Blvd.
Toronto, Ontario, M9W6J6
TEL 416-675-4530
FAX 416-675-0688
http://www.mdsintl.com

MDS Nordion
447 March Road
Ottawa, ON K2K 1X8
Canada
TEL +1 613 592 2790
FAX +1 613 592 6937
http://www.mds.nordion.com/

MDS Sciex
71 Four Valley Drive
Concord, ON L4K 4V8
Canada
http://www.sciex.com

Medinor
Nils Hansens vei 4
Postboks 94 Bryn
0611 Oslo
Norway

Melford Laboratories Ltd.
Bildeston Rd.
Chelsworth
Ipswich, Suffolk, IP7 7LE
UK
TEL (+44) 1449 741178
FAX (+44) 1449 741217

Melles Griot
051 Palomar Airport Rd. #200,
Carlsbad, CA 92009
2TEL (760) 268-5131
TEL (800) 835-2626
FAX (760) 804-0049
E-MAIL sales@irvine.mellesgriot.com
http://www.mellesgriot.com

Melles Griot Ltd.
1 Saint Thomas Place
Cambridgeshire Business Park
Angel Drove, Ely, CB7 4EX
Cambridgeshire
TEL 01353 654500
FAX 01353 654555
http://www.mellesgriot.com

MENZEL-GLAZER
Gerhard Menzel
Glasbearbeitungswerk
Gmbh & Co. KG
Saarbrückener Str. 248
D-38116 Braunschweig
Germany
TEL 0531/59008-0
FAX 0531/509799

Merck
Frankfurter Str. 250
64293 Darmstadt
Germany
TEL +49 6151 72-3000
FAX +49 6151 72-3333
FAX +49 6151 72 7495
E-MAIL catalog@merck.de, AR@merck.de
http://www.merck.de

Merck & Co., Inc.
One Merck Drive
P.O. Box 100
Whitehouse Station, NJ 08889-0100
USA
TEL 908-423-1000

Merck Biosciences
Boulevard Industrial Park
Padge Road
Beeston, Nottingham NG9 2JR
UK
http://www. merckbiosciences.co.uk

Merck Biosciences GmbH (formerly Calbiochem-
    Novabiochem GmbH)
Ober der Roth 4
D-65796 Bad Soden
Germany
FAX 49 (6196) 62361
http://www.calbiochem.com

Merck Biosciences Ltd.*
Boulevard Industrial Park
Padge Road
Beeston, Nottingham, NG9 2JR
UK
TEL 0115 943 0840
FAX 0115 943 0951
http://www.merckbiosciences.co.uk/html/CNUK/
    account_managers.htm

Merck Eurolab
10, rue de la Durance
B.P. 36
67023 Strasbourg Cedex 1
France
TEL +33 3 88 65 80 20
FAX +33 3 88 39 74 41
http://www.merckeurolab.fr

Merck KGaA
D-64271 Darmstadt
Germany
TEL (49)-6151-72-0
FAX (49)-6151-72-2000
E-MAIL catalog@merck.de
http://www.merck.de

Merck Ltd.
Laboratory Supplies
Merck House
Seldown, Poole, Dorset, BH15 1TD, UK
TEL 44-1202-669700
FAX 44-1202-665599
E-MAIL stella.taylor@merck-ltd.co.uk
http://www.merck-ltd.co.uk

Merck Schuchardt OHG
Eduard-Buchner-Str. 14-20
85662 Hohenbrunn
Germany
TEL +49 8102 802-0
FAX +49 8102 802-175
http://www.schuchardt.de

Merck/VWR International GmbH
Hilpertstr. 20a
D-64295 Darmstadt
Germany
E-MAIL darmstadt@de.vwr.com
http://www.vwr.com

Mica House
2A Pretoria Street
Calcutta 700 071
India

Michrom BioResources
1945 Industrial Drive
Auburn, CA 95603

Microbix Biosystems, Inc.
341 Bering Avenue
Toronto, Ontario, M8Z 3A8
Canada
TEL 416-234-1624
FAX 416-234-1626
http://www.microbix.com

MICROM International GmbH
Robert-Bosch-Str. 49
D-69190 Walldorf
TEL +49 6227-836 0
FAX +49 6227-836 111
http://www.microm.de

Micro Video Instruments Inc.
11 Robbie Road
P.O. Box 518
Avon, MA 02322
USA
TEL 1-508-580-0080
FAX 1-508-580-8623
E-MAIL Via the website
http://www.mvi-inc.com/home.htm

Milian SA
Route du Vélodrome 35
CH-1228 Plan-les-Ouates, Geneva
Switzerland
http://www.milian.com

Millipore
290 Concord Rd.
Billerica, MA 01821
USA
TEL 1-978-715-4321
http://www.millipore.com

Millipore AG
Chriesbaumstrasse 6
CH-8604 Volketswil
Switzerland
TEL +41 848 645 645
FAX +41 848 645 644
http://www.millipore.com

Miltenyi Biotech GmbH
Friedrich-Ebert-Str. 68
51429 Bergisch-Gladbach
Germany

Miltex instruments
http://www.ssrsurgical.com

MJ Research, Inc.
5350 Capital Court, #102
Reno, NV 89502
USA
http://www.mjr.com/

MJ RESEARCH, INC.
590 Lincoln Street
Waltham, MA 02451
TEL (617) 923-8000
TEL (888) 735-8437
FAX (617) 923-8080
E-MAIL info@mjr.com
http://www.mjr.com

MJ Research Inc.
149 Grove St.
Watertown, MA 02172

MoBiTec
Wagenstieg 5 D-37077
Göttingen
TEL (+49) 551-371062
FAX (+49) 551-34987

MoBiTec GmbH
Lotzestrasse 22a
37083 Göttingen, Germany
http://www.mobitec.de

Molectron Detector, Inc.
7470 SW Bridgeport Road
Portland, OR 97224
USA
TEL 1-800-366-4340
FAX 1-503-620-8964
E-MAIL info@molectron.com
http://www.molectron.com/index.asp

Molecular Biology Insights
8685 US Highway 24
Cascade, CO 80809-1333
USA
TEL 1-800-747-4362
TEL 1-719-684-7988
FAX 1-719-684-7989
http://www.oligo.net/

Molecular BioProducts, Inc.
9880 Mesa Rim Road
San Diego, CA 92121-2979
USA

Molecular Devices Corporation
1311 Orleans Avenue
Sunnyvale CA 94089-1136
USA
TEL +1-408-747-1700
TEL 800-635-5577
FAX +1-408-747-3601
http://www.moleculardevices.com

Molecular Devices Ltd.
135 Wharfedale Road
Winnersh Triangle
Winnersh, Wokingham RG41 5RB
England

Molecular Dynamics
928 East Arques Avenue
Sunnyvale, CA 94068
USA
TEL 1-800-333-5703
http://www.mdyn.com

Molecular Probes
29851 Willow Creek Road
PO Box 22010
Eugene, OR 97402-0469
USA
TEL 1-800-438-2209
FAX 1-800-438-0228
http://www.probes.com/

Molecular Probes
PoortGebouw
Rijnsburgerweg 10
2333 AA Leiden
The Netherlands
TEL +31 71 52 36 850
FAX +31 71 52 33 419
http://www.probes.nl

Molecular Probes Europe BV
Poortgebouw
Rijnsburgerweg 10
2333 AA Leiden
The Netherlands
TEL +31-71-5233378
FAX +31-71-5233419
http://www.probes.com/

Moss, Inc.
P.O. Box 189
Pasadena, Maryland 21123-0189
USA
TEL +1-800-932-6677
FAX +1-410-768-3971
http://www.mosssubstrates.com

Motion Analysis Corporation
3617 Westwind Blvd.
Santa Rosa, California 95403
TEL 707.579.6500
FAX 707.526.0629
http://www.motionanalysis.com

mouse models of human cancer consortium
    (MMHCC)
http://web.ncifcrf.gov/researchresources/mmhcc/

MP Biomedicals Corporate Headquarters
15 Morgan
Irvine, CA 92618-2005
TEL 800.633.1352
E-MAIL sales@mpbio.com

MSD Sharp & Dohme GmbH
Lindenplatz 1
D-85540 Haar
Germany

MTX Labs, Inc.
8456 Tyco Road
Building D
Vienna, Virginia 22182
USA
TEL 01.703.821.1045
TEL 1.800.848.6474
FAX 01.703.821.1046
Tech Support 703.821.3948
E-MAIL Info@mtxlsi.com

MWG Biotech (Headquarters)
Anzinger Str. 7a
D-85560 Ebersberg
Germany
TEL +49-08092-8289-0
FAX +49-08092-21084
E-MAIL info@mwgdna.com
http://www.mwg-biotech.com/

Nacalai tesque
498 Higashitamaya-cho
Nijo Karasuma, Nakagyo-ku Kyoto 6040855
Japan
TEL +81-75-251-1723
FAX +81-75-251-1762
E-MAIL info.intl@nacalai.co.jp

Nakagyo-ku
Kyoto 604-0855
Japan
TEL 81-75-211-2516
FAX 81-75-231-2455

Nalge (Europe) Limited
Unit 1a, Thorn Business Park
Hereford HR2 6JS
UK
TEL +44 (0) 1432 263933
FAX +44 (0) 1432 376567

Nalge (Europe) Ltd.
Foxwood Court
Rotherwas, Hereford HR2 6JQ
UK
TEL +44-01432-263933
FAX +44-01432-351923

Nalgene
TEL 1-800-625-4327
FAX 585-586-8987
E-MAIL nnitech@nalgenunc.com
http://www.nalgenunc.com/

Nalge Nunc Internacional
75 Panorama Creek Drive
P.O. Box 20365
Rochester, NY 14602-0365
USA
TEL 1-800-625-4327
FAX 585-586-8987
E-MAIL nnics@nalgenunc.com
http://www.nalgenunc.com

Nanoprobes
95 Horse Block Road
Yaphank, NY 11980-9710
USA

Narashige
27-9 Minamikarasuyama 4-chome
Setagaya-Ku, Tokyo
Japan
TEL 81-3-3308-8233
FAX 81-3-3308-2005
E-MAIL sales@narishige.co.jp
http://www.narishige.co.jp/products/electro/
  index2.htm

Narishige International LTD
Unit 7, Willow Business Park
Willow Way, London SE26 4QP
UK
http://www.narishige.co.jp/niusa/index.htm

Narishige International USA, INC.
1710 Hempstead Turnpike
East Meadow, NY 11554
USA
TEL 516-794-8000
TEL 1-800-445-7914 (within the USA)
FAX 516-794-0066
E-MAIL narishige-usa@pb.net
http://www.narishige.co.jp/niusa/index.htm

NASCO—Fort Atkinson
901 Janesville Avenue
P.O. Box 901
Fort Atkinson, WI 53538-0901
TEL 1-800-558-9595
FAX 920-563-8296
http://www.nascofa.com/prod/Home

National Diagnostics U.S.A.
305 Patton Drive
Atlanta, Georgia 30336
USA
http://www.nationaldiagnostics.com

National Instruments Corporation
11500 N Mopac Expwy
Austin, TX 78759-3504
TEL (+01) 512-683-0100
FAX (+01) 512-683-8411
http://www.ni.com

National Scientific Co.
205 East Paletown Road
Quakertown, PA 18951
USA
TEL 215-536-2577
FAX 215-536-5811

NBS Biologicals Ltd.
14 Tower Square
Hungtingdon, Cambs PE29 7TD
England
http://www.nbsbio.co.uk

NEN
PerkinElmer, European Headquarters
Via Tiepolo, 24
20052 Monza
Italy
http://it.Instruments.perkinelmer.com

NEN (New England Nuclear)
Now part of Perkin Elmer Life and Analytical
  Sciences, Inc.
549 Albany Street
Boston MA 02118-2512 (USA)
las.perkinelmer.com/content/corporate/about/
  nenlifescience.html

NeoLab
Rischerstr. 7
69123 Heidelberg

NeoLab MIGGE Laborbedarf-Vetriebs GmbH
Rischerstr. 7
D-69123 Heidelberg
Postbox 10 40 80
Germany
TEL (0)-62 21/84 42-0 switchboard
FAX (0)-62 21/84 42 33
neoLab (0)-62 21/83 32 26 MIGGE
E-MAIL info@neolab.deTEL
http://www.neolab.de

Neomarkers
47790 Westinghouse Drive
Fremont, California 94539
USA

Nerliens
Postboks 2955
Tøyen, 0608 Oslo
Norway
TEL +47 22 66 65 00
FAX +47 22 66 65 01
E-MAIL info@nmas.no
http://www.vwrsp.com

New Brunswick Scientific Co., Inc.
P.O. Box 4005
44 Talmadge Road
Edison, New Jersey 08818-4005
TEL +1 (732) 287-1200
TEL +1 (800) 631-5417
FAX +1 (732) 287-4222
E-MAIL bioinfo@nbsc.com
http://www.nbsc.com/

NEW ENGLAND BIOLABS
32 Tozer Road
Beverly, MA 01915-5599
2TEL (978) 927- 5054
TEL (USA Orders) 1-800-632-5227
http://www.neb.com

New England Nuclear
Contact PerkinElmer Life and Analytical Sciences
   Inc.
549 Albany Street
Boston, MA 02118
USA
TEL 1-800-762-4000
http://las.perkinelmer.com/

New Focus Corporate Offices
2584 Junction Avenue
San Jose, CA 95134
USA
TEL 1-866-NUFOCUS (USA and Canada only)
TEL (408) 919-1500
E-MAIL contact@newfocus.com

New Objective, Inc.
2 Constitution Way
Woburn, MA 01801-1023
USA
TEL (888) 220-2998 U.S.
TEL (781) 933-9560
FAX (781) 933-9564
E-MAIL sales@newobjective.com

Newport Corp.
1791 Deere Ave.
Irvine, CA 92714
USA
TEL 800-222-6440
FAX 714-963-2015
E-MAIL uk@newport.com 253-1680
http://www.newport.com/

Newport Scientific
8246-E Sandy Court
Jessup, Maryland 20794
http://www.newport.com

Nihon Pharmaceutical
Higashicanda 1-9-8
Chiyoda Tokyo 101-0031
Japan
TEL 03-3864-8411
FAX 03-3864-8837

NIKON
1300 Walt Whitman Road
Melville, NY 11747
http://www.nikonusa.com

Nikon DIAPHOT-TMD
Fuji Bldg. 2-3, 3-chrom
Marunouchi Chiyoda-Ku, Tokyo 157
Japan

Nipa Laboratories Inc.
Llantwit Fadre
Pontypridd, Mid Glamorgan CF38 2SN
UK
TEL (+44) 1443 205311
FAX (+44) 1443 207746

Nissui Pharmaceutical Co.
Tokyo, Japan
http://www.nissui-pharm.co.jp
Nitta Gelatin Co. Ltd.
Osaka, Japan
http://www.nitta-gelatin.co.jp

NORTON PERFORMANCE PLASTICS CORP.
Akron, Ohio 44305
USA
http://www.tygon.com/

Nova Biochem
Laufelfingen, Switzerland
http://www.emdbiosciences.com/html/NBC/
    home.html

Novagen
Merck Biosciences Ltd.
Boulevard Industrial Park
Padge Road
Beeston, Nottingham, NG9 2JR
UK
TEL 0800 622935
FAX 0115 943 0951
E-MAIL customer.service@merckbiosciences.co.uk

Novocastra Laboratories Ltd
Balliol Business Park West
Benton Lane, Newcastle upon Tyne, NE12 8EW
UK
TEL +44 (0) 191 215 0567
FAX +44 (0) 191 215 1152

Novo Nordisk
NOVO NORDISK A/S
NOVO ALLE, DK 2880 BAGSVAERD
DENMARK

Novus Biologicals, Inc.
PO Box 802
Littleton, CO, 80160
USA
E-MAIL novus-biological.com

NuAire, Inc.
2100 Fernbrook Lane
Plymouth, MN 55447
USA
http://www.nuaire.com

NUNC
75 Panorama Creek Drive
Rochester, NY 14625
USA
http://www.nalgenunc.com/

Nunc A/S
Box 280
Kamstrup, DK 4000
Roskilde
Denmark
TEL +45-42359065
http://www.nunc.nalgenunc.com

Nunclon
Nunc Gmbh & Co. KG
Postfach 120543
D-65083 Wiesbaden
TEL +49 6111 86740
FAX +49 6111 867474
E-MAIL nunc@nunc.de
http://www.nunc.de

N.V. Mettler-Toledo S.A.
Leuvensesteenweg 384
1932 Zaventem
Belgium
TEL +32 2 334 02 11
FAX +32 2 334 03 34
http://www.mt.com

Oakdale Engineering, Inc.
23 Tomey Road
Oakdale, PA 15071
USA

Office Depot Corporate Support Center
2200 Old Germantown Road
Delray Beach, FL 33445
TEL 1-800-463-3768
http://www.officedepot.com/

Office Max
http://www.officemax.com

Oligos Etc. Inc.
PO Box 727
9775 SW Commerce Circle C-6
Wilsonville, OR 97070

Olympus America
Olympus America Inc.
2 Corporate Center Drive
Melville, NY 11747
USA
TEL 1-800-645-8160
http://www.olympusamerica.com/seg_section/
    seg_home.asp

Olympus Optical Co., Ltd.
Tokyo, Japan
http://www.olympus-global.com/

Omega Optical, Inc.
210 Main Street
Brattleboro, VT 05301
USA
TEL (802) 254-2690
TEL (866) 488-1064
FAX (802) 254-3937
http://www.omegafilters.com

Once/Millpledge Veterinary
Whinleys Estate
Clarborough
Redford, Knotts, DN229A
UK

Oncogene Science from Cedarlane Laboratories Ltd.
5516-8th Line
R.R.2
Hornby, ON L0P 1E0
Canada
TEL 1-800-268-5058

Ophir Optronics Inc.
9 Electronics Avenue
Danvers Industrial Park
Danvers, MA, USA 01923
TEL 1-800-383-0814
FAX 1-978-774-8202
E-MAIL sales@ophiropt.com
http://www.ophiropt.com/

Optical Insights, LLC
1807 Second Street
Suite #100
Santa Fe, NM 87505
2TEL 505-955-1585
International: 001-505-955-1585
http://www.optical-insights.com

Oriel
150 Long Beach Blvd.
Stratford, CT 06615
USA
2TEL (203) 377-8282
FAX (203) 378-2457
http://www.oriel.com

Osmonics
G.A. Murdock Inc.
1200 Division Ave. S.
P.O. Box 465
Madison, SD 57042, USA
http://www.gamurdock.com/gam/out/BRANDS/
    Osmonics.htm

Osram
Hellabrunner Straße 1
81543 München
Germany

Oswel Research Products Ltd.
Unit 2
Winchester Hill Commercial Park
Winchester Hill
Romsey, Hampshire SO51 7UT
England
OVA Production
Farmer: Pär-Erik Wejåker Sörgården,
    Morgongåva
Sweden

Owens polyscience Ltd.
34 Chester Road
Macclesfield, Cheshire, SK11 8DG

Oxoid Limited
Wade Road
Basingstoke, Hampshire, RG24 8PW
UK
http://www.oxoid.com

Oxoid Ltd.
Wade Road
Basingstoke RG24 OPW
England

PAA Laboratories
1 Technine
Guard Avenue
Houndstone Business Park
Yeovil Somerset BA22 8YE
UK
TEL ++44 193 541-1418

PAA Laboratories GmbH
Haidmannweg 9
A-4061 Pasching
Austria
TEL 43 7229 64865
FAX 43 7229 64866

Packard, see PerkinElmer Life and Analytical
   Sciences, Inc.
549 Albany Street
Boston, MA 02118-2512
USA
TEL (+1) 203-925-4602
FAX 203-944-4902
E-MAIL productinfo@perkinelmer.com
http://www.perkinelmer.com

Packard Instruments Co. Inc.
2200 Warrenville rd.
Downers Grove, Illinois 60515
USA
TEL (1) 312-969-6000

Pall Corporation
Europa House
Havant Street
Portsmouth, Hampshire, PO1 3PD
UK
TEL +44 (0) 23 9230 3303
FAX +44 (0) 23 9230 2509
http://www.pall.com/

Panasonic
Matsushita Electric Corporation of America
One Panasonic Way
Secaucus, NJ 07094
http://www.panasonic.com

Panreac química S.A.
Riera de Sant Cugat, 1
E-08110 Montcada i Reixac (Barcelona)
Spain
http://www.panreac.es

Parale Mitsui
Bldg., 8
Higashida-cho
Kawasaki-ku, Kawasaki, Kanagawa, 210-0005
Japan
http://www.nikon.com

Parr Instrument Company
211 53rd Street
Moline, Illinois 61265-9984
TEL 1-800-872-7720
TEL (309) 762-7716
FAX (309) 762-9453

Partec AG
Sonnenweg 7
CH-4144 Arlesheim
Switzerland
TEL 061 72 77 55

PE Applied Biosystems
850 Lincoln Centre Drive
Foster City, CA 94404
USA
TEL 800.327.3002
TEL 650.638.5800
http://www.appliedbiosystems.com

Pelco International
P.O. Box 492477
Redding, CA 96049-2477
U.S.A.
TEL 530-243-2200
TEL 800-237-3526 (Canada)
FAX 530-243-3761
E-MAIL sales@pelcoint.com
http://www.pelcoint.com/

PeproTech EC Ltd.
PeproTech House
29 Margravine Road
London W6 8LL
UK
TEL (0)20 7610 3055
TEL (0)20 7610 3062
FAX (0)20 7610 3430

Perbio GmbH
Adenauerallee 113
53113 Bonn
Germany
TEL 49 228 9125650
FAX 49 228 9125651

Perbio-Science
Knutpunkten 34
SE-252 78 Helsingborg
Sweden
TEL +46 42 26 90 90
FAX +46 42 26 90 98
http://www.perbio.com

Perbio Science (Branch Office)
Industriezone III
Industrielaan 27
B-9320 Erembodegem-Aalst
BELGIUM
TEL +32 53 83 44 04
FAX +32 53 83 76 38
http://www.piercenet.com

Perbio Science UK Ltd.
Century House
High Street
Tattenhall, Cheshire CH3 9RJ

Perkin Elmer
204 Cambridge Science Park
Cambridge CB4 0GZ
UK

PerkinElmer
45 William Street
Wellesley, MA 02481-4078
USA

PerkinElmer Instruments GmbH
Ferdinand-Porsche-Ring 17
63110 Rodgau-Jügesheim
FAX 49 (01803) 929526
http://www.perkin-elmer.de

PerkinElmer Life & Analytical Sciences, Inc.
(New England Nuclear Lifesciences)
549 Albany Street
Boston, MA 02118-2512
TEL 1-800-762-4000
TEL 203-925-4602
FAX 1-203-944-4902
E-MAIL productinfo@perkinelmer.com
http://www.perkinelmer.com/nenlifescience.asp

PerSeptive
850 Lincoln Centre Drive
Foster City, CA 94404
http://www.appliedbiosystems.com/

Perseptive Biosystems
500 Old Connecticut Path
Framingham, MA 01701
USA
TEL 800-899-5858
FAX 508-383-7851

PGC Scientific
7311 Governors Way
Frederick, MD 21704
USA
TEL (301) 620-7777
http://www.pgcsci.com/

Pharmacia
Am Wolfsmantel 46
D-91058 Erlangen
Germany

Pharmacia & Upjohn Pty LtD
15 Brodie Hall Dve
Bentley WA 6102
Australia
http://www.pharmacia.com

Pharmacia Fine Chemicals
Amersham Biosciences Corp.
800 Centennial Ave
P.O. Box 1327
Piscataway, NJ 08855
USA
http://www5.amershambiosciences.com

Pharmingen
1-6800 Kitimat Road
Mississauga ON, L5N 5M1
Canada
TEL 1-888-259-0187

Phillip Harris Scientific
618 Western Avenue
Park Royal, London W3 0TE
UK
TEL (+44) 181 992 5555
FAX (+44) 181 993 8020

Phoenix Flow Systems
11575 Sorrento Valley Road
San Diego, CA, 92121
TEL 1-800-886-FLOW (3569)
TEL (619) 453-5095
FAX (619) 259-5268
http://www.phnxflow.com

Photometrics Roper Scientific
3440 E. Britannia Drive
Suite 100
Tucson, Arizona 85706
Roper Scientific Benelux
Ir. D.S. Tuijnmanweg 10
Triple P Building, 4131 PN Vianen
The Netherlands
http://www.photomet.com

Photometrics Roper Scientific
3440 East Britannia Drive
Tucson, AZ 85706
USA
http://photomet.com

Photon Technology International, Inc.
1009 Lenox Drive
Lawrenceville, NJ 08648
http://www.pti-nj.com

Physik Instrumente (PI) GmbH & Co. KG
Auf der Roemerstrasse
D-76228 Karlsruhe/Palmbach
Germany
TEL (+49) 721 4846-0
FAX (+49) 721 4846-100
E-MAIL info@pi.ws
http://www.pi.ws

Physik Instrumente GmbH
Auf der Romerstrase 1
Karlsruhe null, D76228
GERMANY
TEL 49-7243-604-100
FAX 49-7243-604-145
http://www.physikinstrumente.com/

Pierce
P.O. Box 117
3747 N. Meridian Road
Rockford, IL 61105
USA
TEL +1-800-842-5007
TEL 815-968-8148
FAX 1-800-842-5007
E-MAIL cs@piercenet.com
http://www.piercenet.com

Pierce Chemical Company
Perbio Science UK Ltd.
Century House
High Street
Tattenhall, Cheshire CH3 9RJ
UK
TEL +44 (0)1829 771 744
FAX +44 (0)1829 771 644
http://www.piercenet.com

Polaroid Corporation
Corporate Headquarters
1265 Main Street—
Bldg. W3
Waltham, MA 02451
USA
TEL 1-800-343-5000
http://www.polaroid.com/us/index.jsp

Polymicro Technologies, LLC
18019 N. 25th Avenue
Phoenix, AZ 85023-1200
TEL 602-375-4100
FAX 602-375-4110

Polysciences, Inc.
400 Valley Rd.
Warrington, PA 18976-2522
TEL 800/523-2575
FAX 800/343-3291
E-MAIL info@polysciences.com

PolysciencesEurope
GmbHHandelsstr. 3
D-69214 Eppelheim
Germany
TEL (49)6221-765767
FAX (49)6221-764620
E-MAIL info@polysciences.de
http://www.polysciences.com/shop/

Precision Brand Products Inc.
Downers Grove, IL
USA
http://www.precisionbrand.com

Prime Tech
635 Nakamukaihara Tsuchiura-shi
Ibaraki 300-0841
JAPAN
TEL +81-29-830-4517
FAX +81-29-830-4515
E-MAIL pmm@primetech-jp.com
http://www.primetech-jp.com/english/
    newproduct.htm

Princeton Research Instruments, Inc.
42 Cherry Valley Road
Princeton, NJ 08542
USA

Princeton Scientific Instruments, Inc.
7 Deer Park Drive
Monmouth Junction, New Jersey 08852
USA
TEL +1 732 274 0774
FAX +1 732 274 0775
http://www.prinsci.com/

PROBES
Molecular Probes Europe BV
Poortgebouw, Rijnsburgerweg 10
2333 AA Leiden
The Netherlands
TEL +31-71-5233378
FAX +31-71-5233419
http://www.probes.com

Progen Biotechnik GmbH
Maaßstrasse 30
69123 Heidelberg
Germany
TEL 49 6221 8278-13
FAX 49 6221 8278-23

ProImmune Limited
Oxford BioBusiness Centre
Littlemore Park
Littlemore, Oxford OX4 4SS
UK
TEL +44 1865 405 128
FAX +44 1865 405 123
E-MAIL enquiries@proimmune.com
http://www.proimmune.co.uk

Promega Benelux BV.
Kenauweg 34
PO Box 391
2300 AJ Leiden
THE NETHERLANDS
TEL (31) 71 5324244
FAX (31) 71 5324907

Promega Corporation
2800 Woods Hollow Road
Madison, WI 53711
USA
TEL 606-274-4330
TEL 800-356-9526
FAX 608-277-2601
http://www.promega.com

Promega—from Fisher Scientific
112 Colonnade Road
Nepean ON K2E 7L6
Canada
TEL 1-800-234-7437

Promega GmbH
Schildkrötstr. 15
D-68199 Mannheim
Germany

Promega Ltd.
Delta House
Enterprise Road
Chilworth Research Centre, Southampton
England

Proxeon Bioseparations
Staermosegaardsvej 6
DK 5230 Odense M
Denmark

Purite Ltd.
Bandet Way
Thame, Oxon OX9 3SJ

QA Bio.
240 Eaton Road
San Mateo, CA 94402
USA

Qbiogene, Inc.
2251 Rutherford Road
Carlsbad, CA 92008
USA
TEL 800-424-6101
FAX 760-918-9313
E-MAIL orders@qbiogene.com
http://qbiogene.com

Qbiogene, Europe offices
Parc d'Innovation
BP 50067
67402 ILLKIRCH CEDEX
France
TEL +33 3 88 67 54 25
FAX +33 3 88 67 19 45
E-MAIL eurorders@qbiogene.com
http://qbiogene.com/about/distributors

QIAGEN
28159 Avenue Stanford
Valencia, CA 91355
TEL 800-426-8157
FAX (800) 718-2056
http://www.qiagen.com

QIAGEN
QIAGEN HOUSE
Fleming Way
Crawley West Sussex RH10 9NQ
UK

QIAGEN AG
Auf dem Wolf 39
CH-4052 Basel
Switzerland

QIAGEN GmbH
Max-Volmer-Str 4
40724 Hilden
Germany
http://www.qiagen.com/

QIAGEN Inc.
28159 Avenue Stanford
Valencia, CA 91355
TEL 800-426-8157
TEL 800-DNA-PREP
TEL (800-362-7737)
FAX 800-718-2056

Quantel SA
17, Avenue de l'Atlantique
Z.A. de Courtaboeuf
BP 23-91941 Les Ulis Cedex
France
TEL 33 (1) 69 29 17 00
FAX 33 (1) 69 29 17 29

Quantronix Corp.
45 Adams Ave.
Hauppauge, NY 11788
USA
TEL 800-235-5953

Queen's University Core Facility
Ms. Sook Shin
Queen's University Kingston, ON K7L 3N6
Canada
TEL (613)533-6837

R&D Systems Europe
Abington Science Park
Abington, OX14 3NB
UK
http://www.RnDSystems.com

R&D Systems Inc.
614 McKinley Place N.E.
Minneapolis, MN 55413
TEL 1 800 349 74 75
TEL 1 612 379 29 56
FAX 1 612 656 44 00
E-MAIL info@rndsystems.com
http://www.rndsystems.com

Rainin
7500 Edgewater Dr.
Oakland, CA 94621
TEL (510)564-1600
http://www.rainin.com

RA Lamb
Units 4 & 5 Parkview Industrial Estate
Alder Close
Lottbridge Drove
Eastbourne, East Sussex, BN23 6QE
UK
TEL 01323 737000
FAX 01323 733000
E-MAIL sales@ralamb.com

Rathburn Chemicals Ltd.
Walkerburn, Scotland
EH43 6AU
TEL (44) (0)1 896 870 651
FAX (44) (0)1 896 870 633

Raymond A Lamb LLC.
7304 Vanclaybon Drive
Apex, NC 27502
USA
TEL 919387-1237
FAX 919-387-1736
E-MAIL sales@ralamb.com

Raytest GmbH
Benzstr. 4
D 75339 Straubenhardt
Germany
E-MAIL info@raytest.de
http://www.raytest.de

Reichert Division der Leica Aktiengesellschaft
Hernalser Hauptstr. 219
Postfach 95
A-1171 Vienna
Austria
TEL +43 1 4616 410
FAX +43 1 46 0326

Research Products International Corp.
410 N. Business Center Drive
Mount Prospect, IL 60056-2190
TEL 800 323 9814
http://www.rpicorp.com

Rheodyne LP
PO box 1909
Rohnert Park, CA 94927-1909
USA

Riedel-de-Haen
Sigma-Aldrich Laborchemikalien GmbH
P.O. Box 100262
30918 Seelze
Germany
TEL +49 5137 82 38 0
FAX +49 5137 82 38 0
E-MAIL Riedel@sial.com
http://www.sigma-aldrich.com

Roche
Basel, Switzerland
F. Hoffmann-La Roche Ltd
Group Headquarters
Grenzacherstrasse 124
CH-4070 Basel
Switzerland
TEL +41-61-688 1111
FAX +41-61-691 9391
http://www.roche.com/

Roche (F. Hoffmann-La Roche Ltd)
Diagnostics Division
Grenzacherstrasse 124
CH-4070 Basel
Switzerland
TEL 41-61-688-1111
FAX 41-61-691-9391
http://www.roche.com/

Roche Applied Science
9115 Hague Road
P.O. Box 50414
Indianapolis, IN 46250-0414, USA
TEL (800) 428-5433
FAX (800) 428-2883
http://www.roche-applied-science.com

Roche Diagnostics
Sandhoferstrasse 116
68305 Mannheim
Germany
FAX 49 (621) 759-4083
http://www.roche-diagnostics.com

Roche Diagnostic Corporation
9115 Hague Road
PO Box 50457
Indianapolis, IN 46256
USA
TEL 1-800-428-5433
http://www.roche-diagnostic.com

Roche Diagnostics, S.L.
Molecular Biochemicals
Copérnico, 60
08006 Barcelona
Spain
http://biochem.roche.com

Roche Diagnostics Corporation
201 Boulevard Armand Frappier
Laval, Quebec, H7V 4A2
Canada
TEL 800-363-5887
FAX 800-667-7050
http://www.biochem.roche.com

Roche Diagnostics Ltd.
Bell Lane, Lewes
East Sussex BN7 1LG

Roche Diagnostics Nederland B.V.
Postbus 1007
300 BA Almere
The Netherlands

Roche Molecular Biochemicals
Roche Norge
Division Diagnostics
Pb 6610 Etterstad, N-0607 Oslo
Norway

Roche Pharma (Schweiz) AG
Schönmattstrasse 2
CH-4153 Reinach BL
Switzerland
TEL +41-61-715 4111
FAX +41-61-715 4112
http://www.roche.com/

Roche Products Ltd
40 Broadwater Road
Welwyn Garden City
Hertfordshire, AL7 3AY

Rockland Immunochemicals
P. O. Box 316
Gilbertsville, PA 19525
http://rockland-inc.com

Romil Ltd.
The Source
Convent Drive
Ultra-pure waterbeach, Cambridge GB-CB59QT

Roper Scientific
3660 Quakerbridge Road
Trenton, NJ 08619
USA
TEL 609-587-9797
FAX 609-587-1970
sold by Visitron (CH and Germany) or Universal
    Imaging (UK)
http://www.roperscientific.com

Roper Whitney of Rockford, Inc.
2833 Huffman Blvd.
Rockford, IL 61103
USA
http://66.165.84.245/index2.asp

Rossville
Gold Shield Chemical Company
Hayward, CA 94545

RZPD (German Resource Centre for Genome
    Research GmbH)
Heubnerweg 6
D-14059 Berlin
Germany

Sakura Finetek Europe B.V.
Energieweg 1
3640 AB Mijdrecht
PO Box 80
The Netherlands
TEL +297 280 666
FAX +297 280 373
http://www.bayer.nl

Samco
1050 Arroyo Ave.
San Fernando, CA 91340-1822
USA

Santa Cruz Biotechnology, Inc.
2145 Delaware Ave.
Santa Cruz, CA 95060
TEL 831-457-3800
TEL 800-457-3801
FAX 831-457-3801
E-MAIL scbt@scbt.com
http://www.scbt.com

SARSTEDT AG & Co.
Rommelsdorfer Straße
Postfach 1220
51582 Nümbrecht
Germany
TEL +49 2293 305 0
FAX +49 2293 305 122
E-MAIL info@sarstedt.com
http://www.sarstedt.com

Sarstedt, Inc.
PO Box 468
Newton, NC 28658

SARSTEDT Inc.
6373 Des Grandes Prairies
Montreal PQ H1P 1A5
Canada
TEL 1-888-727-7833

Sartorius Corp.
131 Heartland Blvd.
Edgewood, NY 11717
USA
FAX 1 (631) 254-4253
http://www.sartoriuscorp.com

Sartorius GmbH
Weender Landstr. 94-108
D-37075 Göttingen
Germany
FAX 49 (551) 308-289
http://www.sartorius.de

Savant (now a division of Thermo Electron
    Corporation)
Thermo Electron Corporate Headquarters
81 Wyman Street
Waltham, MA 02454-9046
USA
TEL 1-877- 843-7668
FAX 1-781-622-1207
http://www.thermo.com/

Scanalytics, Inc.
8550 Lee Highway
Suite 400
Fairfax, VA 22031 USA
TEL 1-703-208-2230
FAX 1-703-208-1960
http://www.scanalytics.com/

Scharlau Chimie S.A.
Ctra. de Polinyà a Sentmenat
Km. 8, 2
08181 Sentmenat (Barcelona)
Spain
http://www.sharlau.com

Schering-Plough Co., Ltd.
Osaka, Japan
http://www.schering-plough.co.jp

Schleicher & Schuell
Hahnestrasse 3
D-37586 Dassel
Germany
TEL ++49 5561 791 463
FAX ++49 5561 791 583
http://www.schleicher-schuell.com/icm11be.nsf/
    (html)/FramesetBioScience

Schleicher & Schuell Bioscience Inc.
10 Optical Avenue
Keene, NH 03431

Schleicher & Schuell GmbH
Grimsehlstr. 23
37574 Einbeck
Germany
http://www.schleicher-schuell.com/

Schott Glasware
Bacto Laboratories Pty Ltd
PO Box 295
Liverpool, NSW, Australia 2170
TEL +61 (0)2 9602 5499
FAX +61 (0)2 9601 82
http://www.bacto.com.au

Scientific Industries, Inc.
70 Orville Dr.
Bohemia, NY 11716
USA
http://www.scientificindustries.com

Scientific Laboratory Supplies
Unit 27
Wilford Industrial Estate
Ruddington Lane
Wilford, Nottingham NG11 7EP
UK
TEL (+44) 115 9821111
FAX (+44) 115 9825275

Scios Nova Inc.
820 West Maude Avenue
Sunnyvale, California 94086
USA
http://www.sciosinc.com/

Scripophily.com
P.O. Box 223795
Chantilly, Virginia, 20153

Sefar America Inc.
http://www.sefaramerica.com

SEIKAGAKU
124 Bernard Saint Jean Dr.
East Falmouth, MA 02536-4445
USA
http://www.acciusa.com/seikagaku/index.asp

SEIKAGAKU
Seikagaku Corporation
1-5 Nihonbashi-honcho
2-chome, Chuo-ku, Tokyo 103-0023
Japan
TEL (81)-3-3270-0966
FAX (81)-3-3270-0538

Sequenom Inc.
3595 John Hopkins Court
San Diego, CA 92121-1331

Sera Laboratories International Ltd.
Unit A 1 horsted
Haywards Heath, RH17 7BA (Road Map) West
    Sussex
TEL 01722 790000
http://www.sli-ltd.com

Serologicals Corp.
5655 Spalding Drive
Norcross, GA 30092
USA
TEL 1-679-728-2000
http://www.serologicals.com

Serotec
22 Bankside
Station Approach
Kidlington, Oxford, OX5 1JE

Serva
AL-Labortechnik
Friedmühle 430 A
A—3300 Amstetten
Austria
TEL +43 7472 234 233
FAX +43 7472 234 234
http://www.serva.de

Serva Electrophoresis GmbH
Carl-Benz-Str. 7
D-69115 Heidelberg
P.O.B. 10 52 60
Germany
TEL +49 (0) 6221 13840-0
FAX +49 (0) 6221 13840-10
E-MAIL info@serva.de
http://www.serva.de

Severn Biotech Ltd.
Unit 2
Park Lane
Stourport Rd
Kidderminster, Worcestershire, DY11 6TJ
UK

SGE, Inc.
2007 Kramer Lane
Austin, TX 78758
USA

Shimadzu America/ Shimadzu Scientific
    Instruments, Inc.
7102 Riverwood Drive
Columbia, MD 21046, USA
TEL (410) 381-1227/ 800-477-1227
FAX (410) 381-1222
http://www.ssi.shimadzu.com

Shinjuku Monolith
3-1 Nishi-Shinjuku 2-chome
Shinjuku-ku, Tokyo 163-0914
Japan

Shipley
Corporate and Manufacturing Headquarters
455 Forest Street
Marlborough, MA 01752
TEL 508 481-7950
TEL 800 832-6200
FAX 508 485-9113

Siegfried (USA), Inc.
33 Industrial Park Road
Pennsville, NJ 08070
TEL +1 856 678 3601
FAX +1 856 678 4008
E-MAIL info@siegfried-usa.com

Siegfried Ltd.
Untere Bruehlstrasse 4
CH-4800 Zofingen
Switzerland
TEL +41 62 746 1111
FAX +41 62 746 1103
http://www.siegfried.ch/

Sierra BioSource Inc.
260 Cochrane Circle
Ste E
Morgan Hill, CA 9503

Sigma Aldrich Denmark A/S
Vejlegaardsvej 65B
DK-2665
Vallensbaek Strand
http://www.sigmaaldrich.com

Sigma Aldrich
P.O. Box 14508
St. Louis, MO 63178
http://www.sigma-aldrich.com

SIGMA-ALDRICH Handels GmbH
Favoritner Gewerbering 10
1100 Wien
Austria
TEL +43 1 605 81 10
FAX +43 1 605 81 20
E-MAIL sigma@sigma.co at
http://www.sigmaaldrich.com

Sigma-Aldrich Sweden AB
Stockholm, Sweden
http://www.sigmaaldrich.com/

Sigma-Aldrich
3050 Spruce St.
St. Louis, MO 63103
TEL 314-771-5750
TEL 800-325-3010
TEL 314-771-5765
FAX 800-325-5052
FAX 314-771-5757
E-MAIL custserv@sial.com
http://www.sigma-aldrich.com

Sigma-Aldrich
Norway A/S
Tevlingveien 23
1081 Oslo

Sigma-Aldrich Canada Ltd.
2149 Winston Park Dr.
Oakville, Ontario, L6H 6J8
TEL 800-565-1400
www.sial.com

SIGMA-Aldrich Chemie GmbH
Eschenstr. 5
D-82024 Taufkirchen
Germany
TEL 49 89 65131103
FAX 49 89 65131144
E-MAIL DECustsv@eurnotes.sial.com
http://www.sigma-aldrich.com

Sigma-Aldrich Chimie S.ar.l.
L'Isle d'Abeau Chesnes
B.P. 701
38297 Saint Quentin
Fallavier Cedex, France
TEL 08 00 21 14 08
FAX 08 00 03 10 52
http://www.sigma-aldrich.com

Sigma-Aldrich Co. Ltd.
Fancy Road
Poole, Dorset, BH12 4QH
UK
TEL +44 (0)1202733114
FAX +44 (0)1202715460

Sigma-Aldrich Company Ltd.
The Old Brickyard
New Rd
Gillingham, Dorset SP8 4XT
UK
TEL +44 (0) 1747 833000, 0800 717181
FAX 0800 378785
http://www.sigmaaldrich.com

Sigma-Aldrich Denmark A/S
Vejlegårdsvej 65B
2665 Vallensbæk Strand
Denmark

Sigma Aldrich Germany
Eschenstr. 5
D 82024 Taufkirchen
Germany

Sigma-Aldrich Japan
Tokyo, Japan
E-MAIL sia1jpts@sial.com
http://www.wako-chem.co.jp

Sigma Aldrich Química S.A.
Ronda de Poniente 3
2ª planta
Apartado de Correos 278
Tres Cantos (Madrid)
Spain
http://www.sigmaaldrich.com

Sigma Biosciences
Fancy Road
Poole, Dorset BH17 7NH
UK
TEL (+44) 1202 733114
FAX (+44) 1202 715460

Sigma-Isotec
3858 Benner Rd.
Miamisburg, OH 45342
USA
http://www.isotec.com

Silicon Graphics
1600 Amphitheatre Parkway
Mountain View, CA 94043

Simco Nederland B.V.
Aalsvoort 74
7241 MB Lochem
TEL +31 (0)573 288333
FAX +31 (0)573 257319
E-MAIL info@antistatic.nl

SIS Chemicals
Linden House1
The Square
Pennington, Lymington, Hampshire S0418GN
UK

Skatron Instruments
P.O. Box 8
3401 Lier
Norway
TEL 47-3285-4250
FAX 47-328-54204

Small Parts, Inc.
13980 N.W. 58th Court
P.O. Box 4650
Miami Lakes, FL 33014-0650
http://www.smallparts.com/

Solamere Technology Group
1427 Perry Ave
Salt Lake City, UT 84103
USA

Solis Biodyne
Pikk 14
51013 Tartu
Estonia

Solon Manufacturing Mompany
Meriden Cooper Corporation
PO Box 692
Meriden CT 06450
TEL (203) 237-8448
FAX (203) 238-1314

Sonics & Material Inc.
53 Church Hill Road
Newtown, CT 06470-1614
USA

Soquelec Ltd.
5757 Cavendish Blvd.
Suite 540
Montreal, Quebec W4H 2W8
(Canada)
http://www.soquelec.com

Sorvall
Kendro Laboratory Products GmbH
Wiegelestraße 4
A-1230 Wien
AUSTRIA
TEL +43 1 801 40-0
FAX +43 1 801 40 40
E-MAIL info.at@kendro.spx.com
http://www.sorvall.com

Specialty media
580 Marshall Street
Phillipsburg, NJ 08865
USA
TEL 888-209-8870
TEL 908-213-6555
FAX 908-387-1670
E-MAIL questions@cmt-inc.net
http:///www.specialitymedia.com

Spectra Physics
1335 Terra Bella Avenue
Post Office Box 7013
Mountain View, CA 94039
USA
TEL 650-961-2550
FAX 650-968-5215
E-MAIL service@splasers.com
http://www.spectraphysics.com/

Spectrum Europe B.V. Europe
P.O. Box 3262
NL-4800 DG Breda
The Nederlands
TEL +31 76 5719 419
FAX +31 76 5719 772
http://www.spectrapor.com

Spherotech, Inc.
1840 Industrial Dr.
Suite 270
Libertyville, IL 60048-9467
TEL (800) 368-0822
TEL (847) 680-8922
FAX (847) 680 8927

Stanford Photonics
1032 Elwell Court
Suite 104
Palo Alto, CA 94303
http://www.stanfordphotonics.com/

Stratagene
11011 N. Torrey Pines Road
La Jolla, CA 92037
USA
TEL 1-858-535-5400
http://www.stratagene.com

Stratagene
1834 State Highway 71 West
Cedar Creek, TX 78612
TEL (800) 424-5444
FAX (512) 321-3128
http://www.stratagene.com

Strathmann
Habichthorst 30
22459 Hamburg
Germany

Structure Probe, Inc.
569 East Gay Street
West Chester, PA 19380
USA
TEL 1-800-242-4774
TEL 1-610-436-5400
FAX 1-610-436-5755
E-MAIL spi3spi@2spi.com
http://www.2spi.com/spihome.html

Sumalsa (Suministros de material y aparatos de
    laboratorio S.A.)
Obispo Tajón
18 bajo
50005 Zaragoza
Spain
http://www.thermospectronic.com

Sun Microsystems, Inc.
4150 Network Circle
Santa Clara, CA 95054
USA
TEL 1-800-555-9786
TEL 1-650-960-1300
http://www.sun.com/

Surgipath Medical Industries, Inc.
5205 Rt. 12
Richmond, IL 60071
USA
TEL 800-225-3035
http://www.surgipath.com/

SÜSS MicroTec Lithography GmbH
Schleissheimer Str. 90
D-85748 Garching / Munich
Germany
TEL (+49) [0] (89) 32007-0
FAX (+49) [0] (89) 32007-162

Sutter Instrument Company
51 Digital Drive
Novato, CA 94949
USA
TEL 415.883.0128
FAX 415.883.0572
E-MAIL info@sutter.com
http://www.sutter.com

Swiss Institute of Bioinformatics
Biozentrum—Basel University
Klingelbergstrasse 50-70
4056 Basel
Switzerland
TEL +41 61 267 2042
FAX +41 61 267 2024
E-MAIL admin@isb-sib.ch
http://www.expasy.org/

Taab Laboratories Equipment Ltd.
3 Minerva House
Calleva Ind Park
Aldermaston, Reading, Berks, RG7 4QW
England
TEL +44 (0)1734817775
FAX +44 (0)1734817881

Taconic M&B
Bomholtvej 10
P.O. Box 1079
DK-8680 Ry
Denmark

TaKaRa
Takara Biomedical Group
TAKARA SHUZO CO., LTD.
Seta 3-4-1, Otsu, Shiga 520-2193
Japan

Tecan Austria GmbH
Untersbergstr. 1a
5082 Grödig
Austria

Techne
3 Terri Lane
Suite 10
Burlington, N.J. 08016
TEL 800-225-9243
http://www.techneusa.com

Technologiepark
Emmy-Noether Strasse 10
76131 Karlsruhe
Germany

Technical Video
PO Box 693
Woods Hole MA 02543
USA
TEL 508-563-6377
FAX 508 563-6265

Ted Pella, Inc.
P.O. Box 492477
Redding, CA 96049
TEL 800-237-3526

TeleChem International, Inc.
ArrayIt Division
524 E. Weddell Drive
Suite #3
Sunnyvale, CA 94089, USA
TEL +1-408-744-1331
FAX +1-408-744-1711
http://arrayit.com/

Termo Spectronic (Europe, Middle East and Africa)
Mercers Row
Cambridge CB5 8HY
U.K.

Terumo Neolus
44-1,2 chome
Hatagaya, Shibuya-ku, Tokyo,
Japan
Terumo Neolus, Benelux Sales Division
Researchpark Zone 2
Haasrode Interleuvenlaan 40 B-3001 Leuven
Belgium
TEL 32-16-38-14-02
FAX 32-16-38-15-55
http://www.terumo.co.jp

Tetko Inc.
P.O. Box 346
Lancaster, NY 14086
USA
http://tetko.com/

The MathWorks, Inc.
3 Apple Hill Drive
Natick, MA 01760-2098
USA
TEL 508-647-7000
FAX 508-647-7001

Therma Electron Corporation (formerly Forma
 Scientific)
Headquarters: Informatics and Services
18 Commerce Way
Woburn, MA 01801-1086
United States
TEL +1 781 933 4689 (866 INFOLAB)
FAX +1 781 933 6322
http://www.forma.com/

Thermo Electron Corp.
Via Fisher Scientific Canada
Customer Service Centre
112 Colonnade Rd
Nepean, ON, K2E 7L6
TEL 800-234-7437
http://www.thermo.com

Thermo Electron Informatics (former Thermo
 LabSystem)
St. Georges Court
Hanover Business Park
Altrincham, Cheshire, WA14 5TP
UK
TEL +44 161 942 3000
FAX +44 161 942 3001
http://www.thermo.com/com/cda/home/
 1,1089,,00.html

Thermo Electron Molecular Biology
Sample Preparation
Molecular Biology
450 Fortune Blvd
Milford, MA 01757
USA
TEL +1 508 482 7000
http://www.thermo.com

ThermoFinnigan
355 River Oaks Parkway
San Jose, CA 95134
USA
TEL 408-965-6000

Thermo Hybaid
Contact Thermo Electron Corporation
81 Wyman Street
Waltham, MA 02454
USA
TEL1-877-843-7668
http://www.thermo.com/

Thermo Labsystems Oy
Ratastie 2
P.O.Box 100
Vantaa, 01620
Finland
TEL +358-9-329100
FAX +358-9-32910500
http://www.thermo.com

Thermo Life Sciences—UK
Unit 5
The Ringway Centre
Edison Rd, Basingstoke RG21 6YH
UK
TEL 01256 817282
FAX 01256 817292
http://www.thermo.com/

Thermo Savant / Thermo EC
100 Colin Drive
Holbrook, NY 11741-4306
USA
http://www.thermo.com

Thermo Shandon, Inc.
171 Industry Dr.
Pittsburgh, PA 15275
USA
TEL 412-788-1133
TEL 800-547-7429
FAX 412-788-1138
E-MAIL thermoshandon@thermoshandon.com
http://www.thermoshandon.com

Thomas Scientific
P.O. Box 99
Swedesboro, NJ 08085
U.S.A
TEL 800-345-2100
FAX 856-467-3087
http://www.thomassci.com/contact/index.jsp

Thompson B&SH Co. Ltd.
8148 Ch. Devonshire
Montreal, QC
TEL 514-739-1971

T.I.L.L. Photonics GmbH
Lochhamer Schlag 19
D-82166 Gräfelfing
Germany
http://www.till-photonics.com

Treff
Treff AG
Taastrasse 16
CH-9113 Degersheim

TR Tech/Bex Co. Ltd.
2-22-3-104
Sengoku, Bunkyo-ku, Tokyo 112-0011
JAPAN

Turner BioSystem
845 W. Maude Avenue
Sunnyvale, CA, 94085

UCSF Cell Culture Facility
Box 0528
San Francisco CA 94143
http://www.ccf.ucsf.edu/

Uniblitz by Vincent Associates
803 Linden Avenue
Rochester, NY 14625
USA

United Chemical Technologies, Inc.
2731 Bartram Road
Bristol, PA 19007-6893
TEL (215) 781-9255
TEL 800-541-0559
FAX (215) 785-1226

United States Biochemical Corp.
P.O. Box 22400
Cleveland OH 44122
USA
TEL +1-216-765-5000
FAX +1-216-464-5075

Universal Imaging Corp.
502 Brandywine Parkway
West Chester, PA 19380
USA
TEL 610-344-9410
FAX 610-344-9515
http://www.universal-imaging.com/

Universal Imaging Corporation Limited
PO Box 1192
43 High Street Marlow
Buckinghamshire SL7 1GB
UK
TEL +44 1628 890858
FAX +44 1628 898381
http://www.image1.com

Upchurch Scientific
P.O. Box 1529
619 Oak Street
Oak Harbor, WA 98277
USA
TEL 800-426-0191
FAX 800-359-3460
Upjohn
7171 Portage Rd
Kalamazoo, MI 49001-0199

Upstate Biotechnology, Inc.
10 Old Barn Road
Lake Placid NY 12946
http://www.upstatebiotech.com/.

USA Scientific, Inc.
PO Box 3565
Ocala, FL 34478-3565
TEL 1-800-LAB-TIPS (522-8477)
TEL 352-237-6288
http://www.usascientific.com

USB Corp.
26111 Miles Road
Cleveland, OH 44128
USA
TEL 800 321 9322
FAX 216 464 5075

VACUUBRAND GMBH + CO KG
Postfach 1664
D-97866 Wertheim
Alfred-Zippe-Str. 4
D-97877 Wertheim
TEL +49 (0) 9342 / 808-0
FAX +49 (0) 9342 / 59880
http://www.vacuubrand.de/

Valco Instruments Co. Inc.
P.O. Box 55603
Houston, TX 77255
USA
TEL 800-227-9770
TEL 800-367-8424
FAX 713-688-8106

Valeant Pharmaceuticals (previously ICN)
Valeant Plaza
3300 Hyland Avenue
Costa Mesa, CA 92626
USA
TEL 1.800.548.5100
TEL 1.714.545.0100
FAX 1.714.556.0131
http://www.valeant.com

Vector Laboratories
16 Wulfric Square
Bretton, Peterborough, Cambridgeshire PE3 8RF
England
http://www.vectorlabs.com/

Vector Laboratories
3390 South Service Road
Burlington ON L7N 3J5
Canada
TEL 1-888-629-2121

Vector Laboratories, Inc.
30 Ingold Road
Burlingame, CA 94010
USA
TEL 800 227 6666
TEL 415-697-3600
FAX 415-697-0339
http://www.vectorlabs.com

Vedco Inc.
http://www.vedco.com

Vel, Merck-Belgolabo
Haasrode Researchpark Zone 3
Geldenaaksebaan 464
B-3001 Leuven
Belgium
TEL +32 16 385 011
http://www.vel.be

Verity Software House
PO Box 247
Topsham, ME 04086
E-MAIL verity@vsh.com
http://www.vsh.com

Video Scope International, Ltd.
105 Executive Drive
Dulles, Virginia 20166-9558
TEL (703) 437-5534
FAX (703) 742-8947
E-MAIL info@videoscopeintl.com

VidraFoc S.A.
Badajoz, 50
08005 Barcelona
Spain
http://www.vidrafoc.com

VisiTech International Ltd.
Unit 92 Silverbriar
Sunderland Enterprise Park (East), Sunderland, SR5
    2TQ
UK

Visitron Systems GmbH
Gutenbergstrasse 9
D-82178 Puchheim
Germany
http://visitron.de

VWR
3745 Bayshore Blvd.
Brisbane CA 94005
TEL 1-800-932-5000
http://www.vwr.com

VWR
50 D'Angello Road
Marlboro, MA 01752

VWR
P.O. Box 5015
Bristol, CT 06011

VWR International
1310 Goshen Pkwy
West Chester, PA 19380
TEL 800-932-5000
TEL (610) 431-1700
FAX (610)431-9174
http://www.vwrsp.com

VWR International
2360 Argentia Road
Mississauga, Ontario L5N 5Z7
Canada
TEL 800-932-5000
TEL 905-813-7377
FAX 800-668-6348
http://www.vwrsp.com

VWR International, Inc.
10105 Carroll Canyon Road
San Diego, CA 92131
TEL 858-695-7600

VWR International AS
PO Box 45
Kalbakken, (Kakkelovnskroken, 1), 0901 Oslo
Norway
TEL +47 02290
FAX +47 22 90 00 40
E-MAIL info@no.vwr.com
http://www.vwrsp.com

VWR International GmbH (formerly Merck)
Hilpertstrasse 20a
64295 Darmstadt
Germany

VWR International Holding GmbH
Branch Office Belgium
Woluwedal 28, B-1932
Zaventem
TEL 00322-7115858
FAX 00322-7115920
http://www.vwr.com

VWR International Limited
Hunter Blvd
Magna Park
Lutterewordth, Leicestershire LE17 4XN
UK
TEL 0800 223344
FAX (+44) 01202 665599

VWR International Ltd
Merck House
Poole, Dorset, BH15 1TD
England
TEL 01202 660444
FAX 01202 666856
http://www.bdh.com

VWR Scientific
USA
VWR International
1310 Goshen Parkway
West Chester, PA 19380
USA
http://www.vwr.com

VWR Scientific Corporation
P.O. Box 1002
600C Corporate Court
South Plainfield, NJ 07080
USA

VWR Scientific Products
133 South Center
Suite 700
Morrisville, NC 27560
USA

Wacker-Chemie GmbH
Hauptverwaltung
Hanns-Seidel-Platz 4
81737 München
GERMANY
TEL +49 (0)89 6279-01
FAX +49 (0)89 6279-1770
http://www.wacker.com

Wako Pure Chemicals Industries, Ltd.
1-2,Doshomachi 3-Chome
Chuo-ku, Osaka 540-8605
Japan

Wako Chemicals USA, Inc.
1600 Bellwood Road
Richmond, VA 23237-1326
TEL 804-271-7677
FAX 804-271-7791
TEL 877-714-1924

Waring Laboratory
314 Ella T. Grasso Ave.
Torrington, CT 06790
TEL 1-800-4WARING
TEL (1-800-492-7464)
E-MAIL waring@conair.com

Warner Instrument Corp.
1125 Dixwell Ave.
Hamden, CT06514
USA

Waters Ltd UK
730-740 Centennial Court
Centennial Park
Elstree, Hertsfordshire, WD6 3SZ

Waters/Micromass
34 Maple Street
Milford, MA. 01757
http://www.micromass.se

Watkins and Doncaster
PO Box 5
Cranbrook, Kent TN18 5E2
UK

WeldtiteProducts Ltd.
Harrier Road
Barton-on-Humber, DN18 SRP
UK
TEL 01652 660000
http://www.weldtite.co.uk

Western Region
Sci. Pro. Div.
39899 Balentine Dr. #325
Newark, CA 94560
USA
TEL 800-222-7740

Whatman, Inc.
401 West Morgan Road
Ann Arbor, MI 48108
USA
http://www.whatman.com

Whatman Inc.
9 Bridewell Place
Clifton, NJ 07014
TEL 800-441-6555

Whatman International Ltd.
Whatman House
St Leonard's Road
20/20 Maidstone
Kent, ME16 0LS
UK
TEL +44 (0) 1622 676670
FAX +44 (0) 1622 677011
http://www.whatman.com

Wheaton Science Products
An Alcan Packaging Company
1501 N. 10th Street
Millville, NJ 08332-2093
USA
TEL 800-225-1437
FAX 856-825-1368
TEL 1 856-825-1100
FAX 1 856-825-4568

Wolfram Research (Mathematica)
http://www.wolfram.com/

World Precision Instruments
Astonbury Farm Business Centre
Aston, Stevenage. SG2 7EG.
TEL +44 (0)1438 880025
FAX +44 (0)1438 880026
E-MAIL wpiuk@wpi-europe.com

World Precision Instruments, Inc.
175 Sarasota Center Boulevard
Sarasota, Florida 34240
USA
TEL (941) 371-1003
FAX (941) 377-5428
http://www.wpiinc.com/

Worthington Biochemical Corporation
Pharmacia Biotech Inc.
800 Centennial Ave.
P.O. Box 1327
Piscataway, NJ 08855-1327
USA
TEL 1 800-445-9603 (Canada and the US)
TEL 1 732-942-1660 (International)
E-MAIL Corporate Webmaster@roche.com
http://www.worthington-biochem.com

Wyeth: Fort Dodge Animal Health
http://www.wyeth.com

XENOTEK Engineering (Steve Novotny)
525 North 27th Street
Belleville, IL 62226
USA
TEL 618-235-9110
TEL 800-260-5656
FAX 618-235-9167
FAX 800-859-0898

Yamato Scientific Co. Ltd.
1-6 Honcho 2-Chome
Nihonbashi Chuo-ku
Tokyo103
Japan
TEL +81-3-3231-1124
FAX +81-3-3231-1144
E-MAIL english-website@yamato-net.co.jp
http://www.yamato-net.co.jp

Zeiss
Carl Zeiss Promenade 10
07745 Jena
Germany

Zimmer, Inc.
P.O. Box 708
1800 West Center Street
Warsaw, IN 46581-0708

Zymed Laboratories, Inc.
52 South Linden Avenue, Suite 3
South San Francisco, CA 94080
USA
TEL 800 874 4494
TEL 1 415 871 4494
FAX 1 415 871 4499
http://www.zymed.com/

Index

2to3 program, 4:392
3D (three-dimensional) BM assay, 1:139
18S rRNA, 2:75
293T cells, 1:208–210

A

A23187 (calcium ionophore), 4:49
α-actin, 2:165
α-actinin, 2:173
AAV (adeno-associated virus), 1:457
AbCam, 2:428
ablating UV laser, 3:95
absorption peak, 1:305
acetone, 2:162, 2:163, 2:165, 3:230
    preparation from rabbit skeletal muscle,
        2:173–174
    quality of when making fluorescent
        actin, 3:150
acetonitrile, 4:374, 4:420, 4:423, 4:445, 4:448
acetylation, 1:521
acetylcholinesterase, 1:118
N-acetylglucosamine, 2:220
β-N-acetylhexosaminidase, 2:37
Acholeplasma laidlauii, 1:53
acidification, 2:61
acid phosphatase, 2:74
acousto-optical tunable filters (AOTFs),
    2:318, 3:74
acridine orange (AO) dye
    differential staining of cellular DNA
        and RNA, 1:282, 1:283
    photosensitization of chromosomes
        with, 3:36
    preparation of, 1:328
    use in live cell DNA labeling, 1:306
    use in micronuclei and comet assay,
        1:329
acrylamide/bisacrylamide solution, 1:113,
    1:218–219, 4:105, 4:176, 4:177, 4:185,
    4:198, 4:200, 4:261, 4:332
acrylic resins, see Lowicryl resins
ActA protein, 2:397
actin, 3:137–151
    β-actin, 2:165, 2:168–171
    Ca-actin, 2:173, 2:175
    Ca-G-actin, 2:174–175
    cytoskeleton dynamics, 3:111
    fluorescent speckle microscopy of
        camera electronics, 3:143

CCD chip, 3:142–144
    cooled CCD camera, 3:141–142
    electronically controlled shutter, 3:141
    matching microscope and detector
        resolution, 3:144
    overview, 3:137–140
    pitfalls, 3:150
    principles of, 3:137–140
    procedures, 3:145–150
    software for control of shutter and
        image acquisition, 3:143–144
    upright or inverted epifluorescent
        microscope and optics, 3:141
actin filaments, 2:387, 2:390, 2:393, 3:41
actin in vitro motility assays, 2:387–392
    materials and instrumentation,
        2:387–388
    overview, 2:387
    pitfalls, 2:391–392
    procedures, 2:388–391
        construction of flow cells, 2:388
        overview, 2:388
        preparation of rhodamine-phalloidin-
            labeled actin, 2:388
        preparation of sample for motility
            assay, 2:388–389
        presentation of data, 2:390–391
        recording and quantifying data steps
            and equipment, 2:389–390
actin purification, 2:161–175
    acetone use in, 2:162, 2:165, 2:173–174
    laboratory equipment, 2:173
    materials and instrumentation, 2:161,
        2:165–166, 2:173
        chemicals, 2:161
        gel filtration media, 2:161
    procedures, 2:162, 2:173–175
        conversion of Ca-actin to Mg-actin,
            2:175
        preparation of acetone powder from
            rabbit skeletal muscle, 2:173–174
        preparation of Ca-G-actin from
            acetone powder, 2:174–175
        purification of recombinant b-actin,
            2:168–171
        fermentor culture of yeast expressing
            recombinant actin, 2:169
        purification of recombinant actin
            from yeast, 2:169–171
    solutions, 2:161–162
    tests, 2:162

cosedimentation assays, 2:162
    general polymerization, 2:162
active contour models, 3:98
acylation, 1:523
AD (transcriptional activation domain),
    4:295
AD and DB vectors, 4:299
adaptor protein (AP), 2:51
ADcDNA library, 4:298
adeno-associated virus (AAV), 1:457
adenoviral infection, 2:190
ADF (Array Design File) format, 4:97
adherent cells, 1:16–17
    cryopreservation of, 1:19–20
        media and solutions, 1:19
        notes, 1:19–20
        procedures, 1:19
    subculturing, 1:16–17
        media and solutions, 1:16–17
        notes, 1:17
        procedures, 1:17
        protocols, 1:16
adhesion complexes, 2:18
adhesive forces during microdissection,
    3:342
adipocytes, 1:21, 1:83
AD-ORF, 4:302
AD plasmid, 4:297
ADP-ribosylation factor 1 (Arf1), 2:45
adrenal glands, bovine, purification of
    clathrin-coated vesicles from,
    2:51–56
    cleaning of bovine brain cortices, 2:52
    differential centrifugations, 2:53–54
    homogenization, 2:52–53
    materials and instrumentation, 2:52
    overview, 2:51
    pitfalls, 2:55–56
    procedures, 2:52–54
adrenocorticotropin, 1:512
advanced granulation technology (AGT),
    1:39
A/E (attaching and effacing) lesions,
    2:399
AEP, see affinity electrophoresis (AEP)
AfCS-Nature Signaling Gateway, 2:431
affinity electrophoresis (AEP), 4:197–206,
    4:289
    gel cassette, 4:203
    materials and instrumentation,
        4:197–198